D0138337

REF. 550.3 K968
Kusky, Timothy M.
Encyclopedia of earth
 science

MID-CONTINENT PUBLIC LIBRARY
Raytown Branch
6131 Raytown Road   RT
Raytown, MO 64133

ENCYCLOPEDIA OF
EARTH SCIENCE

ENCYCLOPEDIA OF
EARTH SCIENCE

TIMOTHY KUSKY, Ph.D.

Department of Earth and Atmospheric Sciences,
Saint Louis University

☑®
Facts On File, Inc.

MID-CONTINENT PUBLIC LIBRARY
Raytown Branch
6131 Raytown Road
Raytown, MO 64133
RT

MID-CONTINENT PUBLIC LIBRARY

3 0000 12659776 8

Encyclopedia of Earth Science

Copyright © 2005 by Timothy Kusky, Ph.D.

All rights reserved. No part of this book may be reproduced or utilized in any form or by any
means, electronic or mechanical, including photocopying, recording, or by any information
storage or retrieval systems, without permission in writing from
the publisher. For information contact:

Facts On File, Inc.
132 West 31st Street
New York NY 10001

Library of Congress Cataloging-in-Publication Data

Kusky, Timothy M.
Encyclopedia of earth science / Timothy Kusky.
p. cm.
Includes bibliographical references and index.
ISBN 0-8160-4973-4
1. Earth sciences—Encyclopedias. I. Title.
QE5.K85 2004
550′.3—dc22
2004004389

Facts On File books are available at special discounts when purchased in bulk quantities
for businesses, associations, institutions, or sales promotions. Please call our
Special Sales Department in New York at 212/967-8800 or 800/322-8755.

You can find Facts On File on the World Wide Web at http://www.factsonfile.com

Text design by Joan M. Toro
Cover design by Cathy Rincon
Illustrations by Richard Garratt and Facts On File, Inc.

Printed in the United States of America

VB Hermitage 10 9 8 7 6 5 4 3 2

This book is printed on acid-free paper.

Dedicated to G.V. Rao (1934–2004)

CONTENTS

Acknowledgments
xi

Introduction
xiii

Entries A–Z
1

Feature Essays:
"Coping with Sea-Level Rise in Coastal Cities"
81

"Gaia Hypothesis"
82

"Desertification and Climate Change"
116

"Earthquake Warning Systems"
131

"Loma Prieta Earthquake, 1989"
132

"Mississippi River Basin and the Midwest Floods of
1927 and 1993"
152

"Age of the Earth"
166

"Formation of the Earth and Solar System"
203

"Galveston Island Hurricane, 1900"
214

"Is There Life on Mars?"
262

"Lahars of Nevado del Ruiz, Colombia, 1985"
266

"History of Ocean Exploration"
302

"The World's Oldest Ophiolite"
307

"*Homo sapiens sapiens* and Neandertal Migration and Relations
in the Ice Ages"
338

"Is Your Home Safe from Radon?"
350

"Why Is Seawater Blue?"
379

"Seismology and Earth's Internal Structure"
383

"December 26, 2004: Indian Ocean Earthquake and Tsunami"
437

"Volcanoes and Plate Tectonics"
452

Appendixes:

Appendix I
Periodic Table of the Elements
477

Appendix II
The Geologic Timescale
479

Classification of Species
480

Summary of Solar System Data
480

Evolution of Life and the Atmosphere
480

Index
481

ACKNOWLEDGMENTS

Many people have helped me with different aspects of preparing this encyclopedia. Frank Darmstadt, Executive Editor at Facts On File, reviewed and edited all text and figures in the encyclopedia, providing guidance and consistency throughout. Rose Ganley spent numerous hours as editorial assistant correcting different versions of the text and helping prepare figures, tables, and photographs. Additional assistance in the preparation was provided by Soko Made, Justin Kanoff, and Angela Bond. Many sections of the encyclopedia draw from my own experiences doing scientific research in different parts of the world, and it is not possible to individually thank the hundreds of colleagues whose collaborations and work I have related in this book. Their contributions to the science that allowed the writing of this volume are greatly appreciated. I have tried to reference the most relevant works, or in some cases more recent sources that have more extensive reference lists. Any omissions are unintentional.

Finally, I would especially like to thank Carolyn, my wife, and my children Shoshana and Daniel for their patience during the long hours spent at my desk preparing this book. Without their understanding this work would not have been possible.

INTRODUCTION

The *Encyclopedia of Earth Science* is intended to provide a broad view of some of the most important subjects in the field of earth sciences. The topics covered in the encyclopedia include longer entries on the many broad subdisciplines in the earth sciences (hydrology, structural geology, petrology, isotope geology, geochemistry, geomorphology, atmospheric sciences, climate, and oceanography), along with entries on concepts, theories and hypotheses, places, events, the major periods of geological time, history, people who have made significant contributions to the field, technology and instruments, organizations, and other subjects.

The *Encyclopedia of Earth Science* is intended to be a reference for high school and college students, teachers and professors, scientists, librarians, journalists, general readers, and specialists looking for information outside their specialty. The encyclopedia is extensively illustrated with photographs and other illustrations including line art, graphs, and tables, and contains 19 special essays on topics of interest to society. The work is extensively cross-referenced and indexed to facilitate locating topics of interest.

Entries in the *Encyclopedia of Earth Science* are based on extensive research and review of the scientific literature, ranging from the general science to very specialized fields. Most of the entries include important scientific references and sources listed as "Further Reading" at the end of each section, and the entries are extensively cross-referenced with related entries. Some parts of the encyclopedia draw from my collected field notes, class notes, and files of scientific reprints about selected topics and regions, and I have tried to provide uniformly detailed coverage of most topics at a similar level. Some of the more lengthy entries, however, go into deeper levels on topics considered to be of great importance.

ENTRIES A–Z

A

aa lava Basaltic lava flows with blocky broken surfaces. The term is of Hawaiian origin, its name originating from the sound that a person typically makes when attempting to walk across the lava flow in bare feet. Aa lava flows are typically 10–33 feet (3–10 m) thick and move slowly downhill out of the volcanic vent or fissure, moving a few meters per hour. The rough, broken, blocky surface forms as the outer layer of the moving flow cools, and the interior of the flow remains hot and fluid and continues to move downhill. The movement of the interior of the flow breaks apart the cool, rigid surface, causing it to become a jumbled mass of blocks with angular steps between adjacent blocks. The flow front is typically very steep and may advance into new areas by dropping a continuous supply of recently formed hot, angular blocks in front of the flow, with the internal parts of the flow slowly overriding the mass of broken blocks. These aa lava fronts are rather noisy places, with steam and gas bubbles rising through the hot magma and a continuous clinking of cooled lava blocks rolling down the lava front. Gaps that open in the lava front, top, and sides may temporarily expose the molten lava within, showing the high temperatures inside the flow. Aa flows are therefore hazardous to property and may bulldoze buildings, forests, or anything in their path, and then cause them to burst into flames as the hot magma comes into contact with combustible material. Since these flows move so slowly, they are not considered hazardous to humans.

See also PAHOEHOE LAVA; VOLCANO.

abyssal plains Flat, generally featureless plains that form large areas on the seafloor. In the Atlantic Ocean, abyssal plains form large regions on either side of the Mid-Atlantic Ridge, covering the regions from about 435–620 miles (700–1,000 km), and they are broken occasionally by hills and volcanic islands such as the Bermuda platform, Cape Verde Islands, and the Azores. The deep abyssal areas in the Pacific Ocean are characterized by the presence of more abundant hills or seamounts, which rise up to 0.6 miles (1 km) above the seafloor. Therefore, the deep abyssal region of the Pacific is generally referred to as the abyssal hills instead of the abyssal plains. Approximately 80–85 percent of the Pacific Ocean floor lies close to areas with hills and seamounts, making the abyssal hills the most common landform on the surface of the Earth.

Many of the sediments on the deep seafloor (the abyssal plain) are derived from erosion of the continents and are carried to the deep sea by turbidity currents, wind (e.g., volcanic ash), or released from floating ice. Other sediments, known as deep-sea oozes, include pelagic sediments derived from marine organic activity. When small organisms die, such as diatoms in the ocean, their shells sink to the bottom and over time can create significant accumulations. Calcareous ooze occurs at low to middle latitudes where warm water favors the growth of carbonate-secreting organisms. Calcareous oozes are not found in water that is more than 2.5–3 miles (4–5 km) deep, because this water is under such high pressure that it contains dissolved CO_2 that dissolves carbonate shells. Siliceous ooze is produced by organisms that use silicon to make their shell structure.

See also CONTINENTAL MARGIN.

accretionary wedge Structurally complex parts of subduction zone systems, accretionary wedges are formed on the landward side of the trench by material scraped off from the subducting plate as well as trench fill sediments. They typically have wedge-shaped cross sections and have one of the most complex internal structures of any tectonic element known on Earth. Parts of accretionary wedges are characterized by numerous thin units of rock layers that are repeated by numerous thrust faults, whereas other parts or other wedges are characterized by relatively large semi-coherent or folded packages of rocks. They also host rocks known as tectonic mélanges that are complex mixtures of blocks and thrust slices of many rock types (such as graywacke, basalt, chert, and limestone) typically encased in a matrix of a different rock type (such as shale or serpentinite). Some accretionary wedges contain small blocks or layers of high-pressure low-temperature metamorphic rocks (known as blueschists) that have formed deep within the wedge where pressures are high and temperatures are low because of the insulating effect of the cold subducting plate. These high-pressure rocks were brought to the surface by structural processes.

Accretionary wedges grow by the progressive offscraping of material from the trench and subducting plate, which constantly pushes new material in front of and under the wedge as plate tectonics drives plate convergence. The type and style of material that is offscraped and incorporated into the wedge depends on the type of material near the surface on the subducting plate. Subducting plates with thin veneers of sediment on their surface yield packages in the accretionary wedge dominated by basalt and chert rock types, whereas subducting plates with thick sequences of graywacke sediments yield packages in the accretionary wedge dominated by graywacke. They may also grow by a process known as underplating, where packages (thrust slices of rock from the subducting plate) are added to the base of the accretionary wedge, a process that typically causes folding of the overlying parts of the wedge. The fronts or toes of accretionary wedges are also characterized by material slumping off of the steep slope of the wedge into the trench. This material may then be recycled back into the accretionary wedge, forming even more complex structures. Together, the processes of offscraping and underplating tend to steepen structures and rock layers from an orientation that is near horizontal at the toe of the wedge to near vertical at the back of the wedge.

The accretionary wedges are thought to behave mechanically somewhat as if they were piles of sand bulldozed in front of a plow. They grow a triangular wedge shape that increases its slope until it becomes oversteepened and mechanically unstable, which will then cause the toe of the wedge to advance by thrusting, or the top of the wedge to

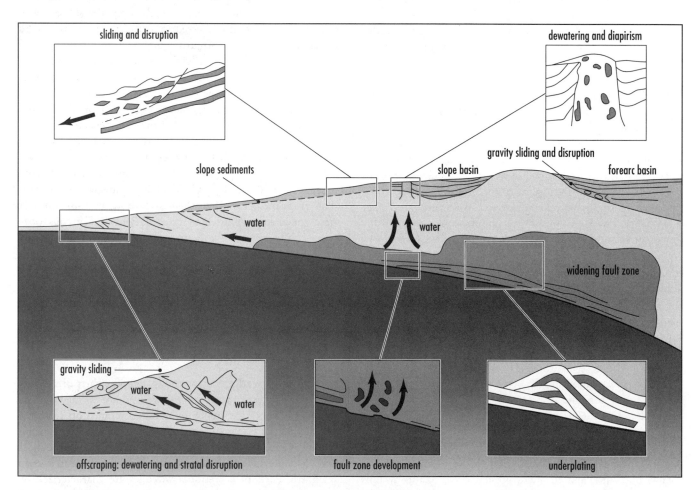

Cross section of typical accretionary wedge showing material being offscraped at the toe of the wedge and underplated beneath the wedge

collapse by normal faulting. Either of these two processes can reduce the slope of the wedge and lead it to become more stable. In addition to finding the evidence for thrust faulting in accretionary wedges, structural geologists have documented many examples of normal faults where the tops of the wedges have collapsed, supporting models of extensional collapse of oversteepened wedges.

Accretionary wedges are forming above nearly every subduction zone on the planet. However, these accretionary wedges presently border open oceans that have not yet closed by plate tectonic processes. Eventually, the movements of the plates and continents will cause the accretionary wedges to become involved in plate collisions that will dramatically change the character of the accretionary wedges. They are typically overprinted by additional shortening, faulting, folding, and high-temperature metamorphism, and intruded by magmas related to arcs and collisions. These later events, coupled with the initial complexity and variety, make identification of accretionary wedges in ancient mountain belts difficult, and prone to uncertainty.

See also CONVERGENT PLATE MARGIN PROCESSES; MÉLANGE; PLATE TECTONICS; STRUCTURAL GEOLOGY.

Further Reading

Kusky, Timothy M., and Dwight C. Bradley. "Kinematics of Mélange Fabrics: Examples and Applications from the McHugh Complex, Kenai Peninsula, Alaska." *Journal of Structural Geology* 21, no. 12 (1999): 1,773–1,796.

Kusky, Timothy M., Dwight C. Bradley, Peter Haeussler, and Susan M. Karl. "Controls on Accretion of Flysch and Mélange Belts at Convergent Margins: Evidence from The Chugach Bay Thrust and Iceworm Mélange, Chugach Terrane, Alaska." *Tectonics* 16, no. 6 (1997): 855–878.

Adirondack Mountains The Adirondack Mountains occupy the core of a domal structure that brings deep-seated Late Proterozoic rocks to the surface and represents a southern extension of the Grenville province of Canada. The Late Proterozoic rocks are unconformably overlain by the Upper Cambrian/Lower Ordovician Potsdam Sandstone, dipping away from the Adirondack dome. The late Cenozoic uplift is shown by the anomalous elevations of the Adirondack Highlands compared with the surrounding regions and the relatively young (Tertiary) drainage patterns. Uplift is still occurring on the order of few millimeters per year.

Five periods of intrusion and two main periods of deformation are recognized in the Adirondacks. The earliest intrusions are the tonalitic and calc-alkaline intrusions that are approximately 1,350–1,250 million years old. These intrusions were followed by the Elzevirian deformation at approximately 1,210–1,160 million years ago. The largest and most significant magmatic event was the emplacement of the anorthosites, mangerites, charnockites, and granites, commonly referred to as the AMCG suite. This suite is thought to have been intruded

about 1,155–1,125 million years ago. This magmatism was followed by two more magmatic events; hornblende granites and leucogranites at approximately 1,100–1,090 million years ago (Hawkeye suite) and 1,070–1,045 million years ago (Lyon Mountain granite), respectively. The most intense metamorphic event was the Ottawan orogeny, which occurred 1,100–1,000 million years ago, with "peak" metamorphism occurring at about 1,050 million years ago.

The Adirondacks are subdivided into two provinces: the Northwest Lowlands and the Highlands, separated by the Carthage-Colton mylonite zone. Each province contains distinct rock types and geologic features, both of which have clear affinities related to the Canadian Grenville province.

The Northwest Lowlands

The Northwest Lowlands are located in the northwest portion of the Adirondack Mountains. On the basis of lithologies, the Lowlands are closely related to the Frontenac terrane of the Canadian metasedimentary belt and are thought to be connected via the Frontenac Arch. The Northwest Lowlands are smaller in area, have lower topographic relief than the Highlands, and are dominated by metasedimentary rocks interlayered with leucocratic gneisses. Both lithologies are metamorphosed to upper amphibolite grade. The metasedimentary rocks are mostly marbles but also contain units of quartzites and mica schists, suggesting a platform sedimentary sequence provenance. The protoliths of the leucocratic gneisses are controversial. Some geologists consider the leucocratic gneisses to be basal rhyolitic and dacitic ash-flow tuff deposits that have been metamorphosed, based on geochemical signatures and the absence of xenoliths in the formations. However, others question this interpretation and suggest that the leucocratic bodies are intrusive in nature, based on crosscutting field evidence and geothermometry. The geothermometry on the leucocratic gneiss yields a temperature of 1,436°F–1,490°F (780°C–810°C). This is an anomalously high metamorphic temperature compared with other rocks in the region, suggesting that they may be igneous crystallization temperatures.

The Highlands

The Highlands are correlative with the central granulite terrain of the Canadian Grenville province. The Green Mountains of Vermont may also be correlative with the Highlands, although other Proterozoic massifs in the northern Appalachians such as the Chain Lakes massif may be exotic to Laurentia. The Highlands are dominated by meta-igneous rocks, including abundant anorthosite bodies. The largest anorthosite intrusion is the Mount Marcy massif located in the east-central Adirondacks; additional anorthosite massifs are the Oregon and Snowy Mountain domes that lie to the south-southwest of Mount Marcy. The anorthosite bodies are part of the suite of rocks known as the AMCG suite; anorthosites, mangerites,

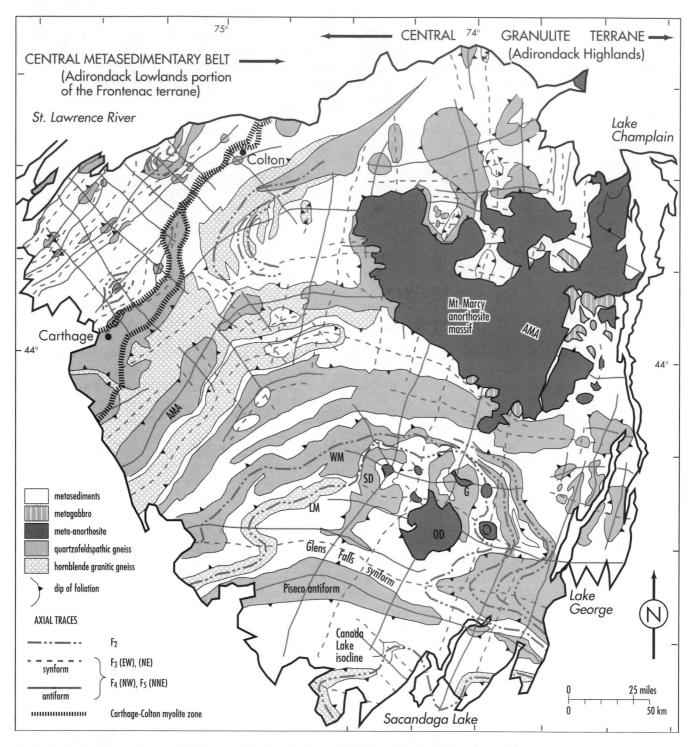

Structural map showing axial traces of folds in the Adirondack Mountains: AMA: Arab Mountain antiform; G: Gore Mountain; LM: Little Moose Mountain synform; OD: Oregon Dome; SD: Snowy Mountain Dome; WM: Wakeley Mountain nappe

charnockites, granitic gneisses. Between the Marcy massif and the Carthage-Colton mylonite zone is an area known as the Central Highlands. Here, the rock types consist of AMCG rocks and hornblende gneisses, both of which exhibit variable amounts of deformation. The Southern Highlands are comprised of granitic gneisses from the AMCG suite with infolded metasedimentary rocks that are strongly deformed. Within the Southeastern Highlands, metasedimentary rocks are found;

these metasedimentary rocks may be correlative with rocks in the Northwest Lowlands. The following sections briefly review the important Highland suites.

TONALITIC SUITE   The tonalitic suite outcrops in the extreme southern Adirondacks where they are highly deformed. These tonalitic rocks are one of the oldest suites in the Adirondacks and have been dated at circa 1.3 billion years. The tonalitic gneiss is thought to be igneous in origin based on the presence of xenoliths from the surrounding rock and the subophitic textures. Strong calc-alkaline trends suggest that these rocks are arc-related; however, this geochemical signature does not differentiate between an island-arc and an Andean arc-type setting. This suite may be correlative with tonalitic rocks in the Green Mountains of Vermont based on age relations and petrographic features. They are also similar in composition with the somewhat younger Elzevirian batholith (1.27–1.23 billion years old) in the central metasedimentary belt. Consequently, the tonalitic suite in the Adirondacks is thought to have been emplaced in the early intraoceanic history of the Elzevirian arc, prior to collision at circa 1,200 million years ago.

AMCG SUITE   The circa 1,555–1,125-million-year-old AMCG suite occurs predominantly in the Adirondack Highlands and central granulite terrain of the Canadian Grenville province. Though highly deformed, the AMCG suite has been characterized as igneous in origin based on the presence of relict igneous textures. Several geologists, pioneered by Jim McLelland, have suggested that the post-collisional delamination of the subcontinental lithospheric mantle generated gabbroic melts that ponded at the mantle-crust boundary. This ponding would have provided a significant source of heat, thereby affecting the lower crust in two ways: it created melts in the lower crust, thus producing a second generation of more felsic magma. This model is supported by the bimodal nature of the AMCG suite. The second effect was weakening of the crust, which provided a conduit for the hot, less dense magmas to ascend to the surface. This hypothetical emplacement model is supported by the AMCG suite's anhydrous nature in conjunction with the shallow crustal levels the magma has invaded.

Large-Scale Structural Features

The structure of the Adirondack Mountains has puzzled geologists for decades. This is due to the polyphase deformation that complexly deformed the region during the Ottawan orogeny (1.1–1.0 billion years ago). In 1936 J. S. Brown was one of the first investigators who recognized that the stratigraphy of the Northwest Lowlands is repeated by a series of folds. Later workers, including Ynvar Isachsen, suggested that there are five sets of large-scale folds that occur throughout the Adirondacks. In addition, rocks of the central and southern Adirondacks are strongly foliated and lineated. The large-scale folds and rock fabrics suggest northwest directed tectonic transport, which is consistent with other kinematic indicators in the rest of the Grenville province.

Even the most generalized geologic maps of the Adirondacks reveal that this region possesses multiple large-scale folds. Delineating the various fold sets is difficult, due to the fold interference patterns, but at least five sets of folds are recognized. The timing of these fold sets has remained obscure, but at least some are related to the Ottawan orogeny. It is also not clear whether these folds formed as a progressive event or as part of distinct events.

Fold nomenclature, i.e., anticline and syncline, is based on structural evidence found in the eastern parts of the Adirondacks. The shapes of igneous plutons and orientation of igneous compositional layering have aided structural geologists to determine fold superposition in this region. The earliest fold set (F_1 folds) are reclined to recumbent folds. Mainly minor, intrafolial F_1 folds have been documented, with rare outcrop-scale examples. The presence of larger F_1 folds is suspected based on rotated foliations associated with F_1 folding in the hinge areas of F_2 folds. Many F_1 folds may have eluded detection because of their extremely large size.

The F_2 folds are the earliest mappable folds in the Adirondacks, an example being the Wakely Mountain nappe. In general the F_2 folds are recumbent to reclined, isoclinal folds. The F_2 folds are coaxial with the F_1 folds and have fold axes that trend northwest to east-west. Both of these fold sets have been suggested to be associated with thrust nappes.

The F_3 folds are large, upright-open folds that trend west-northwest to east-west. Therefore, they are considered coaxial with F_1 and F_2 folds. F_3 folds are best developed in the south-central Adirondack Highlands. Examples of these folds are the Piseco anticline and the Glens Falls syncline. Northwest trending F_4 folds are best developed in the Northwest Lowlands and are rare in the Highlands, except in the southern regions. North-northeast trending F_5 folds are open, upright folds except near Mount Marcy where they become tight. F_5 folds are better developed in the eastern parts of the Adirondacks. Due to the spatial separation of F_4 and F_5 folds, distinguishing relative timing between the two is difficult.

See also GRENVILLE PROVINCE; PROTEROZOIC; STRUCTURAL GEOLOGY; SUPERCONTINENT CYCLE.

Further Reading

Brown, John S. "Structure and Primary Mineralization of the Zinc Mine at Balmat, New York." *Economic Geology* 31, no. 3 (1936): 233–258.

Buddington, Arthur F. "Adirondacks Igneous Rocks and Their Metamorphism." *Geological Society of America Memoir* 7 (1939): 1–354.

Chiarenzelli, Jeffrey R., and Jim M. McLelland. "Age and Regional Relationships of Granitoid Rocks of the Adirondack Highlands." *Journal of Geology* 99 (1991): 571–590.

Corrigan, Dave, and Simon Hanmer. "Anorthosites and Related Granitoids in the Grenville Orogen: A Product of the Convective Thinning of the Lithosphere?" *Geology* 25 (1997): 61–64.

Davidson, Anthony. "Post-collisional A-type Plutonism, Southwest Grenville province: Evidence for a Compressional Setting." *Geological Society of America Abstracts with Programs* 28 (1996): 440.

———. "An Overview of Grenville province Geology, Canadian Shield." In "Geology of the Precambrian Superior and Grenville provinces and Precambrian Fossils in North America," edited by S. B. Lucas and Marc R. St-Onge. *Geological Society of America, Geology of North America* C-1 (1998): 205–270.

Hoffman, Paul F. "Did the Breakout of Laurentia Turn Gondwanaland Inside-Out?" *Science* 252 (1991): 1,409–1,411.

Kusky, Timothy M., and Dave P. Loring. "Structural and U/Pb Chronology of Superimposed Folds, Adirondack Mountains: Implications for the Tectonic Evolution of the Grenville province." *Journal of Geodynamics* 32 (2001): 395–418.

McLelland, Jim M., J. Stephen Daly, and Jonathan M. McLelland. "The Grenville Orogenic Cycle (ca. 1350–1000 Ma): an Adirondack perspective." *In Tectonic Setting and Terrane Accretion in Precambrian Orogens*, edited by Timothy M. Kusky, Ben A. van der Pluijm, Kent C. Condie, and Peter J. Coney. *Tectonophysics* 265 (1996): 1–28.

McLelland, Jim M., and Ynvar W. Isachsen. "Synthesis of Geology of the Adirondack Mountains, New York, And Their Tectonic Setting within the Southwestern Grenville province." In *The Grenville province*, edited by J. M. Moore, A. Davidson, and Alec J. Baer. *Geological Association of Canada Special Paper* 31 (1986): 75–94.

———. "Structural Synthesis of the Southern and Central Adirondacks: A Model for the Adirondacks as a Whole and Plate Tectonics Interpretations." *Geological Society of America Bulletin* 91 (1980): 208–292.

Moores, Eldredge M. "Southwest United States-East Antarctic (SWEAT) Connection: A Hypothesis." *Geology* 19 (1991): 425–428.

Rivers, Toby. "Lithotectonic Elements of the Grenville province: Review and Tectonic Implications." *Precambrian Research* 86 (1997): 117–154.

Rivers, Toby, and Dave Corrigan. "Convergent Margin on Southeastern Laurentia during the Mesoproterozoic: Tectonic Implications." *Canadian Journal of Earth Sciences* 37 (2000): 359–383.

Rivers, Toby, J. Martipole, Charles F. Gower, and Anthony Davidson. "New Tectonic Subdivisions of the Grenville province, Southeast Canadian Shield." *Tectonics* 8 (1989): 63–84.

Afar Depression, Ethiopia One of the world's largest, deepest regions below sea level that is subaerially exposed on the continents, home to some of the earliest known hominid fossils. It is a hot, arid region, where the Awash River drains northward out of the East African rift system, and is evaporated in Lake Abhe before it reaches the sea. It is located in eastern Africa in Ethiopia and Eritrea, between Sudan and Somalia, and across the Red Sea and Gulf of Aden from Yemen. The reason the region is so topographically low is that it is located at a tectonic triple junction, where three main plates are spreading apart, causing regional subsidence. The Arabian plate is moving northeast away from the African plate, and the Somali plate is moving, at a much slower rate, to the southeast away from Africa. The southern Red Sea and north-central Afar Depression form two parallel north-northwest-trending rift basins, separated by the Danakil Horst, related to the separation of Arabia from Africa. Of the two rifts, the Afar depression is exposed at the surface, whereas the Red Sea rift floor is submerged below the sea. The north-central Afar rift is complex, consisting of many grabens and horsts. The Afar Depression merges southward with the northeast-striking Main Ethiopian Rift, and eastward with the east-northeast-striking Gulf of Aden. The Ethiopian Plateau bounds it on the west. Pliocene volcanic rocks of the Afar stratoid series and the Pleistocene to Recent volcanics of the Axial Ranges occupy the floor of the Afar Depression. Miocene to recent detrital and chemical sediments are intercalated with the volcanics in the basins.

The Main Ethiopian and North-Central Afar rifts are part of the continental East African Rift System. These two kinematically distinct rift systems, typical of intracontinental rifting, are at different stages of evolution. In the north and east, the continental rifts meet the oceanic rifts of the Red Sea and the Gulf of Aden, respectively, both of which have propagated into the continent. Seismic refraction and gravity studies indicate that the thickness of the crust in the Main Ethiopian Rift is less than or equal to 18.5 miles (30 km). In Afar the thickness varies from 14 to 16 miles (23–26 km) in the south to 8.5 miles (14 km) in the north. The plateau on both sides of the rift has a crustal thickness of 21.5–27 miles (35–44 km). Rates of separation obtained from geologic and geodetic studies indicate 0.1–0.2 inches (3–6 mm) per year across the northern sector of the Main Ethiopian Rift between the African and Somali plates. The rate of spreading between Africa and Arabia across the North-Central Afar rift is relatively faster, about 0.8 inches (20 mm) per year. Paleomagnetic directions from Cenozoic basalts on the Arabian side of the Gulf of Aden indicate 7 degrees of counterclockwise rotation of the Arabian plate relative to Africa, and clockwise rotations of up to 11 degrees for blocks in eastern Afar. The initiation of extension on both sides of the southernmost Red Sea Rift, Ethiopia, and Yemen appear coeval, with extension starting between 22 million and 29 million years ago.

See also DIVERGENT OR EXTENSIONAL BOUNDARIES; RIFT.

Further Reading

Tesfaye, Sansom, Dave Harding, and Timothy Kusky. "Early Continental Breakup Boundary and Migration of the Afar Triple Junction, Ethiopia." *Geological Society of America Bulletin* 115 (2003): 1,053–1,067.

agate An ornamental, translucent variety of quartz, known for its spectacular colors and patterns. It is extremely fine-grained (or cryptocrystalline), and mixed with layers of opal, which is another variety of colored silica that has combined

Landsat Thematic Mapper image of the area where the Ethiopian rift segment of the East African rift meets the Tendaho rift, an extension of the Red Sea rift, and the Goba'ad rift, an extension of the Gulf of Aden rift system. Note the dramatic change in orientation of fault-controlled ridges and how internal drainages like the Awash River terminate in lakes such as Lake Abhe, where the water evaporates.

with variable amounts of water molecules. Opal is typically iridescent, displaying changes in color when viewed in different light or from different angles. Agate and opal typically form colorful patterns including bands, clouds, or moss-like dendritic patterns indicating that they grew together from silica-rich fluids. Agate is found in vugs in volcanic rocks and is commonly sold at rock and mineral shows as polished slabs of ornamental stone.

See also MINERALOGY.

air pressure   The weight of the air above a given level. This weight produces a force in all directions caused by constantly moving air molecules bumping into each other and objects in the atmosphere. The air molecules in the atmosphere are constantly moving and bumping into each other with each air molecule averaging a remarkable 10 billion collisions per second with other air molecules near the Earth's surface. The density of air molecules is highest near the surface, decreases rapidly upward in the lower 62 miles (100 km) of the atmosphere, then decreases slowly upward to above 310 miles (500 km). Air molecules are pulled toward the Earth by gravity and are therefore more abundant closer to the surface. Pressure, including air pressure, is measured as the force divided by the area over which it acts. The air pressure is greatest near the Earth's surface and decreases with height, because there is a greater number of air molecules near the Earth's surface (the air pressure represents the sum of the total mass of air above a certain point). A one-square-inch column of air extending from sea level to the top of the atmosphere weighs about 14.7 pounds. The typical air pressure at sea level is therefore 14.7 pounds per square inch. It is commonly measured in units of millibars (mb) or hectopascals (hPa), and also in inches of mercury. Standard air pressure in these units equals 1,013.25 mb, 1,013.25 hPa, and 29.92 in of mercury. Air pressure is equal in all directions, unlike some pressures (such as a weight on one's head) that act in one direction. This explains why objects and people are not crushed or deformed by the pressure of the overlying atmosphere.

Air pressure also changes in response to temperature and density, as expressed by the gas law:

Pressure = temperature × density × constant (gas constant, equal to 2.87×10^6 erg/g K).

From this gas law, it is apparent that at the same temperature, air at a higher pressure is denser than air at a lower pressure. Therefore, high-pressure regions of the atmosphere are characterized by denser air, with more molecules of air than areas of low pressure. These pressure changes are caused by wind that moves air molecules into and out of a region. When more air molecules move into an area than move out, the area is called an area of net convergence. Conversely, in areas of low pressure, more air molecules are moving out than in, and the area is one of divergence. If the air density is constant and the temperature changes, the gas law states that at a given atmospheric level, as the temperature increases, the air pressure decreases. Using these relationships, if either the temperature or pressure is known, the other can be calculated.

If the air above a location is heated, it will expand and rise; if air is cooled, it will contract, become denser, and sink closer to the surface. Therefore, the air pressure decreases rapidly with height in the cold column of air because the molecules are packed closely to the surface. In the warm column of air, the air pressure will be higher at any height than in the cold column of air, because the air has expanded and more of the original air molecules are above the specific height than in the cold column. Therefore, warm air masses at height are generally associated with high-pressure systems, whereas cold air aloft is generally associated with low pressure. Heating and cooling of air above a location causes the air pressure to change in that location, causing lateral variation in air pressure across a region. Air will flow from high-pressure areas to low-pressure areas, forming winds.

The daily heating and cooling of air masses by the Sun can in some situations cause the opposite effect, if not overwhelmed by effects of the heating and cooling of the upper atmosphere. Over large continental areas, such as the southwestern United States, the daily heating and cooling cycle is associated with air pressure fall and rise, as expected from the gas law. As the temperature rises in these locations the pressure decreases, then increases again in the night when the temperature falls. Air must flow in and out of a given vertical column on a diurnal basis for these pressure changes to occur, as opposed to having the column rise and fall in response to the temperature changes.

See also ATMOSPHERE.

Aleutian Islands and trench Stretching 1,243 miles (2,000 km) west from the western tip of the Alaskan Peninsula, the Aleutian Islands form a rugged chain of volcanic islands that stretch to the Komandorski Islands near the Kamchatka Peninsula of Russia. The islands form an island arc system above the Pacific plate, which is subducted in the Aleutian trench, a 5-mile (8-km) deep trough ocean-ward of the Aleutian Islands. They are one of the most volcanically active island chains in the world, typically hosting several eruptions per year.

The Aleutians consist of several main island groups, including the Fox Islands closest to the Alaskan mainland, then moving out toward the Bering Sea and Kamchatka to the Andreanof Islands, the Rat Islands, and the Near Islands. The climate of the Aleutians is characterized by nearly constant fog and heavy rains, but generally moderate temperatures. Snow may fall in heavy quantities in the winter months. The islands are almost treeless but have thick grasses, bushes, and sedges, and are inhabited by deer and sheep. The local Inuit population subsists on fishing and hunting.

The first westerner to discover the Aleutians was the Danish explorer Vitus Bering, when employed by Russia in 1741. Russian trappers and traders established settlements on the islands and employed local Inuit to hunt otters, seals, and fox. The Aleutians were purchased by the United States along with the rest of Alaska from Russia in 1867. The only good harbor in the Aleutian is at Dutch Harbor, used as a transshipping port, a gold boomtown, and as a World War II naval base.

See also PLATE TECTONICS.

alluvial fans Fan- or cone-shaped deposits of fluvial gravel, sand, and other material radiating away from a single point source on a mountainside. They represent erosional-depositional systems in which rock material is eroded from mountains and carried by rivers to the foot of the mountains, where it is deposited in the alluvial fans. The apex of an alluvial fan is the point source from which the river system emerges from the mountains and typically breaks into several smaller distributaries forming a braided stream network that frequently shifts in position on the fan, evenly distributing alluvial gravels across the fan with time. The shape of alluvial fans depends on many factors, including tectonic uplift and subsidence, and climatic influences that change the relative river load-discharge balance. If the discharge decreases with time, the river may downcut through part of the fan and emerge partway through the fan surface as a point-source for a new cone. This type of morphology also develops in places where the basin is being uplifted relative to the mountains. In places where the mountains are being uplifted relative to the basin containing the fan, the alluvial fan typically displays several, progressively steeper surfaces toward the fan apex. In many places, several alluvial fans merge together at the foot of a mountain and form a continuous depositional surface known as a bajada, alluvial apron, or alluvial slope.

The surface slope of alluvial fans may be as steep as 10° near the fan apex and typically decreases in the down-fan direction toward the toe of the fan. Most fans have a concave upward profile. The slope of the fan at the apex is typically the same as that of the river emerging from the mountains, showing that deposition on the fans is not controlled by a sudden decrease in gradient along the river profile.

Alluvial fans that form at the outlets of large drainage basins are larger than alluvial fans that form at the outlets of smaller drainage basins. The exact relationships between fan size and drainage basin size is dependent on time, climate, type of rocks in the source terrain in the drainage basin, structure, slope, tectonic setting, and the space available for the fan to grow into.

Alluvial fans are common sights along mountain fronts in arid environments but also form in all other types of climatic conditions. Flow on the fans is typically confined to a single or a few active channels on one part of the fan, and shifts to other parts of the fan in flood events in humid environments or in response to the rare flow events in arid environments. Deposition on the fans is initiated when the flow leaves the confines of the channel, and the flow velocity and depth decrease dramatically. Deposition on the fans may also be induced by water seeping into the porous gravel and sand on the fan surface, which has the effect of decreasing the flow discharge, initiating deposition. In arid environments it is common for the entire flow to seep into the porous fan before it reaches the toe of the fan.

The sedimentary deposits on alluvial fans include fluvial gravels, sands, and overbank muds, as well as debris flow and mudflow deposits on many fans. The debris flows are

Alluvial fan, Death Valley, California. Recent channels are light-colored, whereas older surfaces are coated with a dark desert varnish. *(Photo by Timothy Kusky)*

characterized by large boulders embedded in a fine-grained, typically mud-dominated matrix. These deposits shift laterally across the fan, although the debris and mudflow deposits tend to be confined to channels. The fan surface may exhibit a microtopography related to the different sedimentary facies and deposit types.

The development of fan morphology, the slope, relative aggradation versus downcutting of channels, and the growth or retreat of the toe and apex of the fan are complex phenomena dependent on a number of variables. Foremost among these are the climate, the relative uplift and subsidence of the mountains and valleys, base level in the valleys, and the sediment supply.

See also DESERT; DRAINAGE BASIN; GEOMORPHOLOGY.

Further Reading

Bull, William B. "Alluvial Fans." *Journal of Geologic Education* 16 (1968): 101–106.
Ritter, Dale F., R. Craig Kochel, and Jerry R. Miller. *Process Geomorphology,* 3rd ed. Boston: WCB-McGraw Hill, 1995.

Alps An arcuate mountain system of south central Europe, about 497 miles (800 km) long and 93 miles (150 km) wide, stretching from the French Riviera on the Mediterranean coast, through southeastern France, Switzerland, southwestern Germany, Austria, and Yugoslavia (Serbia). The snow line in the Alps is approximately 8,038 feet (2,450 m), with many peaks above this being permanently snowcapped or hosting glaciers. The longest glacier in the Alps is the Aletsch, but many landforms attest to a greater extent of glaciation in the Pleistocene. These include famous landforms such as the Matterhorn and other horns, aretes, U-shaped valleys, erratics, and moraines.

The Alps were formed by plate collisions related to the closure of the Tethys Ocean in the Oligocene and Miocene, but the rocks record a longer history of deformation and events extending back at least into the Mesozoic. Closure of the Tethys Ocean was complex, involving contraction of the older Permian-Triassic Paleo-Tethys Ocean at the same time that a younger arm of the ocean, the Neo-Tethys, was opening in Triassic and younger times. In the late Triassic, carbonate platforms covered older evaporites, and these platforms began foundering and were buried under deepwater pelagic shales and cherts in the early Jurassic. Cretaceous flysch covered convergent margin foreland basins, along with felsic magmatism and high-grade blueschist facies metamorphism. Continent-continent collision-related events dominate the Eocene-Oligocene, with the formation of giant nappes, thrusts, and deposition of syn-orogenic flysch. Late Tertiary events are dominated by late orogenic uplift, erosion, and deposition of post-orogenic molasse in foreland basins. Deformation continues, mostly related to post-collisional extension.

See also CONVERGENT PLATE MARGIN PROCESSES; PLATE TECTONICS; STRUCTURAL GEOLOGY.

altimeter   An instrument, typically an aneroid barometer, that is used for determining the elevation or height above sea level. Aneroid barometer style altimeters operate by precisely measuring the change in atmospheric pressure, that decreases with increasing height above sea level, since there is less air exerting pressure at a point at higher elevations than at lower elevations. Altimeters need to be calibrated each day at a known elevation, to account for weather-related changes in atmospheric pressure.

Before 1928 there was no possible way for pilots to know how far above the ground they were. The German inventor Paul Kollsman invented the first reliable and accurate barometric altimeter. The altimeter measured altitude by barometric pressure. Pilots still use the barometric altimeter today.

In 1924 Lloyd Espenschied invented the first radio altimeter. In 1938 Bell Labs demonstrated the first radio altimeter. A radio altimeter uses radio signals that bounce off of the ground and back to the receiver in the plane showing pilots the altitude of the aircraft. A radar altimeter works much in the same way except it bounces the signal off of an object in the air thus telling the height of the object above the ground. A laser altimeter can measure the distance from a spacecraft or satellite to a fixed position on Earth. The measurement when compiled with radial orbit knowledge can provide the topography of the Earth.

Altiplano   A large, uplifted plateau in the Bolivian and Peruvian Andes of South America. The plateau has an area of about 65,536 square miles (170,000 km²), and an average elevation of 12,000 feet (3,660 m) above sea level. The Altiplano is a sedimentary basin caught between the mountain ranges of the Cordillera Oriental on the east and the Cordillera Occidental on the west. Lake Titicaca, the largest high-altitude lake in the world, is located at the northern end of the Altiplano.

The Altiplano is a dry region with sparse vegetation, and scattered salt flats. Villagers grow potatoes and grains, and a variety of minerals are extracted from the plateau and surrounding mountain ranges.

See also ANDES.

Amazon River   The world's second longest river, stretching 3,900 miles (6,275 km) from the foothills of the Andes to the Atlantic Ocean. The Amazon begins where the Ucayali and Maranon tributaries merge and drains into the Atlantic near the city of Belem. The Amazon carries the most water and has the largest discharge of any river in the world, averaging 150 feet (45 m) deep. Its drainage basin amounts to about 35 percent of South America, covering 2,500,000 square miles (6,475,000 km²). The Amazon lowlands in Brazil include the largest tropical rainforest in the world. In this region, the Amazon is a muddy, silt-rich river with many channels that wind around numerous islands in a complex maze. The delta region of the Amazon is marked by numerous fluvial islands and distributaries, as the muddy waters of the river get dispersed by strong currents and waves into the Atlantic. A strong tidal bore, up to 12 feet (3.7 m) high runs up to 500 miles (800 km) upstream.

The Amazon River basin occupies a sediment-filled rift basin, between the Precambrian crystalline basement of the Brazil and Guiana Shields. The area hosts economic deposits of gold, manganese, and other metals in the highlands, and detrital gold in lower elevations. Much of the region's economy relies on the lumber industry, with timber, rubber, vegetable oils, Brazil nuts, and medicinal plants sold worldwide.

Spanish commander Vincent Pinzon was probably the first European in 1500 to explore the lower part of the river basin, followed by the Spanish explorer Franciso de Orellana in 1540–41. De Orellana's tales of tall strong female warriors gave the river its name, borrowing from Greek mythology. Further exploration by Pedro Teixeira, Charles Darwin, and Louis Agassiz led to greater understanding of the river's course, peoples, and environment, and settlements did not appear until steamship service began in the middle 1800s.

amber   A yellow or yellowish brown translucent fossil plant resin derived from coniferous trees. It is not a mineral but an organic compound that often encases fossil insects, pollen, and other objects. It is capable of taking on a fine polish and is therefore widely used as an ornamental jewelry piece and is also used for making beads, pipe mouthpieces, or bookshelf oddities. Amber is found in many places, including soils, clays, and lignite beds. It is well known from locations including the shores of the Baltic Sea and parts of the Dominican Republic. Amber contains high concentrations of succinic acid (a crystalline dicarboxylic acid, with the formula HOOCCH$_2$CH$_2$COOH), and has highly variable C:H:O ratios. Amber of Oligocene age seems particularly abundant, although it is known from as old as the Cretaceous and includes all ages since sap-producing trees have proliferated on Earth.

Many species of fossil insects and plants have been identified in amber, particularly from the spectacular amber deposits found along the southeastern shores of the Baltic Sea. There, yellow, brown, orange, and even blue amber is rich in contained fossils, though most of the amber was mined by the end of Roman times. Amber has retained a sort of mystical quality since early times, probably because it has some unusual properties. Amber stays warm whereas minerals often feel cool to the touch, and amber burns giving off a scent of pine sap (from which it is derived). Even more astounding to early people was that when rubbed against wool or silk, amber becomes electrically charged and gives off sparks. This feature led the early Greeks to call amber "electron." Many theories were advanced for the origin of amber, ranging from tears of gods to solidified sunshine. The origin of amber was first appreciated by Pliny the Elder, who, in his famous work *Historia Naturalis* (published in C.E. 77), suggested that amber is derived from plants.

The Romans mined the amber deposits of the Baltic Sea because they thought amber had medicinal qualities that enabled it to ward off fever, tonsillitis, ear infections, and poor eyesight.

Decorative amber has been used for burial rituals and to ward off evil spirits for thousands of years, and in Europe it has been found in graves as old as 10,000 years. Amber was widely transported on the ancient silk roads and in ancient Europe, where figurines, beads, and other decorative items were among the most valuable items in the markets.

Further Reading

Zahl, P. A. "Golden Window on the Past." *National Geographic* 152, no. 3 (1977): 423–435.

American Geological Institute (AGI)

The American Geological Institute (http://www.agiweb.org) was founded in 1948. It plays a major role in strengthening geoscience education and increasing public awareness of the vital role that geosciences play in society. AGI supports its programs and initiatives through sales of its publications and services, royalties, contracts, grants, contributions, and affiliated society dues. AGI's staff provides professional and informational services related to government affairs; earth-science education, outreach, human resources, and scholarships; the bibliographic database GeoRef and its Document Delivery Service; and the monthly newsmagazine *Geotimes* and other publications. The Member Society Council meets twice a year in conjunction with the annual meetings of the American Association of Petroleum Geologists and the Geological Society of America.

American Geophysical Union (AGU)

AGU (http://www.agu.org), a nonprofit scientific organization, was established in 1919 by the National Research Council. AGU is supplying an organizational framework within which geophysicists have created the programs and products needed to advance their science. AGU now stands as a leader in the increasingly interdisciplinary global endeavor that encompasses the geophysical sciences.

AGU's activities are focused on the organization and dissemination of scientific information in the interdisciplinary and international field of geophysics. The geophysical sciences involve four fundamental areas: atmospheric and ocean sciences; solid-Earth sciences; hydrologic sciences; and space sciences. AGU has a broad range of publications and meetings and educational and other activities that support research in the Earth and space sciences.

American Meteorological Association (AMS)

The American Meteorological Society (http://www.ametsoc.org) was founded in 1919. The society's initial publication, the *Bulletin of the American Meteorological Society*, serves as a supplement to the *Monthly Weather Review*, which was initially published by the U.S. Weather Bureau. The role of the AMS is serving the atmospheric and related sciences. The AMS now publishes in print and online nine well-respected scientific journals and an abstract journal. The AMS administers two professional certification programs, the Radio and Television Seal of Approval and the Certified Consulting Meteorologist (CCM) programs, and also offers an array of undergraduate scholarships and graduate fellowships to support students pursuing careers in the atmospheric and related oceanic and hydrologic sciences.

amphibole

A group of dark-colored ferromagnesian silicate minerals with the general chemical formula:

$$A_{2-3} B_5 (Si, Al)_8 O_{22}(OH)_2$$

where A = Mg, Fe^{+2}, Ca, or Na, and B = Mg, Fe^{+3}, Fe^{+2}or Al. Amphiboles contain continuous double chains of cross-linked double silicate tetrahedra. The chains are bound together by cations such as Ca, Mg, and Fe, which satisfy the negative charges of the polymerized tetrahedra. Most amphiboles are monoclinic, but some crystallize in the orthorhombic crystal system. They have good prismatic cleavage intersecting at 56° and 124° and typically form columnar or fibrous prismatic crystals. Amphiboles are very common constituents of metamorphic and igneous rocks and have a chemical composition similar to pyroxenes. Some of the common amphibole minerals include hornblende, tremolite, actinolite, anthophyllite, cummingtonite, riebeckite, and glaucophane.

Amphibole is a fairly common mineral in intermediate to mafic igneous rocks such as granodiorite, diorite, and gabbro, forming up to 25 percent of these rocks in some cases. Since amphibole is a hydrous mineral, it typically forms in igneous environments where water is available. Amphibole is best known, however, as a metamorphic mineral indicative of medium grade pressure-temperature metamorphism of mafic rocks. When basalt, gabbro, or similar rocks are heated to 930°F–1,300°F (500°C–700°C) at 3–10 kilobars pressure (equivalent to 6 to 20-mile or 10 to 30-km depth), the primary mineral assemblage will commonly turn to an assemblage of amphibole+plagioclase feldspar. Many field geologists will call such a rock an "amphibolite," although this term should be reserved for a description of the metamorphic conditions (known as facies) at which these rocks formed.

See also MINERALOGY.

Andes   A 5,000-mile (8,000-km) long mountain range in western South America, running generally parallel to the coast, between the Caribbean coast of Venezuela in the north and Tierra del Fuego in the south. The mountains merge with ranges in Central America and the West Indies in the north, and with ranges in the Falklands and Antarctica in the south. Many snow-covered peaks rise more than 22,000 feet (6,000 m), making the Andes the second largest mountain belt in the world, after the Himalayan chain. The highest range in the Andes is the Aconcagua on the central and northern Argentine-Chile border. The high cold Atacama desert is located in the northern Chile sub-Andean range, and the high Altiplano Plateau is situated along the great bend in the Andes in Bolivia and Peru.

The southern part of South America consists of a series of different terranes added to the margin of Gondwana in the late Proterozoic and early Proterozoic. Subduction and the accretion of oceanic terranes continued through the Paleozoic, forming a 155-mile (250-km) wide accretionary wedge. The Andes formed as a continental margin volcanic arc system on the older accreted terranes, formed above a complex system of subducting plates from the Pacific Ocean. They are geologically young, having been uplifted mainly in the Cretaceous and Tertiary, with active volcanism, uplift, and earthquakes. The specific nature of volcanism, plutonism, earthquakes, and uplift is found to be strongly segmented in the Andes, and related to the nature of the subducting part of the plate, including its dip and age. Regions above places where the subducting plate dips more than 30 degrees have active volcanism, whereas regions above places where the subduction zone is sub-horizontal do not have active volcanoes.

See also CONVERGENT PLATE MARGIN PROCESSES; PLATE TECTONICS.

andesite   A fine-grained, dark-colored intermediate volcanic rock, andesite typically has phenocrysts of zoned sodic plagioclase, and biotite, hornblende, or pyroxene. It has 56–63 percent silica, although basaltic andesites with silica contents down to 52 percent have a composition that is transitional with basalts. Andesite is the extrusive equivalent of

diorite and is characteristic of volcanic belts formed above subduction zones that dip under continents. The name was coined by Buch (1826) for rocks in the Andes Mountains of South America.

Andesite is generally associated with continental margin or Andean-type magmatic arcs built on continental crust above subduction zones. Their composition is thought to reflect a combination of processes from the melting of the mantle wedge above the subducting plate, plus some contamination of the magmas by partial melting of the continental crust beneath the arc.

The average composition of the continental crust is approximately andesitic to dacitic. Many models for the formation and growth of continents therefore invoke the formation of andesitic to dacitic magmas at convergent margins, with the andesitic arcs colliding to form larger continental masses. This is known as the andesite model of continental growth.

See also CONVERGENT PLATE MARGIN PROCESSES; PLATE TECTONICS; VOLCANO.

anticline   Folds in rocks in which a convex upward warp contains older rocks in the center and younger rocks on the sides. They typically occur along with synclines in alternating anticline-syncline pairs forming a fold train. Their geometry is defined by several artificial geometric surfaces, known as the fold axial surface, which divides the fold into two equal limbs, and a fold hinge, parallel to the line of maximum curvature on the folded layers.

The anticlines may be of any size, ranging from microscopic folds of thin layers to large mountain-scale uplifts. Regional parts of mountain ranges that are characterized by generally uplifted rocks in the center are known as anticlinoria.

Anticlines and broad upwarps of strata make particularly good oil and gas traps if the geologic setting is appropriate

Anticline in the Canadian Rockies, near McConnel *(Photo by Timothy Kusky)*

for the formation of oil. Oil and gas tend to migrate upward in geologic structures, and if they find a layer with significant porosity and permeability, the oil and gas may become trapped in the anticlinal structure. The broader and gentler the upwarp, the larger the area that the hydrocarbons may be trapped in, and the larger the oil or gas field. Some famous oil and gas fields that are located in anticlinal structures include those of the Newport-Inglewood trend in California and the Zagros Mountains of Iran.

See also FOLD; STRUCTURAL GEOLOGY.

Appalachians A mountain belt that extends for 1,600 miles (1,000 km) along the east coast of North America, stretching from the St. Lawrence Valley in Quebec, Canada, to Alabama. Many classifications consider the Appalachians to continue through Newfoundland in maritime Canada, and before the Atlantic Ocean opened, the Appalachians were continuous with the Caledonides of Europe. The Appalachians are one of the best-studied mountain ranges in the world, and understanding of their evolution was one of the factors that led to the development and refinement of the paradigm of plate tectonics in the early 1970s.

Rocks that form the Appalachians include those that were deposited on or adjacent to North America and thrust upon the continent during several orogenic events. For the length of the Appalachians, the older continental crust consists of Grenville Province gneisses, deformed and metamorphosed about 1.0 billion years ago during the Grenville orogeny. The Appalachians grew in several stages. After Late Precambrian rifting, the Iapetus Ocean evolved and hosted island arc growth, while a passive margin sequence was deposited on the North American rifted margin in Cambrian-Ordovician times. In the Middle Ordovician, the collision of an island arc terrane with North America marks the Taconic orogeny, followed by the Mid-Devonian Acadian orogeny, which probably represents the collision of North America with Avalonia, off the coast of Gondwana. This orogeny formed huge molassic fan delta complexes of the Catskill Mountains and was followed by strike-slip faulting. The Late Paleozoic Alleghenian orogeny formed striking folds and faults in the southern Appalachians but was dominated by strike-slip faulting in the northern Appalachians. This event appears to be related to the rotation of Africa to close the remaining part of the open ocean in the southern Appalachians. Late Triassic-Jurassic rifting reopened the Appalachians, forming the present Atlantic Ocean.

The history of the Appalachians begins with rifting of the one-billion-year-old Grenville gneisses and the formation of an ocean basin known as Iapetus approximately 800–570 million years ago. Rifting was accompanied by the formation of normal-fault systems and grabens and by the intrusion of swarms of mafic dikes exposed in places in the Appalachians such as in the Long Range dike swarm on Newfoundland's Long Range Peninsula. Rifting was also accompanied by the deposition of sediments, first in rift basins, and then as a Cambrian transgressive sequence that prograded onto the North American craton. This unit is generally known as the Potsdam Sandstone and is well-exposed around the Adirondack dome in northern New York State. Basal parts of the Potsdam sandstone typically consist of a quartz pebble conglomerate and a clean quartzite.

Overlying the basal Cambrian transgressive sandstone is a Cambrian-Ordovician sequence of carbonate rocks deposited on a stable carbonate platform or passive margin, known in the northern Appalachians as the Beekmantown Group. Deposition on the passive margin was abruptly terminated in the Middle Ordovician when the carbonate platform was progressively uplifted above sea level from the east, then migrated to the west, and then suddenly dropped down to water depths too great to continue production of carbonates. In this period, black shales of the Trenton and Black River Groups were deposited, first in the east and then in the west. During this time, a system of normal faults also migrated across the continental margin, active first in the east and then in the west. The next event in the history of the continental margin is deposition of coarser-grained clastic rocks of the Austin Glen and correlative formations, as a migrating clastic wedge, with older rocks in the east and younger ones in the west. Together, these diachronous events represent the first stages of the Taconic orogeny, and they represent a response to the emplacement of the Taconic allochthons on the North American continental margin during Middle Ordovician arc-continent collision.

The Taconic allochthons are a group of Cambrian through Middle Ordovician slates resting allochthonously on the Cambro-Ordovician carbonate platform. These allochthons are very different from the underlying rocks, implying that there have been substantial displacements on the thrust faults beneath the allochthons, probably on the order of 100 miles (160 km). The allochthons structurally overlie wild flysch breccias that are basically submarine slide breccias and mudflows derived from the allochthons.

Eastern sections of the Taconic aged rocks in the Appalachians are more strongly deformed than those in the west. East of the Taconic foreland fold-thrust belts, a chain of uplifted basement with Grenville ages (about one billion years) extends discontinuously from Newfoundland to the Blue Ridge Mountains and includes the Green Mountains of Vermont. These rocks generally mark the edge of the hinterland of the orogen, and the transition into greenschist and higher metamorphic facies. Some of these uplifted basement gneisses are very strongly deformed and metamorphosed, and they contain domal structures known as gneiss domes, with gneisses at the core and strongly deformed and metamorphosed Cambro-Ordovician marbles around their rims. These rocks were deformed at great depths.

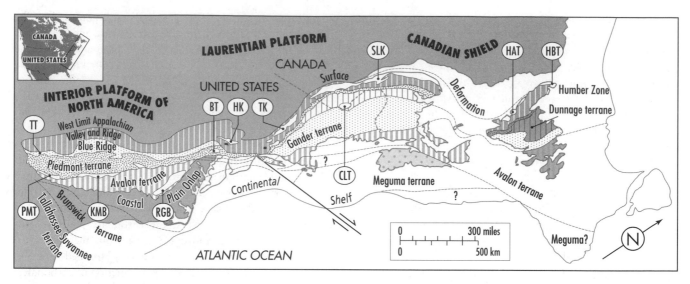

Tectonic map of the Appalachian Mountains showing the distribution of major lithotectonic terranes. Abbreviations as follows: HBT: Hare Bay terrane; HAT: Humber Arm terrane; CLT: Chain Lakes terrane; SLK: St. Lawrence klippe; TK: Taconic klippe; HK: Hamburg klippe; BT: Brunswick terrane; RGB: Raleigh-Goochland belt; KMB: Kings Mountain belt; TT: Talladega terrane; PMT: Pine Mountain terrane (belt)

Also close to the western edge of the orogen is a discontinuous belt of mafic and ultramafic rocks comprising an ophiolite suite, interpreted to be remnants of the ocean floor of the Iapetus Ocean that closed during the Taconic orogeny. Spectacular examples of these ophiolites occur in Newfoundland, including the Bay of Islands ophiolite complex along Newfoundland's western shores.

Further east in the Taconic orogen are rocks of the Bronson Hill anticlinorium or terrane, which are strongly deformed and metamorphosed and have been affected by both the Taconic and Acadian orogenies. These rocks have proven very difficult to map and have been of controversial significance for more than a century. Perhaps the best interpretation is that they represent rocks of the Taconic island arc that collided with North America to produce the Taconic orogeny.

The Piscataquis volcanic arc is a belt of Devonian volcanic rocks that extends from central Massachusetts to the Gaspe Peninsula. These rocks are roughly coextensive with the Ordovician arc of the Bronson Hill anticlinorium and include basalts, andesites, dacites, and rhyolites. Both subaerial volcanics and subaquatic pillow lavas are found in the belt. The Greenville plutonic belt of Maine (including Mount Kathadin) is included in the Piscataquis arc, and interpreted by some workers to be post-Acadian, but is more typical of syn-tectonic arc plutons. The eastern part of the Taconic orogenic belt was also deformed by the Acadian orogeny and contains some younger rocks deposited on top of the eroded Taconic island arc, then deformed in the Acadian orogeny.

The Taconic allochthons turn out to be continental rise sediments that were scraped off the North American continental margin and transported on thrusts for 60–120 miles (100–200 km) during the Taconic arc continent collision. A clastic wedge (Austin Glen and Normanskill Formations) was deposited during emplacement of the allochthons, by their erosion, and spread out laterally in the foreland. As Taconic deformation proceeded, the clastic wedge and underlying carbonates and Grenville basement became involved in the deformation, rotating them, forming the Taconic angular unconformity.

The Acadian orogeny has historically been one of the most poorly understood aspects of the regional geology of the Appalachians. Some of the major problems in interpreting the Acadian orogeny include understanding the nature of pre-Acadian, post-Taconic basins such as the Kearsage–Central Maine basin, Aroostook-Matapedia trough, and the Connecticut Valley–Gaspe trough. The existence and vergence of Acadian subduction zones is debated, and the relative importance of post-Acadian strike-slip movements is not well-constrained.

Examining the regional geology of the northern Appalachians using only the rocks that are younger than the post-Taconic unconformity yields a picture of several distinctive tectonic belts, including different rock types and structures. The North American craton includes Grenville gneisses and Paleozoic carbonates. The foreland basin includes a thick wedge of Devonian synorogenic clastic rocks, such as the Catskill Mountains, that thicken toward the mountain belt. The Green Mountain anticlinorium is a basement thrust slice, and the Connecticut Valley–Gaspe trough is a post-Taconic basin with rapid Silurian subsidence and deposition. The Bronson Hill–Boundary Mountain anticlinorium (Piscataquis volcanic arc) is a Silurian–Mid-Devonian volcanic belt formed along the North American continental margin. The Aroostook-Matapedia trough is a Silurian extensional basin, and the Miramichi massif represents remnants of a high-standing Ordovician (Taconic) arc. The Kearsarge–Central

Maine basin (Merrimack trough) preserves Silurian deepwater sedimentary rocks, preserved in accretionary prisms, and is the most likely site where the Acadian Ocean closed. The Fredrickton trough is a continuation of the Merrimack trough, and the Avalon Composite terrane (coastal volcanic arc) contains Silurian–Early Devonian shallow marine volcanics built upon Precambrian basement of Avalonia.

Synthesizing the geology of these complex belts, the tectonics of the Acadian orogeny in the Appalachian Mountains can be summarized as follows. The Grenville gneisses and some of the accreted Taconic orogen were overlain by a Paleozoic platform sequence, and by mid-Devonian times the region was buried beneath thick clastics of the Acadian foreland basin, best preserved in the Catskill Mountains. Nearly two miles (3 km) of fluvial sediments were deposited in 20 million years, derived from mountains to the east. Molasse and red beds of the Catskills once covered the Adirondack Mountains and pieces are preserved in a diatreme in Montreal, and they are exposed along strike as the Old Red Sandstone in Scotland and on Spitzbergen Island.

The Connecticut Valley–Gaspe trough is a complex basin developed over the Taconic suture and was active from Silurian through Early Devonian. It is an extensional basin containing shallow marine sedimentary rocks and may have formed from oblique strike-slip after the Taconic collision, with subsidence in pull-apart basins. The Aroostook-Matapedia trough is an Ordovician-Silurian turbidite belt, probably a post-Taconic extensional basin, and perhaps a narrow oceanic basin.

The Miramichi massif contains Ordovician arc rocks intruded by Acadian plutons and is part of the Taconic arc that persisted as a high area through Silurian times and became part of the Piscataquis volcanic arc in Silurian-Devonian times. The coastal volcanic arc (Avalon) is exposed in eastern Massachusetts though southern New Brunswick and includes about 5 miles (8 km) of basalt, andesites, rhyolite, and deep and shallow marine sediments. It is a volcanic arc that was built on Precambrian basement that originated in the Avalonian or Gondwana side of the Iapetus Ocean.

The Kearsage–Central Maine basin (Fredericton trough) is the location of a major post-Taconic, pre-Acadian ocean that closed to produce the Acadian orogeny. It contains polydeformed deepwater turbidites and black shales, mostly Silurian. The regional structural plunge results in low grades of metamorphism in Maine, high grades in New Hampshire, Massachusetts, and Connecticut. There are a few dismembered ophiolites present in the belt, structurally incorporated in about 3 miles (5 km) of turbidites.

Volcanic belts on either side of the Merrimack trough are interpreted to be arcs built over contemporaneous subduction zones. In the Late Silurian, the Acadian Ocean basin was sub-

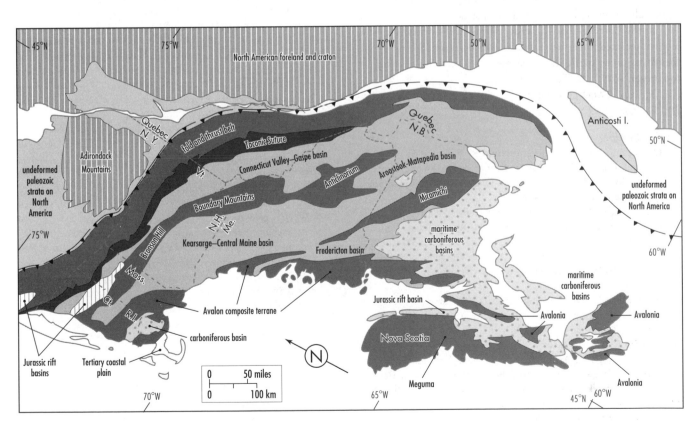

Map of the northern Appalachians showing the main Early Paleozoic tectonic terranes

ducting on both sides, forming accretionary wedges of opposite vergence, and forming the Coastal and Piscataquis volcanic arcs. The Connecticut Valley–Gaspe trough is a zone of active strike-slip faulting and pull-apart basin formation behind the Piscataquis arc. In the Devonian, the accretionary prism complexes collided, and west-directed overthrusting produced a migrating flexural basin of turbidite deposition, including the widespread Seboomook and Littleton Formations. The collision continued until the Late Devonian, then more plutons intruded, and dextral strike-slip faulting continued.

Acadian plutons intrude all over the different tectonic zones and are poorly understood. Some are related to arc magmatism, some to crustal thickening during collision. Late transpression in the Carboniferous includes abundant dextral strike-slip faults, disrupted zones, and formed pull-apart basins with local accumulations of several miles of sediments. About 200 miles (300 km) of dextral strike-slip offsets are estimated to have occurred across the orogen.

The Late Paleozoic Alleghenian orogeny in the Carboniferous and Permian included strong folding and thrusting in the southern Appalachians and formed a fold/thrust belt with a ramp/flat geometry. In the southern Appalachians the foreland was shortened by about 50 percent during this event, with an estimated 120 miles (200 km) of shortening. The rocks highest in the thrust belt have been transported the farthest and are the most allochthonous. At the same time, motions in the northern Appalachians were dominantly dextral strike-slip in nature.

In the Late Triassic–Jurassic, rifting and normal faulting were associated with the formation of many small basins and the intrusion of mafic dike swarms, related to the opening of the present-day Atlantic Ocean.

See also CALEDONIDES; PENOBSCOTTIAN OROGENY; PLATE TECTONICS.

Further Reading

Ala Drake, A., A. K. Sinha, Jo Laird, and R. E. Guy. "The Taconic Orogen." Chapter 3 in "The Geology of North America, vol. A: The Geology of North America, an Overview." *Geological Society of America* (1989): 101–178.

Bird, John M., and John F. Dewey. "Lithosphere Plate-Continental Margin Tectonics and the Evolution of the Appalachian Orogen." *Geological Society of America Bulletin* 81 (1970): 1,031–1,060.

Bradley, Dwight C. "Tectonics of the Acadian Orogeny in New England and Adjacent Canada." *Journal of Geology* 91 (1983): 381–400.

Bradley, Dwight C., and Timothy M. Kusky. "Geologic Methods of Estimating Convergence Rates during Arc-Continent Collision." *Journal of Geology* 94 (1986): 667–681.

Dewey, John F., Michael J. Kennedy, and William S. F. Kidd. "A Geotraverse through the Appalachian of Northern Newfoundland." In *Profiles of Orogenic Belts,* edited by N. Rast and F. M. Delany. *AGU/Geological Society of America,* Geodynamics series 10, 1983.

Hatcher, Robert D., Jr., William A. Thomas, Peter A. Geiser, Arthur W. Snoke, Sharon Mosher, and David V. Wiltschko. "Alleghenian Orogeny." Chapter 5 in "The Geology of North America, vol. A:

The Geology of North America, an Overview." *Geological Society of America* (1989): 233–319.

Kusky, Timothy M., J. Chow, and Samuel A. Bowring. "Age and Origin of the Boil Mountain Ophiolite and Chain Lakes Massif, Maine: Implications for the Penobscottian Orogeny." *Canadian Journal of Earth Sciences* 34, no. 5 (1997): 646–654.

Kusky, Timothy M., and William S. F. Kidd. "Early Silurian Thrust Imbrication of the Northern Exploits Subzone, Central Newfoundland." *Journal of Geodynamics* 22 (1996): 229–265.

Kusky, Timothy M., William S. F. Kidd, and Dwight C. Bradley. "Displacement History of the Northern Arm Fault, and Its Bearing on the Post-Taconic Evolution of North-Central Newfoundland." *Journal of Geodynamics* 7 (1987): 105–133.

Manspeizer, Warren, Jelle Z. de-Boer, John K. Costain, Albert J. Froelich, Cahit Coruh, Paul E. Olsen, Gregory J. McHone, John H. Puffer, and David C. Prowell. "Post-Paleozoic Activity." Chapter 6 in "The Geology of North America, vol. A: The Geology of North America, an Overview." *Geological Society of America* (1989): 319–374.

McKerrow, W. Stuart, and A. M. Ziegler. "Paleozoic Oceans." *Nature* 240 (1972): 92–94.

Neuman, Robert B., A. R. Palmer, and J. Thomas Dutro. "Paleontologic Contributions to Paleozoic Paleographic Reconstructions of the Appalachians." Chapter 7 in "The Geology of North America, vol. A: The Geology of North America, an Overview." *Geological Society of America* (1989): 375–384.

Osberg, Phil, James F. Tull, Peter Robinson, Rudolph Hon, and J. Robert Butler. "The Acadian Orogen." Chapter 4 in "The Geology of North America, vol A: The Geology of North America, an Overview." *Geological Society of America* (1989): 179–232.

Rankin, D. W., Avery Ala Drake, Jr., Lynn Glover III, Richard Goldsmith, Leo M. Hall, D. P. Murray, Nicholas M. Ratcliffe, J. F. Read, Donald T. Secor, Jr., and R. S. Stanley. "Pre-Orogenic Terranes." Chapter 2 in "The Geology of North America, vol. A: The Geology of North America, an Overview." *Geological Society of America* (1989): 7–100.

Rast, Nick. "The Evolution of the Appalachian Chain." Chapter 12 in "The Geology of North America, vol. A: The Geology of North America, an Overview." *Geological Society of America* (1989): 323–348

Rowley, David B., and William S. F. Kidd. "Stratigraphic Relationships and Detrital Composition of the Medial Ordovician Flysch of Western New England: Implications for the Tectonic Evolution of the Taconic Orogeny." *Journal of Geology* 89 (1981): 199–218.

Roy, D., and James W. Skehan. *The Acadian Orogeny.* Geological Society of America Special Paper 275, 1993.

Socci, Anthony D, James W. Skehan, and Geoffrey W. Smith. *Geology of the Composite Avalon Terrane of Southern New England.* The Geological Society of America Special Paper 245, 1990.

Stanley, Rolfe S., and Nicholas M. Ratcliffe. "Tectonic Synthesis of the Taconian Orogeny in Western New England." *Geological Society of America Bulletin* 96 (1985): 1,227–1,250.

aquifer   Any body of permeable rock or regolith saturated with water through which groundwater moves. The term *aquifer* is usually reserved for rock or soil bodies that contain economical quantities of water that are extractable by existing methods. The quality of an aquifer depends on two main quantities, porosity and permeability. Porosity is a measure of the total amount of open void space in the material. Permeability is a term that refers to the ease at which a fluid can move through the open pore spaces, and it depends in part on the size, the shape, and how connected individual pore spaces are in the material. Gravels and sandstone make good aquifers, as do fractured rock bodies. Clay is so impermeable that it makes bad aquifers or even aquicludes, which stop the movement of water.

There are several main types of aquifers. In uniform, permeable rock and soil masses, aquifers will form as a uniform layer below the water table. In these simple situations, wells fill with water simply because they intersect the water table. However, the rocks below the surface are not always homogeneous and uniform, which can result in a complex type of water table known as a perched water table. This results from discontinuous impermeable rock or soil bodies in the subsurface, which create domed pockets of water at elevations higher than the main water table, resting on top of the impermeable layer.

When the upper boundary of the groundwater in an aquifer is the water table, the aquifer is said to be unconfined. In many regions, a saturated permeable layer, typically sandstone, is confined between two impermeable beds, creating a confined aquifer. In these systems, water only enters the system in a small recharge area, and if this is in the mountains, then the aquifer may be under considerable pressure. This is known as an artesian system. Water that escapes the system from the fracture or well reflects the pressure difference between the elevation of the source area and the discharge area (hydraulic gradient), and it rises above the aquifer as an artesian spring, or artesian well. Some of these wells have made fountains that have spewed water 200 feet (60 m) high.

See also FRACTURE ZONE AQUIFERS; GROUNDWATER.

Arabian shield The Arabian shield comprises the core of the Arabian Peninsula, a landmass of near trapezoidal shape bounded by three water bodies. The Red Sea bounds it from the west, the Arabian Sea and the Gulf of Aden from the south, and the Arabian Gulf and Gulf of Oman on the east. The Arabian Peninsula can be classified into two major geological provinces, including the Precambrian Arabian shield and the Phanerozoic cover.

The Precambrian shield is located along the western and central parts of the peninsula. It narrows in the north and the south but widens in the central part of the peninsula. The shield lies between latitudes 12° and 30° north and between longitudes 34° and 47° east. The Arabian shield is considered as part of the Arabian-Nubian shield that was formed in the upper Proterozoic Era and stabilized in the Late Proterozoic around 600 million years ago. The shield has since subsided and been covered by thick deposits of Phanerozoic continen-

tal shelf sediments along the margins of the Tethys Ocean. Later in the Tertiary the Arabian-Nubian shield was rifted into two fragments by the Red Sea rift system.

Phanerozoic cover overlies the eastern side of the Arabian shield unconformably and dips gently toward the east. Parts of the Phanerozoic cover are found overlying parts of the Precambrian shield, such as the Quaternary lava flows of Harrat Rahat in the middle and northern parts of the shield as well as some sandstones, including the Saq, Siq, and Wajeed sandstones in different parts of the shield.

History of Tectonic Models
The Arabian shield includes an assemblage of Middle to Late Proterozoic rocks exposed in the western and central parts of the Arabian Peninsula and overlapped to the north, east, and south by Phanerozoic sedimentary cover rocks. Several parts of the shield are covered by Tertiary and Quaternary lava flows that were extruded concurrently with rifting of the Red Sea. Rocks of the Arabian shield may be divided into assemblages of Middle to Late Proterozoic stratotectonic units, volcano-sedimentary, and associated mafic to intermediate intrusive rocks. These rocks are divided into two major categories, the layered rocks and the intrusive rocks. Researchers variously interpret these assemblages as a result of volcanism and magmatism in ensialic basins or above subduction zones. More recent workers suggested that many of these assemblages belong to late Proterozoic volcanic-arc systems that comprise distinct tectonic units or terranes, recognized following definitions established in the North America cordillera.

Efforts in suggesting models for the evolution of the Arabian shield started in the 1960s. Early workers suggested that the Arabian shield experienced three major orogenies in the Late Proterozoic Era. They also delineated four classes of plutonic rocks that evolved in chemistry from calc-alkaline to peralkaline through time. In the 1970s a great deal of research emerged concerning models of the tectonic evolution of the Arabian shield. Two major models emerged from this work, including mobilistic plate-tectonic models, and a nonmobilistic basement-tectonic model.

The main tenet of the plate-tectonic model is that the evolution of the Arabian shield started and took place in an oceanic environment, with the formation of island arcs over subduction zones in a huge oceanic basin. On the contrary, the basement-tectonic model considers that the evolution of the Arabian shield started by the rifting of an older craton or continent to form intraoceanic basins that became the sites of island arc systems. In both models, late stages of the formation of the Arabian-Nubian shield are marked by the sweeping together and collision of the island arcs systems, obduction of the ophiolites, and cratonization of the entire orogen, forming one craton attached to the African craton. Most subsequent investigators in the 1970s supported one of these two models and tried to gather evidence to support that model.

As more investigations, mapping, and research were carried out in the 1980s and 1990s, a third model invoking microplates and terrane accretion was suggested. This model suggests the existence of an early to mid-Proterozoic (2,000–1,630-million-year-old) craton that was extended, rifted, then dispersed causing the development of basement fragments that were incorporated as allochthonous microplates into younger tectonostratigraphic units. The tectonostratigraphic units included volcanic complexes, ophiolite complexes, and marginal-basin and fore-arc stratotectonic units that accumulated in the intraoceanic to continental-marginal environments that resulted from rifting of the preexisting craton. These rocks, including the older continental fragments, constituted five large and five small tectonostratigraphic terranes that were accreted and swept together between 770 million and 620 million years ago to form a neo-craton on which younger volcano-sedimentary and sedimentary rocks were deposited. Most models developed in the period since the early 1990s represent varieties of these three main classical models, along with a greater appreciation of the role that the formation of the supercontinent of Gondwana played in the formation of the Arabian-Nubian shield.

Geology of the Arabian Shield

Peter Johnson and coworkers have synthesized the geology of the Arabian shield and proposed a general classification of the geology of the Arabian shield that attempts to integrate and resolve the differences between the previous classifications. According to this classification, the layered rocks of the Arabian shield are divided into three main units separated by periods of regional tectonic activity (orogenies). This gives an overall view that the shield was created through three tectonic cycles. These tectonic cycles include early, middle, and late Upper Proterozoic tectonic cycles.

The early Upper Proterozoic tectonic cycle covers the time period older than 800 million years ago and includes the oldest rock groups that formed before and up to the Aqiq orogeny in the south and up to the Tuluhah orogeny in the north. In this general classification, the Aqiq and Tuluhah orogenies are considered as part of one regional tectonic event or orogeny that is given a combined name of the Aqiq-Tuluhah orogeny.

The middle Upper Proterozoic tectonic cycle is considered to have taken place in the period between 800 and 700 million years ago. It includes the Yafikh orogeny in the south and the Ragbah orogeny in the north. These two orogenies were combined together into one regional orogeny named the Yafikh-Ragbah orogeny.

The late Upper Proterozoic tectonic cycle took place in the period between 700 and 650 million years ago. It includes the Bishah orogeny in the south and the Rimmah orogeny in the north. These two orogenies are combined together into one regional orogeny named the Bishah-Rimmah orogeny.

Classification of Rock Units

The layered rocks in the Arabian shield are classified into three major rock units, each of them belonging to one of the three tectonic cycles mentioned above. These major layered rock units are the lower, middle, and upper layered rock units.

The lower layered rock unit covers those rock groups that formed in the early upper Proterozoic tectonic cycle (older than 800 million years ago) and includes rocks with continental affinity. The volcanic rocks that belong to this unit are characterized by basaltic tholeiite compositions and by the domination of basaltic rocks that are older than 800 million years. The rock groups of this unit are located mostly in the southwestern and eastern parts of the shield.

The rock groups of the lower layered unit include rocks that were formed in an island arc environment and that are characterized by basic tholeiitic volcanic rocks (Baish and Bahah Groups) and calc-alkaline rocks (Jeddah Group). These rocks overlie in some places highly metamorphosed rocks of continental origin (Sabia Formation and Hali schists) that are considered to have been brought into the system either from a nearby craton such as the African craton, or from microplates that were rifted from the African plate such as the Afif microplate.

The middle layered rock unit includes the layered rock groups that formed during the middle upper Proterozoic tectonic cycle in the period between 800 and 700 million years ago. The volcanic rocks are predominately intermediate igneous rocks characterized by a calc-alkaline nature. These rocks are found in many parts of the shield, with a greater concentration in the north and northwest, and scattered outcrops in the southern and central parts of the shield.

The upper layered rock unit includes layered rock groups that formed in the late upper Proterozoic tectonic cycle in the period between 700 and 560 million years ago and are predominately calc-alkaline, alkaline intermediate, and acidic rocks. These rock groups are found in the northeastern, central, and eastern parts of the shield.

Intrusive Rocks

The intrusive rocks that cut the Arabian shield are divided into three main groups. These groups are called (from the older to the younger) pre-orogenic, syn-orogenic, and post-orogenic.

The pre-orogenic intrusions are those intrusions that cut through the lower layered rocks unit only and not the other layered rock units. It is considered older than the middle layered rock unit but younger than the lower layered rock unit. These intrusions are characterized by their calcic to calc-alkaline composition. They are dominated by gabbro, diorite, quartz-diorite, trondhjemite, and tonalite. These intrusions are found in the southern, southeastern, and western parts of the shield and coincide with the areas of the lower layered rocks unit. These intrusions are assigned ages between 1,000 and 700 million years old. Geochemical signatures including strontium isotope ratios show that these intrusions were derived from magma that came from the upper mantle.

The syn-orogenic intrusions are those intrusions that cut the lower and the layered rock units as well as the pre-orogenic intrusions, but that do not cut or intrude the upper layered rocks unit. These intrusions are considered older than the upper layered rocks unit and younger than the pre-orogenic intrusions, and they are assigned ages between 700 and 620 million years old. Their chemical composition is closer to the granitic calc-alkaline to alkaline field than the pre-orogenic intrusions. These intrusions include granodiorite, adamalite, monzonite, granite, and alkali granite with lesser amount of gabbro and diorite in comparison with the pre-orogenic intrusions. The general form of these intrusions is batholithic bodies that cover wide areas. They are found mostly in the eastern, northern, and northeastern parts of the Arabian shield. The initial strontium ratio of these intrusions is higher than that of the pre-orogenic intrusions and indicates that these intrusions were derived from a magma that was generated in the lower crust.

Post-orogenic intrusions are intrusions that cut through the three upper Proterozoic layered rocks units as well as the pre- and syn-orogenic intrusions. These are assigned ages between 620 and 550 million years old. They form circular, elliptical, and ring-like bodies that range in chemical composition from alkaline to peralkaline. These intrusions are made mainly of alkaline and peralkaline granites such as riebeckite granite, alkaline syenite, pink granite, biotite granite, monzo-granite, and perthite-biotite granite.

Ringlike bodies and masses of gabbro are also common, and the post-orogenic magmatic suite is bimodal in silica content. These intrusions are found scattered in the Arabian shield, but they are more concentrated in the eastern, northern, and central parts of the shield.

The initial strontium ratio of the post-orogenic intrusions ranges between 0.704 and 0.7211, indicating that these intrusions were derived from a magma that was generated in the lower crust.

Ophiolite Belts

Mafic and ultramafic rocks that comply with the definition of the ophiolite sequence are grouped into six major ophiolitic belts. Four of these belts strike north while the other two belts strike east to northeast. These ophiolite belts include:

1. The Amar-Idsas ophiolite belt
2. Jabal Humayyan–Jabal Sabhah ophiolite belt
3. The Bijadiah-Halaban ophiolite belt
4. Hulayfah-Hamdah "Nabitah" ophiolite belt
5. Bi'r Umq–Jabal Thurwah ophiolite belt
6. Jabal Wasq–Jabal Ess ophiolite belt

These rocks were among other mafic and ultramafic rocks considered as parts of ophiolite sequences, but later only these six belts were considered to comply with the definition of ophiolite sequences. However, the sheeted dike complex of the typical ophiolite sequence is not clear or absent in some of these belts, suggesting that the dikes may have been obscured by metamorphism, regional deformation, and alteration. These belts are considered to represent suture zones, where convergence between plates or island arc systems took place, and are considered as the boundaries between different tectonic terranes in the shield.

Najd Fault System

One of the noticeable structural features of the Arabian shield is the existence of a fault system in a zone 185 miles (300 km) wide with a length of nearly 750 miles (1,200 km) extending from the southeastern to the northwestern parts of the shield. This system was generated just after the end of the Hijaz tectonic cycle, and it was active from 630 to 530 million years, making it the last major event of the Precambrian in the Arabian shield. These faults are left-lateral strike-slip faults with a 150-mile (250-km) cumulative displacement on all faults in the system.

The main rock group that was formed during and after the existence of the Najd fault system is the Ji'balah Group. This group formed in the grabens that were formed by the Najd fault system and are the youngest rock group of the Precambrian Arabian shield. The Ji'balah Group formed between 600 and 570 million years ago. The Ji'balah Group is composed of coarse-grained clastic rocks and volcanic rocks in the lower parts, by stromatolitic and cherty limestone and argillites in the middle parts, and by fine-grained clastic rocks in the upper parts. These rocks were probably deposited in pull-apart basins that developed in extensional bends along the Najd fault system.

Tectonic Evolution of the Arabian Shield

The Arabian shield is divided into five major and numerous smaller terranes separated by four major and many smaller suture zones, many with ophiolites along them. The five major terranes include the Asir, Al-Hijaz, Midyan, Afif, and Ar-Rayn. The first three terranes are interpreted as interoceanic island arc terranes while the Afif terrane is considered continental, and the Ar-Rayn terrane is considered to be probably continental. The four suture zones include the Bi'r Umq, Yanbu, Nabitah, and Al-Amar-Idsas. These suture zones represent the collision and suturing that took place between different tectonic terranes in the Arabian shield. For example, the Bi'r Umq suture zone represents the collision and suturing between two island arc terranes of Al-Hijaz and Asir, while the Yanbu suture zone represents the collision zone between the Midyan and Al-Hijaz island arc terranes. The Nabitah suture zone represents collision and suturing between a continental microplate (Afif) in the east and island arc terranes (Asir and Al-Hijaz) in the west; Al-Amar-Idsas suture zone represents the collision and suturing zone between two continental microplates, Afif and Ar-Rayn.

Five main stages are recognized in the evolution of the Arabian shield, including rifting of the African craton (1,200–950 million years ago), formation of island arcs over oceanic crust (950–715 million years ago), formation of the Arabian shield craton from the convergence and collision of microplates with adjacent continents (715–640 million years ago), continental magmatic activity and tectonic deformation (640–550 million years ago), and epicontinental subsidence (550 million years ago).

Information about the rifting stage (1,200–950 million years ago) is limited but it can be said that the Mozambique belt in the African craton underwent rifting in the time interval between 1,200 million and 950 million years ago. This rifting resulted in the formation of an oceanic basin along the present northeastern side of the African craton. This was a part of the Mozambique Ocean that separated the facing margins of East and West Gondwana. Alternatively there may have been more than one ocean basin, separated by rifted micro-continental plates such as the Afif micro-continental plate.

The island arc formation stage (950–715 million years ago) is characterized by the formation of oceanic island arcs in the oceanic basins formed in the first stage. The stratigraphic records of volcanic and sedimentary rocks in the Asir, Al-Hijaz, and some parts of the Midyan terranes, present rocks with ages between 900 and 800 million years old. These rocks are of mafic or bimodal composition and are considered products of early island arcs, particularly in the Asir terrane. These rocks show mixing or the involvement of rocks and fragments that formed in the previous stage of rifting of the African craton.

The formation of island arc systems did not take place at the same time but rather different arc systems evolved at different times. The Hijaz terrane is considered to be the oldest island arc, formed between 900 million and 800 million years ago. This terrane may have encountered some continental fragments now represented by the Khamis Mushayt Gneiss and Hali Schist, which are considered parts of, or derived from, the old continental crust from the previous stage of rifting.

Later on in this stage (760–715 million years ago), three island arc systems apparently formed simultaneously. These are the Hijaz, Tarib, and Taif island arc systems. These island arc systems evolved and formed three crustal plates including the Asir, Hijaz, and Midyan plates. Later in this stage the Amar Andean arc formed between the Afif plate and Ar-Rayn plate, and it is considered part of the Ar-Rayn plate. Oceanic crustal plateaus may have been involved in the formation of the oceanic crustal plates in this stage.

In the collision stage (715–640 million years ago) the five major terranes that formed in the previous stages were swept together and collisions took place along the four suture zones mentioned above. The collision along these suture zones did not take place at the same time. For example, the collision along the Hijaz and Taif arcs occurred around 715 million years ago, and the collision along the Bir Omq suture zone took place between 700 million and 680 million years ago, while the island arc magmatic activity in the Midyan terrain continued until 600 million years ago. It appears that the collision along the Nabitah suture zone was diachronous along strike. The collision started in the northern part of the Nabitah suture between the Afif and Hijaz terranes at about 680 million to 670 million years ago, and at the same time the southern part of the suture zone was still experiencing subduction. Further collision along the Nabitah suture zone shut off the arc in the south, and the Afif terrain collided with the Asir terrain. As a result, the eastern Afif plate and the western island arc plates of the Hijaz and Asir were completely sutured along the Nabitah orogenic belt by 640 million years ago. In this stage three major magmatic arcs developed, and later on in this stage they were shut off by further collision. These arcs include the Furaih magmatic arc that developed on the northern part of the Nabitah suture zone and on the southeastern part of the Hijaz plate, the Sodah arc that developed on the eastern part of the Afif plate, and an Andean-type arc on the eastern part of the Asir plate.

The Ar-Rayn collisional orogeny along the Amar suture was between the two continental plates of Afif and Ar-Rayn and took longer than any other collisions in the shield (from 700 million to 630 million years ago). Many investigators suggest that the Ar-Rayn terrain is part of a bigger continent (one that extends under the eastern Phanerozoic cover and is exposed in Oman) that collided with or into the Arabian shield from the east and was responsible for the development of Najd left-lateral fault system.

By 640 million years ago the five major terranes had collided with each other forming the four mentioned suture zones and the Arabian shield was stabilized. Since then, the shield behaved as one lithospheric plate until the rifting of the Red Sea. However, orogenic activity inside the Arabian shield continued for a period of about 80 million years after collision, during which the Najd fault system developed as the last tectonic event in the Arabian shield in the late Proterozoic Era.

After development of the Najd fault system, tectonic activity in the Arabian shield ended and the Arabian-Nubian shield subsided and was peneplained, as evidenced by the existence of epicontinental Cambro-Ordovician sandstone covering many parts of the shield in the north and the south. The stratigraphic records of the Phanerozoic cover show that the Arabian shield has been tectonically stable with the exception of ophiolite obduction and collision along the margins of the plate during the closure of the Tethys Sea until rifting of the Red Sea in the Tertiary.

See also CRATONS; KUWAIT; OMAN MOUNTAINS; ZAGROS AND MAKRAN MOUNTAINS.

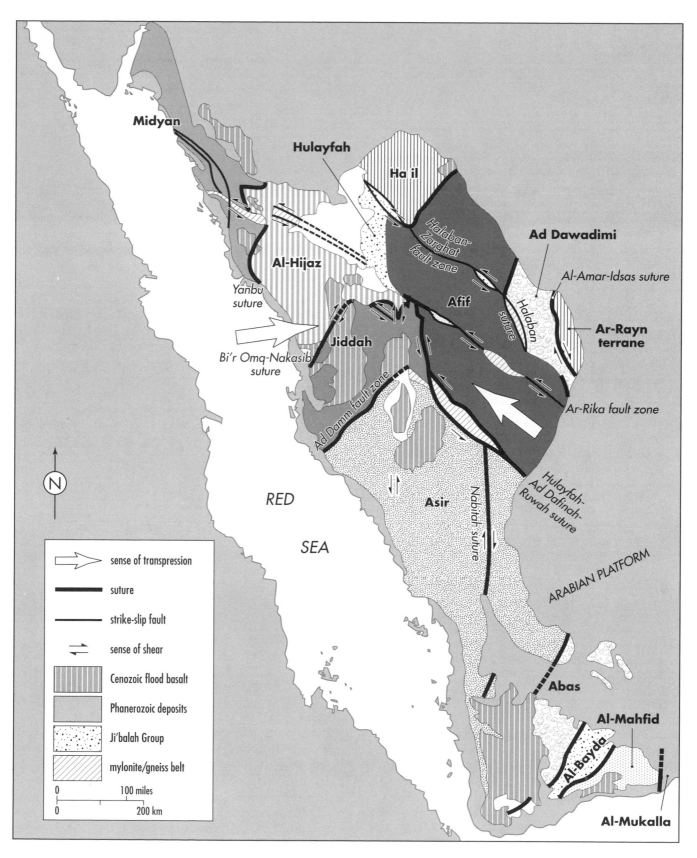

Map showing the tectonic division of the Arabian shield, consisting of amalgamated island arc, accretionary prism, and continental fragments

Further Reading

Abdelsalam, Mohamed G., and Robert J. Stern. "Sutures and Shear Zones in the Arabian-Nubian Shield." *Journal of African Earth Sciences* 23 (1996).

Al-Shanti, A. M. S. *The Geology of the Arabian Shield.* Saudi Arabia: Center for Scientific Publishing, King AbdelAziz University, 1993.

Brown, Glen F., Dwight L. Schmidt, and Curtis A. Huffman, Jr. *Geology of the Arabian Peninsula, Shield Area of Western Saudi Arabia.* United States Geological Survey Professional Paper 560-A, 1989.

Delfour, Jacques. "Geology and Mineral Resources of the Northern Arabian Shield, A Synopsis of BRGM Investigations, 1965–1975." *BRGM Technical Record* TR-03-1 (1983).

Johnson, Peter R. "Post-amalgamation Basins of the NE Arabian Shield and Implications for Neoproterozoic III Tectonism in the Northern East African Orogen." In *Evolution of the East African and Related Orogens, and the Assembly of Gondwana,* edited by Timothy M. Kusky, Mohamed Abdelsalam, Robert Tucker, and Robert Stern. *Precambrian Research,* 2003.

Johnson, Peter R., Erwin Scheibner, and Alan E. Smith. "Basement Fragments, Accreted Tectonostratigraphic Terranes, and Overlap Sequences: Elements in the Tectonic Evolution of the Arabian Shield, Geodynamics Series." *American Geophysical Union* 17 (1987): 324–343.

Kusky, Timothy M., Mohamed Abdelsalam, Robert Tucker, and Robert Stern, eds., *Evolution of The East African and Related Orogens, and the Assembly of Gondwana. Precambrian Research,* 2003.

Kusky, Timothy M., and Mohamed Matsah. "Neoproterozoic Dextral Faulting on the Najd Fault System, Saudi Arabia, Preceded Sinistral Faulting and Escape Tectonics Related to Closure of the Mozambique Ocean." *In Proterozoic East Gondwana: Supercontinent Assembly and Break-up,* edited by M. Yoshida, Brian F. Windley, S. Dasgupta, and C. Powell, 327–361. *Journal of the Geological Society of London,* 2003.

Stern, Robert J. "Arc Assembly and Continental Collision in the Neoproterozoic East African Orogen: Implications for Consolidation of Gondwanaland." *Annual Review of Earth and Planetary Sciences* 22 (1994): 319–351.

Stoeser, Douglas B., and Victor E. Camp. "Pan-African Microplate Accretion of the Arabian Shield." *Geological Society of America Bulletin* 96 (1985): 817–826.

Stoeser, Douglas B., and John S. Stacey. "Evolution, U-Pb Geochronology, and Isotope Geology of the Pan-African Nabitah Orogenic Belt of the Saudi Arabian Shield." In *The Pan-African Belts of Northeast Africa and Adjacent Areas,* edited by S. El Gaby and R. O. Greiling, 227–288. Braunschweig: Friedr Vieweg and Sohn, 1988.

Aral Sea   A large inland sea in southwestern Kazakhstan and northwest Uzbekistan, east of the Caspian Sea. The Aral Sea is fed by the Syr Darya and Amu Darya Rivers that flow from the Hindu Kush and Tien Shan Mountains to the south and is very shallow, attaining a maximum depth of only 220 feet (70 m). In the latter half of the 20th century, the Soviet government diverted much of the water from the Syr Darya and Amu Darya Rivers for irrigation, which has had dramatic effects on the inland sea. In the 1970s the Aral was the fourth largest lake, covering 26,569 square miles (68,000 km^2). It had an average depth of 52.5 feet (16 m) and was the source of about 45,000 tons of carp, perch, and pike fish each year. Since the diversion of the rivers, the Aral has shrunk dramatically, retreating more than 31 miles (50 km) from its previous shore, lowering the average depth to less than 30 feet (9 m), reducing its area to less than 15,376 square miles (40,000 km^2), and destroying the fishing industry in the entire region. Furthermore, since the lake bottom has been exposed, winds have been blowing the salts from the evaporated water around the region, destroying the local farming. The loss of evaporation from the sea has even changed the local climate, reducing rainfall and increasing the temperatures, all of which exacerbate the problems in the region. Disease and famine have followed, devastating the entire central Asian region.

Archean (Archaean)   Earth's first geological era for which there is an extensive rock record, the Archean also preserves evidence for early primitive life forms. The Archean is the second of the four major eras of geological time: the Hadean, Archean, Proterozoic, and Phanerozoic. Some time classification schemes use an alternative division of early time, in which the Hadean, Earth's earliest era, is considered the earliest part of the Archean. The Archean encompasses the one and one-half-billion-year long (Ga = giga année, or 10^9 years) time interval from the end of the Hadean era to the beginning of the Proterozoic era. In most classification schemes, it is divided into three parts, including the Early Archean (4.0–3.5 Ga), the Middle Archean (3.5–3.1 Ga), and the Late Archean, ranging up to 2.5 billion years ago.

The oldest known rocks on Earth are the 4.0-billion-year-old Acasta gneisses from northern Canada that span the Hadean-Archean boundary. Single zircon crystals from the Jack Hills and Mount Narryer in western Australia have been dated to be 4.3–4.1 billion years old. The oldest well-documented and extensive sequence of rocks on Earth is the Isua belt located in western Greenland, estimated to be 3.8 billion years old. Life on Earth originated during the Archean, with the oldest known fossils coming from the 3.5-billion-year-old Apex chert in western Australia, and possible older traces of life found in the 3.8-billion-year-old rocks from Greenland.

Archean and reworked Archean rocks form more than 50 percent of the continental crust and are present on every continent. Most Archean rocks are found in cratons, or as tectonic blocks in younger orogenic belts. Cratons are low-relief tectonically stable parts of the continental crust that form the nuclei of many continents. Shields are the exposed parts of cratons, other parts of which may be covered by younger platformal sedimentary sequences. Archean rocks in cratons and shields are generally divisible into a few basic types. Relatively low-metamorphic grade greenstone belts consist of deformed metavolcanic and metasedimentary rocks. Most Archean plutonic rocks are tonalites, trondhjemites, granodiorites, and granites that intrude or are in structural contact with strongly deformed and metamorphosed sedimentary and volcanic rocks in greenstone belt associations. Together, these rocks form the granitoid-greenstone association that characterizes many Archean cratons. Granite-greenstone terranes are common in parts of the Canadian Shield, South America, South Africa, and Australia. Low-grade cratonic basins are preserved in some places, including southern Africa and parts of Canada. High-grade metamorphic belts are also common in Archean cratons, and these generally include granitic, metasedimentary, and metavolcanic gneisses that were deformed and metamorphosed at middle to deeper crustal levels. Some well-studied Archean high-grade gneiss terranes include the Lewisian and North Atlantic Province, the Limpopo Belt of southern Africa, the Hengshan of North China, and parts of southern India.

The Archean witnessed some of the most dramatic changes to Earth in the history of the planet. During the Hadean, the planet was experiencing frequent impacts of asteroids, some of which were large enough to melt parts of the outer layers of the Earth and vaporize the atmosphere and oceans. Any attempts by life to get a foothold on the planet in the Hadean would have been difficult, and if any organisms were to survive this early bombardment, they would have to have been sheltered in some way from these dramatic changes. Early atmospheres of the Earth were blown away by asteroid and comet impacts and by strong solar winds from an early T-Tauri phase of the Sun's evolution. Free oxygen was either not present or present in much lower concentrations, and the atmosphere evolved slowly to a more oxygenic condition.

The Earth was also producing and losing more heat during the Archean than in younger times, and the patterns, styles, and rates of mantle convection and the surface style of plate tectonics must have reflected these early conditions. Heat was still left over from early accretion, core formation, late impacts, and the decay of some short-lived radioactive isotopes such as ^{129}I. In addition, the main heat-producing radioactive decay series were generating more heat then than now, since more of these elements were present in older half-lives. In particular, ^{235}U, ^{238}U, ^{232}Th, ^{40}K were cumulatively producing two to three times as much heat in the Archean as at present. Since we know from the presence of rocks that formed in the Archean that the planet was not molten then, this heat must have been lost by convection of the mantle. It is possible that the temperatures and geothermal gradients were 10–25 percent hotter in the mantle during the Archean, but most of the extra heat was likely lost by more rapid convection, and by the formation and cooling of oceanic lithosphere in greater volumes. The formation and cooling of oceanic lithosphere is presently the most efficient mechanism of global heat loss through the crust, and it is likely that the most efficient mechanism was even more efficient in times of higher heat production. A highly probable scenario for removing the additional heat is that there were more ridges, producing thicker piles of lava, and moving at faster rates in the Archean as compared with the present. However, there is currently much debate and uncertainty about the partitioning of heat loss among these mechanisms, and it is also possible that changes in mantle viscosity and plate buoyancy would have led to slower plate movements in the Archean as compared with the present.

Archean Granitoid Greenstone Terranes

Archean granitoid-greenstone terranes are one of the most distinctive components of Archean cratons. About 70–80 percent of the Archean crust consists of granitoid materials, most of which are compositionally tonalites and granodiorites. Many of these are intrusive into metamorphosed and deformed volcanic and sedimentary rocks in greenstone belts. Greenstone belts are generally strongly deformed and metamorphosed, linear to irregularly shaped assemblages of volcanic and sedimentary rocks. They derive their name from the green-colored metamorphic minerals chlorite and amphibole, reflecting the typical greenschist to amphibolite facies metamorphism of these belts. Early South African workers preferred to use the name schist belt for this assemblage of rocks, in reference to the generally highly deformed nature of the rocks. Volcanic rocks in greenstone belts most typically include basalt flows, many of which show pillow structures where they are not too intensely deformed, and lesser amounts of ultramafic, intermediate, and felsic rocks. Ultramafic volcanic rocks with quench-textures and high MgO contents, known as komatiite, are much more abundant in Archean greenstone belts than in younger orogenic belts, but they are generally only a minor component of greenstone belts. Some literature leads readers to believe that Archean greenstone belts are dominated by abundant komatiites; however, this is not true. There have been a inordinate number of studies of komatiites in greenstone belts since they are such an unusual and important rock type, but the number of studies does not relate to the abundance of the rock type. Sedimentary rocks in greenstone belts are predominantly graywacke-shale sequences (or their metamorphic equivalents), although conglomerates, carbonates, sandstones, and other sedimentary rocks are found in these belts as well.

Suites of granitoid rock that are now deformed and metamorphosed to granitic gneisses typically intrude the volcanic and sedimentary rocks of the greenstone belts. The deformation of the belts has in many cases obscured the original relationships between many greenstone belts and gneiss terrains. Most of the granitoid rocks appear to intrude the greenstones, but in some belts older groups of granitic gneisses have been identified. In these cases it has been important to determine the original contact relationships between granitic gneisses and greenstone belts, as this relates to the very uncertain tec-

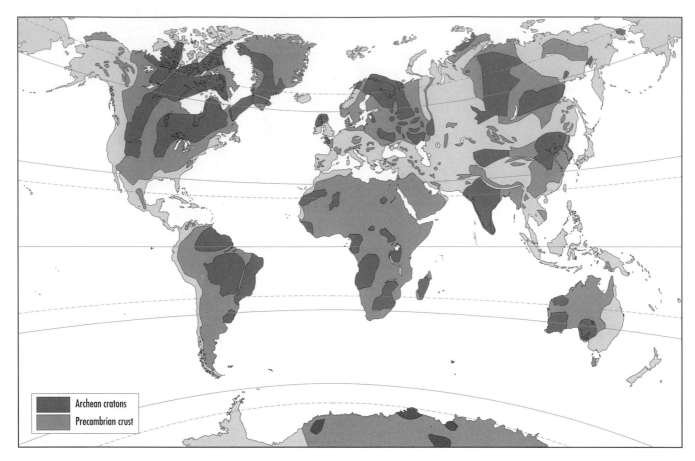

Map of the world showing distribution of Archean, Proterozoic, and Phanerozoic rocks

tonic setting of the Archean greenstones. If contact relationships show that the greenstone belts were deposited unconformably over the granitoid gneisses, then it can be supposed that greenstone belts represent a kind of continental tectonic environment that is unique to the Archean. In contrast, if contact relationships show that the greenstone belts were faulted against or thrust over the granitoid gneisses, then the greenstone belts may be allochthonous (far-traveled) and represent closed ocean basins, island arcs, and other exotic terrains similar to orogenic belts of younger ages.

Prior to the mid-1980s and 1990s, many geologists believed that many if not most greenstone belts were deposited unconformably over the granitoid gneisses, based on a few well-preserved examples at Belingwe, Zimbabwe; Point Lake, Canada; Yellowknife and Cameron River, Canada; Steep Rock Lake, Canada; and in the Yilgarn, western Australia. However, more recent mapping and structural work on these contact relationships has revealed that all of them have large-scale thrust fault contacts between the main greenstone belt assemblages and the granitoid gneisses, and these belts have since been reinterpreted as allochthonous oceanic and island arc deposits similar to those of younger mountain belts.

There seems to be an age-dependent variation in the style of Archean greenstone belts. Belts older than 3.5 billion years have sediments, including chert, banded iron formation, evaporites, and stromatolitic carbonates, indicating shallow water deposition, and contain only very rare conglomerates. They also have more abundant komatiites than younger greenstone belts. Younger greenstone belts seem to contain more intermediate volcanic rocks, such as andesites, and have more deep-water sediments and conglomerates. They also contain banded iron formations, stromatolitic carbonates, and chert. However, since there are so few early Archean greenstone belts preserved, it is difficult to tell if these apparent temporal variations represent real time differences in the style of global tectonics, or if they are a preservational artifact.

Archean Gneiss Terranes
High-grade granitoid gneiss terranes form the second main type of Archean terrane. Examples include the Limpopo belt of southern Africa, the Lewisian of the North Atlantic Province, the Hengshan of North China, and some less-well-documented belts in Siberia and Antarctica. The high-grade gneiss assemblage seems similar in many ways to the lower-

grade greenstone belts, but more strongly deformed and metamorphosed, reflecting burial to 12.5–25 miles (20–40 km) depth. Strongly deformed mylonitic gneisses and partially melted rocks known as migmatites are common, reflecting the high degrees of deformation and metamorphism. Most of the rocks in the high-grade gneiss terranes are metamorphosed sedimentary rocks including sandstones, graywackes, and carbonates, as well as layers of volcanic rocks. Many are thought to be strongly deformed continental margin sequences with greenstone-type assemblages thrust over them, deformed during continent-continent collisions. Most high-grade gneiss terranes have been intruded by several generations of mafic dikes, reflecting crustal extension. These are typically deformed into boudins (thin layers in the gneiss), making them difficult to recognize. Some high-grade gneiss terranes also have large layered mafic/ultramafic intrusions, some of which are related to the mafic dike swarms.

The strong deformation and metamorphism in the Archean high-grade gneiss terranes indicates that they have been in continental crust that has been thickened to double crustal thicknesses of about 50 miles (80 km), and some even

more. This scale of crustal thickening is typically associated with continental collisions and/or thickened plateaus related to Andean style magmatism. High-grade gneiss terrains are therefore typically thought to represent continent-continent collision zones.

Archean Cratonic Basins

A third style of rock association also typifies the Archean but is less common than the previous two associations. The cratonic basin association is characterized by little-deformed and metamorphosed sequences of clastic and carbonate sedimentary rocks, with a few intercalated volcanic horizons. This category of Archean sequence is most well developed on South Africa's Kaapvaal craton and includes the Pongola, Witwatersrand, Ventersdorp, and Transvaal Supergroups. These groups include sequences of quartzites, sandstones, arkoses, carbonates, and volcanic rocks, deposited in shallow marine to lacustrine basins. They are interpreted to represent rift, foreland basin, and shallow marine cratonic epeirogenic sea type deposits, fortuitously preserved although only slightly metamorphosed and deformed. Several shallow-water carbonate

Pillow lavas from the 2.6-billion-year-old Yellowknife greenstone belt, Nunavet *(Photo by Timothy Kusky)*

High-grade metamorphic gneiss from the 3.5-billion-year-old Tokwe gneiss terrain, Zimbabwe *(Photo by Timothy Kusky)*

shelf associations are also preserved, including the 3.0 Ga Steep Rock platform in Canada's Superior Province, and possibly also the 2.5 Ga Hamersley Group in western Australia.

Archean Life

It is clear that life had already been established on Earth by the Early Archean. The geologic setting and origin of life are topics of current intense interest, research, and thought by scientists and theologians. Any models for the origin of life need to explain some observations about early life from Archean rock sequences.

Evidence for early life comes from two separate lines. The first includes remains of organic compounds and chemical signatures of early life, and the other line consists of fossils, microfossils, and microstructures. The best organic evidence for early life comes from kerogens, which are nonsoluble organic compounds or the non-extractable remains of early life that formed at the same time as the sediments that they are found in. Other extractable organic compounds such as amino acids, fats, and sugars may also represent remains of early life, but these substances are very soluble in water and may have entered the rocks much after the deposition of

the sediments. Therefore, most work on the biochemistry of early life has focused on the non-extractable kerogens. Biological activity changes the ratio of some isotopes, most notably $^{13}C/^{12}C$, producing a distinctive biomarker that is similar in Archean through present-day life. Such chemical evidence of early life has been documented in Earth's oldest sedimentary rocks, the 3.8 Ga Isua belt in Greenland.

The earliest known fossils come from the 3.5–3.6 Ga Apex chert of the Pilbara craton in western Australia. Three distinctive types of microfossils have been documented from the Apex chert. These include spheroidal bodies, 5–20 microns in diameter, some of which have been preserved in the apparent act of cell division. These microfossils are similar to some modern cyanobacteria and show most clearly that unicellular life was in existence on Earth by 3.5 Ga. Simple rod-shaped microfossils up to 1 micron long are also found in the Apex chert, and these have shapes and characteristics that are also remarkably similar to modern bacteria. Finally, less-distinctive filamentous structures up to several microns long may also be microfossils, but they are less convincing than the spheroidal and rod-shaped bodies. All of these show, however, that simple, single-celled probably prokaryotic life-forms were present on Earth by 3.5 billion years ago, one billion years after the Earth formed.

Stromatolites are a group of generally dome-shaped or conical mounds, or sheets of finely laminated sediments produced by organic activity. They were most likely produced by cyanobacteria, or algae, that alternately trapped sediment with filaments that protruded above the sediment/water interface and secreted a carbonate layer during times when little sediment was passing to be trapped. Stromatolites produced a distinctive layering by preserving this alternation between sediment trapping and secretion of carbonate layers. Stromatolites are common in the Archean and Proterozoic record and show that life was thriving in many places in shallow water and was not restricted to a few isolated locations. The oldest stromatolites known are in 3.6-billion-year-old sediments from the Pilbara craton of western Australia, with many examples in early, middle, and late Archean rock sequences. Stromatolites seem to have peaked in abundance in the Middle Proterozoic and largely disappeared in the late Proterozoic with the appearance of grazing metazoans.

See also ATMOSPHERE; BANDED IRON FORMATION; CRATONS; HADEAN; LIFE'S ORIGINS AND EARLY EVOLUTION; PHANEROZOIC; PROTEROZOIC; STROMATOLITE.

Further Reading

Burke, Kevin, William S. F. Kidd, and Timothy M. Kusky. "Archean Foreland Basin Tectonics in the Witwatersrand, South Africa." *Tectonics* 5, no. 3 (1986): 439–456.

———. "The Pongola Structure of Southeastern Africa: The World's Oldest Recognized Well-Preserved Rift." *Journal of Geodynamics* 2, no. 1 (1985): 35–50.

———. "Is the Ventersdorp Rift System of Southern Africa Related to a Continental Collision between the Kaapvaal and Zimbabwe Cratons at 2.64 Ga Ago?" *Tectonophysics* 11 (1985): 1–24.

Kusky, Timothy M. "Structural Development of an Archean Orogen, Western Point Lake, Northwest Territories." *Tectonics* 10, no. 4 (1991): 820–841.

———. "Evidence for Archean Ocean Opening and Closing in the Southern Slave Province." *Tectonics* 9, no. 6 (1990): 1,533–1,563.

———. "Accretion of the Archean Slave Province." *Geology* 17 (1989): 63–67.

Kusky, Timothy M., Z. H. Li, Adam Glass, and H. A. Huang. "Archean Ophiolites and Ophiolite Fragments of the North China Craton." *Earth Science Reviews*, in review.

Kusky, Timothy M., and Ali Polat. "Growth of Granite-Greenstone Terranes at Convergent Margins and Stabilization of Archean Cratons." In *Tectonics of Continental Interiors*, edited by Stephen Marshak and Ben van der Pluijm, 43–73. Tectonophysics (1999).

Kusky, Timothy M., and Peter J. Hudleston. "Growth and Demise of an Archean Carbonate Platform, Steep Rock Lake, Ontario Canada." *Canadian Journal of Earth Sciences* 36 (1999): 1–20.

Kusky, Timothy M., and Julian Vearncombe. "Structure of Archean Greenstone Belts." Chap. 3 in *Tectonic Evolution of Greenstone Belts*, edited by Maarten J. de Wit and Lewis D. Ashwal, 95–128. Oxford Monograph on Geology and Geophysics (1997).

Kusky, Timothy M., and Pamela A. Winsky. "Structural Relationships along a Greenstone/Shallow Water Shelf Contact, Belingwe Greenstone Belt, Zimbabwe." *Tectonics* 14, no. 2 (1995): 448–471.

Kusky, Timothy M., and William S. F. Kidd. "Remnants of an Archean Oceanic Plateau, Belingwe Greenstone Belt, Zimbabwe." *Geology* 20, no. 1 (1992): 43–46.

Kusky, Timothy M., ed., *Precambrian Ophiolites and Related Rocks, Developments in Precambrian Geology XX*. Amsterdam: Elsevier Publishers, 2003.

McClendon, John H. "The Origin of Life." *Earth Science Reviews* 47 (1999): 71–93.

Schopf, William J. *Cradle of Life, The Discovery of Earth's Earliest Fossils*. Princeton: Princeton University Press, 1999.

artesian system *See* AQUIFER; ARTESIAN WELL.

artesian well A well that taps confined groundwater. In many regions, a permeable layer, typically sandstone, is confined between two impermeable beds, creating a confined aquifer. In these situations, water only enters the system in a small recharge area, and if this is in the mountains, then the aquifer may be under considerable pressure. This is known as an artesian system. Water that escapes the system from a fracture or well reflects the pressure difference between the elevation of the source area and the discharge area (hydraulic gradient) and rises above the groundwater level in the aquifer as an artesian spring, or artesian well. The water may or may not rise above the ground surface. Some scientists use the term *artesian well* only to refer to wells in which the water does rise above the ground surface. Some of these wells have made fountains that have spewed water 200 feet (60 m) high.

One example of an artesian system is in Florida, where water enters in the recharge area and is released near Miami about 19,000 years later. Other examples are abundant east of the Rocky Mountains of the western United States, in the fens of the United Kingdom, and in some desert oases of Egypt and North Africa.

See also GROUNDWATER.

asbestos A commercial term for a group of silicate minerals that form thin, strong, heat resistant fibers. These minerals include several varieties from the asbestos group, as well as some varieties of amphibole. Asbestos was widely used as a flame retardant in buildings through the middle 1970s, and it is present in millions of buildings in the United States. It was also used in vinyl flooring, ceiling tiles, and roofing material. It is no longer used in construction since it was recognized that asbestos might cause certain types of diseases, including asbestosis (pneumoconiosis), a chronic lung disease. Asbestos particles get lodged in the lungs, and the lung tissue hardens around the particles, decreasing lung capacity. This decreased lung capacity causes the heart to work harder, leading to heart failure and death. Virtually all deaths from asbestosis can be attributed to long-term exposure to asbestos dust in the workplace before environmental regulations governing asbestos were put in place. A less common disease associated with asbestos is mesothelioma, a rare cancer of the lung and stomach linings. Asbestos has become one of the most devastating occupational hazards in U.S. history, costing billions of dollars for cleaning up asbestos in schools, offices, homes, and other buildings. Approximately $3 billion a year are currently spent on asbestos removal in the United States.

Asbestos is actually a group of six related minerals, all with similar physical and chemical properties. Asbestos includes minerals from the amphibole and serpentine groups that are long and needle-shaped, making it easy for them to get lodged in people's lungs. The Occupational Safety and Health Administration (OSHA) defined asbestos as having dimensions of greater than 5 micrometers (0.002 in.) long, with a length to width ratio of at least 3:1. The minerals in the amphibole group included in this definition are grunerite (known also as amosite), reibeckite (crocidolite), anthophyllite, tremolite, and actinolite, while the serpentine group mineral that fits the definition is chrysotile. Almost all of the asbestos used in the United States is chrysotile (known as white asbestos), while about 5 percent of the asbestos used is crocidolite (blue asbestos) and amosite (brown asbestos). There is currently considerable debate among geologists, policy makers, and medical officials on the relative threats from different kinds of asbestos.

In 1972 OSHA and the U.S. government began regulating the acceptable levels of asbestos fibers in the workplace. The Environmental Protection Agency (EPA) agreed, and declared asbestos a Class A carcinogen. The EPA composed the Asbestos Hazard Emergency Response Act, which was signed by President Reagan in 1986. OSHA gradually lowered the acceptable limits from a pre-regulated estimate of greater than 4,000 fibers per cubic inch (1,600 fibers per cm³), to 4 fibers per cubic inch (1.6 particles per cm³) in 1992. Responding to public fears about asbestosis, Congress passed a law requiring that any asbestos-bearing material that appeared to be visibly deteriorating must be removed and replaced by non-asbestos-bearing material. This remarkable ruling has resulted in billions of dollars being spent on asbestos removal, which in many cases may have been unnecessary. The asbestos can only be harmful if it is an airborne particle, and only long-term exposure to high concentrations leads to disease. In some cases it is estimated that the processes of removing the asbestos resulted in the inside air becoming more hazardous than before removal, as the remediation can cause many small particles to become airborne and fall as dust throughout the building.

Asbestos fibers in the environment have led to some serious environmental disasters, as the hazards were not appreciated during early mining operations before the late 1960s. One of the worst cases is the town of Wittenoom, Australia, where crocidolite was mined for 23 years (between 1943 and 1966). The mining was largely unregulated, and asbestos dust filled the air of the mine and the town, where the 20,000 people who lived in Wittenoom breathed the fibers in high concentrations daily. More than 10 percent or 2,300 people who lived in Wittenoom have since died of asbestosis, and the Australian government has condemned the town and is in the process of burying it in deep pits to rid the environment of the hazard.

In the United States, W. R. Grace and Co. in Libby, Montana, afflicted hundreds of people with asbestos-related diseases through mining operations. Vermiculite was mined at Libby from 1963 to 1990 and shipped to Minneapolis to make insulation products, but the vermiculite was mixed with the tremolite (amphibole) variety of asbestos. In 1990 the EPA tested residents of Libby and found that 18 percent of residents who had been there for at least six months had various stages of asbestosis, and that 49 percent of the W. R. Grace mine employees had asbestosis. The mine was closed down, and Libby is now being considered a potential superfund site by the EPA. The problem was not limited to Libby, however, and 24 workers at the processing plant in Minneapolis and one resident who lived near the factory have since died from asbestosis. The EPA and Minnesota's Department of Health are currently assessing the level of exposure of other nearby residents.

See also MINERALOGY.

Further Reading
"Asbestos: Try Not to Panic." *Consumer Reports* (July 1995): 468–469.
Ross, M. "The Health Effects of Mineral Dusts." In *The Environmental Geochemistry of Mineral Deposits; Part A: Processes, Techniques, and Health Issues. Society of Economic Geologists Reviews in Economic Geology* 6A (1999): 339–356.

asteroid Small to nearly planet-sized celestial bodies that are orbiting the Sun, asteroids are mostly found in the 250-million-mile (400-million-km) wide asteroid belt that lies between the orbits of Mars and Jupiter. There are six large gaps in the asteroid belt where virtually no asteroids are found, known as Kirkwood Gaps (after the mathematician Daniel Kirkwood), and they form from effects of Jupiter's gravitational attraction pushing objects out of orbits that are multiples of Jupiter's orbit. They generally have an irregular shape and rotate on their spin axis about once every 4–20 hours. There are at least one million stony and metallic asteroids in the asteroid belt, and some of these occasionally get knocked out of their regular orbit and placed into an Earth-crossing orbit. Some asteroids lie outside the main asteroid belt, and many remain in the same orbit as Jupiter. These asteroids, known as the Trojans, reside at two points of gravitational stability known as Lagrange Points on either side of Jupiter. It is estimated that there are as many Trojans as there are asteroids in the main asteroid belt.

Asteroids are classified based on their composition and include three main classes—P-type or primitive asteroids are rich in carbon and water, and are thought to represent unaltered material left over from the formation of the solar system. They reside primarily in outer parts of the asteroid belt. C-type asteroids are metamorphic and, like the P-type asteroids, contain abundant carbon, but their water has been removed during heating and metamorphism. They reside mainly in the center part of the asteroid belt. S-type asteroids are igneous and represent the most common type of meteorite found on Earth. These asteroids were partially melted and have a composition similar to that of the Earth's mantle. S-type asteroids reside in the inner part of the asteroid belt. Some of these classes are thought by some scientists to represent a planet that broke up during a massive collision early in the history of the solar system. The metallic or iron meteorites would represent the core of this hypothesized destroyed planet, whereas the stony meteorite would represent the mantle and crust. Other scientists believe that the asteroids never

formed a single planet but represent several different planetsimals that never coalesced but have experienced many collisions, forming the metamorphism and partial melting observed in some meteorites.

More than 2,000 asteroids that are larger than a kilometer wide are in Earth-crossing orbits. When asteroids enter the Earth's atmosphere, their outer surface burns up and creates a fiery streak moving across the sky. Asteroids that enter the Earth's atmosphere are known as meteorites. Small meteorites burn up completely before hitting the Earth, whereas larger ones may reach the Earth before burning up.

Asteroids, comets, and meteorites have been objects of fascination, speculation, and fear for most of recorded human history. Early peoples thought that fiery streaks in the sky were omens of ill fortune and sought refuge from their evil powers. Impacts of comets and meteorites with Earth are now recognized as the main cause for several periods of mass extinction on the planet, including termination of the dinosaurs 65 million years ago. Comets and meteorites may

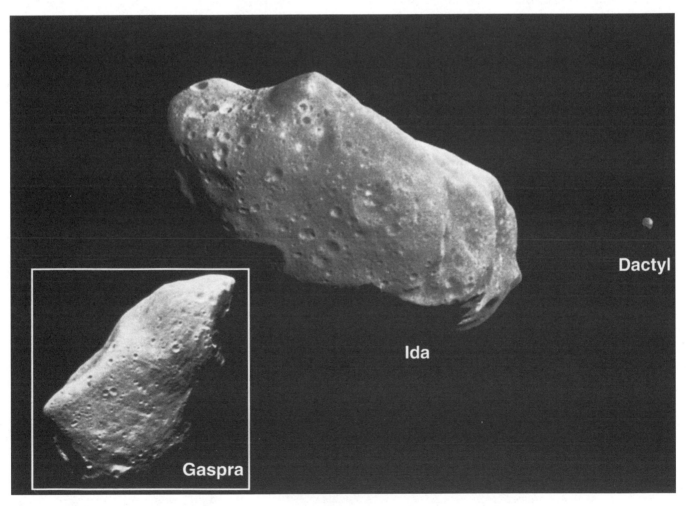

A view of asteroid Ida, with moon Dactyl, and asteroid Gaspra (inset), taken from *Galileo* *(Photo courtesy of NASA)*

also have brought much of the water, air, and perhaps even life to Earth, and the planets themselves coalesced from numerous smaller asteroids, comets, and interplanetary dust. Asteroids, comets, and meteorites were therefore essential for the formation of Earth and life and also responsible for the extinction of many species.

New asteroids are constantly being discovered as our powers of observation become better. The mass of currently known asteroids in the main asteroid belt is about half of that of the Earth's moon, but it was probably much higher in the earlier history of the solar system before many of the asteroids collided with other planets. Some large asteroids were discovered in the last century. For instance, the nearly 620-mile (1,000-km) wide asteroid Ceres was discovered in 1801 by the Italian astronomer Giuseppe Piazzi. The three largest known asteroids are Ceres, Pallas, and Vesta, each several hundred kilometers wide. More than 1,000 asteroids are larger than 20 miles (30 km) wide, and it is estimated that more than a million are more than 2,625 feet (800 m) across. We only know the orbits of about 18,000 of these asteroids.

The Earth has experienced large impacts from meteorites. Several have caused 50 to 90 percent of all species alive on the planet at the time of the impact to go extinct, paving the way for the evolution and diversification of new organisms. The impact of a six-mile (10-km) wide meteorite with the Yucatán Peninsula is now thought to have ended the reign of the dinosaurs. The impact instantly formed a fireball 1,240 miles (2,000 km) across, followed by tsunami hundreds of meters tall. The dust thrown out of the deep crater excavated by the impact plunged the world into a fiery darkness, then months or even years of freezing temperatures. As soon as the dust settled, carbon dioxide released by the impact caused Earth to soar into an intense greenhouse warming. Few species handled these changing environmental stresses well, and 65 percent of all species went extinct.

NASA and other agencies have recently begun to plan possible defenses that the human race might mount against any asteroid or comets on an impact course with Earth. The future of the human race may well depend on increased awareness of how to handle this potential threat. In 1996 an asteroid about 2,300 feet (700 m) across barely missed hitting Earth, speeding past at a distance about equal to the distance to the moon. The sobering reality of this near collision is that the asteroid was not even spotted until a few days before it sped past Earth. What if the object was bigger, or slightly closer? Would it have been stoppable, and if not, what would have been the consequences of its collision with Earth?

See also COMET; METEOR.

Further Reading

Erickson, J. *Asteroids, Comets, and Meteorites: Cosmic Invaders of the Earth*. New York: Facts On File, 2003.

asthenosphere   The layer of the mantle between the lithosphere and mesosphere. Its depth in the Earth ranges from about 155 miles (250 km) to 0 miles below the mid-ocean ridges, and 31–62 miles (50–100 km) below different parts of the continents and oceans. Some old continental cratons have deep roots that extend deeper into the asthenosphere. It is characterized by small amounts (1–10 percent) of partial melt that greatly reduces the strength of the layer and is thought to be the layer across which much of the movement of the plates and vertical isostatic motions are accommodated. The name derives from the Greek for "weak sphere." The asthenosphere is clearly demarcated by S-wave seismic velocities, which show a dramatic drop through the asthenosphere because of the partial melt present in this zone. The asthenosphere is also known therefore as the low velocity zone, and it shows the greatest attenuation of seismic waves anywhere in the Earth.

The asthenosphere is composed of the rock type peridotite, consisting primarily of olivine, with smaller amounts of orthopyroxene, clinopyroxene, and other minerals including spinels such as chromite. The asthenosphere is flowing in response to heat loss in the deep Earth, and there is a current debate on the relative coupling between the flowing asthenosphere and the overlying lithosphere. In some models, the convection in the asthenosphere exerts a considerable mantle drag force on the base of the lithosphere and significantly influences plate motions. In other models, the lithosphere and asthenosphere are thought to be largely uncoupled, with the driving forces for plate tectonics being more related to the balance between the gravitational ridge push force, slab pull force, the slab drag force, transform resistance force, and the subduction resistance force. There is also a current debate on the relationship between upper mantle (asthenosphere) convection, and convection in the mesosphere. Some models have double or several layers of convection, whereas other models purport that the entire mantle is convecting as a single layer.

See also CONVECTION AND THE EARTH'S MANTLE; MANTLE; PLATE TECTONICS.

Atacama Desert   An elevated arid region located in northern Chile, extending over 384 square miles (1,000 km²) south from the border with Peru. The desert is located 2,000 feet (600 m) above sea level and is characterized by numerous dry salt basins (playas), flanked on the east by the Andes and on the west by the Pacific coastal range. The Atacama is one of the driest places on Earth, with no rain ever recorded in many places, and practically no vegetation in the region. Nitrate and copper are mined extensively in the region.

The Atacama is first known to have been crossed by the Spanish conquistador Diego de Almagro in 1537, but was ignored until the mid-19th century, when mining of nitrates in the desert began. However, after World War I, synthetic nitrates were developed, and the region has been experiencing economic decline.

Atlas Mountains A series of mountains and plateaus in northwest Africa extending about 1,500 miles (2,500 km) in southwest Morocco, northern Algeria, and northern Tunisia. The highest peak in the Atlas is Jabel Toubkal at 13,665 feet (4,168 m) in southwest Morocco. The Atlas are dominantly folded sedimentary rocks uplifted in the Jurassic and related to the Alpine system of Europe. The Atlas consists of several ranges separated by fertile lowlands in Morocco, from north to south including the Rif Atlas, Middle Atlas, High Atlas (Grand Atlas), and the Anti Atlas. The Algerian Atlas consists of a series of plateaus including the Tell and Saharan Atlas rimming the Chotts Plateau, then converging in Tunisia. The Atlas form a climatic barrier between the Atlantic and Mediterranean basins and the Sahara, with rainfall falling on north-facing slopes, but arid conditions dominating on the rain shadow south-facing slopes. The Atlas are rich in mineral deposits, including coal, iron, oil, and phosphates. The area is also used extensively for sheep grazing, with farming in the more fertile intermountain basins.

atmosphere Thin sphere around the Earth consisting of the mixture of gases we call air, held in place by gravity. The most abundant gas is nitrogen (78 percent), followed by oxygen (21 percent), argon (0.9 percent), carbon dioxide (0.036 percent), and minor amounts of helium, krypton, neon, and xenon. Atmospheric (or air) pressure is the force per unit area (similar to weight) that the air above a certain point exerts on any object below it. Atmospheric pressure causes most of the volume of the atmosphere to be compressed to 3.4 miles (5.5 km) above the Earth's surface, even though the entire atmosphere is hundreds of kilometers thick.

The atmosphere is always moving, because more of the Sun's heat is received per unit area at the equator than at the poles. The heated air expands and rises to where it spreads out, then it cools and sinks, and gradually returns to the equator. This pattern of global air circulation forms Hadley cells that mix air between the equator and mid-latitudes. Similar circulation cells mix air in middle to high latitudes, and between the poles and high latitudes. The effects of the Earth's rotation modify this simple picture of the atmosphere's circulation. The Coriolis effect causes any freely moving body in the Northern Hemisphere to veer to the right, and toward the left in the Southern Hemisphere. The combination of these effects forms the familiar trade winds, easterlies and westerlies, and doldrums.

The atmosphere is divided into several layers, based mainly on the vertical temperature gradients that vary significantly with height. Atmospheric pressure and air density both decrease more uniformly with height, and therefore they do not serve as a useful way to differentiate the atmospheric layers.

The lower 36,000 feet (11 km) of the atmosphere is known as the troposphere, where the temperature generally decreases gradually, at about 7.0°F per mile (6.4°C per km), with increasing height above the surface. This is because the Sun heats the surface, which in turn warms the lower part of the troposphere. Most of the atmospheric and weather phenomena we are familiar with occur in the troposphere.

Above the troposphere is a boundary region known as the tropopause, marking the transition into the stratosphere that continues to a height of about 31 miles (50 km). The base of the stratosphere contains a region known as an isothermal, where the temperature remains the same with increasing height. The tropopause is generally at higher elevations in the summer than the winter, and it is also the region where the jet streams are located. Jet streams are narrow, streamlike channels of air that flow at high velocities, often exceeding 115 miles per hour (100 knots). Above about 12.5 miles (20 km), the isothermal region gives way to the upper stratosphere where temperatures increase with height, back to near surface temperatures at 31 miles (50 km). The heating of the stratosphere is due to ozone at this level absorbing ultraviolet radiation from the Sun.

The mesosphere lies above the stratosphere, extending between 31 and 53 miles (50–85 km). An isothermal region known as the stratopause separates the stratosphere and mesosphere. The air temperature in the mesosphere decreases dramatically above the stratopause, reaching a low of –130°F (–90°C) at the top of the mesosphere. The mesopause separates the mesosphere from the thermosphere, which is a hot layer where temperatures rise to more than 150°F (80°C). The relatively few oxygen atoms at this level absorb solar energy, heat quickly, and may change dramatically in response to changing solar activity. The thermosphere continues to thin upward, extending to about 311 miles (500 km) above the surface. Above this level, atoms dissociate and are able to shoot outward and escape the gravitational pull of Earth. This far region of the atmosphere is sometimes referred to as the exosphere.

In addition to the temperature-based division of the atmosphere, it is possible to divide the atmosphere into different regions based on their chemical and other properties. Using such a scheme, the lower 46.5–62 miles (75–100 km) of the atmosphere may be referred to as the homosphere, where the atmosphere is well mixed and has a fairly uniform ratio of gases from base to top. In the overlying heterosphere, the denser gases (oxygen, nitrogen) have settled to the base, whereas lighter gases (hydrogen, helium) have risen to greater heights, resulting in chemical differences with height.

The upper parts of the homosphere and the heterosphere contain a large number of electrically charged particles known as ions. This region is known also as the ionosphere, which strongly influences radio transmission and the formation of the aurora borealis and aurora australis.

Atmospheric gases are being produced at approximately the same rate that they are being destroyed or removed from the atmospheric system, although some gases are gradually increasing or decreasing in abundance as described below. Soil bacteria and other biologic agents remove nitrogen from the atmosphere, whereas decay of organic material releases nitrogen back to the atmosphere. However, decaying organic material removes oxygen from the atmosphere by combining with other substances to produce oxides. Animals also remove oxygen from the atmosphere by breathing, whereas oxygen is added back to the atmosphere through photosynthesis.

Water vapor is an extremely important gas in the atmosphere, but it varies greatly in concentration (0–4 percent) from place to place, and from time to time. Water vapor is invisible, and it becomes visible as clouds, fog, ice, and rain when the water molecules coalesce into larger groups. Water forms water vapor gas, liquid, and solid, and constitutes the precipitation that falls to Earth and is the basis for the hydrologic cycle. Water vapor is also a major factor in heat transfer in the atmosphere. A kind of heat known as latent heat is released when water vapor turns into solid ice or liquid water. This heat is a major source of atmospheric energy that is a major contributor to the formation of thunderstorms, hurricanes, and other weather phenomena. Water vapor may also play a longer-term role in atmospheric regulation, as it is a greenhouse gas that absorbs a significant portion of the outgoing radiation from the Earth, causing the atmosphere to warm.

Carbon dioxide, although small in concentration, is another very important gas in the Earth's atmosphere. Carbon dioxide is produced during decay of organic material, from volcanic outgassing, from cow, termite, and other animal emissions, deforestation, and from the burning of fossil fuels. It is taken up by plants during photosynthesis and is also used by many marine organisms for their shells, made of $CaCO_3$ (calcium carbonate). When these organisms (for instance, phytoplankton) die, their shells can sink to the bottom of the ocean and be buried, removing carbon dioxide from the atmospheric system. Like water vapor, carbon dioxide is a greenhouse gas that traps some of the outgoing solar radiation that is reflected from the Earth, causing the atmosphere to warm up. Because carbon dioxide is released by the burning of fossil fuels, its concentration is increasing in the atmosphere as humans consume more fuel. The concentration of CO_2 in the atmosphere has increased by 15 percent since 1958, enough to cause considerable global warming. It is estimated that the concentration of CO_2 will increase by another 35 percent by the end of the 21st century, further enhancing global warming.

Other gases also contribute to the greenhouse effect, notably methane (CH_4), nitrous oxide (NO_2) and chlorofluorocarbons (CFCs). Methane is increasing in concentration in the atmosphere and is produced by the breakdown of organic material by bacteria in rice paddies and other environments, termites, and in the stomachs of cows. NO_2, produced by microbes in the soil, is also increasing in concentration by 1 percent every few years, even though it is destroyed by ultraviolet radiation in the atmosphere. Chlorofluorocarbons have received a large amount of attention since they are long-lived greenhouse gases increasing in atmospheric concentration as a result of human activity. Chlorofluorocarbons trap heat like other greenhouse gases and also destroy ozone (O_3), our protective blanket that shields the Earth from harmful ultraviolet radiation. Chlorofluorocarbons were used widely as refrigerants and as propellants in spray cans. Their use has been largely curtailed, but since they have such a long residence time in the atmosphere, they are still destroying ozone and contributing to global warming and will continue to do so for many years.

Ozone (O_3) is found primarily in the upper atmosphere where free oxygen atoms combine with oxygen molecules (O_2) in the stratosphere. The loss of ozone has been dramatic in recent years, even leading to the formation of "ozone holes" with virtually no ozone present above the Arctic and Antarctic in the fall. There is currently debate about how much of the ozone loss is due to human-induced ozone loss by chlorofluorocarbon production, and how much may be related to natural fluctuations in ozone concentration.

Many other gases and particulate matter play important roles in atmospheric phenomena. For instance, small amounts of sulfur dioxide (SO_2) produced by the burning of fossil fuels mix with water to form sulfuric acid, the main harmful component of acid rain. Acid rain is killing the biota of many natural lake systems, particularly in the northeastern United States, and it is causing a wide range of other environmental problems across the world. Other pollutants are major causes of respiratory problems, environmental degradation, and the major increase in particulate matter in the atmosphere in the past century has increased the hazards and health effects from these atmospheric particles.

Formation and Evolution of the Atmosphere

There is considerable uncertainty about the origin and composition of the Earth's earliest atmosphere. Many models assume that methane and ammonia, instead of nitrogen and carbon dioxide, dominated the planet's early atmosphere as it is presently. The gases that formed the early atmosphere could have come from volcanic outgassing by volcanoes, from extraterrestrial sources (principally cometary impacts), or, most likely, both. It is also likely that comets brought organic molecules to Earth. A very large late impact is thought to have melted outer parts of the Earth, formed the Moon, and blown away the earliest atmosphere. The present atmosphere must therefore represent a later, secondary atmosphere formed after this late impact.

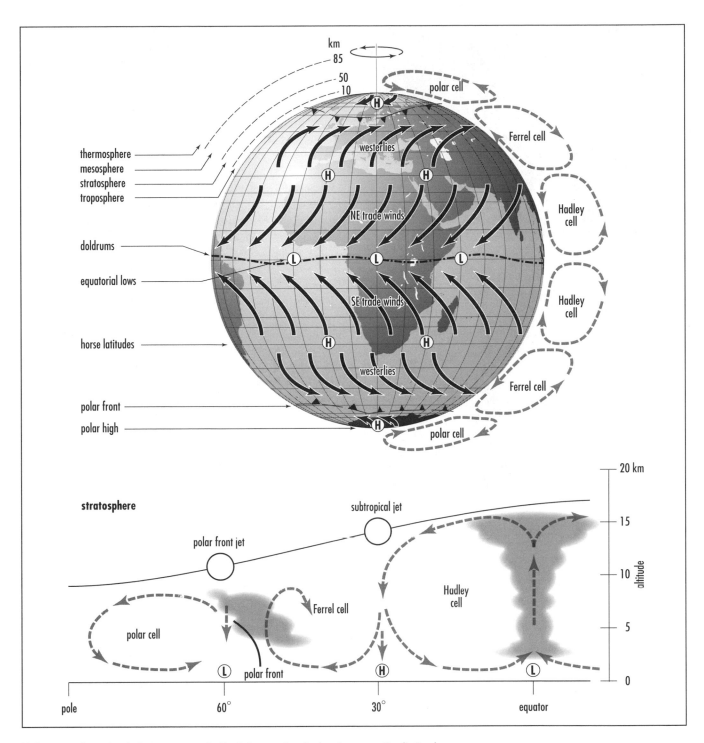

Major atmosphere circulation patterns on the Earth, in map view (top) and cross section (bottom)

During the early Archean, the Sun was only about 70 percent as luminous as it is presently, so the Earth must have experienced a greenhouse warming effect to keep temperatures above the freezing point of water, but below the boiling point. Increased levels of carbon dioxide and ammonia in the early atmosphere could have acted as greenhouse gases, accounting for the remarkable maintenance of global temperatures within the stability field of liquid water, allowing the development of life. Much of the carbon dioxide that was in the early atmosphere is now locked up in deposits of sedi-

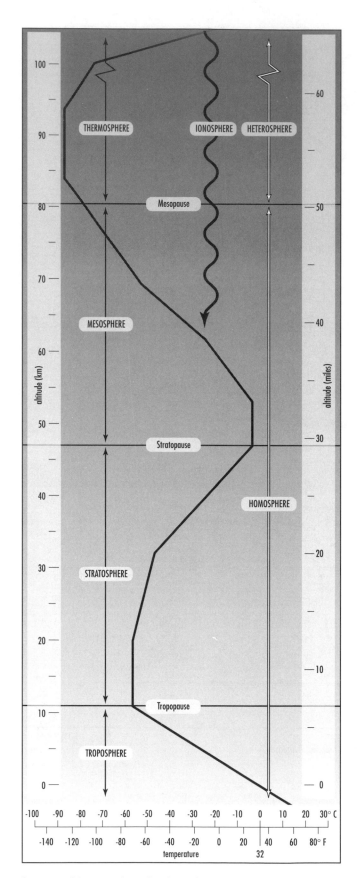

Structure of the atmosphere showing various layers and temperature profile with height

mentary limestone, and in the planet's biomass. The carbon dioxide that shielded the early Earth and kept temperatures in the range suitable for life to evolve now forms the bodies and remains of those very life-forms.

Role of the Atmosphere in Global Climate

Interactions between the atmosphere, hydrosphere, biosphere, and lithosphere control global climate. Global climate represents a balance between the amount of solar radiation received and the amount of this energy that is retained in a given area. The planet receives about 2.4 times as much heat in the equatorial regions as in the polar regions. The atmosphere and oceans respond to this unequal heating by setting up currents and circulation systems that redistribute the heat more equally. These circulation patterns are in turn affected by the ever-changing pattern of the distribution of continents, oceans, and mountain ranges.

The amounts and types of gases in the atmosphere can modify the amount of incoming solar radiation, and hence global temperature. For instance, cloud cover can cause much of the incoming solar radiation to be reflected back to space before being trapped by the lower atmosphere. On the other hand, greenhouse gases allow incoming short-wavelength solar radiation to enter the atmosphere but trap this radiation when it tries to escape in its longer-wavelength reflected form. This causes a buildup of heat in the atmosphere and can lead to a global warming known as the greenhouse effect.

The amount of heat trapped in the atmosphere by greenhouse gases has varied greatly over Earth's history. One of the most important greenhouse gases is carbon dioxide (CO_2). Plants, which release O_2 to the atmosphere, now take up CO_2, by photosynthesis. In the early part of Earth's history (in the Precambrian before plants covered the land surface), photosynthesis did not remove CO_2 from the atmosphere, with the result that CO_2 levels were much higher than at present. Atmospheric CO_2 is also presently taken up by marine organisms that remove it from the ocean surface water (which is in equilibrium with the atmosphere) and use the CO_2 along with calcium to form their shells and mineralized tissue. These organisms make $CaCO_3$ (calcite), which is the main component of limestone, a rock composed largely of the dead remains of marine organisms. Approximately 99 percent of the planet's CO_2 is presently removed from the atmosphere/ocean system because it is locked up in rock deposits of limestone on the continents and on the seafloor. If this amount of CO_2 were released back into the atmosphere, the global temperature would increase dramatically. In the early Precambrian, when this CO_2 was free in the atmosphere, global temperatures averaged about 550°F (290°C).

The atmosphere redistributes heat quickly by forming and redistributing clouds and uncondensed water vapor around the planet along atmospheric circulation cells. Oceans are able to hold and redistribute more heat because of the greater amount of water in the oceans, but they redistribute this heat more slowly than the atmosphere. Surface currents are formed in response to wind patterns, but deep ocean currents that move more of the planet's heat follow courses that are more related to the bathymetry (topography of the seafloor) and the spinning of the Earth than they are related to surface winds.

The balance of incoming and outgoing heat from the Earth has determined the overall temperature of the planet through time. Examination of the geological record has enabled paleoclimatologists to reconstruct periods when the Earth had glacial periods, hot dry periods, hot wet periods, or cold dry periods. In most cases, the Earth has responded to these changes by expanding and contracting its climate belts. Warm periods see an expansion of the warm subtropical belts to high latitudes, and cold periods see an expansion of the cold climates of the poles to low latitudes.

See also AIR PRESSURE; AURORA; CLIMATE; GREENHOUSE EFFECT; WEATHERING.

Further Reading

Ashworth, William, and Charles E. Little. *Encyclopedia of Environmental Studies, New Edition.* New York: Facts On File, 2001.

Bekker, Andrey, H. Dick Holland, P. L. Wang, D. Rumble III, H. J. Stein, J. L. Hannah, L. L. Coetzee, and Nick Beukes. "Dating the Rise of Atmospheric Oxygen." *Nature* 427 (2004): 117–120.

Kasting, James F. "Earth's Early Atmosphere." *Science* 259 (1993): 920–925.

atoll Consisting of circular, elliptical, or semicircular islands made of coral reefs that rise from deep water, atolls surround central lagoons, typically with no internal landmass. Some atolls do have small central islands, and these, as well as parts of the outer circular reef, are in some cases covered by forests. Most atolls range in diameter from half a mile to more than 80 miles (1–130 km) and are most common in the western and central Pacific Ocean basin, and in the Indian Ocean. The outer margin of the semicircular reef on atolls is the most active site of coral growth, since it receives the most nutrients from upwelling waters on the margin of the atoll. On many atolls, coral growth on the outer margin is so intense that the corals form an overhanging ledge from which many blocks of coral break off during storms, forming a huge talus slope at the base of the atoll. Volcanic rocks, some of which lie more than half a mile (1 km) below current sea level, underlay atolls. Since corals can only grow in very shallow water less than 65 feet (20 m) deep, the volcanic islands must have formed near sea level, grown coral, and subsided with time, with the corals growing at the rate that the volcanic islands were sinking.

Charles Darwin proposed such an origin for atolls in 1842 based on his expeditions on the *Beagle* from 1831 to 1836. He suggested that volcanic islands were first formed with their peaks exposed above sea level. At this stage, coral reefs were established as fringing reef complexes around the volcanic island. He suggested that with time the volcanic islands subsided and were eroded, but that the growth of the coral reefs was able to keep up with the subsidence. In this way, as the volcanic islands sank below sea level, the coral reefs continued to grow and eventually formed a ring circling the location of the former volcanic island. When Darwin proposed this theory in 1842 he did not know that ancient eroded volcanic mountains underlay the atolls he studied. More than 100 years later, drilling confirmed his prediction that volcanic rocks would be found beneath the coralline rocks on several atolls.

With the advent of plate tectonics in the 1970s, the cause of the subsidence of the volcanoes became apparent. When oceanic crust is created at mid-ocean ridges, it is typically about 1.7 miles (2.7 km) below sea level. With time, as the oceanic crust moves away from the mid-ocean ridges, it cools and contracts, sinking to about 2.5 miles (4 km) below sea level. In many places on the seafloor, small volcanoes form on the oceanic crust a short time after the main part of the crust forms at the mid-ocean ridge. These volcanoes may stick up above sea level a few hundred meters. As the oceanic crust moves away from the mid-ocean ridges, these volcanoes subside below sea level. If the volcanoes happen to be in the tropics where corals can grow, and if the rate of subsidence is slow enough for the growth of coral to keep up with subsidence, then atolls may form where the volcanic island used to be. If corals do not grow or cannot keep up with subsidence, then the island subsides below sea level and the top of the island gets scoured by wave erosion, forming a flat-topped mountain that continues to subside below sea level. These flat-topped mountains are known as guyots, many of which were mapped during exploration of the seafloor associated with military operations of World War II.

See also CORALS; PLATE TECTONICS; SEAMOUNT.

aurora Aurora Borealis and Aurora Australis are glows that are sometimes visible in the Northern and Southern Hemispheres, respectively. They are informally known as the northern lights and the southern lights. The glows are strongest near the poles and originate in the Van Allen radiation belts, which are regions where high-energy charged particles of the solar wind that travel outward from the Sun are captured by the Earth's magnetic field. The outer Van Allen radiation belt consists mainly of protons, whereas the inner Van Allen belt consists mainly of electrons. At times, electrons spiral down toward Earth near the poles along magnetic field lines and collide with ions in the thermosphere, emitting light in the process. Light in the aurora is emitted between a base level of about 50–65 miles (80–105 km), and an upper level of about 125 miles (200 km) above the Earth's surface.

The solar wind originates when violent collisions between gases in the Sun emit electrons and protons, which escape the gravitational pull of the Sun and travel through space at about 250 miles per second (more than 1 million km/hr) as a plasma known as the solar wind. When these charged particles move close to Earth, they interact with the magnetic field, deforming it in the process. The natural undisturbed state of the Earth's magnetic field is broadly similar to a bar magnet, with magnetic flux lines (of equal magnetic intensity and direction) coming out of the south polar region and returning back into the north magnetic pole. The solar wind deforms this ideal state into a teardrop-shaped configuration known as the magnetosphere. The magnetosphere has a rounded compressed side facing the Sun, and a long tail (magnetotail) on the opposite side that stretches past the orbit of the moon. The magnetosphere shields the Earth from many of the charged particles from the Sun by deflecting them around the edge of the magnetosphere, causing them to flow harmlessly into the outer solar system.

The Sun periodically experiences periods of high activity when many solar flares and sunspots form. During these periods the solar wind is emitted with increased intensity, and the plasma is emitted with greater velocity, in greater density, and with more energy than in its normal state. During these periods of high solar activity the extra energy of the solar wind distorts the magnetosphere and causes more electrons to enter the Van Allen belts, causing increased auroral activity.

When the electrons from the magnetosphere are injected into the upper atmosphere, they collide with atoms and molecules of gases there. The process involves the transfer of energy from the high-energy particle from the magnetosphere to the gas molecule from the atmosphere, which becomes excited and temporarily jumps to a higher energy level. When the gas molecule returns to its normal, regular energy level, it releases radiation energy in the process. Some of this radiation is in the visible spectrum, forming the Aurora Borealis in the Northern Hemisphere and the Aurora Australis in the Southern Hemisphere.

Auroras typically form waving sheets, streaks, and glows of different colors in polar latitudes. The colors originate because different gases in the atmosphere emit different characteristic colors when excited by charged particles from the magnetosphere, and the flickering and draperies are caused by variations in the magnetic field and incoming charged particles. The auroras often form rings around the magnetic poles, being most intense where the magnetic field lines enter and exit the Earth at 60–70° latitude.

Aurora Australis   *See* AURORA.

Aurora Borealis   *See* AURORA.

avalanche   *See* MASS WASTING.

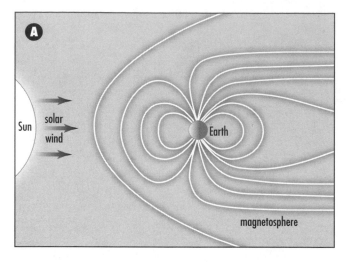

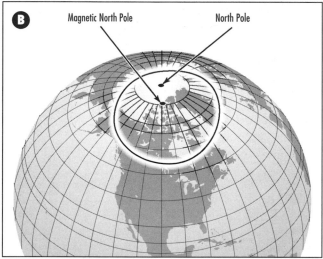

(A) Magnetosphere drawing showing asymmetric shape created by distortion of the Earth's magnetic field by the solar wind; (B) Earth showing typical auroral ring with the greatest intensity of auroral activity about 20–30° from the magnetic pole, where magnetic field lines are most intense

B

badlands An erosional landform that develops in arid to semiarid terranes, characterized by a very high drainage density (77 to 747 miles per square mile), steep slopes, and narrow ridges separating individual stream segments. They may have steep pinnacles, walls, and heavily gullied slopes leading into valleys that branch and bend forming a complex, patternless land surface. Badlands develop on poorly consolidated soils such as clay, silt, and, in some examples, water-soluble minerals including gypsum and anhydrite. The development of badlands is aided by poor drainage, a lack of vegetation, or loss of vegetation on the unconsolidated surface soils. The French fur traders (who called it the *mauvaises terres*) first used the

Badlands, South Dakota *(Photo by Timothy Kusky)*

term for an area in western South Dakota, part of which is now Badlands National Park. The complex land with many valleys and difficult access offered hideouts for bandits during the United States expansion in the west.

Baja California An elongate peninsula extending about 760 miles (1,225 km) south from southern California. It is separated from mainland Mexico by the Gulf of California, and bordered on the west by the Pacific Ocean. It is divided into the Mexican states of Baja California in the north and Baja California Sur in the south. The peninsula is dominated by rugged mountain ranges about 5,000 feet (1,500 m) high, and two large coastal plains on the Pacific side. The area is desolate and arid, but littered with archeological sites, and unique marine and other natural environments. Coastal Baja was explored by the Spaniards in the 1530s, and briefly occupied by American forces during 1853–54 in the Spanish-American War. The population is now more than one million, with most people in coastal cities.

Baltic shield The Baltic (Fennoscandian) shield is an Archean craton that traditionally is divided into three distinct parts. The northern, Lapland-Kola province, mainly consists of several previously dispersed Archean crustal terranes that together with the different Paleoproterozoic belts have been involved in a collisional-type orogeny at 2.0 billion to 1.9 billion years ago. A central, northwest-trending segment known as the Belomorian mobile belt is occupied by assemblages of gneisses and amphibolites. This part of the Baltic shield has experienced two major orogenic periods, in the Neoarchean and Paleoproterozoic. The Neoarchean period included several crust-forming events between 2.9 billion and 2.7 billion years ago that can be interpreted in terms of first subduction-related and later collisional orogeny. In the end of the Paleoproterozoic at 1.9–1.8 billion years ago, strong structural and thermal reworking occurred during an event of crustal stacking and thrusting referred to as the Svecofennian orogeny caused by overthrusting of the Lapland granulite belt onto the Belomorian belt. Although the Svecofennian high-grade metamorphism and folding affected all of the belt, its major Neoarchean crustal structure reveals that early thrust and fold nappes developed by 2.74–2.70 billion years ago. In contrast, the Karelian province displays no isotopic evidence for strong Paleoproterozoic reworking. The Karelian craton forms the core of the shield and largely consists of volcanic and sedimentary rocks (greenstones) and granites/gneisses that formed between 3.2 billion and 2.6 billion years ago and were metamorphosed at low-grade. Local synformal patches of Paleoproterozoic 2.45 billion to 1.9 billion year old volcano-sedimentary rocks unconformably overlie the Karelian basement. To the southwest of the Archean Karelian craton, the Svecofennian domain represents a large portion of Paleoproterozoic crust developed between 2.0 billion and 1.75 billion years ago.

Although tectonic settings of the Karelian Archean greenstone belts are still a matter of debate, there are some indications that subduction-accretion processes similar to modern-day convergent margins operated, at least, since 2.9 billion years ago. However, a large involvement of deep mantle-plume derived oceanic plateaus in Archean crustal growth processes remains questionable in respect to subduction style.

See also KOLA PENINSULA.

Further Reading

Shchipansky, Andrey, Andrei V. Samsonov, E. V. Bibikova, Irena I. Babarina, Alexander N. Konilov, K. A. Krylov, Aleksandr I. Slabunov, and M. M. Bogina. "2.8 Ga Boninite-Hosting Partial Suprasubduction Zone Ophiolite Sequences from the North Karelian Greenstone Belt, NE Baltic Shield, Russia." In *Precambrian Ophiolites and Related Rocks, Developments in Precambrian Geology,* edited by Tim Kusky. Amsterdam: Elsevier, 2004.

banded iron formation A distinctive type of sedimentary rock that formed predominantly during the Precambrian and is the major source of the world's iron reserves. Banded iron formations (BIFs) are a thinly bedded, chemically precipitated, iron-rich rock, with layers of iron ore minerals typically interbedded with thin layers of chert or microcrystalline silica. Many are completely devoid of detrital or clastic sedimentary input. Most banded iron formations formed between 2.6 billion and 1.8 billion years ago, and only a few very small similar types of deposits have been discovered in younger mountain belts. This observation suggests that the conditions necessary to form the BIFs were present on Earth in early (Precambrian) time, but largely disappeared by 1.8 billion years ago. The chemical composition and reduced state of much of the iron of BIFs suggest that they may have formed in an oxygen-poor atmosphere/ocean system, explaining their disappearance around the time that atmospheric oxygen was on the rise. BIFs may also be intimately associated with early biological activity and may preserve the record of the development of life on Earth. The world's oldest BIF is located in the 3.8-billion-year-old Isua belt in southwestern Greenland, and some geologists have suggested that this formation contains chemical signatures that indicate biological activity was involved in its formation.

Banded iron formations can be divided into two main types based on the geometric characteristics of the deposits. Algoma-type BIFs are lenticular bodies that are closely associated with volcanic rocks, typically basalts. Most are several hundred meters to kilometers in scale. In contrast, Superior-type BIFs are very large in scale, many initially covering tens of thousands of square kilometers. Superior-type BIFs are closely associated with shallow marine shelf types of sedimentary rocks including carbonates, quartzites, and shales.

Banded iron formations are also divisible into four types based on their mineralogy. Oxide iron formations contain layers of hematite, magnetite, and chert (or cryptocrystalline silica). Silicate iron formations contain hydrous silicate minerals, including chlorite, amphibole, greenalite, stilpnomelande, and minnesotaite. Carbonate iron formations contain siderite, ferrodolomite, and calcite. Sulfide iron formations contain pyrite.

In addition to being rich in iron, BIFs are ubiquitously silica-rich, indicating that the water from which they precipitated was saturated in silica as well as iron. Other chemical characteristics of BIFs include low alumina and titanium, elements that are generally increased by erosion of the continents. Therefore, BIFs are thought to have been deposited in environments away from any detrital sediment input. Some BIFs, especially the sulfide facies Algoma-type iron formation, have chemical signatures compatible with formation near black smoker types of seafloor hydrothermal vents, whereas others may have been deposited on quiet marine platforms. In particular, many of the Superior-types of deposits have many characteristics of deposition on a shallow shelf, including their association with shallow water sediments, their chemical and mineralogical constituency, and the very thin and laterally continuous nature of their layering. For instance, in the Archean Hamersley Basin of Western Australia, millimeter-thick layers in the BIF can be traced for hundreds of kilometers.

The environments that BIFs formed in and the mechanism responsible for the deposition of the iron and silica in BIFs prior to 1.8 Ga ago is still being debated. Any model must explain the large-scale transport and deposition of iron and silica in thin layers, in some cases over large areas, for a limited time period of Earth's history. Some observations are pertinent. First, to form such thin layers, the iron and silica must have been dissolved in solution. For iron to be in solution, it needs to be in the ferrous (reduced) state, in turn suggesting that the Earth's early oceans and atmosphere had little if any free oxygen and were reducing. The source of the iron and silica is also problematic; it may have come from weathering of continents, or from hydrothermal vents on the seafloor. There is currently evidence to support both ideas for individual and different kinds of BIFs, although the scales seem to be tipped in favor of hydrothermal origins for Algoma-types of deposits, and weathering of continents for Superior-type deposits.

The mechanisms responsible for causing dissolved iron to precipitate from the seawater to form the layers in BIFs have also proven elusive and problematical. It seems likely that changes in pH and acidity of seawater may have induced the iron precipitation, with periods of heavy iron deposition occurring during a steady background rate of silica deposition. Periods of nondeposition of iron would then be marked by deposition of silica layers. Prior to 1.8 Ga the oceans did not have organisms (e.g., diatoms) that removed silica from the oceans to make their shells, so the oceans would have been close to saturated in silica at this time, easing its deposition.

Several models have attempted to bring together the observations and requirements for the formation of BIFs, but none appear completely satisfactory at present. Perhaps there is no unifying model or environment of deposition, and multiple origins are possible. One model calls on alternating periods of evaporation and recharge to a restricted basin (such as a lake or playa), with changes in pH and acidity being induced by the evaporation. This would cause deposition of alternating layers of silica and iron. However, most BIFs do not appear to have been deposited in lakes. Another model calls on biological activity to induce the precipitation of iron, but fossils and other traces of life are generally rare in BIFs, although present in some. In this model, the layers would represent daily or seasonal variations in biological activity. Another model suggests that the layering may have been induced by periodic mixing of an early stratified ocean, where a shallow surface layer may have had some free oxygen resulting from near-surface photosynthesis, and a deeper layer would be made of reducing waters, containing dissolved elements produced at hydrothermal seafloor vents. In this model, precipitation and deposition of iron would occur when deep reducing water upwelled onto continental shelves and mixed with oxidized surface waters. The layers in this model would then represent the seasonal (or other cycle) variation in the strength of the coastal upwelling. This last model seems most capable of explaining features of the Superior-types of deposits, such as those of the Hamersley Basin in Western Australia. Variations in the exhalations of deep-sea vents may be responsible for the layering in the Algoma-type deposits. Other variations in these environments, such as oxidation, acidity, and amount of organic material, may explain the mineralogical differences between different banded iron formations. For instance, sulfide-facies iron formations have high amounts of organic carbon (especially in associated black shales and cherts) and were therefore probably deposited in shallow basins with enhanced biological activity. Carbonate-facies BIFs have lower amounts of organic carbon, and sedimentary structures indicative of shallow water deposition, so these probably were deposited on shallow shelves but further from the sites of major biological activity than the sulfide-facies BIFs. Oxide-facies BIFs have low contents of organic carbon but have a range of sedimentary structures indicating deposition in a variety of environments.

The essential disappearance of banded iron formation from the geological record at 1.8 billion years ago is thought to represent a major transition on the planet from an essentially reducing atmosphere to an oxygenated atmosphere. The exact amounts and rate of change of oxygen dissolved in the atmosphere and oceans would have changed gradually, but the sudden disappearance of BIFs at 1.8 Ga seems to mark the time when the rate of supply of biologically produced oxygen overwhelmed the ability of chemical reactions in the

A banded iron formation, Ungava Peninsula, Canada *(Photo by Timothy Kusky)*

oceans to oxidize and consume the free oxygen. The end of BIFs therefore marks the new dominance of photosynthesis as one of the main factors controlling the composition of the atmosphere and oceans.

See also ATMOSPHERE; PRECAMBRIAN.

Further Reading

Morris, R. C. "Genetic Modeling for Banded Iron Formation of the Hamersley Group, Pilbara Craton, Western Australia." *Precambrian Research* 60 (1993): 243–286.

Simonson, Bruce M. "Sedimentological Constraints on the Origins of Precambrian Banded Iron Formations." *Geological Society of America Bulletin* 96 (1985): 244–252.

barrier beach Narrow elongate deposits of sand situated offshore and generally elongate parallel to the main shoreline. They are typically only elevated slightly above high tide level and separated from the mainland by a lagoon. Barrier islands are slightly broader than barrier beaches and are typically a few to several meters above sea level. Different geomorphic zones in barrier island systems include the beachfront, sand dunes, vegetated zones, and swampy terranes and a lagoon separating it from the mainland. Barrier beaches and islands often form chains of islands known as barrier chains, famous examples of which are located on the southern side of Long Island (New York), forming the Outer Banks of North and South Carolina, and the Lido of Venice (Italy).

The sand that comprises barrier beaches and islands is very mobile, being reworked by storms and waves that strike the shore obliquely, transporting sand along the shore in a process called longshore drift. Individual sand grains in the beach swash and backwash zone may move as much as several hundred meters in a single day. In such an environment, the beach face is continually changing, with tidal inlets and other beach features constantly appearing or disappearing.

Barrier beaches and islands are extremely sensitive environments that can be easily destroyed or modified by human activity disrupting the vegetation and sand surfaces of the different geomorphic zones. Hurricanes and other storms can also disrupt, totally rework, or even remove barrier beaches, making them very hazardous places to live. The United States's most deadly natural disaster occurred in 1900 when a hurricane hit Galveston, Texas, built on a barrier island.

Nearly 7,000 people died when a storm surge and high winds flooded and demolished most of the city.

See also BEACH; HURRICANE.

barrier island *See* BARRIER BEACH.

basic rocks Dark-colored rocks with low silica contents generally falling between 44 and 52 percent. They include one of four groups of rocks classified based on silica content, including ultramafic (less than 44 percent SiO_2), basic, intermediate (53–65 percent SiO_2), and silicic (greater than 65 percent SiO_2). They are generally dark colored, and include basalt, gabbro, and related rocks. Basic rocks are typically rich in iron (Fe) and magnesium (Mg) and are also known as mafic rocks.

See also IGNEOUS ROCKS.

basin A depression in the surface of the Earth or other celestial body. There are many types of basins, including depressed areas with no outlet or with no outlet for deep levels (such as lakes, oceans, seas, and tidal basins), and areas of extreme land subsidence (such as volcanic calderas or sinkholes). In contrast, drainage basins include the total land area that contributes water to a stream. Drainage (river or stream) basins are geographic areas defined by surface slopes and stream networks where all the surface water that falls in the drainage basin flows into that stream system or its tributaries. Groundwater basins are areas where all the groundwater is contained in one system, or flows toward the same surface water basin outlet. Impact basins are circular depressions excavated instantaneously during the impact of a comet or asteroid with the Earth or other planetary surface.

Areas of prolonged subsidence and sediment accumulation are known as sedimentary basins, even though they may not presently be topographically depressed. There are several types of sedimentary basins, classified by their shape and relationships to bordering mountain belts or uplifted areas. Foreland basins are elongate areas on the stable continent sides of orogenic belts, characterized by a gradually deepening, generally wedge-shaped basin, filled by clastic and lesser amounts of carbonate and marine sedimentary deposits. The sediments are coarser grained and of more proximal varieties toward the mountain front, from where they were derived. Foreland basins may be several hundred feet to about 9–12 miles (100s of meters to 15–20 km) deep and filled entirely by sedimentary rocks, and they are therefore good sites for hydrocarbon exploration. Many foreland basins have been overridden by the orogenic belts from where they were derived, producing a foreland fold-thrust belt, and parts of the basin incorporated into the orogen. Many foreland basins show a vertical profile from a basal continental shelf type of assemblage, upward to a graywacke/shale flysch sequence, into an upper conglomerate/sandstone molasse sequence.

Rift basins are elongate depressions in the Earth's surface where the entire thickness of the lithosphere has ruptured in extension. They are typically bounded by normal faults along their long sides, and display rapid lateral variation in sedimentary facies and thicknesses. Rock types deposited in the rift basins include marginal conglomerate, fanglomerate, and alluvial fans, grading basinward into sandstone, shale, and lake evaporite deposits. Volcanic rocks may be intercalated with the sedimentary deposits of rifts and in many cases include a bimodal suite of basalts and rhyolites, some with alkaline chemical characteristics.

Several other less common types of sedimentary basins form in different tectonic settings. For instance, pull-apart rift basins and small foreland basins may form along bends in strike-slip fault systems, and many varieties of rift and foreland basins form in different convergent margin and divergent margin tectonic settings.

See also CONVERGENT PLATE MARGIN PROCESSES; DIVERGENT OR EXTENSIONAL BOUNDARIES; TRANSFORM PLATE MARGIN PROCESSES; DRAINAGE BASIN; FORELAND BASIN; OCEAN BASIN; RIFT; PLATE TECTONICS.

Further Reading
Allen, Philip A., and John R. Allen. *Basin Analysis, Principles and Applications.* Oxford: Blackwell Scientific Publications, 1990.

Basin and Range Province, United States The Basin and Range Province of the United States extends from eastern California to central Utah, and from southern Idaho into the state of Sonora, Mexico. The region is arid, characterized by steep-sided linear mountain ranges separated by generally flat, alluvial fan-bounded basins. The Earth's crust in the entire region has been extended by up to 100 percent of its original width, thinning the crust and forming large extensional faults. The faults are generally north-south trending, bound the uplifted mountain ranges, and accommodate the intervening basin's down-dropping and subsidence, forming the distinctive alternating pattern of linear mountain ranges and basins.

Although there are many faults of different ages in the Basin and Range Province, the extension and crustal stretching that shaped the modern landscape produced mainly normal faults. On the upthrown sides of these faults are enormous mountain ranges that characteristically rise abruptly and steeply, whereas the down-dropped sides create distinct low valleys. Most of the fault planes dip about 60°, and many have throws or displacements of more than 2 miles (3 km).

As the mountains rise, they are subjected to weathering and erosion, being attacked by water, ice, wind, and other erosional agents. These physical and chemical weathering processes are very important in the erosion of the mountains and the formation of sediments in the intervening basins. Extreme climate conditions, ranging from cold winters with heavy snowfall to long hot summers with little or no rainfall, enhance weathering processes, with frost and heat causing large boulders to spall off of cliff faces and fall to the basin floor, where they are further reduced by chemical processes.

The Basin and Range Province is very active tectonically. The northern part of the province is an actively deforming intercontinental plateau lying between stable blocks of the Sierra Nevada and the Colorado Plateau. The Basin and Range Province is experiencing rapid extension and active tectonics along its predominant faults. The province has extended by a factor of about two in the last 20 million years, and extension continues, with ongoing seismic activity and slip along numerous faults distributed across a zone 500 miles (800 km) wide. The internal part of the province is generally free from such deformation, with most occurring along the outer edges of the province. Space geodetic measurements broadly define movement across the province, whereas local surveys have mapped concentrated deformation in several seismically active zones. Determining the detailed pattern of active faulting is important since it defines the current seismic hazard zones, with zones of high-velocity gradients having more frequent damaging earthquakes than regions of low gradient.

batholith   Large igneous intrusion with a surface area greater than 60 square miles (100 km²). Several types of igneous intrusions are produced by magmas (generated from melting rocks in the Earth) and intrude the crust, taking one of several forms. A pluton is a general name for a large cooled igneous intrusive body in the Earth. The specific type of pluton is based on its geometry, size, and relations to the older rocks surrounding the pluton, known as country rock. Concordant plutons have boundaries parallel to layering in the country rock, whereas discordant plutons have boundaries that cut across layering in the country rock. Dikes are tabular but discordant intrusions, and sills are tabular and concordant intrusives. Volcanic necks are conduits connecting a volcano with its underlying magma chamber. A famous example of a volcanic neck is Devils Tower in Wyoming. Some plutons are so large that they have special names.

Batholiths and plutons have different characteristics and relationships to surrounding country rocks, based on the depth at which they intruded and crystallized. Epizonal plutons are shallow and typically have crosscutting relationships with surrounding rocks and tectonic foliations. They may have a metamorphic aureole surrounding them, where the country rocks have been heated by the intrusion and grew new metamorphic minerals in response to the heat and fluids escaping from the batholith. Rings of hard contact metamorphic rocks in the metamorphic aureole surrounding batholiths are known as hornfels rocks. Mesozonal rocks intrude the country rocks at slightly deeper levels than the epizonal plutons, but not as deep as catazonal plutons and batholiths. Catazonal plutons and batholiths tend to have contacts parallel with layering and tectonic foliations in the surrounding country rocks, and they do not show such a large temperature gradient with the country rocks as those from shallower crustal levels. This is because all the rocks are at relatively high temperatures. Catazonal plutons tend to be foliated, especially around their margins and contacts with the country rocks.

Batholiths are derived from deep crustal or deeper melting processes and may be linked to surface volcanic rocks. Batholiths form large parts of the continental crust, are associated with some metallic mineral deposits, and are used for building stones.

See also IGNEOUS ROCKS; PLUTON.

bauxite   Aluminous ore produced by intense humid tropical weathering of limestone, basalt, shale, marl, or virtually any igneous, metamorphic, or sedimentary rock. Intense tropical weathering of the parent rocks removes soluble material, leaving the insoluble alumina, iron, and some silica behind in laterite and bauxite deposits. It is typically a reddish brown, yellow, or whitish rock consisting of amorphous or crystalline hydrous aluminum oxides and hydroxides (gibbsite, boehmite, and diaspore) along with free silica, silt, iron hydroxides, and clay minerals. The bauxite may form concretionary, pisolitic, or oolitic forms, and may form dense or earthy horizons in the laterites. It may be either a residual or transported component of lateritic clay deposits in tropical and subtropical environments, and it is the principal economic ore of aluminum. Several different types of bauxite may be distinguished. Orthobauxites consist of massive red gibbsite and are typically overlain by an iron-rich crust called a hardcap. Metabauxites are white, boehmitic deposits that have pisolitic structures, are poor in iron, and are typically underlain by kaolinitic nodular iron-rich crusts known as ferricrete. They are thought to represent orthobauxites that have been transformed into metabauxites under arid weathering conditions. Cryptobauxites are orthobauxites that form under extremely humid environmental conditions and are overlain by an organic rich soft kaolinite horizon. Bauxites are common in Jamaica and the Caribbean and occur with laterites in India, Africa, and Asia. The term *bauxite* comes from Les Baux de Provence, near Arles in southern France.

See also LATERITE.

beach   An accumulation of sediment exposed to wave action along a coastline. The beach extends from the limit of the low-tide line to the point inland where the vegetation and landforms change to that typical of the surrounding region. Many beaches merge imperceptibly with grasslands and forests, whereas others end abruptly at cliffs or other permanent features. Beaches may occupy bays between headlands, form elongate strips attached (or detached, in the cases of barrier islands) to the mainland, or form spits that project out into the water. Beaches are very dynamic environments

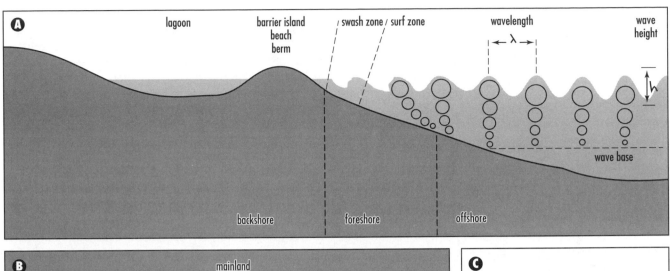

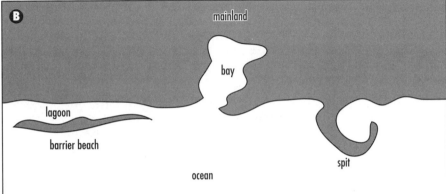

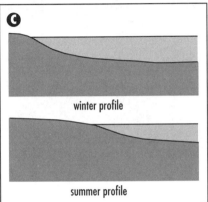

(A) Typical beach profile showing the major elements from the backshore to offshore; (B) different types of beach environments, including barrier beaches, bays, and spits; (C) typical beach profiles in winter and summer showing how large winter storm waves erode the beach and smaller summer waves rebuild the beach

and are always changing, with material being eroded and redeposited constantly from day to day and from season to season. Beaches are typically eroded to thin strips by strong winter storms, and built up considerably during summer, when storms tend to be less intense. The processes controlling this seasonal change are related to the relative amounts of energy in summer and winter storms—summer storms (except for hurricanes) tend to have less energy than winter storms, so they have waves with relatively short wavelengths and heights. These waves gradually push the offshore sands up to the beach face, building the beach throughout the summer. In contrast, winter storms have more energy with longer wavelength, higher amplitude waves. These large waves break on the beach, erode the beach face, and carry the sand seaward, depositing it offshore.

Some beaches are bordered by steep cliffs, many of which are experiencing active erosion. The erosion is a function of waves undercutting the base of the cliffs and over-steepening the slopes, which attempt to recover to the angle of repose by rainwater erosion or slumping from the top of the cliff. This erosion can be dramatic, with many tens of feet removed during single storms. The material that is eroded from the cliffs replenishes the beaches, and without the erosion the beaches would not exist. Coarser materials are left behind as they cannot be transported by the waves or tidal currents, and these typically form a rocky beach with a relatively flat platform known as a wave-cut terrace.

Lagoons, bays, and sounds separate the mainland from barrier islands that are long narrow offshore beaches. Barrier islands are common along the east coast of the United States (e.g., south shore of Long Island; Atlantic City, New Jersey; Outer Banks of North Carolina; and Galveston, Texas).

Further Reading

Dean, C. *Against the Tide: The Battle for America's Beaches*. New York: Columbia University Press, 1999.

Dolan, Robert, Paul J. Godfrey, and William E. Odum. "Man's Impact on the Barrier Islands of North Carolina." *American Scientist* 61 (1973): 152–162.

Kaufman, W., and Orrin H. Pilkey, Jr. *The Beaches Are Moving*. Durham, N.C.: Duke University Press, 1983.

King, C. A. M. *Beaches and Coasts*. London: Edward Arnold Publishers, 1961.

Komar, Paul D., ed. *CRC Handbook of Coastal Processes and Erosion*. Boca Raton, Fla.: CRC Press, 1983.

Williams, Jeffress, Kurt A. Dodd, and Kathleen K. Gohn. *Coasts in Crisis*. U.S. Geological Survey Circular 1075, 1990.

bedding Also known as stratification, bedding is a primary depositional layering in sedimentary rocks that is defined by variations in grain size of clastic grains, mineralogy, rock type, or other distinguishing features. It forms during the movement, sorting, and deposition of sediments by depositing currents, or it may reflect changes in the character of the environment during deposition of different beds. In a sedimentary rock, beds may be stacked in a series of layers of similar or variable thickness, and the overall character and changes within a sedimentary sequence may be described by noting the changes between and within beds at various heights in the sedimentary or stratigraphic pile. Stratification or bedding may form at many different scales, and understanding the character and mechanisms of formation of the strata can lead to a deep understanding of the depositional history of the rock unit or sedimentary basin in which the rocks formed. Major changes in the character of strata or beds can be used to define different formations in a rock sequence and form the basic subdivision for geologic mapping of an area.

Several styles of bedding are more common than others. Uniform beds are those that do not change in character from base to top. Graded bedding refers to the phenomena whereby the largest sized grains are deposited at the base of the bed, and finer grained particles are deposited progressively upward toward the top of the bed. Graded bedding is produced by a flow regime where the depositing current is losing velocity during deposition, dropping first the heavy coarse-grained particles, and then the finer grained particles as the current loses its strength. Cross bedding refers to the characteristic where prominent layers are oriented obliquely, typically at angle of 15–23 degrees, from the main originally horizontal bedding planes. These are produced by air or water currents that deposit sand or other particles in ripples or dunes, with the sedimentary particles forming layers as they slide down and accumulate on the slip surfaces of the dunes and ripples.

Bedding in rock sequences defines patterns that reflect major and minor changes in the sedimentary depositional environments. For instance, when sea levels rise relative to a coastline, a marine transgression results, and typically strata will change from sandstone to shale to limestone in a vertical sequence. Recognizing that this sequence of rock types in a bedded stratigraphic unit represents a deepening upward cycle would lead the geologist to understand the past series of events that led to the formation of these layers.

With the exception of cross-bedding, most beds are deposited horizontally, providing a reference frame for understanding folding and tilting of rocks in orogenic and deformation belts. The amount of tilting of a rock sequence can be estimated by measuring the inclination (called dip) of the bedding planes relative to the horizontal plane.

See also CROSS-BEDDING; SEDIMENTARY ROCKS; STRATIGRAPHY.

Belingwe greenstone belt, Zimbabwe The Archean Belingwe greenstone belt in southern Zimbabwe has proved to be one of the most important Archean terranes for testing models for the early evolution of the Earth and the formation of continents. It has been variously interpreted to contain a continental rift, arc, flood basalt, and structurally emplaced ophiolitic or oceanic plateau rocks. It is a typical Archean greenstone belt, being an elongate belt with abundant metamorphosed mafic rocks and metasediments, deformed and metamorphosed at greenschist to amphibolite grade. The basic structure of the belt is a refolded syncline, although debate has focused on the significance of early folded thrust faults.

The 3.5-billion-year-old Shabani-Tokwe gneiss complex forms most of the terrain east of the belt and underlies part of the greenstone belt. The 2.8–2.9-billion-year-old Mashaba tonalite and Chingezi gneiss are located west of the belt. These gneissic rocks are overlain unconformably by a 2.8-billion-year-old group of volcanic and sedimentary rocks known as the Lower Greenstones or Mtshingwe Group, including the Hokonui, Bend, Brooklands, and Koodoovale Formations. These rocks, and the eastern Shabani-Tokwe gneiss are overlain unconformably by a shallow water sedimentary sequence known as the Manjeri Formation, consisting of quartzites, banded iron formation, graywacke, and shale. A major fault is located at the top of the Manjeri Formation, and the Upper Greenstones structurally overlie the lower rocks being everywhere separated from them by this fault. The significance of this fault, whether a major tectonic contact or a fold accommodation related structure, has been the focus of considerable scientific debate. The 2.7-billion-year-old Upper Greenstones, or the Ngezi Group, includes the ultramafic-komatiitic Reliance Formation, the 6-kilometer-thick tholeiitic pillow lava-dominated Zeederbergs Formation, and the sedimentary Cheshire Formation. All of the units are intruded by the 2.6-billion-year-old Chibi granitic suite.

The Lower Greenstones have been almost universally interpreted to be deposits of a continental rift or rifted arc sequence. However, the tectonic significance of the Manjeri Formation and Upper Greenstones has been debated. The Manjeri Formation is certainly a shallow water sedimentary sequence that rests unconformably over older greenstones and gneisses. Correlated with other shallow-water sedimentary rocks across the southern craton, it may represent the remnants of a passive-margin type of sedimentary sequence.

The top of the Manjeri Formation is marked by a fault, the significance of which has been disputed. Some scientists have suggested that it may be a fault related to the formation of the regional syncline, formed in response to the rocks in the center of the belt being compressed and moving up and out of the syncline. Work on the sense of movement on the fault zone, however, shows that the movement sense is incompatible with such an interpretation, and that the fault is a folded thrust fault that placed the Upper Greenstones over the Manjeri Formation. Therefore, the tectonic setting of the Upper Greenstones is unrelated to the rocks under the thrust fault, and the Upper Greenstones likely were emplaced from

a distant location. The overall sequence of rocks in the Upper Greenstones, including several kilometers of mafic and ultramafic lavas, is very much like rock sequences found in contemporary oceanic plateaus or thick oceanic crust, and such an environment seems most likely for the Upper Greenstones in Belingwe and other nearby greenstone belts of the Zimbabwe craton.

A tectonic model for the evolution of the Zimbabwe craton has been proposed by Timothy Kusky in his 1998 article "Tectonic Setting and Terrane Accretion of the Archean Zimbabwe Craton." An ancient gneiss complex (The Tokwe Terrain) forms the core of the craton and

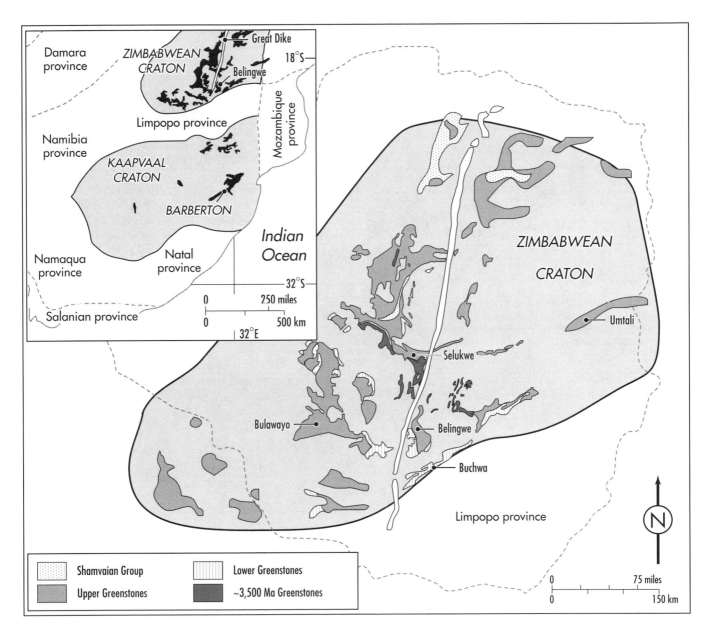

Map of the Zimbabwe craton showing greenstone belts of different ages, including the Belingwe greenstone belt in the southern part of the craton

Photograph of the basal unconformity in the Belingwe greenstone belt. Quartz-pebble conglomerates and sandstone of the Manjeri Formation on the left overlie 3.5-billion-year-old gneiss of the Shabani Complex on the right. *(Photo by Timothy Kusky)*

Further Reading
Kusky, Timothy M. "Tectonic Setting and Terrane Accretion of the Archean Zimbabwe Craton." *Geology* 26 (1998): 163–166.
Kusky, Timothy M., and Pamela A. Winsky. "Structural Relationships along a Greenstone/Shallow Water Shelf Contact, Belingwe Greenstone Belt, Zimbabwe." *Tectonics* 14 (1995): 448–471.
Kusky, Timothy M., and William S. F. Kidd. "Remnants of an Archean Oceanic Plateau, Belingwe Greenstone Belt, Zimbabwe." *Geology* 20 (1992): 43–46.

bench mark   Well-defined, uniformly fixed point on the land's surface that is used for a reference point from which other measurements can be made. It is generally marked by a circular bronze disk with a 3.75-inch (10-cm) diameter embedded firmly in bedrock or other permanent structure. In the United States, bench marks are installed and maintained by the U.S. Coast and Geodetic Survey and by the U.S. Geological Survey. The elevations of many benchmarks were in the past established by a surveying technique called differential leveling. Now it is more common to determine elevations using satellite-based differential global positioning systems. Bench marks are marked on topographic maps using the abbreviation B.M., and they are used for determining elevation and for surveying and construction.

benthic   The benthic environment includes the ocean floor, and the benthos are those organisms that dwell on or near the seafloor. Bottom-dwelling benthos organisms include large plants that grow in shallow water, as well as animals that dwell on the seafloor at all depths.

Many of the sediments on the deep seafloor are derived from erosion of the continents and carried to the deep sea by turbidity currents, carried by wind (e.g., volcanic ash), or released from floating ice. Other sediments, known as deep-sea oozes, include pelagic sediments derived from marine organic activity. When small organisms die, such as diatoms in the ocean, their shells sink to the bottom and over time can make significant accumulations. Calcareous ooze occurs at low to middle latitudes where warm water favors the growth of carbonate-secreting organisms. Calcareous oozes are not found in water that is more than 2.5–3 miles (4–5 km) deep because this water is under such high pressure that it contains a lot of dissolved CO_2, which dissolves carbonate shells. Siliceous ooze is produced by organisms that use silicon to make their shell structure.

The benthic world is amazingly diverse yet parts of the deep seafloor are less explored than the surface of the Moon. Organisms that live in the benthic community generally use one or more of three main strategies for living. Some attach themselves to anchored surfaces and get food by filtering it from the seawater. Other organisms move freely about on the ocean bottom and get their food by predation. Still others burrow or bury themselves in the ocean bottom sediments and get nourishment by digesting and extracting nutrients

extended at 2.9–2.8 billion years ago, forming the Lower Greenstones and related rocks. At 2.7 billion years ago, a continental margin volcanic arc was built on the northwestern side of the craton, while an ocean basin closure on the southeastern margin emplaced the Upper Greenstones over the continental shelf/passive margin sequence preserved in the Manjeri Formation. Folding, intrusion of the Chibi intrusives, and other deformation is related to the closure of the basin in the southeast, and later collision of the Zimbabwe and Kaapvaal cratons by 2.5 billion years ago. The tectonic history of the Belingwe greenstone belt and Zimbabwe craton is therefore very much like the evolution of modern convergent margins, showing that Precambrian plate tectonics was similar to modern style plate tectonics.

See also ARCHEAN; CRATONS; GREENSTONE BELTS; STRUCTURAL GEOLOGY.

from the benthic sediments. All the benthic organisms must compete for living space and food, with other factors including light levels, temperature, salinity, and the nature of the bottom controlling the distribution and diversity of some organisms. Species diversification is related to the stability of the benthic environment. Areas that experience large fluctuations in temperature, salinity, and water agitation tend to have low species diversification, but they may have large numbers of a few different types of organisms. In contrast, stable environments tend to show much greater diversity with a larger number of species present.

There are a large number of different benthic environments. Rocky shore environments in the intertidal zone have a wide range of conditions from alternately wet and dry to always submerged, with wave agitation and predation being important factors. These rocky shore environments tend to show a distinct zonation in benthos, with some organisms inhabiting one narrow niche, and other organisms in others. Barnacles and other organisms that can firmly attach themselves to the bottom do well in wave-agitated environments, whereas certain types of algae prefer areas from slightly above the low tide line to about 33 feet (10 m) deep. The area around the low tide mark tends to be inhabited by abundant organisms, including snails, starfish, crabs, mussels, sea anemones, urchins, and hydroids. Tide pools are highly variable environments that host specialized plants and animals including crustaceans, worms, starfish, snails, and seaweed. The subtidal environment may host lobster, worms, mollusks, and even octopus. Kelp, which are a brown benthic algae, inhabit the subtidal zone in subtropical to subpolar waters. Kelp can grow down to a depth of about 130 feet (40 m), often forming thick underwater forests that may extend along a coast for many kilometers.

Sandy and muddy bottom benthic environments often form at the edges of deltas, sandy beaches, marshes, and estuaries. Many of the world's temperate to tropical coastlines have salt marshes in the intertidal zone, and beds of sea grasses growing just below the low-tide line. Surface dwelling organisms in these environments are known as epifauna, whereas organisms that bury themselves in the bottom sand and mud are called infauna. Many of these organisms obtain nourishment either by filtering seawater that they pump though their digestive system, or by selecting edible particles from the seafloor. Deposit-feeding bivalves such as clams inhabit the area below the low tide mark, whereas other deposit feeders may inhabit the intertidal zone. Other organisms that inhabit these environments include shrimp, snails, oysters, tube-building crustaceans, and hydroids.

Coral reefs are special benthic environments that require warm water greater than 64.4°F (18°C) to survive. Colonial animals secrete calcareous skeletons, placing new active layers on top of the skeletons of dead organisms, and thus build the reef structure. Encrusting red algae, as well as green and red algae, produces the calcareous cement of the coral reefs. The reef hosts a huge variety and number of other organisms, some growing in symbiotic relationships with the reef builders, others seeking shelter or food among the complex reef. Nutrients are brought to the reef by upwelling waters and currents. The currents release more nutrients produced by the reef organisms. Some of the world's most spectacular coral reefs include the Great Barrier Reef off the northeast coast of Australia, reefs along the Red Sea, Indonesia, and in the Caribbean and South Florida.

Unique forms of life were recently discovered deep in the ocean near hot vents located along the mid-ocean ridge system. The organisms that live in these benthic environments are unusual in that they get their energy from chemosynthesis of sulfides exhaled by hot hydrothermal vents, and not from photosynthesis and sunlight. The organisms that live around these vents include tube worms, sulfate-reducing chemosynthetic bacteria, crabs, giant clams, mussels, and fish. The tube worms grow to enormous sizes, some being 10 feet in length and 0.8–1.2 inches wide (3 m long and 2–3 cm wide). Some of the bacteria that live near these vents include the most heat-tolerant (thermophyllic) organisms recognized on the planet, living at temperatures of up to 235°F (113°C). They are thought to be some of the most primitive organisms known, being both chemosynthetic and thermophyllic, and may be related to some of the oldest life-forms that inhabited the Earth.

The deep seafloor away from the mid-ocean ridges and hot vents is also inhabited by many of the main groups of animals that inhabit the shallower continental shelves. However, the number of organisms on the deep seafloor is few, and the size of the animals tends to be much smaller than those found at shallower levels. Some deep-water benthos similar to the hot-vent communities have recently been discovered living near cold vents above accretionary prisms at subduction zones, near hydrocarbon vents on continental shelves, and around decaying whale carcasses.

See also BEACH; BLACK SMOKER CHIMNEYS; CONTINENTAL MARGIN.

benthos *See* BENTHIC.

Big Bang Theory One of several possible theoretical beginning moments of the universe. One of the deepest questions in cosmology, the science that deals with the study and origin of the universe and everything in it, relates to how the universe came into existence. Cosmologists estimate that the universe is 10–20 billion years old and consists of a huge number of stars grouped in galaxies, clusters of galaxies, and superclusters of galaxies, surrounded by vast distances of open space. The universe is thought to be expanding because measurements show that the most distant galaxies, quasars, and most other objects in the universe are moving away from each other and from the center of the universe. The Big

Bang Theory states that the expanding universe originated 10–20 billion years ago in a single explosive event in which the entire universe suddenly exploded out of nothing, reaching a pea-sized supercondensed state with a temperature of 10 billion million million degrees Celsius in one million-million-million-million-million-million-millionth (10^{-36}) of a second after the Big Bang. Some of the fundamental parts of the expanding universe models come from Albert Einstein, who in 1915 proposed the General Theory of Relativity relating how matter and energy warp space-time to produce gravity. When Einstein applied his theory to the universe in 1917, he discovered that gravity would cause the universe to be unstable and collapse, so he proposed adding a cosmological constant as a "fudge factor" to his equations. The cosmological constant added a repulsive force to the General Theory, and this force counterbalanced gravity enabling the universe to continue expanding in his equations. William de Sitter further applied Einstein's General Theory of Relativity to predict that the universe is expanding. In 1927, Georges Lemaitre proposed that the universe originated in a giant explosion of a primeval atom, an event we would now call the Big Bang. In 1929, Edwin Hubble measured the movement of distant galaxies and discovered that galaxies are moving away from each other, expanding the universe as if the universe is being propelled from a big bang. This idea of expansion from an explosion negated the need for Einstein's cosmological constant, which he retracted, referring to it as his biggest blunder. This retraction, however, would later come back to haunt cosmologists.

Also in the 1920s, George Gamow worked with a group of scientists and suggested that elements heavier than hydrogen, specifically helium and lithium, could be produced in thermonuclear reactions during the Big Bang. Later, in 1957, Fred Hoyle, William Fowler, Geoff and Margaret Burbidge showed how hydrogen and helium could be processed in stars to produce heavier elements such as carbon, oxygen, and iron, necessary for life.

The inflationary theory is a modification of the Big Bang Theory and suggests that the universe underwent a period of rapid expansion immediately after the Big Bang. This theory was proposed in 1980 by Alan Guth, and it attempts to explain the present distribution of galaxies, as well as the 3°K cosmic background radiation discovered by Arno Penzias and Robert Wilson in 1965. This uniformly distributed radiation is thought to be a relict left over from the initial explosion of the Big Bang. For many years after the discovery of the cosmic background radiation, astronomers searched for answers to the amount of mass in the universe and to determine how fast the universe was expanding, and how much the gravitational attraction of bodies in the universe was causing the expansion to slow. A relatively high density of matter in the universe would eventually cause it to decelerate and collapse back upon itself, forming a "Big Crunch," and perhaps a new

Big Bang. Cosmologists called this the closed universe model. A low-density universe would expand forever, forming what cosmologists called an open universe. In between these end member models was a "flat" universe, that would expand ever more slowly until it froze in place.

An alternative theory to the Big Bang is known as the steady state theory, in which the universe is thought to exist in a perpetual state with no beginning or end, with matter continuously being created and destroyed. The steady state theory does not adequately account for the cosmic background radiation. For many years cosmologists argued, almost religiously, whether the Big Bang Theory or the steady state theory better explained the origin and fate of the universe. More recently, with the introduction of new high-powered instruments such as the Hubble Space telescope, the Keck Mirror Array, and supercomputers, many cosmology theories have seen a convergence of opinion. A new, so-called standard model of the universe has been advanced and is currently being refined to reflect this convergence of opinion.

In the standard model for the universe, the Big Bang occurred 14 billion years ago and marked the beginning of the universe. The cause and reasons for the Big Bang are not part of the theory but are left for the fields of religion and philosophy. Dr. William Percival of the University of Edinburgh leads a group of standard model cosmologists, and they calculate that the Big Bang occurred 13.89 billion years ago, plus or minus half a billion years. Most of the matter of the universe is proposed to reside in huge invisible clouds of dark matter, thought to contain elementary particles left over from the Big Bang. Galaxies and stars reside in these huge clouds of matter and comprise a mere 4.8 percent of the matter in the universe. The dark matter forms 22.7 percent of the universe, leaving another 72.5 percent of the universe as nonmatter. At the time of the proposal of the standard model, this ambiguous dark matter had yet to be conclusively detected or identified. In 2002 the first-ever atoms of antimatter were captured and analyzed by scientific teams from CERN, the European Organization for Nuclear Research.

Detailed observations of the cosmic background radiation by space-borne platforms such as NASA's COBE (Cosmic Background Explorer) in 1992 revealed faint variations and structure in the background radiation, consistent with an inflationary expanding universe. Blotches and patterns in the background radiation reveal areas that may have been the seeds or spawning grounds for the origin of galaxies and clusters. Detailed measurements of this background radiation have revealed that the universe is best thought of as flat—however, the lack of sufficient observable matter to have a flat universe requires the existence of some invisible dark matter. These observations were further expanded in 2002, when teams working with the DASI (Degree Angular Scale Interferometer) experiment reported directional differences (called polarizations) in the cosmic microwave background radiation dating from 450,000 years after the Big Bang. The astronomers were able to relate these directional differences to forces that led to the formation of galaxies and the overall structure of the universe today. These density differences are quantum effects that effectively seeded the early universe with protogalaxies during the early inflation period, and their observation provides strong support for the standard model for the universe.

Recent measurements have shown that the rate of expansion of the universe seems to be increasing, which has led cosmologists to propose the presence of a dark energy that is presently largely unknown. This dark energy is thought to comprise the remaining 72.5 percent of the universe, and it is analogous to a repulsive force or antimatter. Recognition in 1998 that the universe is expanding at ever increasing rates has toppled questions about open versus closed universe models and has drastically changed perceptions of the fate of the universe. Amazingly, the rate of acceleration of expansion is remarkably consistent with Einstein's abandoned cosmological constant. The expansion seems to be accelerating so fast that eventually the galaxies will be moving apart so fast, they will not be able to see each other and the universe will become dark. Other cosmologists argue that so little is known of dark matter and dark energy that it is difficult to predict how it will act in the future, and the fate of the universe is not determinable from our present observations.

Alan Guth and coworkers have recently proposed modifications of the inflationary universe model. They propose that the initial inflation of the universe, in its first few microseconds, can happen over and over again, forming an endless chain of universes, called multiverses by Dr. Martin Rees of Cambridge University. With these ideas, our 14-billion-year-old universe may be just one of many, with Big Bangs causing inflations of the perhaps infinite other universes. According to the theories of particle physics it takes only about one ounce of primordial starting material to inflate to a universe like our own. The process of growing chains of bubble-like universes through multiple Big Bangs and inflationary events has been termed eternal inflation by Dr. Andrei Linde of Stanford University.

Cosmologists, astronomers, and physicists are searching for a grand unifying theory that is able to link Einstein's General Theory of Relativity with quantum mechanics and new observations of our universe. One attempt at a grand unifying theory is the string theory, in which elementary particles are thought to be analogous to notes being played on strings vibrating in 10- or 11-dimensional space. A newer theory emerging is called M-theory, or Matrix theory, in which various dimensional membranes including universes can interact and collide, setting off Big Bangs and expansions that could continue or alternate indefinitely.

Cosmology and the fate of theories like the Big Bang are undergoing rapid and fundamental changes in understanding,

induced by new technologies, computing abilities, philosophy, and from the asking of new questions about creation of the universe. Although it is tempting to think of current theories as complete, perhaps with a few unanswered questions, history tells us that much can change with a few new observations, questions, or understanding.

biogenic sediment Sediments that are produced directly by the physiological activities of plants or animals. Rocks derived from biogenic sediments include both subaqueous and subaerial varieties. Coal is a biogenic rock, produced by the burial, heating and compression of terrestrial organic remains, including peat that typically forms in swampy environments. Lime muds and shell accumulations produce micritic limestone and shelly limestone, both of which are biogenic rocks. Coral reefs produce a variety of biogenic rocks, including coralline limestone and accumulations of reef organisms that may form packstones and coralline breccias. The benthic environment also hosts biogenic sediments, including calcareous and siliceous oozes that form by the gradual rain of shells of deceased planktonic organisms.

In many sedimentary environments, sediments with biogenic, clastic, or chemical origins are altered in some way by living organisms. Biogenic structures include a variety of markings left by animals, but not the skeletal remains of the organisms themselves. Bioturbation is a process whereby organisms may churn, stir up, or otherwise disrupt previously deposited sediment. Other biogenic trace fossils or ichnofossils include burrows left by worms and organisms that bury themselves beneath the sediment/water interface, tracks or footprints left on a subaqueous or subaerial surface, or borings left in solid surfaces. The types of trace fossils present in a rock can be used to learn about the sedimentary environment of deposition, rates of deposition, and as paleocurrent indicators.

See also BENTHIC; HYDROCARBONS; LIMESTONE; SEDIMENTARY ROCKS.

biosphere The biosphere encompasses the part of the Earth that is inhabited by life, and includes parts of the lithosphere, hydrosphere, and atmosphere. Life evolved more than 3.8 billion years ago and has played an important role in determining the planet's climate and insuring that it does not venture out of the narrow window of parameters that allow life to continue. In this way, the biosphere can be thought of as a self-regulating system that interacts with chemical, erosional, depositional, tectonic, atmospheric, and oceanic processes on the Earth.

Most of the Earth's biosphere uses photosynthesis as its primary source of energy, driven ultimately by energy from the Sun. Plants and many bacteria use photosynthesis as their primary metabolic strategy, whereas other microorganisms and animals rely on photosynthetic organisms as food for their energy and thus use solar energy indirectly. Most of the organisms that rely on solar energy live, by necessity, in the upper parts of the oceans (hydrosphere), lithosphere, and lower atmosphere. Bacteria are the dominant form of life on Earth (comprising about 5×10^{30} cells) and also live in the greatest range of environmental conditions. Some of the important environmental parameters for bacteria include temperature, between $-41°F$ to $235°F$ ($-5°C$ to $113°C$), pH levels from 0 to 11, pressures between a near-vacuum and 1,000 times atmospheric pressure, and from supersaturated salt solutions to distilled water.

Bacteria and other life-forms exist with diminished abundance to several kilometers or more beneath the Earth's surface, deep in the oceans, and some bacterial cells and fungal spores are found in the upper atmosphere. Life in the upper atmosphere is extremely limited by a lack of nutrients and by the lethal levels of solar radiation above the shielding effects of atmospheric ozone.

Soils and sediments in the lithosphere contain abundant microorganisms and invertebrates at shallow levels. Bacteria exist at much deeper levels and are being found in deeper and deeper environments as exploration continues. Bacteria are known to exist to about 2 miles (3.5 km) in pore spaces and cracks in rocks, and deeper in aquifers, oil reservoirs, and salt and mineral mines. Deep microorganisms do not rely on photosynthesis but rather use other geochemical or geothermal energy to drive their metabolic activity.

The hydrosphere and especially the oceans are teeming with life, particularly in the near-surface photic zone environment where sunlight penetrates. At greater depths below the photic zone most life is still driven by energy from the Sun, as organisms rely primarily on food provided by dead organisms that filter down from above. In the benthic environment of the seafloor there may be as many as 10,000 million (10^{10}) bacteria per milliliter of sediment. Bacteria also exist beneath the level that oxygen can penetrate, but the bacteria at these depths are anaerobic, primarily sulfate-reducing varieties. Bacteria are known to exist to greater than 2,789 feet (850 m) beneath the seafloor.

In the 1970s, a new environment for a remarkable group of organisms was discovered on the seafloor along the mid-ocean ridge system, where hot hydrothermal vents spew heated nutrient-rich waters into the benthic realm. In these environments, seawater circulates into the ocean crust where it is heated near oceanic magma chambers. This seawater reacts with the crust and leaches chemical components from the lithosphere, then rises along cracks or conduits to form hot black and white smoker chimneys that spew the nutrient rich waters at temperatures of up to $680°F$ ($350°C$). Life has been detected in these vents at temperatures of up to $235°F$ ($113°C$). The vents are rich in methane, hydrogen sulfide, and dissolved reduced metals such as iron that provide a chemical energy source for primitive bacteria. Some of the bacteria around these vents are sulfate-reducing chemosyn-

thetic thermophyllic organisms, living at high temperatures using only chemical energy and therefore existing independently of photosynthesis. These and other bacteria are locally so great in abundance that they provide the basic food source for other organisms, including spectacular worm communities, crabs, giant clams, and even fish.

See also ATMOSPHERE; BENTHIC; BLACK SMOKER CHIMNEYS; SUPERCONTINENT CYCLE.

biotite A common black, dark brown, or green mineral of the mica group. It is an aluminosilicate with the formula $K(Mg,Fe^{+2})_3(Al,Fe^{+3})Si_3O_{10}(OH)_2$ and is a sheet silicate that forms thin cellophane-like sheets that easily tear off underlying sheets. Biotite is a common constituent of igneous rocks and also forms as a metamorphic mineral especially in aluminous metasedimentary and mafic igneous rocks. It forms common platy minerals in metamorphic gneisses, typically being one of the main minerals that define tectonic foliations, where many biotite grains are aligned in a similar orientation.

Biotite exhibits monoclinic symmetry with pseudo-rhombohedral forms and has well developed basal cleavage. It typically has short tabular or prismatic crystals, exhibits striations on cleavage planes, and exhibits strong pleochroism under the polarizing microscope. Ionic substitution is common, and there are many varieties of biotite. When altered by the addition of water, biotite may exhibit a bronze-colored luster on cleavage surfaces.

See also MINERALOGY; STRUCTURAL GEOLOGY.

bioturbation *See* BIOGENIC SEDIMENT.

black smoker chimneys Hydrothermal vent systems that form near active magmatic systems along the mid-ocean ridge system, approximately 2 miles (3 km) below sea level. They were first discovered by deep submersibles exploring the oceanic ridge system near the Galapagos Islands in 1979, and many other examples have been documented since then, including a number along the Mid-Atlantic Ridge.

Black smokers are hydrothermal vent systems that form when seawater percolates into fractures in the seafloor rocks near the active spreading ridge, where the water gets heated to several hundred degrees Celsius. This hot pressurized water leaches minerals from the oceanic crust and extracts other elements from the nearby magma. The superheated water and brines then rise above the magma chamber in a hydrothermal circulation system and escape at vents on the seafloor, forming the black smoker hydrothermal vents. The vent fluids are typically rich in hydrogen sulfides (H_2S), methane, and dissolved reduced metals such as iron. The brines may escape at temperatures greater than $680°F$ ($360°C$), and when these hot brines come into contact with cold seawater, many of the metals and minerals in solution rise in plumes, since the hot fluids are

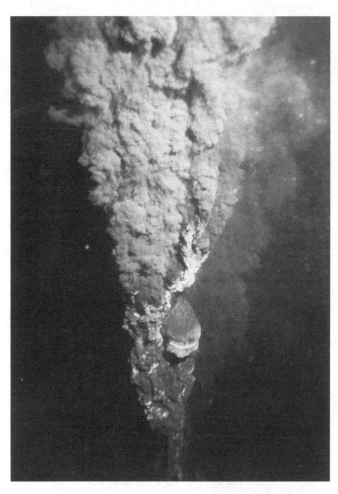

A black smoker chimney from the East Pacific Rise showing sulfide-rich vent fluid escaping into the seawater *(Photo by Robert D. Ballard, courtesy of Woods Hole Oceanographic Institution)*

than 66 feet (20 m) high. Clusters of black smoker chimneys several meters high may occupy the central area of mounds and deposit iron-copper sulfides. White smoker chimneys typically form in a zone around the central mound, depositing iron-zinc sulfides and iron oxides. Some mounds on the seafloor have been drilled to determine their internal structure. The TAG hydrothermal mound on the Mid-Atlantic Ridge is capped by central chimneys made of pyrite, chalcopyrite, and anhydrite, overlying massive pyrite breccia, with anhydrite-pyrite and silica-pyrite rich zones found a few to tens of meters below the surface. Below this, the host basalts are highly silicified, then at greater depths form a network of chloritized breccia. White smoker chimneys rim the central mound, and these are made of pyrite (FeS_2) and sphalerite (ZnS). In addition to the sulfides, oxides, hydroxides, and orthohydroxides, including several percent copper and zinc, the TAG mound contains minor amounts of gold.

Seafloor hydrothermal mounds and particularly the black smoker chimneys host a spectacular community of unique life-forms, found only in these environments. Life-forms include primitive sulfate-reducing thermophyllic bacteria, giant worms, giant clams, crabs, and fish, all living off the chemosynthetic metabolism made possible by the hydrothermal vent systems. Life at the black smokers draws energy from the internal energy of the Earth (not the Sun), via oxidation in a reducing environment. Some of the bacteria living at these vents are the most primitive organisms known on Earth, suggesting that early life may have resembled these chemosynthetic thermophyllic organisms.

Black smoker chimneys and the entire hydrothermal mounds bear striking similarities to volcanogenic massive sulfide (VMS) deposits found in Paleozoic and older ophiolite and arc complexes including the Bay of Islands ophiolite in

more buoyant than the colder seawater. The plumes are typically about 0.6 miles (1 km) high and 25 miles (40 km) wide, and they can be detected by temperature and chemical anomalies, including the presence of primitive ^3He isotopes derived from the mantle. These plumes may be rich in dissolved iron, manganese, copper, lead, zinc, cobalt, and cadmium, which rain out of the plumes, concentrating these elements on the seafloor. Manganese remains suspended in the plumes for several weeks, whereas most of the other metals are precipitated as sulfides (e.g., pyrite, FeS_2; chalcopyrite, $CuFeS_2$; sphalerite, ZnS), oxides (e.g., hematite, Fe_2O_3), orthohydroxides (e.g., goethite, $FeOOH$), or hydroxides (e.g., limonite, $Fe(OH)_3$). A group of related hydrothermal vents that form slightly further from central black smoker vents are known as white smokers, which typically have vent temperatures from 500°F–572°F (260°C–300°C).

On the seafloor along active spreading ridges, the hydrothermal vent systems form mounds that are typically 164–656 feet (50–200 m) in diameter and some are more

Hydrothermal vent on seafloor with tube worms and other biological activity around vent *(Photo courtesy of the U.S. Geological Survey)*

Newfoundland, the Troodos ophiolite in Cyprus, and the Semail ophiolite in Oman. Even older VMS deposits are common in Archean greenstone belts, and these are typically basalt or rhyolite-hosted chalcopyrite, pyrite, sphalerite, copper-zinc-gold deposits that many workers have suggested may be ancient seafloor hydrothermal vents. Interestingly, complete hydrothermal mounds with preserved black and white smoker chimneys have recently been reported from the 2.5-billion-year-old North China craton, in the same belt where the world's oldest well-preserved ophiolite is located.

The tectonic setting for the origin of life on the early Earth is quite controversial. Some favor environments in shallow pools, some favor deep ocean environments where the organisms could get energy from the chemicals coming out of seafloor hydrothermal vents. It is significant that black smoker types of hydrothermal vents have been discovered in Archean ophiolite sequences. The physical conditions at these mid-ocean ridges at more than 2.5 billion years ago permit the inorganic synthesis of amino acids and other prebiotic organic molecules. Some scientists think that the locus of precipitation and synthesis for life might have been in small iron-sulfide globules, such as those that form around black smokers. By examining black smoker chimneys we may be effectively looking at a window into the past, and the origin of life on Earth.

See also BENTHIC; BIOSPHERE; GREENSTONE BELT; OPHIOLITES.

Further Reading
Scott, Steven. "Minerals on Land, Minerals in the Sea." *Geotimes* 47, no. 12 (2002): 19–23.

boudinage   A process where a relatively stiff or rigid layer in a layered rock is ductilely extended parallel to layering into a series of connected or disconnected segments. Softer or more ductile layers of rock that separate the boudinaged layer flow into the spaces created by the extending stiff layer. Under some deformation conditions fibrous minerals or other minerals, typically quartz or calcite, may fill the spaces between the boudins. The degree of contrast in the competence (strength) of the stiff and soft layers determines the shapes of the boudins. Large competence contrasts between the stiff and soft layers produce boudins with sharp edges, whereas small competence contrasts produce rounded

Boudinaged quartz vein in a softer shale and sandstone matrix, Maine *(Photo by Timothy Kusky)*

boudins. The term *boudinage* is from the French *boudin,* for sausage links, which these structures resemble.

Simple boudins often form during folding of layered rock sequences, and in these cases the long axes of the boudins (and the boudin necks) are elongate parallel to the fold axes. Under other conditions, such as flattening perpendicular to layers, with extension in all directions parallel to layers, structures called chocolate block boudins may form. These have roughly equidimensional shapes, with boudin necks in two perpendicular directions.

See also DEFORMATION OF ROCKS; STRUCTURAL GEOLOGY.

Bowen, Norman Levi (1887–1956) Canadian *Petrologist, Geologist* Dr. Norman Levi Bowen was one of the most brilliant igneous petrologists in the 20th century. Although he was born in Ontario, Canada, he spent most of his productive research career at the Geophysical Laboratories of the Carnegie Institute in Washington, D.C. Bowen studied the relationships between plagioclase feldspars and iron-magnesium silicates in crystallizing and melting experiments. Based in these experiments, he derived the continuous and discontinuous reaction series explaining the sequence of crystallization and melting of these minerals in magmas. He also showed how magmatic differentiation by fractional crystallization can result from a granitic melt from an originally basaltic magma through the gradual crystallization of mafic minerals, leaving the felsic melt behind. Similarly, he showed how partial melting of one rock type can result in a melt with a different composition than the original rock, typically forming a more felsic melt than the original rock, and leaving a more mafic residue (or restite) behind. Bowen also worked on reactions between rocks at high temperatures and pressures, and the role of water in magmas. In 1928, N. L. Bowen published his now-classic book, *The Evolution of Igneous Rocks.*

See also IGNEOUS ROCKS.

Further Reading

Bowen, Norman Levi. "Progressive Metamorphism of Siliceous Limestone and Dolomite." *Journal of Geology* 48, no. 3 (1940): 225–274.

———. "Recent High-Temperature Research on Silicates and Its Significance in Igneous Geology." *American Journal of Science* 33 (1937): 1–21.

Bowen, Norman Levi, and John Frank Schairer. "The Problem of the Intrusion of Dunite in the Light of the Olivine Diagram." *International Geological Congress* 1 (1936): 391–396.

brachiopod Solitary marine invertebrate bivalves of the phylum Brachiopoda. The shells of brachiopods may be made of calcium carbonate or chitinophosphate. They range in age from Lower Cambrian to the present. Brachiopods are known as "lamp shells," and they have two clam-like shells joined for protection from predators. They feed from a thin feather-like device called a lophophore that filters the food from the water. Brachiopods attach themselves to hard surfaces or burrow through soft sediments using a pedicle, which is a long fleshy stalk that protrudes from the base of the shell.

The earliest brachiopods are known as inarticulates because they lack teeth and sockets to hold the two sides of the shell together. These oldest brachiopods used strong muscles to keep the shells closed when attacked and are virtually identical to the presently living brachiopod *Lingula*. Later, brachiopods developed sockets that aided in keeping the shell closed and the animal out of reach of predators. Early brachiopods had shells made of chitinophosphate, whereas the later varieties used calcium carbonate and developed teeth, ornaments, ribs, but remained bilaterally symmetric.

See also PALEONTOLOGY.

braided stream A braided stream consists of two or more adjacent but interconnected channels separated by bars or an island. They form in many different settings, including mountain valleys, broad lowland valleys, and in front of glaciers. Braided streams tend to form where there are large variations in the volume of water flowing in the stream and a large amount of sediment is available to be transported during times of high flow. The channels typically branch, separate and reunite, forming a pattern similar to a complex braid. Braided streams have constantly shifting channels, which move as the bars are eroded and redeposited (during large fluctuations in discharge). Most braided streams have highly variable discharge in different seasons, and they carry more load than meandering streams. Braided streams form where the stream load is greater than the stream's capacity to carry the load.

breccia A coarse-grained rock with angular clasts held together by a finer grained matrix. There are sedimentary, igneous, and structural breccias, each of which forms in very different ways. Sedimentary breccias are distinguished from conglomerates that have rounded clasts set in a fine grained matrix. The angular nature of clasts in a sedimentary breccia indicates that the clasts have not been transported far from their source of origin. Many breccias form along the bases of cliffs or steep slopes in talus aprons, and some may be eroded from fault scarps. Breccias may also form above solution cavities such as caves that collapse in karst terrains. The clasts in breccias can often be traced to a nearby source rock, and the geometry of the breccia deposit can be used to help decipher the sedimentary environment in which it formed.

Igneous breccias include both intrusion breccias and volcanic breccias. Intrusion breccias typically form when fluid and gas-rich magmatic rocks intrude country rocks, causing many fragments of the country rock to break off and be incorporated into the magmatic rock. Igneous breccias may also form through a process called stoping, where hot magma causes angular fragments of the country rock to shatter off and drop into the magma chamber. Volcanic breccias form by

A braided stream on the north side of Mount McKinley, Denali National Park, Alaska. Note the wide flood plain that is supplied with gravel by alluvial fans from the sides and glacial outwash from upstream regions. The braided channels are constantly shifting within the wide flood plain. *(Photo by Timothy Kusky)*

several processes, including the slow movement of lava flows that may be partly molten at depth and solid at the surface. Movement of the molten lava causes the overlying lava to fracture and break into many small angular pieces, forming a volcanic breccia. Aa flows that form in basalts, and are common in Hawaii, are one example of a volcanic breccia. Another type of volcanic breccia is known as a hyaloclastite. These are produced by the eruption of pillow lavas underwater. When magma inside the pillows drains out, typically forming lava tubes of new pillows budding off older pillows, the older pillow may collapse inward (implode), forming a breccia made of angular fragments of pillow lava. Subaerial volcanic eruptions can also produce a variety of breccias, including both fragments of cooled magma and fragments ripped off of from older volcanic rocks.

Structural or tectonic breccias are produced along fault zones. When a fault moves, the two adjacent sides grind against each other and often break off pieces of the wall rocks from either side of the fault. These get concentrated in the fault zone, or in damage zones on either side of the main fault zone. If deformation in the fault zone is strong, the breccia may be further broken down into a finer-grained rock called fault gouge, consisting of sand and clay-sized angular

to rounded clasts produced by breaking and grinding of the rocks, a process called cataclasis.

See also FAULT; IGNEOUS ROCKS; SEDIMENTARY ROCKS; STRUCTURAL GEOLOGY.

Brooks Range, Alaska The northernmost part of the Rocky Mountains is known as the Brooks Range, in northern Alaska. The mountain range extends almost 600 miles (1,000 km) in an east-west direction across northern Alaska and is a rugged, barren, windblown and snow-covered uninhabited range. The Brooks Range separates central Alaska's Yukon Basin from the oil-rich northern coastal plain, including the Arctic National Wildlife Refuge, and the northern petroleum reserve.

See also NORTH SLOPE OF ALASKA AND ANWR.

bryozoa Colonial invertebrates (phylum Bryozoa) with a calcareous membrane or chitinous membrane. They form coral-like skeletons with hundreds or thousands of tiny holes but are more closely related to brachiopods than corals. Some bryozoans formed large branching colonies, whereas others formed irregular lumpy masses. Each of the tiny holes in a bryozoan skeleton hosts a small filter-feeding animal that obtains its food through a lophophore similar to those of brachiopods. They are characterized by a U-shaped alimentary canal extending from mouth to anus and are known as moss corals, sea mats, and polyzoans. Bryozoans have a range from Upper Cambrian or Early Ordovician to present.

See also PALEONTOLOGY.

Burke, Kevin C. (1929–) British, American *Tectonicist* Kevin Burke was born in London, England, on November 13, 1929, and lived there till the age of 23, taking bachelor's and doctor's degrees at University College, London. The latter involved field-mapping of crystalline rocks in Galway, Ireland. In 1953 Burke was appointed a lecturer at what is now the University of Ghana, and apart from five years working with the British Geological Survey (1956–61) he spent the next 20 years teaching and doing field-related research at universities in Ghana, Korea, Jamaica, Nigeria, and Canada. In Canada he spent two years working with Tuzo Wilson at the University of Toronto, and in 1973 he joined the Geology Department in SUNY Albany where he spent 10 years working mainly with John Dewey, Bill Kidd, and Celal Sengör on a variety of tectonic problems. This group formulated tectonic models for many of the world's basins and mountain belts and made reconstructions of the continents at various times in Earth history. In 1983 Kevin Burke was appointed professor at the University of Houston, spending most of his efforts between 1983 and 1988 as director of the Lunar and Planetary Institute in Clear Lake. Between 1989 and 1992 Burke worked at the National Research Council in Washington with scientists putting together a major report on the future of the Solid Earth Sciences. Burke has focused many of his efforts on Africa, although he publishes extensively on other parts of the world, especially Asia and the Caribbean. Burke has devoted great efforts to editing journals and to national and international committees. He is the editor of the *Journal of Asian Earth Sciences* and was the president of the Scientific Committee on the Lithosphere of the International Council of Scientific Unions.

Further Reading

Burke, Kevin. "The African Plate." *South African Journal of Geology* 99, no. 4 (1996): 339–409.

———. "Tectonic Evolution of the Caribbean Plate." *Eos, Transactions of the American Geophysical Union* 76, no. 46 (1995).

Burke, Kevin, Duncan S. MacGregor, and N. R. Cameron. "Africa's Petroleum Systems; Four Tectonic 'Aces' in the Past 600 Million Years." *Geological Society of London* Special Publication no. 207 (2003): 21–60.

Burke, Kevin, and J. Tuzo Wilson. "Is the African Plate Stationary?" *Nature* 239 (1972): 387.

Dewey, John F., and Kevin Burke. "Tibetan, Variscan, and Precambrian Basement Reactivation: Products of Continental Collision." *Journal of Geology* 81 (1973): 683–692.

Dewey, John F., and Kevin Burke. "Plume Generated Triple Junctions." *Eos, Transactions of the American Geophysical Union* 54, no. 4 (1973): 406–433.

C

calcite A common trigonal rock-forming mineral, typically forming white, pink, colorless, yellow, or gray crystals with a variety of shapes, including rhombohedrons and clusters of small crystals known as nailhead spar and dogtooth spar. Distinct crystal faces and striations, or cleavage traces, on the crystal faces make many varieties of calcite easily distinguished from quartz. Some of the clear crystals of calcite known as Iceland spar form perfect rhombohedrons that exhibit strong double refraction. When an object is viewed through the clear crystals of Iceland spar, a double image appears through the crystal. Calcite is the main constituent of many carbonate rocks such as limestone, and also occurs as crystals in marble, in gangue mineral in many mineral deposits, as stalactites and stalagmites in cave deposits, and as loose earthy material in chalk deposits.

Calcite has the chemical formula $CaCO_3$, and it is trimorphous with aragonite and vaterite. Aragonite has orthorhombic symmetry, is denser and harder than calcite, has a less distinct cleavage, and typically forms as fibrous aggregates with gypsum in hot springs and in shallow marine muds and coral reefs. Vaterite is a rare hexagonal form of $CaCO_3$.

Most fossil shells are made of calcite, either as an original precipitate from seawater or from aragonite that reverted to the more stable calcite. Calcite also forms a common cement in sedimentary rocks and in many vein and fault systems.

See also MINERALOGY.

caldera Huge basin-shaped semicircular depressions, calderas, like Crater Lake in Oregon, are often many kilometers in diameter, produced when deep magma chambers under a volcano empty out (during an eruption), and the overlying land collapses inward producing a topographic depression. Yellowstone Valley occupies one of the largest calderas in the United States. Many geysers, hot springs, and fumaroles in the valley are related to groundwater circulating to depths, being heated by shallow magma, and mixing with volcanic gases that escape through minor cracks in the crust of the Earth.

Calderas may be relatively stable for tens or hundreds of thousands of years, but often magma reenters the collapsed magma chamber and causes it to rise up or inflate. This forms a resurgent caldera, which can produce a catastrophic volcanic eruption. Some of the largest volcanic eruptions in history have come from resurgent calderas. For instance, 600,000 years ago Yellowstone caldera experienced a resurgent eruption that spewed more than 240 cubic miles (1,000 km³) of volcanic material into the air, covering much of the United States with thick volcanic debris. This is more than 1,000 times the amount of material erupted by Mount Saint Helens in 1980. Other dormant calderas in the United States include Long Valley in California and Crater Lake in Oregon. One of the greatest volcanic eruptions in history was the 1883 explosion from the caldera of Krakatau in the Java-Sumatra Straits. On August 27, 1883, the four-mile (6-km) wide caldera of Krakatoa exploded with a force 5,000 times stronger than the atomic bomb dropped on Hiroshima. Water surged into the collapsed caldera and exploded with a sonic boom that was heard 1,240 miles (2,000 km) away in Australia. Ash covered 270,000 square miles (700,000 km²), and 13 percent of the global sunlight was blocked, lowering global temperatures by several degrees for more than a year. The explosion also generated a huge 130-foot (40-m) high tsunami that killed approximately 36,000 people.

See also VOLCANO.

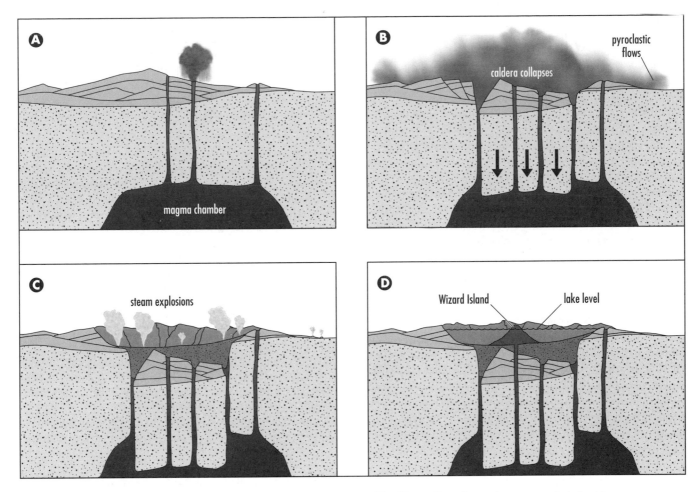

Four-stage schematic evolution of caldera collapse: (A) typical volcanic eruption; (B) caldera collapse by gravity as magma is drained from magma chamber; (C) posteruption fumarole activity; and (D) building of cinder cones

Picture of Crater Lake, Oregon, with the cinder cone of Wizard Island filling caldera produced by collapse of Mount Mazama *(Photo courtesy of H. R. Cornwall, U.S. Geological Survey)*

Caledonides An early Paleozoic orogenic belt in North and East Greenland, Scandinavia, and the northern British Isles. The Caledonides were continuous with the Appalachian Mountains before the opening of the Atlantic Ocean, together extending more than 4,100 miles (6,600 km). The history of the opening and closing of the early Paleozoic Iapetus Ocean and Tornquist Sea is preserved in the Caledonian-Appalachian orogen, which is one of the best known and studied Paleozoic orogenic belts in the world. The name is derived from the Roman name for the part of the British Isles north of the firths of Clyde and Forth, used in modern times for Scotland and the Scottish Highlands.

The Paleozoic Iapetus Ocean separated Laurentia (proto-North America) from Baltica and Avalonia, and the Tornquist Sea separated Baltica from Avalonia. The eastern margin of Laurentia has Neoproterozoic and Cambrian rift basins overlain by Cambro-Ordovician carbonate platforms, representing a rifting to trailing or passive margin sequence developed as the Iapetus Ocean opened. Similarly, Baltica has Neoproterozoic rift basins overlain by Cambro-Ordovician shelf sequences, whereas the Avalonian margin in Germany and Poland records Neoproterozoic volcanism and deformation, overlain by Cambro-Silurian shelf sequences, with an arc accretion event in the Ordovician. Gondwana sequences include Neoproterozoic orogens overlain by Ordovician shelf rocks deformed in the Devonian and Carboniferous. Importantly, faunal assemblages in Laurentia, Baltica, Gondwana, and Avalonia all show very different assemblages, interpreted to reveal a wide ocean between these regions in the early Paleozoic. This conclusion is supported by paleomagnetic data. Middle Ordovician ophiolites and flysch basins on Laurentia and Baltica reflect an arc accretion event in the Middle Ordovician, with probable arc polarity reversal leading to volcanism and thin-skinned thrusting preceding ocean closure in the Silurian.

From these and many other detailed studies, a brief tectonic history of the Appalachian-Caledonide orogen is as follows. Rifting of the Late Proterozoic supercontinent Rodinia at 750–600 million years ago led to the formation of rift to passive margin sequences as Gondwana and Baltica drifted away from Laurentia, forming the wide Iapetus Ocean and the Tornquist Sea. Oceanic arcs collided with each other in the Iapetus in the Cambrian and with the margin of Laurentia and Avalonia (still attached to Gondwana) in the Ordovician. These collisions formed the well-known Taconic orogeny on Laurentia, ophiolite obduction, and the formation of thick foreland basin sequences. Late Ordovician and Silurian volcanism on Laurentia reflects arc polarity reversal and subduction beneath Laurentia and Gondwana, rifting Avalonia from Gondwana and shrinking the Iapetus as ridges were subducted and terranes were transferred from one margin to another. Avalonia and Baltica collided in the Silurian (430–400 million years ago), and Gondwana collided with Avalon and the southern Appalachians by 300 million years

ago, during the Carboniferous Appalachian orogeny. At this time, the southern Rheic Ocean also closed, as preserved in the Variscan orogen in Europe.

See also APPALACHIANS; PLATE TECTONICS; SCOTTISH HIGHLANDS.

Further Reading
Bradley, Dwight C., and Timothy M. Kusky. "Geologic Methods of Estimating Convergence Rates during Arc-Continent Collision." *Journal of Geology* 94 (1986): 667–681.

Gayer, Rodney A., ed. *The Tectonic Evolution of the Appalachian-Caledonide Orogen.* Wisbaden: Vieweg and Sohn, 1985.

Kusky, Timothy M., James S. Chow, and Samuel A. Bowring. "Age and Origin of the Boil Mountain Ophiolite and Chain Lakes Massif, Maine: Implications for the Penobscottian Orogeny." *Canadian Journal of Earth Sciences* 34 (1997): 646–654.

Kusky, Timothy M., and William S. F. Kidd. "Tectonic Implications of Early Silurian Thrust Imbrication of the Northern Exploits Subzone, central Newfoundland." *Journal of Geodynamics* 22, no. 3–4 (1996): 229–265.

Kusky, Timothy M., William S. F. Kidd, and Dwight C. Bradley. "Displacement History of the Northern Arm Fault and Its Bearing on the Post-Taconic Evolution of North Central Newfoundland." *Journal of Geodynamics* 7 (1987): 105–133.

Williams, Hank, ed. "Geology of the Appalachian-Caledonide Orogen in Canada and Greenland, Geological Survey of Canada." *Geology of Canada* 6, 1995.

Cambrian The first geologic period of the Paleozoic Era and the Phanerozoic Eon, beginning at 544 million years ago (Ma), and ending 505 million years ago. It is preceded by the Late Proterozoic Eon and succeeded by the Ordovician Period. The Cambrian System refers to the rocks deposited during this period. The Cambrian is named after Cambria, which was the Roman name for Wales, where the first detailed studies of rocks of this age were completed.

The Cambrian is sometimes referred to as the age of invertebrates, and until this century, the Cambrian was thought to mark the first appearance of life on Earth. As the oldest period of the Paleozoic Era, meaning "ancient life," the Cambrian is now recognized as the short time period in which a relatively simple pre-Paleozoic fauna suddenly diversified in one of the most remarkable events in the history of life. For the 4 billion years prior to the Cambrian explosion, life consisted mainly of simple single-celled organisms, with the exception of the remarkable Late Proterozoic soft-bodied Ediacaran (Vendian) fauna, sporting giant sea creatures that all went extinct by or in the Cambrian and have no counterpart on Earth today. The brief 40-million-year-long Cambrian saw the development of multicelled organisms, as well as species with exoskeletons, including trilobites, brachiopods, arthropods, echinoderms, and crinoids.

At the dawn of the Cambrian most of the world's continents were distributed within 60°N/S of the equator, and many of the continents that now form Asia, Africa, Australia,

Antarctica, and South America were joined together in the supercontinent of Gondwana. These continental fragments had broken off from an older supercontinent (Rodinia) between 700 and 600 million years ago, then joined together in the new configuration of Gondwana, with the final ocean closure of the Mozambique Ocean between East and West Gondwana occurring along the East African Orogen at the Precambrian-Cambrian boundary. Even though the Gondwana supercontinent had only formed at the end of the Proterozoic, it was already breaking up and dispersing different continental fragments by the Cambrian.

Neoproterozoic closure of the Mozambique Ocean sutured East and West Gondwana and intervening arc and continental terranes along the length of the East African Orogen. Much active research in the Earth Sciences is aimed at providing a better understanding of this ancient mountain belt and its relationships to the evolution of crust, climate, and life at the end of Precambrian time and the opening of the Phanerozoic. There have been numerous and rapid changes in our understanding of events related to the assembly of Gondwana. The East African Orogen encompasses the Arabian-Nubian Shield in the north and the Mozambique Belt in the south. These and several other orogenic belts are commonly referred to as Pan-African belts, recognizing that many distinct belts in Africa and other continents experi-

enced deformation, metamorphism, and magmatic activity in the general period of 800–450 Ma. Pan-African tectonic activity in the Mozambique Belt was broadly contemporaneous with magmatism, metamorphism, and deformation in the Arabian-Nubian Shield. The difference in lithology and metamorphic grade between the two belts has been attributed to the difference in the level of exposure, with the Mozambican rocks interpreted as lower crustal equivalents of the rocks in the Arabian-Nubian Shield.

The timing of Gondwana's amalgamation is remarkably coincident with the Cambrian explosion of life, which has focused the research of many scientists on relating global-scale tectonics to biologic and climatic change. It is thought that the dramatic biologic, climatic, and geologic events that mark Earth's transition into the Cambrian might be linked to the distribution of continents and the breakup and reassembly of a supercontinent. The formation and dispersal of supercontinents causes dramatic changes in the Earth's climate and changes the distribution of environmental settings for life to develop within. Plate tectonics and the formation and breakup of the supercontinents of Rodinia and Gondwana set the stage for life to diversify during the Cambrian explosion, bringing life from the primitive forms that dominated the Precambrian to the diverse fauna of the Paleozoic. The breaking apart of supercontinents creates abundant shallow and warm water

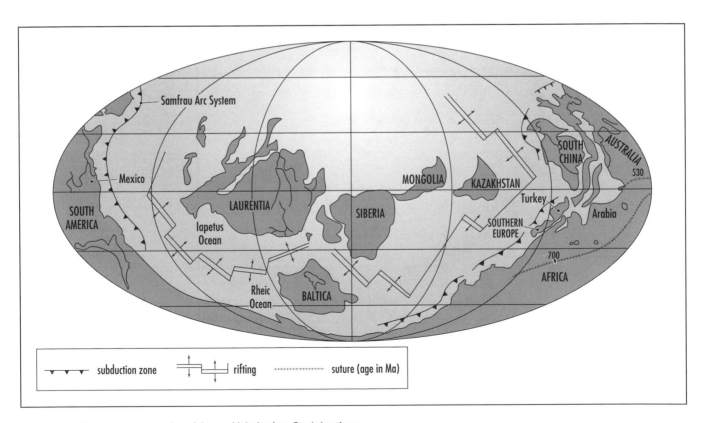

Paleogeographic plate reconstruction of the world during Late Cambrian times

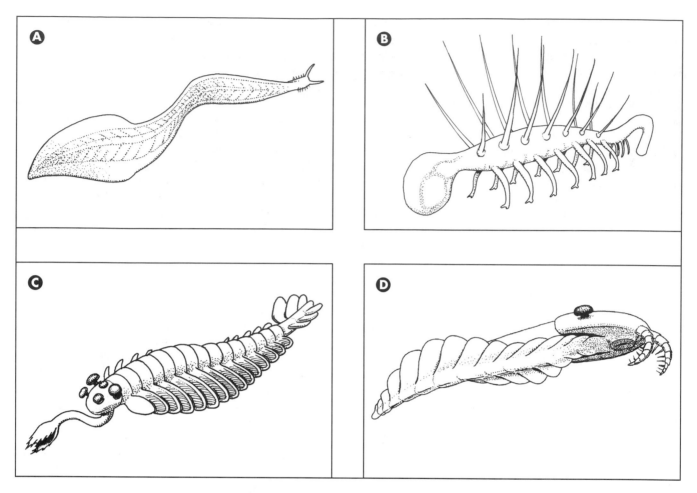

Some of the remarkable fauna from the Burgess Shale, including (A) Pikaia; (B) Hallucigenia; (C) Opabinia; and (D) Anomocaris

inland seas, as well as shallow passive margins along the edges of the rifted fragments. It is thought that as rifting separated continental fragments from Rodinia, they moved across warm oceans, and new life-forms developed on these shallow passive margins and inland seas. When these "continental icebergs" carrying new life collided with the supercontinent of Gondwana, the new life-forms could rapidly expand and diversify, then compete with the next organism brought in by the next continent. This process happened over and over again, with the formation and breakup of the two Late Proterozoic-Cambrian supercontinents of Rodinia and Gondwana.

North America began rifting away from Gondwana as it was forming, with the rifting becoming successful enough to generate rift-type volcanism by 570 million years ago, and an ocean named Iapetus by 500 million years ago. Iapetus saw some convergent activity in the Cambrian, but it was not until the Middle Ordovician that this ocean experienced major contractional events. North and south China had begun rifting off of Gondwana in the Cambrian, as did Kazakhstan, Siberia, and Baltica. The margins of these continental fragments sub-

sided and accommodated the deposition of thick, carbonate passive margin sequences that heralded the rapid development of life, some of which are now petroleum provinces.

A Middle Cambrian sequence of fine-grained turbidites near Calgary, Alberta, Canada, has yielded a truly remarkable group of extremely well-preserved fauna. The Burgess Shale was discovered by Charles D. Walcott in 1909 and is thought to preserve organisms deposited in a lagoon that was buried suddenly in anaerobic muds, resulting in the organisms being so well preserved that even the soft parts show fine detail. Fossils from the Burgess Shale and related rocks have revealed much of what we know about the early life-forms in the Cambrian, and it represents the earliest known fauna. The Burgess Shale has yielded some of the best preserved jellyfish, worms, sponges, brachiopods, trilobites, arthropods, mollusks, and first invertebrate chordates.

One of the important steps for the rapid expansion of life-forms in the Cambrian was the rapid radiation of acritarchs, which are small spores of planktonic algae such as green algae or dinoflagellates. The acritarchs served as the primary source of food and the base of the food chain for the higher animals that developed. Acritarchs appeared first during the Late Proterozoic, but 75 percent of all their taxa became extinct in the Late Proterozoic glaciation. The Late Proterozoic-Cambrian transition also saw the first appearance of trace fossils of soft-bodied organisms, showing that they developed and rapidly diversified in this time period. Traces of worm paths are most common, where they searched for food or burrowed into soft sediments.

Shelly fossils first appeared slightly later in the Cambrian, during the Tommotian, a 15-million-year-long stage added to the base of the Cambrian time scale in the 1970s. Most of the early shelly fossils were small (1–2 mm) conical shells, tubes, plates, or spicules made of calcite or calcium phosphate. They represent different phyla, including mollusks, brachiopods, armored worms, sponges, and archeocyathid reefs. The next major phase of the Cambrian radiation saw calcite shells added to trilobites, enabling their widespread preservation. Trilobites rapidly became abundant, forming about 95 percent of all preserved Cambrian fossils. However, since trilobites form only 10 percent of the Burgess Shale, it is thought that this number is biased in favor of the hard-shelled trilobites over other soft-bodied organisms. Trilobites experienced five major extinctions in the Cambrian, each one followed by an adaptive radiation and expansion of new species into vacant ecological niches. Other arthropods to appear in the Cambrian include crustaceans (lobsters, crabs, shrimp, ostracods, and barnacles) and chelicerates (scorpions, spiders, mites, ticks, and horseshoe crabs).

Mollusks also appeared for the first time at the beginning of the Cambrian, resulting in the first clam (pelecypod) by the end of the Cambrian. Snails (gastropods) also emerged, including those with multiple gas-filled chambers, which includes the cephalopods. Echinoderms with hard skeletons first appeared in the Early Cambrian, and these include starfish (asteroids), brittle stars (ophiuroids), sea urchins (echinoids), and sea cucumbers (holothuroids).

See also GONDWANA; SUPERCONTINENT CYCLE; VENDIAN.

Further Reading

Cowrie, J. W., and M. D. Braiser, eds. *The Precambrian-Cambrian Boundary.* Oxford: Clarendon Press, 1989.

Gould, Steven J. *Wonderful Life: Burgess Shale and the Nature of History.* New York: W. W. Norton, 1989.

Kusky, Timothy M., Mohamed Abdelsalam, Robert Tucker, and Robert Stern, eds. *Evolution of the East African Orogen and the Assembly of Gondwana. Precambrian Research*, 2003.

Lipps, Jere H., and Philip W. Signor, eds. *Origin and Early Evolution of the Metazoa.* New York: Plenum, 1992.

McMenamin, Mark A. S., and Diana L. S. McMenamin. *The Emergence of Animals: The Cambrian Breakthrough.* New York: Columbia University Press, 1989.

Prothero, Donald R., and Robert H. Dott. *Evolution of the Earth*, 6th ed. Boston: McGraw Hill, 2002.

Cape Cod, Massachusetts A sandy cape that protrudes into the Gulf of Maine on the Atlantic Ocean from the east coast of Massachusetts, United States. It is curved like an outstretched and upraised arm, protecting Cape Cod Bay between its inner curve and the mainland. The once fertile fishing grounds of George's Bank are to the southeast, and Nantucket Sound separates the islands of Nantucket and Martha's Vineyard from the Cape. Cape Cod, the islands of Martha's Vineyard and Nantucket, as well as Long Island, are terminal and recessional moraines from the last Pleistocene glaciation, so they contain sand and gravel derived from Canada and New England. Many parts of the Cape have been reworked into giant sand dunes both in the periglacial environment and in more recent times. Cape Cod is a favorite summertime vacation spot for New Englanders, especially those from the Boston area.

Cape Horn and the Strait of Magellan The southern tip of South America has consistently horrid weather with high winds, rain and ice storms, and large sea waves. Southernmost South America is a large island archipelago known as Tierra del Fuego, separated from the mainland by the Strait of Magellan, and the southern tip of which is known as Cape Horn. The Drake passage separates Cape Horn from the northern tip of the Antarctic Peninsula. Tierra del Fuego and the Strait of Magellan were discovered by Magellan in 1520, and settled by Europeans, Argentineans, and Chileans after the discovery of gold in the 1880s. These peoples brought diseases that spread to and killed off all of the indigenous people of the islands.

carbon-14 dating Also known as radiocarbon dating, a technique for determining the age of organic samples that are less than about 35,000 years old. Carbon-14 (^{14}C) is a cosmogenic isotope, generated by the interaction of cosmic rays with matter on the Earth. Cosmic rays enter the Earth's atmosphere and transform common ^{12}C to radioactive ^{14}C, which has a half-life of 5,730 million years. Within about 12 minutes of being struck by cosmic rays in the upper atmosphere, the ^{14}C combines with oxygen to produce radioactive carbon dioxide. ^{14}C is exchanged with living organisms as $^{14}CO_2$, and the isotopic clock for dating is set when the organism dies and stops exchanging $^{14}CO_2$ with the atmosphere. Since the initial $^{14}C/^{12}C$ ratio and the half life ^{14}C are well known, the present $^{14}C/^{12}C$ ratio can accurately determine the age of an appropriate sample. Radiocarbon dating is widely used to date fossil wood, carbon in sediments, charcoal from archaeological sites, corals, foraminifera, and organic material in sediments. Willard F. Libby of the University of Chicago pioneered the carbon-14 dating technique in 1946.

See also GEOCHRONOLOGY.

carbonate A sediment or sedimentary rock containing the carbonate ion (CO_3^{-2}). It may be formed by the organic or inorganic precipitation from aqueous solutions in warm waters. Typical carbonate rock types include limestone and dolostone, most of which are deposited on shallow marine platforms, mounds, or in reefs, where there is a lack of siliciclastic sediment input.

Carbonate platforms form most commonly along subsiding passive margins, where sedimentation or reef growth keeps up with the rate of subsidence, maintaining shallow water depths. The platforms grow laterally by shedding carbonate sediment into the adjacent basin until the water depths are shallow enough for the platform or reef to grow outward over the slope deposits. At times of high eustatic sea level, carbonate muds may cover vast areas of the continents, forming shallow inland seas that can accumulate tens or even hundreds of meters of carbonate sediments. Carbonate reefs may also form fringing complexes around volcanic atolls.

Several types of carbonate platforms are distinguished. Carbonate ramps are gently seaward-sloping surfaces with no marked break between shallow and deep water. These typically have carbonate sands and oolitic deposits formed near the continent and fringing reefs, grading seaward into skeletal sands, muddy sands, and mud. Rimmed shelves have sharp breaks between shallow and deep water, with a high energy facies near the slope/shelf break. The shelves are typically covered by carbonate sands, muddy sands, and muds, with isolated patch reefs. The shelf edge may develop extensive reef complexes (such as the Great Barrier Reef) that are fed by nutrient-rich upwelling water and shed reef debris into the adjacent basin during storms.

Reefs are framework supported carbonate mounds built by carbonate-secreting organisms, whereas carbonate mounds are general positive features on the seafloor built by biological processes. Reefs contain a plethora of organisms that together build a wave-resistant structure and provide shelter for fish and other organisms. The spaces between the framework are typically filled by skeletal debris, which together with the framework becomes cemented together to form a wave-resistant feature that shelters the shelf from high-energy waves. Reef organisms (presently consisting mainly of zooxanthellae) can only survive in the photic zone, so reef growth is restricted to the upper 328 feet (100 m) of the seawater. Carbonate mounds differ from reefs in that they do not have a rigid framework but are formed by sediment-trapping processes such as those of sea grass, algae, and other organisms.

Carbonates, carbonate platforms, and reefs are represented in all geological eras from the Archean to the present, although the organisms that have constructed these similar morphological features have changed through time. Carbonates are the primary source rocks for the world's hydrocarbon deposits.

See also ATOLL; CARBONATE MINERALS; CARBON CYCLE; DOLOMITE; LIMESTONE; PASSIVE MARGIN; REEF; SEDIMENTARY ROCKS.

Further Reading
Friedman, Gerald M., and John E. Sanders. *Principles of Sedimentology.* New York: John Wiley, 1978.
Read, J. Fred. "Carbonate Platform Facies Models." *American Association of Petroleum Geologists Bulletin* 69 (1985): 1–21.
Tucker, Maurice E., and Paul V. Wright. *Carbonate Sedimentology.* Oxford: Blackwell Publications, 1990.

carbonate minerals Minerals that contain the carbonate anion (CO_3^{-2}). The two dominant carbonate minerals are calcite and aragonite, which are pseudomorphs (minerals with the same chemical formula but different crystal structures) with the formula $CaCO_3$. Calcite has a trigonal crystal form, typically forming white, pink, colorless, yellow, or gray crystals with a variety of shapes, including rhombohedrons and clusters of small angular crystals. Distinct crystal faces and striations, or cleavage traces, on the crystal faces make many varieties of calcite easily distinguished from quartz. Aragonite has orthorhombic symmetry, is denser and harder than calcite, has a less distinct cleavage, and typically forms as fibrous aggregates with gypsum in hot springs and in shallow marine muds and coral reefs. Aragonite is less stable than calcite and typically reverts to calcite in rock sequences. Most aragonite is relatively pure, but several varieties are formed by substitution. Dolomite is another common carbonate mineral, made of $(Ca,Mg)CO_3$. In strontianite ($SrCO_3$), the strontium ion substitutes for calcite, and in witherite ($BaCO_3$), barium substitutes for calcium. Other common carbonates include magnesite ($MgCO_3$), rhodochrosite ($MnCO_3$), and siderite (CO_3). Other orthorhombic carbonate minerals include alstonite ($Ca,Ba[CO_3]_2$) and cerussite ($PbCO_3$), which is an alteration product of galena. Vaterite is a rare hexagonal form of $CaCO_3$.

See also CALCITE; CARBONATE; MINERALOGY.

carbonatite Rare igneous rocks that contain more than 50 percent carbonate. Some carbonatites are calcitic (sovite) whereas magnesio-carbonatites (rauhaugites) have dolomite as the primary carbonate mineral. Ferro-carbonatites have calcium-magnesium and iron carbonate minerals. There are more than 350 carbonatite intrusions and volcanoes recognized on Earth, with many of these in the East African rift system. Northern Tanzania hosts the world's most dense carbonatite province, where active eruptions from centers such as Ol Doninyo Lengai produce short-lived bizarre flows with twisted spires, swirls, and oozes made of the extremely fluid carbonatite. The carbonatite lavas flow with a viscosity similar to that of olive oil, the lowest viscosity known for any magma on Earth. Other carbonatites are associated with kimberlites and alkalic rocks and are all located on continental crust. Most carbonatites are very small, being less than 1 square kilometer in area, with the largest known complex being 7.5 square miles (20 km²). They range in age from Proterozoic (2.0 billion years old) to the present with most being younger than 150 million years old.

The chemistry of carbonatites shows that they formed by magmatic processes in the mantle. They have high concentrations of rare elements such as phosphorous and sodium and compositionally grade into kimberlites. Intrusion of both kimberlites and carbonatites causes intense alteration of surrounding country rocks, characterized by the addition of sodium and potassium.

Carbonatites are thought to form by small amounts of partial melting in the upper mantle, although there has been a large amount of controversy about the origin of this volumetrically minor group of igneous rocks. Some models suggested that carbonatites were formed by the remobilization of sedimentary limestones by heat from nearby silicic intrusions, but these ideas have been largely discredited. Carbonatite magmatism tends to occur repeatedly, but not continuously, for hundreds of millions or even billions of years in the same location, suggesting that some unusual aspect of the mantle in these areas leads to the formation of magmas with such rare compositions. It has been noted that carbonatite magmatism in the African plate seems to coincide with distant plate collisions or periods of increased global magmatic activity, although the cause of these apparent trends is unknown.

See also DIVERGENT OR EXTENSIONAL BOUNDARIES; IGNEOUS ROCKS; RIFTS.

Further Reading
Bourne, Joel K. "Ol Doinyo Lengai." *National Geographic* (January 2003).

carbon cycle   The carbon cycle represents a complex series of processes where the element carbon makes a continuous and complex exchange between the atmosphere, hydrosphere, lithosphere and solid Earth, and biosphere. Carbon is one of the fundamental building blocks of Earth, with most life-forms consisting of organic carbon and inorganic carbon dominating the physical environment. The carbon cycle is driven by energy flux from the Sun and plays a major role in regulating the planet's climate.

Several main processes control the flux of carbon on the Earth, and these processes are presently approximately balanced. Assimilation and dissimilation of carbon, by photosynthesis and respiration by life, cycles about 10^{11} metric tons of carbon each year. Some carbon is simply exchanged between systems as carbon dioxide, and other carbon undergoes dissolution or precipitation as carbonate compounds in sedimentary rocks.

Atmospheric carbon forms the long-lived compounds carbon dioxide and methane and the short-lived compound carbon monoxide that has a very short atmospheric residence time. Global temperatures and the amount of carbon (chiefly as CO_2) in the atmosphere are closely correlated, with more CO_2 in the atmosphere resulting in higher temperatures. However, it is yet to be determined if increased carbon flux to the atmosphere from the carbon cycle forces global warming, or if global warming causes an increase in the carbon flux. Since the industrial revolution, humans have increased CO_2 emissions to the atmosphere resulting in measurable global warming, showing that increased carbon flux can control global temperatures.

The oceans represent the largest carbon reservoir on the planet, containing more than 60 times as much carbon as the atmosphere. Dissolved inorganic carbon forms the largest component, followed by the more mobile dissolved organic carbon. The oceans are stratified into three main layers. The well-mixed surface layer is about 246 feet (75 m) thick and overlies the thermocline, which is a stagnant zone characterized by decreasing temperature and increasing density to its base at about 0.6-mile (1-km) depth. Below this lie the deep cold bottom waters where dissolved CO_2 transferred by descending cold saline waters in polar regions may remain trapped for thousands of years. Cold polar waters contain more CO_2 because gases are more soluble in colder water. Some, perhaps large amounts, of this C gets incorporated in gas hydrates, which are solid, ice-like substances made of cases of ice molecules enclosing gas molecules like methane, ethane, butane, propane, carbon dioxide, and hydrogen sulfide. Gas hydrates have recently been recognized as a huge global energy resource, with reserves estimated to be at least twice that of known fossil fuel deposits. However, gas hydrates form at high pressures and cold temperatures, and extracting them from the deep ocean without releasing huge amounts of CO_2 to the atmosphere may be difficult.

Carbon is transferred to the deep ocean by its solubility in seawater, whereas organic activity (photosynthesis) in the oceanic surface layer accounts for 30–40 percent of the global vegetation flux of carbon. About 10 percent of the C that is used in respiration in the upper oceanic layer is precipitated out and sinks to the lower oceanic reservoir.

The majority of Earth's carbon is locked up in sedimentary rocks, primarily limestone and dolostone. This stored carbon reacts with the other reservoirs at a greatly reduced rate (millions and even billions of years) compared with the other mechanisms discussed here. Some cycles of this carbon reservoir are related to the supercontinent cycle and the weathering of carbonate platforms when they are exposed by continental collisions.

The Earth's living biomass, the decaying remains of this biomass (litter), and soil all contain significant C reserves that interact in the global carbon cycle. Huge amounts of carbon are locked in forests, as well as in arctic tundra. Living vegetation contains about the same amount of carbon as is in the

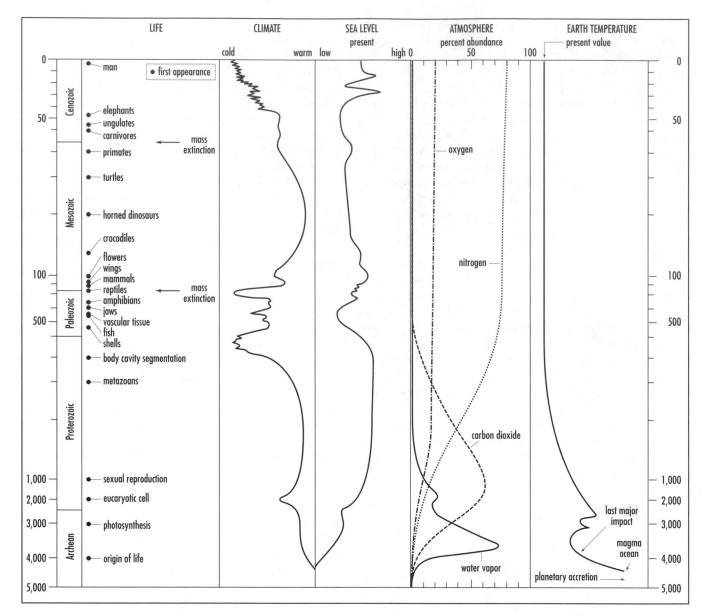

Graphs showing changes in CO₂, O₂, and temperature in the atmosphere with time

atmosphere, whereas the litter or dead biomass contains about twice the amount in the living biomass. It is estimated that land plants absorb 100 gigatons of carbon a year and return about half of this to the atmosphere by respiration. The remainder is transformed to organic carbon and incorporated into plant tissue and soil organic carbon.

Understanding the global carbon cycle is of great importance for predicting and mitigating climate change. Climatologists, geologists, and biologists are just beginning to understand and model the consequences of changes to parts of the system induced by changes in other parts of the system. For instance, a current debate centers on how plants respond to greater atmospheric CO₂. Some models indicate that they

may grow faster under enhanced CO₂, tending to pull more C out of the atmosphere in a planetary self-regulating effect. This is known as the fertilization effect. Many observations and computer models are being performed to investigate the effects of natural and human-induced changes (anthropogenic) to the global carbon cycle, and to better understand what the future may hold for global climates.

See also GAS HYDRATES; GREENHOUSE EFFECT.

Carboniferous   A late Paleozoic geologic period in which the Carboniferous System of rocks was deposited, between 355 million and 285 million years (Ma) ago. The system was named after coal-bearing strata in Wales and has the distinction of being the first formally established stratigraphic system. In the United States, it is customary to use the divisions Mississippian Period (355–320 Ma) and Pennsylvania Period (320–285 Ma), whereas Europeans and the rest of the world refer to the entire interval of time as the Carboniferous Period and divide the rocks deposited in the period into two subsystems, the Upper and Lower, and five series.

The Carboniferous is known as the age of amphibians, or the age of coal. The supercontinent Pangea straddled the equator in the early Carboniferous, with warm climates dominating the southern (Gondwana) and northern (Laurasia) landmasses. In the Lower Carboniferous giant seed ferns and great coal forests spread across much of Gondwana and Laurasia, and most marine fauna that developed in the Lower Paleozoic flourished. Brachiopods, however, declined in number and species. Fusulinid foraminifera appeared for the first time. Primitive amphibians roamed the Lower Carboniferous swamps, along with swarms of insects including giant dragonflies and cockroaches.

In the Early Carboniferous (Mississippian), Gondwana was rotating northward toward the northern Laurentian continent, closing the Rheic Ocean. Continental fragments that now make up much of Asia were rifting from Gondwana, and the west coasts of North and South America were subduction-type convergent margins open to the Panthallassic Ocean. Several arc and other collisions with North America were under way, including the Antler orogeny in the western United States. The Hercynian orogeny in Europe marked the collision between Baltica, southern Europe, and Africa. In the Late Carboniferous (Pennsylvanian), Laurentia and Gondwana had finally collided, forming the single large landmass of Pangea. This collision resulted in the Alleghenian orogeny in the Appalachians of the eastern United States, the Ouachita orogeny in southern United States and South America, and formed the ancestral Rocky Mountains. In Asia, Kazakhstan collided with Siberia, forming the Altai Mountains. Several microcontinents were rifted off the Gondwana continents to be accreted to form much of present-day Asia.

Global climates in the Carboniferous ranged from tropical around much of Laurentia and northern Gondwana, to polar on southern Gondwana, which experienced glaciation in the Pennsylvanian. This widespread glaciation formed in response to Gondwana migrating across the South Pole and is characterized by several advances and retreats and glacial deposits on Africa, Australia, South America, and India. Coal formed at both high and low latitudes in the Pennsylvanian, reflecting the warm climates from easterly trade winds around the closing Rheic Ocean and future opening of the Tethys Ocean. Most of the coal deposits formed in foreland basins associated with continental collisions.

Many sedimentary deposits of Carboniferous age worldwide show development in a repetitive cycle, including accumulation of organic material (vegetation), deposition of carbonates, deposition of clastic sands, and erosion to sea level and soil development. These types of sedimentary deposits have become known as cyclothems and reflect a uniform fluctuation of sea level by 500–650 feet (150–200 m). Analysis of the ages of each cyclothem have led to the recognition that each cycle represents 300,000 years, but the cause of the repetitive cycles remains a mystery. They may be related to cyclical variations in orbital parameters (Milankovitch cycles), or to variations in the intensity of the southern glaciation.

Extinctions in the Late Devonian paved the way for rapid expansion of new marine invertebrate forms in many ecological niches. Radiations in the brachiopods, ammonoids, bryozoans, crinoids, foraminifera, gastropods, pelcypods, and calcareous algae became widespread. Crinoids were particularly abundant in the Mississippian forming dense submarine gardens, along with reefs made of bryozoans and calcareous algae. Fusulinid foraminifera with distinctive coiled forms evolved in the beginning of the Pennsylvanian and serve as a useful index fossil since they evolved so quickly and are abundant in many environments.

Land plants originated in the Devonian and saw additional diversification in the Carboniferous. Chordates, a prominent gymnosperm with long thin leaves, flourished in the Mississippian, whereas conifers appeared in the Late Pennsylvanian. The tropical coal forests of the Pennsylvanian had trees that were more than 98 feet (30 m) tall, including the prominent Lepidodendron and Calamites trees and the seed-bearing Glossopteris shrub that covered much of the cooler parts of Gondwana. Warm climates in the low-latitude coal swamps led to a flourishing fungi flora. The dense vegetation of the Carboniferous led to high levels of atmospheric oxygen, estimated to have comprised about 35 percent of the gases in the atmosphere, compared to present-day levels of 21 percent.

The insects radiated in the Early Pennsylvanian and included the wingless hexapods and the primitive Paleoptera, ancestors of the modern dragonfly and mayfly. A giant Pennsylvanian dragonfly had a wingspan of 24 inches (60 cm) and preyed largely on other insects. Exopterygota, primitive crickets and cockroaches, appeared in the Pennsylvanian. Endopterygota, the folding wing insects including flies and beetles, did not appear until the Permian.

The Carboniferous is famous for the radiation of amphibians. Ten different amphibian families appeared by the end of the Mississippian, living mostly in water and feeding on fish. Eryops and other amphibians of this time resembled crocodiles and include relatives of modern frogs and salamanders. Embolomeres evolved into large (up to 13 feet, or 4 m) eel-like forms with small legs, some living on land and eating insects. Leopospondyls remained in the water, eating mollusks and insects. The earliest known reptile, Westlothiana, evolved from the amphibians in the Late Mississippian by 338 million years ago. The transition from amphibians to reptiles occurred quickly, within a few tens of

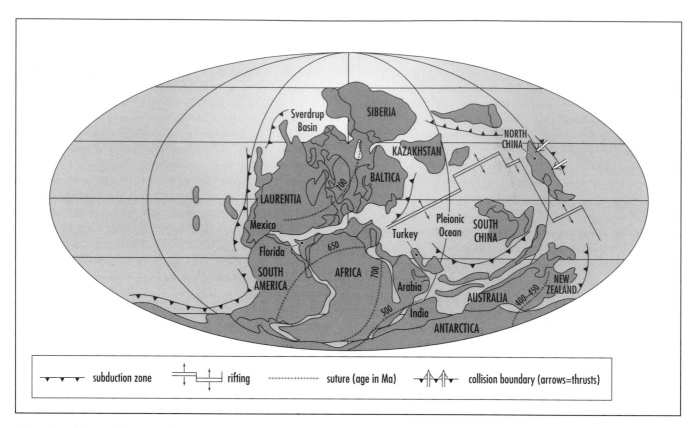

Paleogeographic map of Gondwana in the Carboniferous

millions of years after the origin of amphibians. Amniotes are four-legged animals (tetrapods) that produced eggs similar to the modern bird egg and include reptiles with scales. The rise of amniotes represents a major evolutionary step, since the older amphibians went through an early tadpole stage in which the young are vulnerable to prey. In contrast, the eggs of the amniotes and later reptiles had enough food to provide for the growth of the embryo in a safer environment. Descendants of the amniotes include mammals and birds.

The evolutionary transition between reptiles and mammals is gradual, with more intermediate evolutionary steps known than for any other high-order taxa. Like many other major evolutionary periods in Earth history, this evolutionary step occurred during a supercontinental amalgamation, enabling many species to compete. Many species were intermediate between reptiles and mammals (the so-called mammal-like reptiles), and these dominated the land fauna for about 100 million years until the period of the dinosaurs began in the Permian. Mammal-like reptiles include two orders, the Pelycpsaurs and the Therasids. The mammal-like reptiles had evolved into true mammals by this time, but they did not become dominant until the dinosaurs were killed off at the end of the Cretaceous.

See also MILANKOVITCH CYCLES; PALEOCLIMATOLOGY; PALEOZOIC.

Caspian Sea, Trapped Oceanic Crust?   The Caspian is a large, shallow, and salty inland sea, located between southern Russia, Kazakhstan, Turkmenistan, Iran, and Azerbaijan. It measures 144,444 square miles (373,000 km²), and its surface rests 92 feet (28 m) below sea level. It has a maximum depth of only 3,280 feet (1,000 m) in the south and is very shallow in the north with an average depth of only 16.5 feet (5 m). Thus, changes in the level of the sea result in large changes in the position of the shoreline. These historical changes in the shoreline position are evident in the lowland continuation of the Caspian depression in the Kalymkiya region to the northwest of the sea. More than 75 percent of the water flowing into the Caspian is from the Volga River, which flows in from the north, draining the western side of the Urals and the European plains. Other rivers that flow into the Caspian include the Ural, Emba, Kura, and Temek, but there is no outlet. The Caucasus Mountains strike into the sea on the west, and the Elbruz Mountains line its southern border.

The Caspian is mineral rich, blessed with large oil and gas deposits in several regions, and it is one of the most active exploration areas in the world. It is estimated that the Caspian may hold as much as 200 billion barrels of oil, as much as Iraq and Iran combined. Rich petroleum deposits off the Apseran Peninsula on the west led to the development of Baku, where the Nobels made their fortune at the end of the

19th century. Unfortunately, decades of careless environmental practices associated with state-run oil extraction have led to widespread pollution and contamination, only recently being cleaned up.

The origin of the Caspian depression is somewhat controversial, but many geologists believe that much of the basin is ocean crust trapped during closure of the Tethys Ocean, then deeply buried by sedimentary sequences that host the many petroleum deposits in the area. The sea is also rich in salt deposits and is extensively fished for sturgeon, although the catches have been dramatically declining since the early 1990s. The reasons for the fish decline include loss of spawning grounds, extensive poaching, overfishing, and pollution. A single large (typically up to 15 feet) female Beluga sturgeon may weigh 1,300 pounds and carry 200 pounds of roe, which retailers can sell as caviar for $250,000 in the United States.

cataclasis A process of rock deformation involving the breaking and grinding of preexisting rocks. It amounts to a macroscopic ductile flow produced by grain-scale fracturing, rotation, and sliding of grains to produce smaller fragments. It is typically confined to narrow, tabular-shaped zones known as shear zones or faults, in which the original wall rocks are crushed and ground up to produce fault rocks. Cataclasis or cataclastic flow occurs under relatively low temperature and low to moderate confining pressure conditions in the Earth's crust. The range of cataclastic deformation may be extended to high pressure and temperature conditions by increasing the strain rate, or rate of deformation. High fluid pressures in a deforming rock may aid the cataclastic deformation processes. Cataclasites are distinctive types of fault rocks. Included are coarse-grained fault breccias with visible fragments, fine-grained fault gouge, microbreccias, fractures and microfractures, and even frictional melts known as pseudotachylites. Cataclastic flow may increase the volume of a deforming rock mass since the fracturing creates new open void space. Fluids that flow in the faults commonly deposit vein minerals that often become fractures themselves during continued deformation.

Most cataclastic deformation occurs within a few kilometers of the Earth's surface. However, some minerals behave brittlely under higher temperature conditions and may fracture and deform cataclastically when other minerals are deforming ductilely. For instance, feldspar, garnet, and a few other minerals may break and fracture at relatively high temperatures, typically 930°F–1,830°F (500°C–1,000°C), depending on the strain rate. Most other common minerals, including quartz and calcite, stop behaving brittlely and deform ductilely at 570°F–1,470°F (300°C–800°C). It is often possible to estimate the pressure and temperature conditions of deformation by determining which minerals deformed cataclastically and which minerals deformed ductilely.

See also DEFORMATION OF ROCKS; STRUCTURAL GEOLOGY.

Further Reading
Higgins, M. W. *Cataclastic Rocks.* U.S. Geological Survey Professional Paper 687, 1971.

cathodoluminescence Cold cathodoluminescence is the emission of rays of light from crystalline minerals by excitation with cathode rays. A cold cathode gun with an active current of 620 to 630 mA paired with an ordinary light microscope directs the electrons from a discharge of gas onto the samples. The electrons are generated under vacuum at a cathode and are pulsated toward an anode over a potential difference of 15 kV, before striking the sample. By using a high vacuum, the energy that is imported to electrons in activator ions within the grain causes luminescence.

The cold cathodoluminescence (CL) can measure many properties. It is mostly helpful in determining rock composition and mineral arrangement, revealing growth-zoning cements, separation of quartz overgrowths from detrital quartz, and intensity and localization of fracturing and fracture cements. The CL microscope is very useful for determining and studying the diagenesis of quartz-rich sandstones. It is possible to classify different quartzite samples according to their luminescence.

caves Underground openings and passageways in rock that are larger than individual spaces between the constituent grains of the rock. The term is often reserved for spaces that are large enough for people to enter. Some scientists use the term to describe any rock shelter, including overhanging cliffs. Many caves are small pockets along enlarged or widened cavities, whereas others are huge open underground spaces. The largest cave in the world is the Sarawak Chamber in Borneo, with a volume of 65 million cubic feet. The Majlis Al Jinn (Khoshilat Maqandeli) Cave in Oman is the second largest cave known in the world and is big enough to hold several of the Sultan of Oman's Royal Palaces, with a 747 flying overhead (for a few seconds). Its main chamber is more than 13 million cubic feet in volume, larger than the biggest pyramid at Giza. Other large caves include the world's third, fourth, and fifth largest caves, the Belize Chamber, Salle de la Verna, and the largest "Big Room" of Carlsbad Cavern, which is a large chamber 4,000 feet (1,200 m) long, 625 feet (190 m) wide, and 325 feet (100 m) high. Each of these has a volume of at least 3 million cubic feet. Some caves form networks of linked passages that extend for many miles in length. Mammoth Cave in Kentucky, for instance, has at least 300 miles (485 km) of interconnected passageways. While the caves are forming, water flows through these passageways in underground stream networks.

The formation of caves and sinkholes in karst regions begins with a process of dissolution. Rainwater that filters through soil and rock may work its way into natural fractures or breaks in the rock, and chemical reactions that remove ions from the limestone slowly dissolve and carry away parts of the limestone in solution. Fractures are gradually enlarged, and new passageways are created by groundwater flowing in underground stream networks through the rock. Dissolution of rocks is most effective if the rocks are limestone, and if the water is slightly acidic (acid rain greatly helps cave formation). Carbonic acid (H_2CO_3) in rainwater reacts with the limestone, rapidly (at typical rates of a few millimeters per thousand years) creating open spaces, cave and tunnel systems, and interconnected underground stream networks.

See also KARST.

Cenozoic   The Cenozoic Era marks the emergence of the modern Earth, starting at 66 million years ago and continuing until the present. Also spelled Cainozoic and Kainozoic, the term Cenozoic is taken from the Greek meaning *recent life* and is commonly referred to as the age of the mammals. It is divided into the Tertiary (Paleogene and Neogene) and Quaternary Periods, and the Paleocene, Eocene, Oligocene, Miocene, Pliocene, Pleistocene, and Holocene Epochs.

Modern ecosystems developed in the Cenozoic, with the appearance of mammals, advanced mollusks, birds, modern snakes, frogs, and angiosperms such as grasses and flowering weeds. Mammals developed rapidly and expanded to many different environments. Unlike the terrestrial fauna and flora, the marine biota underwent only minor changes, with the exception of the origin and diversification of whales.

Cretaceous-Tertiary Boundary

The Cenozoic began after a major extinction at the Cretaceous-Tertiary boundary, marking the boundary between the Mesozoic and Cenozoic Eras. This extinction event was probably caused by a large asteroid impact that hit the Yucatán Peninsula near Chicxulub at 66 million years ago. Dinosaurs, ammonites, many marine reptile species, and a large number of marine invertebrates suddenly died off, and the planet lost about 26 percent of all biological families and numerous species. Some organisms were dying off slowly before the dramatic events at the close of the Cretaceous, but a clear sharp event occurred at the end of this time of environmental stress and gradual extinction. Iridium anomalies have been found along most of the clay layers that mark this boundary, considered by many to be the "smoking gun" indicating an impact origin for cause of the extinctions. One-half million tons of iridium are estimated to be in the Cretaceous-Tertiary boundary clay, equivalent to the amount that would be contained in a meteorite with a 6-mile (9.5-km) diameter. Some scientists have argued that volcanic processes within the Earth can produce iridium, and an impact is not necessary to explain the iridium anomaly. However, other rare elements and geochemical anomalies are present along the Cretaceous-Tertiary boundary, supporting the idea that a huge meteorite hit the Earth at this time.

Many features found around and associated with an impact crater on Mexico's Yucatán Peninsula suggest that it is the crater associated with the death of the dinosaurs. The Chicxulub crater is about 66 million years old and lies half-buried beneath the waters of the Gulf of Mexico and half on land. Tsunami deposits of the same age are found in inland Texas, much of the Gulf of Mexico, and the Caribbean, recording a huge tsunami perhaps several hundred feet high that was generated by the impact. The crater is at the center of a huge field of scattered spherules that extends across Central America and through the southern United States. It is a large structure, and is the right age to be the crater that resulted from the impact at the Cretaceous-Tertiary boundary, recording the extinction of the dinosaurs and other families.

The 66-million-year-old Deccan flood basalts, also known as traps, cover a large part of western India and the Seychelles. They are associated with the breakup of India from the Seychelles during the opening of the Indian Ocean. Slightly older flood basalts (90–83 million years old) are associated with the breaking away of Madagascar from India. The volume of the Deccan traps is estimated at 5 million cubic miles (20,841,000 km³), and the volcanics are thought to have been erupted within about 1 million years, starting slightly before the great Cretaceous-Tertiary extinction. Most workers now agree that the gases released during the flood basalt volcanism stressed the global biosphere to such an extent that many marine organisms had gone extinct, and many others were stressed. Then the planet was hit by the massive Chicxulub impactor, causing the massive extinction including the end of the dinosaurs. Faunal extinctions have been correlated with the eruption of the Deccan flood basalts at the Cretaceous-Tertiary (K-T) boundary. There is still considerable debate about the relative significance of flood basalt volcanism and impacts of meteorites for the Cretaceous-Tertiary boundary. However, most scientists would now agree that the global environment was stressed shortly before the K-T boundary by volcanic-induced climate change, and then a huge meteorite hit the Yucatán Peninsula, forming the Chicxulub impact crater, causing the massive K-T boundary extinction and the death of the dinosaurs.

Cenozoic Tectonics and Climate

Cenozoic global tectonic patterns are dominated by the opening of the Atlantic Ocean, closure of the Tethys Ocean and formation of the Alpine–Himalayan Mountain System, and mountain building in western North America. Uplift of mountains and plateaus and the movement of continents severely changed oceanic and atmospheric circulation patterns, changing global climate patterns.

As the North and South Atlantic Oceans opened in the Cretaceous, western North America was experiencing contractional orogenesis. In the Paleocene (66–58 Ma) and

Eocene (58–37 Ma), shallow dipping subduction beneath western North America caused uplift and basin formation in the Rocky Mountains, with arc-type volcanism resuming from later Eocene through late Oligocene (about 40–25 Ma). In the Miocene (starting at 24 Ma), the Basin and Range Province formed through crustal extension, and the formerly convergent margin in California was converted to a strike-slip or transform margin, causing the initial formation of the San Andreas fault.

The Cenozoic saw the final breakup of Pangea and closure of the tropical Tethys Ocean between Eurasia and Africa, Asia, and India and a number of smaller fragments that moved northward from the southern continents. Many fragments of Tethyan ocean floor (ophiolites) were thrust upon the continents during the closure of Tethys, including the Semail ophiolite (Oman), Troodos (Cyprus), and many Alpine bodies. Relative convergence between Europe and Africa, and Asia and Arabia plus India continues to this day and is responsible for the uplift of the Alpine-Himalayan chain of mountains. The uplift of these mountains and the Tibetan Plateau has had important influences on global climate, including changes in the Indian Ocean monsoon and the cutting off of moisture that previously flowed across southern Asia. Vast deserts such as the Gobi were thus born.

The Tertiary began with generally warm climates, and nearly half of the world's oil deposits formed at this time. By the mid-Tertiary (35 Ma) the Earth began cooling again, culminating in the ice house climate of the Pleistocene, with many glacial advances and retreats. The Atlantic Ocean continued to open during the Tertiary, which helped lower global temperatures. The Pleistocene experiences many fluctuations between warm and cold climates, called glacial and interglacial stages (we are now in an interglacial stage). These fluctuations are rapid—for instance, in the past 1.5 million years, the Earth has experienced 10 major and 40 minor periods of glaciation and interglaciation. The most recent glacial period peaked 18,000 years ago when huge ice sheets covered most of Canada and the northern United States, and much of Europe.

The human species developed during the Holocene Epoch (since 10,000 years ago). The Holocene is just part of an extended interglacial period in the planet's current ice house event, raising important questions about how the human race will survive if climate suddenly changes back to a glacial period. Will humans survive? Since 18,000 years ago, the climate has warmed by several degrees, sea level has risen 500 feet (150 m), and atmospheric CO_2 has climbed. Some of the global warming is human induced. One scenario of climate evolution is that global temperatures will rise, causing some of the planet's ice caps to melt, raising global sea level. This higher sea level may increase the Earth's reflectance of solar energy, suddenly plunging the planet into an ice house event and a new glacial advance.

See also CLIMATE CHANGE; PLATE TECTONICS.

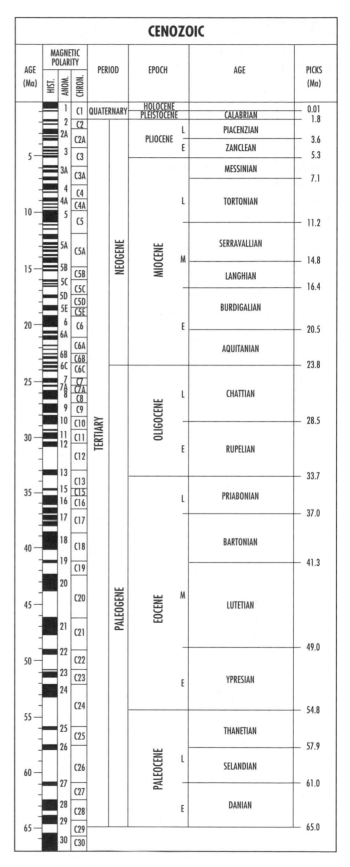

Cenozoic timescale

Further Reading

Pomerol, Charles. *The Cenozoic Era: Tertiary and Quaternary.* Chichester: Ellis Horwood, 1982.

Proterero, Donald, and Robert Dott. *Cenozoic History, Threshold of the Present.* In *Evolution of the Earth,* 6th ed. New York: McGraw Hill, 2002.

Stanley, Steven M. *Earth and Life Through Time.* New York: Freeman, 1986.

Challenger The H.M.S. *Challenger* left dock at Portsmouth, England, on December 21, 1872, on the world's first purely scientific oceanographic expedition. This four-year voyage of the *Challenger,* the accounts of which are published in 50 volumes, changed the course of scientific history. The ship included an interdisciplinary crew of geologists, biologists, chemists, and physicists, who collaborated to map the oceans and collect information about the natural and human environment of the seas. The *Challenger* circumnavigated the globe, discovered and cataloged numerous new species, described and collected samples forming a wealth of geological data, and sounded the depths of the oceans to 26,850 feet (8,185 m), providing the first maps of the seafloor. During the voyage, the ship crossed the oceans many times, visiting every continent including Antarctica. The scientists on board the *Challenger* also described the many indigenous people they encountered in remote parts of the globe. The ship was piloted by Sir Charles Wyville Thomson, a natural history professor from the University of Edinburgh, who died from exhaustion from the rigors of the trip. He was aided by Sir John Murray, a biologist whose observations were extremely influential in establishing the field of marine biology.

The H.M.S. *Challenger* was a 226-foot (69-m) long three-masted square rigger with 2,300 tons of displacement, plus a 1,200-horsepower motor. Since it was a former naval vessel, 16 of the 18 cannons were removed to make room for sampling lines and scientific equipment. The ship was equipped with some of the best labs available for the cruise, including a natural history laboratory for describing biological and geological samples, and a chemistry laboratory. The ship carried a carbonic acid analysis apparatus, a Buchanan water sampler, and a dredge-style bottom sampler.

The ship's crew and scientific team took regular samples and measurements while at sea and stopped at 362 stations. At these stations, they would determine the water depth, collect a bottom sample, collect waters from different depths, do temperature profiles, sample the fauna and flora, collect geological samples, take meteorological observations, and measure currents.

The voyage resulted in a number of major scientific findings. The crew produced the first systematic map of ocean currents and temperatures around the world and mapped the bottom deposits and main contours of the oceans. They discovered the Mid-Atlantic Ridge, a submarine mountain range, and found the deepest point then known on the Earth, the Challenger Deep (26,900 feet; 8,200 m) in the Marianas trough. In addition the scientists on board discovered and described 715 new genera and 4,717 new species of ocean life-forms, and they discovered prodigious life-forms at great depths in the ocean, dispelling earlier claims that the deep oceans were devoid of life.

chemical sediment Sediments deposited directly by precipitation from solution, evaporation, or from deposition of insoluble precipitates. They differ from the other main type of sediments and sedimentary rocks, known as clastic sediments and rocks, which are made of particles broken from another source, transported, and deposited as the new sediment.

Common types of chemical sediments include limestones, dolostones, evaporites, chert, phosphates, and ironstones. Some chemical sediments are produced by a combination of biochemical processes, especially varieties of limestone, dolostone, some types of phosphates, and banded iron formations.

Limestones and dolostones are carbonate rocks composed primarily of calcite and dolomite and produced by a combination of accumulation of dead organisms above the level that calcium carbonate is saturated in seawater (the calcium carbonate compensation depth), and by biochemical and chemical reactions that precipitate carbonate from solution. Evaporites, commonly gypsum ($CaSO_4 \cdot 2 H_2O$), anhydrite ($CaSO_4$), and halite ($NaCl$), form most commonly by the evaporation of seawater or shallow restricted lake basins in arid environments, known as playas. In some of these examples, rainwater flows through older rock deposits, dissolving the evaporite minerals from the older rocks, then reprecipitating the minerals as new evaporites in restricted basins. Chert is a rock composed of nearly pure silica, consisting of microcrystalline or cryptocrystalline quartz. Chert may form from the inorganic or biologic precipitation from seawater. Phosphates commonly form nodules on the seafloor, as layered sediments in shallow water environments, and as guano accumulations, especially from sea birds. Manganese forms nodules on the seafloor, and some are associated with black smoker types of submarine hydrothermal vents. Others are located in shallow water environments. Ironstones and banded iron formations are iron-rich sedimentary rocks that are especially common in Precambrian rocks and may reflect different levels of atmospheric oxygen and iron redox state in the early oceans. These unique formations contain the world's largest reserves of iron ore, but the lack of a modern analog environment makes understanding their origin somewhat enigmatic.

See also BANDED IRON FORMATION; BLACK SMOKER CHIMNEYS; CARBONATE; CHERT.

chemical weathering *See* WEATHERING.

chemosynthesis A metabolic process that differs drastically from photosynthesis, the dominant metabolic process on Earth that relies on energy from the Sun. Chemosynthetic bacteria use hydrogen sulfide produced at deep seafloor hydrothermal vents known as black smoker chimneys as an energy source for their metabolism. The hydrogen sulfide is produced by water that is heated by magma in the crust and reacts with the seafloor basalt, converting sulfate in the seawater to hydrogen sulfide. The primitive bacteria at these seafloor hydrothermal vents, first discovered in 1977, metabolize the hydrogen sulfide as an energy source and become the primary producers in a complex food web around the vents. Other organisms that in turn depend on the chemosynthetic bacteria include giant tube worms, giant clams, lobster, and fish. Some models for the origin of life on Earth suggest that the earliest life on the planet was probably chemosynthetic and heat-loving (thermophyllic).

See also BLACK SMOKER CHIMNEYS; LIFE'S ORIGINS AND EARLY EVOLUTION.

chert A fine-grained chemical sedimentary rock that exhibits concoidal fracture and has been used for stone tools such as arrowheads and knives by prehistoric peoples. It is dense, typically has a dull to semivitreous luster, and has a variety of colors based on types of impurities. The dark gray variety is known as flint, whereas the bright red variety containing iron oxide impurities is known as jasper. Most of the silica in chert consists of very small microcrystalline or cryptocrystalline, less than 30 micron interlocking quartz crystals, but chert may also contain amorphous silica known as opal. Chert may form as diagenetic (replacement) products in limestone, often forming rows or football-shaped nodules. It also forms bedded deposits that may form by accumulation of siliceous tests (skeletons) of planktonic marine organisms, or through chemical and biochemical reactions in seawater.

See also CHEMICAL SEDIMENT.

Chicxulub, Yucatán, Mexico A small town on the northwest side of the Yucatán Peninsula, Mexico, at the center of a 110-mile (177-km) wide meteorite impact crater. Chicxulub is located on the coast, and the impact crater is circular, half on land and half under the shallow sea offshore. A series of concentric rings defined by gravity—and, to a lesser extent, by paleotopography buried under a little over half a mile (1 km) of younger sediments—defines the crater. Much of the underlying rock is limestone that has developed a series of caves (cenotes in local terminology) that follow specific stratigraphic units, also outlining parts of the crater. The Chicxulub crater has been estimated to be 65 million years old, so its age coincides with the mass extinction event associated with the death of the dinosaurs at the Cretaceous/Tertiary boundary. Many models suggest that the impact triggered giant tsunami as high as one thousand feet, and formed a fireball a thousand miles

wide that erupted into the upper atmosphere. Huge earthquakes accompanied the explosion. An ensuing global firestorm released huge amounts of dust and carbon into the atmosphere, blocking the Sun for years. The dust blown into the atmosphere immediately initiated a dark global winter, and as the dust settled months or years later, the extra carbon dioxide in the atmosphere warmed the Earth for many years, forming a greenhouse condition. Many forms of life, including the dinosaurs, could not tolerate these rapid changes and perished.

China's Dabie Shan ultra-high pressure metamorphic belt The Qingling–Dabie-Sulu, or Dabie Shan, is the world's largest ultrahigh-pressure metamorphic belt, containing Triassic (240–220 Ma) eclogite facies metamorphic rocks formed during the collision of the North and South China cratons. Most remarkably, the orogen contains coesite and diamond-bearing eclogite rocks, indicating metamorphic burial to depths of more than 60 miles (100 km). The Dabie Shan metamorphic belt stretches from the Tanlu fault zone between Shanghai and Wuhan, approximately 1,250 miles (2,000 km) to the west northwest of the Qaidam basin north of the Tibetan plateau. The orogen is only 30–60 miles (50–100 km) wide in most places, and it separated the North China craton on the north from the Yangtze craton (also called the South China block) on the south. A small tectonic block or terrane known as the South Qingling is wedged between the North and South China cratons in the orogen and is thought to have collided with the North China craton in the Triassic, before the main collision.

Further Reading
Okay, Aral I., and A. M. Celal Sengör. "Evidence for Intracontinental Thrust-Related Exhumation of the Ultra-High Pressure Rocks in China." *Geology* 20 (1992).
Zhang, Ruth Y., Juhn G. Liou, and Bolin Cong. "Petrogenesis of Garnet-Bearing Ultramafic Rocks and Associated Eclogites in the Su-Lu Ultrahigh P Metamorphic Terrane, Eastern China." *Journal of Metamorphic Geology* 12 (1994).

China's Dongwanzi ophiolite, oldest complete ophiolite The North China, Tarim, and Yangtze cratons form the bulk of Precambrian rocks in China, covering an area of about 15.5 million square miles (4 million km²), or about 40 percent of China. The North China craton is approximately 65.6 million square miles (1.7 million km²) in area and forms a triangular shape covering most of north China, the southeastern part of northeast China, Inner Mongolia, Bohai Bay, Northern Korea, and part of the Yellow Sea regions. It is divided into the eastern and western blocks and the central orogenic belt. The western block and central orogenic belt are separated by the younger Huashan-Lishi fault in the south and by the Datong-Duolun fault in the north. The central orogenic belt and eastern block are locally separated by the younger Xinyang-Kaifen-Jianping fault. Rock formation ages range from 2.7–2.5 Ga in the eastern and western blocks and 2.5 Ga for the central orogenic belt. Detrital zircons from the eastern block have been dated at 3.8–3.5 Ga and are the oldest ages obtained from the North China craton. The average lithospheric thickness ranges from 50 miles (80 km) in the east to 100 miles (160 km) in the west. The difference in thickness between the blocks is reflected in the lack of a thick mantle root in the eastern block.

The Zunhua structural belt is located in the northern part of the central orogenic belt, in the eastern part of the Hebei province, and covers an 81×12.5 square mile (130×20 km^2) area. Numerous mafic and ultramafic boudins have been identified within the Zunhua structural belt, including typical ophiolitic assemblages, which in turn include cumulate ultramafic rocks and pillow lavas, podiform chromite, well-preserved sheeted dikes, pyroxenite, wehrlite, and partly serpentinized harzburgite. These tectonic ultramafic blocks are part of a complex mélange zone found along with the Dongwanzi ophiolite and probably represent deeper-crustal parts of the ophiolite. Podiform chromites are found within harzburgite tectonite and dunite host rocks and contain nodular and orbicular textures. These types of chromite textures are found exclusively in ophiolites and are thought to have formed during partial melting of the flowing upper mantle. Chromites from the ophiolite have been dated by the Re-Os method to be $2,547 \pm 10$ million years old.

The Dongwanzi ophiolite belt is located in the northeast part of the Zunhua structural belt and has been interpreted to represent a large ophiolitic block within the Zunhua ophiolite mélange. The Dongwanzi ophiolite belt is about 31 miles (50 km) long and from three to six miles (5–10 km) wide and preserves the upper or crustal part of the ophiolite suite, with deeper sections of the ophiolite being preserved in related blocks. The ophiolite is broken into three main thrust slices, including the northwest belt, central belt, and the southeast belt. All of the belts are metamorphosed to at least greenschist facies and typically amphibolite facies, with conditions approaching granulite in the west.

A high-temperature shear zone intruded by the 2.4-Ga-old diorite and tonalite marks the base of the ophiolite. Exposed ultramafic rocks along the base of the ophiolite include strongly foliated and lineated dunite and layered harzburgite. Aligned pyroxene crystals and generally strong deformation of serpentinized harzburgite resulted in strongly foliated rock. Harzburgite shows evidence for early high temperature deformation. This unit is interpreted to be part of the lower residual mantle, from which the overlying units were extracted.

The cumulate layer represents the transition zone between the lower ultramafic cumulates and upper mafic assemblages. The lower part of the sequence consists of orthocumulate pyroxenite, dunite, wehrlite, lherzolite and websterite, and olivine gabbro-layered cumulates. Many layers grade from dunite at the base, through wehrlite, and are capped by clinopyroxene. Some unusual ultramafic cumulate rocks in the central belt include hornblende-pyroxenites, hornblendites, and plagioclase-bearing pyroxene hornblendites. Basaltic dikes cut through the cumulates and are similar mineralogically and texturally to dikes in the upper layers.

The gabbro complex of the ophiolite is up to three miles (5 km) thick and grades up from a zone of mixed layered gabbro and ultramafic rocks to one of strongly layered gabbro that is topped by a zone of isotropic gabbro. Thicknesses of individual layers vary from centimeter to meter scale and include clinopyroxene and plagioclase-rich layers. Layered gabbros from the lower central belt alternate between fine-grained layers of pyroxene and metamorphic biotite that are separated by layers of metamorphic biotite intergrown with quartz. Biotite and pyroxene layers show a random orientation of grains. Coarse-grained veins of feldspar and quartz are concentrated along faults and fractures. Plagioclase feldspar shows core replacement and typically has irregular grain boundaries. The gabbro complex of the ophiolite has been dated by the U-Pb method on zircons to be $2,504 \pm 2$ million years old.

The sheeted dike complex is discontinuous over several kilometers. More than 70 percent of the dikes exhibit one-way chilling on their northeast side. Gabbro screens are common throughout the complex and increase in number and thickness downward marking the transition from the dike complex to the fossil magma chamber. In some areas, the gabbro is cut by basaltic-diabase dikes, but in others it cuts through xenoliths of diabase suggesting comagmatic formation.

The upper part of ophiolite consists of altered and deformed pillow basalts, pillow breccias, and interpillow sediments (chert and banded iron formations). Many of the pillows are interbedded with more massive flows and cut by sills; however, some well-preserved pillows show typical lower cuspate and upper lobate boundaries that define stratigraphic younging. Pyroxenes from pillow lavas from the ophiolite have been dated by the Lu-Hf method to be 2.5 billion years old, the same age as estimated for the gabbro and mantle sections.

Prior to the discovery of the Dongwanzi ophiolite, portions of several Archean greenstone belts had been interpreted to contain dismembered or partial ophiolites, but none of these contain the complete ophiolite sequence. Several well-documented dismembered Archean ophiolites have three or four of the main magmatic components of a full ophiolite. Archean greenstone belts have a greater abundance of accreted ophiolitic fragments compared to Phanerozoic orogens, suggesting that thick, relatively buoyant, young Archean oceanic lithosphere may have had a rheological structure favoring delamination of the uppermost parts during subduction and collisional events. The preservation of a complete Archean ophiolite sequence in the North China craton is therefore of great importance for understanding processes of Archean seafloor spreading, as it is the most complete record of this process known to exist.

Despite the apparent abundance of partial dismembered Archean ophiolites, no complete and laterally extensive Archean ophiolites had been previously described from the geologic record, leading some workers to the conclusion that Archean tectonic style was fundamentally different from that of younger times. The presence of a complete Archean ophiolite suggests that Archean and similar younger tectonic environments were not so different, and that seafloor spreading operated as a planetary heat loss mechanism 2.5 billion years ago much as it does today.

See also ARCHEAN; CRATON; OPHIOLITE; PRECAMBRIAN.

Further Reading

Kusky, Timothy M., Jianghai Li, Adam Glass, and Alan Huang. "Archean Ophiolites and Ophiolite Fragments of the North China Craton." *Earth Science Reviews* (2004).

Kusky, Timothy M., Jianghai Li, and Robert T. Tucker. "The Archean Dongwanzi Ophiolite Complex, North China Craton: 2.505 Billion Year Old Oceanic Crust and Mantle." *Science* 292 (2001).

Li, Jianghai, Timothy M. Kusky, and Alan Huang. "Neoarchean Podiform Chromitites and Harzburgite Tectonite in Ophiolitic Melange, North China Craton, Remnants of Archean Oceanic Mantle." *GSA Today* (2002).

China's Sinian fauna   Parts of northeastern and southern China are well known for Late Proterozoic-age sedimentary deposits that host some of the world's most spectacular early animal fossils. One of the best studied sequences is the Doushantuo formation of phosphatic sedimentary rocks exposed in South China's Guizhou province, dated to be 580–600 million years old. The Sinian fauna is therefore older than the well-known Ediacaran metazoan fauna and is currently the oldest known assemblage of multicelled animal fossils on the Earth. The macrofossil assemblages are associated with prokaryotic and eukaryotic microfossils and are remarkable in the well-preserved cellular and tissue structures and even organic kerogen. Many of the fossils are unusual acritarchs, organic walled fossils with peripheral processes such as spines, hairs, and flagellum, that cannot be confidently placed into any living plant or animal group classification. There is current debate about the origin of some of the fossils, whether they may be metazoan embryos, multicelled algae, filamentous bacteria, acritarchs, or phytoplankton.

See also PROTEROZOIC; VENDIAN.

Further Reading

Hongzhen, Wang, Lin Baoyu, and Liu Xiaoliang. "Cnidarian Fossil from the Sinian System of China and Their Stratigraphic Significance." *Paleontographica Americana* 54 (1984).

Jinbao, Chen, Zhang Huimin, Xing Yusheng, and Ma Guogan. "On the Upper Precambrian (Sinian Erathem) in China." *Precambrian Research* 15 (1981).

Chinook winds   A special type of Foehn winds that descend the leeward slopes of mountains. The Chinook is a warm, dry wind that descends along the eastern slope of the Rocky Mountains from New Mexico into Canada, generally affecting a several hundred-kilometer-wide belt in the mountains' foothills and slope. Foehn winds can cause temperatures in the affected area to rise rapidly, sometimes more than 36°F (20°C) in a single hour, with a corresponding drop in humidity.

Chinooks form in the Rocky Mountain region when strong westerly winds blow over a mountain and create a trough of low pressure on the mountain's east slope. This low pressure forces air downslope, which becomes compressed and warmed (by compressional heating) by about 50°F (30°C) per kilometer as it descends.

Chinooks can be predicted when a wall-like bank of clouds is observed forming over the Rockies. These clouds remain stationary, but the winds coming out of the mountains can be quite strong.

See also SANTA ANA WINDS.

chondrite   *See* METEORITE.

chromite   A black to brownish black octahedral crystal of the spinel group exhibiting a submetallic to metallic luster, with the chemical formula $(Fe,Mg)(Cr,Al)_2O_4$. It forms a primary accessory mineral in mafic and ultramafic igneous rocks and may form economically important layers, or pods, in layered intrusions and ophiolitic complexes, and it may also accumulate in detrital deposits derived from erosion of igneous parent rocks. Chromite is isomorphous with magnesiochromite $(MgCr_2O_4)$, and it is the principal ore of chromium. In some deposits, chromite and related minerals (magnesiochromite, magnetite, and hematite) are so concentrated that they make up the bulk of the rock, which may then be referred to as a chromitite.

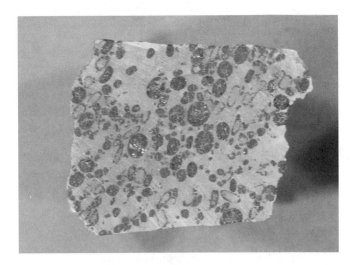

Chromite, North China *(Photo by Timothy Kusky)*

Chromite ores that are associated with continental layered intrusions tend to form stratiform bodies, with several centimeters to a meter-thick layers of chromite interlayered with dunite, pyroxenite, and other types of peridotite. The layers may be continuous for many tens if not hundreds of meters. In contrast, chromite deposits associated with ophiolitic complexes tend to form discontinuous pod-shaped bodies in dunite or harzburgite. These podiform chromites frequently show uniform geological characteristics, including: (1) lensoidal geometry, distributed along foliation; (2) rootless folds with chromite bands; (3) unique magmatic textures and structures, such as dunite envelopes for the chromite ores, and nodular and orbicular textures of chromites; and (4) strong plastic deformation associated with harzburgites. The origin of the podiform chromitites is attributed to melt-rock reaction or dynamic magmatism within melt channels in the upper mantle. The oldest known podiform chromite deposits are associated with China's 2.5-billion-year-old Dongwanzi ophiolite complex.

See also CHINA'S DONGWANZI OPHIOLITE; OPHIOLITE.

Further Reading

Edwards, S. J., J. A. Pearce, and J. Freeman. "New Insights Concerning the Influence of Water during the Formation of Podiform Chromitite." In "Ophiolites and Oceanic Crust," edited by Y. Dilek, E. Moores, D. Elthon, and A. Nicolas. *Geological Society of America Special Paper* 349 (2000): 139–147.

Kusky, Timothy M., Jianghai H. Li, and Robert T. Tucker. "The Dongwanzi Ophiolite: Complete Archean Ophiolite with Extensive Sheeted Dike Complex, North China Craton." *Science* 292 (2001): 1,142–1,145.

Lago, Bernard L., Michel Rabinowicz, and Adolphe Nicolas. "Podiform Chromitite Ore Bodies: a Genetic Model." *Journal of Petrology* 23 (1982): 103–125.

Leblanc, Marc, and Adolphe Nicolas. "Ophiolitic Chromitites." *International Geology Reviews* 34 (1992): 653–686.

Li, Jianghai H., Timothy M. Kusky, and Xiongnan Huang. "Neoarchean Podiform Chromitites and Harzburgite Tectonite in Ophiolitic Melange, North China Craton, Remnants of Archean Oceanic Mantle." *GSA Today* 12, no. 7 (2002): 4–11.

Melcher, Frank, Walter Grum, Tatiana V. Thalhammer, and Oskar A. R. Thalhammer. "The Giant Chromite Deposits at Kempirsai, Urals; Constraints from Trace Element (PGE, REE) and Isotope Data." *Mineralium Deposita* 34 (1999): 250–272.

Nicolas, Adolphe, and H. A. Azri. *Chromite-Rich and Chromite-Poor Ophiolites: the Oman Case. In Ophiolite Genesis and Evolution of the Oceanic Lithosphere,* edited by T. J. Peters, A. Nicolas, and R. G. Coleman, 261–274. Boston: Kluwer Academic Publishers, 1991.

Stowe, Clive W. "Compositions and Tectonic Settings of Chromite Deposits through Time." *Economic Geology* 89 (1994): 528–546.

Thayer, T. P. "Gravity Differentiation and Magmatic Re-emplacement of Podiform Chromite Deposits." *Economic Geology Monograph,* no. 4 (1969): 132–146.

Zhou, Meifu, Paul T. Robinson, John Malpas, and Li Zijin. "Podiform Chromitites in the Luobusa Ophiolite (South Tibet): Implications for Melt-Rock Interaction and Chromite Segregation in the Upper Mantle." *Journal of Petrology* 37 (1996): 3–21.

Chugach Mountains, Alaska Southern Alaska is underlain by a complex amalgam of accreted terranes, including the Wrangellia superterrane (consisting of the peninsular, Wrangellia, and Alexander terranes), and farther outboard, the Chugach–Prince William superterrane. During much of the Mesozoic, the two superterranes formed a magmatic arc and accretionary wedge, respectively, above a circum-Pacific subduction zone. The Border Ranges fault forms the boundary between the Wrangellia and Chugach–Prince William superterranes; it initiated as a subduction thrust but has been reactivated in various places as a strike-slip or normal fault. On the Kenai Peninsula the Chugach terrane contains two major units. Farther inboard lies the McHugh complex, composed mainly of basalt, chert, argillite, and graywacke, as well as several large ultramafic massifs. Radiolarians from McHugh cherts throughout south-central Alaska range in age from Ladinian (middle Triassic) to Albian-Aptian (mid-Cretaceous). The interval during which the McHugh complex formed by subduction-accretion is not well known, but probably spanned most of the Jurassic and Cretaceous. The McHugh has been thrust seaward on the Eagle River/Chugach Bay fault over a relatively coherent tract of trench turbidites assigned to the Upper Cretaceous Valdez Group. After the protracted episode of subduction-accretion that built the Chugach terrane, the accretionary wedge was cut by near-trench intrusive rocks, assigned to the Sanak-Baranof plutonic belt, probably related to ridge subduction.

The McHugh complex of south-central Alaska and its lateral equivalent, the Uyak complex of Kodiak, are part of the Mesozoic/Cenozoic subduction complex of the Chugach terrane. Despite its vast extent and its potential value in reconstructing the tectonics of the Pacific realm, the McHugh has not been very extensively studied, especially compared to similar tracts such as the Franciscan of California or the Shimanto of Japan. The evolution of the McHugh and its equivalents can be broken down into three broad, somewhat overlapping phases: (1) origin of igneous and sedimentary protoliths; (2) incorporation into the subduction complex ("accretion"), and attendant deformation and metamorphism; and (3) younger deformations.

Few fossil ages have been reported from the McHugh complex with the best paleontological control now available from the Seldovia quadrangle. At several places, mostly in Kachemak Bay, radiolarian chert depositionally overlies pillow basalt. Precise radiolarian age calls show that the base of the chert varies in age from Ladinan to Albian-Aptian. Other chert sections, which are typically fault-bounded and have no stratigraphic context, also range from Ladinian to Albian. Graywacke depositionally overlying chert has yielded Early Jurassic (Pleinsbachian) radiolarians. These ages are readily explained by a stratigraphic model in which the McHugh basalts were formed by seafloor spreading, the overlying cherts were deposited on the ocean floor as it was conveyed toward a trench, and the argillite and graywacke record deposition in the trench, just prior to subduction-accretion. The timing of the subduction-accretion is not well known but probably spanned most of the Jurassic and Cretaceous.

Limestones within the McHugh complex are of two categories. A limestone clast in McHugh conglomerate has yielded conodonts with a possible age range of late Meramecian to early Morrowan (Late Mississippian to Early Pennsylvanian). This clast could have been shed from the Strelna formation of the Wrangellia terrane. Most of the dated limestones, however, are tectonic blocks—typically occurring as severely extended strings of boudins—that have yielded Permian fusulinids or conodonts. Both the fusulinids and conodonts are of shallow-water, tropical, Tethyan affinity; the fusulinids are quite distinct from those of Wrangellia. The limestone blocks might represent the tops of seamounts that were decapitated at the subduction zone. If so, some of the ocean floor that was offscraped to form the McHugh complex must have formed in the Paleozoic.

The seaward part of the Chugach terrane is underlain by the Valdez group of Late Cretaceous (Campanian? to Maastrichtian) age. In the Kenai Peninsula, it includes medium- and thin-bedded graywacke turbidites, black argillite, and minor pebble to cobble conglomerate. These strata were probably deposited in a deep-sea trench and accreted shortly thereafter. Most of the Valdez group consists of relatively coherent strata, deformed into regional-scale tight to isoclinal folds, cut by a slaty cleavage. The McHugh complex and Valdez group are juxtaposed along a thrust, which in the area of Turnagain Arm has been called the Eagle River fault, and on the Kenai Peninsula is known as the Chugach Bay thrust. Beneath this thrust is a mélange of partially to thoroughly disrupted Valdez group turbidites. This monomict mélange, which is quite distinct from the polymict mélanges of the McHugh complex, can be traced for many kilometers in the footwall of the Eagle River thrust and its along-strike equivalents.

In early Tertiary time, the Chugach accretionary wedge was cut by near-trench intrusive rocks forming the Sanak-Baranof plutonic belt. The near-trench magmatic pulse migrated 1,367 miles (2,200 km) along the continental margin, from about 65–63 Ma at Sanak Island in the west to about 50 Ma at Baranof Island in the east. The Paleogene near-trench magmatism was related to subduction of the Kula-Farallon spreading center.

Mesozoic and Cenozoic rocks of the accretionary wedge of south-central Alaska are cut by abundant late brittle faults. Along Turnagain Arm near Anchorage, four sets of late faults are present: a conjugate pair of east-northeast-striking dextral and northwest-striking sinistral strike-slip faults; north-northeast-striking thrusts; and less abundant west-northwest-strik-

Glaciated terrain typical of Chugach Mountains *(Photo by Timothy Kusky)*

ing normal faults. All four fault sets are characterized by quartz ± calcite ± chlorite fibrous slickenside surfaces and appear to be approximately coeval. The thrust- and strike-slip faults together resulted in subhorizontal shortening perpendicular to strike, consistent with an accretionary wedge setting. Motion on the normal faults resulted in strike-parallel extension of uncertain tectonic significance. Some of the late brittle faults host gold-quartz veins, which are the same age as nearby near-trench intrusive rocks. By implication, the brittle faulting and gold mineralization are probably related to ridge subduction.

Scattered fault-bounded ultramafic-mafic complexes in southern Alaska stretch 600 miles (1,000 km) from Kodiak Island in the south to the Chugach Mountains in the north. These generally consist of dunite +/- chromite, wehrlite, clinopyroxenite, and websterite, which grade upward into gabbronorites. These rocks are intruded by quartz diorite, tonalite, and granodiorite. Because of general field and petrographic similarities, these bodies are generally regarded as having a similar origin and are named the Border Ranges ultramafic-mafic complex (BRUMC). The BRUMC includes six bodies on Kodiak and Afognak Islands, plus several on the Kenai Peninsula (including Red Mountain, parts of the Kachemak terrane (now obsolete, but the ultramafic complex is renamed the Halibut Cove ultramafic massif), and other smaller bodies. In the northern Chugach Mountains the BRUMC includes the Eklutna, Wolverine, Nelchina, and Tonsina complexes and the Klanelneechena complex in the central Chugach Mountains.

Some models for the BRUMC suggest that all these bodies represent cumulates formed at the base of an intraoceanic arc sequence and were cogenetic with volcanic rocks now preserved on the southern edge of the Wrangellian composite terrane located in the Talkeetna Mountains. However, some of the ultramafic massifs on the southern Kenai Peninsula are probably not related to this arc but represent deep oceanic material accreted in the trench. The ultramafic massifs on the Kenai Peninsula appear to be part of a dismembered assemblage that includes the ultramafic cumulates at the base, gabbroic-basalt rocks in the center, and basalt-chert packages in the upper structural slices. The ultramafic massifs may represent pieces of an oceanic plate subducted beneath the Chugach terrane, with fragments offscraped and accreted during the subduction process. There are several possibilities as to what the oceanic plate may have been, including "normal" oceanic lithosphere, an oceanic plateau, or an immature arc. Alternatively, the ultramafic/mafic massifs may represent a forearc or "suprasubduction zone" ophiolite, formed seaward of the incipient Talkeetna (Wrangellia) arc during a period of forearc extension.

See also CONVERGENT PLATE MARGIN PROCESSES.

Further Reading

Bradley, Dwight C., Timothy M. Kusky, Peter Haeussler, D. C. Rowley, Richard Goldfarb, and S. Nelson. "Geologic Signature of Early Ridge Subduction in the Accretionary Wedge, Forearc Basin, and Magmatic Arc of South-Central Alaska." In "Geology of a Transpressional Orogen Developed During a Ridge-Trench Interaction Along the North Pacific Margin," edited by Virginia B. Sisson, Sarah M. Roeske, and Terry L. Pavlis. *Geological Society of America* Special Paper 371 (2003): 19–50.

Bradley, Dwight C., Timothy M. Kusky, Peter Haeussler, S. M. Karl, and D. Thomas Donley. Geologic Map of the Seldovia Quadrangle, U.S. Geological Survey Open File Report 99-18, scale 1:250,000, with marginal notes, 1999. Also available as an Internet publication: http://wrgis.wr.usgs.gov/open-file/of 99-18/

Burns, L. E. "The Border Ranges Ultramafic and Mafic Complex, South-Central Alaska: Cumulate Fractionates of Island Arc Volcanics." *Canadian Journal of Earth Science,* v. 22 (1985): 1,020–1,038.

Clark, Sandra H. B. "The McHugh Complex of South-Central Alaska." U.S. Geological Society Bulletin 1372-D (1973): D1–11.

Connelly, W. "Uyak Complex, Kodiak Islands, Alaska—A Cretaceous Subduction Complex." *Geological Society of America Bulletin,* v. 89 (1978): 755–769.

Cowan, Darrel S. "Structural Styles in Mesozoic and Cenozoic Mélanges in the Western Cordillera of North America." *Geological Society of America Bulletin,* v. 96 (1985): 451–462.

Cowan, Darrel S., and R. F. Boss. "Tectonic Framework of the Southwestern Kenai Peninsula, Alaska." *Geological Society of America Bulletin,* v. 89 (1978): 155–158.

DeBari, Susan M., and R. G. Coleman. "Examination of the Deep Levels of an Island Arc: Evidence from the Tonsina Ultramafic-Mafic Assemblage, Tonsina, Alaska." *Journal of Geophysical Research,* v. 94 (1989): 4,373–4,391.

Guild, Philip W. "Chromite Deposits of the Kenai Peninsula, Alaska." *U.S. Geological Society Bulletin* 931-G (1942): 139–175.

Hudson, Travis. "Calc-alkaline Plutonism along the Pacific Rim of Southern Alaska: Circum-Pacific Terranes." *Geological Society of America Memoir* 159 (1983): 159–169.

Kusky, Timothy M., Dwight C. Bradley, D. Thomas Donley, D. C. Rowley, and Peter Haeussler. "Controls on Intrusion of Near-Trench Magmas of the Sanak-Baranof Belt, Alaska, during Paleogene Ridge Subduction, and Consequences for Forearc Evolution." In "Geology of a Transpressional Orogen Developed During a Ridge-Trench Interaction Along the North Pacific Margin," edited by Virginia B. Sisson, Sarah M. Roeske, and Terry L. Pavlis. *Geological Society of America Special Paper* 371 (2003): 269–292.

Kusky, Timothy M., and C. Young. "Emplacement of the Resurrection Peninsula Ophiolite in the Southern Alaska Forearc during a Ridge-Trench Encounter." *Journal of Geophysical Research,* v. 104, no. B12 (1999): 29,025–29,054.

Kusky, Timothy M., Dwight C. Bradley, Peter Haeussler, and S. Karl. "Controls on Accretion of Flysch and Mélange Belts at Convergent Margins: Evidence from The Chugach Bay Thrust and Iceworm Mélange, Chugach Terrane, Alaska." *Tectonics,* v. 16, no. 6 (1997): 855–878.

Kusky, Timothy M., Dwight C. Bradley, and Peter Haeussler. "Progressive Deformation of the Chugach Accretionary Complex, Alaska, during a Paleogene Ridge-Trench Encounter." *Journal of Structural Geology,* v. 19, no. 2 (1997): 139–157.

Moffit, Fred H. "Geology of the Prince William Sound Region, Alaska." *U.S. Geological Survey Bulletin* 989-E (1954): 225–310.

Pavlis, Terry L., and Virginia B. Sisson. "Structural History of the Chugach Metamorphic Complex in the Tana River Region, Eastern Alaska: A Record of Eocene Ridge Subduction." *Geological Society of America Bulletin* 107 (1995): 1,333–1,355.

Plafker, George, and H. C. Berg. "Overview of the Geology and Tectonic Evolution of Alaska." In *The Geology of Alaska, Decade of North American Geology, G-1,* edited by G. Plafker and H. C. Berg. *Geological Society of America* (1994): 389–449.

Plafker, George, James C. Moore, and G. R. Winkler. "Geology of the Southern Alaska Margin." In *The Geology of Alaska, Decade of North American Geology, G-1,* edited by G. Plafker and H. C. Berg. *Geological Society of America* (1994): 989–1,022.

Sisson, Virginia B., Lincoln S. Hollister, and Tullis C. Onstott. "Petrologic and Age Constraints on the Origin of a Low-Pressure/High Temperature Metamorphic Complex, Southern Alaska." *Journal Geophysical Research,* v. 94, no. B4 (1989): 4,392–4,410.

Sisson, Virginia B., and Lincoln S. Hollister. "Low-Pressure Facies Series Metamorphism in an Accretionary Sedimentary Prism, Southern Alaska." *Geology,* v. 16 (1988): 358–361.

Toth, M. I. "Petrology, Geochemistry, and Origin of the Red Mountain Ultramafic Body near Seldovia, Alaska." *U.S. Geological Society* Open-file Report (1981): 81–514.

Tysdal, Russell G., J. E. Case, G. R. Winkler, and Sandra H. C. Clark. "Sheeted Dikes, Gabbro, and Pillow Basalt in Flysch of Coastal Southern Alaska." *Geology,* v. 5 (1977): 377–383.

Tysdal, Russell G., and George Plafker. "Age and Continuity of the Valdez Group, Southern Alaska." In *Changes in Stratigraphic Nomenclature by the United States Geological Survey, 1977,* edited by Norman F. Sohl and W. B. Wright. *U.S. Geological Society Bulletin* 1457-A (1978): A120–A124.

cirque Glacially formed bowl-shaped hollows that open downstream and are bounded upstream by a steep wall. Frost wedging, glacial plucking, and abrasion all work to excavate cirques from previously rounded mountaintops. Many cirques

Active and abandoned cirques. (A) Aerial view of cirque, Eastern Chugach Mountains, Alaska, bounded by knife-edged ridges known as aretes. Note the line of crevasses marking the place where the glacier is moving over a ledge and out of the cirque. (B) View of cirque feeding large icefield and valley glaciers. Harding Icefield, Kenai Peninsula, Alaska. (C) Mountaintop cirque that appears to be retreating from the large, crevasse-dominated valley glacier in the foreground. Harding Icefield, Alaska. (D) Abandoned cirque showing classic bowl-shaped profile and bounding aretes. Lower Kenai Peninsula, Alaska. *(Photos by Timothy Kusky)*

contain small lakes called tarns, which are blocked by small ridges at the base of the cirque. Cirques continue to grow during glaciation, and where two cirques form on opposite sides of a mountain, a ridge known as an arete forms. Where three cirques meet, a steep-sided mountain forms, known as a horn. The Matterhorn of the Swiss Alps is an example of a glacially carved horn.

See also GLACIER.

clastic rocks Sedimentary rocks that are composed primarily of fragments of older rocks and minerals that have been broken off from their parent rock, transported, and redeposited. The word *clastic* is derived from the Greek word *klastos*, meaning broken. The grain size is quite variable and may include coarse-grained breccias and conglomerates, generally deposited close to their source or in high-energy environments, sandstones, siltstone, muds, and their metamorphic equivalents. Clastic rocks are classified based on their texture,

composition of clasts and matrix, and on their grain size. Grain size classifications are based on divisions known as the Wentworth scale, named after Chester K. Wentworth (1891–1969). The smallest particles in the Wentworth classification are clay particles, with a diameter of less than 1/256 millimeter (.00015 inch). Silt particles range from clay-sized to 1/16 millimeter (.0025 inch), sand from 1/16 to 2 millimeters (.0025–0.079 inch), pebbles from 2 to 64 millimeters (.079–2.5 inch), cobbles from 64 to 256 millimeters (2.5–10 inches), and boulders, larger than 256 millimeters.

See also FACIES; SEDIMENTARY ROCKS.

clay A fine-grained sediment with particles less than 0.000015 inch (1/256 mm) in diameter. Clays display colloidal behavior as they are easily suspended in a fluid and may not precipitate until different particles stick together to form larger particles. Clay minerals include a number of loosely defined amorphous or finely crystalline hydrous aluminosilicates that form monoclinic lattices with two to three repeating layers. Most clay minerals form by the chemical alteration or weathering of silicate minerals such as feldspar, pyroxene, and amphibole, and they may also be abundant in the alteration zones around mineral deposits. Clay minerals typically accumulate in soils, clay deposits, and shales. The most common clay minerals include kaolinite, montmorillonite, and illite.

Clay has been used for making pottery and other artifacts since Neolithic times, because it is easily shaped when wet, and baking clay items in a furnace transforms them into rigid, impermeable vessels. However, some clays are hazardous in certain circumstances. For instance, a group of clays known as expansive clays expand by up to 400 percent when water is added to their environment. The expansion of these clays is powerful and has the force to crumble bridges, foundations, and tall buildings slowly. Damage from expanding clays is one of the most costly of all natural hazards in the United States, causing billions of dollars in damage every year.

In areas with expansive clays it is important to know the shrink/swell potential of a soil before construction begins in an area. The shrink/swell potential is a measure of a soil's ability to add or lose water at a molecular level. Expansive clays add layers of water molecules between the plates of clay minerals (made of silica, aluminum, and oxygen), loosely bonding the water in the mineral structure. Most expansive clays are rich in montmorillonite, a clay mineral that can expand up to 15 times its normal dry size. Most soils do not expand more than 25 to 50 percent of their dry volume, but an expansion of 3 percent is considered hazardous.

Damage from shrinking/swelling clays is mostly to bridges, foundations, and roadways, all of which may crack and move during expansion. Regions with pronounced wet and dry seasons tend to have a greater problem with expansive clays than regions with more uniform precipitation distributed throughout the year. This is because the adhered water content of the clays changes less in regions where the soil moisture remains more constant. Some damage from shrinking and swelling soils can be limited, especially around homes. Trees that are growing near foundations can cause soil shrinkage during dry seasons and expansion in wet seasons. These dangers of shrinking/swelling clays can be avoided by not planting trees too closely to homes and other structures. Local topography and drainage details also influence the site-specific shrink/well potential. Buildings should not be placed in areas with poor drainage as water may accumulate there and lead to increased soil expansion. Local drainage may be modified to allow runoff away from building sites, reducing the hazards associated with soil expansion.

In general, soils that are rich in clay minerals and organic material tend to have low strength, low permeability, high compressibility, and the greatest shrink/swell potential. These types of soils should be avoided for construction projects, where possible. If they cannot be avoided, steps must be taken to accommodate these undesirable traits into the building construction. Sand and gravel-rich soils pose much less danger than clay and organic-rich soils. These soils are well drained, strong, have low compressibility and sensitivity, and low shrink/swell potential.

See also WEATHERING.

cleavage A structural property of certain rocks to split along closely spaced secondary penetrative features such as aligned microcracks and tectonic foliations. Cleavage is a type of foliation defined by penetrative or zonal alignments of mica and quartz grains and domains, or by the preferred alignment of microcracks. It may form discrete zones, domains, be defined by anastomosing dissolution surfaces, or by microfractures, with the character of the cleavage being determined by the original rock type and the conditions of deformation and metamorphism. It is best-developed in mica-rich rocks such as slate. Cleavage often shows a regular geometric relationship with folds in regionally deformed rocks, suggesting that the folds and cleavage both form during the same regional contraction event. The cleavage is typically axial planar to folds, being aligned to the fold axial surface. In complexly deformed terrains there may be cleavages and foliations of several generations, forming more complex fold/cleavage relationships.

Minerals may exhibit a different type of cleavage. Mineral cleavage is the property of minerals to split along smooth, crystallographically defined planes that are parallel to possible crystal faces in the mineral. The number of cleavage faces in a crystal is determined by the crystallographic symmetry of the crystal lattice. The cleavage in minerals forms because atomic bonds are weaker across these planes than in other directions, leading them to be planes of weakness along which the mineral breaks. Cleavage faces in minerals are typically smooth and glistening but may also have striations on them.

See also DEFORMATION OF ROCKS; MINERALOGY; STRUCTURAL GEOLOGY.

Further Reading

Borradaile, Graham J., M. Brian Bayly, and Chris M. Powell. *Atlas of Deformational and Metamorphic Rock Fabrics.* Berlin: Springer-Verlag, 1982.

Gray, Dave R. "Morphological Classification of Crenulation Cleavage." *Journal of Geology* 85 (1977): 229–235.

climate The average weather of a place or area, and its variability over a period of years. The term *climate* is derived from the Greek word *klima,* meaning inclination and referring specifically to the angle of inclination of the Sun's rays, a

Rock cleavage, Maine. The cleavage forms vertically dipping planes and is axial planar to and at right angles to beds in this photo. *(Photo by Timothy Kusky)*

function of latitude. The average temperature, precipitation, cloudiness, and windiness of an area determine a region's climate. Factors that influence climate include latitude; proximity to oceans or other large bodies of water that could moderate the climate; topography, which influences prevailing winds and may block precipitation; and altitude. All of these factors are linked together in the climate system of any region on the Earth. The global climate is influenced by many other factors. The rotation of the Earth and latitudinal position determine where a place is located with respect to global atmospheric and oceanic circulation currents. Chemical interactions between seawater and magma significantly change the amount of carbon dioxide in the oceans and atmosphere and may change global temperatures. Pollution from humans also changes the amount of greenhouse gases in the atmosphere, which may be contributing to global warming. Climatology is the field of science that is concerned with climate, including both present-day and ancient climates. Climatologists study a variety of problems, ranging from the classification and effects of present-day climates through to the study of ancient rocks to determine ancient climates and their relationship to plate tectonics. An especially important

field being actively studied by climatologists is global climate change, with many studies focused on the effects that human activities have had and will have on global climate. For many of these models it is necessary to use powerful supercomputers and to construct computer models known as global circulation models. These models input various parameters at thousands or millions of grid points on a model Earth and see how changing one or more variables (e.g., CO_2 emissions) will affect the others.

Classifications of climate must account for the average, extremes, and frequencies of the different meteorological elements. There are many different ways to classify climate, and most modern classifications are based on the early work of the German climatologist Wladimir Koppen. His classification (initially published in 1900) was based on the types of vegetation in an area, assuming that vegetation tended to reflect the average and extreme meteorological changes in an area. He divided the planet into different zones such as deserts, tropics, rainforest, tundra, etc. In 1928 a Norwegian meteorologist named Tor Bergeron modified Koppen's classification to include the types of air masses that move through an area, and how they influence the vegetation patterns. The

British meteorologist George Hadley reached another fundamental understanding of the factors that influence global climate in the 18th century. Hadley proposed a simple, convective type of circulation in the atmosphere, in which heating by the Sun causes the air to rise near the equators and move poleward, where the air sinks back to the near surface, then returning to the equatorial regions. We now recognize a slightly more complex situation, in that there are three main convecting atmospheric cells in each hemisphere, named Hadley, Ferrel, and Polar cells. These play very important roles in the distribution of different climate zones, as moist or rainy regions are located, in the Tropics and at temperate latitudes, where the atmospheric cells are upwelling and release water. Deserts and dry areas are located around zones where the convecting cells downwell, bringing descending dry air into these regions.

The rotation of the Earth sets up systems of prevailing winds that modify the global convective atmospheric (and oceanic) circulation patterns. The spinning of the Earth sets up latitude-dependent airflow patterns, including the trade winds and westerlies. In addition, uneven heating of the Earth over land and ocean regions causes regional airflow patterns such as rising air over hot continents that must be replenished by air flowing in from the sides. The Coriolis force is a result of the rotation of the Earth, and it causes any moving air mass in the Northern Hemisphere to be deflected to the right, and masses in the Southern Hemisphere to be deflected to the left. These types of patterns tend to persist for long periods of time and move large masses of air around the planet, redistributing heat and moisture and regulating the climate of any region.

Temperature is a major factor in the climate of any area, and this is largely determined by latitude. Polar regions see huge changes in temperature between winter and summer months, largely a function of the wide variations in amount of incoming solar radiation and length of days. The proximity to large bodies of water such as oceans influences temperature, as water heats up and cools down much slower than land surfaces. Proximity to water therefore moderates temperature fluctuations. Altitude also influences temperature, with temperature decreasing with height.

Climate may change in cyclical or long-term trends, as influenced by changes in solar radiation, orbital variations of the Earth, amount of greenhouse gases in the atmosphere, or through other phenomena such as the El Niño or La Niña.

See also ATMOSPHERE; CLIMATE CHANGE; EL NIÑO; PLATE TECTONICS.

climate change   Earth's climate changes on many different timescales, ranging from tens of millions of years to decadal and even shorter timescale variations. In the last 2.5 billion years, several periods of glaciation have been identified, separated by periods of mild climate similar to that of today.

Other periods are marked by global hothouse type conditions, when the Earth had a very hot and wet climate, approaching that of Venus. These dramatic climate changes are caused by a number of different factors that exert their influence on different timescales. One of the variables is the amount of incoming solar radiation, and this changes in response to several astronomical effects such as orbital tilt, eccentricity, and wobble. Changes in the incoming solar radiation in response to changes in orbital variations produce cyclical variations known as Milankovitch cycles. Another variable is the amount of heat that is retained by the atmosphere and ocean, or the balance between the incoming and outgoing heat. A third variable is the distribution of landmasses on the planet. Shifting continents can influence the patterns of ocean circulation and heat distribution, and placing a large continent on one of the poles can cause ice to build up on that continent, increasing the amount of heat reflected back to space and lowering global temperatures in a positive feedback mechanism.

Shorter term climate variations include those that operate on periods of thousands of years, and shorter, less regular decadal scale variations. Both of these relatively short-period variations are of most concern to humans, and considerable effort is being expended to understand their causes and to estimate the consequences of the current climate changes the planet is experiencing. Great research efforts are being expended to understand the climate history of the last million years and to help predict the future.

Variations in formation and circulation of ocean currents may be traced some thousands of years to decadal scale variations in climate. Cold water forms in the Arctic and Weddell Seas. This cold salty water is denser than other water in the ocean, so it sinks to the bottom and gets ponded behind seafloor topographic ridges, periodically spilling over into other parts of the oceans. The formation and redistribution of North Atlantic cold bottom water accounts for about 30 percent of the solar energy budget input to the Arctic Ocean every year. Eventually, this cold bottom water works its way to the Indian and Pacific Oceans where it upwells, gets heated, and returns to the North Atlantic. This cycle of water circulation on the globe is known as thermohaline circulation. Recent research on the thermohaline circulation system has shown a correlation between changes in this system and climate change. Presently, the age of bottom water in the equatorial Pacific is 1,600 years, and in the Atlantic it is 350 years. Glacial stages in the North Atlantic have been correlated with the presence of older cold bottom waters, approximately twice the age of the water today. This suggests that the thermohaline circulation system was only half as effective at recycling water during recent glacial stages, with less cold bottom water being produced during the glacial periods. These changes in production of cold bottom water may in turn be driven by changes in the North American ice sheet, perhaps itself driven by 23,000-year orbital (Milankovitch) cycles. It is thought that a growth in the ice sheet would cause the polar front to shift southward, decreasing the inflow of cold saline surface water into the system required for efficient thermohaline circulation. Several periods of glaciation in the past 14,500 years (known as the Dryas) are thought to have been caused by sudden, even catastrophic injections of glacial meltwater into the North Atlantic, which would decrease the salinity and hence density of the surface water. This in turn would prohibit the surface water from sinking to the deep ocean, inducing another glacial interval.

Shorter term decadal variations in climate in the past million years are indicated by so-called Heinrich Events, defined as specific intervals in the sedimentary record showing ice-rafted debris in the North Atlantic. These periods of exceptionally large iceberg discharges reflect decadal scale sea surface and atmospheric cooling. They are related to thickening of the North American ice sheet, followed by ice stream surges, associated with the discharge of the icebergs. These events flood the surface waters with low-salinity freshwater, leading to a decrease in flux to the cold bottom waters, and hence a short period global cooling.

Changes in the thermohaline circulation rigor have also been related to other global climate changes. Droughts in the Sahel and elsewhere are correlated with periods of ineffective or reduced thermohaline circulation, because this reduces the amount of water drawn into the North Atlantic, in turn cooling surface waters and reducing the amount of evaporation. Reduced thermohaline circulation also reduces the amount of water that upwells in the equatorial regions, in turn decreasing the amount of moisture transferred to the atmosphere, reducing precipitation at high latitudes.

Atmospheric levels of greenhouse gases such as CO_2 and atmospheric temperatures show a correlation to variations in the thermohaline circulation patterns and production of cold bottom waters. CO_2 is dissolved in warm surface water and transported to cold surface water, which acts as a sink for the

Coping with Sea-Level Rise in Coastal Cities

People have built villages, towns, cities, and industrial sites near the sea for thousands of years. The coastal setting offers beauty and convenience but also may bring disaster with coastal storms, tsunami, and invading armies. Coastal communities are currently experiencing the early stages of a new incursion, that of the sea itself, as global sea levels slowly and inexorably rise.

Sea-level rises and falls by hundreds of feet over periods of millions of years have forced the position of the coastline to move inland and seaward by many tens of miles over long time periods. The causes of sea-level rise and fall are complex, including growth and melting of glaciers with global warming, changes in the volume of the mid-ocean ridges, thermal expansion of water, and other complex interactions of the distribution of the continental landmass in mountains and plains during periods of orogenic and anorogenic activity. Most people do not think that changes over these time frames will affect their lives, but a sea-level rise of even a foot or two, which is possible over periods of tens of years, can cause extensive flooding, increased severity of storms, and landward retreat of the shoreline. Sea-level rise is rapidly becoming one of the major global hazards that the human race is going to have to deal with in the next century, since most of the world's population lives near the coast in the reach of the rising waters. Cities may become submerged and farmlands covered by shallow, salty seas. An enormous amount of planning is needed, as soon as possible, to begin to deal with this growing threat. The current rate of rise of an inch or so every 10 years seems insignificant, but it will have truly enormous consequences. When sea-level rises, beaches try to maintain their equilibrium profile, moving each beach element landward. A sea-level rise of one inch is generally equated with a landward shift of beach elements of more than four feet. Most sandy beaches worldwide are retreating landward at rates of 20 inches–3 feet per year, consistent with sea-level rise of an inch every 10 years. If the glacial ice caps on Antarctica begin to melt faster, the sea-level rise will be much more dramatic.

What effect will rising sea levels have on the world's cities and low-lying areas? Many of the world's large cities, including New York, London, Houston, Los Angeles, Washington D.C., Cairo, Shanghai, Brussels, and Calcutta have large areas located within a few feet of sea level. If sea levels rise a few feet, many of the streets in these cities will be underwater, not to mention basements, subway lines, and other underground facilities. Imagine Venice-like conditions in New York! If sea levels rise much more, many of the farmlands of the midwest United States, North Africa, Mesopotamia, northern Europe, Siberia, and eastern China will be submerged in shallow seas. These areas are not only populated but serve as some of the most fertile farmlands in the world. Thus, large sea-level rise will at best displace or more likely simply eliminate the world's best agricultural lands, necessary for sustaining global population levels.

What can be done to prepare for sea-level rise? Some lessons can be learned from the Netherlands, where the Dutch have built numerous dikes to keep the sea out of low-lying areas, at costs of billions of dollars. If the United States had to build such barriers around the coastlines of low-lying areas, the cost would be unbearable and would amount to one of the largest construction projects ever undertaken. Humans are contributing to global warming, which in turn is probably contributing to enhanced melting of the glaciers and ice caps. Although it is too late to stop much of the warming and melting, it may not be too late to stop the warming before it is catastrophic and the ice caps melt, raising sea levels by hundreds of feet. In any case, it is time that governments, planners, and scientists begin to make more sophisticated plans for action during times of rising sea levels.

Gaia Hypothesis

For billions of years the Earth has maintained its temperature and atmospheric composition in a narrow range that has permitted life to exist on the surface. Many scientists have suggested that this remarkable trait of the planet is a result of life adapting to conditions that happen to exist and evolve on the planet. An alternative idea has emerged that the planet behaves as some kind of self-regulating organism that invokes a series of positive and negative feedback mechanisms to maintain conditions within the narrow window in which life can exist. In this scenario, organisms and their environment evolve together as a single coupled system, regulating the atmospheric chemistry and composition to the need of the system. Dr. James Lovelock, an atmospheric chemist at Green College in Oxford, U.K., pioneered this second idea, known as the Gaia hypothesis. However, the idea of a living planet dates back at least to Sir Isaac Newton.

How does the Gaia hypothesis work? The atmosphere is chemically unstable, yet it has maintained conditions conducive to life for billions of years even despite a 30 percent increase in solar luminosity since the Early Precambrian. The basic tenet of the hypothesis is that organisms, particularly microorganisms, are able to regulate the atmospheric chemistry and hence temperature to keep conditions suitable for their development. Although this tenet has been widely criticized, some of the regulating mechanisms have been found to exist, lending credence to the possibility that Gaia may work. Biogeochemical cycles of nutrients including iodine and sulfur have been identified, with increases in the nutrient supply from land to ocean leading to increased biological production and increased emissions to the atmosphere. Increased production decreases the flux of nutrients from the oceans to the land, in turn decreasing the nutrient supply, biological production, and emissions to the atmosphere.

As climate warms, rainfall increases, and the weathering of calcium-silicate rocks increases. The free calcium ions released during weathering combine with atmospheric carbon dioxide to produce carbonate sediments, effectively removing the greenhouse gas carbon dioxide from the atmosphere. This reduces global temperatures in another self-regulating process. An additional feedback mechanism was discovered between ocean algae and climate. Ocean algae produce dimethyl sulfide gas, which oxidizes in the atmosphere to produce nuclei for cloud condensation. The more dimethyl sulfide that algae produce, the more clouds form, lowering temperatures and lowering algal production of dimethyl sulfide in a self-regulating process.

That the Earth and its organisms have maintained conditions conducive for life for 4 billion years is clear. However, at times the Earth has experienced global icehouse and global hothouse conditions, where the conditions extend beyond the normal range. Lovelock relates these brief intervals of Earth history to fevers in an organism, and he notes that the planet has always recovered. Life has evolved dramatically on Earth in the past 4 billion years, but this is compatible with Gaia. Living organisms can both evolve with and adapt to their environment, responding to changing climates by regulating or buffering changes to keep conditions within limits that are tolerable to life on the planet as a whole. However, there are certainly limits, and the planet has never experienced organisms such as humans that continually emit huge quantities of harmful industrial gases into the atmosphere. It is possible that the planet, or Gaia, will respond by making conditions on Earth uninhabitable for humans, saving the other species on the planet. As time goes on, in about a billion years the Sun will expand and eventually burn all the water and atmosphere off the planet, making it virtually uninhabitable. By then humans may have solved the problem of where to move to and developed the means to move global populations to a new planet.

CO_2. During times of decreased flow from cold, high-latitude surface water to the deep ocean reservoir, CO_2 can build up in the cold polar waters, removing it from the atmosphere and decreasing global temperatures. In contrast, when the thermohaline circulation is vigorous, cold oxygen-rich surface waters downwell and dissolve buried CO_2 and even carbonates, releasing this CO_2 to the atmosphere and increasing global temperatures.

The present-day ice sheet in Antarctica grew in the Middle Miocene, related to active thermohaline circulation that caused prolific upwelling of warm water that put more moisture in the atmosphere, falling as snow on the cold southern continent. The growth of the southern ice sheet increased the global atmospheric temperature gradients, which in turn increased the desertification of midlatitude continental regions. The increased temperature gradient also induced stronger oceanic circulation, including upwelling, and removal of CO_2 from the atmosphere, lowering global temperatures, and bringing on late Neogene glaciations.

Major volcanic eruptions inject huge amounts of dust into the troposphere and stratosphere, where it may remain for several years, reducing incoming solar radiation and resulting in short-term global cooling. For instance, the eruption of Tambora volcano in Indonesia in 1815 resulted in global cooling and the year without a summer in Europe. The location of the eruption is important, as equatorial eruptions may result in global cooling, whereas high-latitude eruptions may only cool one hemisphere.

It is clear that human activities are changing the global climate, primarily through the introduction of greenhouse gases such as CO_2 into the atmosphere, while cutting down tropical rain forests that act as sinks for the CO_2 and put oxygen back into the atmosphere. The time scale of observation of these human, also called anthropogenic, changes is short but the effect is clear, with a nearly one degree change in global temperature measured for the past few decades. The increase in temperature will lead to more water vapor in the atmosphere, and since water vapor is also a greenhouse gas, this will lead to a further increase in temperature. Many computer-based climate models are attempting to predict how much global temperatures will rise as a consequence of our anthropogenic influences, and what effects this temperature rise will have on melting of the ice sheets (which could be catastrophic), sea-level rise (perhaps tens of meters or more), and runaway greenhouse temperature rise (which is possible).

Climate changes are difficult to measure, partly because the instrumental and observational records go back only a couple of hundred years in Europe. From these records, global temperatures have risen by about one degree since 1890, most notably in 1890–1940, and again since 1970. This variation, however, is small compared with some of the other variations induced by natural causes, and some scientists argue that it is difficult to separate anthropogenic effects from the background natural variations. Rainfall patterns have also changed in the past 50 years, with declining rainfall totals over low latitudes in the Northern Hemisphere, especially in the Sahel, which has experienced major droughts and famine. However, high-latitude precipitation has increased in the same time period. These patterns all relate to a general warming and shifting of the global climate zones to the north.

See also DESERT; GLACIER; GREENHOUSE EFFECT; ICE AGE; MILANKOVITCH CYCLE; SEA-LEVEL RISE.

clinometer An instrument used in surveying for measuring angles of elevation, slope, or incline, also know as an inclinometer. Clinometers are measuring devices used to measure the angle of a line of sight above or below the horizontal. The height of an object can be determined both by using the clinometer and by measuring the distance to the object. By calculating a vertical angle and a distance, users can perform simple trigonometry to figure out the height of an object.

Clinometers are fairly simple to use. A clear line of sight is needed to measure the angle of an object, and it is common practice to make more than one reading and take the average for a more precise result.

Cloos, Hans (1885–1951) German *Geologist* Hans Cloos was born in Madeburg, Germany. After secondary school, he studied architecture but quickly became interested in geology. He continued to study geology in Bonn and Jena, then he moved to Freiburg in Breisgau where he received his doctorate. From 1909 to 1931 he worked on applied geology in South Africa, where he looked at the granite massifs in the Erongo Mountains, and in Java, where he worked as a petroleum geologist and studied the active volcanoes and their structures. In 1919 he became a professor of geology and petrology at the University of Breslau and developed the field of granite tectonics. This discipline is the reconstruction of dynamics of movement and of emplacement of a mobilized pluton from its internal structure. This idea came from his studies of the granite massifs in Silesia. Cloos discovered that

granite had clearly oriented features obtained during or directly after intrusion, including linear or laminar flow, textures of solidifying magma, regular relationship of joints and vein systems to flow textures, and cleavage in granite. He also developed the field of granite and magmatic tectonics as well as continuing to look at other tectonic problems, such as jointing and cleavage as typical types of deformation of solid rocks. He extensively used wet clay to demonstrate rift valley formations, and his many trips to Africa and North America helped to develop his point of view.

Cloud, Preston (1912–1991) American *Historical Geologist, Geobiologist* Preston Cloud was an eminent geobiologist and paleontologist, who contributed important observations and interpretations that led to greater understanding of the evolution of the atmosphere, oceans, and crust of the Earth and, most important, to understanding the evolution of life on the planet. Preston Cloud was educated at Yale University and received his Ph.D. in 1940 for a study of Paleozoic brachiopods. From there, he moved to Missouri School of Mines in Rolla, but then returned to Yale University. In 1965 Cloud moved to the University of California, Los Angeles, then in 1968 he moved to the Santa Barbara campus. In 1979 Cloud retired, but he remained active in publishing books on life on the planet and was also active on campus. Preston Cloud emphasized complex interrelationships between biological, chemical, and physical processes throughout Earth history. His work expanded beyond the realm of rocks and fossils, and he wrote about the limits of the planet for sustaining the exploding human population, and he recognized that limited material, food, and energy resources with the expanding activities of humans could lead the planet into disaster. Preston Cloud was elected a member of the Academy of Sciences and was an active member for 30 years. The Preston Cloud Laboratory at the University of California, Santa Barbara, is dedicated to the study of pre-Phanerozoic life on Earth.

Further Reading

Cloud, Preston. "Life, Time, History and Earth Resources." *Terra Cognita* 8 (1988): 211.
———. "Aspects of Proterozoic Biogeology." *Geological Society of America Memoir* 161 (1983): 245–251.
———. "A Working Model of the Primitive Earth." *American Journal of Science* 272 (1972): 537–548.

clouds Visible masses of water droplets or ice crystals suspended in the lower atmosphere, generally confined to the troposphere. The water droplets and ice crystals condense from water vapor around small dust, pollen, salt, ice, or pollution particles that aggregate into cloud formations, classified according to their shape and height in the atmosphere. Luke Howard, an English naturalist, suggested the classification system still widely used today in 1803. He suggested Latin names based on 10 genera, then broken into species. In 1887

Abercromby and Hildebrandsson further divided the clouds into high, middle, and low-level types, as well as clouds that form over significant vertical distances. The basic types of clouds include the heaped cumulus, layered stratus, and wispy cirrus. If rain is falling from a cloud the term nimbus is added, as in cumulonimbus, the common thunderhead cloud.

High clouds form above 19,685 feet (6,000 m) and are generally found at mid to low latitudes. The air at this elevation is cold and dry, so the clouds consist of ice crystals and appear white to the observer at the ground except at sunrise and sunset. The most common high clouds are the cirrus, thin, wispy clouds typically blown into thin horsehair-like streamers by high winds. Most cirrus clouds are blown from west to east by prevailing high-level winds and are a sign of good weather. Cirrocumulus clouds are small, white, puffy clouds that sometimes line up in ripple-like rows and at other times form individually. Their appearance over large parts of the sky is often described as a mackeral sky, because of the resemblance to fish scales. Cirrostratus are thin, sheet-like clouds that typically cover the entire sky. They are so thin that the Sun, Moon, and some stars can be seen through them. They are composed of ice crystals, and light that refracts through these clouds often forms a halo or sun dogs. These high clouds often form in front of an advancing storm and typically foretell of rain or snow in 12–24 hours.

Middle clouds form between 6,560 feet and 22,965 feet (2,000 m and 7,000 m), generally in middle latitudes. They are composed mostly of water droplets, with some ice crystals in some cases. Altocumulus clouds are gray, puffy masses that often roll out in waves, with some parts appearing darker than others. Altocumulus are usually less than 0.62 mile (1 km) thick. They form with rising air currents at cloud level, and morning appearance often predicts thunderstorms by the late afternoon. Altostratus are thin blue-gray clouds that often cover the entire sky, and the sun may shine dimly through, appearing as a faint irregular disk. Altostratus clouds often form in front of storms that bring regional steady rain.

Low clouds have bases that may form below 6,650 feet (2,000 m) and are usually composed entirely of water droplets. In cold weather they may contain ice and snow. Nimbostratus are the dark gray rather uniform-looking clouds associated with steady light to moderate rainfall. Rain from the nimbostratus clouds often causes the air to become saturated with water, and a group of thin ragged clouds that move rapidly with the wind may form. These are known as stratus fractus, or scud clouds. Stratocumulus clouds are low, lumpy-looking clouds that form rows or other patterns, with clear sky visible between the cloud rows. The sun may form brilliant streaming rays known as crepuscular rays through these clouds. Stratus clouds have a uniform gray appearance and may cover the sky, resembling fog but not touching the ground. They are common around the seashore especially in summer months.

Some clouds form over a significant range of atmospheric levels. Cumulus are flat-bottomed puffy clouds with irregular, domal, or towering tops. Their bases may be lower than 3,280 feet (1,000 m). On warm summer days, small cumulus clouds may form in the morning and develop significant vertical growth by the afternoon, forming a towering cumulus or cumulus congestus cloud. These may continue to develop further into the giant cumulonimbus, giant thunderheads with bases that may be as low as a few hundred meters, and tops extending to more than 39,370 feet (12,000 m) in the tropopause. Cumulonimbus clouds release tremendous amounts of energy in the atmosphere and may be associated with high winds, vertical updrafts and downdrafts, lightning, and tornadoes. The lower parts of these giant clouds are made of water droplets, the middle parts may be both water and ice, whereas the tops may be made entirely of ice crystals.

There are many types of unusual clouds that form in different situations. Plieus clouds may form over rising cumulus tops, looking like a halo or fog around the cloud peak. Banner clouds form over and downwind of high mountaintops, sometimes resembling steam coming out of a volcano. Lenticular clouds form wave-like figures from high winds moving over mountains and may form elongate pancake-like shapes. Unusual, and even scary-looking, mammatus clouds form bulging bag-like sacks underneath some cumulonimbus clouds, forming when the sinking air is cooler than the surrounding air. Mammatus-like clouds may also form underneath clouds of volcanic ash. Finally, jet airplanes produce condensation trails, produced when water vapor from the jet's exhaust mixes with the cold air, which becomes suddenly saturated with water and forms ice crystals. Pollution particles from the exhaust may provide the nuclei for the ice. In dry conditions condensation, or contrails, will evaporate quickly. However, in more humid conditions the contrails may persist as cirrus-like clouds. With the growing numbers of jet flights in the past few decades, contrails have rapidly become a significant source of cloudiness, contributing to the global weather and perhaps climate.

Clouds have a large influence on the Earth's climate. They are highly reflective and reflect short wavelength radiation from the Sun back to space, cooling the planet. However, since they are composed of water they also stop the longer wavelength radiation escaping in a greenhouse effect. Together these two apparently opposing effects of clouds strongly influence the climate of the Earth. In general, the low and middle level clouds cool the Earth whereas abundant high clouds tend to warm the Earth with the greenhouse effect.

See also ATMOSPHERE; CLIMATE.

Further Reading

Schaefer, Vincent J., and John A. Day. *A Field Guide to the Atmosphere, A Peterson Field Guide.* Boston: Houghton Mifflin, 1981.

coal A combustible rock that contains more than 50 percent (by weight) carbonaceous material formed by the compaction and induration of plant remains, coal is the most abundant fossil fuel. It is a black sedimentary rock that consists chiefly of decomposed plant matter, with less than 40 percent inorganic material. Coal is classified according to its rank and impurities. With increasing temperature and pressure, coal increases in rank and carbon content. Most coal was formed in ancient swamps, where stagnant oxygen-deficient water prevented rapid decay and allowed burial and trapping of organic matter. In addition, anaerobic bacteria in these environments attack the organic matter, releasing more oxygen and forming peat. Peat is a porous mass of organic matter that still preserves recognizable twigs and other plant parts. Peat contains about 50 percent carbon and burns readily when dried. With increasing temperature and pressure, peat is transformed into lignite, bituminous coal, and eventually anthracite. Anthracite contains more than 90 percent carbon, and it is much shinier, brighter, and harder than bituminous coal and lignite.

See also HYDROCARBON.

coastal downwelling A phenomenon where winds moving parallel to the coast cause the ocean surface waters to move toward the shoreline, necessitating a corresponding deeper flow from the coast below the shoreward-moving coastal water. Ocean surface currents generally move at right angles to the dominant wind and move at a velocity of about 2 percent of the winds. Therefore, onshore winds cause currents to move parallel to the coasts, whereas alongshore winds set up currents that move toward or away from the shore.

See also COASTAL UPWELLING; EKMAN SPIRALS; OCEAN CURRENTS; OCEANOGRAPHY.

coastal upwelling Caused by surface winds that blow parallel to the coast, forming ocean surface movements at 90° to the direction of surface winds in Ekman spirals. In many cases, the upper few tens of meters of surface waters move away from the shoreline forcing a corresponding upwelling of water from depth to replace the water that has moved offshore. This is known as coastal upwelling. Upwelling is most common on the eastern sides of ocean basins where the surface layer is thin and near capes and other irregularities in the coastline. Upwelling also occurs away from the coasts along the equator, where surface waters diverge because of the change in sign of the Coriolis force across the equator. Water from depth upwells to replace the displaced surface water.

Zones of coastal and other upwelling, where the water comes from more than 325 feet (100 m) depth, are typically very productive organically, with abundant marine organisms, including plants and fish. This is because upwelling coastal waters are rich in nutrients that suddenly become available to benthic and planktonic photic zone organisms.

See also COASTAL DOWNWELLING; EKMAN SPIRALS; OCEAN CURRENTS; OCEANOGRAPHY.

comet Wanderers of the solar system, occasionally appearing as bright objects that move across the night sky, growing long brilliant tails that have mystified people for thousands of years. They are thought to be made of stony inner cores covered by icy outer layers, often described as dirty snowballs because small rock fragments may be mixed with the icy outer layers. Most are only a few kilometers to tens of kilometers across. Comets represent primitive material formed during the early accretion of the outer solar system, and as such they are analogous to asteroids, which represent primitive material that formed in the early inner solar system. Most comets are located in one of two prominent belts in the outer solar system. These are known as the Kuiper belt, outside the orbit of Neptune, and the Oort cloud, forming the outer reaches of the solar system.

The Kuiper belt contains many comets outside the orbit of Neptune, and the objects Pluto and its large moon Charon are considered by many to represent large comets. Most comets reside in the Oort cloud, a spherical cloud located a light-year (approximately 6 trillion miles) from the Sun. There are thought to be a trillion comets in the Oort cloud, with a total mass of about 40 Earths. Occasionally, gravitational interactions between comets in the Oort cloud will deflect one from this far region and cause it to have an orbit that intersects the inner solar system. When these comets pass though the inner solar system, heat from the Sun vaporizes some of the ice that forms outer layers of the comet, and the solar wind causes this vapor tail (consisting of both gas and dust) to become elongated in a direction pointing directly away from the Sun. These tails may be hundreds of millions of kilometers long, and some are clearly visible from Earth. The gravity of objects in the solar system (such as Jupiter and the Sun) can further distort the orbits of these errant comets, causing them to obtain regular elliptical orbits circling the

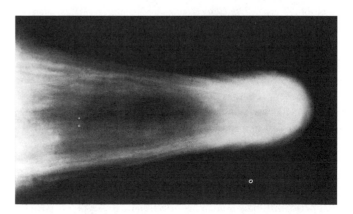

The head of Comet Halley in 1986, on its closest approach to Earth *(Courtesy of NASA)*

Sun with a predictable period. Such comets include the famous Halley's comet, Hale-Bopp, and others.

Comets have occasionally hit the Earth and other planets, forming impact craters, and very likely brought many of the volatile elements, water, and the atmosphere to our planet. Some scientists believe that comets may also have brought primitive organic molecules or even life to Earth. A cometary impact with Earth is thought to have caused a huge explosion and fireball over Tunguska, Siberia, in 1908. The explosion from this event leveled 1,158 square miles (3,000 km²) of forest and sent shock waves around the planet.

There are approximately 10 million comets that have orbits that cross the paths that the planets take around the Sun. Some of these cross the Earth's orbit, leaving a trail of dust particles behind. When the Earth crosses these cometary dust trails, the dust enters the upper atmosphere and burns up, forming spectacular meteorite showers. Halley's comet has left a large trail of dust particles in its wake during its numerous orbits around the Sun. The Earth crosses this trail two times a year, producing the Eta Aquarid meteor shower in early May, and the Oronoid meteor shower in mid–late May. Other meteor showers including the Perseids, Leonids, and Geminids, probably also originating from dust trails from comets, but these comets have probably lost their icy outer layers and have yet to be identified.

It is estimated that 25,000 small comets enter the Earth's atmosphere every day, or about several million per year. These objects are mostly less than 30 feet (10 m) wide but carry 100 tons of water. Cumulatively over the age of the Earth, these minicomets would have added a volume of water to the atmosphere equivalent to the present-day oceans. These comets also carry organic molecules and provide one plausible mechanism to bring complex organic molecules to Earth that could have helped the development of life.

See also ASTEROID; METEOR; LIFE'S ORIGINS AND EARLY EVOLUTION.

Further Reading
Erickson, J. *Asteroids, Comets and Meteorites, Cosmic Invaders of the Earth.* New York: Facts On File, 2003.

compass   An instrument that indicates the whole circle bearing from the magnetic meridian to a particular line of sight. The magnetic compass consists of a needle that aligns itself with the Earth's magnetic flux, and with some type of index that allows for a numeric value for the calculation of bearing.

A compass can be used for many things. The most common application is for navigation. People are able to navigate throughout the world simply by using a compass and map. The accuracy of a compass is dependent on other local magnetic influences such as man-made objects or natural abnormalities such as local geology. The compass itself does not really point to true north, but the compass is attracted by magnetic force that varies in different parts of the world, and it is constantly changing. For example, when you read north on a compass, you are reading the direction toward the magnetic north pole. To offset this phenomena, calculated declination values are used to convert the compass reading to a usable map reading. Since the magnetic flux changes through time, it is necessary to replace older maps with newer maps to insure accurate and precise up-to-date declination values.

concretion   Generally hard, compact, ellipsoidal mineral masses in a sedimentary or volcanic rock, with a composition different from the host rock. Most have regular smooth shapes, but some have irregular or odd-shaped boundaries. Concretions commonly are made of carbonate minerals in a sandstone matrix or chert in limestone. Other common concretions include pyrite, iron oxides, and gypsum. Most concretions are several centimeters to a few decimeters in diameter, but some are as large as a few meters across. Concretions are thought to form from precipitation from aqueous solutions, and they often precipitate on a nucleus made of an unusual mineral, fossil, or grain.

See also DIAGENESIS.

cone of depression   A type of depression in the water table formed by overpumping from a well. Wells are normally drilled to penetrate well below the water table to allow for seasonal rises and falls of the water table related to changes in the balance between recharge and discharge from the aquifer. If a water well is used to pump water out of the aquifer faster than it can be recharged, then the water table level around the well will be drawn down to a lower level, approaching that of the intake pipe on the well. If the aquifer is characterized by uniform material in terms of porosity and permeability, then this area of drawdown will form a cone in which the water table is depressed around the well. This in turn creates a steeper groundwater slope and steeper hydraulic gradient toward the well, causing the groundwater to flow faster toward the well intake pipe.

The size of a cone of depression around a well depends on the permeability of the ground, and the amount and rate that groundwater is withdrawn. Faster and larger amounts of withdrawal create larger cones of depression, and the cones tend to have steeper sides in areas with lower permeability. The formation of cones of depression is a result of the overdraft of the aquifer, defined as when more water is taken out of the aquifer than is replaced by recharge. Regional drawdown and overdraft can cause the water table to be lowered significantly on a regional scale, causing wells to go dry. In coastal areas drawdown of the aquifer can lead to seawater intrusion in which the seawater gradually moves into the aquifer, replacing the freshwater with unusable salty water.

See also AQUIFER; GROUNDWATER.

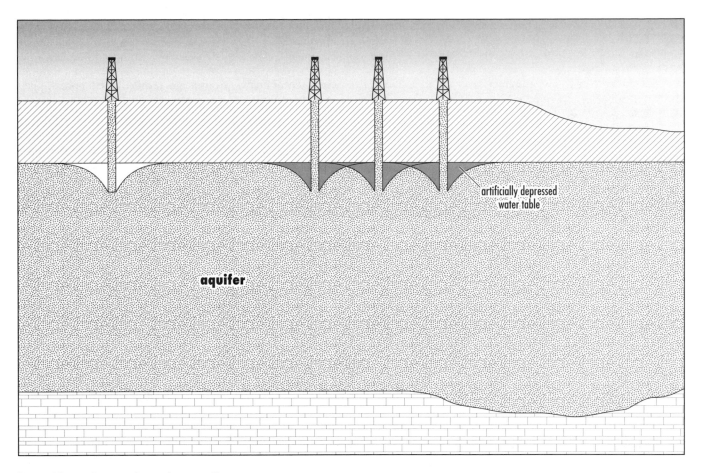

Cones of depression around groundwater wells

conglomerate A clastic sedimentary rock characterized by rounded or subangular fragments that are larger than 0.079 inch (2 mm) in diameter, typically set in a finer grained matrix of sand or silt-sized particles. It is the lithified or hardened equivalent of gravel. The matrix and grains are held together by a cementing agent such as calcium carbonate, silica, iron oxide, or clay. The clasts may be made of a single type of parent rock, in a monomictic conglomerate, or of several types of parent rock, in a polymictic conglomerate. The amount of rounding of clasts can in many cases be related to the amount that the clasts have been transported from their source, with greater transport distances equated with more rounding. Conglomerates may be classified by the types of clasts present, the type of cement, and, less commonly, by the inferred environment of deposition.

Many conglomerates are deposited in channels in fluvial (stream) systems, whereas others are deposited at the bases of submarine slopes, and still others are formed by storms ripping up shallow shelf muds and depositing the fragments in deeper water. Thick conglomeratic sequences are an indication of the erosion of high topography, and the sequence of erosion and exposure of different crustal levels can sometimes be learned by examining the types of clasts throughout the deposit. During mountain-building events, a sequence of conglomerates, sandstones, and shales known as flysch is typically deposited during active deformation, followed by a sequence of conglomerates and stream deposits known as molasse as the mountains get eroded.

See also BRECCIA; CLASTIC ROCKS; FLYSCH; MOLASSE; SEDIMENTARY ROCKS.

conodont An extinct group of small, tooth-like phosphatic marine fossil fragments that range in age from Cambrian through Upper Triassic. Their range was widespread and they diversified quickly, making them ideal index fossils that can help paleontologists identify the age of specific strata. One hundred and forty different conodont zones are recognized from the Paleozoic through Triassic, making them one of the widest applicable biostratigraphic markers in the fossil record. They grew by successively forming layers of the mineral apatite around a nucleus, and most fossil fragments were apparently interior parts of bilaterally symmetric animals that were embedded in some secretionary tissue. The conodont animal was apparently a several-centimeter-long worm-like animal with tail fins, and most of the preserved fossil elements are from the hard feeding apparatus located in the head of this animal.

Conodont fossils are common from benthic (bottom dwelling) and pelagic (free swimming) deposits, but rare from shallow water paleoenvironments, and absent from the deep ocean and basin deposits. Their fossils are most commonly from nutrient-rich continental shelf environments in warm, equatorial paleolatitudes.

With burial and metamorphism, conodonts have been shown to change colors in a remarkably consistent way that can quickly tell the geologist the maximum temperature the rock that the fossil is enclosed within has reached. This amazing property has proved extremely useful to the petroleum industry, as hydrocarbons are only formed and preserved within a narrow temperature window. Simply determining the color of a conodont is often enough to tell the explo-

ration geologist whether or not it is worthwhile to search for petroleum in a specific area any further.

See also PALEONTOLOGY.

continental crust   Continental crust covers about 34.7 percent of the Earth's surface, whereas exposed continents cover only 29.22 percent of the surface, with the discrepancy accounted for in the portions of continents that lie underwater on the continental shelves. Its lateral boundaries are defined by the slope break between continental shelves and slopes, and its vertical extent is defined by a jump in seismic velocities to 4.7–5.0 miles per second (7.6–8.0 km/s) at the Mohorovicic discontinuity. The continental crust ranges in thickness from about 12.5 miles to about 37 miles (20–60 km), with an average thickness of 24 miles (39 km). The continents are divided into orogens, made of linear belts of concentrated deformation, and cratons, making the stable, typically older interiors of the continents.

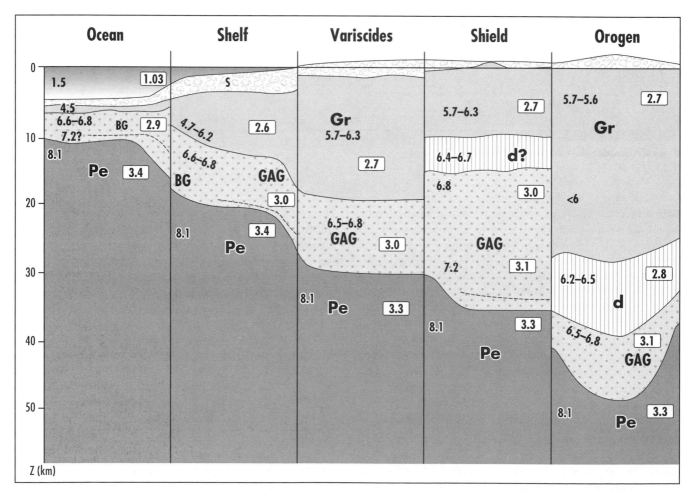

General crustal structure of different provinces as determined by seismology. Numbers in boxes represent densities in grams/cm³, and other numbers represent the seismic velocity of P waves in kilometers per second. BG: basalt-gabbro; GAG: amphibolite and granulite; Pe: peridotite; d: diorite; S: sediments; Gr: granitic-gneissic upper crust

Most of the continental crust is now preserved in Archean cratons, which form the cores of many continents. They are composed of ancient rocks that have been stable for billions of years since the Archean. Cratons generally have low heat flow, few if any earthquakes, no volcanism, and many are overlain by flat lying shallow water sedimentary sequences. Continental shields are places where the cratonic crust is exposed at the surface, whereas continental platforms are places where the cratonic rocks are overlain by shallow water sedimentary rocks, presently exposed at the surface.

Orogens and orogenic belts are elongate regions that are eroded mountain ranges and typically have abundant folds and faults. Young orogens are mountainous and include such familiar mountain ranges as the Rockies, Alps, and the slightly older Appalachians. Many Archean cratons are welded together by Proterozoic and younger orogens. In fact, many Archean cratons can be divided into smaller belts that represent fragments of the planet's oldest orogenic belts.

Orogens have been added to the edges of the continental shield and cratons through processes of mountain building related to plate tectonics. Mountain belts are of three basic types, including fold and thrust belts, volcanic mountain ranges, and fault-block ranges. Fold and thrust belts are contractional mountain belts, formed where two tectonic plates collided, forming great thrust faults, folds, metamorphic rocks, and volcanic rocks. Detailed mapping of the structure in the belts can enable geologists to reconstruct their history and essentially pull them apart. It is found that many of the rocks in fold and thrust belt types of mountain ranges were deposited on the bottom of the ocean, continental rises, slopes, shelves, or on ocean margin deltas. When the two plates collide, many of the sediments get scraped off and deformed, forming the mountain belts. Thus, fold and thrust mountain belts mark places where oceans have closed.

Volcanic mountain ranges include places such as Japan's Fuji, and Mount St. Helens in the Cascades of the western United States. These mountain ranges are not formed primarily by deformation but by volcanism associated with subduction and plate tectonics.

Fault-block mountains, such as the Basin and Range Province of the western United States, are formed by the extension or pulling apart of the continental crust, forming elongate valleys separated by tilted fault-bounded mountain ranges.

Every rock type known on Earth is found on the continents, so averaging techniques must be used to determine the overall composition of the crust. It is estimated that continental crust has a composition equivalent to andesite (or granodiorite) and is enriched in incompatible trace elements, which are the elements that do not easily fit into lattices of most minerals and tend to get concentrated in magmas.

The continents exhibit a broadly layered seismic structure that is different from place to place, and different in orogens, cratons, and parts of the crust with different ages. In shields, the upper layer typically may be made of a few hundred meters of sedimentary rocks underlain by generally granitic types of material with seismic velocities of 3.5 to 3.9 miles per second (5.7–6.3 km/s) to depths of a few to 10 kilometers, then a layer with seismic velocities of 3.9–4.2 miles per second (6.4–6.7 km/s). The lower crust is thought to be made of layered amphibolite and granulite with velocities of 4.2–4.5 miles per second (6.8–7.2 km/s). Orogens tend to have thicker low-velocity upper layers, and a lower-velocity lower crust.

There is considerable debate and uncertainty about the timing and processes responsible for the growth of the continental crust from the mantle. Most scientists agree that most of the growth occurred early in Earth history since more than half of the continental crust is Archean in age, and about 80 percent is Precambrian. Some debate centers on whether early tectonic processes were similar to those currently operating, or if they were considerably different. The amount of current growth and how much crust is being recycled back into the mantle are currently constrained. Most petrological models for the origin of the crust require that it be derived by a process including partial melting from the mantle, but simple mantle melting produces melts that are not as chemically evolved as the crust. Therefore, the crust is probably derived through a multistage process, most likely including early melts derived from seafloor spreading and island arc magmatism, with later melts derived during collision of the arcs with other arcs and continents. Other models seek to explain the difference by calling on early higher temperatures leading to more evolved melts.

See also CRATONS; GREENSTONE BELTS; OROGENY.

Further Reading

Meissner, Rolf. *The Continental Crust: A Geophysical Approach.* International Geophysics Series, vol. 34. Orlando: Academic Press, 1986.

Rudnick, Roberta L. "Making Continental Crust." *Nature* 378 (1995): 571–578.

Taylor, Stuart Ross, and Scott M. McLennan. *The Continental Crust: Its Composition and Evolution, An Examination of the Geochemical Record Preserved in Sedimentary Rocks.* Oxford: Blackwell Scientific Publications, 1985.

continental drift A theory that may be thought of as the precursor to plate tectonics. It was proposed most clearly by Alfred Wegener in 1912 and states that the continents are relatively light objects that are floating and moving freely across a substratum of oceanic crust. The theory was largely discredited because it lacked a driving mechanism, and it seemed implausible if not physically impossible to most geologists and geophysicists at the time. However, many of the ideas of continental drift were later incorporated into the paradigm of plate tectonics.

Early geologists recognized many of the major tectonic features of the continents and oceans. Cratons are very old and stable portions of the continents that have been inactive since the Precambrian and typically have subdued topography including gentle arches and basins. Orogenic belts are long, narrow belts of structurally disrupted and metamorphosed rocks, typified (when active) by volcanoes, earthquakes, and folding of strata. Abyssal plains are stable, flat parts of the deep oceanic floor, whereas oceanic ridges are mountain ranges beneath the sea with active volcanoes, earthquakes, and high heat flow. In order to explain the large-scale tectonic features of the Earth, early geologists proposed many hypotheses, including popular ideas that the Earth was either expanding or shrinking, forming ocean basins and mountain ranges. In 1910–25, Alfred Wegener published a series of works including his 1912 treatise on *The Origin of Continents and Oceans*. He proposed that the continents were drifting about the surface of the planet, and that they once fit together to form one great supercontinent, known as Pangea. To fit the coastlines of the different continental masses together to form his reconstruction of Pangea, Wegener defined the continent/ocean transition as the outer edge of the continental shelves. The continental reconstruction proposed by Wegener showed remarkably good fits between coastlines on opposing sides of ocean basins, such as the Brazilian Highlands of South America fitting into the Niger delta region of Africa. Wegener was a meteorologist,

and since he was not formally trained as a geologist, few scientists at the time believed him, although we now know that he was largely correct.

Most continental areas lie approximately 985 feet (300 m) above sea level, and if we extrapolate present erosion rates back in time, we find that continents would be eroded to sea level in 10–15 million years. This observation led to the application of the principle of isostasy to explain the elevation of the continents. Isostasy, which is essentially Archimedes' Principle, states that continents and high topography are buoyed up by thick continental roots floating in a denser mantle, much like icebergs floating in water. The principle of isostasy states that the elevation of any large segment of crust is directly proportional to the thickness of the crust. Significantly, geologists working in Scandinavia noticed that areas that had recently been glaciated were rising quickly relative to sea level, and they equated this observation with the principle of isostatic rebound. Isostatic rebound is accommodated by the flow of mantle material within the zone of low viscosity beneath the continental crust, to compensate the rising topography. These observations revealed that mantle material can flow at rates of several centimeters per year.

In *The Origin of Continents and Oceans,* Wegener was able to take all the continents and fit them back together to form a Permian supercontinent, known as Pangea (or all land). Wegener also used indicators of paleoclimate, such as locations of ancient deserts and glacial ice sheets, and distributions of certain plant and animal species, to support his ideas. Wegener's ideas were supported by a famous South African geologist, Alexander L. Du Toit, who, in 1921, matched the stratigraphy and structure across the Pangea landmass. Du Toit found the same plants, such as the Glossopterous fauna, across Africa and South America. He also documented similar reptiles and even earthworms across narrow belts of Wegener's Pangea, supporting the concept of continental drift.

Even with evidence such as the matching of geological belts across Pangea, most geologists and geophysicists doubted the idea, since it lacked a driving mechanism and it seemed mechanically impossible for relatively soft continental crust to plow through the much stronger oceans. Early attempts at finding a mechanism were implausible and included ideas such as tides pushing the continents. Because of the lack of credible driving mechanisms, continental drift encountered stiff resistance from the geologic community, as few could understand how continents could plow through the mantle.

In 1928 Arthur Holmes suggested a driving mechanism for moving the continents. He proposed that heat produced by radioactive decay caused thermal convection in the mantle, and that the laterally flowing mantle dragged the continents with the convection cells. He reasoned that if the mantle can flow to allow isostatic rebound following glaciation, then maybe it can flow laterally as well. The acceptance

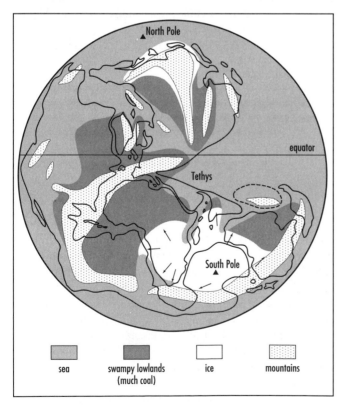

Modification of Alfred Wegener's reconstruction of Pangea

of thermal convection as a driving mechanism for continental drift represented the foundation of modern plate tectonics. In the 1950s and 1960s, the paleomagnetic data was collected from many continents and argued strongly that the continents had indeed been shifting, both with respect to the magnetic pole and also with respect to each other. When seafloor spreading and subduction of oceanic crust beneath island arcs was recognized in the 1960s, the model of continental drift was modified to become the new plate tectonic paradigm that revolutionized and unified many previously diverse fields of the Earth Sciences.

See also PLATE TECTONICS.

continental drift and plate tectonics: historical development of theories The science of tectonics began with Niels Stensens's (alias Nicolaus Steno) publication in 1669 of strain analyses and reconstructions of the pre-deformational configuration of Tuscany, northern Italy. The 18th and early 19th centuries were dominated by views that deformation and uplift of mountain belts were caused by magmatic intrusions and volcanic eruptions and/or the downhill sliding of rock massifs, for example, James Hutton in 1795, Sir James Hall in 1815, Baron Leopold von Buch in 1824, de Saussure in 1796, Gillet-Laumont in 1977, and Studer in 1851–53. In 1829 and 1852 Leonce Elie de Beaumont published an alternative view in which the deformation and uplift of mountain belts was attributed to thermal contraction of a cooling Earth, and this view was popular until the mid-20th century. Mountain belts were viewed as products of crustal shortening by the Rogers brothers (1843) and by James Dwight Dana (1866, 1873), but these scientists still attributed the shortening to global thermal contraction and explosive volcanism (A. M. Celal Sengör, 1982).

Classical theories of orogenesis were upheaved in 1875 with the publication of Eduard Suess's *Die Entstehung der Alpen*. Suess synthesized structural and stratigraphic relationships in mountain belts of the world and emphasized the controlling role of crustal shortening. Although still following the scheme of global contraction, Suess noticed that the global orientation of mountain belts did not obey any simple geometry as expected from uniform global contraction but seemed to reflect an interaction between older rigid massifs and a progression of contractional events on the margins of the massifs. All these models assumed local contraction and did not consider any large-scale translations of rock massifs. In 1884 Marcel Bertrand reinterpreted geological relations in the Alpine Glaurus in terms of a "nappe," which required large horizontal movements, around 25 miles (40 km). Although the nappe theory was not officially recognized until 1903 at the International Geological Congress in Vienna, the publication of this paper marks the beginning of widespread recognition of significant translations in rock massifs. In Suess's last volume of *Das Antlitz der Erde* (1909), he admit-

ted that the amount of shortening observed in mountain belts was greater than could be explained by global cooling and contraction, and he suggested that perhaps other translations have occurred in response to tidal forces and the rotation of the planet. Suess argued on geological grounds for a former unity between Africa and India.

The idea of stable plate interiors can be traced back to Leopold Kober's *Der Bau der Erde* (1921), in which he defined kratogens as areas of low or no mobility, and orogens as areas of high mobility. About this same time (1908, 1916), Emily Argand, clearly of the mobilist school, had documented a continuous Mesozoic through Cenozoic kinematic evolution of the Alpine Orogen. In 1915 and 1929 Alfred Wegener published *Die Entstehung der Kontinente* and *Die Entstehung der Kontinente und Ozeane*. Wegener argued strongly for large horizontal motions between cratons made of sial, and using such data as the match of restored coastlines and paleontological data, he founded the theory of continental drift. The most notable of these early mobilists were Alex du Toit, Reginald A. Daly, Arthur Holmes, Solomon Calvi, and Ishan Ketin (A. M. Celal Sengör, 1982). Alfred Wegener (1915) and Alex du Toit (1927) matched positions of Precambrian cratons and younger fold belts and sedimentary basins between Africa and South America to support continental drift, and it was not until much later, in the 1960s, that continental drift theory developed into a true, multidisciplinary science. It was the geophysical exploration of the seafloor and documentation of rock paleomagnetism that provided the strongest evidence for seafloor spreading and continental drift and led to the theory of plate tectonics (J. Tuzo Wilson, 1965).

Relative plate motions are now corroborated by a plethora of different types of data, including but not limited to the qualitative assertions of the early pioneers. Geological similarities between continents with similar strata and structures continue to offer some of the strongest evidence supporting continental drift. Precise computer-aided geological correlations continue to this day to be one of the most powerful tools for reconstructing past configurations of the continents.

Paleontological and biological data have supported the continental drift hypothesis, with copious early arguments by Arldt (1917), Jaworski (1921), and Alfred Wegener (1929). Since then, paleontological research in tectonics has established paleobiogeographic provinces and has led to the establishment of biostratigraphic timescales (e.g., Truswell, 1981), used for dating geological events. Paleontological data have been directly used to infer plate movements in a few cases. Nordeng (1959, 1963) and Vologdin (1961, 1963) described heliotropic growth of stromatolite columns toward the equator and noted that the apparently changing directions of the equator through time could be interpreted in terms of a wandering pole (or shifting continents). William McKerrow and Cocks (1976) used paleobiogeography to infer the rate of closure of the Iapetus Ocean; Runnegar (1977) used the distribution of cold water circumpolar faunas to infer the drift of Gondwana over the poles. Dwight Bradley and Timothy Kusky (1986) used graptolite biostratigraphy to infer convergence rates and directions for the Taconic orogeny.

Tectonic syntheses based on paleoclimatology have exploited relationships between certain types of deposits, which are most likely to form between specific latitudes. For instance, tillites associated with glaciation tend to form in high latitudes, whereas evaporites and carbonate reefs have a preference for growing within 30°N/S of the equator (e.g., Briden and Irving, 1964). Thus, evaporites, tillites, red beds, coal deposits, and phosphorites all give some indication of paleolatitude and can be used with other data for plate reconstructions (e.g., Frakes, 1981; Ziegler, 1981).

Further Reading

Argand, Emily. "Sur l'Arc des Alpine Occidentales." *Ecologic Geol. Helvet.,* 14 (1916): 145–191.

———. *Carte Géologique du Massif de la Dent Blanche (Moitié Septentrionele),* au 1:50,000. Commission Géologique Suisse, Carte Special no. 52, 1908.

Ardlt, T. *Handbuch der Palaogeographie.* Leipzig: 1917.

de Beaumont, L. E. *Notice sur les Systems des Montagnes.* Paris: 1852.

Bertrand, M. A., "Rapports de Structures des Alpine de Glaris et du Basin Houller du Nord." *Bulletin of the Geological Society of France,* 3ème Sér., 12 (1884): 318–330.

Bradley, Dwight C., and Timothy M. Kusky. "Geologic Evidence for Rate of Plate Convergence during the Taconic Arc-Continent Collision." *Journal of Geology* 94 (1986): 667–681.

Briden, J. C., and E. Irving. "Paleolatitude Spectra of Sedimentary Paleoclimate Indicators." In *Problems in Paleoclimatology,* edited by Alan E. M. Nairn, 199–244. New York: Wiley Interscience, 1964.

von Buch, L. C. "Ueber Geognostische Erscheinungen im Fassathal." v. *Leonard's Mineralogisches Taschenbuch für das Jahr,* 1824 (1933): 396–437.

Dana, James Dwight. "On Some Results of the Earth's Contraction from Cooling, Including a Discussion of the Origin of Mountains and the Nature of the Earth's Interior." *American Journal of Science* 5, series 3 (1873): 423–443.

———. "On Some Results of the Earth's Contraction from Cooling, Including a Discussion of the Origin of Mountains and the Nature of the Earth's Interior." *American Journal of Science* 6, series 3 (1873): 6–14, 104–115, 161–171.

———. "Observations on the Origin of Some of the Earth's Features." *American Journal of Science* 42, series 2 (1866): 205–211, 252–253.

du Toit, Andrew L. "A Geological Comparison of South America with South Africa." *Carnegie Institution of Washington,* pub. 381 (1927): 1–157.

Frakes, Larry A. "Late Paleozoic Paleoclimatology." In *Paleoreconstruction of the Continents,* edited by Michael W. McElhinny and Daniel A. Valencio, 39–44. American Geophysical Union/Geological Society of America Geodynamics Series 2, 1981.

Gillet-Laumont, F. P. N. "Observations Géologiques sur le Gisement et la Forme des Replis Successif que l'on Remarque dans Cer-

taines Couches de Substances Minérales, et Particulirement de Mines Bouille; Suivies de Conjectures sur Leur Origine," *J. des Mines,* no. 69 (1799): 449–454.

Hall, Sir J. "On the Vertical Position and the Convolutions of Certain Strata and their Relation with Granite." *Transaction of the Royal Society of Edinberg* 7 (1815): 79–85.

Hutton, James. *Theory of the Earth with Proofs and Illustrations.* Messrs. Cadell, Junior Davies, London, and William Creech, Edinburgh. 1795.

Jaworsky. "Das Alter des südatlantishen Beckens." *Geologische Rundschau* 20 (1921): 60–74.

Kober, Leopold. *Der Bau der Erde,* 2nd ed. Berlin: Gebrüder Borntrager, 1921.

McKerrow, W. S., and L. R. M. Cocks. "Progressive Faunal Migration across the Iapetus Ocean." *Nature* 263 (1976): 304–306.

Nordeng, S. C. "Precambrian Stromatolites as Indicators of Polar Shift." In *Polar Wandering and Continental Drift,* edited by A. C. Munyan, 131–139. Society of Economic Paleontology Special Publication 10, 1963.

———. "Possible Use of Precambrian Calcareous Algae Colonies as Indicators of Polar Shifts." *Minnesota Center Contribution to the Institute of Lake Superior Geology,* 5th Annual Meeting (1955): 9.

Roger, W. B., and H. D. Rogers. "On the Physical Structure of the Appalachian Chain, as Exemplifying the Laws Which Have Regulated the Elevation of Great Mountain Chains Generally." *Association of American Geologists Report* (1843): 474–531.

Runnegar, B. "Marine Fossil Invertebrates of Gondwanaland, Paleogeographic Implications." *Fourth Gondwana Symposium, Calcutta* (1977): 1–25.

Sengör, A. M. Celal. "Eduard Suess' Relations to the pre–1950 Schools of Thought in Global Tectonics." *Geol. Rundsch.* 71 (1982): 381–420.

———. "Classical Theories of Orogenesis." In *Orogeny,* edited by A. Miyashoro, K. Aki, and A. M. Celal Sengör, 1–48. New York: Wiley, 1982.

Studer, B. *Geologie der Schweiz.* Wien: W. Braumüller, 1851–1853.

Suess, Edward. *Über unterbrochene Gebirgsfaulting.* Sitzber k. Akad: Wiess. 94, Part I (1886): 111–117.

———. *Das Antlitz der Erde.* Wien: Freytag, 1885–1909.

———. *Die Entstehung der Alpen.* Wien: W. Braumüller, 1875.

Truswell, E. M. "Pre-Cenozoic Palynology and Continental Movements." In *Paleoreconstruction of the Continents,* edited by M. W. McElhinny and D. A. Valencio, 13–26. American Geophysical Union/Geological Society of America Geodynamics Series 2, 1981.

Vologdin, A. G. "Stromatolites and Phototropism." *Dokl. Akad. Nauk.,* S.S.S.R. 151 (1963): 683–686.

———. "Eye Witness of the Migration of the Poles." *Prioroda* 11 (1961): 102–103.

Wegener, Alfred. *Die Enstehung der Kontinente und Ozean.* Fried. Vieweg und Sohn, Braunschweig, 1929.

———. *Die Enstehung der Kontinente.* Fried. Vieweg und Sohn, Braunschweig. 1915.

Wilson, J. T. "A New Class of Faults and Their Bearing on Continental Drift." *Nature* 207 (1965): 343–347.

Ziegler, A. M. "Paleozoic Paleogeography." In *Paleoreconstruction of the Continents,* edited by M. W. McElhinny and D. A. Valencio, 31–38. American Geophysical Union/Geological Society of America Geodynamics Series 2, 1981.

continental margin   The transition zone between thick buoyant continental crust and the thin dense submerged oceanic crust. There are several different types, depending on the tectonic setting. Passive, trailing, or Atlantic-type margins form where an extensional boundary evolves into an ocean basin, and new oceanic crust is added to the center of the basin between originally facing continental margins. These margins were heated and thermally elevated during rifting and gradually cool and thermally subside for several tens of millions of years, slowly accumulating thick sequences of relatively flat sediments, forming continental shelves. These shelves are succeeded seaward by continental slopes and rises. The ocean/continent boundary is typically drawn at the shelf/slope break on these Atlantic-type margins, where water depths average a couple of hundred meters. Passive margins do not mark plate boundaries but rim most parts of many oceans, including the Atlantic and Indian, and form around most of Antarctica and Australia. Young immature passive margins are beginning to form along the Red Sea.

Convergent, leading, or Pacific-type margins form at convergent plate boundaries. They are characterized by active deformation, seismicity, and volcanism, and some have thick belts of rocks known as accretionary prisms that are scraped off of a subducting plate and added to the overriding continental plate. Convergent margins may have a deep-sea trench up to seven miles (11 km) deep marking the boundary between the continental and oceanic plates. These trenches form where the oceanic plate is bending and plunging deep into the mantle. Abundant folds and faults in the rocks characterize convergent margins. Other convergent margins are characterized by old eroded bedrock near the margin, exposed by a process of sediment erosion where the edge of the continent is eroded and drawn down into the trench.

A third type of continental margin forms along transform or transcurrent plate boundaries. These are characterized by abundant seismicity and deformation, and volcanism is limited to certain restricted areas. Deformation along transform margins tends to be divided into different types depending on the orientation of bends in the main plate boundary fault. Constraining bends form where the shape of the boundary restricts motion on the fault and are characterized by strong folding, faulting, and uplift. The Transverse Ranges of southern California form a good example of a restraining bend. Sedimentary basins and subsidence characterize bends in the opposite direction, where the shape of the fault causes extension in areas where parts of the fault diverge during movement. Volcanic rocks form in some of these basins. The Gulf of California and Salton trough have formed in areas of extension along a transform margin in southern California.

See also CONVERGENT PLATE MARGIN PROCESSES; DIVERGENT OR EXTENSIONAL BOUNDARIES; PLATE TECTONICS; TRANSFORM PLATE MARGIN PROCESSES.

continental shield   *See* SHIELD.

convection and the Earth's mantle   The main heat transfer mechanism in the Earth's mantle is convection. It is a thermally driven process in which heating at depth causes material to expand and become less dense, causing it to rise while being replaced by complementary cool material that sinks. This moves heat from depth to the surface in a very efficient cycle since the material that rises gives off heat as it rises and cools, and the material that sinks gets heated only to eventually rise again. Convection is the most important mechanism by which the Earth is losing heat, with other mechanisms including conduction, radiation, and advection. However, many of these mechanisms work together in the plate tectonic cycle. Mantle convection brings heat from deep in the mantle to the surface where the heat released forms magmas that generate the oceanic crust. The mid-ocean ridge axis is the site of active hydrothermal circulation and heat loss, forming black smoker chimneys and other vents. As the crust and lithosphere move away from the mid-ocean ridges, it cools by conduction, gradually subsiding (according to the square root of its age) from about 1.5–2.5 miles (2.5–4.0 km) below sea level. Heat loss by mantle convection is therefore the main driving mechanism of plate tectonics, and the moving plates can be thought of as the conductively cooling boundary layer for large-scale mantle convection systems.

The heat being transferred to the surface by convection is produced by decay of radioactive heat-producing isotopes such as U^{235}, Th^{232}, and K^{40}, remnant heat from early heat-producing isotopes such as I^{129}, remnant heat from accretion of the Earth, heat released during core formation, and heat released during impacts of meteorites and asteroids. Very early in the history of the planet at least part of the mantle was molten, and the Earth has been cooling by convection ever since. It is difficult to estimate how much the mantle has cooled with time, but reasonable estimates suggest that the mantle may have been up to a couple of hundred degrees hotter in the earliest Archean.

The rate of mantle convection is dependent on the ability of the material to flow. The resistance to flow is a quantity measured as viscosity, defined as the ratio of shear stress to strain rate. Fluids with high viscosity are more resistant to flow than materials with low viscosity. The present viscosity of the mantle is estimated to be 10^{20}–10^{21} Pascal seconds (Pa/s) in the upper mantle, and 10^{21}–10^{23} Pa/s in the lower mantle, which are sufficient to allow the mantle to convect and complete an overturn cycle once every 100 million years. The viscosity of the mantle is temperature dependent, so it is possible that in early Earth history the mantle may have been able to flow and convectively overturn much more quickly, making convection an even more efficient process and speeding the rate of plate tectonic processes.

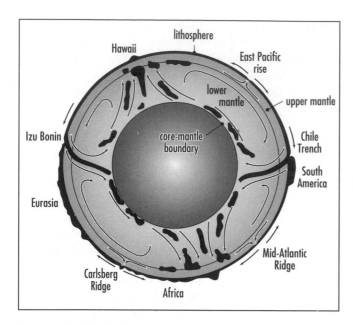

Cross section of the Earth showing possible modes of mantle convection

There is currently an ongoing debate and research about the style of mantle convection in the Earth. The upper mantle is relatively heterogeneous and extends to a depth of 416 miles (670 km), where there is a pronounced increase in seismic velocities. The lower mantle is more homogeneous and extends to the D" region at 1,678 miles (2,700 km), marking the transition into the liquid outer core. One school of mantle convection thought suggests that the entire mantle, including both the upper and lower parts, is convecting as one unit. Another school of thought posits that the mantle convection is divided into two layers, with the lower mantle convecting separately from the upper mantle. A variety of these models, presently held by the majority of geophysicists, is that there is two-layer convection, but that subducting slabs are able to penetrate the 670-kilometer discontinuity from above, and that mantle plumes that rise from the D" region are able to penetrate the 670-kilometer discontinuity from below.

The shapes of mantle convection cells include many possible forms that are reflected to a first order by the distribution of subduction zones and mid-ocean ridge systems. The subduction zones mark regions of downwelling, whereas the ridge system marks broad regions of upwelling. Material is upwelling in a broad planiform cell beneath the Atlantic and Indian Oceans, and downwelling in the circum-Pacific subduction zones. There is thought to be a large plume-like "superswell" beneath part of the Pacific that feeds the planiform East Pacific rise. Mantle plumes that come from the deep mantle punctuate this broad pattern of upper mantle convection, and their plume tails must be distorted by flow in the convecting upper mantle.

The pattern of mantle convection deep in geological time is uncertain. Some periods such as the Cretaceous seem to have had much more rigorous mantle convection and surface volcanism. More or different types or rates of mantle convection may have helped to allow the early Earth to lose heat more efficiently. Some computer models allow periods of convection dominated by plumes, and others dominated by overturning planiform cells similar to the present Earth. Some models suggest cyclic relationships, with slabs pooling at the 670-kilometer discontinuity, then suddenly all sinking into the lower mantle, causing a huge mantle overturn event. Further research is needed on linking the preserved record of mantle convection in the deformed continents to help interpret the past history of convection.

See also BLACK SMOKER CHIMNEYS; DIVERGENT OR EXTENSIONAL BOUNDARIES; PLATE TECTONICS.

Further Reading

Schubert, Gerald, Donald L. Turcotte, and Peter Olson. *Mantle Convection in the Earth and Planets.* Cambridge: Cambridge University Press, 2001.

Turcotte, Donald L., and Gerald Schubert. *Geodynamics,* 2nd ed. Cambridge: Cambridge University Press, 2002.

convergent plate margin processes Structural, igneous, metamorphic, sedimentological processes that occur in the region affected by forces associated with the convergence of two or more plates. Convergent plate boundaries are of two fundamental types, subduction zones and collision zones. Subduction zones are of two basic types, the first of which being where oceanic lithosphere of one plate descends beneath another oceanic plate, such as in the Philippines and Marianas of the southwest Pacific. The second type of subduction zone forms where an oceanic plate descends beneath a continental upper plate, such as in the Andes of South America. The southern Alaska convergent margin is particularly interesting, as it records a transition from an ocean/continent convergent boundary to an ocean/ocean convergent boundary in the Aleutians.

Arcs have several different geomorphic zones defined largely on their topographic and structural expressions. The active arc is the topographic high with volcanoes, and the backarc region stretches from the active arc away from the trench, and it may end in an older rifted arc or continent. The forearc basin is a generally flat topographic basin with shallow to deepwater sediments, typically deposited over older accreted sediments and ophiolitic or continental basement. The accretionary prism includes uplifted strongly deformed rocks that were scraped off the downgoing oceanic plate on a series of faults. The trench may be several to 10 or more kilometers below the average level of the seafloor in the region and marks the boundary between the overriding and underthrusting plate. The outer trench slope is the region from the trench to the top of the flexed oceanic crust that forms a few-hundred-meter-high topographic rise known as the forebulge on the downgoing plate.

Trench floors are triangular-shaped in profile and typically partly to completely filled with graywacke-shale turbidite sediments derived from erosion of the accretionary wedge. They may also be transported by currents along the trench axis for large distances, up to hundreds or even thousands of kilometers from their ultimate source in uplifted mountains in the convergent orogen. Flysch is a term that applies to rapidly deposited deep marine synorogenic clastic rocks that are generally turbidites. Trenches are also characterized by chaotic deposits known as olistostromes that typically have clasts or blocks of one rock type, such as limestone or sandstone, mixed with a muddy or shaly matrix. These are interpreted as slump or giant submarine landslide deposits. They are common in trenches because of the oversteepening of slopes in the wedge. Sediments that get accreted may also include pelagic sediments that were initially deposited on the subducting plate, such as red clay, siliceous ooze, chert, manganiferous chert, calcareous ooze, and windblown dust.

The sediments are deposited as flat-lying turbidite packages, then gradually incorporated into the accretionary wedge complex through folding and the propagation of faults through the trench sediments. Subduction accretion is a process that accretes sediments deposited on the underriding plate onto the base of the overriding plate. It causes the rotation and uplift of the accretionary prism, which is a broadly steady-state process that continues as long as sediment-laden trench deposits are thrust deeper into the trench. Typically new faults will form and propagate beneath older ones, rotating the old faults and structures to steeper attitudes as new material is added to the toe and base of the accretionary wedge. This process increases the size of the overriding accretionary wedge and causes a seaward-younging in the age of deformation.

Parts of the oceanic basement to the subducting slab are sometimes scraped off and incorporated into the accretionary prisms. These tectonic slivers typically consist of fault-bounded slices of basalt, gabbro, and ultramafic rocks, and rarely, partial or even complete ophiolite sequences can be recognized. These ophiolitic slivers are often parts of highly deformed belts of rock known as mélanges. Mélanges are mixtures of many different rock types typically including blocks of oceanic basement or limestone in muddy, shaly, serpentinic, or even a cherty matrix. Mélanges are formed by tectonic mixing of the many different types of rocks found in the forearc, and they are among the hallmarks of convergent boundaries.

There are major differences in processes that occur at Andean-style v. Marianas-style arc systems. Andean-type arcs have shallow trenches, less than 3.7 miles (6 km) deep, whereas Marianas-type arcs typically have deep trenches reaching 6.8 miles (11 km) in depth. Most Andean-type arcs subduct young oceanic crust and have very shallow-dipping subduction zones, whereas Marianas-type arcs subduct old

oceanic crust and have steeply dipping Benioff zones. Andean arcs have back arc regions dominated by foreland (retroarc) fold thrust belts and sedimentary basins, whereas Marianas-type arcs typically have back arc basins, often with active seafloor spreading. Andean arcs have thick crust, up to 43.5 miles (70 km), and big earthquakes in the overriding plate, while Marianas-type arcs have thin crust, typically only 12.5 miles (20 km), and have big earthquakes in the underriding plate. Andean arcs have only rare volcanoes, and these have magmas rich in SiO_2 such as rhyolites and andesites. Plutonic rocks are more common, and the basement is continental crust. Marianas-type arcs have many volcanoes that erupt lava low in silica content, typically basalt, and are built on oceanic crust.

Many arcs are transitional between the Andean or continental margin-types and the oceanic or Marianas-types, and some arcs have large amounts of strike-slip motion. The causes of these variation have been investigated, and it has been determined that the rate of convergence has little effect, but the relative motion directions and the age of the subducted oceanic crust seem to have the biggest effects. In particular, old oceanic crust tends to sink to the point where it has a near-vertical dip, rolling back through the viscous mantle and dragging the arc and forearc regions of overlying Marianas-type arcs with it. This process contributes to the formation of back arc basins.

Much of the variation in processes that occur in convergent margin arcs can be attributed to the relative convergence vectors between the overriding and underriding plates. In this kinematic approach to modeling convergent margin processes, the underriding plate may converge at any angle with the overriding plate, which itself moves toward or away from the trench. Since the active arc is a surface expression of the 68-mile (110-km) isobath on the subducted slab, the arc will always stay 110 kilometers above this zone. The arc therefore separates two parts of the overriding plate that may move independently, including the frontal arc sliver between the arc and trench, that is kinematically linked to the downgoing plate, and the main part of the overriding plate. Different relative angles of convergence between the overriding and underriding plate determine whether or not an arc will have strike-slip motions, and the amount that the subducting slab rolls back (which is age-dependent) determines whether the frontal arc sliver rifts from the arc and causes a back arc basin to open or not. This model helps to explain why some arcs are extensional with big back arc basins, others have strike-slip dominated systems, and others are purely compressional arcs. Convergent margins also show changes in these vectors and consequent geologic processes with time, often switching from one regime to the other quickly with changes in the parameters of the subducting plate.

The thermal and fluid structure of arcs is dominated by effects of the downgoing slab, which is much cooler than the surrounding mantle and serves to cool the forearc. Fluids released from the slab as it descends past 110 kilometers aid partial melting in the overlying mantle and form the magmas that form the arc on the overriding plate. This broad thermal structure of arcs results in the formation of paired metamorphic belts, where the metamorphism in the trench environment grades from cold and low-pressure at the surface to cold and high-pressure at depth, whereas the arc records low and high-pressure high-temperature metamorphic facies series. One of the distinctive rock associations of trench environments is the formation of the unusual high-pressure low-temperature blueschist facies rocks in paleosubduction zones. The presence of index minerals glaucophane (a sodic amphibole), jadeite (a sodic pyroxene), and lawsonite (Ca-zeolite) indicate low temperatures extended to depths of 20–30 kilometers (7–10 kilobars). Since these minerals are unstable at high temperatures, their presence indicates they formed in a low temperature environment, and the cooling effects of the subducting plate offer the only known environment to maintain such cool temperatures at depth in the Earth.

Forearc basins may include several-kilometer-thick accumulations of sediments that were deposited in response to subsidence induced by tectonic loading or thermal cooling of forearcs built on oceanic lithosphere. The Great Valley of California is a forearc basin that formed on oceanic forearc crust preserved in ophiolitic fragments found in central California, and Cook Inlet in Alaska is an active forearc basin formed in front of the Aleutian and Alaska range volcanic arc.

The rocks in the active arcs typically include several different facies. Volcanic rocks may include subaerial flows, tuffs, welded tuffs, volcaniclastic conglomerate, sandstone, and pelagic rocks. Debris flows from volcanic flanks are common, and there may be abundant and thick accumulations of ash deposited by winds and dropped by Plinian and other eruption columns. Volcanic rocks in arcs include mainly calc-alkaline series, showing an early iron enrichment in the melt, typically including basalts, andesites, dacites, and rhyolites. Immature island arcs are strongly biased toward eruption at the mafic end of the spectrum and may also include tholeiitic basalts, picrites, and other volcanic and intrusive series. More mature continental arcs erupt more felsic rocks and may include large caldera complexes.

Back arc or marginal basins form behind extensional arcs or may include pieces of oceanic crust that were trapped by the formation of a new arc on the edge of an oceanic plate. Many extensional back arcs are found in the southwest Pacific, whereas the Bering Sea between Alaska and Kamchatka is thought to be a piece of oceanic crust trapped during the formation of the Aleutian chain. Extensional back arc basins may have oceanic crust generated by seafloor spreading, and these systems very much resemble the spreading centers found at divergent plate boundaries. However, the geochemical signature of some of the lavas shows some subtle

and some not-so-subtle differences, with water and volatiles being more important in the generation of magmas in back arc supra-subduction zone environments.

Compressional arcs such as the Andes have tall mountains, reaching heights of more than 24,000 feet (7,315 m) over broad areas. They have rare or no volcanism but much plutonism and typically have shallow dipping slabs beneath them. They have thick continental crust with large compressional earthquakes, and show a foreland-style retroarc basin in the back arc region. Some compressional arc segments do not have accretionary forearcs but exhibit subduction erosion during which material is eroded and scraped off the overriding plate and dragged down into the subduction zone. The Andes show some remarkable along-strike variations in processes and tectonic style, with sharp boundaries between different segments. These variations seem to be related to what is being subducted and plate motion vectors. In areas where the downgoing slab has steep dips, the overriding plate has volcanic rocks; in areas of shallow subduction there is no volcanism.

Collisions

Collisions are the final products of subduction. There are several general varieties of collisions. They may be between island arcs and continents, such as the Ordovician Taconic orogeny in eastern North America, or they may be between passive margins on one continent and an Andean margin on another. More rarely, collisions between two convergent margins occur above two oppositely dipping subduction zones, with a contemporary example extant in the Molucca Sea of Indonesia. Finally, collisions may be between two continents, such as the ongoing India/Asia collision that is affecting much of Asia.

Arc/continent collisions are the simplest of collisional events. As an arc approaches a continent, the continental margin is flexed downward by the weight of the arc, much like a ruler pushed down over the edge of a desk. The flexure induces a bulge a few hundred kilometers wide in front of the active collision zone, and this bulge migrates in front of the collision as a few-hundred-meter-high broad topographic swell. As the arc terrane rides up onto the continent, the thick sediments in the continental rise are typically scraped off and progressively added to the accretionary prism, with the oldest thrust faults being the ones closest to the arc, and progressively younger thrust faults along the base of the prism. Many forearc regions have ophiolitic basement, and these ophiolites get thrust upon the continents during collision events and are preserved in many arc/continent collisional orogens. The accretionary wedge grows and begins to shed olistostromes into the foredeep basin between the arc and continent, along with flysch and distal black shales. These three main facies migrate in front of the moving arc/accretionary complex at a rate equal to the convergence rate and drown any shallow water carbonate deposition. After the arc

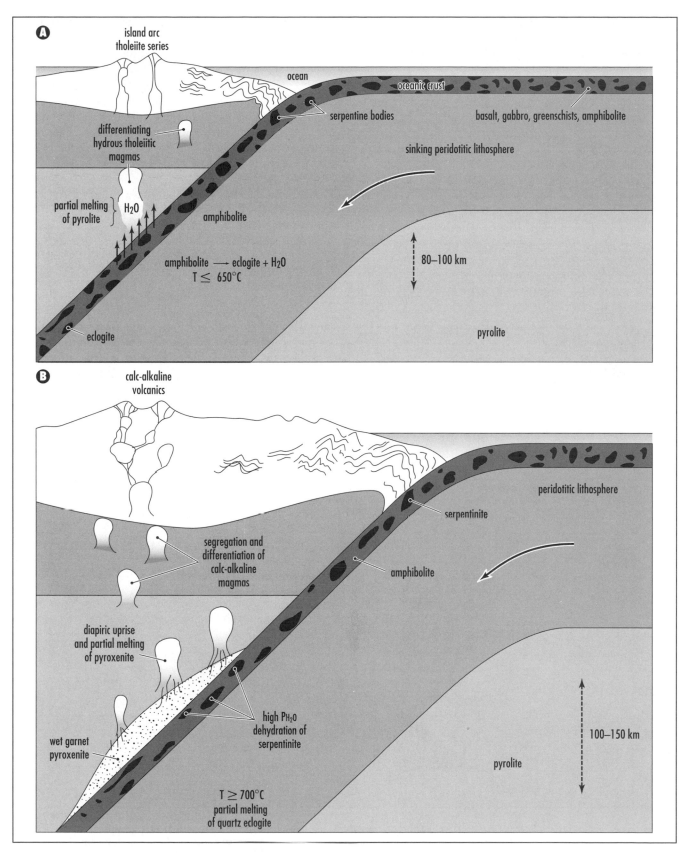

Physiography and geology of arcs: (A) Pacific-type; (B) Andean-type

terrane rides up the continental rise, slope, and shelf, it grinds to a halt when isostatic (buoyancy) effects do not allow continued convergence. At this stage, a new subduction zone may be initiated behind the collided arc, allowing convergence to continue between the two plates.

Continent/continent collisions are the most dramatic of collisional events, with the current example of the convergence and collision of Africa, Arabia, and India with Europe and Asia, affecting much of the continental landmass of the world. Continental collisions are associated with thickening of continental crust and the formation of high mountains, and deformation distributed over wide areas. The convergence between India and Asia dramatically slowed about 38 million years ago, probably associated with initial stages of that collision between 25 million and 40 million years ago. The collision has resulted in the uplift of the Himalayan mountain chain, the Tibetan plateau, and formed a wide zone of deformation that extends well into Siberia and includes much of southeast Asia. Since the collision, there has been 2–2.4 inches per year (5–6 cm/yr) of convergence between India and Asia, meaning that a minimum of 775 miles (1,250 km) has had to be accommodated in the collision zone. This convergence has been accommodated in several ways. Two

large faults between India and Asia, the Main Central thrust and the Main Boundary thrust are estimated to have 250 and 120 miles (400 and 200 km) of displacement on them, so they are able to account for less than half of the displacement. Folds of the crust and general shortening and thickening of the lithosphere may account for some of the convergence, but not a large amount. It appears that much of the convergence was accommodated by underthrusting of the Indian plate beneath Tibet, and by strike-slip faulting moving or extruding parts of Asia out of the collision zone toward the southwest Pacific.

The Tibetan Plateau and Himalayan mountain chain are about 375 miles (600 km) wide, with the crust beneath the region being about 45 miles (70 km) thick, twice that of normal continental crust. This has led to years of scientific investigation about the mechanism of thickening. Some models and data suggest that India is thrust under Asia for 600 kilometers, whereas other models and data suggest that the region has thickened by thrusting at the edges and plane strain in the center. In either case, the base of the Tibetan crust has been heated to the extent that partial melts are beginning to form, and high heat flow in some rifts on the plateau is associated with the intrusions at depth. The intru-

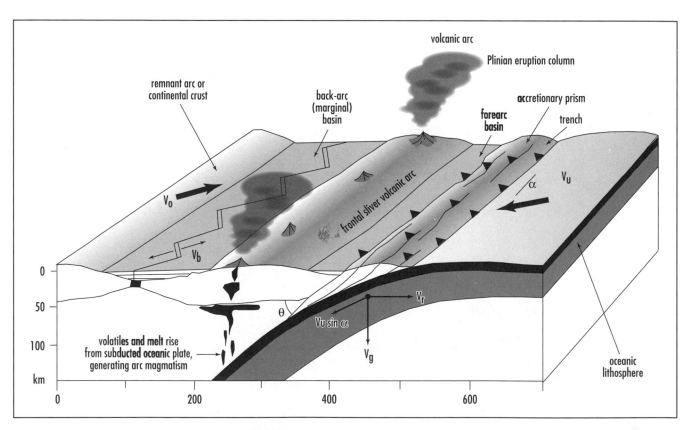

Relative motion vectors in arcs. Changes in relative motions can produce drastically different arc geology. V_u = velocity of underriding plate; V_o = velocity of overriding plate; V_b = slip vector between overriding and underriding plates; V_g = velocity of sinking; V_r = velocity of rollback. (Note that $V_u \sin \alpha$ = velocity of downdip component of subduction, and $V_r = V_g \cot\theta$).

sions are weakening the base of the crust which is starting to collapse under the weight of the overlying mountains, and the entire plateau is on the verge of undergoing extension.

The collisional process is resulting in the formation of a layered differentiated lower continental crust in Tibet with granitic melts forming a layer that has been extracted from a granulitic residue, along with strong deformation. These processes are not readily observable 30 miles (50 km) beneath Tibet, but they are preserved in many very old (generally Precambrian) high-grade gneiss terranes around the world that are thought to have formed in continental collision zones.

Continent/continent collision zones tend to have major effects on global plate motions. Convergence that used to be accommodated between the two continents must be transferred elsewhere on the planet, since all plate motions must sum to zero on the planet. Therefore, continental collisions typically cause events elsewhere, such as the formation of new subduction zones and a global reorganization of plate motions.

See also ACCRETIONARY WEDGE; CONTINENTAL CRUST; CONVERGENT PLATE MARGIN PROCESSES; DEFORMATION OF ROCKS; METAMORPHISM; PLATE TECTONICS; PLUTON; STRUCTURAL GEOLOGY; VOLCANO.

Further Reading
Moores, Eldridge, and Robert Twiss. *Tectonics.* New York: W.H. Freeman and Company, 1995.

corals A group of invertebrate marine fossils of the phylum Cnidaria, characterized by radial symmetry and a lack of cells organized into organs. They are related to jellyfish, hydroids, and sea anemones, all of which possess stinging cells. Corals are the best preserved of this phylum because they secrete a hard calcareous skeleton. The animal is basically a simple sac with a central mouth, surrounded by tentacles, that leads to a closed stomach. Cnidarians are passive predators, catching food that wanders by in their tentacles. Corals and other cnidarians produce alternating generations of two body forms. Medusa are forms that reproduce sexually to form polyps that are asexual forms from which the medusa may bud off. The corals belong to a subclass of the anthozoan cnidarians known as the Zooantharia. The jellyfish belongs to the Scyphozoa class, and the Hydrozoa class includes both fresh and saltwater cnidaria dominated by the polyp stage.

Corals can live in a range of conditions from shallow tidal pools to 19,700 feet (6,000 m) depth. They have a cylindrical or conical skeleton secreted by the polyp stage organism, who lives in the upper exposed part of the structure. The skeleton is characterized by radial ridges known as septa that join the skeleton's outer wall (known as the theca), and may have flat floors that were periodically secreted by the polyp.

Corals range from the Early Ordovician Tabulata forms, joined in the Middle Ordovician by the rugose corals. They both experienced a major extinction in the Late Devonian, from which the rugose forms recovered stronger. Both forms became extinct in the Early Triassic and were replaced by modern coral forms known as Scerlactina, which apparently arose independently from different soft-bodied organisms.

Most corals grow in colonial communities and form reefs that provide numerous advantages including shelter for larvae and young stages. Modern corals can survive only in shallow waters that range in temperature from 77°F to 84°F (25°C to 29°C), at depths of less than 300 feet (90 m). Reefs have various forms including fringing, barrier, and atoll reefs.

See also CARBONATE; PLATFORM; REEF.

Cordillera *See* ROCKY MOUNTAINS.

Coriolis effect The Coriolis effect or force produces a deflection of moving objects and currents to the right in the Northern Hemisphere, and to the left in the Southern Hemisphere. The force ranges from zero at the equator to a maximum at the poles. The force can be understood by considering the rotation of the Earth from a position above the poles. Looking down from above the north pole, the rotation appears to be counterclockwise, and from above the south pole it appears to be clockwise. Points on the poles simply turn around once every 24 hours, whereas points that lie on the equator must speed through space at 1,000 miles per hour (1,600 km/hr).

As air and water move poleward from the equatorial regions they bring the higher velocity they acquired closer to the equator. The slower speeds of rotation of the Earth under the moving air and water causes these fluids to move a greater distance per unit of time than the underlying Earth, resulting in a deflection to the right in the Northern Hemisphere, and to the left in the Southern Hemisphere. Likewise, air and water that are moving from the poles to the equator will be moving slower than the underlying Earth, causing the Earth to move more per unit of time than the air or water. This again causes a deflection to the right in the Northern Hemisphere, and to the left in the Southern Hemisphere.

See also ATMOSPHERE; THERMOHALINE CIRCULATION.

cosmology *See* BIG BANG THEORY.

cratons Large areas of relatively thick continental crust that have been stable for long periods of geological time, generally since the Archean. Most cratons are characterized by low heat flow, no or few earthquakes, and many have a thick mantle root or tectosphere that is relatively cold and refractory, having had a basaltic melt extracted from it during the Archean.

Understanding the origin of stable continental cratons hinges upon recognizing which processes change the volume and composition of continental crust with time, and how and when juvenile crust evolved into stable continental crust. The evidence from the preserved record suggests that the continental landmass has been growing since the early Archean, although the relative rates and mechanisms of crustal recycling and crustal growth are not well known and have been the focus of considerable geological debate. The oldest rocks known on the planet are the circa 4.0 Ga Acasta gneisses from the Anton terrane of the Slave Province. The Acasta gneisses are chemically evolved and show trace and Rare Earth Element (REE) patterns similar to rocks formed in modern supra-subduction zone settings. Furthermore, the 3.8-billion-year-old Isua sequence from Greenland, the oldest known sedimentary sequence, is an accretionary complex. A few circa 4.2 Ga zircon grains have been found, but it is not clear if these were ever parts of large continental landmasses. Approximately half of the present mass of continental crust was extracted from the mantle during the Archean.

Exposed portions of Archean cratons are broadly divisible into two main categories. The first are the "granite-greenstone" terranes, containing variably deformed assemblages of mafic volcanic/plutonic rocks, metasedimentary sequences, remnants of older quartzo-feldspathic gneissic rocks, and abundant late granitoids. The second main class of preserved Archean lithosphere is found in the high-grade quartzo-feldspathic gneiss terranes. Relatively little deformed and metamorphosed cratonic cover sequences are found over and within both types of Archean terrain, but they are especially abundant on southern Africa's Kaapvaal craton. Also included in this category are some thick and laterally extensive carbonate platforms similar in aspect to Phanerozoic carbonate platforms, indicating that parts of the Archean lithosphere were stable, thermally subsiding platforms.

Although the rate of continental growth is a matter of geological debate, most geological data indicates that the continental crust has grown by accretionary and magmatic processes taking place at convergent plate boundaries since the early Archean. Arc-like trace element characteristics of continental crust suggest that subduction zone magmatism has played an important role in the generation of the continental crust. Convergent margin accretionary processes that contribute to the growth of the continental crust can be divided into five major groups: (1) oceanic plateau accretion; (2) oceanic island arc accretion; (3) normal ocean crust (mid-ocean ridge) accretion/ophiolite obduction; (4) back arc basin accretion; and (5) arc-trench migration/Turkic-type orogeny accretion. These early accretionary processes are typically followed by intrusion of late stage anatectic granites, late gravitational collapse, and late strike-slip faulting. Together, these processes release volatiles from the lower crust and mantle and help to stabilize young accreted crust and form stable continents.

Juvenile Island Arc Accretion

Many Archean granite-greenstone terranes are interpreted as juvenile island arc sequences that grew above subduction zones and later amalgamated during collisional orogenesis to form new continental crust. The island arc model for the origin of the continental crust is supported by geochemical studies, which show that the crust has a bulk composition similar to arcs. Island arcs are extremely complex systems that may exhibit episodes of distinctly different tectonics, including accretion of ophiolite fragments, oceanic plateaux, intra-arc extension with formation and preservation of back arc, and intra-arc basins. Many juvenile arcs evolve into mature island arcs in which the magmatic front has migrated through its own accretionary wedge, and many evolve into continental margin arcs after they collide with other crustal fragments or continental nuclei.

Although accretion of immature oceanic arcs appears to have been a major mechanism of crustal growth in Archean orogens, it has been argued that oceanic arc-accretion alone is insufficient to account for the rapid crustal growth in Precambrian shields. Furthermore, most oceanic arcs are characterized by mafic composition, whereas the continental crust is andesitic in composition.

Ophiolite Accretion

Ophiolites are a distinctive association of allochthonous rocks interpreted to form in a variety of plate tectonic settings such as oceanic spreading centers, back arc basins, forearcs, arcs, and other extensional magmatic settings including those in association with plumes. A complete ophiolite grades downward from pelagic sediments into a mafic volcanic complex that is generally made of mostly pillow basalts, underlain by a sheeted dike complex. These are underlain by gabbros exhibiting cumulus textures, then tectonized peridotite, resting above a thrust fault that marks the contact with underlying rock sequences. The term *ophiolite* refers to this distinctive rock association and should not be used in a purely generic way to refer to allochthonous oceanic lithosphere rocks formed at mid-ocean ridges.

Very few complete Phanerozoic-like ophiolite sequences have been recognized in Archean greenstone belts. However, the original definition of ophiolites includes "dismembered," "partial," and "metamorphosed" varieties, and many Archean greenstone belts contain two or more parts of the full ophiolite sequence. Archean oceanic crust was possibly thicker than Proterozoic and Phanerozoic counterparts, resulting in accretion predominantly of the upper section (basaltic) of oceanic crust. The crustal thickness of Archean oceanic crust may in fact have resembled modern oceanic plateaus. If this were the case, complete Phanerozoic-like MORB (Mid-Ocean Ridge Basalt)-type ophiolite sequences would have been very unlikely to be accreted or obducted during Archean orogenies. In contrast, only the upper, pillow lava-dominated sections would likely be accreted.

Portions of several Archean greenstone belts have been interpreted to contain dismembered or partial ophiolites.

Accretion of MORB-type ophiolites has been proposed as a mechanism of continental growth in a number of Archean, Proterozoic, and Phanerozoic orogens. Several suspected Archean ophiolites have been particularly well-documented. One of the most disputed is the circa 3.5 Ga Jamestown ophiolite in the Barberton greenstone belt of the Kaapvaal craton of southern Africa—a 1.8-mile (3-km) thick sequence including a basal peridotite tectonite unit with chemical and textural affinities to Alpine-type peridotites, overlain by an intrusive-extrusive igneous sequence, and capped by a chert-shale sequence. This partial ophiolite is pervasively hydrothermally altered and shows chemical evidence for interaction with seawater with high heat and fluid fluxes. SiO_2 and MgO alteration and black smoker-like mineralization is common, with some hydrothermal vents traceable into banded iron formations, and subaerial mudpool structures. These features led Maarten de Wit and others in 1992 to suggest that this ophiolite formed in a shallow sea and was locally subaerial, analogous to the Reykjanges ridge of Iceland. In this sense, Archean oceanic lithosphere may have looked very much like younger oceanic plateaux lithosphere.

Several partial or dismembered ophiolites have been described from the Slave Province of northern Canada. A fault-bounded sequence on Point Lake grades downward from shales and chemical sediments (umbers) into several kilometers of pillow lavas intruded by dikes and sills, locally into multiple dike/sill complexes, then into isotropic and cumulate-textured layered gabbro. The base of this partial Archean ophiolite is marked by a one-kilometer thick shear zone composed predominantly of mafic and ultramafic mylonites, with less-deformed domains including dunite, websterite, wherlite, serpentinite, and anorthosite. Synorogenic conglomerates and sandstones were deposited in several small foredeep basins and are interbedded with mugearitic lavas (and associated dikes), all deposited/intruded in a foreland basin setting.

A complete but dismembered and metamorphosed 2.5-billion-year-old ophiolite complex has been described from the North China craton. This ophiolite has structurally complex pillow lavas, mafic flows, breccia, and chert overlying a mixed dike and gabbro section that grades down into layered gabbro, cumulate ultramafics, and mantle peridotites. High-temperature mantle fabrics and ophiolitic mantle podiform chromitites have also been documented from the Dongwanzi ophiolite, and it has ophiolitic mélange intruded by arc magmas.

Dismembered ophiolites appear to be a widespread component of greenstone belts in Archean cratons, and many of these apparently formed as the upper parts of Archean oceanic crust. Most of these are interpreted to have been accreted within forearc and intra-arc tectonic settings. The observation that Archean greenstone belts have such an abundance of accreted ophiolitic fragments compared with Phanerozoic orogens suggests that thick, relatively buoyant, young Archean oceanic lithosphere may have had a rheological structure favoring delamination of the uppermost parts during subduction and collisional events.

Oceanic Plateaux Accretion

Oceanic plateaux are thicker than normal oceanic crust formed at mid-ocean ridges: they are more buoyant and relatively unsubductable, forming potential sources of accreted oceanic material to the continental crust at convergent plate boundaries. Accretion of oceanic plateaux has been proposed as a mechanism of crustal growth in a number of orogenic belts, including Archean, Proterozoic, and Phanerozoic examples. Oceanic plateaux are interpreted to form from plumes or plume heads that come from the lower mantle (D") or the 415-mile (670-km) discontinuity, and they may occur either within the interior of plates, or interact with the upper mantle convective/magmatic system and occur along mid-ocean ridges. Oceanic plateaux may be sites of komatiite formation preserved in Phanerozoic through Archean mountain belts, based on a correlation of allochthonous komatiites and high-MgO lavas of Gorgona Island, Curaçao, and in the Romeral fault zone, with the Cretaceous Caribbean oceanic plateau.

Portions of several komatiite-bearing Archean greenstone belts have been interpreted as pieces of dismembered Archean oceanic plateaux. For instance, parts of several greenstone belts in the southern Zimbabwe craton are allochthonous and show a similar magmatic sequence including a lower komatiitic unit overlain by several kilometers of tholeiitic pillow basalts. These may represent a circa 2.7 Ga oceanic plateau dismembered during a collision between the passive margin sequence developed on the southern margin of the Zimbabwe craton and an exotic crustal fragment preserved south of the suture-like Umtali line.

The accretion of oceanic plateaux and normal oceanic crust in arc environments may cause a back-stepping of the subduction zone. As the accretionary complex grows, it is overprinted by calc-alkaline magmatism as the arc migrates through the former subduction complex. Further magmatic and structural events can be caused by late ridge subduction and strike-slip segmentation of the arc. Average geochemical compositions of the continental crust, however, are not consistent with ocean plateau accretion alone.

Parts of many Archean, Proterozoic, and Phanerozoic greenstone belts interpreted as oceanic plateau fragments are overprinted by arc magmatism, suggesting that they either formed the basement of intra-oceanic island arcs, or they have been intruded by arc magmas following their accretion. Perhaps the upper and lower continental crusts have grown through the accretion of oceanic island arcs and ocean plateaus, respectively. Accreted oceanic plateaux may form a significant component of the continental crust, although most are structurally disrupted and overprinted by arc magmatism.

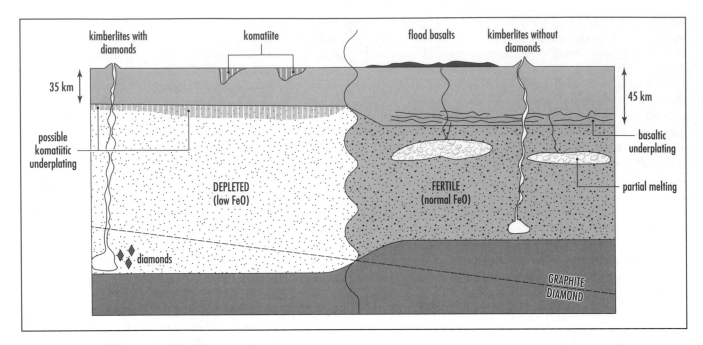

Idealized cross section of a craton showing thick mantle root

Back Arc Basin Accretion

The formation, closure, and preservation of back arc basin sequences has proven to be a popular model for the evolution of some greenstone belts. Paradoxically, the dominance of buoyant subduction styles in the Archean should have led to dominantly compressional arc systems, but many workers suggest back arc basins (which form in extensional arcs) as a modern analog for Archean greenstone belts.

Arc-Trench Migration and Accretionary Orogens; A Paradigm for the Archean

Turkic-type accretionary orogens are large, subcontinent size accretionary complexes built on one or two of the colliding continents before collision, through which magmatic arc axes have migrated, and are later displaced by strike-slip faulting. These accretionary wedges are typically built of belts of flysch, disrupted flysch, and mélange, and accreted ophiolites, plateaux, and juvenile island arcs. A. M. Celal Sengör and Boris Natal (1996) review the geology of several Phanerozoic and Precambrian orogens and conclude that Turkic or accretionary-type orogeny is one of the principal builders of continental crust with time. The record of Archean granite-greenstone terranes typically shows important early accretionary phases followed by intrusion by arc magmatism, possibly related to the migration of magmatic fronts through large accretionary complexes. In examples like the Superior Province, many subparallel belts of accreted material are located between continental fragments that are separated by many hundreds of kilometers and thus may represent large accretionary complexes that formed prior to a "Turkic-type" collision. Late-stage strike-slip faulting is important in these Archean orogens, as in the Altaids and Nipponides, and may be partly responsible for the complexity and repetition of belts of similar character across these orogens.

Turkic or accretionary-type of orogeny provides a good paradigm for continental growth. These orogenic belts possess very large sutures (up to several hundred kilometers wide) characterized by subduction-accretion complexes and arc-derived granitoid intrusions, similar to the Circum-Pacific accreted terranes (e.g., Alaska, Japan). These subduction-accretion complexes are composed of tectonically juxtaposed fragments of island arcs, back arc basins, ocean islands/plateaux, trench turbidites, and microcontinents. Turkic or accretionary-type orogens may also experience late-stage extension associated with gravitational collapse of the orogen, especially in association with late collisional events that thicken the crust in the internal parts of the orogen. In the Archean, slightly higher mantle temperatures may have reduced the possible height that mountains would have reached before the strength of deep-seated rocks was exceeded, so that extensional collapse would have occurred at crustal thickness lower than those of the younger geological record. Another important feature of these orogens is the common occurrence of orogen parallel strike-slip fault systems, resulting in lateral stacking and bifurcating lithological domains. In these respects, the accretionary-type orogeny may be considered as a unified accretionary model for the growth of the continental crust.

Late Stage Granites and Cratonization

Archean cratons are ubiquitously intruded by late to post-kinematic granitoid plutons, which may play a role in or be the result of some process that has led to the stabilization or "cratonization" of these terranes and their preservation as continental crust. Most cratons also have a thick mantle root or tectosphere, characterized by a refractory composition (depleted in a basaltic component), relatively cold temperatures, high flexural rigidity, and high shear wave velocities.

Outward growth and accretion in granite-greenstone terranes provides a framework for the successive underplating of the lower parts of depleted slabs of oceanic lithosphere, particularly if some of the upper sections of oceanic crust are off-scraped and accreted, to be preserved as greenstone belts, or eroded to form belts of graywacke turbidites. These underplated slabs of depleted oceanic lithosphere will be cold and compositionally buoyant compared with surrounding asthenosphere (providing that the basalt is offscraped and not subducted and converted to eclogite) and may contribute to the formation of cratonic roots. One of the major differences between Archean and younger accretionary orogens is that Archean subducted slabs were dominantly buoyant, whereas younger slabs were not. This may be a result of the changing igneous stratigraphy of oceanic lithosphere, resulting from a reduction in heat flow with time, perhaps explaining why Archean cratons have thick roots and are relatively undeformable compared with their younger counterparts. Geometric aspects of underplating these slabs predict that they will trap supra-subduction mantle wedges of more fertile and hydrated mantle, from which later generations of basalt can be generated.

Many granites in Archean terranes appear to be associated with crustal thickening and anatexis during late stages of collision. However, some late-stage granitoids may be a direct result of decompressional melting associated with upper-crustal extensional collapse of Archean orogens thickened beyond their limit to support thick crustal sections, as determined by the strength of deep-seated rocks. Decompressional melting generates basaltic melts from the trapped wedges of fertile mantle, which intrude and partially melt the lower crust. The melts assimilate lower crust, become more silicic in composition, and migrate upward to solidify in the mid to upper crust, as the late to post-kinematic granitoid suite. In this model, the tectosphere (or mantle root) becomes less dense (compositionally buoyant) and colder than surrounding asthenosphere, making it a stable cratonic root that shields the crust from further deformation.

Late-stage strike-slip faults that cut many Archean cratons may also play an important role in craton stabilization. Specifically the steep shear zones may provide conduits for massive fluid remobilization and escape from the subcontinental lithospheric mantle, which would both stabilize the cratonic roots of the craton and initiate large-scale granite emplacement into the mid and upper crust.

See also CHINA'S DONGWANZI OPHIOLITE; GREENSTONE BELTS; OROGEN.

Further Reading

de Wit, Maarten J., Chris Roering, Robert J. Hart, Richard A. Armstrong, C. E. J. de Ronde, Rod W. E. Green, Marian Tredoux, Ellie Peberdy, and Roger A. Hart. "Formation of an Archean Continent." *Nature* 357 (1992): 553–562.

Kusky, Timothy M. "Collapse of Archean Orogens and the Generation of Late- to Post-Kinematic Granitoids." *Geology* 21 (1993): 925–928.

Kusky, Timothy M., Jianghai Li, and Robert T. Tucker. "The Archean Dongwanzi Ophiolite Complex, North China Craton: 2.505 Billion Year Old Oceanic Crust and Mantle." *Science* 292 (2001): 1,142–1,145.

Kusky, Timothy M., and Ali Polat. "Growth of Granite-Greenstone Terranes at Convergent Margins and Stabilization of Archean Cratons." *Tectonophysics* 305 (1999): 43–73.

Kusky, Timothy M., ed., *Precambrian Ophiolites and Related Rocks, Developments in Precambrian Geology XX*. Amsterdam: Elsevier Publishers, 2003.

Li, Jianghai H., Timothy M. Kusky, and Xiongnan Huang. "Neoarchean Podiform Chromitites and Harzburgite Tectonite in Ophiolitic Melange, North China Craton, Remnants of Archean Oceanic Mantle." *GSA Today* 12, no. 7 (2002): 4–11.

Sengör, A. M. Celal, and Boris A. Natal. "Turkic-type Orogeny and Its Role in the Making of the Continental Crust." *Annual Review of Earth and Planetary Sciences* 24 (1996): 263–337.

creep The imperceptible slow downslope flowing movement of regolith (soil plus rock fragments) under the influence of gravity. It involves the very slow plastic deformation of the regolith, as well as repeated microfracturing of bedrock at nearly imperceptible rates. Creep occurs throughout the upper parts of the regolith, and there is no single surface along which slip has occurred. Creep rates range from a

The effects of creep. Regolith including gravel and boulders is slowly creeping downhill, dragging the tops of the steeply dipping beds forming a fold near the surface. *(Photo by Timothy Kusky)*

fraction of an inch per year up to about two inches per year (5 cm/yr) on steep slopes. Creep accounts for leaning telephone poles, fences, and many of the cracks in sidewalks and roads. Although creep is slow and not very spectacular, it is one of the most important mechanisms of mass wasting, and it accounts for the greatest total volume of material moved downhill in any given year. One of the most common creep mechanisms is through frost heaving. Creep through frost heaving is extremely effective at moving rocks, soil, and regolith downhill because when the ground freezes, ice crystals form and grow, pushing rocks upward perpendicular to the surface. As the ice melts in the freeze-thaw cycle, gravity takes over and the pebble or rock moves vertically downward, ending up a fraction of an inch downhill from where it started. Creep can also be initiated by other mechanisms of surface expansion and contraction, such as warming and cooling, or the expansion and contraction of clay minerals with changes in moisture levels.

In a related phenomenon, the freeze-thaw cycle can push rocks upward through the soil profile, as revealed by farmers' fields in New England and other northern climates, where the fields seem to grow boulders. The fields are cleared of rocks, and years later, the same fields are filled with numerous boulders at the surface. In these cases, the freezing forms ice crystals below the boulders that push them upward, and during the thaw cycle, the ice around the edges of the boulder melt first, and mud and soil seep down into the crack, and find their way beneath the boulder. This process, repeated over years, is able to lift boulders to the surface, keeping the northern farmer busy.

The operation of the freeze-thaw cycle makes rates of creep faster on steep slopes than on gentle slopes, and faster with more water, and greater numbers of freeze-thaw cycles. Rates of creep of up to half an inch per year are common.

See also MASS WASTING.

Cretaceous   The youngest of the three periods of the Mesozoic, during which rocks of the Cretaceous System were deposited. It ranges from 144 million years (Ma) ago until 66.4 Ma, and it is divided into the Early and Late Epochs, and 12 ages. The name derives from the Latin *creta* for chalk, in reference to the chalky terrain of England of this age.

Pangea was dispersing during the Cretaceous, and the volume of ridges plus apparently the rate of seafloor spreading were dramatically increased. The consequential displacement of seawater caused global sea levels to rise, so the Late Cretaceous was marked by high sea levels and the deposition of shallow water limestones in many epicontinental seas around the world. On the North American craton, the Zuni Sequence was deposited across wide parts of the craton during this transgression. Increased magmatic activity in the Cretaceous may reflect more rapid mantle convection or melting, as marked by a number of igneous events worldwide. The

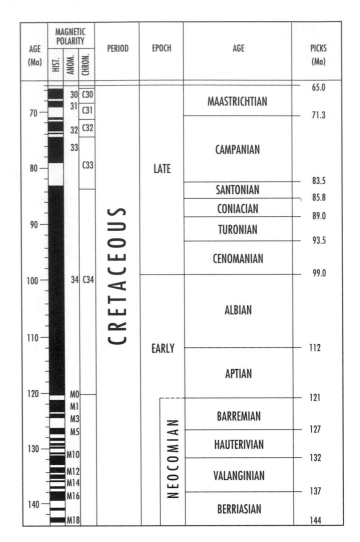

Cretaceous timescale

South American Cordillera and the western United States saw unusual amounts of intrusive and volcanic activity. The giant flood basalt provinces of Parana in South America and the Deccan of India were formed, and kimberlite pipes punctured the lithosphere of south Africa and Greenland. The dispersal of Pangea was associated with the opening of the Atlantic Ocean. Africa rotated counterclockwise away from South America, closing the Tethys Ocean in the process of opening the Atlantic. The closure of Tethys was associated with the emplacement of many ophiolites onto continents, including the giant Oman (Semail) ophiolite that was thrust to the south onto the Arabian continental margin.

Cretaceous sedimentary patterns suggest that the climate was warming through the period and was more varied and seasonal than in the earlier Mesozoic. The famous Cretaceous chalks were formed by the accumulation of tests (exoskeletons or external skeletons) of calcareous marine algae known as coccoliths, which thrived in the shallow warm seas. The chalks are in many places interbedded with fossiliferous limestones with abundant brachiopods and rudist coral fragments.

Life on the Cretaceous continents saw the development of the angiosperms, which became the planet's dominant flora by the middle of the period. Invertebrate and vertebrate species were abundant, and they included many species of dinosaurs, giant flying Pterosaurs, and giant marine reptiles. Dinosaurs occupied many different geological niches, and fossil dinosaurs are found on most continents, including herbivores, carnivores, and omnivores. Birds had appeared, including both flying and swimming varieties. Mammals remained small, but their diversity increased. Many life-forms began a dramatic and progressive disappearance toward the end of the period. These marine and land extinctions seem to be a result of a combination of events, including climate change, exhalations from the massive volcanism in the Indian Deccan South American Parana flood basalt provinces, coupled with an impact of a 6-mile (10-km) wide meteorite that hit the Yucatán Peninsula of Mexico. The extinctions were not all sudden—many of the dinosaur and other genera had gone extinct probably from climate stresses before the meteorite hit the Yucatán Peninsula. When the impact occurred, a thousand-mile-wide fireball erupted into the upper atmosphere, and tsunami hundreds or thousands of feet high washed across the Caribbean, southern North America, and much of the Atlantic. Huge earthquakes accompanied the explosion. The dust blown into the atmosphere immediately initiated a dark global winter, and as the dust settled months or years later, the extra carbon dioxide in the atmosphere warmed the Earth for many years, forming a greenhouse condition. Many forms of life could not tolerate these rapid changes and perished. The end Cretaceous extinction, commonly referred to as the K-T event, is one of the most significant mass extinction events known in the history of life.

See also FLOOD BASALT; KIMBERLITE; PANGEA.

crinoid Disk-shaped marine echinoderms with calcareous skeletons that form branching stems and columns. They are commonly known as sea lilies, even though they are animals and not plants. They have an age range of Ordovician to the present but are much more common in the fossil record than in the present. Crinoids formed rose garden-like communities in Paleozoic shallow seas and form a common fossil in many shallow water limestones. The fossils typically are dismembered, forming "life-saver" or "cheerio-like" rings, or small stem fragments. The branching, flower-like tops are rarely preserved.

cross-bedding Also known as cross-stratification; a primary sedimentary structure characterized by an arrangement of strata that are inclined to the main stratification. Cross-stratification is commonly referred to as cross-bedding, although it is formally divided into cross-lamination for inclined layers that are less than 0.4 inch (1 cm) thick, and cross-bedding for inclined layers that are thicker than 1 centimeter. Cross-beds are referred to as sets if they are part of a single related group, and cosets if they are from a different higher or lower layer in the deposit. Cross-beds and cross-laminations are produced by the migration of bed forms produced by water or air currents including current and wave ripples, and sand dunes. The largest cross-beds are produced by wind-blown sand dunes.

cross section Vertical or inclined diagrammatic slices through the upper part of the crust, showing how structures, rock units, and other features vary with depth. Cross sections are extremely useful because geologic maps provide only a two-dimensional view of three-dimensional structures. Cross sections are generally highly interpretative, based on knowledge only of the surface geology. Data from drill holes, or knowledge of the local stratigraphy, however, may help in constructing cross sections. Most cross sections are drawn perpendicular to the local or regional strike so that the dip of structures is accurately portrayed. Cross sections oriented obliquely to strike will show a dip less than the true dip of the structure.

In areas of complex structure, several parallel cross sections may be shown as fence diagrams, or block diagrams may be drawn to give a three-dimensional view of the structure. Many cross sections, particularly in tectonics, are shown with the vertical scale exaggerated relative to the horizontal scale. This has the unpleasant effect of distorting the way structures appear, gives erroneous impressions of the nature of forces involved, and can change the relative thickness of units with different dips.

See also STRUCTURAL GEOLOGY.

crust Thin, low-density rock material making up the outer layer of the solid Earth, ranging in thickness from about 3 miles (5 km) and less near the mid-ocean ridges, to more than 50 miles (70 km) beneath the tallest mountain ranges. This is followed inward by the mantle, a solid rocky layer extending to 1,802 miles (2,900 km). The outer core is a molten metallic layer extending to 3,169-mile (5,100-km) depth, and the inner core is a solid metallic layer extending to 3,958 miles (6,370 km).

The temperature increases with depth with a gradient of 55°F (30°C) per kilometer in the crust and upper mantle, and with a much smaller gradient deeper within the Earth. The heat of the Earth comes from residual heat trapped from initial accretion, radioactive decay, latent heat of crystallization of outer core, and dissipation of tidal energy of the Sun-Earth-Moon system. Heat flows out of the interior of the Earth toward the surface, through convection cells in the outer core and mantle. The top of the mantle and the crust is

Cross-bedding, Utah *(Photo by Timothy Kusky)*

a relatively cold and rigid boundary layer called the lithosphere and is about 65 miles (100 km) thick. Heat escapes through the lithosphere largely by conduction, transport of heat in igneous melts, and in convection cells of water through mid-ocean ridges.

The Earth's crust is divisible broadly into continental crust of granodioritic composition and oceanic crust of basaltic composition. Continents comprise 29.22 percent of the surface, whereas 34.7 percent is underlain by continental crust, with continental crust under continental shelves accounting for the difference. The continents are in turn divided into orogens, made of linear belts of concentrated deformation, and cratons, the stable, typically older interiors of the continents.

The distribution of surface elevation is strongly bimodal, as reflected in the hypsometric diagrams. Continental freeboard is the difference in elevation between the continents and ocean floor and results from difference in thickness and density between continental and oceanic crust, tectonic activity, erosion, sea level, and strength of continental rocks.

See also CONTINENTAL CRUST; CRATONS; LITHOSPHERE; OCEAN BASIN.

crustacean Arthropods characterized by joints and segmented bodies with an exoskeleton, with two pairs of antennae on their heads. Most crustaceans live or lived in the marine environment, although some varieties also live in freshwater and subaerial settings. They are second in abundance of individuals only to insects. They range from Early Cambrian to the present, with modern examples including shrimp, lobster, barnacles, and ostracods.

crystal Homogeneous solid structures composed of chemical elements or compounds, having a regularly repeating arrangement of atoms. All minerals are solids, and in all minerals the atoms are arranged in a very regular geometric form that is unique to that mineral. Every mineral of that species has an identical crystalline structure. It is this regular structure that gives each mineral its characteristic color, chemistry, hardness, and crystal form.

Many minerals may not have a well-developed external crystal form, but they still must have a regularly repeating crystal lattice composed of the constituent atoms. Crystal forms and the internal atomic lattice exhibit symmetry, which may be of several different varieties. There are four main types of crystal symmetry. Mirror plane symmetry is the simplest, in which the crystal can be divided by an imaginary mirror plane that would divide the crystal into two halves that are mirror-images of one another. Crystals may have symmetry about an axis that runs through the center of the crystal, in which the crystal lattice would be rotated into an identical configuration two, three, four, or, six times in a 360° circuit. These symmetry systems are known as diads, triads, tetrads, and hexads. A more complex form of symmetry is known as roto inversion, characterized by a rotational operation followed by inversion of the lattice across its center point. Finally, there is a simple inversion across the center of the crystal leading to a crystal face dia-metrically opposite to every other crystal face. Using various combinations of these symmetry operations, all crystals are found to belong to one of seven crystal systems. These include cubic, tetragonal, orthorhombic, monoclinic, triclinic, hexagonal, and trigonal.

Crystals may appear to be close to perfect but always have millions of atomic defects. These include vacancies in the crystal lattice, various types of defects in the arrangement of the atoms and lattice, and replacement of one type of atom or ion by another with a similar charge and size.

See also MINERALOGY.

cyclone   *See* HURRICANE.

D

dacite A fine-grained intermediate volcanic rock that is the extrusive equivalent of granodiorite. It is similar in composition and mineralogy to andesite, except that it has more quartz and less calcic plagioclase than andesite. Dacites are erupted in convergent margin settings including island arcs, and in greater abundance in continental margin or Andean-style arcs. They are typically associated with more silicic volcanic rocks such as rhyolite.

See also CONVERGENT PLATE MARGIN PROCESSES; IGNEOUS ROCKS.

Dana, James Dwight (1813–1895) *American Geologist* In terms of enduring scientific achievement, James Dwight Dana is one of Yale's most notable scientific figures. His contributions to geology, mineralogy, and zoology formed the basis of classification systems still in use today by scientists in these fields of study. Dana was educated at Yale, where he received scientific training from Benjamin Silliman, the prominent scientist and founder of *The American Journal of Science*. In 1836 Dana was invited to be a scientific participant of the United States Exploring Expedition, due to sail to the South Seas in 1838. Originally invited on the expedition as its geologist, Dana assumed the role of zoologist after the departure of James Couthouy in 1840. Dana produced two important monographs based on his study of animals collected by the exploring expedition. These monographs, one on corals and anemones and the other on crustaceans, were extraordinary for their sheer size, scope, and detail. Virtually no modern coral or crustacean researcher can undertake significant systematic research without encountering the legacy left by James Dana.

Further Reading

Dana, James Dwight. "On the Areas of Subsidence in the Pacific as Indicated by the Distribution of Coral Islands." *American Journal of Science* 45 (1943): 131–135.

———. "On Certain Parallel Relations between the Classes of Vertebrates, and on the Bearing of These Relations on the Question of the Distinctive Features of the Reptilian Birds." *American Journal of Science* 36 (1863): 315–321.

———. "A New Mineralogical Nomenclature." *Annals of the Lyceum of Natural History of New York* 8 (1837): 9–34.

Darcy's Law A formula that describes the flow of fluids through porous media, assuming laminar flow and that the effects of inertia are negligible. The law states that the velocity of the flow is proportional to the pressure gradient multiplied by the ratio of permeability times density, divided by the viscosity of the fluid. Mathematically, Darcy's Law may be expressed simply as $V=kS$, where V is the velocity of the flow, k is the coefficient with the same units as velocity, and S is the slope of the hydraulic gradient.

See also HYDROLOGY.

Darwin, Charles (1809–1882) *British Natural Historian, Geologist, Evolutionist* Charles Darwin was a British naturalist who is well known for his theories on evolution and natural selection. He was born in Shrewsbury, England, and for five years of his life he traveled on the H.M.S. *Beagle* to several places including South America, where he discovered fossils of extinct animals that appeared to be similar to those of modern-day animals. In the Galapagos Islands he discovered several variations of plants and animals that were similar to those in South America. When he returned to London he studied all his findings from around the world and derived the following theories: (a) evolution did occur; (b) the evolutionary change was gradual, requiring up to 3 million years to occur; (c) the main factor that contributed to evolution was natural selection; and (d) all the species of life that are present today came from a single unique life-form. Darwin's theory states that within each species, nature randomly selects which animal or plant will survive and which will die out. This is done based on how adaptable the species is to its surrounding environment. His theories were published in his book entitled *On the Origin of Species by Means of Natural Selection, or the Preservation of Favored Races in the Struggle for Life.* His book caused great controversy, especially with religious schools of thought. It was said that his theories went against the teachings of the church. Darwin did not talk about religious views in his works, and was in fact a very religious man, but other scientists after him have used his work as a basis to their own theories. Darwin continued to work in the fields of botany, geology, and zoology until his death.

Further Reading
Darwin, Charles. *The Origin of Species by Means of Natural Selection, or the Preservation of Favoured Races in the Struggle for Life,* 6th ed., edited by C. and W. Irvine. New York: Frederick Ungar Publishing Co., 1956.
———. *Voyage of a Naturalist, or Journal of Researches into the Natural History and Geology of the Countries Visited during the Voyage of H. M. S. Beagle round the World, under the Command of Capt. Fitz Roy, R. N.* New York: Harper and Brothers, 1846.

Dead Sea The lowest point on the Earth is the surface of the Dead Sea, with an elevation of 1,400 feet (420 m) below sea level. The sea has an area of about 390 square miles (1,010 km^2) and extends about 45 miles (70 km) north-south through the Jordan Valley along the Israel-Jordan border, between Ghor in the north and Wadi Arabah in the south. Steep rock cliffs bound the sea on the east and west sides, rising 2,500–4,000 feet (762–1,220 m) and creating dramatic topographic relief. Fed only by the Jordan River and a number of small intermittent streams, which are used extensively for domestic drinking and irrigation, the surface of the Dead Sea is currently dropping by 1.5–2.0 inches per year (3.8–5 cm/yr). Combined effects of tectonic subsidence and surface water overuse have led to more than 20 feet (6 m) of subsidence over 15 years. However, the Israeli and Jordanian governments have plans to cooperate on building a pipeline to bring water from the Gulf of Aqaba into the Dead Sea to replenish its dwindling water supply. The water in the Dead Sea is among the saltiest on Earth, and mineral salts such as potash and bromine are commercially extracted from its shorelines.

The Dead Sea is a salt lake formed in a tectonic depression known as a pull-apart basin, formed by extension between mismatched strands of the Dead Sea transform fault. One fault strand extends from the Gulf of Aqaba to the eastern side of the Dead Sea and steps to the west across the sea, extending to the north along the west side. Sinistral (left-lateral) motion along a left-step in a strike-slip fault produces extension in the region of the fault step, creating the topographic low that the Dead Sea is located within.

See also TRANSFORM PLATE MARGIN PROCESSES.

debris avalanche *See* MASS WASTING.

Deccan flood basalts, India The end of the Cretaceous Period at 66.4 million years ago was marked by a global mass extinction event made most famous by the death of the dinosaurs. One of the major events on the planet that contributed to the mass extinction event was the eruption of massive quantities of mafic lava through fissures in western India, forming the Deccan traps or flood basalt province. Latest Cretaceous through Eocene basalts cover more than 315 square miles (510,000 km^2) in western and central India, reaching more than 1.2 miles (2,000 m) in thickness near Mombai (Bombay). Estimates of the original size of the flood basalt province range up to 600,000 square miles (1.5 million km^2), making this one of the largest volcanic provinces on Earth. The volume of preserved basalts is roughly 12,275 cubic miles (512,000 km^3), compared with 1 cubic kilometer of material erupted during the 1980 Mount St. Helens eruption. The basalt plateau is made of at least 48 separate flows with some intervening pyroclastic layers, and locally tens to hundreds of feet of interbedded sediments. Tholeiitic basalts are by far the dominant lava type, along with minor alkaline olivine basalt, picritic basalt, mugearite, and volcanic glass (obsidian). Ultramafic to acidic plutons and a ring dike complex are reportedly associated with the traps, and minor trachyte, andesite, and granophyre are also reported. The variations in the composition of the lavas can be explained by fractional crystallization of a tholeiitic parent magma with sinking and separation of early-formed heavy minerals such as pyroxene.

See also CRETACEOUS; FLOOD BASALT; MASS EXTINCTIONS.

Further Reading
Rampino, Michael R., and Richard B. Stothers. "Flood Basalt Volcanism during the Past 250 Million Years." *Science* 241 (1988).

declination The horizontal angle at any location on the Earth that measures the difference between the magnetic north pole and the Earth's true or rotational north pole. It is one of the magnetic elements used to describe the planet's magnetic field lines, the other element being the inclination, which measures the inclination of the magnetic field lines, measured in the vertical plane.

On average the Earth's magnetic and rotational poles are coincident, but the magnetic poles migrate around the rotational poles, with the magnetic pole currently located approximately 20° from the rotational pole in the Canadian Arctic islands. Navigators must adjust their magnetic compasses to account for the declination at any point in order to account for this deviation of the magnetic pole from the rotational pole. Maps showing the declination are used for this purpose. Topographic maps show the magnetic declination and the rate and direction of change with time.

Decollement, Newfoundland *(Photo by Timothy Kusky)*

decollement A type of detachment fault system that is characterized by different styles of deformation above and below the fault plane. They are typically associated with large-scale thrusting and folding of the thrust sheet overlying the fault, with the rocks underneath the fault remaining relatively undeformed. Decollement faults originate as relatively flat lying but may become rotated to steeper attitudes in progressive or younger deformation events. Beautifully exposed decollement faults are known from the American and Canadian Rockies, the Alps, and many other mountain belts around the world.

See also STRUCTURAL GEOLOGY.

deformation of rocks Deformation of rocks is measured by three components: strain, rotation, and translation. Strain measures the change in shape and size of a rock, rotation measures the change in orientation of a reference frame in the rock, and translation measures how far the reference frame has moved between the initial and final states of deformation.

The movement of the lithospheric plates causes rocks to deform, forming mountain belts and great fault systems like the San Andreas. To describe how rocks are deformed, we use the terms *stress* and *strain*. Stress is a measure of force per unit area, and it is a property that has directions of maximum, minimum, and intermediate values. Strain is a term used to describe the changes in the shape and size of an object, and it is a result of stress.

There are three basic ways in which a solid can deform. The first is known as elastic deformation, which is a reversible deformation, like a stretching rubber band or the rocks next to a fault that bend and then suddenly snap back in place during an earthquake. Most rocks can only undergo a small amount of elastic deformation before they suffer permanent, nonreversible, nonelastic strain. Elastic deformation obeys Hooke's Law, which simply states that for elastic deformation, a plot of stress v. strain yields a straight line. In other words, strain is linearly proportional to the applied stress. So, for elastic deformation, the stressed solid returns to its original size and shape after the stress is removed.

Solids may deform through fracturing and grinding processes during brittle failure, or by flowing during ductile deformation processes. Fractures form when solids are strained beyond the elastic limit and the rock breaks, and they are permanent, or irreversible, strains. Ductile deformation is also irreversible, but the rock changes shape by flowing, much like toothpaste being squeezed out of a tube.

When compressed, rocks first experience elastic deformation, then as stress is increased they hit the yield point, at which point ductile flow begins, and eventually the rock may rupture. Many variables determine why some rocks deform by brittle failure and others by ductile deformation. These variables include temperature, pressure, time, strain rate, and composition. The higher the temperature of the rock during deformation, the weaker and less brittle the rock will be. High temperature therefore favors ductile deformation mechanisms. High pressures increase the strength of the rock, leading to a loss of brittleness. High pressures therefore hinder fracture formation. Time is also a very important factor determining which type of deformation mechanism may operate. Fast deformation favors the formation of brittle structures, whereas slow deformation favors ductile deformation mechanisms. Strain rate is a measure of how much deformation (strain) occurs over a given time interval. Finally, the composition of the rock is also important in determining what type of deformation will occur. Some minerals (like quartz) are relatively strong, whereas others (such as calcite) are weak. Strong minerals or rocks may deform by brittle mechanisms under the same (pressure, temperature) conditions that weak minerals or rocks deform by ductile flow. Water reduces the strength of virtually all minerals and rocks, therefore, the presence of even a small amount of water can significantly affect the type of deformation that occurs.

Bending of Rocks

The bending or warping of rocks is referred to as folding. Monoclines, folds in which both sides are horizontal, often form over deeper faults. Anticlines are upward-pointing arches that have the oldest rocks in the center, and synclines are downward-pointing arches, with the oldest rocks on the outside edges of the structure. There are many other geometric varieties of folds, but most are variations of these basic types. The fold hinge is the region of maximum curvature on the fold, whereas the limbs are the regions between the fold hinges. Folds may be further classified based on how tight the hinges are, which can be measured by the angle between individual fold limbs. Gentle folds have interlimb angles between 180° and 120°, open folds have interlimb angles between 120° and 70°, close folds between 70° and 30°, tight folds have interlimb angles of less than 30°, and isoclinal folds have interlimb angles of 0°. Folds may be symmetrical, with similar lengths of both fold limbs, or asymmetrical in which one limb is shorter than the other limb. Fold geometry may also be described by using the orientation of an imaginary surface (known as the axial surface), which divides the fold limbs into two symmetric parts, and the orientation of the fold hinge. Folds with vertical axial surfaces and subhorizontal hinges are known as upright gently plunging folds, whereas folds with horizontal hinges and axial surface are said to be recumbent.

Breaking of Rocks

Brittle deformation results in the breaking of rock along fractures. Joints are fractures along which no movement has occurred. These may be tectonic structures, formed in response to regional stresses, or formed by other processes such as cooling of igneous rocks. Columnar joints are common in igneous rocks, forming six-sided columns when the magma cools and shrinks.

Fractures along which relative displacement has occurred are known as faults. Most faults are inclined surfaces, and we call the block of rock above the fault the hanging wall, and the block beneath the fault the footwall, after old mining terms. Faults are classified according to the dip of the fault, and the direction of relative movement across the fault. Normal faults are faults along which the hanging wall has moved down rela-

Recumbent syncline developed in slates of the Taconic allochthon near Lake Bomoseen, Vermont *(Photo by Timothy Kusky)*

(A) Thrust fault in Swiss Alps, placing relatively flat beds over folded beds below. The fold was formed during the faulting. (B) Keystone thrust fault near Las Vegas, Nevada, placing Cambrian (black) rocks over younger Tertiary rocks in light tones. *(Photos by Timothy Kusky)*

tive to the footwall. Reverse faults are faults along which the hanging wall moves up relative to the footwall. Thrust faults are a special class of reverse faults that dip less than 45°. Strike-slip faults are steeply dipping (nearly vertical) faults along which the principal movement is horizontal. The sense of movement on strike-slip faults may be right lateral or left lateral, determined by standing on one block and describing whether the block across has moved to the right or to the left.

Regional Deformation of Rocks

Deformation of rocks occurs at a variety of scales, from the atomic to the scale of continents and entire tectonic plates. Deformation at the continental to plate-scale produces some distinctive regional structures. Cratons are large stable blocks of ancient rocks that have been stable for a long time (since 2.5 billion years ago). Cratons form the cores of many continents and represent continental crust that was formed in the Archean Era. Most are characterized by thick continental

roots made of cold mantle rocks, by a lack of earthquakes, and by low heat flow.

Orogens, or orogenic belts, are elongate regions that represent eroded mountain ranges, and they typically form belts around older cratons. They are characterized by abundant folds and faults and typically show shortening and repetition of the rock units by 20–80 percent. Young orogens are mountainous—for instance, the Rocky Mountains have many high peaks, and the slightly older Appalachians have lower peaks.

Continental shields are places where ancient cratons and mountain belts are exposed at the surface, whereas continental platforms are places where younger, generally flat-lying sedimentary rocks overlie the older shield. Many orogens contain large portions of crust that have been added to the edges of the continental shield through a mountain building process related to plate tectonics. Mountain belts may be subdivided into three basic types: fold and thrust belts, volcanic mountain chains, and fault-block ranges.

Fold and Thrust Belts

Fold and thrust mountain chains are contractional features, formed when two tectonic plates collide, forming great thrust faults and folding metamorphic rocks and volcanic rocks. By examining and mapping the structure in the belts we can reconstruct their history and essentially pull them back apart, in the reverse of the sequence in which they formed. By reconstructing the history of mountain belts in this way, we find that many of the rocks in the belts were deposited on the bottom of the ocean, or on the ocean margin deltas, and continental shelves, slopes, and rises. When the two plates collide, many of the sediments get scraped off and deformed, forming the mountain belts, thus fold and thrust mountain belts mark places where oceans have closed.

The Appalachians of eastern North America represent a fold and thrust mountain range. They show a detachment surface, or decollement, folds, and thrust faults. The sedimentary rocks in the mountain belt are like those now off the coast, so the Appalachians are interpreted to represent a place where an old ocean has closed.

Volcanic Mountain Ranges

Volcanic mountain ranges represent thick segments of crust that formed by addition of thick piles of volcanic rocks, generally above a subduction zone. Examples of volcanic mountain chains include the Aleutians of Alaska, the Fossa Magna of Japan (including Mount Fuji), and the Cascades of western United States (including Mount Saint Helens). These mountain belts are not formed primarily by deformation but by volcanism associated with subduction and plate tectonics. However, many do have folds and faults, showing that there is overlap between fold and thrust types of mountain chains and volcanic ranges.

Fault-Block Mountains

Fault-block mountains are generally formed by extension of the continental crust. The best examples include the Basin and Range Province of the western United States, and parts of the East African rift system, including the Ethiopian Afar. These mountain belts are formed by the extension or pulling apart of the continental crust, forming basins between individual tilted fault-block mountains. These types of ranges are associated with thinning of the continental crust, and some have active volcanism as well as active extensional deformation.

See also OROGENY; STRUCTURAL GEOLOGY.

Further Reading

Hatcher, Robert D. *Structural Geology, Principles, Concepts, and Problems,* 2nd ed. Englewood Cliffs, N.J.: Prentice Hall, 1995.

van der Pluijm, Ben A., and Stephen Marshak. *Earth Structure, An Introduction to Structural Geology and Tectonics.* Boston: WCB-McGraw Hill, 1997.

deltas Low flat deposits of alluvium at the mouths of streams and rivers that form broad triangular or irregular shaped areas that extend into bays, oceans, or lakes. They are typically crossed by many distributaries from the main river and may extend for a considerable distance underwater. When a stream enters the relatively still water of a lake or the ocean, its velocity and its capacity to hold sediment drop suddenly. Thus, the stream dumps its sediment load here, and the resulting deposit is known as a delta. The term *delta* was first used for these deposits by Herodotus in the 5th century B.C.E., for the triangular-shaped alluvial deposits at the mouth of the Nile River. The stream first drops the coarsest material, then progressively finer material further out, forming a distinctive sedimentary deposit. In a study of several small deltas in ancient Lake Bonneville, Grove Karl Gilbert in 1890 recognized that the deposition of finer-grained material further away from the shoreline also resulted in a distinctive vertical sequence in delta deposits. The resulting foreset layer is thus graded from coarse nearshore to fine offshore. The bottomset layer consists of the finest material, deposited far out. As this material continues to build outward, the stream must extend its length and forms new deposits, known as topset layers, on top of all this. Topset beds may include a variety of sub-environments, both subaqueous and subaerial, formed as the delta progrades seaward.

Most of the world's large rivers such as the Mississippi, the Nile, and the Ganges, have built enormous deltas at their mouths, yet all of these are different in detail. Deltas may have various shapes and sizes or may even be completely removed, depending on the relative amounts of sediment deposited by the stream, the erosive power of waves and tides, the climate, and the tectonic stability of the coastal region. The distributaries and main channel of the rivers forming deltas typically move to find the shortest route to the sea, resulting in the shifting of the active locus of deposition on deltas. Inactive areas, which may form lobes or just parts of the delta, typically subside and are reworked by tidal currents and waves. High-constructive deltas form where the fluvial transport dominates the energy balance on the delta. These deltas are typically elongate, such as the modern delta at the mouth of the Mississippi, which has the shape of a bird's foot, or they may be lobate in shape, such as the older Holocene lobes of the Mississippi that have now largely subsided below sea level.

High-destructive deltas form where the tidal and wave energy is high and much of the fluvial sediment gets reworked before it is finally deposited. In wave-dominated high-destructive deltas, sediment typically accumulates as arcuate barriers near the mouth of the river. Examples of wave-dominated deltas include the Nile and the Rhone deltas. In tide-dominated high-destructive deltas, tides rework the sediment into linear bars that radiate from the mouth of the river, with sands on the outer part of the delta sheltering a lower-energy area of mud and silt deposition inland from the segmented bars. Examples of tide-dominated deltas include the Ganges, and Kikari and Fly River deltas in the Gulf of Papua, New Guinea. Other rivers drain into the sea in places where the tidal and wave current is so strong that these systems completely overwhelm the fluvial deposition, removing most of the delta. The Orinoco River in South America has had its sediment deposits transported southward along the South American coast, with no real delta formed at the mouth of the river.

Where a coarse sediment load of an alluvial fan dumps its load in a delta, the deposit is known as a fan-delta. Braid-deltas are formed when braided streams meet local base level and deposit their coarse-grained load.

Deltas create unique and diverse environments where freshwater and saltwater ecosystems meet, and swamps, beaches, and shallow marine settings are highly varied. Deltas also form some of the world's greatest hydrocarbon fields, as the muds and carbonates make good source rocks and the sands make excellent trap rocks.

See also CONTINENTAL MARGIN.

dendrites Irregular branching patterns produced on a surface by a mineral, or in a mineral by a foreign mineral. They are commonly formed by manganese oxide on fracture or other surfaces and are also common in some kinds of agates.

dendritic drainage Characterized by random branching of streams in almost all directions, with smaller tributaries feeding into larger channels. The pattern resembles a branching tree. Streams are arranged in an orderly fashion in drainage basins, known as the stream order. Smallest segments lack tributaries and are known as first order streams, second order streams form where two first order streams converge, third order streams form where two second order streams converge, and so on. Several categories of streams reflect different geologic histories. A consequent stream is one whose course is determined by the direction of the slope of the land. A subsequent stream is one whose course has become adjusted so that it occupies a belt of weak rock or another geologic structure. An antecedent stream is one that has maintained its course across topography that is being uplifted by tectonic forces, crossing high ridges. Superposed streams are those whose course was laid down in overlying strata, onto unlike strata below. Dendritic drainage patterns form on horizontal or beveled sediments, or uniformly resistant crystalline igneous and metamorphic rocks. The drainage patterns form when the terrain attains a gentle slope.

See also DRAINAGE BASIN.

Further Reading

Gregory, K. J., and D. E. Walling. *Drainage Basin Form and Process.* New York: Halsted/John Wiley, 1978.
Ritter, Dale F., R. Craig Kochel, and Jerry R. Miller. *Process Geomorphology,* 3rd ed. Boston: WCB-McGraw Hill Publishers, 1995.

dendrochronology The study of annual growth rings on trees for dating the recent geological past. This field is closely related to dendroclimatology, which is the study of the sizes and relative patterns of tree growth rings to yield information about past climates. Tree rings are most clearly developed in species from temperate forests but not well formed in tropical regions where seasonal fluctuations are not as great. Most annual tree rings consist of two parts—early wood, consisting of widely spaced thin-walled cells, followed by late wood, consisting of thinly-spaced thick-walled cells. The changes in relative width and density of the rings for an individual species are related to changes in climate such as soil moisture, sunlight, precipitation, and temperature, and they will also reflect unusual events such as fires or severe drought stress.

The longest dendrochronology record goes back 9,000 years, using species such as the bristlecone pine, found in the southwestern United States, and oak and spruce species from Europe. To extend the record from a particular tree, it is possible to correlate rings between individuals that lived at different times in the same microenvironment close to the same location.

See also PALEOCLIMATOLOGY.

desalination A group of water treatment processes that remove salt from water. It is becoming increasingly more important as freshwater supplies dwindle and population grows on the planet, yet desalination is exorbitantly expensive and cannot be afforded by many countries. Only 6 percent of the water on the planet is freshwater, and of this 27 percent is locked up in glaciers and another 72 percent is groundwater. The remaining 1 percent of the freshwater on Earth is becoming rapidly polluted and unusable for human consumption.

There are a number of different processes that can accomplish desalination of salty water, whether it comes from the oceans or the ground. These are divided broadly into thermal processes, membrane processes, and minor techniques such as freezing, membrane distillation, and solar humidification. All existing desalination technologies require energy input to work, and they end up separating a clear fraction or stream of water from a stream enriched in concentrated salt that must be disposed of, typically by returning it to the sea.

Thermal distillation processes produce about half of the desalted water in the world. In this process, saltwater is heated or boiled to produce vapor that is then condensed to collect freshwater. There are many varieties of this technique, including processes that reduce the pressure and boiling temperature of water to effectively cause flash vaporization, using less energy than simply boiling the water. The multistage flash distillation process is the most widely used around the world. In this technique, steam is condensed on banks of tubes that carry chemically treated seawater through a series of vessels known as brine heaters with progressively lower pressures, and this freshwater is gathered for use. A technique known as multi-effect distillation has been used for industrial purposes for many years. Multi-effect distillation uses a series of vessels with reduced ambient pressure for condensation and evaporation, and it operates at lower temperatures than multistage flash distillation. Saltwater is generally preheated and then sprayed on hot evaporator tubes to promote rapid boiling and evaporation. The vapor and steam are then collected and condensed on cold surfaces, whereas the concentrated brines are run off. Vapor compression condensation is often used in combination with other processes or by itself for small-scale operations. Water is boiled, and the steam is ejected and mechanically compressed to collect freshwater.

Membrane processes operate on the principle of membranes being able to selectively separate salts from water. Reverse osmosis, commonly used in the United States, is a pressure driven process in which water is pressed through a membrane, leaving the salts behind. Electrodialysis uses electrical potential, driven by voltage, to selectively move salts through a membrane, leaving freshwater behind. Electrodialysis operates on the principle that most salts are ionic and carry an electrical charge, so they can be driven to migrate toward electrodes with the opposite charge. Membranes are built that allow passage of only certain types of ions, typically either positively (cation) or negatively (anion) charged. Direct current sources with positive and negative charge are placed on either side of the vessel, with a series of alternate cation and anion selective membranes placed in the vessel. Salty water is pumped through the vessel, the salt ions migrate through the membranes to the pole with the opposite charge, and freshwater is gathered from the other end of the vessel. Reverse osmosis only appeared technologically feasible in the 1970s. The main energy required for this process is for applying the pressure to force the water through the membrane. The salty feed water is preprocessed to remove suspended solids and chemically treated to prevent microbial growth and precipitation. As the water is forced through the membrane, a portion of the salty feed water must be discharged from the process to prevent the precipitation of supersaturated salts. Presently membranes are made of hollow fibers or spiral wound. Improvements in energy recovery and membrane technology have decreased the cost of reverse osmosis, and this trend may continue, particularly with the use of new nanofiltration membranes that can soften water in the filtration process by selectively removing Ca^{+2} and Mg^{+2} ions.

Several other processes have been less successful in desalination. These include freezing, which naturally excludes salts from the ice crystals. Membrane distillation uses a combination of membrane and distillation processes, which can operate at low temperature differentials but require large fluxes of saltwater. Solar humidification was used in World War II for desalination stills in life rafts, but these are not particularly efficient because they require large solar collection areas, have a high capital cost, and are vulnerable to weather-related damage.

Further Reading

Buros, O. K. *The ABCs of Desalting*. International Desalination Association. Topsfield, Mass.: Topsfield, 2000.
United Nations. *Non-conventional Water Resource Use in Developing Countries*. United Nations Publication no. E.87 II.A.20, 1987.

desert The driest places on Earth, deserts by definition receive less than one inch (250 mm) of rain per year. At present about 30 percent of the global landmass is desert, and the United States has about 10 percent desert areas. With changing global climate patterns and shifting climate zones, much more of the planet is in danger of becoming desert.

Most deserts are hot, with the highest recorded temperature on record being 136°F (58°C) in the Libyan Desert. With high temperatures, the evaporation rate is also high, and, in most cases, deserts are able to evaporate more than the amount of precipitation that falls as rain. Many deserts are capable of evaporating 20 times the amount of rain that falls, and some places, like much of the northern Sahara, are capable of evaporating 200–300 times the amount of rain that falls in rare storms. Deserts are also famous for large variations in the daily temperature, sometimes changing as much as 50°F–70°F (28°C–39°C) between day and night (called a diurnal cycle). These large temperature variations can be enough to shatter boulders. Deserts are also windy places and are prone to sandstorms and dust storms. The winds arise primarily because the heat of the day causes warm air to rise and expand, and other air must rush in to take its place. Airflow directions shift frequently between day and night (in response to the large temperature difference between day and night) and between any nearby water bodies, which tend to remain at a constant temperature over a 24-hour period.

There are many different types of deserts located in all different parts of the world. Some deserts are associated with patterns of global air circulation, and others form because they are in continental interiors far from any sources of moisture. Deserts may form on the "back" or leeward side of mountain ranges, where downwelling air is typically dry, or they may form along coasts where cold upwelling ocean currents lower the air temperature and decrease its ability to hold moisture. Deserts may also form in polar regions, where extremely dry and cold air has the ability to evaporate (or sublimate) much more moisture than falls as snow in any given year. There are parts of Antarctica that have not had any significant ice or snow cover for thousands of years.

Deserts have a distinctive set of landforms and hazards associated with these landforms. The most famous desert landform is a sand dune, which is a mobile accumulation of sand that shifts in response to wind. Some of the hazards in deserts are associated with sand and dust carried by the wind. Dust eroded from deserts can be carried around the globe and is a significant factor in global climate and sedimentation. Some sandstorms can be so fierce that they can remove the paint from cars, or the skin from an unprotected person. Other hazards in deserts are associated with flash floods, debris flows, avalanches, extreme heat, and extreme temperature fluctuations.

Droughts are different from deserts—a drought is a prolonged lack of rainfall in a region that typically gets more rainfall. If a desert normally gets a small amount of rainfall, and it still is getting little rainfall, then it is not experiencing a drought. In contrast, a different area that receives more rainfall than the desert may be experiencing a drought if it normally receives significantly more rainfall than it does at present. A drought-plagued area may become a desert if the drought is prolonged. Droughts can cause widespread famine, loss of vegetation, loss of life, and eventual death or mass migrations of entire populations.

Droughts may lead to conversion of previously productive lands to desert, a process called desertification. Desertification may occur if the land is stressed prior to or during the drought, typically from poor agricultural practices, overuse of ground and surface water resources, and overpopulation. Global climate goes through several different variations that can cause belts of aridity to shift back and forth with time. The Sahel region of Africa has experienced some of the more

Desertification and Climate Change

Deserts cyclically expand and contract, reflecting global environmental changes. Many civilizations on the planet are thought to have met their demise because of desertification of the lands they inhabited, and their inability to move with the shifting climate zones. Desertification is defined as the degradation of formerly productive land, and it is a complex process involving many causes, including climate change and misuse of the land. Climates may change, and land use on desert fringes may make fragile ecosystems more susceptible to becoming desert. Among civilizations thought to have been lost to the sands of encroaching deserts are several Indian cultures of the American southwest such as the Anasazi, and many peoples of the Sahel, where up to 250,000 people are thought to have perished in droughts in the late 1960s. Expanding deserts are associated with shifts in other global climate belts, and these shifts too are thought to have contributed to the downfall of several societies, including the Mycenaean civilization of Greece and Crete, the Mill Creek Indians of North America, and the Viking colony in Greenland. Many deserts are presently expanding into previously productive lands creating enormous drought and famine conditions. For instance, Ethiopia, Sudan, and other countries in the horn of Africa have suffered immensely in the past few decades with the expansion of the Sahara desert into their farmlands.

Desertification is the invasion of a desert into nondesert areas, and it is an increasing problem in the southwestern United States, in part due to human activities. Most notably, water is being moved in huge quantities into California, and people are moving into desert areas in vast numbers, all seeking water from limited groundwater supplies. This decreases water supply, vegetation, and land productivity, with the result being that about 10 percent of the lands in this country have been converted to desert in the last 100 years, while nearly 40 percent are well on the way. Desertification is also a major global problem, costing hundreds of billions of dollars per year. China estimates that the Gobi Desert alone is expanding at a rate of 950 square miles per year (2,460 km²/yr), an alarming increase since the 1950s when the desert was expanding at less than 400 square miles per year (1,035 km²/yr). The expansion of the Gobi is estimated to cost $6.7 billion a year in China and affects the livelihood of more than 400 million people through decreased crop yields and forced migrations of people from formerly productive areas to cities.

Desertification is beginning to drastically alter the distribution of agriculture and wealth on the globe. If deserts continue to expand, within a couple of hundred years the wheat belt of the central United States could be displaced to Canada, the sub-Saharan Sahel might become part of the Sahara, and the Gobi Desert may expand out of the Alashan Plateau and Mongolian Plateau.

Desertification is a multistage process, beginning with drought, crop and vegetation loss, and then establishment of a desert landscape. Drought alone does not cause desertification, but misuse of the land during drought greatly increases the chances of a stressed ecosystem reverting to desert. Desertification is associated with a number of other symptoms, including destruction of native and planted vegetation, accelerated and high rates of soil erosion, reduction of surface and groundwater resources, increased saltiness of remaining water supplies, and famine. Desertification can be accelerated by human-induced water use, population growth, and settlement in areas that do not have the water resources to sustain the exogenous population. Eastern California, Nevada, Arizona, and New Mexico all are experiencing problems related to rapidly increasing populations settling into regions with scarce water supplies, leading to desertification of fragile ecosystems.

Drought often presages the expansion of desert environments, and regions like Africa's sub-Saharan Sahel have experienced periods of drought and desert expansion and contraction, several times in the past few tens of thousands of years. At present much of the Sahara is expanding southward, and peoples of the Sahel have suffered immensely.

Droughts may begin imperceptibly, with seasonal rains often not appearing on schedule. Farmers and herdsmen may be waiting for the rains to water their freshly planted fields and to water their flocks, but the rains do not appear. Local water sources such as streams, rivers, and springs may begin to dry up until soon only deep wells are able to extract water out of groundwater aquifers. This is typically not enough to sustain crops and livestock, so they begin to be slaughtered or die of starvation and dehydration. Crops do not grow, and natural vegetation begins to dry up and die. Brushfires often come next, wiping away the dry brush. Soon people start to become weak, and they cannot manage to walk out of the affected areas, so they stay and the weak, elderly, and young of the population may die off. Famine and disease may follow, killing even more people.

One recent example of this is highlighted by the Sudan, where years of drought have exacerbated political and religious unrest, and opposing parties raid Red Cross relief supplies, sabotage the other side's attempts at establishing aid and agriculture, and the people suffer. The Sudan is in the sub-Saharan Sahel, which is a large region between about 14° and 18° N latitude, characterized by scrubby grasslands, getting on average 14–23 inches (36–58 cm) of rain per year. In the late 1960s this amount of rainfall had fallen to about half of its historical average, and the 25 million people of the Sahel began suffering. One of the unpleasant aspects of human nature is that slow-moving, long-lasting disasters like drought tend to bring out the worst in many people. War and corruption often strike drought-plagued regions once relief and foreign aid begins to bring outside food sources into regions. This food may not be enough to feed the whole population, so factions break off and try to take care of their own people. By 1975 about 200,000 people had died, millions of herd animals were dead, and crops and the very structure of society in many Sahel countries were ruined. Children were born brain-damaged because of malnutrition and dehydration, and corruption had set in. Since then the region has been plagued with continued sporadic drought, but the infrastructure of the region has not returned and the people continue to suffer.

severe droughts in recent times. The Middle East and parts of the desert southwest of the United Sates are overpopulated and the environment is stressed. If major droughts occur in these regions, major famines could result and the land may be permanently desertified.

Location and Formation of Deserts

More than 35 percent of the land area on the planet is arid or semiarid, and these deserts form an interesting pattern on the globe that reveals clues about how they form. There are six main categories of desert, based on their geographic location with respect to continental margins, oceans, and mountains.

Trade Wind or Hadley Cell Deserts

Many of the world's largest and most famous deserts are located in two belts between 15° and 30°N and S latitude. Included in this group of deserts are the Sahara, the world's largest desert, and the Libyan Desert of North Africa. Other members of this group include the Syrian Desert, Rub a'Khali (Empty Quarter), and Great Sand Desert of Arabia; the Dasht-i-Kavir, Lut, and Sind of southwest Asia; the Thar Desert of Pakistan; and the Mojave and Sonoran Deserts of the United States. In the Southern Hemisphere, deserts that fall into this group include the Kalahari Desert of Africa and the Great Sandy Desert of Australia, and this effect contributes to the formation of the Atacama Desert (South America), one of the world's driest places.

The location of these deserts is controlled by a large-scale atmospheric circulation pattern driven by energy from the Sun. The Sun heats equatorial regions more than high-latitude areas, which causes large-scale atmospheric upwelling near the equator. As this air rises, it becomes less dense and can hold less moisture, which helps form large thunderstorms in equatorial areas. This drier air then moves away from the equator at high altitudes, cooling and drying more as it moves, until it eventually forms two circum-global downwelling belts between 15 and 30°N and S latitude. This cold downwelling air is dry and has the ability to hold much more water than it has brought with it on its circuit from the equator. These belts of circulating air are known as Hadley Cells and are responsible for the formation of many of the world's largest, driest deserts. As this air completes its circuit back to the equator, it forms dry winds that heat up as they move toward the equator. The dry winds dissipate existing cloud cover and allow more sunlight to reach the surface, which then warms even more.

Deserts formed by global circulation patterns are particularly sensitive to changes in global climate, and seemingly small changes in the global circulation may lead to catastrophic expansion or contraction of some of the world's largest deserts. For instance, the sub-Saharan Sahel has experienced several episodes of expansion and contraction of the Sahara, displacing or killing millions of people in this vicious cycle. When deserts expand, croplands are dried up, and livestock and people can not find enough water to survive. Desert expansion is the underlying cause of some of the world's most severe famines.

Continental Interior/Mid-Latitude Deserts

Some places on Earth are so far from ocean moisture sources that by the time weather systems reach them, most of the moisture they carry has already fallen. This effect is worsened if the weather systems have to rise over mountains or plateaus to reach these areas, because cloud systems typically lose moisture as they rise over mountains. These remote areas therefore have little chance of receiving significant rainfalls. The most significant deserts in this category are the Taklimakan-Gobi region of China, resting south of the Mongolian steppe on the Alashan Plateau, and the Karakum of western Asia. The Gobi is the world's northernmost desert, and it is characterized by 1,000-foot (305-m) high sand dunes made of coarser-than-normal sand and gravel, built up layer-by-layer by material moved and deposited by the wind. It is a desolate region, conquered successively by Genghis Khan, warriors of the Ming Dynasty, then the People's Army of China. The sands are still littered with remains of many of these battles, such as the abandoned city of Khara Khoto. In 1372 Ming Dynasty warriors conquered this walled city by cutting off its water supply consisting of the Black River, waiting, then massacring any remaining people in the city.

Rainshadow Deserts

A third type of desert is found on the leeward (or back) side of some large mountain ranges, such as the sub-Andean Patagonian Gran Chaco and Pampas of Argentina, Paraguay, and Bolivia. A similar effect is partly responsible for the formation of the Mojave and Sonoran deserts of the United States. These deserts form because as moist air masses move toward the mountain ranges they must rise to move over the ranges. As the air rises it cools, and cold air can hold less moisture than warm air. The clouds thus drop much of their moisture on the windward side of the mountains, explaining why places like the western Cascades and western Sierras of the United States are extremely wet, as are the western Andes in Peru. However the eastern lee sides (or back sides) of these mountains are extremely dry. The reason for this is that as the air rose over the fronts or windward sides of the mountains, it dropped its moisture as it rose. As the same air descends on the lee side of the mountains it gets warmer and is able to hold more moisture than it has left in the clouds. The result it that the air is dry and it rarely rains. This explains why places like the eastern sub-Andean region of

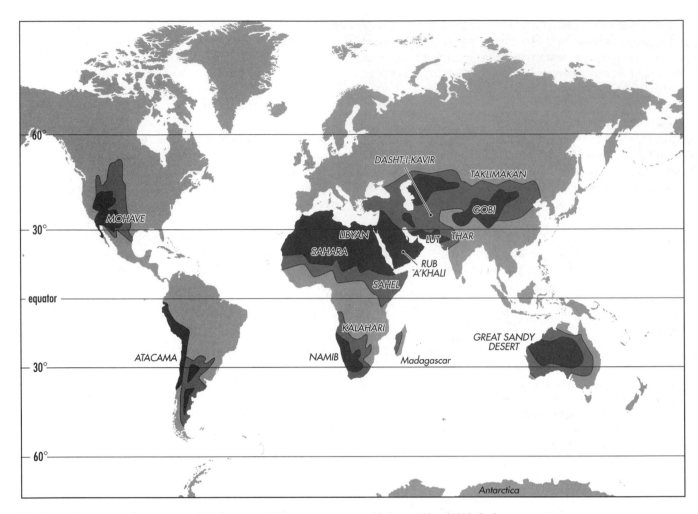

World map showing locations of deserts. Note how most deserts are concentrated between 20° and 40° latitude.

South America and the Sonoran and Mojave deserts of the western United States are extremely dry.

Rainshadow deserts tend to be mountainous because of the way they form, and they are associated with a number of mass wasting hazards such as landslides, debris flows, and avalanches. Occasional rainstorms that make it over the blocking mountain ranges can drop moisture in the highlands, leading to flash floods coming out of mountain canyons into the plains or intermountain basins on the lee side of the mountains.

Coastal Deserts

There are some deserts that are located along coastlines, where intuition would seem to indicate that moisture should be plentiful. However, the driest place on Earth is the Atacama Desert, located along the coast of Peru and Chile. The Namib Desert of southern Africa is another coastal desert, which is known legendarily as the Skeleton Coast, because it is so dry that many of the large animals that roam out of the

more humid interior climate zones perish there, leaving their bones sticking out of the blowing sands.

How do these coastal deserts form adjacent to such large bodies of water? The answer lies in the ocean currents, for in these places cold water is upwelling from the deep ocean, which cools the atmosphere. The effect is similar to rainshadow deserts, where cold air can hold less moisture, and the result is a lack of rain.

Monsoon Deserts

In some places on the planet seasonal variations in wind systems bring alternating dry and wet seasons. The Indian Ocean is famous for its monsoonal rains in the summer as the southeast trade winds bring moist air on shore. However, as the moisture moves across India it loses moisture and must rise to cross the Aravalli Mountain Range. The Thar Desert of Pakistan and the Rajasthan Desert of India are located on the lee side of these mountains and do not generally receive this seasonal moisture.

Polar Deserts

A final class of deserts is the polar desert, found principally in the Dry Valleys and other parts of Antarctica, parts of Greenland, and northern Canada. Approximately 3 million square miles (7.8 million km²) on Earth consist of polar desert environments. In these places, cold downwelling air lacks moisture, and the air is so dry that the evaporation potential is much greater than the precipitation. Temperatures do not exceed 50°F (10°C) in the warmest months, and precipitation is less than one inch per year. There are places in the Dry Valleys of Antarctica that have not been covered in ice for thousands of years.

Polar deserts are generally not covered in sand dunes but are marked by gravel plains or barren bedrock. Polar deserts may also have landforms shaped by frost wedging, where alternating freeze-thaw cycles allow small amounts of water to seep into cracks and other openings in rocks. When the water freezes it expands, pushing large blocks of rock away from the main mountain mass. In polar deserts and other regions affected by frost wedging large talus slopes may form adjacent to mountain fronts, and these are prone to frequent rock falls from frost wedging.

See also SAND DUNES; SAND SEA.

Further Reading

Abrahams, Athol D., and Anthony J. Parsons. *Geomorphology of Desert Environments*. London: Chapman and Hall, 1994.

Bagnold, Ralph A. *The Physics of Blown Sand and Desert Dunes.* London: Methuen, 1941.

McKee, E. D., ed., *A Study of Global Sand Seas*. U.S. Geological Survey Professional Paper 1052, 1979.

Walker, A. S. "Deserts: Geology and Resources." *U.S. Geological Survey Publication* 60 (1996): 421–577.

Webster, D. "Alashan, China's Unknown Gobi." *National Geographic* (2002): 48–75.

Devonian

Devonian   The fourth geological period in the Paleozoic Era, ranging from 408 million to 360 million years ago. It is named after exposures in Devonshire in southwest England and was first described in detail by Adam Sedgwick and Roderick I. Murchison in 1839. The Devonian is divided into three series and seven stages based on its marine fauna.

Devonian rocks are known from all continents and reflect the distribution of the continents grouped into a large remnant Gondwanan fragment in the Southern Hemisphere, and parts of Laurasia (North America and Europe), Angaraland (Siberia), China, and Kazakhstania in the Northern Hemisphere. The eastern coast of North America and adjacent Europe experienced the Acadian orogeny, formed in response to subduction and eventual collision between Avalonian fragments and ultimately Africa with Laurasia. Other orogenies affected North China, Kazakhstania, and other fragments. These mountain building events shed large clastic wedges, including the Catskill delta in North America and the Old Red sandstone in the British Isles.

The Devonian experienced several global (eustatic) sea level changes and had times of glaciation. There appeared to be a strong climatic gradation with tropical and monsoonal conditions in equatorial regions and cold water conditions in more polar regions.

Marine life in the Devonian was prolific, with brachiopods reaching their peak. Rugose and tabulate corals, stromatoporoids, and algae built carbonate reefs in many parts of the world, including North America, China, Europe, North Africa, and Australia. Crinoids, trilobites, ostracods, and a variety of bivalves lived around the reefs and in other shallow water environments, whereas calcareous foraminifera and large ammonites proliferated in the pelagic realm. The pelagic conodonts peaked in the Devonian, and their great variety, widespread distribution, and rapid changes make them useful biostratigraphic markers and form the basis for much of the biostratigraphic division of Devonian time. Bony fish evolved in the Devonian and evolved into tetrapod amphibia by the end of the period.

The land was inhabited by primitive plants in the Early Devonian, but by the middle of the period great swampy forests with giant fern trees (Archaeopteris) and spore-bearing organs populated the land. Insects, including some flying varieties, were found in these swamps.

The end of the Devonian saw widespread mass extinction of some marine animal communities, including brachiopods, trilobites, conodonts and corals. The cause of this extinction is not well known, with models including cooling caused by a southern glaciation, or a meteorite impact.

See also APPALACHIANS; PALEOZOIC.

Dewey, John F.

Dewey, John F. (1937–) British *Tectonicist, Structural Geologist*   John F. Dewey received his Ph.D. from the University of London in 1960. He is a distinguished professor and a member of the National Academy of Sciences. John Dewey was one of the early scientists to pioneer the development of the plate tectonic paradigm, offering explanations of the Appalachian mountain belt in terms of plate tectonics and later expanding his studies to a global scale. He presented plate tectonic concepts in a kinematic framework, clearly describing many phenomena for the first time. His basic interests and knowledge are in structural geology and tectonics from the small-scale materials science of deformed rocks to the large-scale origin of topography and structures. Ongoing field-based research is on the rock fabrics and structures of transpression and transtension, especially in California, New Zealand, Norway, Ireland, and Newfoundland. Evolving interests are in the neotectonics of California and Nevada in the relationship between faulting, topography, and sediment provenance, yield, and distribution. Derivative interests are in the geohazard of volcanoes, earthquakes, and landslides.

Further Reading
Bird, John M., and John F. Dewey. "Lithosphere Plate-Continental Margin Tectonics and the Evolution of the Appalachian Orogen." *Geological Society of America Bulletin* 81 (1970): 1,031–1,059.
Dewey, John F., and John M. Bird. "Mountain Belts and the New Global Tectonics." *Journal of Geophysical Research* 75 (1970): 2,625–2,647.
Dewey, John F., Michael J. Kennedy, and William S. F. Kidd. "A Geotraverse through the Appalachians of Northern Newfoundland." *Geodynamics Series* 10 (1983): 205–241.

diagenesis A group of physical and chemical processes that affect sediments from the time when they are deposited until deformation and metamorphism begin to set in. It therefore occurs at low temperature (T) and pressure (P) conditions, with its upper PT limit defined as when the first metamorphic minerals appear. Diagenesis typically changes the sediment from a loose unconsolidated state to a rock that is cemented, lithified, or indurated.

The style of diagenetic changes in a sediment are controlled by several factors other than pressure and temperature, including grain size, rate of deposition, composition of the sediment, environment of deposition, nature of pore fluids, porosity and permeability, and the types of surrounding rocks. One of the most important diagenetic processes is dewatering, or the expulsion of water from the pore spaces by the weight of overlying, newly deposited sediments. These waters may escape to the surface or enter other nearby more porous sediments, where they can precipitate or dissolve soluble minerals. Compaction and reduction of the thickness of the sedimentary pile result from dewatering of the sediments. For instance, many muds may contain 80 percent water when they are deposited, and compaction is able to rearrange the packing of the constituent mineral grains to reduce the water-filled pore spaces to about 10 percent of the rock. This process results in the clay minerals being aligned, forming a bedding-plane parallel layering known as fissility. Organic sediments also experience large amounts of compaction during dewatering, whereas other types of sediments including sands may experience only limited compaction. Sands typically are deposited with about 50 percent porosity and may retain about 30 percent even after deep burial. The porosity of sandstone is reduced by the pressing of small grains into the pore spaces between larger grains and the addition of cement.

Chemical processes during diagenesis are largely controlled by the nature of the pore fluids. Fluids may dissolve or more commonly add material to the pore spaces in the sediment, increasing or decreasing pore space, respectively. These chemical changes may occur in the marine realm, or in the continental realm with freshwater in the pore spaces. Chemical diagenetic processes tend to be more effective at the higher PT end of the diagenetic spectrum when minerals are more reactive and soluble.

Organic material experiences special types of diagenesis, as bacteria aid in the breakdown of the organic sediments to form kerogen and release methane and carbon dioxide gas. At higher diagenetic temperatures, kerogen breaks down to yield oil and liquid gas. Humus and peat are progressively changed into soft brown coal, hard brown coal, and then bituminous coal during a diagenetic process referred to as coalification. This increases the carbon content of the coal and releases methane gas in the process.

Most sandstones and coarse-grained siliciclastic sediments experience few visible changes during diagenesis, but they may experience the breakdown of feldspars to clay minerals and see an overall reduction in pore spaces. The pore spaces in sandstones may become filled with cements such as calcite, quartz, or other minerals. Cements may form at several times in the diagenetic process. Carbonates are very susceptible to diagenetic changes and typically see early and late cements, and many are altered by processes such as replacement by silica, dolomitization, and the transformation of aragonite to calcite. Carbonates typically show an interaction of physical and chemical processes, with the weight of overlying sediments forming pressure solution surfaces known as stylolites, where grains are dissolved against each other. These stylolites are typically crinkly or wavy surfaces oriented parallel to bedding. The material that is dissolved along the stylolites then is taken in solution and expelled from the system or more commonly reprecipitated as calcite or quartz veins, often at high angles to the stylolites reflecting the stresses induced by the weight of the overburden.

See also COAL; HYDROCARBON; STRUCTURAL GEOLOGY.

diamond Stable crystalline forms of pure carbon that crystallize as isometric tetrahedral crystals. Diamonds form only at high pressures in cool locations in the Earth's mantle and thus are restricted in their genesis to places in the subcontinental mantle where these conditions exist, at 90–125 miles (150–200 km) depth. Most diamonds that make their way back to the surface are brought up from these great depths by rare explosive volcanic eruptions known as kimberlites. More rarely small diamonds, often as inclusions in other minerals, are exposed along major thrust faults that have exposed rocks from the mantle.

Diamonds are the hardest substance known and are widely used as a gemstone. Uncut varieties may show many different crystal shapes and many show striated crystal faces. They have concoidal fracture, have a greasy luster, and may be clear, yellow, red, orange, green, blue, brown, or even black. Triangular depressions are common on some crystals, and others may form elongate or even pear-shaped forms. Diamonds have been found in alluvial deposits such as gravel, and some mines have been located by tracing the source of the gravel back to the kimberlite where the diamonds were brought to the surface. Some diamond-mining operations such as those of the Vaal River, South Africa (discovered in 1867), proceeded for many years before it was recognized that the source was in nearby kimberlites.

Dating the age of formation of small mineral inclusions in diamonds has yielded some very important results. All dated diamonds from the mantle appear to be Precambrian in age, with one type being up to 3.2 billion years old, and another type being 1.0–1.6 billion years old. Since diamonds form at high pressures and low temperatures, their very existence shows that the temperature deep in the Earth beneath the continents in the Precambrian was not much hotter than it is today. The diamonds were stored deep beneath the continents for billions of years before being erupted in the kimberlite pipes.

See also KIMBERLITE; PRECAMBRIAN.

diamond anvil press and multi-anvil press The diamond anvil press and multi-anvil press are scientific equipment that re-create the high pressures found deep within and at the center of the Earth and in other celestial bodies. They provide scientists with indirect ways to study the conditions and mineral phases present deep in the Earth. The diamond anvil press is the smaller of the two and is only able to put small amounts of material under about 100 GPa of pressure, or roughly about the pressure found at the center of the Earth. The multi-anvil press is a larger press and can accommodate more material but can produce far less pressure (about 11 GPa) compared to the diamond anvil press. The diamond anvil press can experimentally produce greater amounts of pressure, but the multi-anvil press is more widely used because it can accommodate greater amounts of material that can be studied after the experiment. Many of the diamond and multi-anvil presses can also pass electric current through the samples that reproduce higher temperatures found inside the Earth.

Mineral physicists mainly use olivine, a silicate mineral, to represent the upper mantle and also iron and nickel combinations to represent the core in the high pressure experiments. The diamond anvil press and the multi-anvil press have allowed scientists to learn about how the core and the mantle separated. They have also been used to study the phases and stabilities of minerals at high pressure, earthquakes that occur deep inside the Earth, and the properties of other materials at high pressure and temperatures.

diapir Intrusive bodies of relatively buoyant rocks that produce an upwarp, a fold, or penetrate through surrounding denser rocks. They may be made of sedimentary, igneous, or even metamorphic rocks.

In sedimentary sequences, diapirs are typically made of salt or shale that became mobilized and buoyant when buried under thick piles of younger strata. Most salts and shales of this category were initially formed as sedimentary layers but intruded upward to form diapirs when the weight of overlying rocks became sufficient to mobilize the salt and shale deposits. The salts typically become buoyant relative to the surrounding rocks when they reach a burial depth of about 1,450–3,250 feet (450–1,000 m), but it takes some triggering mechanism (such as shaking on a fault) to initiate the movement of the salt, referred to as diapirism. In a few places, such as the Great Kavir in the Zagros Mountains of Iran, salt diapirs have emerged at the surface and have formed great salt glaciers. Salt diapirs move by ductile flow forming complex and strongly attenuated folds. Externally, most salt diapirs form mushroom shapes with thin necks and expanded heads. These are referred to as salt domes. Some salt diapirs have moved very far, even tens of kilometers from their area of deposition, forming large sheets that are said to be allochthonous (far traveled). Many other shapes of salt and shale diapirs are known. Salt and shale diapirs typically form folds and make permeability barriers that form economically significant petroleum traps, so a great deal of seismic exploration for hydrocarbons has been done around salt domes and diapirs.

Diapirs may also be composed of igneous rocks that intrude surrounding country rocks. Igneous diapirs may be isolated bodies or internal parts of plutons. Some form shapes similar to salt and shale diapirs, but internal structures are more difficult to recognize because they do not have as many marker horizons as sedimentary salts and shales. Some metamorphic gneiss domes may also rise diapirically, although this is disputed because similar patterns can be formed by fold interference patterns in gneiss terrains.

See also CONTINENTAL CRUST; PLUTON; STRUCTURAL GEOLOGY; ZAGROS AND MAKRAN MOUNTAINS.

diatom Microscopic single-celled plants found in the oceans and freshwater bodies. They have shells (also known as tests) made of silica that form thick accumulations known as siliceous oozes in quiet marine environments. Many diatoms have distinctive markings on their tests known as frustules, and many varieties are used as index fossils to determine the age of ancient deposits.

diatreme *See* KIMBERLITE.

digital elevation models A digital elevation model (DEM) is a digital file made up of topographic elevations for known ground positions at uniform horizontal intervals. The U.S. Geological Survey is responsible for five separate digital elevation models. The five models are all kept in the same matter of data, but the intervals, geographic references, coverage areas, and accuracies are different. The models include the following:

1. 7.5 Minute DEM 30 × 30 meter data intervals
2. 1-Degree DEM 3 × 3 arc second data intervals

3. 2-Arc-Second DEM 2 × 2 arc second data intervals
4. 15-Minute Alaska DEM 2 × 3 arc second data intervals
5. 7.5-Minute Alaska DEM 1 × 2 arc second data intervals

The DEMs are used for three-dimensional graphics that display certain terrain characteristics. They provide scientists with a computer application that allows them to view the terrain in great detail. DEMs have also been used for many nongraphic applications such as the administration and exploration of natural energy resources. The U.S. Geological Survey is currently planning to convert all DEMs to Spatial Data Transfer Standard (SDTS), which is a widely used universal system used to transfer data between separate and different computer operating systems.

disconformity   *See* UNCONFORMITY.

divergent or extensional boundaries   The world's longest mountain chain is the mid-ocean ridge system, extending 25,000 miles (40,000 km) around the planet, representing places where two plates are diverging, and new material is upwelling from the mantle to form new oceanic crust and lithosphere. These mid-ocean ridge systems are mature extensional boundaries, many of which began as immature extensional boundaries in continents, known as continental rifts. Some continental rift systems are linked to the world rift system in the oceans and are actively breaking continents into pieces. An example is the Red Sea–East African rift system. Other continental rifts are accommodating small amounts of extension in the crust and may never evolve into oceanic rifts. Examples of where this type of rifting occurs on a large scale include the Basin and Range Province of the western United States and Lake Baikal in Siberia.

Divergent Plate Boundaries in Continents
Rifts are elongate depressions formed where the entire thickness of the lithosphere has ruptured in extension. These are places where the continent is beginning to break apart and, if successful, may form a new ocean basin. The general geomorphic feature that initially forms is known as a rift valley. Rift valleys have steep, fault-bounded sides, with rift shoulders that typically tilt slightly away from the rift valley floor. Drainage systems tend to be localized internal systems, with streams forming on the steep sides of the rift, flowing along the rift axis, and draining into deep narrow lakes within the rift. If the rift is in an arid environment, such as much of East Africa, the drainage may have no outlet and the water will evaporate before it can reach the sea. Such evaporation leaves distinctive deposits of salts and evaporite minerals, one of the hallmark deposits of continental rift settings. Other types of deposits in rifts include lake sediments in rift centers and conglomerates derived from rocks exposed along the rift shoulders. These sediments may be interleaved with volcanic rocks, typically alkaline in character and bimodal in silica content (i.e., basalts and rhyolites).

Modes of Extension
There are three main end-member models for the mechanisms of extension and subsidence in continental rifts. These are the pure shear (McKenzie) model, the simple shear (Wernicke) model, and the dike injection (Royden and Sclater) model.

In the pure shear model, the lithosphere thins symmetrically about the rift axis, with the base of the lithosphere (defined by the 2,425°F (1,330°C) isotherm) rising to 10–20 miles (15–30 km) below the surface. This causes high heat flow and high geothermal gradients in rifts and is consistent with many gravity measurements that suggest an excess mass at depth (this would correspond to the denser asthenosphere near the surface). Stretching mechanisms in the pure shear model include brittle accommodation of stretching on normal faults near the surface. At about 4 miles (7 km) depth, the brittle ductile transition occurs, and extension below this depth is accommodated by shear on mylonite and ductile shear zones.

In the simple shear model, an asymmetric detachment fault penetrates thickness of the lithosphere, dipping a few degrees forming a system of asymmetric structures across the rift. A series of rotated fault blocks may form where the detachment is close to the surface, whereas the opposite side of the rift (where the lithosphere experiences the most thinning) may be dominated by volcanics. Thermal effects of the lithospheric thinning typically dome the detachment fault up. This model explains differences on either side of rifts, such as faulted and volcanic margins now on opposite continental margins (conjugate margins) of former rifts that have evolved into oceans. These asymmetric detachments have been observed in seismic reflection profiles.

The dike injection model for rifts suggests that a large number of dense mafic dikes (with basaltic composition) intrude the continental lithosphere in rifts, causing the lithosphere to become denser and to subside. This mechanism does not really explain most aspects of rifts, but it may contribute to the total amount of subsidence in the other two models.

In all of these models for initial extension of the rift, initial geothermal gradients are raised, and the isotherms become elevated and compressed beneath the rift axis. After the initial stretching and subsidence phases, the rift either becomes inactive or evolves into a mid-ocean ridge system. In the latter case, the initial shoulders of the rift become passive continental margins. Failed rifts and passive continental margins both enter a second, slower phase of subsidence related to the gradual recovery of the isotherms to their deeper, prerifting levels. This process takes about 60 million years and typically forms a broad basin over the initial rifts, characterized by no active faults, no volcanism, and rare lakes. The transition from initial stretching with coarse clastic sediments

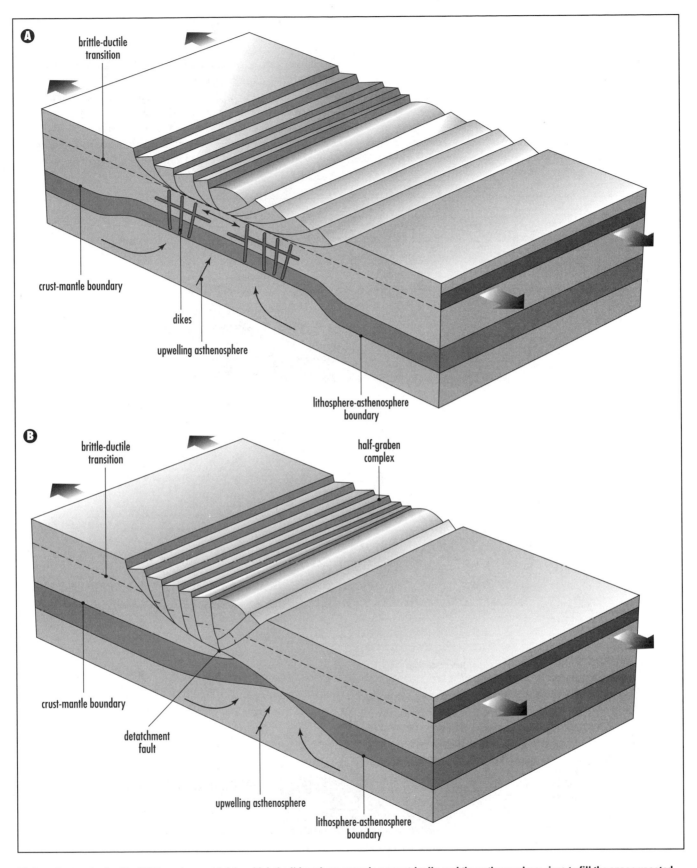

Modes of extension in rifts. (A) Pure shear model, in which the lithosphere extends symmetrically, and the asthenosphere rises to fill the space vacated by the extending lithosphere. (B) Simple shear or asymmetric rifting, where a shallow-dipping detachment fault penetrates the thickness of the lithosphere, and the asthenosphere rises asymmetrically on the side of the rift where the fault enters the asthenosphere. Faulting patterns are also asymmetric, with different styles on either side of the rift.

and volcanics to the thermal subsidence phase is commonly called the "rift to drift" transition on passive margins.

Divergent Margins in the Oceans:
The Mid-Ocean Ridge System

Some continental rifts may evolve into mid-ocean ridge spreading centers. The world's best example of where this transition can be observed is in the Ethiopian Afar, where the East African continental rift system meets juvenile oceanic spreading centers in the Red Sea and Gulf of Aden. Three plate boundaries meet in a wide plate boundary zone in the Afar, including the African/Arabian boundary (Red Sea spreading center), the Arabian/Somalian boundary (Gulf of Aden spreading center), and the African/Somalian boundary

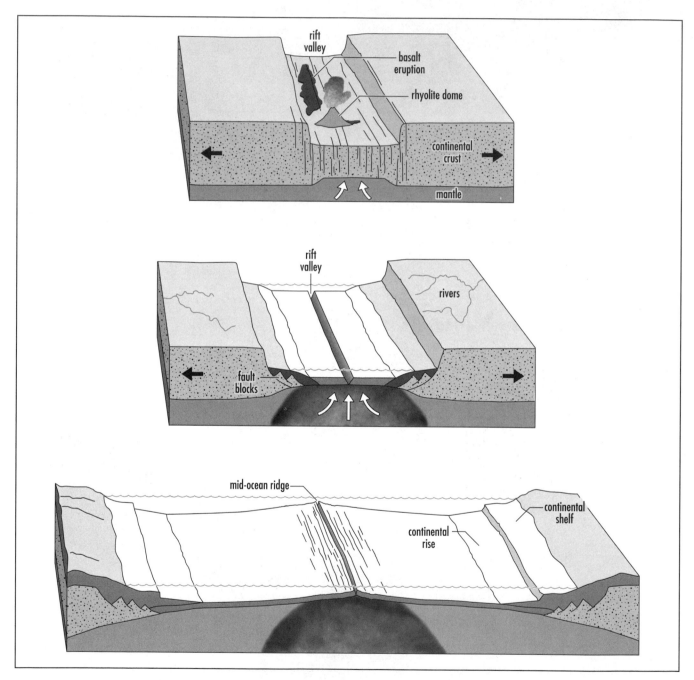

Figure showing simplified three-stage evolution of divergent margins. Young rift valley stage like that in the East African rift system has steep rift shoulders and basaltic and rhyolitic volcanoes. Young ocean stage, similar to the modern Red Sea, has seafloor spreading and steep rift shoulders. Mature ocean stage is like the modern Atlantic ocean, with thick passive margin sequences developed on continental edges around wide ocean basin.

(East African rift). The boundary is a complex system known as an RRR (rift-rift-rift) triple junction. The triple junction has many complex extensional structures, with most of the Afar near sea level, and isolated blocks of continental crust such as the Danakil horst isolated from the rest of the continental crust by normal faults.

The Red Sea has a juvenile spreading center similar in some aspects to the spreading center in the middle of the Atlantic Ocean. Geologists recognize two main classes of oceanic spreading centers, based on geomorphology and topography. These types are found to be related to spreading rate, with slow spreading rates, 0.2–0.8 inch per year (0.5–2 cm/yr), on Atlantic-type ridges, and faster rates, generally 1.5–3.5 inches per year (4–9 cm/yr), on Pacific-type ridges.

Atlantic-type ridges are characterized by a broad, 900–2,000-mile (1,500–3,000-km) wide swell in which the seafloor rises 0.6–1.8 miles (1–3 km) from abyssal plains at 2.5 miles (4.0 km) below sea level to about 1.7 miles (2.8 km) below sea level along the ridge axis. Slopes on the ridge are generally less than 1°. Slow or Atlantic-type ridges have a median rift, typically about 20 miles (30 km) wide at the top to 0.6–2.5 miles (1–4 km) wide at the bottom of the 1-kilometer deep medial rift. Many constructional volcanoes are located along the base and inner wall of the medial rift. Rugged topography and many faults forming a strongly block-faulted slope characterize the central part of Atlantic-type ridges.

Pacific-type ridges are generally 1,250–2,500 miles (2,000–4,000 km) wide and rise 1.2–1.8 miles (2–3 km) above the abyssal plains, with 0.1° slopes. Pacific-type ridges have no median valley but have many shallow earthquakes, high heat flow, and low gravity in the center of the ridge, suggesting that magma may be present at shallow levels beneath the surface. Pacific-type ridges have much smoother flanks than Atlantic-type ridges.

The high topography of both types of ridges shows that they are isostatically compensated, that is, they are underlain by low-density material and are floating on this hot substrate. New magma upwells beneath the ridges and forms small magma chambers along the ridge axis. The magma in these chambers crystallizes to form the rocks of the oceanic crust that gets added (in approximately equal proportions), to both diverging plates. The crust formed at the ridges is young, hot, and relatively light, so it floats on the hot underlying asthenosphere. As the crust ages and moves away from the ridge it becomes thicker and denser, and subsides, explaining the topographic profile of the ridges. The rate of thermal subsidence is the same for fast- and slow-spreading ridges (a function of the square root of the age of the crust), explaining why slow-spreading ridges are narrower than fast-spreading ridges.

The centers of the mid-ocean ridges are characterized by abundant volcanoes, with vast outpourings of basaltic lava. The lavas are typically bulbous-shaped forms called pillows, as well as tubes and other more massive flows. The ridge axes are also characterized by very high heat flow, with many thermal vents marking places where seawater has infiltrated the oceanic crust, made its way to deeper levels where it is heated by coming close to the magma, then risen again to vent on the seafloor. Many of these vents precipitate sulfide and other minerals in copious quantities, forming chimneys called "black smokers" that may be many meters tall. These chimneys have high-temperature metal- and nutrient-rich water flowing out of them (at temperatures of several hundred degrees Celsius), with the metals precipitating once the temperature drops upon contact with the cold seawater outside the vent. These systems may cover parts of the oceanic crust with layers of sulfide minerals. Unusual primitive communities of sulfide-reducing bacteria, tube worms, and crabs have been found near several black smoker vents along mid-ocean ridges. Many scientists believe that similar settings may have played an important role in the early appearance and evolution of life on the planet.

Seismic refraction studies in the 1940s and 1950s established that the oceanic crust exhibits seismic layering that is similar in many places in the oceans. Seismic layer 1 consists of sediments, layer 2 is interpreted to be a layer of basalt 0.6–1.5 miles (1–2.5 km) thick, and layer 3 is approximately 4 miles (6 km) thick and interpreted to be crystal cumulates, underlain by the mantle. Some ridges and transform faults expose deeper levels of the oceanic lithosphere, which can be shown to typically include a diabase dike complex, thick sections of gabbro, and ultramafic cumulates. In some places, rocks of the mantle are exposed, typically consisting of depleted harzburgite tectonites. Much of the detailed information about the deep structure of oceanic crust comes from the study of ophiolites, which are interpreted to be tectonically emplaced on-land equivalents of oceanic crust. Studies of ophiolites have confirmed the general structure of the oceanic crust as inferred from the seismic reflection and refraction studies and limited drilling. Numerous detailed studies of ophiolites have allowed unprecedented detail about the structure and chemistry of inferred oceanic crust and lithosphere to be completed, and as many variations as similarities have been discovered. The causes of these variations are numerous, including differences in spreading rate, magma supply, temperature, depth of melting, tectonic setting (arc, forearc, back arc, mid-ocean ridge, etc.), and the presence or absence of water. However, the ocean floor is still largely unexplored, and we know more about many other planetary surfaces than we know about the ocean floor of the Earth.

See also OPHIOLITES; PLATE TECTONICS.

doldrums Light winds in the region near the equator, or more technically the region near the equator characterized by light shifting winds and low pressure. The weather in this region tends to be monotonous, hence the common expression "down in the doldrums." Air is warmed in this region and rises to form huge cumulus clouds and convective hot towers that release tremendous amounts of latent heat, making the air more buoyant, and driving the atmospheric Hadley Cells.

See also ATMOSPHERE; HADLEY CELL.

dolomite A common carbonate mineral with the formula $(Ca,Mg)CO^3$, and the word is commonly applied to a rock made of more than 50 percent of the mineral dolomite, even though this rock type should technically be termed dolostone. The rock and the Dolomite Mountains of northern Italy are both named after Deodat de Dolomieu, who identified them in 1791. Dolomites are more common in Precambrian carbonate rocks than in younger sequence, perhaps related to higher concentrations of CO^2 in the atmosphere then, as compared to younger times.

Dolomite is often found interbedded with or replacing calcite, reflecting different modes of formation. It may be precipitated directly from lake or shallow marine lagoonal water and may also precipitate as a primary cement in pores between grains in deep marine water. Most dolomite (dolostone) is probably a replacement of earlier precipitated limestone aided by magnesium-rich fluids. Conditions favorable for the replacement of calcium carbonate by dolostone are found in brackish water where fresh and saltwater mix, and in brines (salty water) where organisms remove the calcium carbonate leaving a magnesium-rich fluid behind to crystallize the dolomite. Magnesium-rich waters can also be expelled from a sedimentary sequence during burial, and these fluids can alter a limestone sequence to dolostone.

See also SEDIMENTARY ROCKS.

dolostone *See* DOLOMITE.

Doppler radar Doppler is the principle discovered by Austrian scientist Christian Doppler in 1842 that describes the characteristics of sound waves or radio waves that travel from a source and are reflected by a medium back toward the source. Today the advanced radar known as Doppler radar is named after the Austrian scientist; it is mainly used by meteorologists to forecast the weather and also as a real-time picture of advancing weather systems.

Modern Doppler radars operate much like regular radar but also provide radial velocity images, which are useful to provide the users a three-dimensional image of the object medium. Doppler waves that are reflected by objects moving away from the source provide a lower frequency, while objects moving toward the source return at a higher frequency, providing three-dimensional radial velocities. Doppler radar provides scientists with a way of viewing large-scale objects such as hurricanes that have different internal velocities and vectors.

drainage basin The total area that contributes water to a stream is called a drainage basin, and the line that divides different drainage basins is known as a divide (such as the continental divide) or interfluve. Drainage basins are the primary landscape units or systems concerned with the collection and movement of water and sediment into streams and river channels. Drainage basins consist of a number of interrelated systems that work together to control the distribution and flow of water within the basin. Hillslope processes, bedrock and surficial geology, vegetation, climate, and many other systems all interact in complex ways that determine where streams will form and how much water and sediment they will transport. A drainage basin's hydrologic dynamics can be analyzed by considering these systems along with how much water enters the basin through precipitation, and how much leaves the basin in the discharge of the main trunk channel. Streams are arranged in an orderly fashion in drainage basins, with progressively smaller channels branching away from the main trunk channel. Stream channels are ordered and numbered according to this systematic branching. The smallest segments lack tributaries and are known as first-order streams; second-order streams form where two first-order streams converge, third-order streams form where two second-order streams converge, and so on.

Streams within drainage basins develop characteristic branching patterns that reflect, to some degree, the underlying bedrock geology, structure, and rock types. Dendritic or randomly branching patterns form on horizontal strata or on rocks with uniform erosional resistance. Parallel drainage patterns develop on steeply dipping strata, or on areas with systems of parallel faults or other landforms. Trellis drainage patterns consist of parallel main stream channels intersected at nearly right angles by tributaries, in turn fed by tributaries parallel to the main channels. Trellis drainage patterns reflect significant structural control and typically form where eroded edges of alternating soft and hard layers are tilted, as in folded mountains or uplifted coastal strata. Rectangular drainage patterns form a regular rectangular grid on the surface and typically form in areas where the bedrock is strongly faulted or jointed. Radial and annular patterns develop on domes including volcanoes and other roughly circular uplifts. Other, more complex patterns are possible in more complex situations.

Several categories of streams in drainage basins reflect different geologic histories—a consequent stream is one whose course is determined by the direction of the slope of the land. A subsequent stream is one whose course has become adjusted so that it occupies a belt of weak rock or another geologic structure. An antecedent stream is one that has maintained its course across topography that is being uplifted by tectonic forces; these cross high ridges. Superposed streams are those whose courses were laid down in overlying strata onto unlike strata below. Stream capture occurs when headland erosion diverts one stream and its drainage into another drainage basin.

See also GEOMORPHOLOGY.

Further Reading

Leopold, Luna B. *A View of the River*. Cambridge: Harvard University Press, 1994.

Leopold, Luna B., and M. Gordan Wolman. *River Channel Patterns—Braided, Meandering, and Straight*. U.S. Geological Survey Professional Paper 282-B, 1957.

Maddock, Thomas, Jr. "A Primer on Floodplain Dynamics." *Journal of Soil and Water*, 1976.

Parsons, Anthony J., and Athol D. Abrahams. *Overland Flow—Hydraulics and Erosion Mechanics*. London: UCL Press Ltd., University College, 1992.

Ritter, Dale F., R. Craig Kochel, and Jerry R. Miller. *Process Geomorphology*, 3rd ed. Boston: WCB/McGraw Hill, 1995.

Rosgen, David. *Applied River Morphology*. Pasoga Springs, Colo.: Wildland Hydrology, 1996.

Schumm, Stanley A. *The Fluvial System*. New York: Wiley–Interscience, 1977.

drainage system   *See* DRAINAGE BASIN.

Dry Valleys, Antarctica   The Dry Valleys are the largest area on Antarctica not covered by ice. Approximately 98 percent of the continent is covered by ice, but the Dry Valleys, located near McMurdo Sound on the side of the continent closest to New Zealand, have a cold desert climate and receive only 4 inches (10 cm) of precipitation per year, overwhelmingly in the form of snow. The Dry Valleys are one of the coldest, driest places on Earth and are used by researchers from NASA as an analog for conditions on Mars. There is no vegetation in the Dry Valleys, although a number of unusual microbes live in the frozen soils and form cyanobacterial mats in places. In the Southern Hemisphere summer, glaciers in the surrounding Transantarctic Mountains release significant quantities of meltwater so that streams and lakes form over the thick permafrost in the valleys.

dune   *See* SAND DUNES.

dunite   An ultramafic plutonic rock composed primarily of the mineral olivine. It may have less than 10 percent orthopyroxene, clinopyroxene, hornblende, chromite, and other minor constituents. Most dunites are cumulates formed by the settling of dense olivine crystals in a peridotitic melt. They are commonly found in cumulate sequences in ophiolites, from the bases of island arcs, and in continental intrusions. The rock is named after Dun Mountain in New Zealand, where it was first recognized by Ferdinand von Hochstetter in 1864.

See also IGNEOUS ROCKS; ISLAND ARC; OPHIOLITES.

dust devils   Small tornado-like swirls or spinning vortices that are commonly observed on hot days in desert areas. Most have diameters of a few meters and rise less than 325 feet (100 m), but a few can rise many hundred meters and contain damaging winds of 80 miles per hour (129 km/hr). They are known (and feared) as jifn, or evil spirits in much of Arabia, and as willy-willys in Australia. Dust devils form when the Sun's energy on clear, hot days heats the dry surface so much that the air directly above the surface becomes unstable, setting up low-level convection. Winds that blow into the area to replace the rising air can be deflected around topographic irregularities, causing the rising air to rotate and form the dust devils. Dust devils are not as damaging or long-lived as tornadoes, and differ from them in that they form by spinning air rising from the surface, whereas tornadoes form by spinning air descending from a thunderstorm cloud.

dust storms   Dust storms may form in desert areas when strong winds that blow across desert regions pick up dust made of silt and clay particles and transport them thousands of kilometers from their source. For instance, dust from China

Dust storm blowing down the Gulf of Suez, Egypt *(Photos by Timothy Kusky)*

is found in Hawaii, and the Sahara commonly drops dust in Europe. This dust is a nuisance, has a significant influence on global climate, and has at times (as in the dust bowl days of the 1930s) been known to nearly block out the Sun.

Loess is a name for silt and clay deposited by wind. It forms a uniform blanket that covers hills and valleys at many altitudes, which distinguishes it from deposits of streams. In Shaanxi Province, China, an earthquake that killed 830,000 people in 1556 had such a high death toll in part because the people in the region built their homes out of loess. The loess formed an easily excavated material that hundreds of thousands of villagers cut homes into, essentially living in caves. When the earthquake struck, the loess proved to be a poor building material and large-scale collapse of the fine-grained loess was directly responsible for most of the high death toll.

Recently, it has been recognized that windblown dust contributes significantly to global climate. Dust storms that come out of the Sahara can be carried around the world and can partially block some of the Sun's radiation. The dust particles may also act as small nuclei for raindrops to form around, perhaps acting as a natural cloud-seeding phenomenon. One interesting point to ponder is that as global warming increases global temperatures, the amount and intensity of storms increase, and some of the world's deserts expand. Dust storms may serve to reduce global temperatures and increase precipitation. Might the formation of dust storms represent some kind of self-regulating mechanism, whereby the Earth moderates its own climate?

See also DESERT; LOESS.

Du Toit, Alexander Logie (1878–1948) South African Geologist

Alexander Du Toit is known as "the world's greatest field geologist." He was born near Cape Town and went to school at a local diocesan college. He graduated from South Africa College and then spent two years studying mining engineering at the Royal Technical College in Glasgow, and geology at the Royal College of Science in London. In 1901 he was a lecturer at the Royal Technical College and at the University of Glasgow. He returned to South Africa in 1903, joining the Geological Commission of the Cape of Good Hope, and spent the next several years constantly in the field doing geological mapping. This time in his life was the foundation for his extensive understanding and unrivaled knowledge of South African geology. During his first season he worked with Arthur W. Rogers in the western Karoo where they established the stratigraphy of the Lower and Middle Karoo System. They also recorded the systematic phase changes in the Karoo and Cape Systems. Along with these studies they mapped the dolerite intrusives, their acid phases, and their metamorphic aureoles. Throughout the years, Du Toit worked in many locations, including the Stormberg area, and the Karoo coal deposits near the Indian Ocean. He was very interested in geomorphology and hydrogeology. The most significant factor to his work was the theory of continental drift. He was the first to realize that the southern continents had once formed the supercontinent of Gondwana that was distinctly different from the northern supercontinent Laurasia. Du Toit received many honors and awards. He was the president of the Geological Society of South Africa, a corresponding member of the Geological Society of America, and a member of the Royal Society of London.

Further Reading

Du Toit, Alexander L. "The Origin of the Amphibole Asbestos Deposits of South Africa." *Transactions of the Geological Society of South Africa* 48 (1946): 161–206.
———. "The Continental Displacement Hypothesis as Viewed by Du Toit." *American Journal of Science* 17 (1929): 179–183.

E

Earth The third planet from the center of our solar system, located between Venus and Mars at a distance of 93 million miles (150×10^6 km) from the Sun. It has a mean radius of 3,960 miles (6,371 km), a surface area of 5.101×10^8 km^2, and an average density of 5.5 grams per cubic centimeter. It is one of the terrestrial planets (Mercury, Venus, Earth, and Mars), composed of solid rock, with silicate minerals being the most abundant in the outer layers and a dense iron-nickel alloy forming the core material.

The Earth and eight other planets condensed from a solar nebula about 5 billion years ago. In this process a swirling cloud of hot dust, gas, and protoplanets collided with each other, eventually forming the main planets. The accretion of the Earth was a high-temperature process that allowed melting of the early Earth and segregation of the heavier metallic elements such as iron (Fe) and nickel (Ni) to sink to the core, and for the lighter rocky elements to float upward. This process led to the differentiation of the Earth into several different concentric shells of contrasting density and composition and was the main control on the large-scale structure of the Earth today.

The main shells of the Earth include the crust, a light outer shell 3–43 miles (5–70 km) thick. This is followed inward by the mantle, a solid rocky layer extending to 1,802 miles (2,900 km). The outer core is a molten metallic layer extending to 3,170 miles (5,100 km) depth, and the inner core is a solid metallic layer extending to 3,958 miles (6,370 km). With the recognition of plate tectonics in the 1960s, geologists recognized that the outer parts of the Earth were also divided into several zones that had very different mechanical properties. It was recognized that the outer shell of the Earth was divided into many different rigid plates all moving with respect to each other, and some of them carrying continents in continental drift. This outer rigid layer became

known as the lithosphere and is 45–95 miles (75–150 km) thick. The lithosphere is essentially floating on a denser, but partially molten layer of rock in the upper mantle known as the asthenosphere (or weak sphere). It is the weakness of this layer that allows the plates on the surface of the Earth to move about.

The most basic division of the Earth's surface shows that it is divided into continents and ocean basins, with oceans occupying about 60 percent of the surface and continents 40 percent. Mountains are elevated portions of the continents. Shorelines are where the land meets the sea. Continental shelves are broad to narrow areas underlain by continental crust, covered by shallow water. Continental slopes are steep drop-offs from the edge of a shelf to the deep ocean basin, and the continental rise is where the slope flattens to merge with the deep ocean abyssal plains. Ocean ridge systems are subaquatic mountain ranges where new ocean crust is being created by seafloor spreading. Mountain belts on the Earth are of two basic types. Orogenic belts are linear chains of mountains, largely on the continents, that contain highly deformed, contorted rocks that represent places where lithospheric plates have collided or slid past one another. The mid-ocean ridge system is a 40,000-mile (65,000-km) long mountain ridge that represents vast outpourings of young lava on the ocean floor and places where new oceanic crust is being generated by plate tectonics. After it is formed, it moves away from the ridge crests, and new magmatic plates fill the space created by the plates drifting apart. The oceanic basins also contain long, linear, deep ocean trenches that are up to several kilometers deeper than the surrounding ocean floor and locally reach depths of 7 miles (14 km) below the sea surface. These represent places where the oceanic crust is sinking back into the mantle of the earth, completing the plate tectonic cycle for oceanic crust.

Image of Earth, with the Moon in the background. During its flight, the *Galileo* spacecraft returned images of the Earth and Moon. Separate images of the Earth and Moon were combined to generate this view. The *Galileo* spacecraft took the images in 1992 on its way to explore the Jupiter system in 1995–97. The image shows a partial view of the Earth centered on the Pacific Ocean about latitude 20° south. The west coast of South America can be observed as well as the Caribbean; swirling white cloud patterns indicate storms in the southeast Pacific. The distinct, bright ray crater at the bottom of the Moon is the tycho impact basin. The lunar dark areas are lava rock–filled impact basins. The image contains same-scale and relative color/albedo images of the Earth and Moon. False colors via use of the 1-micron filter as red, 727-nm filter as green, and violet filter as blue. The *Galileo* project is managed for NASA's Office of Space Science by the Jet Propulsion Laboratory. *(Courtesy of NASA)*

External layers of the Earth include the hydrosphere, consisting of the oceans, lakes, streams, and the atmosphere. The air/water interface is very active, for here erosion breaks rocks down into loose debris called the regolith.

The hydrosphere is a dynamic mass of liquid, continuously on the move. It includes all the water in oceans, lakes, streams, glaciers, and groundwater, although most water is in the oceans. The hydrologic cycle describes changes, both long and short term, in the Earth's hydrosphere. It is powered by heat from the Sun, which causes evaporation and transpiration. This water then moves in the atmosphere and precipitates as rain or snow, which then drains off in streams, evaporates, or moves as groundwater, eventually to begin the cycle over and over again.

The atmosphere is the sphere around the Earth consisting of the mixture of gases we call air. It is hundreds of kilometers thick and is always moving, because more of the Sun's heat is received per unit area at the equator than at the poles. The heated air expands and rises to where it spreads out, cools and sinks, and gradually returns to the equator. The effects of the Earth's rotation modify this simple picture of the atmosphere's circulation. The Coriolis effect causes any freely moving body in the Northern Hemisphere to veer to the right, and toward the left in the Southern Hemisphere.

The biosphere is the totality of Earth's living matter and partially decomposed dead plants and animals. The biosphere is made up largely of the elements carbon, hydrogen, and oxygen. When these organic elements decay they may become part of the regolith and be returned through geological processes back to the lithosphere, atmosphere, or hydrosphere.

See also ATMOSPHERE; BIOSPHERE; LITHOSPHERE; MANTLE.

earthflow   *See* MASS WASTING.

earthquakes   Sudden movements of the ground, earthquakes may be induced by movement along faults, volcanic eruptions, landslides, bomb blasts, and other triggering mechanisms. Earthquakes can be extremely devastating and costly events, in some cases killing tens or even hundreds of thousands of people and leveling entire cities in a matter of a few tens of seconds. A single earthquake may release energy equivalent to hundreds or thousands of nuclear blasts, and may cost billions of dollars in damage, not to mention the toll in human suffering. Earthquakes are also associated with secondary hazards, such as tsunami, landslides, fire, famine, and disease that also exert their toll on humans.

The lithosphere (or outer rigid shell) of the Earth is broken into 7 large tectonic plates and many other smaller plates, each moving relative to the others. Most of the earthquakes in the world happen where two of these plates meet and are moving past each other, such as in southern California. Most really big earthquakes occur at boundaries where the plates are moving toward each other (as in Alaska), or sliding past one another (as in southern California). Smaller earthquakes occur where the plates are moving apart, such as along mid-oceanic ridges where new magma rises and forms ocean spreading centers.

The area that gets the most earthquakes in the conterminous United States is southern California along the San Andreas Fault, where the Pacific plate is sliding north relative to the North American plate. In this area, the motion is characterized as a "stick-slip" type of sliding, where the two plates stick to each other along the plate boundary as the two plates slowly move past each other, and stresses rise over tens or hundreds of years. Eventually the stresses along the boundary rise so high that the strength of the rocks is exceeded, and the rocks suddenly break, causing the two plates to

Earthquake Warning Systems

Saving lives during earthquakes is critically dependent on three major issues: knowing the hazards, planning accordingly, and receiving adequate warning that an earthquake is occurring. During earthquakes seismic waves travel outward from the epicenter, traveling at up to several miles per second. For large earthquakes, significant damage may be inflicted on structures tens or even hundreds of miles from the epicenter, several or several tens of seconds after the earthquake first strikes in the epicentral region. Typically the most destructive surface waves travel more slowly than P and S waves, and at distances of more than a few miles, the time difference can be significant. Earthquake engineers and urban planners are utilizing these basic physical realities to devise and implement some extremely sophisticated earthquake warning systems for places like southern California and Japan. Seismographs are being linked to sophisticated computer systems that quickly analyze the magnitude of an earthquake and determine if it is going to be destructive enough to merit a warning to a large region. If a warning is issued that a large earthquake is occurring, the systems use satellite and computer networks to send a warning to the surrounding areas to immediately take a prescribed set of actions to reduce the damage, injury, and death from the earthquake. For instance, trains may be automatically stopped before they derail, sirens may sound so that people can take shelter, nuclear plants can be shut down, and gas lines can be blocked. These warning systems may be able to alert residents or occupants of part of the region that a severe earthquake has just occurred in another part of the region, and that they have several or several tens of seconds to take cover. The thought is that if structures are adequately constructed, and if people have an earthquake readiness plan already implemented, they will know how and where to take immediate cover when the warning whistles are sounded, and that this type of system may be able to save numerous lives.

The effectiveness of earthquake warning systems depends on how adequately the plans for such an event were made. Ground shaking causes much of the damage during earthquakes, and the amount of shaking is dependent on the type of soil, bedrock, the geometry or focal mechanism of the earthquake, and how local geologic factors focus the energy to specific sites. Geologists are able to map the different soil and shaking hazard potentials and build a computer-based database that is useful for emergency response. For instance, a type of map known as a shake map may be rapidly generated for specific earthquakes, showing how much shaking might have been experienced in different areas across a region. If the types of buildings and their susceptibility to shaking are known, the consequences of earthquakes in specific neighborhoods can be predicted. Emergency responders can then immediately go to the areas that likely received the most damage, saving lives and helping the most injured before responding to less severely hit areas.

dramatically move (slip) up to a few meters in a few seconds. This sudden motion of previously stuck segments along a fault plane is an earthquake. The severity of the earthquake is determined by how large an area breaks in the earthquake, how far it moves, how deep within the Earth the break occurs, and the length of time that the broken or slipped area along the fault takes to move. The elastic rebound theory states that recoverable (also known as elastic) stresses build up in a material until a specific level or breaking point is reached. When the breaking point or level is attained, the material suddenly breaks, and the stresses are released in an earthquake. In the case of earthquakes, rows of fruit trees, fences, roads, and railroad lines that became gradually bent across an active fault line as the stresses built up are typically noticeably offset across faults that have experienced an earthquake. When the earthquake occurs, the rocks snap along the fault, and the bent rows of trees, fences, or roads/rail-line become straight again, but displaced across the fault.

Some areas away from active plate boundaries are also occasionally prone to earthquakes. Even though earthquakes in these areas are uncommon, they can be very destructive. Boston, Massachusetts; Charleston, South Carolina; and New Madrid, Missouri (near St. Louis), have been sites of particularly severe earthquakes. For instance, in 1811 and 1812 three large earthquakes with magnitudes of 7.3, 7.5, and 7.8 were centered in New Madrid and shook nearly the entire United States, causing widespread destruction. Most buildings near the origin of the earthquake were toppled, and several deaths were reported (the region had a population of only 1,000 at the time but is now densely populated). Damage to buildings was reported from as far away as Boston and Canada, where chimneys toppled, plaster cracked, and church bells were set to ringing by the shaking of the ground.

Many earthquakes in the past have been incredibly destructive, killing hundreds of thousands of people, like the ones in Armenia, Iran, and Mexico City in recent years. Some earthquakes have killed nearly a million people, such as one in

The 10 Worst Earthquakes in Terms of Loss of Life

Place	Year	Deaths	Estimated Magnitude
Shanxi, China	1556	830,000	
Calcutta, India	1737	300,000	
T'ang Shan, China	1976	242,000	m. 7.8
Gansu, China	1920	180,000	m. 8.6
Messina, Italy	1908	160,000	
Tokyo, Japan	1923	143,000	
Beijing, China	1731	100,000	
Chihli, China	1290	100,000	
Naples, Italy	1693	93,000	
Gansu, China	1932	70,000	

1556 in China that killed 800,000–900,000 people. One in Calcutta, India, in 1737 killed about 300,000 people.

Origins of Earthquakes

Earthquakes can originate from sudden motion along a fault, from a volcanic eruption, or from bomb blasts. Not every fault is associated with active earthquakes; in fact, most faults are no longer active but were active at some time in the geologic past. Of the faults that are active, only some are characterized as being particularly prone to earthquakes. Some faults are slippery, and the two blocks on either side just slide by each other passively without producing major earthquakes. In other cases, however, the blocks stick to each other and deform like a rubber band until they reach a certain point where they suddenly snap, releasing energy in an earthquake event.

Rocks and materials are said to behave in a brittle way when they respond to built-up tectonic pressures by cracking, breaking, or fracturing. Earthquakes represent a sudden brittle response to built-up stress and are almost universally activated in the upper few kilometers of the Earth. Deeper than this, the pressure and temperature are so high that the rocks simply deform like silly putty, do not snap, and are said to behave in a ductile manner.

An earthquake originates in one place and then spreads out. The focus is the point in the Earth where the earthquake energy is first released. The epicenter is the point on the Earth's surface that lies vertically above the focus.

When big earthquakes occur, the surface of the Earth actually forms into waves that move across the surface, just as in the ocean. These waves can be pretty spectacular and also extremely destructive. When an earthquake occurs, these seismic waves move out in all directions, just like sound waves, or ripples that move across water after a stone is thrown in a still pond. After the seismic waves have passed through the ground, the ground returns to its original shape, although buildings and other human constructions are commonly destroyed.

During an earthquake, these waves can either radiate underground from the focus—called body waves—or aboveground from the epicenter—called surface waves. The body waves travel through the whole body of the Earth and move faster than surface waves, though surface waves cause most of the destruction associated with earthquakes because they briefly change the shape of the surface of the Earth when they pass. There are two types of body waves: P, or compressional waves, and S, or secondary waves. P-waves deform material through a change in volume and density, and these can pass

Loma Prieta Earthquake, 1989

The Santa Cruz and San Francisco areas were hit by a moderate-sized earthquake (magnitude 7.1) at 5:04 P.M. on Tuesday, October 17, 1989, during a World Series game played in San Francisco. As a result, 67 people died, 3,757 people were injured, and 12,000 were left homeless. A television audience of tens of millions of people watched as the earthquake struck just before the beginning of game three, and the news coverage that followed was unprecedented for capturing an earthquake as it happened.

The earthquake was caused by a 26-mile (42-km) long rupture in a segment of the San Andreas Fault near Loma Prieta peak in the Santa Cruz Mountains south of San Francisco. The segment of the fault that ruptured was the southern part of the same segment that ruptured in the 1906 earthquake, but this rupture occurred at greater depths than the earlier quake. The actual rupturing lasted only 11 seconds, during which time the western (Pacific) plate slid almost six feet (1.9 meters) to the northwest, and parts of the Santa Cruz Mountains were uplifted by up to four feet (1.3 meters). The rupture propagated at 1.24 miles per second (2 km/sec) and was a relatively short-duration earthquake for one of this magnitude. Had it been much longer, the damage would have been much more extensive. As it was, the damage totals amounted to more than $6 billion.

The actual fault plane did not rupture the surface, although many cracks appeared and slumps formed along steep slopes. The Loma Prieta earthquake had been predicted by seismologists because the segment of the fault that slipped had a noticeable paucity of seismic events since the 1906 earthquake and was iden-

tified as a seismic gap with a high potential for slipping and causing a significant earthquake. The magnitude 7.1 event and the numerous aftershocks filled in this seismic gap, and the potential for large earthquakes along this segment of the San Andreas Fault is now significantly lower, since the built-up strain energy was released during the quake. There are, however, other seismic gaps along the San Andreas Fault in heavily populated areas, such as near Los Angeles, that should be monitored closely.

The amount of ground motion associated with an earthquake not only increases with the magnitude of the earthquake but also depends on the nature of the substratum. In general, loose, unconsolidated fill tends to shake more than solid bedrock. This was dramatically illustrated by the Loma Prieta earthquake, where areas built on solid rock near the source of the earthquake vibrated the least (and saw the least destruction), and areas several tens of miles away built on loose clays vibrated the most. Much of the Bay area is built on loose clays and mud, including the Nimitz freeway, which collapsed during the event. The area that saw the worst destruction associated with ground shaking was the Marina district of downtown San Francisco. Even though this area is located far from the earthquake epicenter, it is built on loose unconsolidated landfill, which shook severely during the earthquake, causing many buildings to collapse, and gas lines to rupture, which initiated fires. More than twice as much damage from ground shaking during the Loma Prieta earthquake was reported from areas over loose fill or mud than from areas built over solid bedrock. Similar effects were reported from the 1985 earthquake in Mexico City, which is built largely on old lakebed deposits.

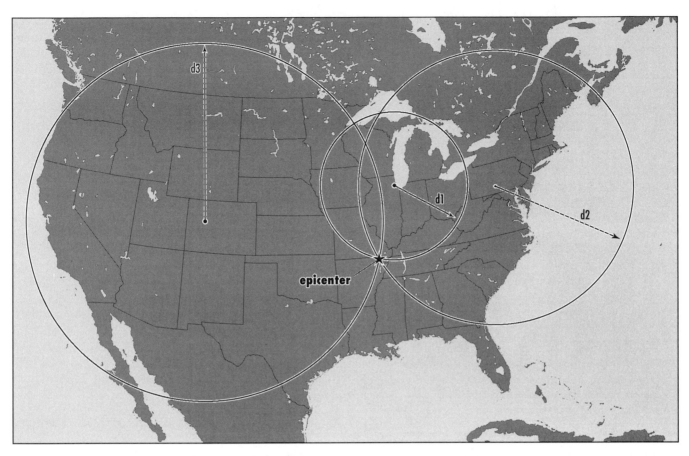

A method of locating earthquake epicenters is by calculating the distance to the source from three different seismic stations. The distance to the epicenter is calculated using the time difference between the first arrivals of P and S waves. The unique place where the three distance circles intersect is the location of the epicenter.

through solids, liquids, and gases. The kind of movement associated with passage of a P-wave is a back and forth type of motion. Compressional (P) waves move with high velocity, about four miles per second (6 km/sec), and are thus the first to be recorded by seismographs. This is why they are called primary (P) waves. P-waves cause a lot of damage because they temporarily change the area and volume of ground that humans built things on or modified in ways that require the ground to keep its original shape, area, and volume. When the ground suddenly changes its volume by expanding and contracting, many of these constructions break. For instance, if a gas pipeline is buried in the ground, it may rupture or explode when a P-wave passes because of its inability to change its shape along with the Earth. It is common for fires and explosions originating from broken pipelines to accompany earthquakes.

The second type of body waves are known as shear waves (S) or secondary waves, because they change the shape of a material but not its volume. Shear waves can only be transmitted by solids. Shear waves move material at right angles to the direction of wave travel, and thus they consist of an alternating series of sideways motions. Holding a jump rope at one end on the ground and moving it rapidly back and forth can simulate this kind of motion. Waves form at the end being held and move the rope sideways as they move toward the loose end of the rope. A typical shear-wave velocity is 2 miles per second (3.5 km/s). These kinds of waves may be responsible for knocking buildings off foundations when they pass, since their rapid sideways or back and forth motion is often not met by buildings. The effect is much like pulling a tablecloth out from under a set table—if done rapidly, the building (as is the case for the table setting) may be left relatively intact, but detached from its foundation.

Surface waves can also be extremely destructive during an earthquake. These have complicated types of twisting and circular motions, much like the circular motions you might feel while swimming in waves out past the surf zone at the beach. Surface waves travel slower than either type of body waves, but because of their complicated types of motion they often cause the most damage. This is a good thing to remember during an earthquake, because if you realize that the body waves have just passed your location, you may have a brief period of no shaking to get outside before the very destructive surface waves hit and cause even more destruction.

Measuring Earthquakes

How are the vibrations of an earthquake measured? Geologists use seismographs, which display Earth movements by means of an ink-filled stylus on a continuously turning roll of

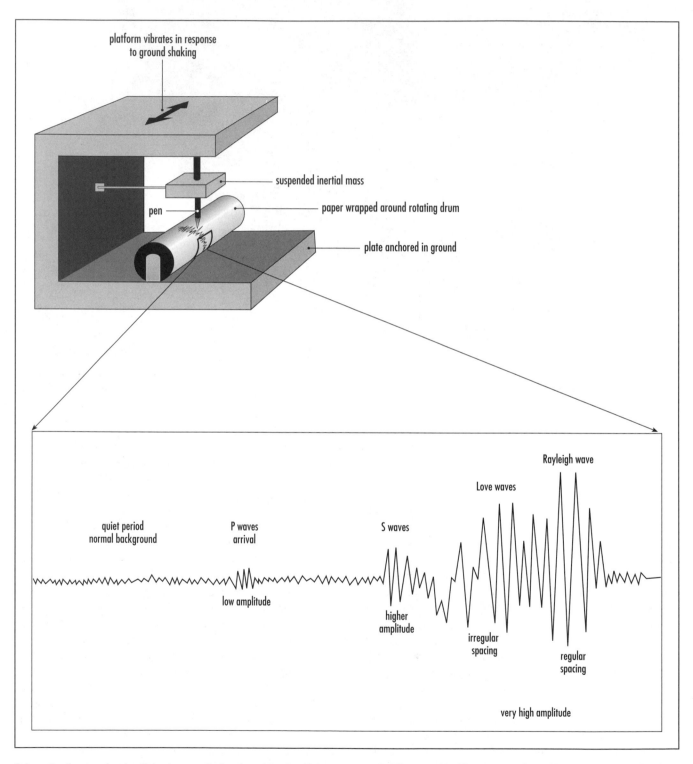

platform vibrates in response to ground shaking

suspended inertial mass

pen

paper wrapped around rotating drum

plate anchored in ground

quiet period normal background

P waves arrival

low amplitude

S waves

higher amplitude

Love waves

Rayleigh wave

irregular spacing

regular spacing

very high amplitude

Schematic diagram of an inertial seismograph showing a large inertial mass suspended from a spring. The mass remains stationary as the ground and paper wrapped around the rotating drum move back and forth during an earthquake, creating the seismogram.

graph paper. When the ground shakes, the needle wiggles and leaves a characteristic zigzag line on the paper. Many seismograph records clearly show the arrival of P and S body waves, followed by the surface waves.

Seismographs are built using a few simple principles. To measure the shaking of the Earth during a quake, the point of reference must be free from shaking, ideally on a hovering platform. However, since building perpetually hovering platforms is impractical, engineers have designed an instrument known as an inertial seismograph. These make use of the principle of *inertia*, which is the resistance of a large mass to sudden movement. When a heavy weight is hung from a string or thin spring, the string can be shaken and the big heavy weight will remain stationary. Using an inertial seismograph, the ink-filled stylus is attached to the heavy weight, and remains stationary during an earthquake. The continuously turning graph paper is attached to the ground, and moves back and forth during the quake, resulting in the zigzag trace of the record of the earthquake motion on the graph paper.

Seismographs are used in series—some set up as pendulums and some others as springs, to measure ground motion in many directions. Engineers have made seismographs that can record motions as small as 1 hundred-millionth of an inch, about equivalent to being able to detect the ground motion caused by a car driving by several blocks away. The ground motions recorded by seismographs are very distinctive, and geologists who study them have methods of distinguishing between earthquakes produced along faults, earthquake swarms associated with magma moving into volcanoes, and even between explosions from different types of construction and nuclear blasts. Interpreting seismograph traces has therefore become an important aspect of nuclear test ban treaty verification.

Earthquake Magnitude
Earthquakes vary greatly in intensity, from undetectable up to ones that kill millions of people and wreak total destruction. For instance, one severe earthquake in December of 2003 killed approximately 50,000 people in Iran, yet several thousand earthquakes that do no damage occur every day throughout the world. The energy released in large earthquakes is enormous; up to hundreds of times more powerful than large atomic blasts. Strong earthquakes may produce ground accelerations greater than the force of gravity, enough to uproot trees, or send projectiles right through buildings, trees, or anything else in their path. Earthquake magnitudes are most commonly measured using the Richter scale.

The Richter scale gives an idea of the amount of energy released during an earthquake and is based on the amplitudes (half the height from wave-base to wave-crest) of seismic waves at a distance of 61 miles (100 km) from the epicenter. The Richter scale magnitude of an earthquake is calculated using the zigzag trace produced on a seismograph, once the epicenter has been located by comparing signals

Modified Mercalli Intensity Scale Compared with Richter Magnitude

Mercalli Intensity	Richter Magnitude	Description
I–II	2	Not felt by most people.
III	3	Felt by some people indoors, especially on high floors.
IV–V	4	Noticed by most people. Hanging objects swing. Dishes rattle.
VI–VII	5	All people feel. Some building damage (especially to masonry), waves on ponds.
VII–VIII	6	Difficult to stand, people scared or panicked. Difficult to steer cars. Moderate damage to buildings.
IX–X	7	Major damage, general panic of public. Most masonry and frame structures destroyed. Underground pipes broken. Large landslides.
XI–XII	8 and higher	Near total destruction.

from several different, widely separated seismographs. The Richter scale is logarithmic, where each step of one corresponds to a tenfold increase in amplitude. This is necessary because the energy of earthquakes changes by factors of more than a hundred million.

The energy released during earthquakes changes even more rapidly with each increase in the Richter scale, because the number of high amplitude waves increases with bigger earthquakes and also because the energy released is according to the square of the amplitude. Thus, it turns out in the end that an increase of one on the Richter scale corresponds to a 30 times increase in energy released. The largest earthquakes so far recorded are the 9.2 Alaskan earthquake of 1964, the 9.5 Chilean earthquake of 1960, and the 9.0–9.2 Sumatra earthquake of 2004, each of which released the energy equivalent to approximately 10,000 nuclear bombs the size of the one dropped on Hiroshima.

Before the development of modern inertial seismographs, earthquake intensity was commonly measured using the modified Mercalli intensity scale. This scale, named after Father Giuseppe Mercalli, was developed in the late 1800s and measures the amount of vibration people remember feeling for low-magnitude earthquakes and the amount of damage to buildings in high-magnitude events. The table compares the Richter and modified Mercalli scales. One of the disadvantages of the Mercalli scale is that it is not corrected for distance from the epicenter. Therefore, people near the source of the earthquake may measure the earthquake as a IX or X, whereas people further from the epicenter might only record a I or II event. However, the modified Mercalli scale has proven very useful for estimating the magnitudes of historical earthquakes that occurred before the development of modern seismographs, since the Mercalli magnitude can be estimated from historical records.

See also PLATE TECTONICS.

Further Reading
Bolt, Bruce A. *Earthquakes,* 4th ed. New York: W.H. Freeman, 1999.
Coburn, Andrew, and Robin Spence. *Earthquake Protection.* Chichester, England: Wiley, 1992.
Erickson, J. *Quakes, Eruptions, and Other Geologic Cataclysms.* New York: Facts On File, 2001.
Logorio, H. *Earthquakes: An Architect's Guide to Non-Structural Seismic Hazards.* New York: Wiley, 1991.
Reiter, L. *Earthquake Hazard Analysis.* New York: Columbia University Press, 1990.
Verney, P. *The Earthquake Handbook.* New York: Paddington Press, 1979.
Wallace, R. E., ed. "The San Andreas Fault System." *Geological Survey Professional Paper 1515,* 1990.

echinoderm Marine benthic, and rarely pelagic, vertebrate animals that have five-fold radial symmetry and an endoskeleton made of spiny plates of calcite. The living echinoderms are divided into five classes with about 6,000 species, including common forms such as starfish, brittle stars, sea urchins, sea cucumbers, and crinoids. A rich fossil record of echinoderms extends back to the Cambrian with about 20 classes recognized.

The wide distribution and evolutionary changes in the echinoderms makes them moderately useful index fossils, but many disarticulate when they die, limiting their utility. They show a record characterized by the sudden appearance of new forms, without any intermediate forms preserved. An initial radiation occurred from the Early Cambrian to Middle Ordovician that led to a stable period in which many stemmed forms, especially crinoids, were prolific until an extinction event at the Permian-Triassic boundary. Many of the echinoderms, including the blastoids, became extinct here, though the sea cucumbers, starfish, sea urchins, and brittle stars survived, but they did not flourish again until the early Mesozoic.

Further Reading
Nichols, Dave. *Echinoderms.* London: Hutchison, 1962.

eclogite A dense, coarse-grained granular green rock consisting of the minerals garnet (almandine-pyrope varieties) and sodic pyroxene (omphacite). Minor amounts of quartz, rutile, orthopyroxene, white mica, and kyanite may also be present. It has the chemical composition of basalt and gabbro, and many examples of eclogite may have formed by the progressive high-grade metamorphism of mafic rocks.

Eclogite facies refers to a group of metamorphic mineral assemblages where mafic rocks are represented by the assemblage omphacite plus garnet. Mineral and phase equilibrium studies show that this assemblage forms at high pressures (more than 10 kilobars, or greater than about 30 kilometers), and some eclogites are known to have formed at more than 30 kilobars pressure, equivalent to more than 62 miles (100 km) depth.

Eclogites are uncommon rocks at the surface because they must be brought to the surface by some unusual process. They are found as inclusion in kimberlites or basalts, bands and lenses in high-grade gneiss terrains, as blocks associated with blueschists, and along some fault zones.

See also METAMORPHISM.

ecosystem An ecological unit that encompasses the total aspect of the physical and biological environment of an area, and the connections between the various parts. It is an integrated unit consisting of a community of living organisms, affected by various factors such as temperature, humidity, light, soil, food supply, and interactions with other organisms. Changes in any one part of an ecosystem are likely to result in changes in the other parts. Relationships between organisms in an ecosystem depend on changes in the energy input and flow and nutrient flux within the system. The term was coined in 1935 by British ecologist Arthur Tansley (1871–1955).

Ekman spirals The turning of water with depth as a result of the Coriolis force. The spirals form because each (infinitesimally thin) layer of the ocean water exerts a frictional drag on the layer below, so that as the top layer moves, the layers below move slightly less with each depth increment. Because the Coriolis force causes moving objects to deflect to the right in the Northern Hemisphere and to the left in the Southern Hemisphere, each successively deeper layer will also be slightly deflected to the right or left of the moving layer above. These effects cause moving water on the surface to be succeeded with depth by progressively slowing and turning particle paths. The Ekman spirals typically extend to about 325 feet (100 m), where the water is actually moving in the opposite direction to that of the surface water that caused the initial flow. The movement of water by Eckman spirals causes a net transport of water to the right of the direction of surface water in the Northern Hemisphere, and to the left of the direction of surface winds in the Southern Hemisphere. This phenomenon is known as Ekman transport.

See also COASTAL DOWNWELLING; COASTAL UPWELLING; CORIOLIS EFFECT.

El-Baz, Farouk (1938–) Egyptian/American *Geologist* Dr. Farouk El-Baz is known for pioneering work in the applications of space photography to the understanding of arid terrain, particularly the location of groundwater resources. Based on the analysis of space photographs, his recommendations have resulted in the discovery of groundwater resources in the Sinai Peninsula, the Western Desert of Egypt, and in arid terrains in northern Somalia and the Red Sea Province of Eastern Sudan. During the past 20 years, he contributed to interdisciplinary field investigations in all major deserts of the world. In 2003 his research objectives included applications of remote sensing technology to the fields of archaeology, geography, and geology.

Between 1967 and 1972, Dr. El-Baz participated in the Apollo program as supervisor of Lunar Science Planning at Bellcomm, Inc., of Bell Telephone Laboratories in Washington, D.C. During these six years, he was secretary of the Site Selection Committee for the Apollo lunar landings, chairman of the Astronaut Training Group, and principal investigator for Visual Observations and Photography. From 1973 to 1983, he established and directed the Center for Earth and Planetary Studies at the National Air and Space Museum, Smithsonian Institution, Washington, D.C. In 1975 Dr. El-Baz was selected by NASA to be principal investigator for earth observations and photography on the Apollo-Soyuz Test Project. This was the first joint American-Soviet space mission. From 1982 to 1986 he was vice president for international development and for science and technology at Itek Optical Systems of Lexington, Massachusetts.

Dr. El-Baz served on the steering committee of earth sciences of the Smithsonian Institution, the arid and semi-arid research needs panel of the National Science Foundation, the Advisory Committee on Extraterrestrial Features of the U.S. Board of Geographic Names, and the Lunar Nomenclature Group of the International Astronomical Union. In 1979, after the United States and China normalized relations, he coordinated the first visit by U.S. scientists to the desert regions of northwestern China. In 1985 he was elected fellow of the Third World Academy of Sciences and represents the academy at the Non-Governmental Organizations Unit of the Economic and Social Council of the United Nations. He also served as science adviser (1978–81) to Anwar Sadat, former president of Egypt.

Dr. Farouk El-Baz is research professor and director of the Center for Remote Sensing at Boston University. He received a B.Sc. (1958) in chemistry and geology from Ain Shams University, Cairo, Egypt, an M.S. (1961) in geology from the Missouri School of Mines and Metallurgy, Rolla, Missouri, and his Ph.D. (1964) in geology from the University of Missouri, after performing research at the Massachusetts Institute of Technology, Cambridge, Massachusetts (1962–63). He taught geology at Egypt's Assiut University 1958 to 1960, and at the University of Heidelberg in Germany between 1964 and 1966. In 1989 Dr. El-Baz received an honorary doctor of science degree from the New England College, Henniker, New Hampshire.

Dr. El-Baz is president of the Arab Society of Desert Research and the recipient of numerous honors and awards, including: NASA's Apollo Achievement Award, Exceptional Scientific Achievement Medal, and Special Recognition Award; the University of Missouri Alumni Achievement Award for Extraordinary Scientific Accomplishments; the Certificate of Merit of the World Aerospace Education Organization; and the Arab Republic of Egypt Order of Merit–First Class. He also received the 1989 Outstanding Achievement Award of the Egyptian American Organization, the 1991 Golden Door Award of the International Institute of Boston, and the 1992 Award for Public Understanding of Science and Technology of the American Association for the Advancement of Science. In 1995 he received the Award for Outstanding Contributions to Science and Space Technology of the American-Arab Anti-Discrimination Committee and the Achievement Award of the Egyptian American Professional Society. He also received the 1996 Michael T. Halbouty Human Needs Award of the American Association of Petroleum Geologists. In 1999 the Geological Society of America established "The Farouk El-Baz Award for Desert Research," to annually encourage and reward arid land studies.

Further Reading

El-Baz, Farouk. "Origin and Evolution of the Desert." *Interdisciplinary Science Reviews* 13 (1988): 331–347.
———. "Gifts of the Desert." *Archaeology* 54 (2001): 42–45.
———. "Sand Accumulation and Groundwater in the Eastern Sahara." *Episodes 21, International Union of Geological Sciences* (1998): 147–151.

electron microprobe   An electron microprobe is an instrument that actively bombards samples of a material with high-energy electrons. It is basically an electron microscope that is able to perform important scientific experiments such as X-ray microanalysis to measure atomic density and elemental abundances. It is necessary to cover the specimen with a thin carbon or metallic coating to prevent the buildup of electric charges on the sample.

Most of the elements on the periodic table can be analyzed using the electron microprobe. The electron microprobe is useful because it can analyze small areas, and that enables scientists to compare that small sample area to the rest of the sample being analyzed. Scientists use the electron microprobe in a wide range of academic fields such as engineering, geology, biology, chemistry, and physics.

electron microscope (TEM and SEM)   There are two main types of electron microscopes. The most widely used electron microscope is the TEM (transmission electron microscope) and the other is the SEM (scanning electron microscope). In an electron microscope a ray of electrons is passed from a cathode heated by an electric current toward the sample. The higher the electric voltage, the greater the resolution. The SEM differs from the TEM in that the SEM primarily uses the same technology as television. The resolution is less than that of the TEM but the focus area of the TEM is much larger than that of the SEM.

One of the main drawbacks of both the TEM and SEM is their inability to scan and analyze living material. Also, it is necessary for most material to be saturated with heavy metals that enable better contrast in both microscopes.

El Niño and the Southern Oscillation (ENSO)

El-Niño–Southern Oscillation is the name given to one of the better-known variations in global atmospheric circulation patterns. Global oceanic and atmospheric circulation patterns undergo frequent shifts that affect large parts of the globe, particularly those arid and semiarid parts affected by Hadley Cell circulation. It is now understood that fluctuations in global circulation can account for natural disasters including the Dust Bowl days of the 1930s in the midwestern United States. Similar global climate fluctuations may explain the drought, famine, and desertification of parts of the Sahel, and the great famines of Ethiopia and Sudan in the 1970s and 1980s.

The secondary air circulation phenomenon known as the El-Niño–Southern Oscillation can also have profound influences on the development of drought conditions and desertification of stressed lands. Hadley Cells migrate north and south with summer and winter, shifting the locations of the most intense heating. There are several zonal oceanic-atmospheric feedback systems that influence global climate, but the most influential is that of the Austral-Asian system. In normal Northern Hemisphere summers, the location of the most intense heating in Austral-Asia shifts from equatorial regions to the Indian subcontinent along with the start of the Indian monsoon. Air is drawn onto the subcontinent, where it rises and moves outward to Africa and the central Pacific. In Northern Hemisphere winters, the location of this intense heating shifts to Indonesia and Australia, where an intense low-pressure system develops over this mainly maritime region. Air is sucked in and moves upward and flows back out at tropospheric levels to the east Pacific. High pressure develops off the coast of Peru in both situations, because cold upwelling water off the coast here causes the air to cool, inducing atmospheric downwelling. The pressure gradient set up causes easterly trade winds to blow from the coast of Peru across the Pacific to the region of heating, causing warm water to pile up in the Coral Sea off the northeast coast of Australia. This also causes sea level to be slightly depressed off the coast of Peru, and more cold water upwells from below to replace the lost water. This positive feedback mechanism is rather stable—it enhances the global circulation, as more cold water upwelling off Peru induces more atmospheric downwelling, and more warm water piling up in Indonesia and off the coast of Australia causes atmospheric upwelling in this region.

This stable linked atmospheric and oceanic circulation breaks down and becomes unstable every two to seven years, probably from some inherent chaotic behavior in the system. At these times, the Indonesian-Australian heating center migrates eastward, and the buildup of warm water in the western Pacific is no longer held back by winds blowing westward across the Pacific. This causes the elevated warm water mass to collapse and move eastward across the Pacific, where it typically appears off the coast of Peru by the end of December. The El-Niño–Southern Oscillation (ENSO) events occur when this warming is particularly strong, with temperatures increasing by 40°F–43°F (22°C–24°C) and remaining high for several months. This phenomenon is also associated with a reversal of the atmospheric circulation around the Pacific such that the dry downwelling air is located over Australia and Indonesia, and the warm upwelling air is located over the eastern Pacific and western South America.

The arrival of El Niño is not good news in Peru, since it causes the normally cold upwelling and nutrient rich water to sink to great depths, and the fish either must migrate to better feeding locations or die. The fishing industry collapses at these times, as does the fertilizer industry that relies on the guano normally produced by birds (which eat fish and anchovies) that also die during El Niño events. The normally cold dry air is replaced with warm moist air, and the normally dry or desert regions of coastal Peru receive torrential rains with associated floods, landslides, death, and destruction. Shoreline erosion is accelerated in El Niño events, because the warm water mass that moved in from across the Pacific raises sea levels by 4–25 inches (10–60 cm), enough to cause significant damage.

The end of ENSO events also leads to abnormal conditions, in that they seem to turn on the "normal" type of circulation in a much stronger way than is normal. The cold upwelling water returns off Peru with such a ferocity that it may move northward, flooding a 1°–2° band around the equator in the central Pacific ocean with water that is as cold as 68°F (20°C). This phenomenon is known as La Niña ("the girl" in Spanish).

The alternation between ENSO, La Niña, and normal ocean-atmospheric circulation has profound effects on global climate and the migration of different climate belts on yearly to decadal timescales, and it is thought to account for about a third of all the variability in global rainfall. ENSO events may cause flooding in the western Andes and southern California, and a lack of rainfall in other parts of South America including Venezuela, northeastern Brazil, and southern Peru. It may change the climate, causing droughts in Africa, Indonesia, India, and Australia, and is thought to have caused the failure of the Indian monsoon in 1899 that resulted in regional famine with the deaths of millions of people. Recently, the seven-year cycle of floods on the Nile has been linked to ENSO events, and famine and desertification in the Sahel, Ethiopia, and Sudan can be attributed to these changes in global circulation as well.

See also CLIMATE; LA NIÑA.

Further Reading

Ahrens, C. D. *Meteorology Today, An Introduction to Weather, Climate, and the Environment,* 6th ed. Pacific Grove, Calif.: Brooks/Cole, 2000.

Eocene The middle epoch of the Paleogene (Lower Tertiary) Period, and the rock series deposited during this time interval. The Eocene ranges from 57.8 million years ago to 36.6 million years ago and is divided from base to top into the Ypresian, Lutetian, Bartonian, and Priabonian ages. The Eocene epoch was named by Charles Lyell in 1833 for mollusk-bearing strata in the Paris basin.

The Tethys Ocean was undergoing final closure during the Eocene, and the main Alpine orogeny occurred near the end of the epoch. Global climates were warm in the Eocene and shallow water benthic nummulites and other large foraminifera became very abundant. Mammals became the dominant tetrapods on the land, and whales were common in the seas about midway through the epoch. Continental flora included broad-leafed trees in low latitude forests, and conifers at high latitudes.

See also TERTIARY.

eolian Meaning "of the wind" (after Aeolus, Greek god of the winds), eolian refers to sediments deposited by wind. Loess is fine-grained windblown silt and dust that covers surfaces and forms thick deposits in some parts of the world, such as Shanxi Province in China. Sand dunes and other forms are moved by the wind and form extensive dune terrains, sand sheets, and sand seas in parts of many deserts in the world.

When wind blows across a surface it creates turbulence that exerts a lifting force on loose, unconsolidated sediment. With increasing wind strengths, the air currents are able to lift and transport larger sedimentary grains, which then bump into and dislodge other grains, causing large-scale movement of sediment by the wind in a process called saltation. When these particles hit surfaces they may abrade or deflate the surface.

Wind plays a significant role in the evolution of desert landscapes. Wind erodes in two basic ways. Deflation is a process whereby wind picks up and removes material from an area, resulting in a reduction in the land surface. Abrasion is a different process that occurs when sand and other sizes of particles impact each other. Exposed surfaces in deserts are subjected to frequent abrasion, which is akin to sandblasting.

Yardangs are elongate streamlined wind-eroded ridges, which resemble an overturned ship's hull sticking out of the water. These unusual features are formed by abrasion, by the long-term sandblasting along specific corridors. The sandblasting leaves erosionally resistant ridges but removes the softer material, which itself will contribute to sandblasting in

Sand dune, Persian Gulf *(Photo by Timothy Kusky)*

the downwind direction and eventually contribute to the formation of sand, silt, and dust deposits.

Deflation is important on a large scale in places where there is no vegetation, and in some places the wind has excavated large basins known as deflation basins. Deflation basins are common in the United States from Texas to Canada as elongate (several-kilometer long) depressions, typically only 3–10 feet (1–3 m) deep. However, in places like the Sahara, deflation basins may be as much as several hundred feet deep.

Deflation by wind can only move small particles away from the source, since the size of the particle that can be lifted is limited by the strength of the wind, which rarely exceeds a few tens of miles per hour. Deflation therefore leaves boulders, cobbles, and other large particles behind. These get concentrated on the surface of deflation basins and other surfaces in deserts, leaving a surface concentrated in boulders known as desert pavement. Desert pavements represent a long-term stable desert surface, and they are not particularly hazardous. However, when the desert pavement is broken, for instance, by being driven across the coarse cobbles, and pebbles get pushed beneath the surface, the underlying sands get exposed to wind action again. Driving across a desert pavement can raise a considerable amount of sand and dust, and if many vehicles drive across the surface then it can be destroyed, and the whole surface becomes active.

Wind moves sand by saltation, in arched paths in a series of bounces or jumps. Wind typically sorts different sizes of sedimentary particles, forming elongate small ridges known as sand ripples, very similar to ripples found in streams. Sand dunes are larger than ripples, up to 1,500 feet (457 m) high, composed of mounds or ridges of sand deposited by wind. Sand dunes are locally very important in deserts, and wind is one of the most important processes in shaping deserts worldwide. Shifting sands are one of the most severe geologic hazards of deserts. In many deserts and desert border areas, the sands are moving into inhabited areas, covering farmlands, villages, and other useful land with thick accumulations of sand. This is a global problem, as deserts are currently expanding worldwide.

See also DESERT; LOESS; SAND DUNES.

eon A large division of geologic time. It is the longest geologic time unit, being an order of magnitude above the era division. Geologic time is divided into four eons, from oldest to youngest, including the Hadean, Archean, Proterozoic, and Phanerozoic.

epeirogeny Vertical motions of the crust and lithosphere, with particular reference to the cratons. Most cratons have been remarkably stable for billions of years since they formed, apparently recording only vertical motions in contrast to mountain or orogenic belts that record large-scale horizontal motions. G. K. Gilbert introduced the term in 1890 to describe and attempt to explain large-scale features of the Earth such as basins and plateaus. Gilbert suggested that epeirogenic movements have created much of the recent topography of cratons and younger eroded orogenic belts, although most of these features and effects are now explained by the plate tectonics paradigm.

epicenter *See* EARTHQUAKES.

epicontinental sea Inland seas, or seas built on the continental shelves. They are generally warm and shallow seas, usually less than 1,000 feet (300 m) deep, that flood many continental areas during sea-level high stands, and have in geologic history been places of great biomass explosions and species diversification. They form often during supercontinent dispersal when large volumes of seawater are displaced onto the continents by voluminous young new oceanic ridges. Shallow epicontinental seas covered much of the continents after the breakup of the supercontinents of Gondwana and Pangea.

See also PLATE TECTONICS; SUPERCONTINENT CYCLE.

epoch A geologic unit of time longer than an age and shorter than a period. A corresponding series of rocks is deposited during an epoch; for instance, the Eocene series was deposited during the Eocene epoch. Sometimes the term is applied informally to refer to an interval of time characterized by a specific condition, such as the glacial epoch, or the global hothouse epoch.

era An interval of geologic time that is the next longest after an eon, and the corresponding *erathem* of rocks deposited during this time interval. For instance the Phanerozoic Eon is divided in the Paleozoic, Mesozoic, and Cenozoic Eras. Some divisions of Precambrian time place the Archean and Proterozoic as eras in the Precambrian Eon, preceded by the Hadean Eon, whereas others divide the Proterozoic Eon into the Paleoproterozoic, Mesoproterozoic, and Neoproterozoic Eras, and the Archean Eon into the Early, Middle and Late Eras.

erosion Erosion encompasses a group of processes that cause Earth material to be loosened, dissolved, abraded, or worn away and moved from one place to another. These processes include weathering, dissolution, corrosion, and transportation. There are two main categories of weathering: physical and chemical processes. Physical processes break down bedrock by mechanical action of agents such as moving water, wind, freeze-thaw cycles, glacial action, forces of crystallization of ice and other minerals, and biological interactions with bedrock such as penetration by roots. Chemical weathering includes the chemical breakdown of bedrock in aqueous solutions. Erosion occurs when the products of weathering are loosened and transported from their origin to another place, most typically by water, wind, or glaciers.

Water is an extremely effective erosional agent, especially when it falls as rain and runs across the surface in finger-sized tracks called rivulets, and when it runs in organized streams and rivers. Water begins to erode as soon as raindrops hit a surface—the raindrop impact moves particles of rock and soil, breaking it free from the surface and setting it in motion. During heavy rains, the runoff is divided into overland flow and stream flow. Overland flow is the movement of runoff in broad sheets. Overland flow usually occurs through short distances before it concentrates into discrete channels as stream flow. Erosion performed by overland flow is known as sheet erosion. Stream flow is the flow of surface water in a well-defined channel. Vegetative cover strongly influences the erosive power of overland flow by water. Plants that offer thicker ground cover and have extensive root systems prevent erosion much more than thin plants and those crops that leave exposed barren soil between rows of crops. Ground cover between that found in a true desert and in a savanna grassland tends to be eroded the fastest, while tropical rainforests offer the best land cover to protect from erosion. The leaves and branches break the force of the falling raindrops, and the roots form an interlocking network that holds soil in place.

Under normal flow regimes streams attain a kind of equilibrium, eroding material from one bank and depositing on another. Small floods may add material to overbank and floodplain areas, typically depositing layers of silt and mud over wide areas. However, during high-volume floods, streams may become highly erosive, even removing entire floodplains that may have taken centuries to accumulate. The most severely erosive floods are found in confined channels with high flow, such as where mountain canyons have formed downstream of many small tributaries that have experienced a large rainfall event. Other severely erosive floods have resulted from dam failures, and in the geological past from the release of large volumes of water from ice-dammed lakes about 12,000 years ago. The erosive power of these floodwaters dramatically increases when they reach a velocity known as supercritical flow, at which time they are able to cut through alluvium like butter and even erode bedrock channels. Luckily, supercritical flow can not be sustained for long periods of time, as the effect of increasing the channel size causes the flow to self-regulate and become subcritical.

Cavitation in streams can also cause severe erosion. Cavitation occurs when the stream's velocity is so high that the

Soil erosion, Madagascar *(Photo by Timothy Kusky)*

vapor pressure of water is exceeded and bubbles begin to form on rigid surfaces. These bubbles alternately form and then collapse with tremendous pressure, and they form an extremely effective erosive agent. Cavitation is visible on some dam spillways, where bubbles form during floods and high discharge events, but it is different from the more common, and significantly less erosive phenomenon of air entrapment by turbulence, which accounts for most air bubbles observed in white-water streams.

Wind is an important but less effective erosional agent than water. It is most important in desert or dry environments, with exposed soil or poor regolith. Glaciers are powerful agents of erosion and are thought to have removed hundreds of meters from the continental surfaces during the last ice ages. Glaciers carve deep valleys into mountain ranges and transport eroded sediments on, within, and in front of glaciers in meltwater stream systems. Glaciers that have layers of water along their bases, known as warm-based glaciers, are more effective erosional agents than cold-based glaciers that do not have any liquid water near their bases. Cold-based glaciers are known from Antarctica.

Mass wasting is considered an erosional process in most definitions, whereas others recognize that mass wasting significantly denudes the surface but classify these sudden events separately. Mass wasting includes the transportation of material from one place to another, so it is included here with erosional processes. Most mass-wasting processes are related to landslides, debris flows, and rock slides and can significantly reduce the elevation of a region, typically occurring in cycles with intervals ranging from tens to tens of thousands of years.

Humans are drastically altering the planet's landscape, leading to enhanced rates of erosion. Cutting down forests has caused severe soil erosion in Madagascar, South America, the United States, and many other parts of the world. Many other changes are difficult to quantify. Urbanization reduces erosion in some places but enhances it elsewhere. Damming of rivers decreases the local gradient, slowing erosion in upland areas but prevents replenishment of the land in downstream areas. Agriculture and the construction of levees have changed the balance of floodplains. Although difficult to quantify, it is estimated that human activities in the past couple of centuries have increased erosion rates on average from five times to 100 times previous levels.

See also DESERT; GLACIER; MASS WASTING; WEATHERING.

esker Long, skinny, steep-walled sinuous deposits of irregularly stratified sand and gravel deposited by subglacial or englacial streams. The streams may have been located between the glacier and rock walls or bases, on top of the glacier, or in tunnels within the glacier. Eskers are typically a few to 650 feet (1–200 m) high and range in length from about 300 feet to 310 miles (100 m–500 km). They typically show branching patterns reflecting where streams flowed together into a single

Esker, Northwest Territories, Canada *(Photo by Timothy Kusky)*

channel. Beaded or lobate eskers are a rare geomorphic type of esker, forming where ice flowed into a lake.

See also GLACIER.

Eskola, Pentti (1883–1964) Finnish *Geologist* Pentti Eskola was one of the first Finnish geologists who applied physicochemical ideas to the study of metamorphism. He laid down the foundations for later studies in metamorphic petrology. Eskola was born in Lellainen and became a chemist at the University of Helsinki before specializing in petrology. In the early 1920s he worked in Norway and in Washington, D.C., in the United States. He then taught at Helsinki from 1924 to 1953. Throughout his life Eskola was interested in the study of metamorphic rocks, taking early interest in the Precambrian rocks of England. Relying heavily on Scandinavian studies, he wanted to define the changing pressure and temperature conditions under which metamorphic rocks were formed. His approach allowed for the comparison of rocks of widely differing compositions in respect of the pressure and temperature under which they had originated.

Further Reading
Eskola, Pentti. "Glimpses of the Geology of Finland." *Manchester Geological Association Journal* 2 (1950): 61–79.
———. "The Nature of Metasomatism in the Processes of Granitization." *International Geological Conference* (1950): 5–13.

estuary   A transitional environment between rivers and the sea where freshwater mixes with seawater and is influenced by tides. Most were formed when sea levels were lower, and rivers carved out deep valleys that are now flooded by water. They are typically bordered by tidal wetlands and are sensitive ecological zones that are prone to disturbances by pollution, storms, and overuse by people.

See also CONTINENTAL MARGIN.

eutrophication   A process in which lakes or other water bodies become enriched in nutrients necessary for plant growth, leading to an increase in primary productivity. The process typically leads to the prolific growth of algae that are in turn necessary for fish and animal life. It is most common in shallow lakes, and the process is usually slow, in some cases being related to the infilling of the lake or the erosion of the lake outlet. Deposits from eutrophic lakes tend to be rich in rapidly decaying organic matter. In contrast to natural eutrophication, human-induced eutrophication can occur rapidly and is often caused by the dumping of sewage, fertilizers, or other nutrient-rich substances into lakes, or an increase in the lake's temperature due to discharge of industrial or power plant cooling waters.

See also ECOSYSTEM.

euxinic   Euxinic environments are characterized by restricted circulation with stagnant, typically anaerobic conditions. Euxinic basins include some fiords, estuaries with ridges or sills near their outlets, lagoons, restricted seas, and narrow foredeep basins restricted by tectonic thrusting. Deposition in euxinic basins typically forms hydrogen-sulfide–rich mudstones and black organic-rich shales, and pyrite-rich or graphitic shales. These are colloquially known as "stinky black shales" by field geologists in mountain belts around the world.

evaporite   Evaporite sediments include salts precipitated from aqueous solutions, typically associated with the evaporation of desert lake basins known as playas, or the evaporation of ocean waters trapped in restricted marine basins associated with tectonic movements and sea level changes.

Evaporation in salt pan, Death Valley, California *(Photo by Timothy Kusky)*

Evaporite salts in dried up salt pan, Death Valley, California *(Photo by Timothy Kusky)*

They are also associated with sabkha environments along some coastlines such as along the southern side of the Persian (Arabian) Gulf, where seawater is drawn inland by capillary action and evaporates, leaving salt deposits on the surface.

Evaporites are typically associated with continental breakup and the initial stages of the formation of ocean basins. For instance, the opening of the south Atlantic Ocean about 110 million years ago is associated with the formation of up to 3,280 feet (1 km) of salts north of the Walvis–Rio Grande Ridge. This ridge probably acted as a barrier that episodically (during short sea-level rises) let seawater spill into the opening Atlantic Ocean, where it would evaporate in the narrow rift basin. It would take a column about 18.5 miles (30 km) thick of ocean water to form the salt deposits in the south Atlantic, suggesting that water spilled over the ridge many times during the opening of the basin. The evaporite forming stage in the opening of the Atlantic probably lasted about 3 million years, perhaps involving as many as 350 individual spills of seawater into the restricted basin. Salts that form during the opening of ocean basins are economically important because when they get buried under thick piles of passive continental margin sedi-

ments, the salts typically become mobilized and intrude overlying sediments as salt diapirs, forming salt domes and other oil traps exploited by the petroleum industry.

Salts may also form during ocean closure, with examples known from the Messinian (Late Miocene) of the Mediterranean region. In this case thick deposits of salt with concentric compositional zones reflect progressive evaporation of shrinking basins, when water spilled out of the Black Sea and Atlantic into a restricted Mediterranean basin. So-called closing salts are also known from the Hercenian orogen north of the Caspian Sea, and in the European Permian Zechstein basin in the foreland of the collision.

As seawater progressively evaporates, a sequence of different salts forms from the concentrated brines. Typically, anhydrite ($CaSO_4$) is followed by halite ($NaCl$), which forms the bulk of the salt deposits. There are a variety of other salts that may form depending on the environment, composition of the water being evaporated, and when new water is added to the brine solution and whether or not it partly dissolves existing salts.

See also DIAPIR; PLAYA; RIFT; SEDIMENTARY ROCKS.

Everglades A marshy low-lying tropical to subtropical area in southern Florida, formerly occupying approximately 5,000 square miles (13,000 km²) but significantly reduced in area by draining and building in past decades. The region receives about 60 inches (152 cm) of rain per year. Much of the area, including a large part of Florida Bay, is now part of Everglades National Park, ensuring its preservation and protection from further land development. The park has also been designated a World Heritage Site and an International Biosphere Preserve. The vegetation and ecosystems in the Everglades are unique, including coastal mangrove forests, cyprus swamps, pinelands, saw grasses, water patches, island-like masses of vegetation known as hammocks, and marine and estuarine environments. Everglades Park animal life includes numerous wading and other bird species, including the wood stork, egrets, herons, and spoonbills, and it is the only place in the world where alligators and crocodiles live together.

The Everglades are underlain by Miocene (6-million-year-old) limestone with a well-developed karst network, so huge quantities of groundwater move through subterranean caverns and passageways and emerge as freshwater springs in other places. Many sinkholes have opened in southern Florida as a result of lowering of the groundwater table. The limestone is overlain by thick layers of decayed organic material commonly referred to as black muck, accumulated over millions of years. In other places, a shallow marine deposit known as the Miami Oolite overlies the older limestone, formed during the interglacial stages of the Pleistocene glacial periods about 100,000 years ago. During glacial periods sea level was about 300 feet lower than at present, but during the interglacials, sea level was up to 100 feet higher than at present. The Miami Oolite consists of round oolites of calcium carbonate, formed as warm shallow water saturated in calcium carbonate precipitated concentric layers of carbonate around small shell and other fragments as they rolled in tidal currents.

The Seminole Indians inhabited the Everglades for centuries, with contact made by several colonial expeditions in the 1500s. In the late 1830s, U.S. forces engaged the Seminoles in military operations that included draining of large tracts of land. A huge fire caused by the over-draining followed in 1939, leading to studies that concluded that the southern parts of the Everglades were uninhabitable, and to establishment of Everglades National Park in 1947. Further development in the Big Cyprus Swamp part of the Everglades and the construction of retaining walls in the 1960s on the south shore of Lake Okeechobee further disrupted the natural flow of water. Ecosystem restoration efforts began in earnest in 1972, when the Florida legislature passed several environmental and growth management laws including the Land Conservation Act. In 1983 an ambitious program called "Save Our Everglades" was launched, with the goal of restoring the ecosystem to its pre-1900 condition.

See also GROUNDWATER; KARST; TROPICS.

evolution Usually regarded as a slow, gradual process that describes how life has changed with time on the Earth, ranging from simple single-celled organisms to the complex biosphere on the planet today. However, it is better defined as a sustained change in the genetic makeup of populations over a period of generations leading to a new species. The field of evolution was pioneered by Jean-Baptiste Lamarck (1809) and Charles Darwin in his *On the Origin of Species* (1859) and *The Descent of Man* (1871), and it is a multidisciplinary science incorporating geology, paleontology, biology, and, with neo-Darwinism, geneticists.

Darwin sailed on the H.M.S. *Beagle* (1831–36) when he made numerous observations of life and fossils from around the world, leading to the development of his theory of natural selection, in which races with favorable traits stand a better chance for survival. The main tenets of his theory are that species reproduce more than necessary, but populations tend to remain stable since there is a constant struggle for food and space, and only the fittest survive. Darwin proposed that the traits that led to some individuals surviving are passed on to their descendants, hence propagating the favorable traits. However, Darwin did not have a good explanation about why some individuals would have favorable traits that others would not, and this evidence would not come until much later with the field of genetics and the recognition that mutations can cause changes in character traits. Sequential passing down to younger generations of mutation-induced changes in character traits can lead to changes in the species, and eventually the evolution of new species. Darwin's process of natural selection worked by eliminating the less successful forms of species, favoring the others that had favorable mutations.

More modern variations on evolution recognize two major styles of change. Macroevolution describes changes above the species level and the origin of major groups, whereas microevolution is concerned with changes below the species level and the development of new species. Another major change in understanding evolution in the past century concerns the rate of evolutionary changes as preserved in the fossil record. Darwin thought that evolution was gradual, with one species gradually changing into a new species, but the fossil record shows only a few examples of this gradual change (with notable examples including changes in trilobites in the Ordovician, and changes in horses in the Cenozoic). Nearly all species in the fossil record exist with little change for long durations of geologic time, then suddenly disappear, and are suddenly replaced by new species. In other cases, new species suddenly appear without the disappearance of other species. Some of the apparent rapid change was initially regarded as an artifact of an incomplete record, but many examples of complete records show that these rapid changes are real. A new paradigm of evolution named punctuated equilibrium, advanced in the 1970s by Steven J. Gould and Niles Eldredge, explains these sudden evolutionary changes. Physical or geographic isolation of some member of a species, expected during supercontinent breakup, can separate and decimate the environment of a species and effectively isolate some members of a species in conditions in which they can change. This small group may have a mutation that favors their new environment, letting them survive. When supercontinents collide, many species that never encountered each other must compete for the same food and space, and only the strongest will survive, leading to extinction of the others.

In other cases, major environmental catastrophes such as meteorite impacts and flood basalt eruptions can cause extreme changes to the planetary environment, causing mass extinction. Relatively minor or threatened species that survive can suddenly find themselves with traits that favor their explosion into new niches and their dominance in the fossil record.

See also MASS EXTINCTIONS; PALEONTOLOGY; SUPERCONTINENT CYCLE.

Further Reading

McKinney, Michael L. *Evolution of Life, Processes, Patterns, and Prospects*. Englewood Cliffs, N.J.: Prentice Hall, 1993.

Stanley, Steven M. *Earth and Life Through Time*. New York: W.H. Freeman and Co., 1986.

exfoliation   A weathering process where successive shells of rock spall off the face of a mountain, like the skin of an onion. The shells may be a few millimeters to several meters thick. It is caused by sliding of rock sheets along joints that form by differential stresses within a rock, typically generated by chemical weathering or by cooling and uplift of the rock mass from depth. For instance, weathering of feldspar to clay causes an increase in volume. This volume increase is accommodated through the generation of fractures parallel to the surface, and the formation of some pop-up structures where two slabs of rock pop up to form a triangular gap with a few centimeters uplift off the surface. These pop-up structures often break off, initiating the sliding of big slabs of rock off the mountain. Exfoliation typically forms rounded, dome-shaped mountains known as exfoliation domes.

See also WEATHERING.

extratropical cyclones   Also known as wave cyclones—hurricane-strength storms that form in middle and high latitudes at all times of the year. Examples of these strong storms include the famous "nor'easters" of New England in the United States; storms along the east slopes of the Rockies and in the Gulf of Mexico, and smaller hurricane-strength storms that form in arctic regions. These storms develop along polar fronts that form semi-continuous boundaries between cold polar air and warm subtropical air. Troughs of low pressure can develop along these polar fronts, and winds that blow in opposite directions to the north and south of the low set up a cyclonic (counterclockwise in the Northern Hemisphere) wind shear that can cause a wave-like kink to develop in the front. This kink is an incipient cyclone and includes (in the Northern Hemisphere) a cold front that pushes southward and counterclockwise, and a warm front that spins counterclockwise and moves to the north. A comma-shaped band of precipitation develops around a central low that develops where the cold and warm fronts meet, and the whole system will migrate east or northeast along the polar front, driven by high-altitude steering winds.

The energy for extratropical cyclones to develop and intensify comes from warm air rising and cold air sinking, transferring potential energy into kinetic energy. Condensation also provides extra energy as latent heat. These storms can intensify rapidly and are especially strong when the cold front overtakes the warm front, occluding the system. The point where the cold front, warm front, and occluded front meet is known as a triple point and is often the site of the formation of a new secondary low-pressure system to the east or southeast of the main front. This new secondary low often develops into a new cyclonic system and moves east or northeastward, and it may become the stronger of the two lows. In the case of New England's "nor'easters," the secondary lows typically develop off the coast of the Carolinas or Virginia, then rapidly intensify as they move up the coast, bringing cyclonic winds and moisture in from the northeast off the Atlantic Ocean.

See also HURRICANE.

extrusive rocks   *See* VOLCANO.

facies Refers to the appearance and characteristics of a rock, with implications for its mode of origin. The term usually refers to sedimentary rocks with descriptions of the rock's lithology and inferred environment of deposition. Other less common usages of the term include metamorphic facies, identified by the mineral assemblages that form in specific rock types under a set range in pressure and temperature conditions, and biofacies, referring to a local assemblage of living or fossil organisms. The term has also been applied, somewhat unsuccessfully, to differentiate igneous and even strain variations in different terrains.

If we travel around the planet we will see that different types of sedimentary rocks are deposited in different places. Within individual systems, such as the Mississippi River Delta, or the Gulf of Mexico, there are also lateral changes, such as differences in grain size, or the type of sedimentary structures. Sedimentary facies refers to a distinctive group of characteristics within a body of sediment that differs from those elsewhere in the same body.

Most sediments change laterally as a result of changes in original depositional environment. An interesting and useful concept, known as Walther's Law, states that the lateral changes in sedimentary facies can also be found in the vertical succession of strata in the system, because of the lateral migration of sedimentary systems. Thus, changes upward in the stratigraphy also reflect changes laterally.

Sedimentary facies in nonmarine environments include stream facies, lake facies, glacial facies, and eolian (or wind-dominated) facies. Within each of these environments, different sub-facies can be described and differentiated from each other. Streams are the principal transporting agent for moving sediments over land. Stream sediments are known as alluvium, and the sedimentary environment is known as fluvial. There are many different sedimentary facies in fluvial systems, and these include conglomerates and sand deposited in the stream channel and fine-grained silt and mud deposited on the floodplain or alluvial plain. Lake sediments are known as lacustrine, with different facies including the lakeshore and the lake bottom. Lakeshore deposits include gravel and sand beaches (for big lakes) and deltas, whereas lake bottom environments include finely laminated clays and silts and vary under the right conditions. Glaciers are great movers of sediments. Sediments deposited directly by glaciers are typically poorly sorted conglomerates mixed with clay, and many fragments may be angular. Many sediments from glaciers are reworked by meltwater, however, and deposited in streams and lakes in front of the glacier. Sediments deposited by wind are referred to as eolian deposits. Since air is less dense than water and can hold less material in suspension, deposits from wind systems tend to be fine-grained. Sand and sand dunes are typical eolian deposits, with large-scale cross-laminations in the sand.

Some of the world's thickest sedimentary deposits are located on the continental shelves, and these are of considerable economic importance because they also host the world's largest petroleum reserves. The continental shelves are divided into many different sedimentary environments and facies. Nearshore environments include estuaries, deltas, beaches, and shallow marine continental shelves. Many of the sediments transported by rivers are deposited in estuaries, which are semi-enclosed bodies of water in which freshwater and seawater mix. In many cases, estuaries are slowly subsiding, and they get filled with thick sedimentary deposits such as muds, dolomitic or limestone muds, silts, and storm deposits. Deltas are formed where streams and rivers meet the ocean and drop their loads because of the reduced flow velocity. Deltas are complex sedimentary systems, with coarse stream channels, fine-grained interchannel and overbank sediments, and a gradation into deepwater deposits. Beaches contain the coarse fraction of material deposited at the oceanfront by rivers and sea cliff erosion. Quartz is typically very abundant, because of its resistance to weathering and its abundance in the crust. Beach sands tend to be well-rounded, as does anything else such as beach glass, because of the continuous abrasion caused by the waves dragging the particles back and forth.

Deep marine facies include pelagic muds, siliceous oozes, and windblown dust that accumulates on the seafloor. The character of deep marine deposits has changed dramatically with time since many of the present deep-sea sediments are produced by the accumulation of the skeletal remains of organisms that did not exist in the Precambrian.

Facies analysis is a very important tool for reconstructing ancient paleoenvironments and depositional settings. When the facies of a rock are understood, facies analysis becomes an important exploration tool and may help find new locations for economically significant hydrocarbon reserves or help understand the tectonic and depositional history of a region.

See also SEDIMENTARY ROCKS.

Further Reading

Reading, Harold G. *Sedimentary Environments: Processes, Facies, and Stratigraphy*, 3rd ed. Oxford: Blackwell Scientific Publication, 1996.

Falkland Plateau

A shallow-water shelf extending 1,200 miles (2,000 km) eastward from Tierra del Fuego on the southern tip of South America past South Georgia Island. The plateau includes the Falkland Islands 300 miles (480 km) east of the coast of South America and is bounded on the south by the Scotia Ridge and on the north by the Agulhas-Falkland fracture zone. The Falkland Islands include two main islands (East and West Falkland) and about 200 small islands and are administered by the British but also claimed by Argentina, with the capital at Stanley. The islands are stark rocky outposts, plagued by severe cold rains and wind, but they have abundant seals and whales in surrounding waters. The highest elevation is 2,315 feet (705 m) on Mount Adam. Thick peat deposits support a sheep-farming community among the dominantly Scottish and Welsh population.

The Falkland Plateau formed as a remnant of the southern tip of Africa that remained attached to South America during the breakup of Gondwana and the movement of South America away from Africa. The Agulhas-Falkland fracture zone extends to the tip of Africa and represents the transform along which divergence of the two continents occurred. Numerous Mesozoic rift basins on the plateau are the site of intensive oil exploration. The geology of the Falklands was first described by Charles Darwin from his expedition on the *Beagle* in 1833, and reported in 1846, and later pioneering studies were completed by Johan G. Andersson in 1916.

Precambrian granite, schist, and gneiss are found on the southwest part of Western Falkland Island, probably correlated with the Nama of South Africa. The Precambrian basement is overlain by a generally flat to gently tilted Paleozoic sequence including 1.7 miles (3,000 m) of Devonian quartzite, sandstone, and shale. A 2.2-mile (3,500-m) thick Permo-Triassic sequence unconformably overlies the Paleozoic sequence and includes tillites and varves indicating glacial influence. These rocks are cut by Triassic-Jurassic dioritic to diabasic dikes and sills related to the Karoo and Parana flood basalts. Quaternary interglacial deposits are overlain by diamictites and long lobes of gravel interpreted as mudflows deposited in a periglacial environment.

The Falklands are folded into a series of northwest-southeast trending folds that intensify to the south and swing to east-west on the east of the plateau.

Further Reading

Andersson, J. G. "Geology of the Falkland Islands." *Wissenschaftliche Ergebnisse der Schwedischen Sudpolar-Expedition*, 1901–1903, 1916.

Darwin, Charles. "Geology of the Falkland Islands." *Quarterly Journal of the Geological Society* 2 (1846).

fault   Discrete surfaces or planes across which two bodies of rock have moved or slid past each other. They are disconti-

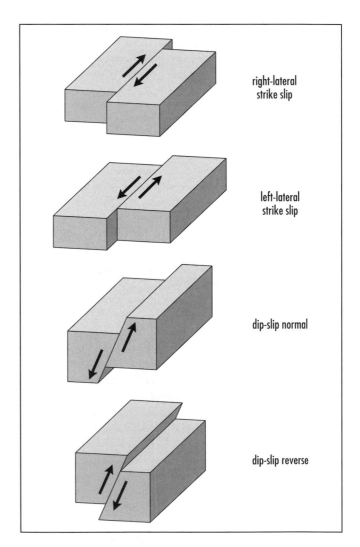

Definition of different types of faults

Fault surface, Newfoundland *(Photo by Timothy Kusky)*

nuities, and they contrast with deformation by homogeneous strain, folding, or buckling. Faults form at relatively high crustal levels and low temperatures and are characterized by brittle deformation. They are sites of rock crushing, grinding, and fluid circulation. Faulting usually occurs in discrete events or earthquakes, or by slower creep. Because the rocks are easily crushed, faults tend to erode, forming gullies, ravines, or valleys with no exposed rock, across which there is a discontinuity in the geology.

Faults may be recognized by the presence of one or several structural elements. They may cause structural or lithological discontinuity, fault zone deformation, deformation of the land surface (scarps, offset features); they may be associated with fault-related sediments and basins, or stratigraphic repetition, omission, or discontinuity. Recent fault traces in tectonically active areas produce ground breaks or other recognizable features associated with the fault trace. The most prominent of these features are the fault scarps, where part of the land's surface is displaced relative to a part that used to be at the same level. Lines of vegetation may be associated with increased water permeability along the fault, and they may form ridges and depressions as different rock slivers are uplifted or down-sagged between bifurcating fault segments. Notches and sad-

dles in ridges in mountainous areas may mark the trace of a fault, since fault zones typically contain rocks that are crushed and easily eroded. Offset stream channels are common along strike-slip faults such as the San Andreas.

Naming faults is based on the orientation of the fault planes and the direction of slip or displacement on the fault. Faults may slip parallel to dip (dip-slip faults), parallel to strike (strike-slip faults), or obliquely. The area above a fault is the hanging wall, whereas the block below the fault is the footwall. Steeply dipping faults are known as high-angle faults, whereas shallow dipping faults are known as low-angle faults. Contractional faults shorten or contract a bed when they move, whereas extensional faults extend or lengthen beds when they move. The most common classification is into normal, thrust (or reverse), and strike-slip faults. Normal faults are dip-slip faults in which the hanging wall drops relative to the footwall. Thrust faults are dip-slip faults in which the hanging wall moves up relative to the footwall. Strike-slip faults are steeply dipping faults that have horizontal slips parallel to the fault trace.

Faulting often creates topographic differentials across the fault, creating opportunities for sedimentation. Faults may form scarps with breccias and olistostromes deposited along their edges. Some sedimentary basins have faulting occurring throughout deposition, forming growth faults, characterized by curving surfaces with thicker layers of sedimentary strata deposited on the downthrown side of the fault.

Fault zones are dominated by brittle structures and solution transfer structures (stylolites, veins), with minor intracrystalline ductile deformation. They normally form under brittle conditions in the upper 10 kilometers in the crust, but transitions do exist with ductile shear zones and mylonites. Frictional sliding along faults typically produces polished and striated surfaces known as slickensides. The striations on the surface are known as slickenlines and are formed parallel to the direction of movement on the fault. The steps on the slickenside surface face in the direction of movement of the opposite block. Slickenlines may be either grooves, scratched into the surface during movement, or fibers grown within small spaces that opened within the fault plane during movement. Fault breccias are angular fragments of rocks formed within fault zones by the crushing of rocks from either side of the fault. Fault gouge is a fine, sandy to clay-like ground-up rock powder formed by intense grinding and milling of the adjacent rocks. Pseudotachylite is an extremely fine-grained glassy rock typically found as dikes and intrusive veins along faults, formed by frictional melting of rocks along the fault. Various types of vein systems are common along faults because of the increased fluid flow along the faults. Mylonite is a ductile shear zone rock, formed at deeper crustal levels than the faults, although there is a transition zone where they both may occur.

See also BRECCIA; MYLONITE; STRUCTURAL GEOLOGY.

faunal succession   Faunal succession, or the Law of Faunal Succession, refers to the sequence of life-forms observed in the geological record. The law states that fossil organisms (fauna and flora) succeed one another in a definite and recognizable order, and that each formation on the Earth preserves a different portion of the history of life than the formations above and below it. The succession of life records a sequence of changes in life on Earth that have been used to demonstrate that evolution records a series of one-way changes in organisms in that once a species disappears from the record, it never appears again.

See also EVOLUTION; PALEONTOLOGY.

Federal Emergency Management Agency (FEMA)   The nation's premier agency that deals with emergency management and preparation and issues warnings and evacuation orders when disasters appear imminent. FEMA maintains a Web site (http://www.fema.gov) that is updated at least daily and includes information on hurricanes, floods and national flood insurance, fires, and disaster prevention, preparation, and emergency management. The Web site also contains information on costs of disasters, maps, and directions on how to do business with FEMA. The agency is divided into national and regional sites. Their main headquarters is at: FEMA, 500 C Street, SW, Washington D.C. 20472.

feldspar   A general name for a group of related framework silicate minerals that form the most abundant constituent in igneous rocks and form about 60 percent of the Earth's crust. They are common in most types of igneous rocks, except for the ultramafic suite, and form major constituents of schists, gneisses, migmatites, pegmatites, and in arenaceous sediments. Most feldspars are white or clear to translucent, and have a hardness of 6 on Moh's hardness scale.

Most feldspars are members of a Ternary system with three compositional end members. The compositions $NaAlSi_3O_8$, $KAlSi_3O_8$, and $CaAl_2Si_2O_8$ are referred to as sodium, potassium, and calcium feldspar. There is a continuous compositional range between the potassium and sodium feldspars, with the intermediate varieties known as alkali feldspars, including sanidine, anorthoclase, and albite. Likewise, the continuous compositional series of feldspars between sodium and calcium feldspar are known as the plagioclase feldspars, with compositional variants including albite, oligoclase, andesine, labradorite, bytownite, and anorthite. Celsian feldspar has the composition $BaAl_2Si_2O_8$.

Feldspars exhibit a variety of crystal forms and structural states. Calcium feldspar (anorthite) and sodium feldspar are triclinic, whereas potassium feldspar is monoclinic. Several of the feldspars crystallize in different structural states and may have different lattice obliquities depending on the temperature of crystallization.

See also MINERALOGY.

Finger Lakes   A group of 11 long, narrow glacial lakes in western New York near Rochester, Syracuse, and Ithaca. The lakes are elongate in a north-south direction reflecting the direction of glacial movement. The largest of the lakes, Seneca and Cayuga, are each more than 35 miles (56 km) long. Geologically, the area is underlain by Devonian limestone and shale, including many richly fossiliferous beds. The Finger Lake region is famous for grapes and outdoor recreation.

fiord   Long, narrow, U-shaped glacial valleys that are flooded by arms of the sea. Many are bordered by several hundred-meter-high steep-walled rocky cliffs along mountainous coasts, such as those found in Scandinavia, Alaska, British Colombia, Patagonia, Greenland, Baffin Island, Iceland, Ellesmere Island, New Zealand, and Antarctica. They typically have a shallow sill or bedrock bench at their mouth and become deeper inland. Fiords form as glacially excavated valleys that later become flooded by the sea after the glacier melts. Fiord is the English spelling of the Norwegian word *fjord*.

See also GLACIER.

McCarty Fiord, Alaska *(Photo by Timothy Kusky)*

First Law of Thermodynamics *See* THERMODYNAMICS.

fission Splitting of heavy atomic nuclei into two or more fragments, accompanied by the release of two or three neutrons and a large amount of nuclear energy. This process occurs naturally and spontaneously in nuclei of U^{235}, which is the main fuel used in nuclear reactors. In nuclear accelerators, this process is achieved by bombarding atomic nuclei with neutrons that become unstable and split, releasing additional neutrons in a series of reactions known as a continuous chain reaction. The minimum amount of fissile material that can undergo a continuous chain reaction is known as the critical mass. This process must be controlled or a nuclear explosion will result. In nuclear reactors the neutrons released from U^{235} are used to set up further reactions in a controlled chain reaction, creating large amounts of harnessable nuclear energy.

See also RADIOACTIVE DECAY.

flood There are several kinds of floods, including those associated with hurricanes and tidal surges in coastal areas, those caused by rare large thunderstorms in mountains and canyon territory, and those caused by prolonged rains over large drainage basins.

Flash floods result from short periods of heavy rainfall and are common near warm oceans, along steep mountain fronts that are in the path of moist winds, and in areas prone to thunderstorms. They are well-known in the mountain and canyon lands of the southwest desert in the United States and many other parts of the world. Some of the heaviest rainfalls in the United States have occurred along the Balcones escarpment in southeastern Texas. Atmospheric instability in this area often forms along the boundary between dry desert air masses to the northwest and warm moist air masses rising up the escarpment from the Gulf of Mexico to the south and east. Up to 20 inches (50 cm) of rain have fallen along the

Balcones escarpment in as little as three hours from this weather situation. The Balcones escarpment also seems to trap tropical hurricane rains, such as those from Hurricane Alice, which dumped more than 40 inches (100 cm) of rain on the escarpment in 1954. The resulting flood waters were 65 feet (20 m) deep, one of the largest floods ever recorded in Texas. Approximately 25 percent of the catastrophic flash flooding events in the United States have occurred along the Balcones escarpment. On a slightly longer time scale, tropical hurricanes, cyclones, and monsoonal rains may dump several feet of rain over periods of a few days to a few weeks, resulting in fast (but not quite flash) flooding.

Flash floods typically occur in localized areas where mountains cause atmospheric upwelling leading to the development of huge convective thunderstorms that can pour several inches of rain per hour onto a mountainous terrain, which focuses the water into steep walled canyons. The result can be frightening, with flood waters raging down canyons as steep, thundering walls of water that crash into and wash away all in their paths. Flash floods can severely erode the landscape in arid and sparsely vegetated regions but do much less to change the landscape in more humid, heavily vegetated areas.

Many of the canyons in mountainous regions have fairly large parts of their drainage basins upriver. Sometimes the storm that produces a flash flood with a wall of water may be located so far away that people in the canyon do not even know that it is raining somewhere, or that they are in immediate and grave danger.

The severity of a flash flood is determined by a number of factors other than the amount of rainfall. The shape of the drainage basin is important, because it determines how quickly rainfall from different parts of the basin converge at specific points. The soil moisture and previous rain history are also important, as are the amounts of vegetation, urbanization, and slope.

The national record for the highest, single-day rainfall is held by the south Texas region, when Hurricane Claudette dumped 43 inches (110 cm) of rain on the Houston area in 1979. The region was hit again by devastating floods during June 8–10, 2001, when an early-season tropical storm suddenly grew off the coast of Galveston and dumped 28–35 inches (70–89 cm) of rain on Houston and surrounding regions. The floods were among the worst in Houston's history, leaving 17,000 people homeless and 22 dead. More than 30,000 laboratory animals died in local hospital and research labs, and the many university and hospital research labs experienced hundreds of millions of dollars in damage. Fifty million dollars were set aside to buy out the properties of homeowners who had built on particularly hazardous flood plains. Total damages have exceeded $5 billion. The standing water left behind by the floods became breeding grounds for disease-bearing mosquitoes, and the humidity led to a dra-

Mississippi River Basin and the Midwest Floods of 1927 and 1993

The Mississippi River is the largest river basin in the United States and the third largest river basin in the world. The river basin is the site of frequent floods that can be devastating because of the millions of people that live there. All of the 11 major tributaries of the Mississippi River have also experienced major floods, including events that have at least quadrupled the normal river discharge in 1883, 1892, 1903, 1909, 1927, 1973, and 1993.

Floods along the Mississippi River in the 1700s and 1800s prompted the formation of the Mississippi River Commission, which oversaw the construction of high levees along much of the length of the river from New Orleans to Iowa. These levees were designed to hold the river in its banks by increasing their natural heights. By the year 1926 more than 1,800 miles (2,900 km) of levees had been constructed, many of them more than 20 feet (6 m) tall. The levees gave people a false sense of security against the floodwaters of the mighty Mississippi and restricted the channel, causing floods to rise more quickly and forcing the water to flow faster.

Many weeks of rain in the late fall of 1926 followed by high winter snow melts in the upper Mississippi River basin caused the river to rise to alarming heights by the spring of 1927. Residents all along the Mississippi were worried, and they began strengthening and heightening the levees and dikes along the river, in the hopes of averting disaster. The crest of water was moving through the upper Midwest and had reached central Mississippi, and the rains continued. In April levees began collapsing along the river sending torrents of water over thousands of acres of farmland, destroying homes and livestock and leaving 50,000 people homeless.

One of the worst-hit areas was Washington County, Mississippi, where an intense late April storm dumped an incredible 15 inches (38 cm) of rain in 18 hours, causing additional levees along the river to collapse. One of the most notable was the collapse of the Mounds Landing levee, which caused a 10-foot deep lobe of water to cover the Washington County town of Greenville on April 22. The river reached 50 miles (80.5 km) in width and flooded approximately one million acres, washing away an estimated 2,200 buildings in Washington County alone. Hundreds of people perished while they were trying to keep the levees from collapsing and were washed away in the deluge. The floodwaters remained high for more than two months, and people were forced to leave the area (if they could afford to) or live in refugee camps on the levees, which were crowded and unsanitary. An estimated 1,000 people perished in the floods of 1927, some from the initial flood and more from famine and disease in the unsanitary conditions in the months that followed.

Once again, in 1972 the waters began rising along the Mississippi, with most tributaries and reservoirs filled by the end of the summer. The rains continued through the winter of 1972–73, and the snowpack thickened over the northern part of the Mississippi basin. The combined snowmelts and continued rains caused the river to reach flood levels at St. Louis in early March, before the snow had even finished melting. Heavy rain continued throughout the Mississippi basin, and the river continued to rise through April and May, spilling into fields and low-lying areas. The Mississippi was so high that it rose to more than 50 feet (15 m) above its average levels for much of the lower river basin, and these river heights caused many of the smaller tributaries to back up until they too were at this height. The floodwaters rose to levels not seen for 200 years. At Baton Rouge, the river nearly broke through its banks and established a new course to the Gulf of Mexico, which would have left New Orleans without a river. The floodwaters began peaking in late April, causing 30,000 people to be evacuated in St. Louis by April 28, and close to 70,000 people were evacuated throughout the region. The river remained at record heights throughout the lower drainage basin through late June. Damage estimates exceeded $750 million (1973 dollars).

In the late summer of 1993 the Mississippi River and its tributaries in the upper basin rose to levels not seen in more than 130 years. The discharge at St. Louis was measured at more than one million cubic feet per second. The weather situation that led to these floods was remarkably similar to that of the floods of 1927 and 1973, only worse. High winter snowmelts were followed by heavy summer rainfalls caused by a low-pressure trough that stalled over the Midwest because it was blocked by a stationary high-pressure ridge that formed over the East Coast of the United States. The low-pressure system drew moist air from the Gulf of Mexico that met the cold air from the eastern high-pressure ridge, initiating heavy rains for much of the summer. The rivers continued to rise until August, when they reached unprecedented flood heights. The discharge of the Mississippi was the highest recorded, and the height of the water was even greater because all the levees that had been built restricted the water from spreading laterally and thus caused the water to rise more rapidly than it would have without the levees in place. More than two-thirds of all the levees in the Upper Mississippi River basin were breached, overtopped, or damaged by the floods of 1993. Forty-eight people died in the 1993 floods, and 50,000 homes were damaged or destroyed. Total damage costs are estimated at more than $20 billion.

The examples of the floods of 1927 and 1993 on the Mississippi reveal the dangers of building extensive levee systems along rivers. Levees adversely affect the natural processes of the river and may actually make floods worse. The first effect they have is to confine the river to a narrow channel, causing the water to rise faster than if it were able to spread across its floodplain. Additionally, since the water can no longer flow across the floodplain it cannot seep into the ground as effectively, and a large amount of water that would normally be absorbed by the ground now must flow through the confined river channel. The floods are therefore larger because of the levees. A third hazard of levees is associated with their failure. When a levee breaks, it does so with the force of hundreds or thousands of acres of elevated river water pushing it from behind. The force of the water that broke through the Mounds Landing Levee in the 1927 flood is estimated to be equivalent to the force of water flowing over Niagara Falls. If the levees were not in place, the water would have risen gradually and would have been much less catastrophic when it eventually came into the farmlands and towns along the Mississippi River basin.

matic increase in the release of mold spores that cause allergies in some people, and some of which are toxic.

The Cherrapunji region in southern India at the base of the Himalayan Mountains has received the world's highest rainfalls. Moist air masses from the Bay of Bengal move toward Cherrapunji, where they begin to rise over the high Himalayan Mountains. This produces a strong orographic effect, where the air mass can not hold as much moisture as it rises and cools, so heavy rains result. Cherrapunji has received as much as 30 feet (9 m) of rain in a single month (July 1861) and more than 75 feet (23 m) of rain for all of 1861.

A final type of flood occurs in areas where rivers freeze over. The annual spring breakup can cause severe floods, initiated when blocks of ice get jammed behind islands, bridges, or along bends in rivers. These ice dams can create severe floods, causing the high spring waters to rise quickly, bringing the ice-cold waters into low-lying villages. When ice dams break up, the force of the rapidly moving ice is sometimes enough to cause severe damage, knocking out bridges, roads, and homes. Ice-dam floods are fairly common in parts of New England, including New Hampshire, Vermont, and Maine.

See also DRAINAGE BASIN; RIVER SYSTEM; URBANIZATION AND FLASH FLOODING.

Further Reading

Baker, Victor R. "Stream-Channel Responses to Floods, with Examples from Central Texas." *Geological Society of America Bulletin* 88 (1977): 1,057–1,071.

Belt, Charles B., Jr. "The 1973 Flood and Man's Constriction of the Mississippi River." *Science* 189 (1975): 681–684.

Junk, Wolfgang J., Peter B. Bayley, and Richard E. Sparks. "The Flood Pulse Concept in River-Floodplain Systems." *Canadian Special Publication Fisheries and Aquatic Sciences* 106 (1989): 110–127.

Maddock, Thomas, Jr. "A Primer on Floodplain Dynamics." *Journal of Soil and Water Conservation* 31 (1976): 44–47.

flood basalt Deposits include vast plateaus of basalts that cover large provinces of some continents and are also known as continental flood basalts, plateau basalts, large igneous provinces, and traps. They have a tholeiitic basalt composition, but some show chemical evidence of minor contamination by continental crust. They are similar to the anomalously thick and topographically high seafloor known as oceanic plateaus, and some volcanic rifted passive margins. At several times in the past several hundred million years, these vast outpourings of lava have accumulated forming thick piles of basalt, representing the largest known volcanic episodes on the planet. These piles of volcanic rock represent times when the Earth moved more material and energy from its interior than during intervals between the massive volcanic events. Such large amounts of volcanism also released large amounts of volcanic gases into the atmosphere, with serious implications for global temperatures and climate, and may have contributed to some global mass extinctions.

The largest continental flood basalt province in the United States is the Columbia River flood basalt in Washington, Oregon, and Idaho. The Columbia River flood basalt province is 6–17 million years old and contains an estimated 1,250 cubic miles (4,900 km³) of basalt. Individual lava flows erupted through fissures or cracks in the crust, then flowed laterally across the plain for up to 400 miles (645 km).

The 66-million-year-old Deccan flood basalts, also known as traps, cover a large part of western India and the Seychelles. They are associated with the breakup of India from the Seychelles during the opening of the Indian Ocean. Slightly older flood basalts (90–83 million years old) are associated with the breakaway of Madagascar from India. The volume of the Deccan traps is estimated at 5 million cubic miles (2,008,000 km³), and the volcanics are thought to have been erupted in about one million years, starting slightly before the great Cretaceous-Tertiary extinction. Most workers now agree that the gases released during the flood basalt volcanism stressed the global biosphere to such an extent that many marine organisms became extinct, and many others were stressed. Then the planet was hit by the massive Chicxulub impact, causing the massive extinction including the end of the dinosaurs.

The breakup of East Africa along the East African rift system and the Red Sea is associated with large amounts of Cenozoic (less than 30 million years old) continental flood basalts. Some of the older volcanic fields are located in East Africa in the Afar region of Ethiopia, south into Kenya and Uganda, and north across the Red Sea and Gulf of Aden into Yemen and Saudi Arabia. These volcanic piles are overlain by younger (less than 15-million-year-old) flood basalts that extend both farther south into Tanzania and farther north through central Arabia, where they are known as Harrats, and into Syria, Israel, Lebanon, and Jordan.

An older volcanic province also associated with the breakup of a continent is known as the North Atlantic Igneous Province. It formed along with the breakup of the North Atlantic Ocean at 62–55 million years ago and includes both onshore and offshore volcanic flows and intrusions in Greenland, Iceland, and the northern British Isles, including most of the Rockall Plateau and Faeroes Islands. In the South Atlantic, similar 129–134-million-year-old flood basalts were split by the opening of the ocean and now are comprised of two parts. In Brazil the flood lavas are known as the Parana basalts, and in Namibia and Angola of West Africa as the Etendeka basalts.

These breakup basalts are transitional to submarine flood basalts that form oceanic plateaus. The Caribbean Ocean floor represents one of the best examples of an oceanic plateau, with other major examples including the Ontong-Java Plateau, Manihiki Plateau, Hess Rise, Shatsky Rise, and Mid Pacific Mountains. All of these oceanic plateaus contain between six and 25-mile (10–40-km) thick piles of volcanic

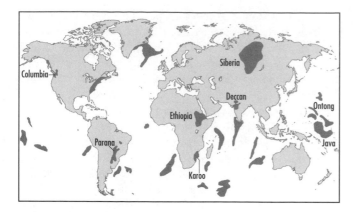

Map of world showing distribution of flood basalts

and subvolcanic rocks representing huge outpourings of lava. The Caribbean seafloor preserves five to 13-mile (8–21-km) thick oceanic crust formed before about 85 million years ago in the eastern Pacific Ocean. This unusually thick ocean floor was transported eastward by plate tectonics, where pieces of the seafloor collided with South America as it passed into the Atlantic Ocean. Pieces of the Caribbean oceanic crust are now preserved in Colombia, Ecuador, Panama, Hispaniola, and Cuba, and some scientists estimate that the Caribbean oceanic plateau may have once been twice its present size. In either case, it represents a vast outpouring of lava that would have been associated with significant outgassing with possible consequences for global climate and evolution.

The western Pacific Ocean basin contains several large oceanic plateaus, including the 20-mile (32-km) thick crust of the Alaskan-sized Ontong-Java Plateau, which is the largest outpouring of volcanic rocks on the planet. It apparently formed in two intervals, at 122 million and 90 million years ago, entirely within the ocean, and represents magma that rose in a plume from deep in the mantle and erupted on the seafloor. It is estimated that the volume of magma erupted in the first event was equivalent to that of all the magma being erupted at mid-ocean ridges at the present time. Sea levels rose by more than 30 feet (9 m) in response to this volcanic outpouring. The gases released during these eruptions are estimated to have raised average global temperatures by 23°F (13°C).

See also IGNEOUS ROCKS; MASS EXTINCTIONS; OCEANIC PLATEAU.

floodplain Generally flat or low-lying areas that are adjacent and run parallel to river channels and are covered by water during flood stages of the river. The floodplain of a river is built by alluvium carried by the river and deposited in overbank environments, forming layers of silt, clay, and sand. Overbank deposits are typically cut by many narrow elongate channels filled by sands and gravels, marking places where the river formerly flowed and meandered away during the course of river evolution. Active and buried channels are typically separated from the floodplain deposits by a sandy or gravelly levee deposit, formed during flood stages of the river. These form because the velocity of floodwater decreases rapidly as it moves out of the channel, causing the current to drop heavy coarse-grained material near the river, forming a levee. Floodplains are also found around some lake basins that experience flood stages.

Floodplains are being increasingly built upon, creating potential and real hazards during floods. Floodplains are characterized by fertile soils and make excellent farmlands, which are nourished by yearly, decadal, and centurial floods, whereas buildings, towns, and cities have a much more difficult time dealing with periodic flooding.

See also FLOOD; GEOMORPHOLOGY; RIVER SYSTEM.

fluvial Meaning "of the river," referring to different and diverse aspects of rivers. The term may be used to refer to the environment around rivers and streams or, more commonly, to refer to sediments deposited by a stream or river system.

Sediments deposited by rivers tend to become finer-grained and more rounded with increasing transport distance from the eroded source terrain, typically an uplifted mountain range. The sediments also tend to become more enriched in the more stable chemical components such as quartz and micas and depleted in chemically vulnerable particles such as feldspars.

Stream channels are rarely straight, and the velocity of flow changes in different places. Friction makes the flow slower on the bottom and sides of the channel, and the bends in the river make the zone of fastest flow swing from side to side. The character of channels changes in different settings because of difference in slope, discharge, and load. Straight channels are very rare, and those that do occur have many properties of curving streams. The thalweg is a line connecting the deepest parts of the channel. In straight segments, the thalweg typically meanders from side to side of the stream. In places where the thalweg is on one side of the channel, a bar may form on the other side. A bar (for example, a sand bar) is a deposit of alluvium in a stream. Most streams move through a series of bends known as meanders. Meanders are always migrating across the floodplain by the process of the deposition of the point bar deposits and the erosion of the bank on the opposite side of the stream with the fastest flow. The erosion typically occurs through slumping of the stream bank. Meanders typically migrate back and forth, and also down-valley at a slow rate. If the downstream portion of a meander encounters a slowly erodable rock, the upstream part may catch up and cut off the meander. This forms an oxbow lake, which is an elongate and curved lake formed from the former stream channel. Braided streams consist of two or more adjacent but interconnected channels separated by bars or islands. Braided streams have constantly shifting channels, which move as the bars are eroded and redeposited, during large fluctuations in discharge. Most braided streams have highly variable discharge in different seasons, and they carry more load than meandering streams.

River and stream channel deposits tend to be composed of sands and gravelly sands that exhibit large-scale three-dimensional ripples, shown in cross section as cross-bedding. These cross-bedded sands are common around the inner bends of channels and mark the former positions of point bars, and they are commonly interbedded with planar bedded sands marking flood stage deposits and gravelly sands deposited during higher flood stages. The tops of channel deposits may be marked by finer-grained sands with small-scale ripples and mud drapes forming flaser-bedding, interbedded with muds, and grading up into overbank flood-plain deposits. This upward-fining sequence is characteristic of fluvial deposits, especially those of meandering streams. In contrast, braided stream deposits show less order and are characteristically dominated by bed load material such as gravel and sand. They include imbricated gravels, gravels deposited in shallow scours, and horizontally bedded sands and gravels deposited in bars.

Fluvial channel deposits form a variety of geometric patterns on a more regional scale. Shoestring sands form an anastomosing pattern of river channels enclosed in overbank shales and muds, formed by meandering and anastomosing river channels. Sheets and wedges of fluvial sediments form in front of uplifted mountain chains in foreland and rift basins and may pass basinward into deltaic or shallow marine sediments and mountainward into alluvial fan deposits. The type of tectonic setting for a basin may be deduced by changes or migration of different fluvial facies with stratigraphic height. Fluvial sediments are widely exploited for hydrocarbon deposits and also are known for placer deposits of gold and other valuable minerals.

See also CLASTIC ROCKS; DRAINAGE BASIN; FLOOD; RIVER SYSTEM.

flysch A synorogenic clastic deposit typically marked by interbedded shales and sandstones. The term was first used for rocks deposited in the Alps in the Cretaceous-Tertiary, before the main erosional event that shed coarser-grained conglomerates known as molasse. Sedimentary structures in flysch typically include a series of graded and cross-laminated layers in sands forming Bouma sequences, indicating that the sands were deposited by turbidity currents. Flysch is typically deposited in foreland basins and forms regionally extensive clastic wedges, underlain by distal black shales, and overlain by fluvial deposits and conglomerates of fluvial origin.

See also MOLASSE; OROGENY; TURBIDITE.

Foehn winds *See* CHINOOK WINDS.

fog A cloud at the Earth's surface formed by the condensation of water vapor from the atmosphere. The fog forms as atmospheric moisture increases and condensation occurs on numerous nuclei, reducing visibility to less than 0.62 miles (1 km).

Fog that forms in dirty city air tends to be thicker than fog that forms at the same atmospheric moisture count over oceans. This is because city air has abundant nuclei that grow more but smaller water droplets than the air over oceans, which typically has fewer nuclei and produces fewer but larger water droplets. Fog that forms in polluted air may be acidic and harmful to human health, particularly if the water droplets combine with sulfur and nitrogen oxides.

One of two main mechanisms is usually responsible for the formation of fog. Fog may form by condensation when the air is cooled past its saturation point (dew point), or it may form by continuous evaporation and mixing of vapor into the air. Radiational cooling of the air near the ground can lower the temperature of the surface layer below the dew point, forming radiation or ground fog. This type of fog forms best on clear nights when a layer of moist air near the surface is overlain by a layer of dry air, forming an atmospheric inversion. Radiation fog forms commonly in the late fall and winter, when nights are longest and cooling of the surface layer lasts the longest. Low winds also help the formation of radiation fog, as the wind promotes interaction of the surface air with the rapidly cooling ground, promoting faster cooling, but strong winds would mix the moist surface layer with dry air aloft, preventing fog from forming. Since the fog is heavy, it typically collects in valleys and low-lying areas. During the day the fog dissipates (it does not burn off) when the ground and low level air warms, causing the water droplets to evaporate.

Advection fog is formed when warm moist air moves over a cold surface, causing the air to cool below the dew point, initiating condensation. This type of fog is common along coastlines and is especially common in central California in the summer. Here, warm ocean water moves over cold upwelling ocean water near the coast, produced by westerly winds, causing the surface air to cool below the dew point and forming fog. Production of advection fog is enhanced by winds that produce rolling clouds of fog moving inland during summer months. Advection fog is also common along headlands of other coasts where converging warm air is cooled by the surface and forced to rise by the convergence.

Upslope fog forms as moist air flows up a mountain or other slope, cooling below the dew point. This type of fog is common around some mountain ranges including the Rockies and also forms around isolated mountains such as volcanoes where the fog may look like a small eruption in progress. A final type of fog is known as evaporation fog, evaporation-mixing fog, or steam fog. This is produced when cold air moves over warm water. This type of fog may commonly be seen forming above warm lakes, rivers, and other water bodies on cold autumn mornings. Steam fog may form over a warm wet surface on a warm summer day, such as after a brief shower drops water on a hot surface. The water quickly evaporates and mixes with the air above, disappearing quickly.

See also ATMOSPHERE; CLOUDS.

fold   Curves or bends in a planar structure such as bedding, foliation, or cleavage, and one of the most common tectonic structures in the crust of the Earth. They generally form by buckling or bending processes in lithified rocks, but some may also form during slumping of soft or wet sediments before they are lithified. The geometry of folds depends on the rheology of the rock, rheological or competence contrasts between layers, and the conditions of deformation, including temperature, pressure, and strain rate.

Folds are described using their orientation, shape, and type of layering being folded. Antiforms are downward closing structures, and synforms are upward closing structures. To portray the geometry of a folded sequence, the shape of a single folded surface needs to be described, as does the form of a sequence of folded layers, the form of series of folds involving a single surface, and the shape of a single folded layer in profile. The fold hinge line is the line of maximum curvature, the limb is the area between hinges, and the fold axis is the imaginary line that when moved parallel to itself generates a cylindrical surface parallel to the folded layer. The fold axial surface (or plane) is the surface containing the hinge lines of successive folds in a folded series of layers. Upright folds have steeply dipping axial surfaces, whereas gently inclined to recumbent folds have shallowly dipping axial surfaces. The plunge of the hinge line determines whether folds are further classified as gently, moderately, or steeply plunging.

A fold train is a group of linked folds of comparable dimensions. It is described by imaginary surfaces called enveloping surfaces that are the limiting surface between which the fold oscillates. A median surface passes through the inflection lines between successive folds. The fold order describes the wavelengths of folds, with the largest being first order, second largest being the second order, etc. A fold system may contain several superimposed fold trains of different dimensions, resulting in a complex pattern of upward closing and downward closing fold structures.

Basin and dome (Type 1) fold interference pattern, Scotland   *(Photo by Timothy Kusky)*

Historically, folds have been described around two geometric models, the parallel fold and the similar fold. In a parallel fold the thickness measured orthogonally across the layers is constant throughout the fold, whereas the similar fold shows considerable variation in layer thickness, with thinning on the limbs. Concentric folds are parallel folds with nearly constant curvature and circular boundary layers, and these are common in high structural levels of fold belts. Tightness is a measure of the interlimb angle between the tangents of folded surfaces measured at the inflexion points. Gentle folds have interlimb angles of 180–120°, open folds have angles of 120–70°, close folds have 70–30°, and tight and isoclinal folds have interlimb angles between 30–10–0°. Elasticas are unusual folds with wide hinge zones and interlimb angles of less than 0°. The fold shape can be further described by characterizing the hinge shape as rounded, angular, or very angular, and the limb shape as planar or curved.

Many folds are asymmetric and appear to be tilted in one direction. Symmetric folds have their axial planes as axes of symmetry and have axial surfaces perpendicular to enveloping surfaces, whereas asymmetric folds do not. The asymmetry can appear to be either "S" or "Z" shaped when viewed in the direction of plunge. A fold's vergence is the direction toward which it is overturned and can be sinistral (S) or dextral (Z), or overturned toward a specific geographic direction. Asymmetric folds form in several ways. They often form as minor folds on the limbs of major folds and verge toward the antiformal axial surface, and away from synformal axes. Knowing a fold's vergence can help mapping locations of larger antiforms and synforms. Asymmetric folds also form in shear zones and during regional tectonic transport, and their vergence can often be used to determine the direction of tectonic movement.

Fold facing is the direction perpendicular to the fold axis along the axial plane, toward the younger beds. Cleavage facing is the direction normal to the bedding plane intersection along the cleavage plane and toward the younger beds, and fault facing is the direction normal to the bedding plane intersection along the fault plane and toward the younger beds. Proper analysis and use of facing data in the field can help locate regions of major fold culminations, determine whether specific units were inverted or right way up prior to a late folding event, aid in the description and analysis of fold geometry, locate regions of refolding, and indicate the direction of tectonic transport.

Many geologic terrains have experienced multiple folding events, and the rocks show complex patterns reflecting this complex refolding. A generation is a group of structures believed to occupy the same position on a relative timescale. The correlation of generations is similar to stratigraphy, in that small-scale structures are used to indicate large-scale structures. Fold style and orientation are not particularly reliable criteria for identifying folds of common origin, and their use may lead to misinterpretation of both geometry and history. Fold generations can be meaningfully defined in terms of overprinting where one fold can be observed to fold geometric elements of earlier fold generations. Where overprinting relationships are not observable, folds should be grouped into style groups instead of generations.

Two main kinds of fold generations are distinguished. The first may be from two differently oriented strains, separated widely or closely in time, and the second may form during one continuous deformation but reflect a change in the material properties through deformation and metamorphism. In the absence of overprinting relationships, fold generations can be correlated using the trend and plunge of hinge lines and the strike and dip of axial surfaces. However, the map projection of hinge lines (fold axes) should not be used for correlations because two very different fold groups can have the same axis orientation. In addition, second generation folds can have different orientations depending on which limb of the original folds they developed on, due to original asymmetry. Furthermore, variations can develop depending on the rock type the folds are developed in.

The correlation of different generations by fold style is often done using the fold profile shape, details of the shape (such as limb tightness), and the type of any axial planar fabric such as cleavage. In many orogenic belts worldwide, fold style changes regularly with different generations. First-generation structures include primary layering (sedimentary beds, igneous layers). Second-generation structures include early folds that are typically tight to isoclinal and exhibit a slaty cleavage or schistosity. Third and later generations are typically more open, have more pointed hinges, a spaced cleavage, or an axial planar cleavage. Crenulations are almost never first-generation structures, because to get crenulations a pre-fold schistosity needs to be present. Kink folds, or chevron folds, have no axial planar foliation.

Folds of similar wavelength and different orientations may interfere in complex ways, forming fold interference patterns. Interpretation of fold interference patterns is typically based on two-dimensional outcrop patterns of refolded units, but accurate interpretation of the geometries of all fold generations requires a three-dimensional analysis. The style of early folds and the angular relationships between fold hinge lines and axial surfaces determines the outcrop patterns of the superimposed folds. Fold interference patterns are classified into several types, which form through different angular relationships between axes of first folds, poles to axial planes of first and second folds, and the "shearing direction" of second folds. Type I interference produces basins and domes, Type II patterns form crescents, and Type III patterns are represented by hooks. To properly interpret refold patterns, it is necessary to measure and analyze a large number of orientation data (fold hinges, axial surfaces) from carefully selected map domains. A caution, however, is needed, as in constric-

Crescents, hooks, and basin and dome types of fold interference patterns, Scotland *(Photo by Timothy Kusky)*

tional strain fields (such as those found in regions between several diapirs) shortening may occur in all directions, and structural forms resembling those produced by successive overprinting may be produced.

See also CLEAVAGE; DEFORMATION OF ROCKS; OROGENY; STRAIN ANALYSIS; STRUCTURAL GEOLOGY.

Further Reading

Borradaile, Graham J., M. Brian Bayly, and Chris M. Powell. *Atlas of Deformational and Metamorphic Rock Fabrics.* Berlin: Springer-Verlag, 1982.

Ramsay, John G., and Martin I. Huber. *The Techniques of Modern Structural Geology,* Vol. 2, *Folds and Fractures.* London: Academic Press, 1987.

footwall   *See* FAULT.

foreland basin   Wedge-shaped sedimentary basins that form on the continent-ward side of fold-thrust belts, filling the topographic depression created by the weight of the mountain belt. Most foreland basins have asymmetric, broadly wedge-shaped profiles with the deeper side toward the mountain range, and a flexural bulge developed about 90 miles (150 km) from the deformation front. The Indo-Gangetic plain on the south side of the Himalayan Mountains is an example of an active foreland basin, whereas some ancient examples include the Cretaceous Canadian Rockies Alberta foreland basin, the Cenozoic Flysch basins of the Alps, and the Ordovician and Devonian clastic wedges in the Appalachian foreland basins. Foreland basins are characterized by asymmetric subsidence, with greater amounts near the thrust front. Typical amounts of sudsidence fall in the range of about 0.6 miles (1 km) every 2–5 million years.

Deformation such as folding, thrust faulting, and repetition of stratigraphic units may affect foreland basins near the transition to the mountain front. These types of foreland basins appear to have formed largely by the flexure of the lithosphere under the weight of the mountain range, with the space created by the flexure filled in by sediments eroded from the uplifted mountains. Sedimentary facies typically grade from fluvial/alluvial systems near the mountains to shallow marine clastic environments farther away from the mountains, with typical deposition of flysch sequences by turbidity currents. These deposits may be succeeded laterally by distal black shales, then shallow water carbonates over a cross-strike distance of several hundred kilometers. There is also often a progressive zonation of structural features across the foreland basin, with contractional deformation (folds and faults) affecting the region near the mountain front, and normal faulting affecting the area on the flexural bulge a few tens to hundreds of kilometers from the deformation front. Sedimentary facies and structural zones all may migrate toward the continent in collisional foreland basins.

A second variety of foreland basins is found on the continent-ward side of non-collisional mountain belts such as the Andes, and these are sometimes referred to as retroarc foreland basins. They differ from the collisional foreland basins described above in that the mountain ranges are not advancing on the foreland, and the basin subsidence is a response to the weight of the mountains, added primarily by magmatism.

Another variety of foreland basins is known as extensional foreland basins and include features such as impactogens and aulacogens, which are extensional basins that form at high angles to the mountain front. Impactogens form during the convergence, whereas aulacogens are reactivated rifts that formed during earlier ocean opening. Many of these basins have earlier structural histories, including formation as a rift at a high angle to an ocean margin. These rifts are naturally oriented at high angles to the mountain ranges when the oceans close and become sites of enhanced subsidence, sedimentation, and locally additional extension. The Rhine graben in front of the Alpine collision of Europe is a well-known example of an aulacogen.

See also CONVERGENT PLATE MARGIN PROCESSES; PLATE TECTONICS.

Further Reading

Allen, P. A., and J. R. Allen. *Basin Analysis, Principles and Applications*. Oxford: Blackwell Scientific Publications, 1990.
Bradley, Dwight C., and Timothy M. Kusky. "Geologic Methods of Estimating Convergence Rates during Arc-Continent Collision." *Journal of Geology* 94 (1986): 667–681.

foreshock   See EARTHQUAKES.

formation   A fundamental lithostratigraphic unit of intermediate rank, characterized by similar lithologic units or rock types, or other distinguishing features. They may range in thickness from about a meter to several thousands of meters. Formations are the only formal lithostratigraphic units defined on the basis of lithology. They must be mappable, distinguishable from other formations, have defined boundaries and relationships to other formations, and be the product of similar conditions. Some formations are made of a singular lithological type, whereas others may be made of a repetition of several different types, such as cyclic alternations between sandstone, shale, and limestone. Formations may represent short or long intervals of geologic time. They may be subdivided into members or grouped into groups and are named after prominent geographic features in the area in which they are defined.

See also STRATIGRAPHY.

fossil   Any remains, traces, or imprints of any plants or animals that lived on the Earth. These remains of past life include body fossils, the preserved record of hard or soft body parts, and trace fossils, which record traces of biological activity such as footprints, tracks, and burrows. The oldest body fossils known are 3.4-billion-year-old remnants of early bacteria, whereas chemical traces of life may extend back to 3.8 billion years.

The conditions that lead to fossilization are so rare that it is estimated that only 10 percent of all species that have ever existed are preserved in the fossil record. The record of life and evolution is therefore very incomplete. In order to be preserved, life-forms become mineralized after they die, with organic tissues typically being replaced by calcite, quartz, or other minerals during burial and diagenesis. Fossils are relatively common in shallow marine carbonate rocks where organisms that produced calcium carbonate shells are preserved in a carbonate matrix.

See also EVOLUTION; PALEONTOLOGY.

fracture   A general name for a break in a rock or other body that may or may not have any observable displacement. Fractures include joints, faults, and cracks formed under brittle deformation conditions and are a form of permanent (nonelastic) strain. Brittle deformation processes generally involve the growth of fractures or sliding along existing fractures. Frictional sliding involves the sliding on preexisting fracture surfaces, whereas cataclastic flow includes grain-scale fracturing and frictional sliding producing macroscopic ductile flow over a band of finite width. Tensile cracking involves the propagation of cracks into unfractured material under tensile stress perpendicular to the maximum compres-

Fracture/joint, China *(Photo by Timothy Kusky)*

Conjugate joints developed in a massive felsic volcanic rock, Alaska
(Photo by Timothy Kusky)

sive stress, whereas shear rupture refers to the initiation of fracture at an angle to the maximum principal stress.

Fractures may propagate in one of three principal modes. Mode I refers to fracture growth by incremental extension perpendicular to the plane of the fracture at the tip. Mode II propagation is where the fracture grows by incremental shear parallel to the plane of the fracture at the tip, in the direction of fracture propagation. Mode III is when the fracture grows by incremental shear parallel to the plane of the fracture at the tip, perpendicular to the direction of propagation.

Joints are fractures with no observable displacement parallel to the fracture surface. They generally occur in subparallel joint sets, and several sets often occur together in a consistent geometric pattern forming a joint system. Joints are sometimes classified into extension joints or conjugate sets of shear joints, a subdivision based on the angular relationships between joints. Most joints are continuous for only short distances, but in many regions master joints may run for very long distances and control geomorphology or form

Plumose structure formed on joint surface *(Photo by Timothy Kusky)*

air photo lineaments. Microfractures or joints are visible only under the microscope and only affect a single grain.

Many joints are contained within individual beds and have a characteristic joint spacing, measured perpendicular to the joints. This is determined by the relative strength of individual beds or rock types, the thickness of the jointed layer, and structural position, and is very important for determining the porosity and permeability of the unit. In many regions, fractures control groundwater flow, the location of aquifers, and the migration and storage of petroleum and gas.

Joints and fractures are found in all kinds of environments and may form by a variety of mechanisms. Desiccation cracks and columnar joints form by the contraction of materials. Bedding plane fissility, characterized by fracturing parallel to bedding, may be produced by mineral changes during diagenesis that lead to volume changes in the layer. Unloading joints form by stress release, such as during uplift, ice sheet withdrawal, or quarrying operations. Exfoliation joints and domes may form by mineral changes, including volumetric changes during weathering, or by diurnal temperature variations. Most joints have tectonic origins, typically forming in response to the last phase of tectonic movements in an area. Other joints seem to be related to regional doming, folding, and faulting.

Many fractures and joints exhibit striated or ridged surfaces known as plumose structures, since they vaguely resemble feathers. Plumose structures develop in response to local variations in propagation velocity and the stress field. The origin is the point at which the fracture originated, the mist is the small ridging on the surface, and the plume axis is the line that starts at the axis and from which individual barbs propagate. The twist hackle is the steps at the edge of the fracture plane along which the fracture has split into a set of smaller fractures.

E. M. Anderson elegantly explained the geometry and orientation of some fracture sets in a now classic work published in 1951. In Andersonian theory, the attitude of a fracture plane tells a lot about the orientation of the stress field that operated when the fracture formed. Fractures are assumed to form as shear fractures in a conjugate set, with the maximum compressive stress bisecting an acute (60°) angle between the two fractures. In most situations, the surface of the Earth may be the maximum, minimum, or intermediate principal stress, since the surface can transmit no shear stress. If the maximum compressive stress is vertical, two fracture sets will form, each dipping 60° toward each other and intersecting along a horizontal line parallel to the intermediate stress. If the intermediate stress is vertical, two vertical fractures will form, with the maximum compressive stress bisecting the acute angle between the fractures. If the least compressive stress is vertical, two gently dipping fractures will form, and their intersection will be parallel to the intermediate principal stress.

Other interpretations of fractures and joints include modifications of Andersonian geometries that include volume changes and deviations of principal stresses from the vertical. Many joints show relationships to regional structures such as folds, with some developing parallel to the axial surfaces of folds and others crossing axial surfaces. Other features on joint surfaces may be used to interpret their mode of formation. For instance, plumose structures typically indicate Mode I or extensional types of formation, whereas the development of fault striations (known as slickensides) indicate Mode II or Mode III propagation. Observations of these surface features, the fractures' relationships to bedding, structures such as folds and faults, and their regional orientation and distribution can lead to a clear understanding of their origin and significance.

See also DEFORMATION OF ROCKS; STRUCTURAL GEOLOGY.

Further Reading
Anderson, E. M. *The Dynamics of Faulting.* London: Oliver and Boyd, 1951.
Pollard, David D., and Aydin Atilla. "Progress in Understanding Jointing over the Past Century." *Geological Society of America* 100 (1988): 1,181–1,204.
Ramsay, John G., and Martin I. Huber. *The Techniques of Modern Structural Geology,* Vol. 2, *Folds and Fractures.* London: Academic Press, 1987.

fracture zone aquifers Faults and fractures develop at various scales from faults that cross continents to fractures that are only visible microscopically. These discontinuities in the rock fabric are located and oriented according to the internal properties of the rock and the external stresses imposed on it. Fractures at various scales represent zones of increased porosity and permeability. They may form networks and, therefore, are able to store and carry vast amounts of water.

The concept of fracture zone aquifers explains the behavior of groundwater in large fault-controlled watersheds. Fault zones in this case serve as collectors and transmitters of water from one or more recharge zones with surface and subsurface flow strongly controlled by regional tectonism.

Both the yield and quality of water in these zones are usually higher than average wells in any type of rock. High-grade water for such a region would be 250 gallons per minute or greater. In addition, the total dissolved solids measured in the water from such high-yielding wells will be lower than the average for the region.

The fracture zone aquifer concept looks at the variations in groundwater flow as influenced by secondary porosity over an entire watershed. It attempts to integrate data on a basin in an effort to describe the unique effects of secondary porosity on the processes of groundwater flow, infiltration, transmissivity, and storage.

The concept includes variations in precipitation over the catchment area. One example is orographic effects wherein the mountainous terrain precipitation is substantially greater than at lower elevations. The rainfall is collected over a large catchment area, which contains zones with high permeability because of intense bedrock fracturing associated with major fault zones. The multitude of fractures within these highly permeable zones "funnel" the water into other fracture zones down gradient. These funnels may be in a network hundreds of square kilometers in area.

The fault and fracture zones serve as conduits for groundwater and often act as channelways for surface flow. Intersections form rectilinear drainage patterns that are sometimes exposed on the surface but are also represented below the surface and converge down gradient. In some regions, these rectilinear patterns are not always visible on the surface due to vegetation and sediment cover. The convergence of these groundwater conduits increases the amount of water available as recharge. The increased permeability, water volume, and ratio of water to minerals within these fault/fracture zones help to maintain the quality of water supply. These channels occur in fractured, nonporous media (crystalline rocks) as well as in fractured, porous media (sandstone, limestone).

At some point in the groundwater course, after convergence, the gradient decreases. The sediment cover over the major fracture zone becomes thicker and acts as a water storage unit with primary porosity. The major fracture zone acts as both a transmitter of water along conduits and a water storage basin along connected zones with secondary (and/or primary) porosity. Groundwater within this layer or lens often flows at accelerated rates. The result can be a pressurization of groundwater both in the fracture zone and in the surrounding material. Rapid flow in the conduit may be replenished almost instantaneously from precipitation. The surrounding materials are replenished more slowly but also release the water more slowly and serve as a storage unit to replenish the conduit between precipitation events.

Once the zones are saturated, any extra water that flows into them will overflow, if an exit is available. In a large area watershed, it is likely that this water flows along subsurface channelways under pressure until some form of exit is found in the confining environment. Substantial amounts of groundwater may flow along an extension of the main fault zone controlling the watershed and may vent at submarine extensions of the fault zone forming coastal or offshore freshwater springs.

The concept of fracture zone aquifers is particularly applicable to areas underlain by crystalline rocks and where these rocks have undergone a multiple deformational history that includes extensional tectonics. It is especially applicable in areas where recharge is possible from seasonal and/or sporadic rainfall on mountainous regions adjacent to flat desert areas.

Fracture zone aquifers are distinguished from horizontal aquifers in that: (a) they drain numerous streams in extensive areas and many extend for tens of kilometers in length; (b) they constitute conduits to mountainous regions where the recharge potential from rainfall is high; (c) some may connect several horizontal aquifers and thereby increase the volume of accumulated water; (d) because the source of the water is at higher elevations, the artesian pressure at the groundwater level may be high; and (e) they are usually missed by conventional drilling because the water is often at the depth of hundreds of meters.

The characteristics of fracture zone aquifers make them an excellent source of groundwater in arid and semiarid environments. Fracture zone aquifers are located by seeking major faults. The latter are usually clearly displayed in images obtained from Earth orbit, because they are emphasized by drainage. Thus, the first step in evaluating the groundwater potential of any region is to study the structures displayed in satellite images to map the faults, fractures, and lineaments. Such a map is then compared to a drainage map showing wadi locations. The combination of many wadis and major fractures indicates a larger potential for groundwater storage. Furthermore, the intersection between major faults would increase both porosity and permeability, and hence, the water collection capacity.

It must be recognized that groundwater resources in arid and semiarid lands are scarce and must be properly used and thoughtfully managed. Most of these resources are "fossil," having accumulated under wet climates during the geological past. The present rates of recharge from the occasional rainfall are not enough to replenish the aquifers. Therefore, the resources must be used sparingly without exceeding the optimum pumping rates for each water well field.

See also GROUNDWATER; STRUCTURAL GEOLOGY.

Further Reading

Bisson, Robert A., and Farouk El-Baz. "Megawatersheds Exploration Model." Proceedings of the 23rd International Symposium on Remote Sensing of Environmental ERIM I, Ann Arbor, Mich., 1990.

El-Baz, Farouk. "Utilizing Satellite Images for Groundwater Exploration in Fracture Zone Aquifers." International Conference on Water Resources Management in Arid Countries, Ministry of Water Resources, Muscat, Oman, v. 2, 1995.

Gale, J. E. "Assessing the Permeability Characteristics of Fractured Rock." In *Recent Trends in Hydrogeology,* edited by T. N. Narasimhan. *Geological Society of America Special Paper* 189 (1982).

Kusky, Timothy M., and Farouk El-Baz. "Structural and Tectonic Evolution of the Sinai Peninsula, Using Landsat Data: Implications for Groundwater Exploration." *Egyptian Journal of Remote Sensing* 1 (1999).

National Academy of Sciences. *Rock Fractures and Fluid Flow: Contemporary Understanding and Applications.* Washington, D.C.: National Academy Press, 1996.

Wright, E. P., and W. G. Burgess. "The Hydrogeology of Crystalline Basement Aquifers in Africa." *Geological Society of London Special Publication* 66 (1992).

freezing rain *See* PRECIPITATION.

fumarole Vents from which gases and vapors are emitted. They typically form in active volcanic terrains where late-stage fluids and gases escape from the magma along fractures or other conduits. They may also form where groundwater seeps

Fumarole, Mount St. Helens

deeply into the crust, gets heated by magmatic or other heat, and returns to the surface to escape at fumarole vents. Geysers are types of fumaroles, as are seafloor hydrothermal vents.

See also BLACK SMOKER CHIMNEYS; GEYSER.

fusion The combination of two light nuclei to form a heavy nucleus, along with the release of a large amount of energy. The atomic bomb combined nuclei of hydrogen to form helium resulting in the sudden release of huge quantities of energy. Nuclear fusion is the main process that powers stars. It operates at very high temperatures and pressures, where the atomic nuclei can approach each other at high enough velocities to overcome the repulsive force between similarly charged atoms. Once at close range the strong nuclear force fuses the nuclei together, releasing large amounts of energy. Significant amounts of research are being done on attempts to find ways to use nuclear fusion to produce energy without causing nuclear explosions.

See also RADIOACTIVE DECAY.

G

gabbro A coarse-grained mafic plutonic rock composed of calcic plagioclase (labradorite or bytwonite) and clinopyroxene (augite), with or without olivine and orthopyroxene. Common accessory minerals include apatite, magnetite, and ilmenite. It is the intrusive equivalent of basalt, named after the town of Gabbro in Tuscany, Italy.

Gabbro is a common constituent of oceanic crust, where it crystallizes from magma chambers that form beneath oceanic spreading centers. Oceanic gabbros have been dredged from the seafloor and studied extensively in ophiolite complexes that represent fragments of oceanic crust thrust onto continents during tectonic collisions. Gabbro from these complexes is quite variable in texture and composition. High-level ophiolitic gabbros are typically homogeneous or isotropic but grade downward into compositionally layered varieties where variations in the amounts of plagioclase and pyroxene result in a pronounced layering to the rock. These layers are thought to represent layers that accumulated along the sides and base of the magma chamber.

Many continental plutons also have gabbroic components or are made predominantly of gabbro. Some continental intrusions form large layered mafic-ultramafic layered complexes that have large quantities of gabbro. Examples of these complexes include the Skaergaard intrusion, Greenland; the Bushveld, South Africa; and the Muskox, Canada.

See also IGNEOUS ROCKS; OPHIOLITES.

Gaia hypothesis Formulated in the 1970s by British atmospheric chemist James Lovelock; proposes that Earth's atmosphere, hydrosphere, geosphere, and biosphere interact as a self-regulating system that maintains conditions necessary for life to survive. In this view the Earth acts as if it is a giant self-regulating organism in which life creates changes in one system to accommodate changes in another, in order to keep conditions on Earth within the narrow limits that allow life to continue.

The average temperature on the Earth has been maintained between 50°F and 86°F (10°C and 30°C) for the past 3.5 billion years, despite the fact that the solar energy received by the Earth has increased by 40–330 percent since the Hadean. The temperature balance has been regulated by changes in the abundance of atmospheric greenhouse gases, controlled largely by volcanic degassing and the reduction of CO_2 by photosynthetic life. A slight increase or decrease in CO_2 and other greenhouse gases could cause runaway greenhouse or icehouse global climates, yet life has been able to maintain the exact balance necessary to guarantee its survival.

The presence of certain gases such as ammonia at critical levels in the atmosphere for maintaining soil pH near 8, the optimal level for sustaining life, is critical for maintaining atmospheric oxygen levels. This critical balance is unusual, as methane is essentially absent from the atmospheres of Venus and Mars where life does not exist. The salinity of the oceans has been maintained at around 3.4 percent, in the narrow range required for marine life, reflecting a critical balance between terrestrial weathering, evaporation, and precipitation.

The exact mechanisms that cause the Earth to maintain these critical balances necessary for life are not well known. However, as solar luminosity increases, the additional energy received by the Earth is balanced by the amount of energy radiated back to space. This can be accomplished by changes in the surface reflectance (albedo) through changes in the amount of ice cover, types of plants, and cloud cover. It is increasingly recognized that changes in one Earth system produce corresponding changes in other systems in self-regulation processes known as homeostasis. Critical for Gaia are the links between organisms and the physical environment, such that many proponents of the theory regard the planet as one giant superorganism.

See also ATMOSPHERE; CLIMATE CHANGE; GREENHOUSE EFFECT; SUPERCONTINENT CYCLE; and feature essay on page 82.

Galveston Island A barrier island off the coast of southeast Texas across from Galveston Bay, an inlet to the Gulf of Mexico. It is part of a string of barrier islands that runs along the Gulf of Mexico from the Mississippi delta in New Orleans to Matagorda and Padre Islands near the border with Mexico. A deep, dredged ship channel extends through the inlet allowing ocean vessels access to the Port of Houston, but the island is still a major port and includes major oil refineries and shipbuilding facilities and has many grain elevators, machine shops, compresses, fishing fleets and processing plants, and chemical plants. The island is also a resort and vacation spot frequented by Texans.

The island is only a few feet above sea level and is prone to be struck by hurricanes, including the nation's worst natural disaster, the hurricane of 1900 in which approximately 10,000 people perished. After this devastating event a giant seawall was built, but the island suffered additional hurricane damage in 1961, and again in 2001.

See also BEACH; HURRICANE.

garnet A group of brittle, transparent to subtransparent cubic minerals with a vitreous luster, no cleavage, and a variety of colors, including red, pink, brown, black, green, yellow, and white. It typically forms distinctive isometric crystals in metamorphic rocks such as schist, gneiss, and eclogite, and it may also occur as an accessory mineral in igneous rocks or as detrital grains in sedimentary deposits. It is used as a semiprecious gemstone and as an industrial abrasive. Garnets are aluminosilicate minerals with the chemical formula $A_3B_2(SiO_4)_3$, where A = Ca, Mg, Fe^{+2}, and Mn^{+2}, B = Al, Fe^{+2}, Mn^{+3}, V^{+3}, and Cr. Varieties include pyrope ($Mg_3Al_2Si_3O_{12}$), almandine ($Fe_3^{+2}Al_2Si_3O_{12}$), spessartine ($Mn_3Al_2Si_3O_{12}$), grossular ($Ca_3Al_2Si_3O_{12}$), andradite ($Ca_3(Fe^{+3},Ti)_2Si_3O_{12}$), uvarovite ($Ca_3Cr_2Si_3O_{12}$), and hydrogrossular ($Ca_3Al_2Si_2O_8)(SiO_4)_{1-m}(OH)_{4m}$.

Different varieties of garnets are more common in certain rock types and tectonic settings. Pyrope is the most common garnet in ultramafic rocks such as mica peridotites, kimberlites, and serpentinites. Eclogites typically have garnets with compositions between pyrope and almandine. Almandine is the most common garnet in regional metamorphic mica schists from the amphibolite through granulite grades. Spessartine or spessartine-almandine is commonly found in granitic pegmatites, whereas grossular garnet forms during regional and thermal (contact) metamorphism of calcareous rocks such as limestone. Andradite is also a common contact metamorphic rock, forming in impure calcareous hosts, and

hydrogrossular garnet forms in altered gabbros, rodingites, and marls. The most rare garnet mineral is uvarovite, found in serpentinites, in association with chromite, and less commonly in contact metamorphic aureoles.

gas hydrates Gas hydrates or clathrates are solid, ice-like water-gas mixtures that form at cold temperatures (40°F–43°F, or 4°F–6°C) and pressures above 50 atmospheres. They form on deep marine continental margins and in polar continental regions, often below the seafloor. The gas component is typically methane but may also contain ethane, propane, butane, carbon dioxide, or hydrogen sulfide, with the gas occurring inside rigid cages of water molecules. The methane is formed by anaerobic bacterial degradation of organic material.

It is estimated that gas hydrates may contain twice the amount of carbon known from all fossil fuel deposits on the planet, and as such they represent a huge, virtually untapped potential source of energy. However, the gases expand by more than 150 times the volume of the hydrates, they are located deep in the ocean, and methane is a significant greenhouse gas. Therefore there are significant technical problems to overcome before gas hydrates are widely mined as an energy source.

See also HYDROCARBON.

gastropods The largest class of the phylum Mollusca, presently containing approximately 1,650 genera but ranges back to the Cambrian. They occur in a variety of environments including marine water, freshwater, and terrestrial settings. Most gastropods have a coiled shell and move via a muscular foot. Snails are probably the most familiar gastropods, consisting of a helically coiled shell from which the head and foot protrude.

Fossil and living gastropods are classified by morphology, with many aspects reflected in the shell. Cambrian gastropods had shells coiled in one plane, whereas others show conical forms. Later forms exhibit torsionally coiled bodies, with the back of the body rotated so that it lies above the head. Gastropods are classified into three subclasses. The Prosobranchia are the most common, show full torsion, and are found in marine, freshwater, and terrestrial varieties. Opisthobranchia lack shells or have them concealed in the mantle, and Pulmonata have spiral shells that have been converted to lungs, since they are terrestrial varieties.

Geiger counter Geiger counters are instruments that detect and measure ionizing radiation that is emitted from radioactive sources. The instrument contains a gas-filled tube that has an electric current passing through it. When any radiation is passed through this tube a gas is discharged inside the tube, and it causes the gas to be a conductor that is measured by the Geiger counter, which emits an audible clicking. The gas-filled glass tube acts like a simple electrode, and radiation simply completes the circuit, enabling the Geiger counter to detect alpha and beta rays that are given off by radioactive material.

See also RADIOACTIVE DECAY.

general circulation models   Attempts to explain the general or average way that the atmosphere and oceans circulate and redistribute heat and moisture. These models do not account for daily variations of the circulation patterns but look at seasonal and longer term averages. Many general circulation models attempt to answer fundamental questions about the motion of the atmosphere, what drives this motion, and what are its consequences. They may start with input such as the latitudinal distribution of heating in different seasons, and try to determine how the atmosphere moves heat from the equator toward the poles, and also consider the return flow of cold air from the poles to the equator. Other inputs to the models would include the distribution of continents and their uneven heating effects, the influences of topography on atmospheric circulation, the rotation of the Earth, and interactions between the oceans and atmosphere.

See also ATMOSPHERE; EL NIÑO; HADLEY CELLS; LA NIÑA.

geobarometry   The study or determination of the pressure conditions under which a rock or mineral formed. It typically involves pressure-sensitive reactions between different minerals or relative partitioning of different elements between mineral pairs. Many mineral assemblages are sensitive to changes in both temperature and pressure, so it is common to discuss these changes with respect to both, under the term thermobarometry.

Mineral reactions are typically limited by kinetic factors such as temperature, which in the Earth increases with pressure, so rocks tend to preserve chemical or mineralogical evidence for reactions that occurred at high temperatures and pressures. Some minerals or mineral assemblages are only stable above certain pressure-temperature conditions, forming good thermobarometers. Other thermobarometers rely on the partitioning of elements such as Fe and Mg, controlled by exchange reactions between different minerals that change with different temperatures and pressures.

See also GEOTHERMOMETRY; METAMORPHISM.

geochemistry   The study of the distribution and amounts of elements in minerals, rocks, ore bodies, rock units, soils, the Earth, atmosphere, and, by some accounts, other celestial bodies. It includes study and analysis of the movement of chemical elements, the properties of minerals as related to their distribution and concentrations of specific elements, and the classification of rocks based on their chemical composition.

The field of geochemistry started with the discovery of 31 chemical elements by Antoine Lavoisier in 1789, with the first mention of the word by Christian F. Shonbein in 1813. In 1884, the U.S. Geological Survey established a laboratory to investigate the chemistry of the planet, and appointed F. W. Clarke as the head of laboratory. Since then, the U.S. Geological Survey has been one of the world's leaders in the collection and analysis of geochemical data. In 1904 the Carnegie Institution in Washington, D.C., established the Geophysical Laboratory, which tested physical and chemical properties of minerals and rocks. The Vernadsky Institute in Russia had a similar charge, and both institutions spearheaded a revolution in technologies applied to analyzing the composition of rock materials, leading to the proposition of the concepts of chemical equilibrium, disequilibrium, and the amassing of huge databases encompassing the chemistry of rocks of the world. Victor M. Goldschmidt from the University of Oslo applied the phase rule, explaining metamorphic changes in terms of chemical equilibrium.

geochronology   The study of time with respect to Earth history, including both absolute and relative dating systems as well as correlation methods. Absolute dating systems include a variety of geochronometers such as radioactive decay series in specific isotopic systems that yield a numerical value for the age of a sample. Relative dating schemes include cross-cutting features and discontinuities such as igneous dikes and unconformities, with the younger units being the cross-cutting features or those overlying the unconformity.

During the 19th and early 20th centuries, geochronologic techniques were very crude. Many ages were estimated by the supposed rate of deposition of rocks and correlation of units with unconformities with other, more complete sequences. With the development of radioactive dating it became possible to refine precise or absolute ages for specific rock units. Radiometric dating operates on the principle that certain atoms and isotopes are unstable. These unstable atoms tend to decay into stable ones by emitting a particle or several particles. Alpha particles have a positive charge and consist of two protons and two neutrons. Beta particles are physically equivalent to electrons or positrons. These emissions are known as radioactivity. The time it takes for half of a given amount of a radioactive element to decay to a stable one is known as the half-life. By matching the proportion of original unstable isotope to stable decay product, and knowing the half-life of that element, one can thus deduce the age of the rock. The precise ratios of parent to daughter isotopes are measured in an instrument known as a mass spectrometer.

Radiocarbon or carbon-14 dating techniques were developed by Willard F. Libby (1908–80) at the University of Chicago in 1946. This discovery represented a major breakthrough in dating organic materials and is now widely used by archaeologists, Quaternary geologists, oceanographers, hydrologists, atmospheric scientists, and paleoclimatologists. Cosmic rays entering Earth's atmosphere transform regular carbon

Age of the Earth

Why do geologists say that the Earth is 4.6 billion years old? For many hundreds of years, most people in European, Western, and other cultures believed the Earth to be about 6,000 years old, based on interpretations of passages in the Torah and Old Testament. However, based on the principles of uniformitarianism outlined by James Hutton and Charles Lyell, geologists in the late 1700s and 1800s began to understand the immensity of time required to form the geologic units and structures on the planet and argued for a much greater antiquity of the planet. When Charles Darwin advanced his ideas about evolution of species, he added his voice to those calling for tens to hundreds of millions of years required to explain the natural history of the planet and its biota. In 1846 the physicist Lord Kelvin joined the argument, but he advocated an even more ancient Earth. He noted that the temperature increased with depth, and he assumed that this heat was acquired during the initial accretion and formation of the planet, and has been escaping slowly ever since. Using heat flow equations Kelvin calculated that the Earth must be 20–30 million years old. However, Kelvin assumed that there were no new inputs of heat to the planet since it formed, and he did not know about radioactivity and heat produced by radioactive decay. In 1896 Madame Curie, working in the labs of Henri Becquerel in France, exposed film to uranium in a light-tight container and found that the film became exposed by a kind of radiation that was invisible to the eye. Soon, many elements were found to have isotopes, or nuclei of the same element with different amounts of neutron in the nucleus. Some isotopes are unstable and decay from one state to another, releasing radioactivity. Radioactive decay occurs at a very specific and fixed average rate that is characteristic of any given isotope. In 1903 Pierre Curie and Albert Laborde recognized that radioactive decay releases heat, a discovery that was immediately used by geologists to reconcile geologic evidence of uniformitarianism with Lord Kelvin's calculated age of the Earth.

In 1905 Ernest Rutherford suggested that the constant rate of decay of radioactive isotopes could be used to date minerals and rocks. Because radioactivity happens at a statistically regular rate for each isotope, it can be used to date rocks. For each isotope an average rate of decay is defined by the time that it takes half of the sample to decay from its parent to daughter product, a time known as the half-life of the isotope. Thus, to date a rock we need to know the ratio of the parent to daughter isotopes and simply multiply by the decay rate of the parent. Half-life is best thought of as the time it takes for half of any size sample to decay, since radioactive decay is a nonlinear exponential process.

The rate of decay of each isotope determines which isotopic systems can be used to date rocks of certain ages. Also, the isotopes must occur naturally in the type of rock being assessed and the daughter products must be present only from decay of the parent isotope. Some of the most accurate geochronologic clocks are made by comparing the ratios of daughter products from two different decay schemes—since both daughters are only present as a result of decay from their parents, and their ratios provide special highly sensitive clocks.

Isotopes and their decay products provide the most powerful way to determine the age of the Earth. Most elements formed during thermonuclear reactions in pre-solar system stars that experienced supernovae explosions. The main constraints we have on the age of the Earth are that it must be younger than 6–7 billion years, because it still contains elements such as K-40, with a half-life of 1.25 billion years. If the Earth were any older, all of the parent product would have decayed. Isotopic ages represent the time that that particular element-isotope system got incorporated in a mineral structure. Since isotopes have been decaying since they were incorporated, the oldest age from an Earth rock gives a minimum age of the Earth. So far, the oldest known rock is the 4.03-billion-year-old Acasta gneiss of the Slave Province in northwest Canada, and the oldest mineral is a 4.2-billion-year-old zircon from western Australia. From these data, we can infer that the Earth is between 4.2 and 6 billion years old.

The crust on the Moon is 4.2–4.5 billion years old, and the Earth, Moon, and meteorites all formed when the solar system formed. The U-Pb isotopic system is one of the most useful for determining the age of the Earth, although many other systems give identical results. Some meteorites contain lead, but no U or Th parents. Since the proportions of the various lead isotopes have remained fixed since they formed, their relative proportions can be used to measure the primordial lead ratios in the early Earth. Then, by looking at the ratios of the four lead isotopes in rocks on Earth from various ages, we can extrapolate back to when they had the same primordial lead ratio. These types of estimates give an age of 4.6–4.7 billion years for the Earth, and 4.3–4.6 billion years for meteorites. So, the best estimate for the age of the Earth is 4.6 billion years, a teenager in the universe.

(C^{12}) to radioactive carbon (C^{14}). Within about 12 minutes of being struck by cosmic rays in the upper atmosphere, the carbon-14 combines with oxygen to become carbon dioxide that has carbon-14. It then diffuses through the atmosphere and is absorbed by vegetation (plants need carbon dioxide in order to make sugar by photosynthesis). Every living thing has carbon in it. While it is alive, each plant or animal exchanges carbon dioxide with the air. Animals also feed on the vegetation and absorb its carbon dioxide. At death, the carbon-14 is no longer exchanged with the atmosphere but continues to decay in the material. Theoretically, analysis of this carbon-14 can reveal the date when the object once lived by the percent of carbon-14 atoms still remaining in the object. The radiocarbon method has subsequently evolved into one of the most powerful techniques to date late Pleistocene and Holocene artifacts and geologic events up to about 50,000 years old.

Uranium, thorium, and lead isotopes form a variety of geochronometers using different parent/daughter pairs. Uranium 238 decays to lead 206 with a half-life of 4.5 billion years. Uranium 235 decays to lead 207 with a half-life of 0.7 billion years, and thorium 232 decays to lead 208 with a half-life of 14.1 billion years. Uranium, thorium, and lead are generally found together in mixtures and each one decays into several daughter products (including radium) before turning into lead. The ^{230}Th/^{234}U disequilibrium method is one of the most commonly used uranium-series techniques. This method is based on the fact that uranium is much more soluble than thorium, so materials such as corals, mollusks, calcic soils, bones, carbonates, cave deposits, and fault zones are enriched in uranium with respect to thorium. This method can be used to date features as old as Precambrian.

Uranium-lead dating also uses the known original abundance of isotopes of uranium and the known decay rates of parents to daughter isotopes. This technique is useful for dating rocks up to billions of years old. All naturally occurring uranium contains ^{238}U and ^{235}U in the ratio of 137.7:1. ^{238}U decays to ^{206}Pb with a half-life of 4,510 Ma through a process of eight alpha-decay steps and six beta-decay steps. ^{235}U decays to ^{207}Pb (with a half-life of 713 Ma) by a similar series of stages that involves seven alpha-decay steps and four beta-decay steps. Uranium-lead dating techniques were initially applied to uranium minerals such as uraninite and pitchblende, but these are rare, so very precise methods of measuring isotopic ratios in other minerals with only trace amounts of uranium and lead (zircon, sphene) were developed. The amount of radiogenic lead in all these methods must be distinguished from naturally occurring lead, and this is calculated using their abundance with ^{204}Pb, which is stable. After measuring the ratios of each isotope relative to ^{204}Pb, the ratios of ^{235}U/^{207}Pb and ^{238}U/^{206}Pb should give the same age for the sample, and a plot with each system plotted on one axis shows each age. If the two ages agree, the ages will plot on a curve known as concordia, which tracks the evolution of these ratios in the Earth v. time. Ages that plot on concordia are said to be concordant. However, in many cases the ages determined by the two ratios are different and they plot off the concordia curve. This occurs when the system has been heated or otherwise disturbed during its history, causing a loss of some of the lead daughter isotopes. Because ^{207}Pb and ^{206}Pb are chemically identical, they are usually lost in the same proportions.

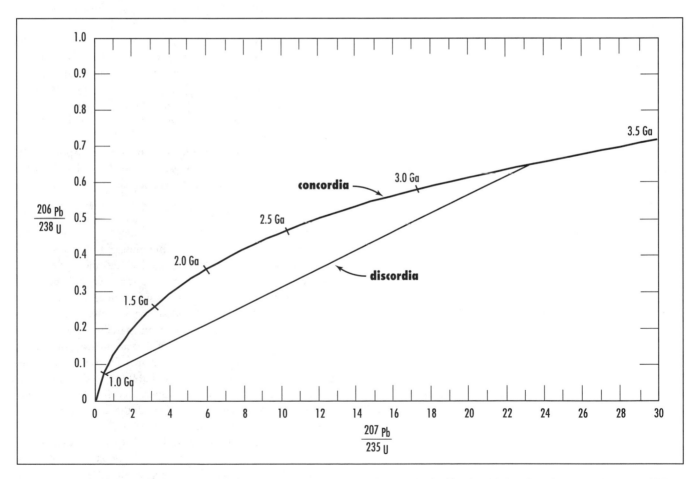

Concordia diagram showing the concordia curve that traces the evolution of the ^{206}Pb/^{238}U v. ^{207}Pb/^{235}U ratio with time, from the present back to 3.5 billion years (Ga) ago. The discordia curve shows the path that the ratio would follow if the rock example used crystallized at 3.2 billion years ago and lost lead (for example, through metamorphism) at 1.0 billion years ago.

The thorium-lead dating technique is similar to the uranium-lead technique and uses the decay from ^{232}Th to ^{208}Pb (with $6He^4$), with a half-life of 13,900 years. Minerals used for this method include sphene, zircon, monazite, apatite, and other rare U-Th minerals. The ratio of $^{208}Pb/^{232}Th$ is comparable with $^{207}Pb/^{235}U$. This method is not totally reliable and is usually employed in conjunction with other methods. In most cases, the results are discordant, showing a loss of lead from the system. The Th-Pb method can also be interpreted by means of isochron diagrams.

Potassium-argon dating is based on the decay of radioactive potassium into calcium and argon gas at a specific rate and is accomplished by measuring the relative abundances of ^{40}K and ^{40}Ar in a sample. The technique is potentially useful for dating samples as old as four billion years. Potassium is one of the most abundant elements in the Earth's crust (2.4 percent by mass). One out of every 100 potassium atoms is radioactive ^{40}K, with 19 protons and 21 neutrons. If one of the protons is hit by a beta particle, it can be converted into a neutron. With 18 protons and 22 neutrons, the atom becomes ^{40}Ar, an inert gas. For every 100 ^{40}K atoms that decay, 11 become ^{40}Ar.

By comparing the proportion of ^{40}K to ^{40}Ar in a sample, and knowing the decay rate of ^{40}K, the age of the sample can be estimated. The technique works well in some cases, but it is unreliable in samples that have been heated or recrystallized after formation. Since it is a gas, ^{40}Ar can easily migrate in and out of potassium-bearing rocks, changing the ratio between parent and daughter.

Fission track dating is used to determine the thermal age of a sample, the time lapsed since the last significant heating event (typically above 215°F, or 102°C). Fission tracks are paths of radiation damage made by nuclear particles released by spontaneous fission or radioactive decay of ^{238}U. Fission tracks are created at a constant rate in uranium bearing minerals, so by determining the density of tracks present it is possible to determine the amount of time that has passed since the tracks began to form in the mineral. Fission track dating is used for determining the thermal ages of samples between about 100,000 and 1,000,000 years old, and it is also used for estimating the uplift and erosional history of areas, by recording when specific points cooled past 102°C.

Thermoluminescence is a chronometric dating method based on the fact that some minerals, when heated, give off a flash of light. The intensity of the light is proportional to the amount of radiation the sample has been exposed to and the length of time since the sample was heated. Luminesence is caused by heating a substance and thus liberating electrons trapped in its crystal defects. The phenomenon is used as a dating technique, especially for pottery. The number of trapped electrons is assumed to be related to the quantity of ionizing radiation to which the specimen has been exposed since firing, because the crystal defects are caused by ionizing radiation, and therefore related to the sample's age. Thus, by measuring the amount of light emitted on heating, an estimate of the age of the sample is obtained.

There are a number of other isotopic systems that are used for geochronology but they are less commonly used or less reliable than the methods described above. Geochronologists also incorporate relative and correlation dating techniques, such as stratigraphic correlation of dated units, to explore the wider implications of ages of dated units. A paleomagnetic timescale has been constructed for the past 180 million years, and in many situations it is now possible to determine the age of a particular part of a stratigraphic column or location on the seafloor by knowing which geomagnetic period the position is located within. Finally, geochronologists use structural cross-cutting relationships to determine which parts of a succession are older than or younger than a dated sample. Eventually the geochronologist is able to put together a temporal history of a rock terrane by dating several samples and combining these ages with cross-cutting observations and correlation with other units.

See also CARBON-14 DATING; DENDROCHRONOLOGY; PALEOMAGNETISM; RADIOACTIVE DECAY; STRATIGRAPHY.

geode Hollow or partly hollow generally subspherical bodies, with inward pointing drusy crystals that grew into an open space. Most geodes form in limestones or volcanic rocks, and fewer form in shales and other rocks, and have an outer shell of chalcedony or fine-grained silica. The crystals inside geodes are often perfectly formed and may exhibit brilliant colors depending on the chemistry of fluids that precipitated the crystals and the impurities that became incorporated in the crystal structure. Typical crystals in geodes include quartz, calcite, barite, and various sulfide minerals.

See also DIAGENESIS.

geodesy The study of the size and shape of the Earth, its gravitational field, and the determination of the precise locations of points on the surface. It also includes the study of the temporal variations in the shape of the planet and the location of points on the surface as a result of tides, rotation, and plate tectonic movements. Geodetic measurements rely heavily on positional measurements from satellite-based global positioning systems (GPS), gravity measurements, and radar altimetry measurements over the oceans. The science of measuring the size of the Earth probably started with Erastothenes in ancient Greece, who measured the distance from Alexandria to Aswan and calculated the curvature of the Earth from his measurements.

One branch of geodesy deals with the measurement of the Earth's gravity field and the geoid, the surface of equal gravitational potential. The geoid has a roughly elliptical shape that is slightly flattened at the poles as a result of the planet's rotation and is approximated by a reference ellipsoid.

Variations in the height of the geoid from the reference ellipsoid are expressed as the geoid height, in many cases reaching tens of meters. These variations reflect variations in the mass distribution within the Earth, and smaller, temporary variations may result from tides or winds changing the mass distribution of the oceans.

Geodetic measurements must use some reference frame, typically an astronomical or celestial, or an inertial reference frame. Many geodetic measurements are between different points on the surface, and these terrestrial measurements are useful for determinations of surface deformation such as motion along faults. Regional geodetic measurements rely on the art of triangulation, first developed by the Dutch scientist Gemma Frisius in the 16th century. Triangulation uses precise measurements of the angles and distances between different points in a network or grid to determine the changes in the shape of the grid with time, and hence the deformation of the surface.

Space geodesy uses satellite positioning techniques where GPS satellites emit microwave signals encoded with information about the position of the satellite, and the precise time at which the signal left the satellite. The distance to the GPS receiver on the surface can therefore be determined, and by using several GPS satellite signals the precise position on the surface of the Earth can be determined. The accuracy of GPS positions can be within a few meters (or less), and can be improved by a technique known as differential GPS in which a satellite receiver at a known position is coupled with and emits signals to a roving receiver. Further, multi-receiver interferometric and kinematic GPS techniques can improve positional measurements to the submillimeter level. The improved precision for these methods has greatly improved observations of surface deformation needed to predict earthquakes and volcanic eruptions and has aided precise navigation, surveying, and guidance systems.

See also GEOID; GLOBAL POSITIONING SYSTEM.

geodynamics The branch of geophysical science that deals with forces and processes in the interior of the Earth. It typically involves the macroscopic analysis of forces associated with a process and may include mathematical or numerical modeling. Geodynamics is a quantitative science closely related to geophysics, tectonics, and structural geology with problems including assessment of the forces associated with mantle convection, plate tectonics, heat flow, mountain building, erosion, volcanism, fluid flow, and other phenomena. The aim of many studies in geodynamics is to assess the relationships between different processes, such as to determine the influence of mantle convection on plate movements, or to assess plate motions in one area with deformation in another. It is contrasted with many other types of geological studies, which tend to be either static, analyzing only present and past states, or kinematic, analyz-

ing the history of motions without a quantitative assessment of the forces involved.

See also GEOPHYSICS; PLATE TECTONICS.

Geographic Information Systems (GIS) Geographic Information Systems (GIS) are computer application programs that organize and link information in a way that enables the user to manipulate that information in a useful way. They typically integrate a database management system with a graphics display that shows links between different types of data. For instance, a GIS may show relationships between geological units, ore deposits, and transportation networks. GIS allows users to layer information over other information that is already in the database. A GIS database allows the storage of information to a particular geographical area no matter what that information may be. GIS is used widely in science, industry, business, and government to sort out the pertinent information for the particular user from the GIS database.

geoid The surface of equipotential gravity equivalent with sea level and extending through the continents on the Earth is known as the geoid, and it is often referred to as the figure of the Earth. It is theoretically everywhere perpendicular to the direction of gravity (the plumb line) and is used as a reference surface for geodetic measurements. If the Earth were spherically symmetric and not spinning, the gravitational equipotential surfaces would consist of a series of concentric shells with increasing potential energy extending away from the Earth, much like a ball having greater potential energy the higher it is raised. However, since the Earth is not perfectly spherical (it is a flattened oblate spheroid) and it is spinning, the gravitational potential is modified so that it is an oblate spheroid with its major axis 0.3 percent longer than the minor axis. A best-fit surface to this spheroid is used by geodeticists, cartographers, and surveyors, but in many places the actual geoid departs from this simple model shape. Nonuniform distributions of topography and mass with depth cause variations in the gravitational attraction, known as geoid anomalies. Areas of extra mass, such as mountains or dense rocks at depth, cause positive geoid anomalies known as geoid highs, whereas mass deficits cause geoid lows. The geoid is measured using a variety of techniques, including direct measurements of the gravity field on the surface, tracking of satellite positions (and deflections due to gravity), and satellite-based laser altimetry that can measure the height of the sea surface to the sub-centimeter level. Variations in the height of the geoid are from up to tens of meters to even more than 100 meters.

geological hazard Geological hazards take many shapes and forms, from earthquakes and volcanic eruptions to the slow downhill creep of material on a hillside and the expansion of clay minerals in wet seasons. Natural geologic processes are in constant operation on the planet. These processes are considered hazardous when they go to extremes and interfere with the normal activities of society. For instance, the surface of the Earth is constantly moving through plate tectonics, yet we do not notice this process until sections of the surface move suddenly and cause an earthquake.

The Earth is a naturally dynamic and hazardous world, with volcanic eruptions spewing lava and ash, earthquakes pushing up mountains and shaking Earth's surface, and tsunami that sweep across ocean basins at hundred of miles per hour, rising in huge waves on distant shores. Mountains may suddenly collapse, burying entire villages, and slopes are gradually creeping downhill moving everything built on them. Storms sweep coastlines and remove millions of tons of sand from one place and deposit it in another in single days. Large parts of the globe are turning into desert, and glaciers that once advanced are rapidly retreating. Sea level is beginning to rise faster than previously imagined. All of these natural phenomena are expected consequences of the way the planet works, and as scientists better understand these geological processes, they are able to better predict when and where natural geologic hazards could become disasters and take preventive measures.

The slow but steady movement of tectonic plates on the surface of the Earth is the cause of many geologic hazards, either directly or indirectly. Plate tectonics controls the distribution of earthquakes and the location of volcanoes and causes mountains to be uplifted. Other hazards are related to Earth's surface processes, including floods of rivers, coastal erosion, and changing climate zones. Many of Earth's surface processes are parts of natural cycles on the Earth, but they are considered hazardous to humans because we have not adequately understood the cycles before building on exposed coastlines and in areas prone to shifting climate zones. A third group of geologic hazards is related to materials, such as clay minerals that dramatically expand when wetted, and sinkholes that develop in limestones. Still other hazards are extraterrestrial in origin, such as the occasional impact of meteorites and asteroids with Earth. The exponentially growing human population on Earth worsens the effect of most of these hazards. Species on the planet are now experiencing a mass extinction event, the severity of which has not been seen since the extinction event 66 million years ago that killed the dinosaurs and many of the other species alive on Earth at the time.

Many geologic hazards are the direct consequence of plate tectonics, associated with the motion of individual blocks of the rigid outer shell of the Earth. With so much energy loss accommodated by plate tectonics, we can expect that plate tectonics is one of the major energy sources for natural disasters and hazards. Most of the earthquakes on the planet are directly associated with plate boundaries, and these sometimes devastating earthquakes account for much of the motion between the plates. Single earthquakes have killed tens and even hundreds of thousands of people, such as the 1976 Tangshan earthquake in China that killed a quarter million people. Earthquakes also cause enormous financial and insurance losses; for instance, the 1994 Northridge earthquake in California caused more than $14 billion in losses. Most of the world's volcanoes are also associated with plate boundaries. Thousands of volcanic vents are located along the mid-ocean ridge system, and most of the volume of magma produced on the Earth is erupted through these volcanoes. However, volcanism associated with the mid-ocean ridge system is rarely explosive, hazardous, or even noticed by humans. In contrast, volcanoes situated above subduction zones at convergent boundaries are capable of producing tremendous explosive eruptions, with great devastation of local regions. Volcanic eruptions and associated phenomena have killed tens of thousands of people in this century, including the massive mudslides at Nevada del Ruiz in Colombia that killed 23,000 in 1985. Some of the larger volcanic eruptions cover huge parts of the globe with volcanic ash and are capable of changing the global climate. Plate tectonics is also responsible for uplifting the world's mountain belts, which are associated with their own sets of hazards, particularly landslides and other mass wasting phenomena.

Some geologic hazards are associated with steep slopes, and the effects of gravity moving material down these steep slopes to places where people live. Landslides and the slow downhill movement of earth material occasionally kill thousands of people in large disasters, such as when parts of a mountain collapsed in 1970 in the Peruvian Andes and buried a village several tens of miles away, killing 60,000 people. More typically, downhill movements are more localized and destroy individual homes, neighborhoods, roads, or bridges. Some downslope processes are very slow and involve the gradual, inch-by-inch creeping of soil and other earth material downhill, taking everything with it during its slide. This process of creep is one of the most costly of natural hazards, costing U.S. taxpayers billions of dollars per year.

Many other geological hazards are driven by energy from the Sun and reflect the interaction of the hydrosphere, lithosphere, atmosphere, and biosphere. Heavy or prolonged rains can cause river systems to overflow, flooding low-lying areas and destroying towns, farmlands, and even changing the courses of major rivers. There are several types of floods, including flash floods in mountainous areas or regional floods in large river valleys such as the great floods of the Mississippi and Missouri Rivers in 1993. Coastal regions may also experience floods, sometimes the result of typhoons, hurricanes, or coastal storms that bring high tides, storm surges, heavy rains, and deadly winds. Coastal storms may cause large amounts of coastal erosion, including cliff retreat, beach and dune migration, and the opening of new tidal inlets and closing of old inlets. These are all normal beach processes but have become hazardous since so many people have migrated into beachfront homes. Hurricane Andrew caused more than $19 billion of damage to the southern United States in 1992.

Deserts and dry regions are associated with their own set of natural geologic hazards. Blowing winds and shifting sands make agriculture difficult, and deserts have a very limited capacity to support large populations. Some of the greatest disasters in human history have been caused by droughts, some associated with the expansion of desert regions into areas that previously received significant rainfall and supported large populations dependent on agriculture. In this century, the sub-Saharan Sahel region of Africa has been hit with drought disaster several times, affecting millions of people and animals.

Desertification is but one possible manifestation of global climate change. The Earth has fluctuated in climate extremes, from hot and dry to cold and dry or cold and wet, and has experienced several periods when much of the land's surface was covered by glaciers. Glaciers have their own set of local-scale hazards that affect those living or traveling on or near their ice—crevasses can be deadly if fallen into, glacial melt-water streams can change in discharge so quickly that encampments on their banks can be washed away without a trace, and icebergs present hazards to shipping lanes. Glaciers may significantly reflect subtle changes in global climate—when glaciers are retreating, climate may be warming and becoming drier. When glaciers advance, the global climate may be getting colder and wetter. Glaciers have advanced and retreated over northern North America several times in the past 100,000 years. We are currently in an interglacial episode, and we may see the start of the return of the continental glaciers over the next few hundred or thousand years.

Geologic materials themselves can be hazardous. Asbestos, a common mineral, is being removed from thousands of buildings in the nation because of the perceived threat that certain types of airborne asbestos fibers present to human health. In some cases (for certain types of asbestos fibers), this threat is real. In other cases, the asbestos would be safer if it were left where it is rather than disturbing it and making the particles airborne. Natural radioactive decay is releasing harmful gases including radon that creep into our homes, schools, and offices, and causing cancer in numerous cases every year. This hazard is easily mitigated, and simple monitoring and ventilation can prevent many health problems. Other materials can be hazardous even though they seem inert. For instance, some clay minerals expand by hundreds of percent when wetted. These expansive clays rest under many foundations, bridges, and highways and cause billions of dollars of damage every year in the United States.

Sinkholes have swallowed homes and businesses in Florida and other locations in recent years. Sinkhole collapse and other subsidence hazards are more important than many people realize. Some large parts of southern California near Los Angeles have sunk tens of feet in response to pumping of groundwater and oil out of underground reservoirs. Other developments above former mining areas have begun sinking into collapsed mine tunnels. Coastline areas that are experiencing subsidence have the added risk of having the ocean rise into former living space. Coastal subsidence coupled with gradual sea-level rise is rapidly becoming one of the major global hazards that the human race is going to have to deal with in the next century, since most of the world's population lives near the coast in the reach of the rising waters. Cities may become submerged and farmlands covered by shallow salty seas. An enormous amount of planning is needed, as soon as possible, to begin to deal with this growing threat.

Occasionally in the Earth's history the planet has been hit with asteroids and meteorites from outer space, and these have completely devastated the biosphere and climate system. Many of the mass extinctions in the geologic record are now thought to have been triggered, at least in part, by large impacts from outer space. For instance, the extinction of the dinosaurs and a huge percent of other species on Earth 66 million years ago is thought to have been caused by a combination of massive volcanism from a flood basalt province preserved in India, coupled with an impact with a six-mile (10-km) wide meteorite that hit the Yucatán Peninsula of Mexico. When the impact occurred, a 1,000-mile (1,610-km) wide fireball erupted into the upper atmosphere, tsunami hundreds or thousands of feet high washed across the Caribbean, southern North America, and across much of the Atlantic, and huge earthquakes accompanied the explosion. The dust blown into the atmosphere immediately initiated a dark global winter, and as the dust settled months or years later, the extra carbon dioxide in the atmosphere warmed the Earth for many years, forming a greenhouse condition. Many forms of life could not tolerate these rapid changes and perished. Similar impacts have occurred at several times in the Earth's history and have had a profound influence on the extinction and development of life on Earth.

The human population is growing at an alarming rate, with the population of the planet currently doubling every 50 years. At this rate, there will only be a 3-foot by 1-foot space for every person on Earth in 800 years. Our unprecedented population growth has put such a stress on other species that we are driving a new mass extinction on the planet. We do not know the details of the relationships between different species, and many fear that destroying so many other life-forms may contribute to our own demise. In response to the population explosion, people are moving into hazardous locations including shorelines, riverbanks, along steep-sloped mountains, and along the flanks of volcanoes. Populations that grow too large to be supported by the environment usually suffer some catastrophe, disease, famine, or other mechanism that limits growth, and we as a species need to find ways to limit our growth to sustainable rates. Our very survival on the planet depends on our ability to maintain these limits.

Advances in science and engineering in recent decades have dramatically changed the way we view natural hazards. In the past, we viewed destructive natural phenomena (including earthquakes, volcanic eruptions, floods, landslides and tsunamis) as unavoidable and unpredictable. Our society's attention to basic scientific research has changed that view dramatically, and we are now able to make general predictions of when, where, and how severe such destructive natural events may be, reducing their consequences significantly. We are therefore able to plan evacuations, strengthen buildings, and make detailed plans of what needs to be done in natural disasters to such a degree that the costs of these natural geological hazards have been greatly reduced. This greater understanding has come with increased governmental responsibility. In the past, society placed little blame on government for the consequences of natural disasters. For instance, nearly 10,000 people perished in a hurricane that hit Galveston, Texas, on September 8, 1900, yet since there were no warning systems in place, no one was to blame. In 2001, 2 feet (0.6 m) of rain with consequent severe flooding hit the same area, and nobody perished. However, billions of dollars worth of insurance claims were filed. Now things are different, and few disasters go without blame being placed on public officials, engineers, or planners. Our extensive warning systems, building codes, and understanding have certainly prevented the loss of thousands of lives, yet they also have given us a false sense of security. When an earthquake or other disaster strikes, we expect our homes to be safe, yet they are only built to be safe to a certain level of shaking. When a natural geological hazard exceeds the expected level, a natural disaster with great destruction may result, and we blame the government for not anticipating the event. However, our planning and construction efforts are only designed to meet certain levels of force for earthquakes and other hazards, and planning for the rare stronger events would be exorbitantly expensive.

Geologic hazards can be extremely costly, in terms of price and human casualties. With growing population and wealth, the cost of natural disasters has grown as well. The amount of property damage measured in dollars has doubled or tripled every decade, with individual disasters sometimes costing tens of billions of dollars. A recent report (2000) to the Congressional Natural Hazards Caucus estimated the costs of some recent disasters; Hurricane Andrew in 1992 cost $23 billion, the 1993 Midwest floods cost $21 billion, and the 1994 Northridge earthquake cost $45 billion. In contrast, the entire first Persian Gulf war cost the United States and its allies $65 billion. That the costs of natural geologic hazards are now similar to the costs of warfare demonstrates the importance of understanding their causes and potential effects.

See also ASBESTOS; HURRICANE; MASS WASTING; PLATE TECTONICS; RADON.

Further Reading
Abbott, P. L. *Natural Disasters*, 3rd ed. Boston: McGraw Hill, 2002.
Bryant, E. A. *Natural Hazards*. Cambridge: Cambridge University Press, 1993.
Erickson, J. *Quakes, Eruptions, and Other Geologic Cataclysms: Revealing the Earth's Hazards*. The Living Earth Series. New York: Facts On File, 2001.
Griggs, Gary B., and J. A. Gilchrist. *Geologic Hazards, Resources, and Environmental Planning*. Belmont, Calif.: Wadsworth Publishing Co., 1983.
Kusky, Timothy M. *Geologic Hazards: A Sourcebook*. Westport, Conn.: Greenwood Press, 2003.
Murck, Barbara W., Brian J. Skinner, and Stephen C. Porter. *Dangerous Earth: An Introduction to Geologic Hazards*. New York: John Wiley and Sons, 1997.

Geological Society of America (GSA) The GSA (http://www.geosociety.org) was established in 1888 and provides access to elements that are essential to the professional growth of earth scientists at all levels of expertise and from all sectors: academic, government, business, and industry. Its main mission is to advance the geosciences, to enhance the professional growth of its members, and to promote the geosciences in the service of humankind. The Geological Society of America publishes several important journals and magazines, including the *Bulletin of the Geological Society of America, Geology,* and *GSA Today.*

geomagnetic reversals *See* GEOMAGNETISM; PALEOMAGNETISM.

geomagnetism The Earth has a magnetic field that is generated within the core of the planet. The field is generally approximated as a dipole, with north and south poles, and magnetic field lines that emerge from the Earth at the south pole and reenter at the north pole. The field is characterized at each place on the planet by an inclination and a declination. The inclination is a measure of how steeply inclined the field lines are with respect to the surface, with low inclinations near the surface, and steep inclinations near the poles. The declination measures the apparent angle between the rotational north pole and the magnetic north pole.

The magnetic field originates in the liquid outer core of the Earth and is thought to result from electrical currents generated by convective motions of the iron-nickel alloy that the outer core is made from. The formation of the magnetic field by motion of the outer core is known as the geodynamo theory, pioneered by Walter M. Elsasser of Johns Hopkins University in the 1940s. The basic principle for the generation of the field is that the dynamo converts mechanical energy from the motion of the liquid outer core, which is an electrical conductor, into electromagnetic energy of the magnetic field. The convective motion of the outer core, maintained by thermal and gravitational forces, is necessary to

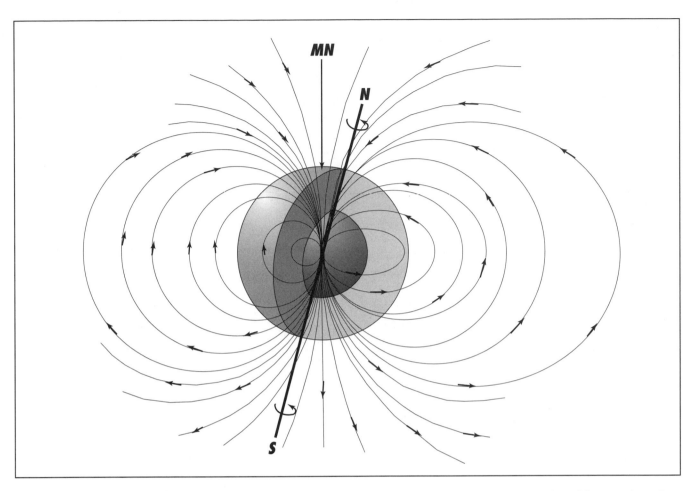

Magnetic field lines of Earth approximate the shape of the field produced by a bar magnet. Magnetic field lines point upward out of the magnetic south pole, form imaginary elliptical belts of equal intensity around the Earth, and plunge back into the Earth at the magnetic north pole. Note how the magnetic poles are not coincident with the rotational poles. The orientation of the magnetic field at any point on the Earth can be expressed as an inclination (plunge into the Earth) and a declination (angular distance between the magnetic and rotational north poles).

maintain the field. If the convection stopped or if the outer core solidified, the magnetic field would cease to be generated. Secular variations in the magnetic field have been well documented by examination of the paleomagnetic record in the seafloor, lava flows, and sediments. Every few thousand years the magnetic field changes intensity and reverses, with the north and south poles abruptly flipping.

See also PALEOMAGNETISM.

geomorphology The description, classification, and study of the physical properties and the origin of the landforms of the Earth's surface. Most studies in geomorphology include an analysis of the development of landforms and their relationships to underlying structures, and how the surface has interacted with other Earth systems such as the hydrosphere, cryosphere, and atmosphere. Geomorphologists have become increasingly concerned with the study of global climate change, and with the development of specific landforms asso-

ciated with active deformation and tectonics. These relatively new fields of global change and active tectonic geomorphology represent a significant movement away from classical geomorphology, concerned mostly with the evolutionary development of landforms.

Geomorphology includes many different processes that operate on the surface of the planet, so the geomorphologist needs to integrate hydrology, climate, sedimentology, geology, forestry, pedology, and many other sciences. The decomposition of bedrock and the development of soils are important for geomorphologists; they relate to global climate systems and may have importance for local engineering problems such as determining slope stability for construction sites. Other geomorphologists may study the development and evolution of drainage basins, along with the analysis of fluvial landforms such as floodplains, terraces, and deltas. Desert geomorphology is concerned with the development of desert landforms and climate, whereas glacial geomorphology analyzes the causes and effects of the movement of glaciers. Coastal processes including erosion, deposition, and longshore movement of sand are treated by coastal geomorphologists, whereas the development of various other types of landforms, such as karst, alpine, the seafloor, and other terrains are treated by other specialists.

Further Reading
Ritter, Dale, R. Craig Kochel, and Jerry Miller. *Process Geomorphology*, 3rd ed. Boston: WCB-McGraw Hill, 1995.

geophysics The study of the Earth by quantitative physical methods, with different divisions including solid-Earth geophysics, atmospheric and hydrospheric geophysics, and solar-terrestrial physics. There are many specialties in the field, including seismology, tectonics, geomagnetics, gravity, atmospheric science, ocean physics, and many others. Geophysics also includes the description and study of the origin and evolution of the major Earth systems, including the core, mantle, and crust of the continents and oceans.

geostrophic currents Currents in the ocean or atmosphere in which the horizontal pressure is balanced by the equal but opposite Coriolis force. These currents are not affected by friction, and they flow to the right of the pressure gradient force along pressure isobars in the Northern Hemisphere atmosphere and oceans, and to the left in the Southern Hemisphere. In the oceans, geostrophic currents are also known as contour currents, since they follow the bathymetric contours on the seafloor, flowing clockwise in the Northern Hemisphere and counterclockwise in the Southern Hemisphere. Downslope currents such as turbidity currents deposit most of the sediments on continental slopes. Since geostrophic or contour currents flow along the bathymetric contours, they rework bottom sediments at right angles to the currents that deposited the sediments. Their work is therefore detectable by examination of paleocurrent indicators that swing from downslope to slope-parallel movement vectors at the top of turbidite and other slope deposits.

See also ATMOSPHERE; CONTINENTAL MARGIN.

geothermal energy Temperatures in the Earth generally increase downward at 68°F–212°F (20°C–100°C) per kilometer, following the geothermal gradient for the region. Some regions near active volcanic vents have even higher geothermal gradients and are typified by abundant fumaroles and hot springs. These systems are usually set up when rising magma heats groundwater in cracks and pore spaces in rocks, and this heated water rises to the surface. Water from the sides of the system then moves in to replace the water that rose into hot springs, fumaroles, and geysers, and a natural hydrothermal circulation system is set up. The best natural hydrothermal systems are found in places where there are porous rocks and a heat source such as young magma.

Geothermal energy may be tapped by drilling wells, frequently up to several kilometers deep, into the natural geothermal systems. Geothermal wells that penetrate these systems commonly encounter water and less commonly steam at temperatures exceeding 572°F (300°C). Since water boils at 212°F (100°C) at atmospheric pressure, and higher temperatures at higher pressures (300°C at a one-kilometer depth), the water can be induced to boil by reducing the pressure by bringing it toward the surface in pipes. For a geothermal well to be efficient, the temperatures at depth should be greater than 392°F (200°C). Turbines attached to generators are attached to the tops of wells, requiring about two kilograms of steam per second to generate each megawatt of electricity.

Some countries are using geothermal energy to produce large amounts of electricity or heated water, with China, Hungary, Iceland, Italy, Japan, Mexico, and New Zealand leading the list. Still, the use of geothermal energy amounts to a very small but growing amount of the total electrical energy used by industrialized nations around the world.

geothermometry One of the aims of metamorphic petrology is to determine the temperature at which minerals in the rock grew and, ultimately, to determine the history of temperatures through which rock has passed. Such knowledge is important for understanding the Earth's temperature structure and the effects of temperature on physical and chemical processes, including volcanic and metamorphic processes. Geothermometry is often aimed at determining the pressure-temperature-time history of a specific rock mass, so that its mineral or hydrocarbon potential can be assessed, since specific minerals and hydrocarbon deposits are only formed and preserved in very limited ranges of temperature. Other geothermometry studies are aimed at understanding tectonic and metamorphic processes, since the temperature history of rock masses is expected to be different in different tectonic settings.

See also HYDROCARBON; METAMORPHISM; PLATE TECTONICS.

geyser A type of spring in which hot water or steam sporadically or episodically erupts as jets from an opening in the surface, in some cases creating a tower of water many tens of meters high. Geysers are often marked on the surface by a cone of siliceous sinter and other minerals that precipitated from the hot water, known as geyser cones. Many also have deposits of thermophyllic (heat-loving) bacteria that can form layers and mounds in stromatolitic buildups. Geysers form where water in pore spaces and cracks in bedrock gets heated by an underlying igneous intrusion or generally hot rock, causing it to boil and erupt, then the lost water is replaced by other water that comes in from the side of the system. In this way, a circulation system is set up that in some cases is quite regular with a predictable period between eruptions. The most famous geyser in the world is Old Faithful in Yellowstone National Park, Wyoming.

See also GEOTHERMAL ENERGY.

Gilbert, Grove K. (1843–1918) American *Geologist, Geomorphologist* Grove Gilbert is well known for his concept of graded streams. This concept maintained that streams always make channels and slopes for themselves either by cutting down their beds or by building them up with sediment so that over a period of time these channels will transport exactly the load delivered into them from above. Gilbert also explained the structure of the Great Basin as a result of extension, that is, individual "basin ranges" are the eroded upper parts of tilted blocks, which were displaced along faults as "comparatively rigid bodies of strata." After 1862, Gilbert received a number of awards and worked with numerous geologists on several geological surveys. When he studied the Henry Mountains (1875–76) he was the first to establish the fact that intrusive bodies are capable of deforming the host rock. He insisted that the Earth's crust is "as plastic in great masses as wax is in small." Unfortunately he exaggerated the fluidity of magma. Gilbert was conscientious in giving credit to those who deserved it but did not pay attention when it was given to him. From 1889 to 1892, he was the chief geologist of the U.S. Geological Survey.

Further Reading

Gilbert, G. K. "Report on the Geology of the Henry Mountains—Reprint of 1880 Paper." *Earth Science* 31 (1976): 68–74.

glacial system *See* GLACIER.

glacier Any permanent body of ice (recrystallized snow) that shows evidence of gravitational movement. Glaciers are an integral part of the cryosphere, which is that portion of the planet where temperatures are so low that water exists primarily in the frozen state. Most glaciers are presently found in the polar regions and at high altitudes. However, at several times in Earth history glaciers have advanced deeply into midlatitudes and the climate of the entire planet was different. Some models suggest that at one time the entire surface of the Earth may have been covered in ice, a state referred to as the "Snowball Earth."

Glaciers are dynamic systems, always moving under the influence of gravity and changing drastically in response to changing global climate systems. Thus, changes in glaciers may reflect coming changes in the environment. There are several types of glaciers. Mountain glaciers form in high elevations and are confined by surrounding topography, such as valleys. These include cirque glaciers, valley glaciers, and fiord glaciers. Piedmont glaciers are fed by mountain glaciers but terminate on open slopes beyond the mountains. Some piedmont and valley glaciers flow into open water, bays, or fiords, and are known as tidewater glaciers. Ice caps form dome-shaped bodies of ice and snow over mountains and flow radially outward. Ice sheets are huge, continent-sized masses of ice that presently cover Greenland and Antarctica and are the largest glaciers on Earth. Ice sheets contain about 95 percent of all the glacier ice on the planet. If global warming were to continue to melt the ice sheets, sea level would rise by 230 feet (66 m). A polar ice sheet covers Antarctica, consisting of two parts that meet along the Transantarctic Mountains. It shows ice shelves (thick glacial ice that floats on the sea), which form many icebergs by calving, which move northward into shipping lanes of the Southern Hemisphere.

Polar glaciers form where the mean average temperature lies below freezing, and these glaciers have little or no seasonal melting because they are always below freezing. Other glaciers, called temperate glaciers, have seasonal melting periods, where the temperature throughout the glacier may be at the pressure melting point (when the ice can melt at that pressure and both ice and water coexist). All glaciers form above the snow line, which is the lower limit at which snow remains year-round, located at sea level in polar regions, and at 5,000–6,000 feet (1,525–1,830 m) at the equator (Mount Kilimanjaro in Tanzania has glaciers, although these are melting rapidly).

Glaciers represent sensitive indicators of climate change and global warming, shrinking in times of warming, and expanding in times of cooling. Glaciers may be thought of as the "canaries in the coal mine" for climate change.

The Earth has experienced at least three major periods of long-term frigid climate and ice ages, interspersed with periods of warm climate. The earliest well-documented ice age is the period of the "Snowball Earth" in the Late Proterozoic, although there is evidence of several even earlier glaciations. Beginning about 350 million years ago, the Late Paleozoic saw another ice age lasting about 100 million years. The planet entered the present ice age about 55 million years ago. The underlying causes of these different glaciations are varied and include anomalies in the distribution of continents and oceans and associated currents, variations in the amount of incoming solar radiation, and changes in the atmospheric balance between the amount of incoming and outgoing solar radiation.

Formation of Glaciers

Glaciers form mainly by the accumulation and compaction of snow and are deformed by flow under the influence of gravity. When snow falls it is very porous, and with time the pore spaces close by precipitation and compaction. When snow first falls, it has a density about 1/10 that of ice; after a year or more, the density is transitional between snow and ice, and it is called firn. After several years, the ice has a density of 0.9 g/cm^3, and it flows under the force of gravity. At this point, glaciers are considered to be metamorphic rocks, composed of the mineral ice.

Retreating valley glacier (Wosnesinski glacier), Kenai Peninsula, Alaska. The glacier has retreated about 3–6 miles (5–10 km) in 90 years, leaving a glacial outwash and till-filled, U-shaped valley where it once flowed. The glacier is fed from the Harding Icefield in the background. *(Photo by Timothy Kusky)*

The mass and volume of glaciers are constantly changing in response to the seasons and to global climate changes. The mass balance of a glacier is determined by the relative amounts of accumulation and ablation (mass loss through melting and evaporation or calving). Some years see a mass gain leading to glacial advance, whereas some periods have a mass loss and a glacial retreat (the glacial front or terminus shows these effects).

Glaciers have two main zones, best observed at the end of the summer ablation period. The zone of accumulation is found in the upper parts of the glacier and is still covered by the remnants of the previous winter's snow. The zone of ablation is below this and is characterized by older dirtier ice,

McCarty tidewater glacier, Kenai Fiords National Park, Alaska. Note how the main trunk glacier is fed by many smaller valley glaciers and how its terminus is floating on the waters of McCarty Fiord. Many icebergs are calving off the front of the glacier. *(Photo by Timothy Kusky)*

from which the previous winter's snow has melted. An equilibrium line, marked by where the amount of new snow exactly equals the amount that melts that year separates these two zones.

Movement of Glaciers

When glacial ice gets thick enough, it begins to flow and deform under the influence of gravity. The thickness of the ice must be great enough to overcome the internal forces that resist movement, which depend on the temperature of the glacier. The thickness at which a glacier starts flowing also depends on how steep the slope it is on is. Thin glaciers can move on steep slopes, whereas to move across flat surfaces, glaciers must become very thick. The flow is by the process of creep, or the deformation of individual mineral grains. This creep leads to the preferential orientation of mineral (ice) grains, forming foliations and lineations, much the same way as in other metamorphic rocks.

Some glaciers develop a layer of meltwater at their base, allowing basal sliding and surging to occur. Where glaciers flow over ridges, cliffs, or steep slopes, their upper surface fails by cracking, forming large deep crevasses (these can be up to 200 feet deep). A thin blanket of snow can cover these crevasses, making for very dangerous conditions.

Ice in the central parts of valley glaciers moves faster than it does at the sides, because of frictional drag against the valley walls on the side of the glacier. Similarly, a profile with depth of the glacier would show that it moves the slowest along its base, and faster internally and along its upper surface. When a glacier surges, it may temporarily move as fast along its base as it does in the center and top. This is because during surges, the glacier is essentially riding on a cushion of meltwater along the glacial base, and frictional resistance is reduced during surge events. During meltwater-enhanced surges, glaciers may advance by as much as several kilometers in a year. Events like this may happen in response to climate changes.

Calving refers to a process in which icebergs break off from the fronts of tidewater glaciers or ice shelves. Typically, the glacier will crack with a loud noise that sounds like an explosion, and then a large chunk of ice will splash into the water, detaching from the glacier. This process allows glaciers to retreat rapidly.

Glaciation and Glacial Landforms

Glaciation is the modification of the land's surface by the action of glacial ice. When glaciers move over the land's surface, they plow up the soils, abrade and file down the bedrock, carry and transport the sedimentary load, steepen valleys, then leave thick deposits of glacial debris during retreat.

In glaciated mountains, a distinctive suite of landforms forms from glacial action. Glacial striations are scratches on the surface of bedrock, formed when a glacier drags boulders across the bedrock surface. Roche moutonnée and other asymmetrical landforms are made when the glacier plucks pieces of bedrock from a surface and carries them away. The step faces in the direction of transport. Cirques are bowl-shaped hollows that open downstream and are bounded upstream by a steep wall. Frost wedging, glacial plucking, and abrasion all work to excavate cirques from previously rounded mountaintops. Many cirques contain small lakes called tarns, which are blocked by small ridges at the base of the cirque. Cirques continue to grow during glaciation, and where two cirques form on opposite sides of a mountain, a ridge known as an arete forms. Where three cirques meet, a steep-sided mountain forms, known as a horn. The Matterhorn of the Swiss Alps is an example of a glacial carved horn.

Valleys that have been glaciated have a characteristic U-shaped profile, with tributary streams entering above the base of the valley, often as waterfalls. In contrast, streams generate V-shaped valleys. Fiords are deeply indented glaciated valleys that are partly filled by the sea. In many places that were formerly overlain by glaciers, elongate streamlined forms known as drumlins occur. These are both depositional features (composed of debris) and erosional (composed of bedrock).

Glacial Transport

Glaciers transport enormous amounts of rock debris, including some large boulders, gravel, sand, and fine silt, called till. The glacier may carry this at its base, on its surface, or internally. Glacial deposits are characteristically poorly sorted or non-sorted, with large boulders next to fine silt. Most of a glacier's load is concentrated along its base and sides, because in these places plucking and abrasion are most effective.

Active ice deposits till as a variety of moraines, which are ridge-like accumulations of drift deposited on the margin of a glacier. A terminal moraine represents the farthest point of travel of the glacier's terminus. Glacial debris left on the sides of glaciers forms lateral moraines, whereas where two glaciers meet, their moraines merge and are known as a medial moraine.

Rock flour is a general name for the deposits at the base of glaciers, where they are produced by crushing and grinding by the glacier to make fine silt and sand. Glacial drift is a general term for all sediment deposited directly by glaciers, or by glacial meltwater in streams, lakes, and the sea. Glacial marine drift is sediment deposited on the seafloor from floating ice shelves or bergs and may include many isolated pebbles or boulders that were initially trapped in glaciers on land, then floated in icebergs that calved off from tidewater glaciers. The rocks melted out while over open water and fell into the sediment on the bottom of the sea. These isolated stones are called dropstones and are often one of the hallmark signs of ancient glaciations in rock layers that geologists find in the rock record. Stratified drift is deposited by meltwater and may

include a range of sizes, deposited in different fluvial or lacustrine environments.

Glacial erratics are glacially deposited rock fragments with compositions different from underlying rocks. In many cases the erratics are composed of rock types that do not occur in the area they are resting in but are only found hundreds or even thousands of miles away. Many glacial erratics in the northern part of the United States can be shown to have come from parts of Canada. Some clever geologists have used glacial erratics to help them find mines or rare minerals that they have located in an isolated erratic—they have used their knowledge of glacial geology to trace the boulders back to their sources following the orientation of glacial striations in underlying rocks. Recently, diamond mines were discovered in northern Canada (Nunavut) by tracing diamonds found in glacial till back to their source region.

Sediment deposited by streams washing out of glacial moraines is known as outwash and is typically deposited by braided streams. Many of these form on broad plains known as outwash plains. When glaciers retreat, the load is diminished, and a series of outwash terraces may form.

See also ICE AGE.

Further Reading

Alley, Richard B., and Michael L. Bender. "Greenland Ice Cores: Frozen in Time." *Scientific American,* February issue, 1998.
Dawson, A. G. *Ice Age Earth*. London: Routledge, 1992.
Erickson, J. "Glacial Geology: How Ice Shapes the Land." Changing Earth Series. New York: Facts On File, 1996.
Schneider, D. "The Rising Seas." *Scientific American,* March issue, 1997.
Stone, G. "Exploring Antarctica's Islands of Ice." *National Geographic,* December issue (2001): 36–51.

global positioning system Global positioning systems, commonly referred to by their acronym GPS, were developed by the U.S. Department of Defense to provide the U.S. military with a superior tool for navigation, viable at any arbitrary point around the world. As of 2002, the U.S. Department of Defense has paid more than $12 billion for the development and maintenance of the GPS program, which has matured a great deal since its conception in the 1960s.

The current configuration of the global positioning system includes three main components: the GPS satellites, the control segment, and the GPS receivers. Working together, these components provide users of GPS devices with their precise location on the Earth's surface, along with other basic information of substantial use such as time, altitude, and direction. Key in understanding how GPS functions is the understanding of these components, and how they interrelate with one another.

GPS satellites, dubbed the Navstar satellites, form the core functionality of the global positioning system. Navstar satellites are equipped with an atomic clock and radio equipment to broadcast a unique signal, called a "pseudo-random code," as well as ephemeris data. This signal serves to identify one satellite from the other and provide GPS receivers accurate information regarding the exact location of the satellite. These satellites follow a particular orbit around the Earth, and the sum of these orbits is called a constellation. The GPS Navstar satellite constellation is configured in such a way that at any point on the Earth's surface, the user of a GPS receiver should be able to detect signals from at least six Navstar satellites. It is very important to the proper functioning of GPS that the precise configuration of the constellation be maintained.

Geosynchronous satellite orbit maintenance is performed by the control segment (or satellite control centers), with stations located in Hawaii, Ascension Island, Diego Garcia, Kwajalein, and Colorado Springs. Should any satellite fall slightly in altitude or deviate from its correct path, the control segment will take corrective actions to restore precise constellation integrity.

Lastly, the component most visible to all who use GPS devices is the GPS receiver. It is important to understand that GPS receivers are single-direction asynchronous communication devices, meaning that the GPS receiver does not broadcast any information to the Navstar satellites, only receives signals from them. Recent years have seen the miniaturization and mass proliferation of GPS receivers. GPS devices are now so small that they can be found in many other hybrid devices such as cellular phones, some radios, and personal desk accessories. They are also standard on many vehicles for land, sea, and air travel. Accuracy of consumer GPS receivers is typically no better than nine feet, but advanced GPS receivers can measure location to less than a centimeter.

GPS resolves a location on the surface of the Earth through a process called trilaterating, which involves determining the distances to the Navstar satellites and the GPS receiver. In order to do this, two things must be true. First, the locations of the Navstar satellites must be known, and second, there must be a mechanism for precision time measurements. For very precise measurements, there must be a system to reconcile error caused by various phenomena.

GPS receivers are programmed to calculate the location of all the Navstar satellites at any given time. A combination of the GPS receiver's internal clock and trilateration signal reconciliation, performed by the GPS receiver, allow a very precise timing mechanism to be established.

When a GPS receiver attempts to locate itself on the surface of the planet, it receives signals from the Navstar satellites. As mentioned previously, these signals are intricate and unique. By measuring the time offset between the GPS receiver's internal pseudo-random code generator, and the pseudo-random code signal received from the Navstar satellites, the GPS receiver can use the simple distance equation to calculate the distance to the Navstar satellite.

To accurately locate a point on the Earth's surface, at least three distances need to be measured. One measurement is only enough to place the GPS receiver within a three-dimensional arc. Two measurements can place the GPS receiver within a circle. Three measurements place the GPS receiver on one of two points. One possible point location will usually be floating in space or traveling at some absurd velocity, and so the GPS receiver eliminates this point as a possibility, thus resolving the GPS receiver's location on the surface of the planet. A fourth measurement will also allow the correct point to be located, as well as provide necessary geometry data to synchronize the GPS receiver's internal clock to the Navstar satellite's clock.

Depending on the quality of the GPS device, error correction may also be performed when calculating location. Errors arise from many sources. Atmospheric conditions in the ionosphere and troposphere cause impurities in the simple distance equation by altering the speed of light. Weather modeling can help calculate the difference between the ideal speed of light and the likely speed of light as it travels through the atmosphere. Calculations based on the corrected speed of light then yield more accurate results.

As weather conditions rarely fit models, however, other techniques such as dual frequency measurements, where two different signals are compared to calculate actual speed of the pseudo-code signal, will be used to reduce atmospheric error.

Ground interference, such as multipath error, which arises from signals bouncing off objects on the Earth's surface, can be detected and rejected in favor of direct signals via very complicated signal selection algorithms.

Still another source of error can be generated by the Navstar satellites being slightly out of position. Even a few meters from the calculated position can throw off a high-precision measurement.

Geometric error can be reduced by using satellites that are far apart, rather than close together, easing certain geometric constraints.

Sometimes precision down to the centimeter is needed. Only advanced GPS receivers can produce precise and accurate measurements at this level. Advanced GPS receivers utilize one of several techniques to more precisely pinpoint locations on the Earth's surface, mostly by reducing error or using comparative signal techniques.

One such technique is called differential GPS, which involves two GPS receivers. One receiver monitors variations in satellite signals and relates this information to the second receiver. With this information, the second receiver is then able to more accurately determine its location through better error correction.

Another method involves using the signal carrier-phase as a timing mechanism for the GPS receiver. As the signal carrier is a higher frequency than the pseudo-random code it carries, carrier signals can be used to more accurately synchronize timers.

Lastly, a geostationary satellite can be used as a relay station for transmission of differential corrections and GPS satellite data. This is called augmented GPS and is the basic idea behind the new WAAS (Wide Area Augmentation System) system of additional satellites installed in North America. The system encompasses 25 ground-monitoring stations and two geostationary WAAS satellites that allow for better error correction. This sort of GPS is necessary for aviation, particularly in landing sequences.

gneiss A coarse-grained strongly foliated metamorphic rock in which bands or layers of granular minerals such as quartz and feldspar generally alternate with bands or layers of flaky or elongate minerals such as mica or amphibole. There are many varieties of gneiss, distinguished by composition and texture. Gneiss derived from igneous rocks is generally known as orthogneiss, whereas gneiss derived from sedimentary rocks is known as paragneiss. However it is often difficult to determine the origin or protolith of the gneiss, so it is recommended that textural and compositional terms be used to describe a gneiss, instead of terms that purport to know the rock's protolith. Augen gneiss has lenticular spots, generally made of quartz and feldspar that resemble eyes in a strongly foliated matrix. Mylonitic gneiss is strongly foliated and shows evidence of high shear strains. Straight gneiss has parallel layers that extend over large distances, resembling a sedimentary layering. Gneiss is generally formed from regional metamorphic and tectonic processes.

See also METAMORPHISM; PLATE TECTONICS.

Gobi Desert One of the world's great deserts, the Gobi is located in central Asia encompassing more than 500,000 square miles (1,295,000 km²) in Mongolia and northern China. The desert covers the region from the Great Khingan Mountains northwest of Beijing to the Tien Shan north of Tibet, but the desert is expanding at an alarming rate, threatening the livelihood of tens of thousands of farmers and nomadic sheepherders every year. Every spring dust from the Gobi covers eastern China, Korea, and Japan and may extend at times around the globe. Northwesterly winds have removed almost all the soil from land in the Gobi, depositing it as thick loess in eastern China. Most of the Gobi is situated on a high plateau resting 3,000–5,000 feet (900–1,500 m) above sea level, and it contains numerous alkaline sabkhas and sandy plains in the west. Regions in the Gobi include abundant steppes, high mountains, forests, and sandy plains. The Gobi has yielded many archaeological, paleontological, and geological finds, including early stone implements, dinosaur eggs, and mineral deposits and precious stones including turquoise and jasper.

See also DESERT.

gold A soft, yellow, native metallic element widely used for jewelry, ornaments, and as the international standard for world finance. Along with silver and copper, gold has been sought after by our human ancestors for seven or eight thousand years as a valuable and malleable decorative metal; since it was found in its native form, it is highly malleable and has desirable aesthetic properties.

Gold usually occurs as a native metal in deposits and typically has some silver mixed with it by atomic substitution. Deposits may be lode or placer type, with lode gold including primary deposits in hard rock (typically metamorphic or igneous), and placer deposits being secondary, found in stream gravels or soils. Lode gold deposits are formed by hot aqueous solutions that move through the crust and are typically associated with the marginal zone or aureoles of plutons. Other lode gold deposits seem to be associated with regional metamorphic fluids, or meteoric waters that circulate deeply and get heated, deposit gold where they cool or react chemically with wall rocks of different composition. Many of the world's famous gold mines are associated with quartz-pyrite-gold veins in faults, including the Mother Lode in California, the Bendigo and Ballarat lodes of Australia, and the ancient workings of the Pharos in the Southeastern Desert of Egypt. Lode gold deposits that form under low pressures and low to intermediate temperatures (302°F–662°F, or 150°C–350°C) are classified as epithermal deposits and tend to be associated with convergent margins.

Gold is chemically unreactive, so it persists through weathering and transportation, and gets concentrated in soils and as heavy minerals in stream placer deposits. Placer gold was the sought-after treasure in the great gold rushes of the Fairbanks gold district and Yukon territories of Canada, where placer and lode deposits are still being discovered and mined. One of the biggest ancient placer deposits in the world is in the 2.6–2.7-billion-year-old Witwatersrand basin of southern Africa, which has supplied more than 40 percent of the gold ever mined in the world.

Further Reading
Goldfarb, Richard J., David I. Groves, and S. Gardoll. "Orogenic Gold and Geologic Time: A Global Synthesis." *Ore Geology Reviews* 18 (2001): 1–75.

Goldschmidt, Victor Moritz (1888–1942) Swiss *Chemist, Geochemist, Mineralogist* Victor Goldschmidt studied chemistry, mineralogy, and geology at the University of Christiania. His work was greatly influenced by the Norwegian petrologist and mineralogist W. C. Bogger and also by earth scientists Paul von Groth and Friedrich Becke. He received his doctorate in 1911 and became a full professor and director of the mineralogical institute of the University of Christiania. His doctoral thesis talked about the factors governing the mineral associations in contact-metamorphic rocks and was based upon the samples he had collected in southern Norway. In later years he became a professor in the Faculty of Natural

Sciences at Gottingen and head of its mineral institute. He began geochemical investigations on the noble and alkali metals, the siderophilic and lithophilic elements. He produced a model of the Earth and was able to show how these different elements and metals were accumulated in various geological domains on the basis of their charges, sizes, and the polarizabilities of their ions. Goldschmidt is one of the pioneers in geochemistry who gave explanations of the composition of the environment.

Further Reading
Goldschmidt, V. M. "The Distribution of the Chemical Elements." *Royal Institution Library of Science, Earth Science* 3 (1971): 219–233.
———. "On the Problems of Mineralogy." *Journal of the Washington Academy of Sciences* 51 (1961): 69–76.

Gondwana (Gondwanaland)

The Late Proterozoic–Late Paleozoic supercontinent of the Southern Hemisphere, named by Eduard Suess after the Gondwana System of southern India. The name Gondwana means "land of the Gonds" (an ancient tribe in southern India) so the more common rendition of the name Gondwanaland for the southern supercontinent is technically improper, meaning "land of the land of the Gonds." It includes the present continents of Africa, South America, Australia, Arabia, India, Antarctica, and many smaller fragments. Most of these continental masses amalgamated in the latest Precambrian during closure of the Mozambique and several other ocean basins and persisted as a supercontinent until they joined with the northern continents in the Carboniferous to form the supercontinent of Pangea.

The different fragments of Gondwana are matched with others using alignment between belts of similar-aged deformation, metamorphism, and mineralization, as well as common faunal, floral, and paleoclimatic belts. The formation and breakup of Gondwana is associated with one of the most remarkable explosions of new life-forms in the history of the planet, the change from simple single-celled organisms and soft-bodied fauna to complex, multicelled organisms. The formation and dispersal of supercontinents strongly influences global climate and the availability of different environmental niches for biological development, linking plate tectonic and biological processes.

Since the early 1990s there has been an emerging consensus that Gondwana formed near the end of the Neoproterozoic from the fragmented pieces of an older supercontinent, Rodinia, itself assembled near the end of the Grenville cycle (~1100 Ma). The now standard model of Gondwana's assembly begins with the separation of East Gondwana (Australia, Antarctica, India, and Madagascar) from the western margin of Laurentia, and the fan-like aggregation of East and West Gondwana. The proposed assembly closed several ocean basins, including the very large Mozambique Ocean, and turned the constituents of Rodinia inside out, such that the

external or "passive" margins in Madagascar and elsewhere became collisional margins in latest Precambrian and earliest Cambrian time.

The notion of a single, short-lived collision between East and West Gondwana is an oversimplification, since geologic relations suggest that at least three major ocean basins closed during the assembly of Gondwana (Pharusian, Mozambique, Adamastor), and published geochronology demonstrates that assembly was a protracted affair. There is currently a large amount of research being done to understand these relationships. For example, an alternative two-stage model for closure of the Mozambique Ocean has been recently advanced that ascribes an older "East African" orogeny (~680 Ma) to collision between Greater India (i.e., India–Tibet–Seychelles–Madagascar–Enderby Land) and the cojoined Congo and Kalahari Cratons. This was followed by a younger "Kuunga" event (~550 Ma) that represents the collision of Australia–East Antarctica with proto-Gondwana, thus completing Gondwana's assembly near the end of the Neoproterozoic.

See also NEOPROTEROZOIC; SUPERCONTINENT CYCLE.

Further Reading
Dalziel, Ian W. D. "Global Paleotectonics; Reconstructing a Credible Supercontinent." *Geological Society of America, Abstracts with Programs* 31, no. 7, A-316, 1999.
de Wit, Maarten J., Margaret Jeffry, Hugh Bergh, and Louis Nicolaysen. *Geological Map of Sectors of Gondwana Reconstructed to Their Disposition at ~150 Ma.* American Association of Petroleum Geologists, Tulsa, Scale 1:10,000,000, 1988.
Hoffman, Paul F. "Did the Breakout of Laurentia Turn Gondwana Inside-out?" *Science* 252 (1991): 1,409–1,412.
Kusky, Timothy M., Mohamed Abdelsalam, Robert Tucker, and Robert Stern, eds., *Evolution of the East African and Related Orogens, and the Assembly of Gondwana.* Amsterdam: Precambrian Research, 2003.

GPS *See* GLOBAL POSITIONING SYSTEM.

Grabau, Amadeus William

(1870–1946) American *Geologist, Paleontologist* Amadeus Grabau was a great contributor to systematic paleontology and stratigraphic geology, and he was also a respected professor and writer. He spent half of his professional life in the United States and the last 25 years in China. Grabau studied at the Massachusetts Institute of Technology and went on to receive his M.S. and D.Sc. at Harvard. He became a professor in paleontology at Colombia University. In the first 20 years of his career, he was the leading scientist in paleontology, stratigraphy, and sedimentary petrology. The greatest effect of his scientific work has been his contributions to the principles of paleoecology and to the genetic aspect of sedimentary paleontology. His stratigraphic work was influential. Not only did it bring about a more developed understanding of the subject but it was the source of understanding Earth movements. The concepts involved in his polar control theory, pulsation theory, and the separation of Pangea allowed for the imaginative syntheses of geologic evidence. He received numerous awards and was a member of the following institutes: the Geological Society of America, the New York Academy of Science, and the Geological Society of China. He was also an honorary member of the Peking Society of Natural History, the China Institute of Mining and Metallurgy, the Academia Sinica, and the Academia Peipinensis.

Further Reading

Grabau, Amadeus W. "The Polar-Control Theory of Earth Development." *Association of Chinese and American Eng.,* J 18 (1937): 202–223.

———. "Fundamental Concepts in Geology and Their Bearing on Chinese Stratigraphy." *Geological Society of China Bulletin* 16 (1937): 127–176.

———. "Revised Classification of the Palaeozoic Systems in the Light of the Pulsation Theory." *Geological Society of China Bulletin* 15 (1936): 23–51.

graben An elongate, fault-bounded topographically depressed region of the Earth's crust, generally associated with rifting. Graben are common in regions of continental rifting, in many cases being bounded by a large normal fault on one side, and a tilted fault block on the opposing side. The crests of slow-spreading oceanic ridges are marked by axial graben, with topographic relief approaching one kilometer. Other graben are associated with volcanism, with parts of magma chamber roofs collapsing into the chamber, forming elongate depressions on the surface.

See also RIFT; STRUCTURAL GEOLOGY.

Grand Canyon The Great Gorge of the Colorado River, known as the Grand Canyon, is a one-mile (1.5-km) deep, 4–18-mile (6.5–29-km) wide, 217-mile (350-km) long gorge in the Colorado Plateau of northwestern Arizona. The canyon was created by more than 8 million years of erosion as the Colorado Plateau has been uplifted by tectonic forces, and it exposed more than 1.7 billion years' worth of history of the region in its stratified rock layers and deeper metamorphic and igneous rocks. The canyon was set aside as a national monument in 1908, and 673,575 acres of the canyon were designated a national park in 1919. The 198,280-acre primitive Grand Canyon National Monument presently adjoins the national park on the west.

The canyon was first explored by boat by a party led by John W. Powell in 1869, after being described by Spanish explorer Garcia Lopez de Cardenas in 1540. Before that, the canyon was inhabited by hundreds of Indian pueblos, many of which are still visible on the lower canyon walls and rim.

From top to base, the canyon exposes a stratigraphic sequence ranging from the Kaibab Limestone, Toroweap Formation, Coconino Sandstone, Hermit Shale, Supai Formation, Redwall Limestone, Muav Limestone, Bright Angel Shale, Tapeats Sandstone, and Vishnu Schist. Most of the rocks are sedimentary except for the Precambrian schists and intrusions of Zoroaster Granite. Two major unconformities are present, one above the Vishnu Schist and other rocks of the Precambrian sequence, and one, known as the Great Unconformity, beneath the Tapeats Sandstone.

The Kaibab Limestone is the highest unit in the canyon stratigraphy, and it is composed of limestone with minor shale and sandstone in its upper parts. Fossils include crinoids, corals, brachiopods, and other shellfish. This formation is known as the giant bathtub ring around the edge of the canyon, because of its resemblance to rings found around many older ceramic tubs. The white color of the formation is caused by the fossilized white sponges, which have been replaced by silica, forming chert. The Kaibab Limestone is therefore strong and erosionally resistant and forms steep cliffs around the edge of the canyon.

The Torweap Formation is a dark yellow to gray limestone underlying the Kaibab, and is in turn underlain by the Coconino Sandstone. This unit consists of light-colored fossil sand dunes with large-scale cross-bedding. Fossil footprints and burrows of invertebrates are common. The Hermit Shale underlies the Coconino Sandstone and consists of soft, rust-colored shale that erodes easily, forming a slope instead of a cliff. This formation contains abundant fossilized plants and footprints of amphibians and reptiles.

The Supai Formation is a red shale formed in a deltaic environment, preserving terrestrial plant fossils and amphibian tracks, with more marine conditions toward the west. This is underlain by the marine Redwall Limestone, a hard cliff-forming brown limestone that is stained red by hematite that has washed down from the overlying red shale. Fossils include marine brachiopods, snails, fish, corals, and trilobites. The sparsely fossiliferous Muav Limestone underlies the Redwall Limestone.

The Bright Angel Shale is composed of a colorful variety of mudstone and shale with fossils of brachiopods, trilobites, and other marine organisms. It is underlain by the Tapeats Sandstone, a cliff-forming rippled brown sandstone, with occasional fossils of trilobites.

Underlying the Great Unconformity at the base of the Tapeats sandstone is the Precambrian sequence, consisting of the Vishnu Schist and intrusive Zoroaster Granite on the southwest side of the canyon. On the northeast side of the canyon, the Vishnu Schist is overlain by another Precambrian sequence, including a variety of sandstones, shales, mudstones, stromatolitic limestone, basaltic lava, and quartzite.

granite A coarse-grained igneous plutonic rock with visible quartz, potassium, and plagioclase feldspar, and dark minerals such as biotite or amphibole, is generally known as granite, but the IUGS (International Union of Geological Scientists) define granite more exactly as a plutonic rock with 10–50 percent quartz, and the ratio of alkali to total feldspar in the range of 65–90 percent.

Granites and related rocks are abundant in the continental crust and may be generated either by the melting of preexisting rocks, or, in lesser quantities, by the differentiation by fractional crystallization of basaltic magma. Many granites are associated with convergent margin or Andean-style magmatic arcs and include such large plutons and batholiths as those of the Sierra Nevada batholith, Coast Range batholith, and many others along the American Cordillera. Granites are also a major component of Archean cratons and granite greenstone terranes.

Many building stones are granitic, since they tend to be strong, durable, nonporous, and exhibit many color and textural varieties. Granite often forms rounded hills with large round or oblong boulders scattered over the hillside. Many of these forms are related to weathering along several typically perpendicular joint sets, where water infiltrates and reacts with the rock along the joint planes. As the joints define cubes in three dimensions, large blocks get weathered out and eventually get rounded as the corners weather faster than the other parts of the joint surface. Granite also commonly forms exfoliation domes, in which large sheets of rock weather off and slide down mountainsides and inselbergs, isolated steep-sided hills that remain on a more weathered plain. Many granites weather ultimately to flat or gently rolling plains covered by erosional detritus including cobbles, boulders, and granitic gravels.

See also CRATONS; EXFOLIATION; IGNEOUS ROCKS; PLUTON.

granulite A coarse-grained metamorphic rock formed at high temperatures and pressures. Granulites generally lack a strong foliation because dehydration reactions remove micas and other hydrous minerals that are platy, and typically

Granulite from Kabuldurga, southern India. Orthopyroxene-bearing granulite forms the dark-colored replacement bands following shear zones in older gneiss. (lens cap for scale) *(Photo by Timothy Kusky)*

define the foliation in other metamorphic rocks. Many types of granulite (such as charnockites) are orthopyroxene-bearing, with large orthopyroxene grains typically overgrowing older primary minerals.

The origin of granulites has been controversial, with abundant evidence indicating that they are residual rocks that have experienced incipient partial melting, and other evidence indicating that the rocks have been permeated by carbon-dioxide–rich fluids during high temperature–high pressure metamorphism.

See also METAMORPHISM.

graptolite Sticklike or serrated colonial marine organisms of the class Graptolithina (phylum Hemichordata) that lived from the Cambrian to Pennsylvanian. Most are less than an inch in length, but they range from 0.2 inch to 3.3 feet (5 mm–1 m) in length. They typically appear as faint sticklike serrated carbon films on bedding surfaces. Since graptolites are relatively common and abundant in Paleozoic shales and show rapid morphological changes, they form useful index fossils for biostratigraphic correlation. They are most common in rocks deposited on the outer continental shelves, and most probably lived in the planktonic near-surface environment, with fewer deepwater varieties preserved.

Reconstruction of some of the best preserved graptolites has revealed details of the structure of the colonies. Graptolites apparently reproduced by asexual budding, with new individuals producing new overlapping sawblade-like arms known as thecae, with overlapping branches known as stipes. The stipes formed the frontal parts of a feeding apparatus attached to a zooid, which was probably part of a larger animal.

There are several different orders of graptolites, classified on the basis of their mode of life. Bush-like forms that were originally attached to the seafloor are known as Dendroidea and were common from Cambrian through Pennsylvanian, whereas Graptoloidean were planktonic forms, abundant from the Ordovician through Early Devonian. Several other orders were much less common, with most exhibiting encrusting types of morphologies.

See also PALEONTOLOGY.

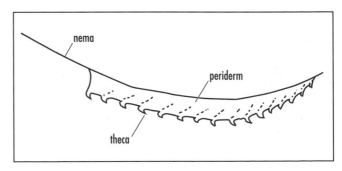

Graptolite morphology showing serrated edge (theca) and stem (nema)

gravel A natural accumulation of rounded sedimentary particles, with the majority being larger than sand (.08 inch, or 2 mm). The particles may include granules, pebbles, cobbles, and boulders in increasing size, and the gravel may be named according to the size or composition of its most abundant particles or by its matrix. When consolidated, gravels are known as conglomerates. Conglomerates as gravels are said to be oligomictic if they are composed of one type of clast, and polymictic if they are composed of clasts of many different types.

Gravels are commonly found in streams and on beaches and cover some alluvial plains and alluvial fans in arid regions. They are deposited in places where fast currents lose velocity and can no longer transport them. Many gravel deposits are only temporarily in place, such as those in streams, and may only move during flood events.

Reefs and other carbonate environments produce coarse-grained limestone or coral conglomerates and breccias, but these types of rocks are usually classified as limestone-breccias, carbonate conglomerates, and other types of carbonates. Some unusual types of gravel and conglomerate include tillites, coarse glacially derived gravel and conglomerate typically in a fine-grained matrix, and diamictite, a nongenetic name for a non-sorted, non-calcareous terrigenous sedimentary rock composed of sand and larger-sized particles in a fine-grained matrix.

See also SEDIMENTARY ROCKS.

gravimeter A gravimeter is an instrument that measures and determines the specific gravity of a substance or measures variations in the gravitational field of the Earth. Common gravimeters often are made with a simple mass-spring assembly. By hanging a mass on the spring we can account for how much gravity is affecting the mass by how far the spring has stretched. Simple gravimeters are easy to operate but lack the precision needed to view the minute changes in the gravitational field of the Earth. Modern gravimeters are much more accurate and are employed in the field to study the gravitational field of the Earth but are much more complicated and employ special springs known as "zero springs."

gravity anomaly The difference between the observed value of gravity at a point and the theoretically calculated value of gravity at that point, based on a simple gravity model. The value of gravity at a point reflects the distribution of mass and rock units at depth, as well as topography. The average gravitational attraction on the surface is 32 feet per seconds squared (9.8 m/s^2), with one gravity unit (g.u.) being equivalent to one ten-millionth of this value. Another older unit of measure, the milligal, is equivalent to 10 gravity units. The range in gravity on the Earth's surface at sea level is about 50,000 g.u., 32.09–32.15 feet per seconds squared (9.78–9.83 m/s^2). A person would weigh slightly more at the equator than at the poles because the Earth has a slightly larger radius at the equator than at the poles.

Geologically significant variations in gravity are typically only a few tenths of a gravity unit, so instruments to measure gravity anomalies must be very sensitive. Some gravity surveys are done using closely to widely spaced gravity meters on the surface, whereas others are done using observations of the perturbations of orbits of satellites.

The determination of gravity anomalies involves subtracting the effects of the overall gravity field of the Earth, accomplished by removing the gravity field at sea level (geoid), leaving an elevation-dependent gravity measurement. This measurement reflects a lower gravitational attraction with height and distance from the center of the Earth, as well as an increase in gravity caused by the gravitational pull of the material between the point and sea level. The free-air gravity anomaly is a correction to the measured gravity calculated using only the elevation of the point and the radius and mass of the Earth. A second correction depends on the shape and density of rock masses at depth and is known as the Bouguer gravity anomaly. Sometimes a third correction is applied to gravity measurements, known as the isostatic correction. This applies when a load such as a mountain, sedimentary basin, or other mass is supported by mass deficiencies at depth, much like an iceberg floating lower in the water. However, there are several different mechanisms of possible isostatic compensation, and it is often difficult to know which mechanisms are important on different scales. Therefore, this correction is often not applied.

Different geological bodies are typically associated with different magnitudes and types of gravity anomalies. Belts of oceanic crust thrust on continents (ophiolites) represent unusually dense material and are associated with positive gravity anomalies of up to several thousand g.u.. Likewise, dense massive sulfide metallic ore bodies are unusually dense and are also associated with positive gravity anomalies. Salt domes, oceanic trenches, and mountain ranges all represent an increase in the amount of low-density material in the crustal column and are therefore associated with increasingly negative gravity anomalies, with negative values of up to 6,000 g.u. associated with the highest mountains on Earth, the Himalayan chain.

See also GEOID; ISOSTACY.

graywacke A clastic sedimentary rock, sandstone, with generally coarse and angular grains of quartz and feldspar set in a compact clay-rich matrix. It may have clasts of many other minerals and rock fragments and typically has a dark or light gray color. Graywacke often occurs as laterally continuous beds interbedded with shale or slate in thick flysch sequences, and many graywacke beds exhibit sedimentary structures known as Bouma sequences, indicating deposition by submarine turbidity currents. Sedimentary structures in Bouma sequences typically include a graded base to a bed, which may show scour marks or other markings such as sole marks or load casts on the bottom of the bed. This lower part of the bed grades up through an interval of parallel laminations, and then an interval with ripple cross-laminations. In a complete Bouma sequence, these are followed upward by an upper layer with parallel laminations, and then a pelitic top to the bed.

The immaturity of graywacke sediments is generally interpreted to reflect that they were eroded, transported, and buried so quickly that chemical weathering did not have a chance to break down the feldspars and separate the clays from the quartz grains. These conditions are met in an actively deforming orogenic belt, and graywackes and flysch belts are typically associated with the erosion of active or recently formed orogenic belts. The sediments are typically eroded in the mountains, transported by streams, and deposited in submarine foreland basins in front of the deforming orogen.

See also FLYSCH; SANDSTONE; SEDIMENTARY ROCKS.

Great Barrier Reef The largest coral reef in the world, the Great Barrier Reef forms a 1,250-mile (2,010-km) long breakwater in the Coral Sea along the northeast coast of Queensland, Australia. The reef has been designated a World Heritage area, the world's largest such site. The reef is comprised of several individual reef complexes, including 2,800 individual reefs stretching from the Swain reefs in the south to the Warrier reefs along the southern coast of Papua New Guinea. Many reef types are recognized, including fringing reefs, flat platform reefs, and elongate ribbon reefs. The reef complexes are separated from the mainland of Queensland by a shallow lagoon ranging 10–100 miles (16–160 km) wide.

There are more than 400 types of coral known on the Great Barrier Reef, as well as 1,500 species of fish, 400 species of sponges, and 4,000 types of mollusk, making it one of the world's richest sites in terms of faunal diversity. Additionally, the reefs hosts animals including numerous sea anemones, worms, crustaceans, and echinoderms, and an endangered mammal known as a dugong. Sea turtles feed on abundant algae and sea grass, and the reef is frequented by humpback whales that migrate from Antarctic waters to have babies in warm waters. Hundreds of bird species have breeding colonies in the islands and cays among the reefs, and these birds include beautiful herons, pelicans, osprey, eagles, and shearwaters.

The reefs also hide dozens of shipwrecks and have numerous archaeological sites of significance to the Aboriginal and Torres Strait Islander peoples.

See also CARBONATE; REEF.

Great Dike, Zimbabwe A 310-mile (500-km) long, 2.5-billion-year-old mafic and ultramafic dike that cuts across the Zimbabwe craton in a north-northeasterly direction. The main dike is associated with many smaller but parallel satel-lite dikes. The Great Dike is one of the most prominent mafic dikes on the planet, clearly visible from space and on satellite imagery. The dike is actually comprised of a number of smaller layered mafic-ultramafic intrusions, some of which exhibit igneous layering. Since the dike is essentially undeformed in most places, its age of 2.5 billion years shows that most deformation in the Zimbabwe craton ceased before then. Using this observation, plus the orientation of the stresses that must have allowed extension perpendicular to the dike margins, several authors have noted that the dike probably intruded as a result of the collision of the Zimbabwe and Kaapvaal cratons to the south, just prior to 2.5 billion years ago. Parts of the dike are mined for chromite and platinum group elements.

Further Reading

Burke, Kevin, William S. F. Kidd, and Timothy M. Kusky. "Archean Foreland Basin Tectonics in the Witwatersrand, South Africa." *Tectonics* 5 (1986).
———. "Is the Ventersdorp Rift System of Southern Africa Related to a Continental Collision between the Kaapvaal and Zimbabwe Cratons at 2.64 Ga Ago?" *Tectonophysics* 11 (1985).

Great Lakes, North America A group of five large interconnected freshwater lakes located along the border between the north-central United States and Canada. From west to east the lakes include Lake Superior, Lake Michigan, Lake Huron, Lake Erie, and Lake Ontario, which feeds the St. Lawrence River that flows to the North Atlantic Ocean. Niagara Falls is located between Lakes Erie and Ontario. Cumulatively, the lakes form the largest body of freshwater in the world with a combined surface area of 95,000 square miles (246,050 km^2). The lakes formed at the end of the Pleistocene glaciation when large ice-carved depressions were filled with glacial meltwater as the glaciers retreated.

The first Europeans to settle in the area were the French fur traders, including Etienne Brulé in 1612 and Samuel de Champlain, who explored Lakes Huron and Ontario. Robert LaSalle explored Lakes Erie and Michigan in 1679. Many years of dispute over land ownership between the French, British, Americans, and native tribes ensued until the War of 1812 was concluded. The region was then rapidly settled by Americans, and the Erie Canal connected Lake Erie with the Mohawk and Hudson Rivers in 1825, greatly increasing development of the entire area.

In a general sense the Great Lakes occupy the geological boundary between the Precambrian Canadian shield on the north and the Paleozoic North American passive margin and foreland basin sequences on the south. The Precambrian basement areas include rich deposits of iron ore, especially around Lake Superior. Together with an abundance of coal, these resources led the region to be developed as a major steel-producing area, which continues to this day.

See also CRATONS; PRECAMBRIAN.

Great Salt Lake, Utah The Great Salt Lake of Utah is a terminal lake, one that has no outlet to the ocean or a larger body of water. Since the lake is in an arid area, evaporation is intense, causing dissolved minerals in the water to become concentrated. As freshwater enters the lake it evaporates, leaving salts behind, causing the lake waters to be about eight times as salty as seawater.

The lake is situated at the base of a flat valley west of the steep Wasatch Mountains, and the lake's shoreline positions change dramatically in response to changes in rainfall, snowmelt, and inflow into the lake. The lake is about 70 miles (113 km) long, 30 miles (48 km) wide, but only 40 feet (12m) deep, so small changes in the water depth cause large changes in the position of the shoreline. Since 1982 rainfall increased in the area and the lake began to rise, covering highways and other urbanized areas around Salt Lake City. Lake levels peaked in 1986–87 and began to fall to more typical levels since then. The Bonneville salt flats west of the Great Salt Lake attest to a once much larger lake and are now used as a place into which to pump extra water from the Great Salt Lake when lake levels get too high. There, the water evaporates leaving additional salt deposits behind.

Great Slave Lake, Nunavet The Great Slave Lake, named after the Dene Slavey Indians, covers 10,980 square miles (28,44 km²) in Nunavet, formerly the Northwest Territories of Canada. The lake is 300 miles (483 km) long and between 12 and 68 miles (19–110 km) wide, and it is the deepest lake in North America at 2,015 feet (614.6 m) and the sixth deepest in the world. The first European to discover the lake was Samuel Hearne in 1771, and he named it Athapuscow Lake. The Mackenzie River, the longest in Canada—2,651 miles (4,241 km)—has its source in the western end of the Great Slave Lake. The lake is located along several geological boundaries, some marked by faults, explaining its great depths. The south and western ends of the lake are covering Devonian clastic rocks derived from the mountains to the west, whereas the northern and eastern ends of the lake cover Precambrian basement. The east arm of the lake follows a Proterozoic rift basin, where spectacular stromatolitic reefs are preserved, whereas the northern side of the lake exposes Archean rocks of the Slave province. The McDonald fault scarp strikes northeast along the east arm of the lake.

Further Reading

Henderson, John B. "Geology of the Yellowknife—Hearne Lake Area, District of MacKenzie: Segment Across an Archean Basin." *Geological Survey of Canada Memoir* 414 (1985).

Hoffman, Paul F. "Continental Transform Tectonics: Great Slave Lake Shear Zone (ca. 1.9 Ga), Northwest Canada." *Geology* 15 (1987): 785–788.

Kusky, Timothy M. "Accretion of the Archean Slave Province." *Geology* 17 (1989): 63–67.

Aerial view of springtime ice breakup on the Great Slave Lake showing huge blocks of ice shuffling around to form extensional, compressional, and strike-slip zones between the giant slabs. *(Photo by Timothy Kusky)*

greenhouse effect The greenhouse effect refers to time periods when the Earth is abnormally warm in response to the atmosphere trapping incoming solar radiation. Global climate represents a balance between the amount of solar radiation received and the amount of this energy that is retained in a given area. The planet receives about 2.4 times as much heat in the equatorial regions as in the polar regions. The atmosphere and oceans respond to this unequal heating by setting up currents and circulation systems that redistribute the heat more equally. These circulation patterns are in turn affected by the ever-changing pattern of the distribution of continents, oceans, and mountain ranges.

The amounts and types of gases in the atmosphere can modify the amount of incoming solar radiation. For instance, cloud cover can cause much of the incoming solar radiation to be reflected back to space before being trapped by the lower atmosphere. On the other hand, certain types of gases (known as greenhouse gases) allow incoming short-wave-

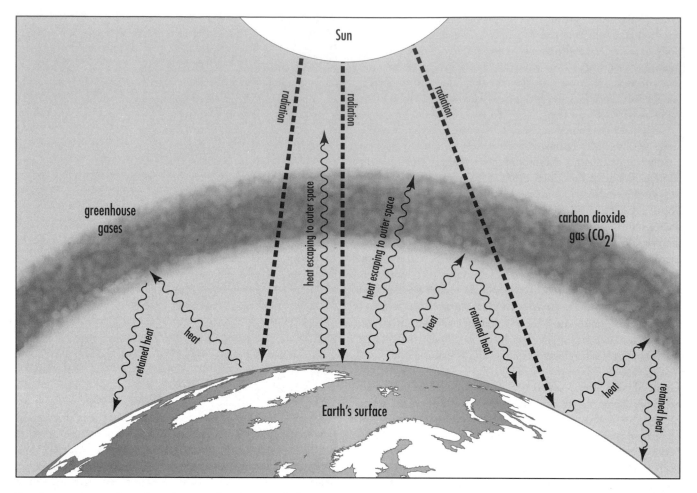

Showing greenhouse effect, where short-wavelength solar radiation penetrates Earth's atmosphere, but reflected longer wavelength radiation is trapped by greenhouse gases, warming the atmosphere.

length solar radiation to enter the atmosphere but trap this radiation when it tries to escape in its longer wavelength reflected form. This causes a buildup of heat in the atmosphere and can lead to a global warming known as the greenhouse effect.

The amount of heat trapped in the atmosphere by greenhouse gases has varied greatly over Earth's history. One of the most important greenhouse gases is carbon dioxide (CO_2) that plants (which release O_2 to the atmosphere) now take up by photosynthesis. In the early part of Earth history (in the Precambrian), before plants covered the land surface, photosynthesis did not remove CO_2 from the atmosphere, so CO_2 levels were much higher than at present. Atmospheric CO_2 is also presently taken up by marine organisms, which remove it from the ocean surface water (which is in equilibrium with the atmosphere) and use the CO_2 along with calcium to form their shells and mineralized tissue. These organisms make $CaCO_3$ (calcite), the main component of limestone, a rock composed largely of the dead remains of marine organisms. Approximately 99 percent of the planet's CO_2 is presently

removed from the atmosphere/ocean system, because it has been locked up in rock deposits of limestone on the continents and on the seafloor. If this amount of CO_2 were released back into the atmosphere, the global temperature would increase dramatically. In the early Precambrian, when this CO_2 was free in the atmosphere, global temperatures averaged about 550°F (290°C).

The atmosphere redistributes heat quickly by forming and redistributing clouds and uncondensed water vapor around the planet along atmospheric circulation cells. Oceans are able to hold and redistribute more heat because of the greater amount of water in the oceans, but they redistribute this heat more slowly than the atmosphere. Surface currents are formed in response to wind patterns, but deep ocean currents (which move more of the planet's heat) follow courses that are more related to the bathymetry (topography of the seafloor) and the spinning of the Earth than they are related to surface winds.

The balance of incoming and outgoing heat from the Earth has determined the overall temperature of the planet through time. Examination of the geological record has enabled paleoclimatologists to reconstruct periods when the Earth had glacial periods, hot and dry periods, hot and wet periods, or cold and dry periods. In most cases the Earth has responded to these changes by expanding and contracting its climate belts. Warm periods see an expansion of the warm subtropical belts to high latitudes, and cold periods see an expansion of the cold climates of the poles to low latitudes.

See also ATMOSPHERE.

greenschist Metamorphic rocks are commonly referred to in the field by informal names that reflect some distinctive physical or mineralogical aspect of the rock. A greenschist is an informal name for a schistose, generally dark green rock (on fresh surfaces), whose color is due to the presence of the metamorphic minerals chlorite, actinolite, epidote, or serpentine. In more formal usage, a greenschist is a schistose rock that has been metamorphosed to the greenschist facies, typically in the range of 570°F–930°F (300°C–500°C), at low to medium pressures. It is a typical low-grade regional metamorphic rock. The greenschist facies is defined in metamorphosed basic (mafic) rocks by the mineral assemblage albite + epidote + chlorite + actinolite. In metamorphosed pelitic rocks, the greenschist facies assemblage typically includes chlorite + white mica + biotite + chloritoid.

See also GREENSTONE BELTS; METAMORPHISM.

greenstone *See* GREENSTONE BELTS.

greenstone belts Elongate accumulations of generally mafic volcanic and plutonic rocks, typically associated with immature graywacke types of sedimentary rocks, banded iron formations, and less commonly carbonates and mature sedimentary rocks. Most greenstone belts are Archean or at least Precambrian in age, although similar sequences are known from orogenic belts of all ages. Most greenstone belts are metamorphosed to greenschist through amphibolite facies and intruded by a variety of granitoid rocks. Older quartzofeldspathic gneisses are found associated with some greenstone belts, although most of these are in fault contact with the greenstones.

Until recently, few complete Phanerozoic-like ophiolite sequences were recognized in Archean greenstone belts, leading some workers to the conclusion that no Archean ophiolites or oceanic crustal fragments are preserved. These ideas were recently challenged by research documenting partial dismembered ophiolites in several greenstone belts, and a complete ophiolite sequence in the North China craton. Archean oceanic crust was possibly thicker than Proterozoic and Phanerozoic counterparts, resulting in accretion predominantly of the upper basaltic section of oceanic crust. The crustal thickness of Archean oceanic crust may in fact have resembled modern oceanic plateaux. If this were the case, complete

Phanerozoic-like Mid-Ocean Ridge Basalt (MORB)–type ophiolite sequences would have been very unlikely to be accreted or obducted during Archean orogenies. In contrast, only the upper, pillow lava-dominated sections would likely be accreted. Archean greenstone belts have an abundance of accreted ophiolitic fragments compared to Phanerozoic orogens, suggesting that thick, relatively buoyant, young Archean oceanic lithosphere may have had a rheological structure favoring delamination of the uppermost parts during subduction and collisional events.

Greenstone belts display a wide variety of shapes and sizes and are distributed asymmetrically across Archean cratons, in a manner reminiscent of tectonostratigraphic zonations in Phanerozoic orogens. For instance, the Yilgarn craton has mostly granitic gneisses in the southwest, mostly circa 2.9 Ga greenstones throughout the central craton, and circa 2.7 Ga greenstones in the east. The Slave Province contains remnants of a circa 4.2–2.9 Ga gneissic terrain in the western part of the province, dominantly mafic greenstone belts in the center, and circa 2.68 Ga mixed mafic, intermediate, and felsic calc-alkaline volcanic rocks in the eastern part of the province. Other cratons are also asymmetric in this respect; for example, the Zimbabwe craton has mostly granitic rocks in the east, and more greenstones in the west. The Superior Province contains numerous subparallel belts, up to thousands of kilometers long, which are distinct from each other, but similar in scale and lithology to "terranes" in Phanerozoic orogens. These distributions of rock types are analogous to asymmetric tectonostratigraphic zonations, which are products of plate tectonics in younger orogenic belts, and emphasize that greenstone belts are perhaps only parts of once larger orogenic systems.

There are three significantly different end-member regional outcrop patterns reflecting the distribution of greenstone belts within cratons. These include broad domal granitoids with interdomal greenstones, broad greenstone terrains with internally bifurcating lithological domains and irregular granitoid contacts, and long, narrow, and straight greenstone belts. The first pattern includes mostly granitoid domes with synformal greenstone belts, which result from either interference folding or domal/diapiric granitoids. The second pattern includes many of the terrains with thrust belt patterns, including much of the Yilgarn Province in Western Australia and the Slave Province of Canada. Contacts with granitoids are typically intrusive. The third pattern includes transpressional belts dominated by late strike-slip shear zones along one or more sides of the belt. Granite-greenstone contacts are typically a fault or shear zone.

Greenstone Belt Geometry

Geophysical surveys have shown that greenstone belts are mostly shallow to intermediate in depth, some have flat or irregular bases, and many are intruded by granitic rocks.

They are not steep synclinal keels. Gravity models consistently indicate that greenstone belts rarely extend to greater than six miles (10 km) in depth, and seismic reflection studies show that the steeply dipping structures characteristic of most greenstone belts disappear into a horizontally layered mid to lower crustal structure. Seismic reflection surveys have also proven useful at demonstrating that boundaries between different "belts" in granite-greenstone terrains are in some cases marked by large-scale crustal discontinuities most easily interpreted as sutures or major strike-slip faults.

Just as greenstone belts are distributed asymmetrically on cratons, many have asymmetrical distributions of rock types and structural vergence within them, and in this respect are very much like younger orogenic belts. For example, the eastern Norseman-Wiluna belt in the Yilgarn craton contains a structurally disrupted and complex association of tholeiites and calc-alkaline volcanic rocks, whereas the western Norseman-Wiluna belt contains disrupted tholeiites and komatiites. In other belts, it is typical to find juxtaposed rocks from different crustal levels, and facies that were originally laterally separated. One of the long-held myths about the structure of greenstone belts is that they simply represent steep synclinal keels of supracrustal rocks squeezed between diapiric granitoids. Where studied in detail, there is a complete lack of continuity of strata from either side of the supposed syncline, and the structure is much more complex than the pinched-synform model predicts. The structure and stratigraphy of greenstone belts will only be unraveled when "stratigraphic" methods of mapping are abandoned, and techniques commonly applied to gneissic terrains are used for mapping greenstone belts. Greenstone belts should be divided into structural domains, defined by structural style, metamorphic history, distinct lithological associations, and age groupings where these data are available.

One of the most remarkable features of Archean greenstone belts is that structural and stratigraphic dips are in most cases very steep to vertical. These ubiquitously steep dips are evidence for the intense tectonism that these belts have experienced, although mechanisms of steepening may be different in different examples. Some belts, including the central Slave Province in Canada and Norseman-Wiluna belt of Western Australia, appear to have been steepened by imbricate thrust stacking, with successive offscraping of thrust sheets steepening rocks toward the hinterland of the thrust belt, whereas greenstone belt rocks on the margins of plutons and batholiths have commonly been further steepened by the intrusions. Examples of this mechanism are found in the Pilbara and northern Zimbabwe cratons. Late homogeneous strain appears to be an important steepening mechanism in other examples, such as in the Theespruit area of the Barberton Belt. Tight to isoclinal upright folding, common in most greenstone belts, and fold interference patterns are responsible for other steep dips. In still other cases, rotations incurred in strike-slip fault systems (e.g., Norseman-Wiluna belt, Superior Province) and on listric normal fault systems (e.g., Quadrilatero Ferrifero, Sao Francisco craton) and may have caused local steepening of greenstone belt rocks. These are the types of structures seen throughout Phanerozoic orogenic belts.

Structural vs. Stratigraphic Thickness of Greenstone Belts

Many studies of the "stratigraphy" of greenstone belts have assumed that thick successions of metasedimentary and metavolcanic rocks occur without structural repetition, and that they have undergone relatively small amounts of deformation. As fossil control is virtually nonexistent in these rocks, stratigraphic correlations are based on gross similarities of lithology and poorly constrained isotopic dates. In pre-1980 studies it was common to construct homoclinal stratigraphic columns that were 6–12 miles (10–20 km) or more thick, but recent advances in the recognition of usually thin fault zones, and precise U-Pb ages documenting older-over-younger stratigraphies makes reevaluation of these thicknesses necessary. It is rare to have intact stratigraphic sections that are more than a couple of kilometers thick in greenstone belts, and further mapping needs to be structural, based on defining domains of like structure, lithology, and age, rather than lithological, attempting to correlate multiply deformed rocks across large distances.

An observation of utmost importance for interpreting the significance of supposed thick stratigraphic sections in greenstone belts is that there is an apparent lack of correlation between metamorphic grade and inferred thicknesses of the stratigraphic pile. If the purported 10–20-kilometer-thick sequences were real stratigraphic thicknesses, an increase in metamorphic grade would be detectable with inferred increase in depth. Because this is not observed, the thicknesses must be tectonic and thus reflect stratigraphic repetition in an environment such as a thrust belt or accretionary prism, where stratigraphic units can be stacked end-on-end, with no increase in metamorphic grade in what would be interpreted as stratigraphically downward. Other mechanisms by which apparent stratigraphic thicknesses may be increased are folding, erosion through listric normal fault blocks, and progressive migration of depositional centers.

Greenstone-Gneiss Contact Relationships

An important problem in many greenstone belt studies is determining the original structural relationships between greenstone belts and older gneiss terrains. In pre-1990 studies the significance of early thrusting along thin "slide" zones went unrecognized, leading to a widespread view that many greenstone belts simply rest autochthonously over older gneisses, or that the older gneisses intruded the greenstone belt. While this may be the case in a few examples, it is difficult to demonstrate, and the "classic areas" in which such relationships were supposedly clearly demonstrable have recently been shown to contain significant early thrust faults between greenstones and older sedimentary rocks that rest unconformably over older gneisses. Such is the case at Steep Rock Lake in the Superior Province, at Point Lake and Cameron River in the Slave Province, in the Theespruit type section of the Barberton greenstone belt, at Belingwe, Zimbabwe, and in the Norseman-Wiluna Belt.

The Norseman-Wiluna, Cameron River, and Point Lake greenstone belts contain up to 1,640-feet (500-m) wide dynamothermal aureoles at their bases. The aureoles contain upper-amphibolite facies assemblages, in contrast to greenschist and lower amphibolite facies assemblages in the rest of the greenstone belts. The aureoles have mylonitic, gneissic, and schistose fabrics parallel to the upper contacts with the greenstone belts and are locally partially melted forming granitic anatectites. The aureole at the base of the Norseman-Wiluna belt is an early shear zone structure related to the juxtaposition of the greenstone belt with older gneisses, and it is overprinted by greenschist facies metamorphic fabrics and two episodes of regional folding, the second of which is the main regional deformation event and is associated with a strong cleavage. A late-spaced cleavage is associated with upright folds. On Cameron River and Point Lake in the Slave Province, the aureoles represent early thrust zones related to the tectonic emplacement of the greenstone belts over the gneisses. Amphibolite-facies mylonites were derived through deformation of mafic and ultramafic rocks at the bases of the greenstone belts, which are largely at greenschist facies. The broad field-scale relationships in these cases are also similar to those found in dynamothermal aureoles attached to the bases of many obducted ophiolites.

Structural Elements of Greenstone Belts

The earliest structures found in greenstone belts are those that formed while the rocks were being deposited. In most greenstone belts, the mafic volcanic/plutonic section is older than the clastic sedimentary section, so penecontemporaneous structures are older in the magmatic rocks than in the sedimentary rocks. Unequivocal evidence for large-scale deformation of the magmatic rocks of greenstone belts during their deposition is lacking. Structures of this generation typically include broken pillow lavas grading into breccias, and possible slump folds and faults in interpillow sedimentary horizons.

From the Jamestown ophiolite complex within the Barberton greenstone belt, Maarten de Wit describes two types of early extensional shear zones that were active prior to regional contractional deformation. The first are low-angle normal faults located along the lower contacts of the ophiolite-related cherts (so-called Middle Marker) and cause extensive brecciation and alteration of adjacent simatic rocks. The faults and adjacent cherts are cut by subvertical mafic rocks of the Onverwacht Group, showing that these faults were active early, during the formation of the ophi-

olitic Onverwacht Group. The second type of early faults in the Barberton greenstone belt occur in both the plutonic and the extrusive igneous parts of the Jamestown ophiolite, and they may represent steepened extensions (root zones) of the higher-level extensional faults, or they may represent transform faults. Other possible examples of early extensional faults have been described from the Cameron River and Yellowknife greenstone belts in the Slave Province, and from the Proterozoic Purtuniq ophiolite in the Cape Smith Belt of the Ungava Orogen. In the Purtuniq ophiolite, early sinuous shear zones locally separate sheeted dikes from mafic schists, causing rotation of the dike complex. In other places, dikes intrude this contact, showing that the shear zones are early features. Although not explicitly interpreted in this way, these shear zones may be related to block faulting in the region of the paleoridge axis.

Detailed mapping in a number of greenstone belts has revealed early thrust faults and associated recumbent folds. Most do not have any associated regional metamorphic fabric or axial planar cleavage, making their identification difficult without very detailed structural mapping. Examples are known from the Zimbabwe and Kaapvaal cratons, the Yilgarn craton, the Pilbara craton, the Slave Province, and the Superior Province. In these cases, it is apparent that early thrust faulting and recumbent folding are responsible for the overall distribution of rock types in the greenstone belts and also account for what were, in some cases, previously interpreted as enormously thick stratigraphic sequences. In some cases, early thrust faults are responsible for juxtaposing greenstone sequences with older gneissic terrains. Few of these early thrust faults are easy to detect; they occupy thin poorly exposed structural intervals within the greenstone belts, some are parallel to internal stratigraphy for many kilometers, and most have been reoriented by later structures. In many greenstone belts there are numerous layer-parallel fault zones, but their origin is unclear because they are not associated with any proved stratigraphic repetition or omission. Although it is possible that these faults are thrust, strike-slip, or normal faults, the evidence so far accumulated in the few well-mapped examples supports the interpretation that they are early thrust faults. In some cases, late intrusive rocks have utilized the zone of structural weakness provided by the early thrusts for their intrusion.

Emplacement of the early thrust and fold nappes typically is not associated with any penetrative fabric development or any regional metamorphic recrystallization, making recognition of these early structures even more difficult. Delineation of early thrusts depends critically on very detailed fieldwork with particular attention paid to structural facing, vergence, and patterns of lineations. Determining the sense of tectonic transport of early nappes is critical for tectonic interpretation, but it is also one of the most elusive goals because of the weak development of critical lineations, and reorientation of the earliest fabric elements by younger structures. Kinematic studies of early "slide" zones have received far less attention than they deserve in Archean greenstone belts.

Early penecontemporaneous structures within the younger metasedimentary sections appear in general to be related to thrust stacking of the mafic volcanic/plutonic section, as shown best by the deformed flysch and molasse of the Pietersburg, South Africa, and Point Lake, Northwest Territories, greenstone belts.

Folds are in many cases the most obvious outcrop to map-scale structures in greenstone belts. Several phases of folding are typical, and fold interference patterns are commonplace. Some greenstone belts show a progression from early recumbent folds (associated with thrust/nappe tectonics), through two or more phases of tight to isoclinal upright folds, which are associated with the most obvious mesoscopic and microscopic fabric elements and metamorphic mineral growth. One or more generations of late open folds or broad crustal arches, with associated crenulation cleavages also affect many greenstone belts. Fold interference patterns most typically reflect the geometry of F_2 and F_3 structures because they have similar amplitudes and wavelengths; F_1 recumbent folds are best recognized by reversals in younging directions, or downward-facing F_2 and F_3 folds.

In many examples, the relationships of individual fold generations to tectonic events is poorly understood, and the orientations of causal stresses are poorly constrained, largely because of uncertainties associated with correctly unraveling superimposed folding events. The relative importance of "horizontal" v. "vertical" tectonics has been debated, in part because it is difficult to distinguish granite-cored domes produced by the interference of different generations of folds from domes produced by diapirism. Two generations of upright folds with similar amplitudes and wavelengths will produce a dome and basin fold interference pattern that very much resembles a diapir pattern.

Many granite-greenstone terrains are cut by late-stage strike-slip faults, some of which are reactivated structures that may have been active at other times during the history of individual greenstone belts. Strike-slip dominated structural styles are known from the Superior Province, the Yilgarn craton, the Pilbara craton, and southern Africa.

Large-scale lineaments of the Norseman-Wiluna Belt near Kalgoorlie and Kambalda show late reverse motion related to regional oblique compression. However, their length, often greater than 62 miles (100 km), and consistent indicators of a sinistral component of motion suggest that they are dominantly strike-slip structures. These corridor bounding structures are not just late cross-cutting faults but lateral ramps, present through the deformation history that bound domains within which unique structures are developed. These major faults may represent reactivation of terrane boundaries, since they separate zones of contrasting stratigraphy and structure.

The Superior Province consists of a number of fault-bounded sub-provinces containing rocks of different lithological associations, ages, structural and metamorphic histories. The greenstone terrains consist of several types, including tholeiitic-komatiitic lava, tholeiitic to calc-alkaline complexes, and shoshonitic/alkalic volcanic rocks with associated fluvial deposits. Belts of variably metamorphosed volcanogenic turbidites, interpreted as accretionary prisms, separate the older individual volcanic belts. A general southward younging of central parts of the Superior Province, together with the contemporaneity of deformation events along strike within individual belts suggests that the Superior Province represents an amalgam of oceanic crust and plateaux, island arcs, continental margin arcs, and accretionary prisms, brought together by dextral oblique subduction, which formed the major sub-province–bounding strike-slip faults. Continued late orogenic strike-slip motion on some of the faults localized alkalic volcanic and fluvial sedimentary sequences in pull-apart basins on some of these strike-slip faults.

In the Vermillion district of the southern Superior Province, the deformation history begins with early nappe-style structures, which are overprinted by the "main" fabric elements related to dextral transpression. This sequence of fabric development is interpreted to reflect dextral-oblique accretion of island arcs and microcontinents of the southern Superior Province. A combination of north-south shortening together with dextral simple shear led to the juxtaposition of zones with constrictional and flattening strains. The constrictional strains in this area were previously interpreted to be a result of "squeezing" between batholiths, necessitating reevaluation of similar theories for the origin of prolate strains in numerous other greenstone belts.

In some cases, strike-slip faults have played an integral role in the localization and generation of greenstone belts in pull-apart basins. Several strike-slip fault systems associated with the formation of greenstone belts in pull-apart basins are known from the Pilbara craton. The Lalla Rookh and circa 2,950 Ma Whim Creek belts are interpreted as "second-cycle greenstones," because they were deposited in strike-slip related pull-apart basins that formed in already complexly deformed and metamorphosed circa 3,500–3,300 Ma rocks of the Pilbara craton. Thus, although these fault systems form an integral part of the structural evolution of the Pilbara craton, they postdate events related to initial formation of the granite-greenstone terrain. One problem that may be addressed by future studies of greenstone belt structure is understanding the nature of the transition from brittle to ductile strains in pull-apart regions along strike-slip fault systems, such as those of the Pilbara.

Late stage major ductile transcurrent shear zones cut many cratons, but few have well constrained kinematic or metamorphic histories. An exception is the 186-mile (300-km) long Koolyanobbing shear zone in the Southern Cross Province of the Yilgarn craton. The Koolyanobbing shear is a four- to nine-mile (6–15-km) wide zone with a gradation from foliated granitoid, through protomylonite, mylonite, to ultramylonite, from the edge to the center of the shear zone. Shallowly plunging lineations and a variety of kinematic indicators show that the shear zone is a major sinistral fault, but regional relationships suggest that it does not represent a major crustal boundary or suture. Fault fabrics both overprint and appear coeval with late stages in the development of the regional metamorphic pattern, suggesting that the shear zone was active around 2.7–2.65 Ga.

Late Extensional Collapse, Diapirism, Denudation

It is widely recognized that in Phanerozoic orogens, late stages of orogenic development are characterized by extensional collapse of structurally over-thickened crust. In granite-greenstone terrains, late stages of orogenesis are characterized by the intrusion of abundant, locally diapiric granitic magmas, with chemical signatures indicative of crustal anatexis. The intrusion of late granitic plutons in greenstone terrains may be related to rapid uplift and crustal melting accompanying extension in the upper crust. Early plutons of the tonalite-trondhjemite-gabbro-granodiorite suite are generated in an island arc setting and are in turn intruded by continental margin arc magmas after the primitive arcs collide and form larger continental fragments. When plate collision causes further crustal thickening, adiabatic melting produces thin diapiric plutons that rise partway through the crust but will crystallize before rising very far, as they do not contain enough heat to melt their way through the crust. The crustal sections in these collisional orogens gravitationally collapse when the strength of quartz and olivine can no longer support the topography. Decompression in the upper mantle and lower crust related to upper crustal extension generates significant quantities of basaltic melts. These basaltic melts rise and partially melt the lower or middle crust, becoming more silicic by assimilating crustal material. The hybrid magmas thus formed intrude the middle and upper crust, forming the late to post-kinematic granitoid suite so common in Archean granite-greenstone terrains. If the time interval between crustal thickening and gravitational collapse is short, then magmas related to decompressional melting may quickly rise up the partially solidified crystal/mush pathways provided by the earlier plutons generated during the crustal thickening phases of orogenesis. Such temporal and spatial relationships easily account for the common occurrence of composite and compositionally zoned plutons in Precambrian and younger orogenic belts.

See also ARCHEAN; CONTINENTAL CRUST; CRATONS.

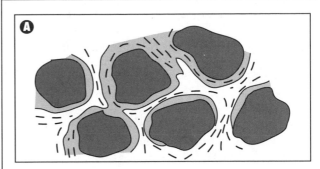

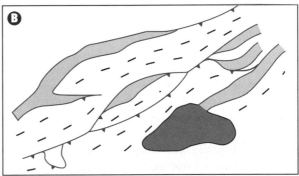

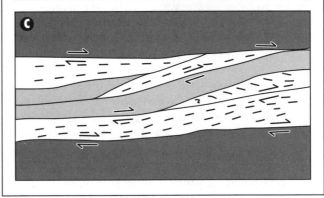

Greenstone belt geometry: (A) granitoids; (B) terrains; (C) belts

Further Reading

Kusky, Timothy M., Jianghai Li, and Robert T. Tucker. "The Archean Dongwanzi Ophiolite Complex, North China Craton: 2.505 Billion Year Old Oceanic Crust and Mantle." *Science* 292 (2001): 1,142–1,145.

Kusky, Timothy M., and Peter J. Hudleston. "Growth and Demise of an Archean Carbonate Platform, Steep Rock Lake, Ontario Canada." *Canadian Journal of Earth Sciences* 36 (1999): 1–20.

Kusky, Timothy M., and Julian Vearncombe. "Structure of Archean Greenstone Belts." In *Tectonic Evolution of Greenstone Belts*, edited by Maarten J. de Wit and Lewis D. Ashwal, 95–128. Oxford Monograph on Geology and Geophysics, 1997.

Kusky, Timothy M., ed., *Precambrian Ophiolites and Related Rocks, Developments in Precambrian Geology XX*. Amsterdam: Elsevier Publishers, 2003.

Li, Jianghai H., Timothy M. Kusky, and Xiongnan Huang. "Neoarchean Podiform Chromitites and Harzburgite Tectonite in

Ophiolitic Melange, North China Craton, Remnants of Archean Oceanic Mantle." *GSA Today* 12, no. 7 (2002): 4–11, plus cover.

Grenville province The Grenville province is the youngest region of the Canadian shield; it is outboard of the Labrador, New Quebec, Superior, Penokean, and Yavapai-Mazatzal provinces. It is the last part of the Canadian shield to experience a major deformational event, this being the Grenville orogeny, which was responsible for complexly deforming the entire region. The Grenville province has an aerial extent of approximately 600,000 square miles (1,000,000 km²). The subterranean extent of Grenville rocks, however, is much greater in area. Phanerozoic rocks cover their exposure from New York State down the length of the Appalachian Mountains and into Texas.

The Grenville province formed on the margin of Laurentia in the middle to late Proterozoic. The rocks throughout the province represent a basement and platform sedimentary sequence that was intruded by igneous rocks. Subsequent to this intrusive event in the late Proterozoic, the entire region underwent high-grade metamorphism and was complexly deformed. However, prior to this high-grade metamorphic event, the rocks of the Grenville province experienced multiple pulses of metamorphism and deformation, including the Elsonian (1,600–1,250-million-year-old) and the Elzevirian (1,250–1,200-million-year-old) orogenies. The Ottawan orogeny was the last and most intense in the Grenville province, culminating 1.1 billion years ago and overprinting much of the earlier tectonic history. This has made it difficult for geologists to describe the earlier orogenies and also to determine the tectonic evolution of the Grenville province. For these reasons, the term *Ottawan orogeny* is usually used synonymously with the term *Grenville orogeny*.

The Grenville province is subdivided into numerous subprovinces, including the central gneiss belt (CGB), central metasedimentary belt (CMB), central granulite terrane (CGT), and one major structural feature: the Grenville front.

The CGB is located in the western part of the Grenville province and contains some of the oldest rocks found in the province. The majority of the rocks are 1.8–1.6-billion-year-old gneisses intruded by 1.5–1.4-billion-year-old granitic and monzonitic plutons. Both the metasedimentary and the igneous rocks of the CGB are metamorphosed from upper amphibolite and locally granulite facies. The CGB is bounded by the Grenville front to the northwest and lies in tectonic contact with the central metasedimentary belt to the southeast. The dominant structural trend is northeast but changes to the northwest near Georgian Bay. The CGB has been divided into smaller terranes including the Nipissing, Algonquin, Tomiko, and Parry Sound, based on lithology, metamorphic grade and structures, namely, shear zones. These terranes are considered to be mainly parautochthonous terranes. The shear zones that separate the various terranes contain kinematic indicators that suggest northwest-directed tectonic transport, and tectonic transport is thought to have occurred between 1.18 billion and 1.03 billion years ago.

The Nipissing terrane is located in the western portion of the central gneiss belt. Part of the Nipissing terrane occupies a region known as the Grenville front tectonic zone (GFTZ), an area that lies within 30 kilometers of the Grenville front. The lithologies here are strongly deformed with northeast-striking foliations and zones of cataclasis and moderately plunging southeast lineations. The heterogeneous gneisses of the Nipissing terrane fall into two categories: Archean and Lower Proterozoic migmatitic gneisses that are likely reworked units of the Southern and Superior provinces and Middle Proterozoic metasedimentary gneiss. These rocks were intruded by 1.7 billion- and 1.45-billion-year-old granitic plutonic rocks, both of which are less deformed than the host rocks. Postdating this intrusive event, the region underwent high-grade metamorphism, experiencing temperatures of 1,200°F–1,280°F (650°C–750°C) and pressures of 8.0–8.5 kilobars.

The Tomiko terrane is located in the extreme northwestern portion of the central gneiss belt. The most striking aspect of the Tomiko terrane is the relative abundance of metasedimentary rocks, but it also contains metamorphosed granitic rocks that are Middle Proterozoic in age. The Tomiko terrane is allochthonous with respect to the Nipissing terrane. Evidence to support this is the distinct detrital zircon population in the Tomiko metaquartzites, dated at 1,687 million years old. This is in sharp contrast to the metaquartzites of the Nipissing terrane, where the detrital zircons are Archean to Lower Proterozoic in age. This suggests that the Nipissing terrane was already adjacent to the Superior province at the time of the Nipissing quartzite formation. Further evidence for the allochthonous nature of the Tomiko terrane is the presence of iron formations in the Tomiko terrane, which are not present elsewhere in the CGB. The metamorphic conditions experienced by the Tomiko terrane are temperatures of less than 1,290°F (700°C) and pressures of 6.0–8.0 kilobars.

The Algonquin terrane is the largest terrane in the CGB and consists of numerous domains. The rocks in this terrane are meta-igneous quartzo-feldspathic gneisses and supracrustal gneisses. Generally, the foliations strike northeast and dip to the southeast; down-dip stretching lineations are common. The southern domains have been interpreted as thrust sheets with a clear polarity of southeasterly dips and the entire Algonquin terrane may be parautochthonous. The metamorphic temperatures and pressures range from 1,240°F–1,520°F (670°C–825°C) and 7.9–9.9 kilobars, respectively.

The Parry Sound terrane is the most studied terrane in the CGB. It is located in the south-central portion of the CGB and contains large volumes of mafic rock, marble, and anorthosite.

The age of the Parry Sound terrane ranges from 1,425 million to 1,350 million years. Both the lithologies and the age of the Parry Sound terrane are different from the rest of the CGB. Therefore, it is not surprising that this terrane is considered as allochthonous and overlying the parautochthonous Algonquin domains. In fact, since the Parry Sound terrane is completely surrounded by the Algonquin terrane, structurally it is considered a klippe. The metamorphic conditions reached by the Parry Sound terrane are in the range of 1,200°F–1,470°F (650°C–800°C) and 8.0–11.0 kilobars.

The CMB has a long history of geologic investigation. One of the reasons is the abundance of metasedimentary rocks which makes it a prime target for locating ore deposits. The CMB was originally named the Grenville series by Sir William Logan in 1863 for an assemblage of rocks near the village of Grenville, Quebec, and is the source of the name for the entire Grenville province. Later, the Grenville series was raised to supergroup status, but presently the Grenville Supergroup is a term limited to a continuous sequence of rocks within the CMB.

The CMB contains Middle Proterozoic metasediments that were subsequently intruded by syn, late, and post-tectonic granites. The time of deposition is estimated to have been from 1.3 billion to 1.1 billion years ago, with the bulk of the material having been deposited before 1.25 billion years ago. After their deposition, the rocks of the CMB underwent deformation and metamorphism from the Elzevirian orogeny 1.19–1.06 billion years ago. The effects of the Elzevirian orogeny were all but wiped out by the later Ottawan orogeny, which deformed and metamorphosed the rocks to middle-upper amphibolite facies. The CMB contains five distinct terranes: Bancroft, Elzevir, Mazinaw, Sharbot Lake, and Frontenac. The Frontenac is correlative with the Adirondack Lowlands.

The Bancroft terrane is located in the northwestern portion of the CMB. The Bancroft is dominated by marbles but also contains nepheline-bearing gneiss and granodioritic orthogneiss metamorphosed to middle through upper amphibolite facies. The Bancroft terrane contains complex structures, such as marble breccias and high strain zones. The orthogneiss occurs in thin structural sheets suggesting that it may occur in thrust-nappe complexes. The thrust sheets generally dip to the southeast with dips increasing toward the dip direction. Rocks of the Bancroft terrane possess a well-developed stretching lineation that also plunges in the southeast direction. Both of these structural orientations suggest northwest-directed tectonic transport.

The Elzevir terrane is located in the central portion of the CMB. It is known for containing the classic Grenville Supergroup. The Elzevir is composed of 1.30–1.25-billion-year-old metavolcanics and metasediments, intruded by 1.27-billion-year-old tonalitic plutons ranging in composition from gabbro to syenite. The largest of these calc-alkaline bodies is the Elzevirian batholith. The calc-alkaline signature of the batholith

suggests that it may have been generated in an arc-type setting. The Elzevir terrane also contains metamorphic depressions. These are areas of lower metamorphic grade, such as greenschist to lower amphibolite facies. These depressions may be related to the region's polyphase deformation history, and in contrast to surrounding high-grade terranes, they contain sedimentary structures enabling the application of stratigraphic principles in order to determine superposition.

The Mazinaw terrane was once mapped as part of the Elzevir terrane and it also contains some of the classic Grenville Supergroup marbles and the Flinton Group. The rocks encountered here are marbles, calc-alkalic metavolcanic and clastic metasedimentary rocks. The Flinton Group is derived from the weathering of plutonic and metamorphic rocks found in the Frontenac terrane. Furthermore, the complex structural style of the Mazinaw terrane is similar to the Frontenac and the Adirondack Lowlands.

The Sharbot Lake terrane was once mapped as part of the Frontenac terrane but is now considered a separate terrane. The Sharbot Lake principally contains marbles and metavolcanic rocks intruded by intermediate and mafic plutonic rocks and may represent a strongly deformed and metamorphosed carbonate basin. Metamorphic grade ranges from greenschist to lower amphibolite. The lithologies, metamorphic grade, and lack of exposed basement rocks to the Sharbot Lake terrane imply that these rocks may be correlative with the Elzevir terrane.

The Frontenac terrane is located in the southeastern portion of the CMB. This terrane extends into the Northwest Lowlands of the Adirondack Mountains. The Frontenac terrane is composed of marble with pelitic gneisses and quartzites. The relative abundances of the gneisses and quartzites increase toward the southeast, while the relative abundances of metavolcanic rocks and tonalitic plutons decrease in the same direction. A trend also exists in the metamorphic grade from northwest to southeast. In the northwest, the metamorphic grade ranges from lower amphibolite to upper amphibolite-granulite facies, but then decreases in the southeast to amphibolite facies. Rock attitudes also change, dipping southeast in the northwest, to vertical in the central part, to the northwest in the Northwest Lowlands.

Throughout the CMB, large-scale folds are present. These folds indicate crustal shortening. More important, however, is the recognition of main structural breaks that lie both parallel to and within the CMB. The structural breaks are marked by narrow zones of highly attenuated rocks, such as mylonites. The Robertson Lake mylonite zone (RLMZ) is one such structural break and lies between the Sharbot Lake terrane and the Mazinaw terrane. The RLMZ has been interpreted as a low angle thrust fault and also as a normal fault caused by unroofing.

To the east of the central metasedimentary belt lies the central granulite terrane. These two subprovinces are separated by the Chibougamau-Gatineau Lineament (CGL), which is a wide mylonite zone. The CGL is well defined on aeromagnetic maps suggesting that it is a crustal-scale feature. The CGL roughly trends northeast-southwest, where it ranges from a few meters to seven kilometers wide. The CGL may be correlative with the Carthage-Colton mylonite zone in the Adirondack Mountains of New York State.

The central granulite terrane (CGT) was originally named by Wynne-Edwards in 1972. It is located in the central and southeastern portion of the Grenville province and is correlative with the Adirondack Highlands. The CGT is often referred to as the core zone of the Grenville orogen and is where the majority of the Grenvillian plutonic activity occurred. This subprovince underwent high-grade metamorphism with paleotemperatures ranging up to 1,470°F (800°C) and paleopressures up to 9.0 kilobars. In order to explain these high pressures and temperatures, a double thickening of the crust is required. For this reason, it has been suggested that the Grenville province represents a continent/continent collision zone.

The most abundant rock constituent of the central granulite belt is anorthosite. The larger anorthosite bodies are termed massifs, such as the Morin massif. The anorthosites along with a whole suite of rocks, known as AMCG (anorthosite, mangerite, charnokite, granite) suite, are thought to have intruded at approximately 1,159–1,126 million years ago based on U-Pb zircon analysis. These dates are in agreement with U-Pb zircon ages of the AMCG rocks in the Adirondack Highlands (1,160–1,125 million years old). This places their intrusion as postdepositional with the sediments of the CMB and before the Ottawan orogeny. The anorthosites were emplaced at shallow levels somewhere between the Grenville supergroup and the underlying basement. A major tectonic event, such as continental collision, must have occurred in order to produce the high paleotemperatures and paleopressures recorded in the anorthosites.

For the most part, the Grenville front (GF) marks the northwestern limit of Grenville deformation and truncates older provinces and structures. The zone is approximately 1,200 miles (2,000 km) long and is dominated by northwest-directed reverse faulting that has been recognized since the 1950s. The GF is recognized by faults, shear zones, and metamorphic discontinuities. Faults, foliations, and lineations dip steeply to the southeast. Interpretation of the GF has changed with time. In the 1960s, with the advent of the theory of plate tectonics, the GF was immediately interpreted as a suture. This suggestion was refuted because Archean-age rocks of the Superior craton continue south across the GF, implying that the suture should lie to the southeast of the GF. It is possible that the suture is reworked somewhere in the Appalachian orogen. There are still several unresolved questions about the tectonic nature of the GF, considering that the GF marks the limit of Grenvillian deformation: (1) the adjacent foreland to the northwest contains no evidence of supracrustal assemblages associated with the Grenville orogen, (2) the zone lacks Grenville age intrusives that are prevalent to the southeast, and (3) the front divides older rocks from a belt of gneisses that appear to be their reworked equivalents.

Tectonic Evolution of the Grenville Province

The tectonic framework of the Grenville province is a topic of considerable debate. Many theories and models have been proposed, although there is no one universally accepted model. Nevertheless, there are some aspects of the tectonic framework researchers do agree upon, in particular that the Grenville province represents a collisional boundary. This model is supported by seismic data and the granulite facies metamorphism, both of which suggest that the crust was doubly thickened during peak deformation and metamorphism.

There are a few ways to thicken the crust: thrusting, volcanism, plutonism, and homogeneous shortening. One or a combination of these mechanisms must have occurred in the late Proterozoic to produce granulite facies metamorphism in the Grenville province. Two models account for similar large tectonic crustal thickening presently occurring on the Earth's surface.

The first model is fashioned after the Andean-type margin. This model suggests that relatively warm, buoyant oceanic crust is subducted under continental crust. This model has several implications. The oceanic crust subducted underneath the South American plate is relatively young. Therefore, it has not had a sufficient time interval to cool and become dense. The relative low density of the young oceanic crust resists subduction. Consequently, the oceanic crust subducts at a relatively shallow angle. A shallow subduction angle creates a compressional stress regime throughout the margin. This has the effect of crustal shortening accommodated by fore-arc frontal thrusts. The subducting oceanic plate also induces plutonism and volcanism that adds to the crustal thickening process.

The second model is based on the Himalayan orogen. This model thickens the crust by a continent-continent collision. This model is somewhat similar to the Andean model, except that the warm, buoyant oceanic crust is replaced by continental crust. The subducting continental crust resists subduction due to its buoyancy, causing the subducting continental crust to get tucked under the overriding continental crust. The underriding crust never subducts down into the asthenosphere but rather underplates the overriding continental crust; hence, crustal thickening. It can be quickly seen that the Andean model may be the predecessor to the Himalayan model. Therefore, it is fair to conclude that a combination of the two models may have worked together to produce the Grenville orogen in the Proterozoic.

A simplistic tectonic model for the Grenville attempts to explain the broad-scale tectonic processes that may account for the large-scale features. An arc-continent collision was followed by a continent-continent collision in the late Proterozoic, probably involving southeastward-directed subduction for the continent-continent collision. Consistent kinematics in domain boundary shear zones (in the CGB) preserve an overall northwesterly direction of tectonic transport, consistent with northwestward stacking of crustal slices.

The calc-alkaline trends of the Elzevirian batholith suggest that this is an island arc-type batholith. Thus, the Elzevir terrane was probably an island arc before it collided with North America. The Elzevirian age metamorphism resulted from the collision of the CMB and the CGT. Ultimately, the southeastward subduction along the western CMB margin resulted in a continent-continent collision with the CGB.

Plate reconstructions for the Late Proterozoic are currently an area of active investigation. Recent research, in the form of geochronology, comparative geology, stratigraphy, and paleomagnetism, have provided a wealth of new information that has proven useful in correlating rocks on a global scale. It is these correlations that geologists use to determine the temporal and spatial plate configurations for the Late Proterozoic. Such plate reconstructions have been a new source of insight in the study of the Grenville province.

Advances in geochronology have been the greatest contributor in helping to correlate rocks globally. Field mapping in previously unmapped areas and improved techniques in paleomagnetic determination further help to narrow possible plate configurations. With this knowledge, geologists take present-day continents and strip away their margins—more precisely, all post-Grenvillian age rocks—and try to piece together the cratons that may once have been conjugate margins.

In 1991 researchers including Paul Hoffman, Eldridge Moores, and Ian Dalziel proposed that a supercontinent existed in the Late Proterozoic. This supercontinent has been named Rodinia and was formed by the amalgamation of Laurentia (North America and Greenland), Gondwana (Africa, Antarctica, Arabia, Australia, India, and South America), Baltica, and Siberia. The joining of these plates resulted in collisional events along the Laurentian margins. It is these orogenic events in the Late Proterozoic that are thought to produce the Grenvillian belts found throughout the world.

Most Late Proterozoic plate reconstructions place the Canadian Grenville province and Amazonian and Congo cratons in close proximity. Therefore, Amazonia and Congo were the probable Late Proterozoic continental colliders with the eastern margin of Laurentia, resulting in the Ottawan orogeny. Evidence supporting this correlation includes the similar Neodynium (Nd) model ages of 1.4 billion years of the Grenvillian belts found on the Amazonian and Congo cratons, the same as the Laurentian Grenville province.

Plate reconstructions for the Late Proterozoic are not absolute. There is no hard evidence, such as hot spot tracks and oceanic magnetic reversal data, to determine plate motions for the Proterozoic as there is for the Mesozoic and Cenozoic. Furthermore, definitive sutures, such as ophiolite sequences and blueschist facies terrains, are deformed and few, making it difficult to determine the exact location of the Grenvillian suture, that would strongly demonstrate a collisional margin. This may be due to the expansive time interval that has ensued, later orogenic events, rifting events, and erosion, all of which help to alter and destroy the geologic record.

Most tectonic models for the Grenville province are broadly similar for the late stages of the evolution of this orogenic belt but differ widely in the early stages. The earliest record of arc magmatism in the central metasedimentary belt comes from the Elzevir terrane or composite arc belt, where circa 1,350–1,225-million-year-old magmatism is interpreted to represent one or more arc/back arc basin complexes. The Adirondack Lowlands terrane may have been continuous with the Frontenac terrane, which together formed the trailing margin of the Elzevirian arc. Isotopic ages for the Frontenac terrane fall in the range of 1,480–1,380 million years, and between 1,450 million and 1,300 million years for the entire central metasedimentary belt, suggesting that the Elzevirian arc is largely a juvenile terrane. The Elzevirian arc is thought to have collided offshore with other components of the composite arc belt by 1,220 million years ago, because of widespread northwestward-directed deformation and tectonic repetition in the central metasedimentary belt at that time. Following amalgamation, subduction is interpreted by some to have stepped southeastward to lie outboard of the composite arc and dipped westward beneath a newly developed active margin. This generated a suite of circa 1,207-million-year-old calc-alkaline plutons (Antwerp-Rossie suite) and 1,214 ± 21–million-year-old dacitic volcaniclastics, metapelites, and diorite-tonalitic plutons. Other models suggest that the Adirondack Highlands and Frontenac/Adirondack Lowlands terranes remained separated until 1,170–1,150 million years ago, when the Frontenac and Sharbot Lakes domains were metamorphosed and intruded by plutons.

The Adirondack Highlands–Green Mountains block is regarded by many workers as a single arc complex, based on abundant circa 1,350–1,250-million-year-old calc-alkaline tonalitic to granodioritic plutons in both areas. The Adirondack Highlands–Green Mountains block may have been continuous with the Elzevirian arc as well, forming one large composite arc complex. Neodymium model ages for the Adirondack Mountains–Green Mountain block fall in the range of 1,450–1,350 Ma, suggesting that this arc complex was juvenile, without significant reworking of older material.

Collision of the Adirondack Highlands–Green Mountain block with Laurentia occurred between the intrusion of the circa 1,207-million-year-old Antwerp-Rossie arc magmas, and formation of the 1,172-million-year-old Rockport-Hyde-School-Wellesley-Wells intrusive suite. This inference is based on the observation that peak metamorphic conditions preced-

ed intrusion of the 1,180–1,150-million-year-old intrusive suite in the Frontenac terrane. Also, metamorphic zircon and monazite (presumably dating the collision) from the central

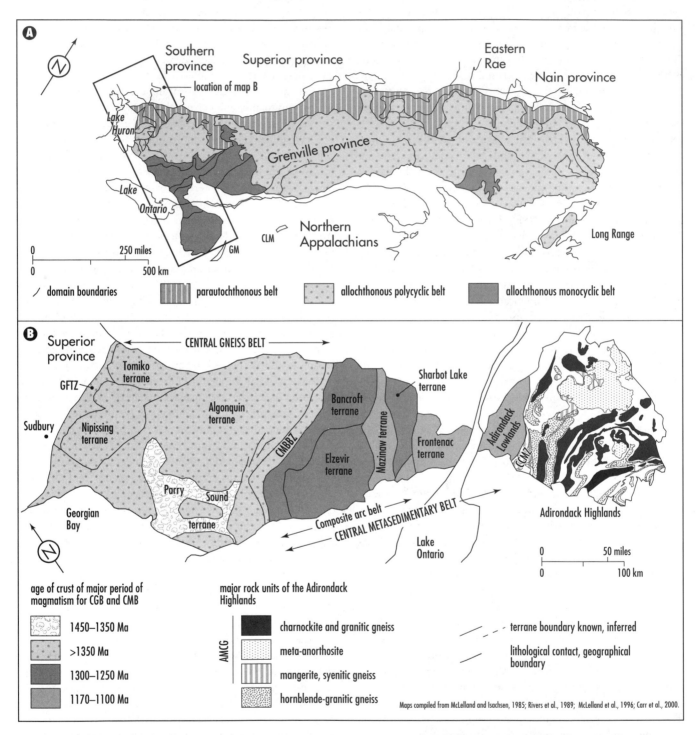

(A) Tectonic subdivisions of the Grenville province according to the classification of Toby Rivers (1989) and Carr et al. (2000) showing also the older domain boundaries of Wynne-Edwards (1972) and others. **(B)** Terranes and shear zones of the central gneiss belt (CGB) and the central metasedimentary belt (CMB) and major geological features of the Adirondack Highlands. Abbreviations as follows: CCMZ: Carthage-Colton mylonite zone; CLM: Chain Lakes massif; CMBBZ: central metasedimentary belt boundary zone; GFTZ: Grenville front tectonic zone; GM: Green Mountains metasedimentary belt fall in the range of 1,190–1,180 million years. The Carthage-Colton mylonite zone may represent a cryptic suture marking the broad boundary along which the Adirondack Highlands–Green Mountain block is juxtaposed with Laurentia from a collision that emplaced the Lowlands over the Highlands. Possible early localized delamination beneath the collision zone may have elevated crustal temperatures and generated crustal melts of the circa 1,172-million-year-old Rockport and Hyde School granites, and we add the Wells leucocratic gneiss to this group. However, the present geometry with relatively low-grade rocks of the Lowlands, juxtaposed with high-grade rocks of the Highlands, suggests that the present structure is an extensional fault that may have reactivated an older structure.

The circa 1,172-million-year-old collisional granites (Rockport, Hyde School gneiss, Wellesley, Wells) are largely syntectonic and we suggest that emplacement of these magmas may have slightly preceded formation of large-scale recumbent nappes including the F_1 fold documented here. These large nappes may be responsible for complex map patterns and repetition of units in the CMB and CGT. High-temperature deformation of monzonites in the Robertson Lake shear zone took place at circa 1,162 million years ago and demonstrated that deformation continued for at least 10 million years after intrusion of the 1,172-million-year-old magmatic suite. Deformation had apparently terminated by 1,160 million years ago, however, as shown by the 1,161–1,157-million-year-old Kingston dikes and Frontenac suite plutons, which cross-cut Elzevirian fabrics and cut the Robertson Lake shear zone.

The widespread monzonitic, syenitic, and granitic plutons (AMCG suite) that intruded the Frontenac terrane in the period of 1,180–1,150 million years ago swept eastward across the orogen forming the AMCG suite in the Highlands at 1,155–1,125 million years ago. It has been suggested that separation of the subcontinental lithospheric mantle that started around 1,180–1,160 million years ago may have proceeded to large-scale delamination beneath the orogen. This would have exposed the base of the crust to hot asthenosphere, causing melting and triggering the formation of the AMCG suite. We suggest that the 1,165-million-year-old metagabbro units are related to this widespread melting and intrusive event in the Adirondacks.

The culminating Ottawan orogeny from circa 1,100–1,020 million years ago in the Adirondacks and Grenville orogen is widely thought to result from the collision of Laurentia with another major craton, probably Amazonia. This collision is one of many associated with the global amalgamation of continents to form the supercontinent of Rodinia. The event is associated with large-scale thrusting, high-grade metamorphism, recumbent folding, and intrusion of a second generation of crustal melts associated with orogenic collapse. The putative suture (Carthage-Colton mylonite zone) between the accreted Adirondack Highlands–Green Mountain block and Laurentia was reactivated as an extensional shear zone in this

Complex fold in strongly deformed and metamorphosed gneiss in the Parry Sound region of the Grenville province, Canada *(Photo by Timothy Kusky)*

event, partly accommodating the orogenic collapse and exhumation of deep-seated rocks in the Adirondack Highlands. The relative timing of igneous events and folding in the Adirondacks has shown that the F_2 and F_3 folding events in the southern part of the Highlands postdated 1,165 million and predated 1,052 million years ago, demonstrating that these folds, and later generations of structures, are related to the Ottawan orogeny. The Ottawan orogeny in this area is therefore marked by the formation of early recumbent fold nappes, overprinted by upright folds.

The regional chronology and overprinting history of folding related to the Ottawan orogeny are generally poorly known. In 1939 Buddington noted isoclinal folds dated circa 1,149 million years old in the Hermon granite gneiss in the Adirondack Highlands, and very large granulite facies fold nappes have been emplaced throughout the Adirondack region. These folds refold an older isoclinal fold generation, so are F_2 folds, and are related to this regional event. The youngest rocks that show widespread development of fabrics attributed to the Ottawan orogeny are the circa 1,100–1,090-

million-year-old Hawkeye suite, that show "peak" conditions of ~ 800°C at 20–25 km depth. These conditions existed from about 1,050 million through approximately 1,013 million years ago. Older thrust faults along the CMB boundary zone were reactivated at about 1,080–1,050 million years ago. The latter parts of the Ottawan orogeny (1,045–1,020 million years ago) are marked by extensional collapse of the orogen, with low-angle normal faults accommodating much of this deformation. Crustal melts associated with orogenic collapse are widespread.

See also ADIRONDACK MOUNTAINS; CONVERGENT PLATE MARGIN PROCESSES; STRUCTURAL GEOLOGY.

Further Reading

Borg, Scott G., and Don J. DePaolo. "Laurentia, Australia and Antarctica as a Late Proterozoic Supercontinent: Constraints from Isotopic Mapping." *Geology* 22 (1994): 307–310.

Carr, S. D., Mike R. Easton, Rebecca A. Jamieson, and Nick G. Culshaw. "Geologic Transect across the Grenville Orogen of Ontario and New York." *Canadian Journal of Earth Science* 37 (2000): 193–216.

Corrigan, David, and Simon Hanmer. "Anorthosites and Related Granitoids in the Grenville Orogen: A Product of the Convective Thinning of the Lithosphere?" *Geology* 25 (1997): 61–64.

Culotta, Raymond C., T. Pratt, and Jack Oliver. "A Tale of Two Orogens: COCORP's Deep Seismic Surveys of the Grenville Province in the Eastern United States Midcontinent." *Geology* 18 (1990): 646–649.

Dalziel, Ian W. D. "Neoproterozoic-Paleozoic Geography and Tectonics: Review, Hypothesis, Environmental Speculation." *Geological Society of America* 109 (1997): 16–42.

———. "Pacific Margins of Laurentia and East Antarctica-Australia as a Conjugate Rift Pair: Evidence and Implications for an Eocambrian Supercontinent." *Geology* 19 (1991): 598–601.

Davidson, Anthony. "An Overview of Grenville Province Geology, Canadian Shield." In "Geology of the Precambrian Superior and Grenville Provinces and Precambrian Fossils in North America," edited by S. B. Lucas and M. R. St-Onge. *Geological Society of America, Geology of North America* C-1 (1998): 205–270.

———. "A Review of the Grenville Orogen in Its North American Type Area." *Journal of Australian Geology and Geophysics* 16 (1995): 3–24.

Hoffman, Paul F. "Did the Breakout of Laurentia Turn Gondwanaland Inside-Out?" *Science* 252 (1991): 1,409–1,411.

Kusky, Timothy M., and Dave P. Loring. "Structural and U/Pb Chronology of Superimposed Folds, Adirondack Mountains: Implications for the Tectonic Evolution of the Grenville Province." *Journal of Geodynamics*, v. 32 (2001): 395–418.

McLelland, Jim M., J. Stephen Daly, and Jonathan M. McLelland. "The Grenville Orogenic Cycle (ca. 1350–1000 Ma): an Adirondack Perspective." In "Tectonic Setting and Terrane Accretion in Precambrian Orogens," edited by Timothy M. Kusky, Ben A. van der Pluijm, Kent Condie, and Peter Coney. *Tectonophysics* 265 (1996): 1–28.

Moores, Eldredge M. "Southwest United States–East Antarctic (SWEAT) Connection: A Hypothesis." *Geology* 19 (1991): 425–428.

Rivers, Toby. "Lithotectonic Elements of the Grenville Province: Review and Tectonic Implications." *Precambrian Research* 86 (1997): 117–154.

Rivers, Toby, and David Corrigan. "Convergent Margin on Southeastern Laurentia during the Mesoproterozoic: Tectonic Implications." *Canadian Journal of Earth Science* 37 (2000): 359–383.

Rivers, Toby, J. Martipole, Charles F. Gower, and Anthony Davidson. "New Tectonic Subdivisions of the Grenville Province, Southeast Canadian Shield." *Tectonics* 8 (1989): 63–84.

Wynne-Edwards, H. R. "The Grenville Province. Variations in Tectonic Styles in Canada." In "Variations in Tectonic Styles Canada," edited by R. A. Price and R. J. W. Douglas. *Geological Society of Canada* Special Paper 11 (1972): 263–334.

groin People are modifying the shoreline environment on a massive scale with the construction of new homes, resorts, and structures that attempt to reduce or prevent erosion along the beach. These modifications have been changing the dynamics of the beach in drastic ways, and most often they result in degradation of the beach. In many cases, obstacles are constructed that disrupt the transportation of sand along the beach in longshore drift. This causes sand to build up at some locations, and to be removed from other locations further along the beach. Some of the worst culprits are groins, or walls of rock, concrete, or wood built at right angles to the shoreline, designed to trap sand from longshore drift and replenish a beach. Groins stop the longshore drift, causing the sand to accumulate on the updrift side and to be removed from the downdrift side. Groins also set up conditions favorable for the formation of rip tides, which tend to take sand (and unsuspecting swimmers) offshore, out of the longshore drift system. The result of groin construction is typically a few triangular areas of sand next to the rocky protrusions, along what was once a continuous beach. Little or no sand will remain in the areas on the downdrift sides of the groins. Therefore, when groins are constructed it usually becomes necessary to begin an expensive program of artificial replenishment of beach sands to fill in the areas that were eroded by the new pattern of longshore drift set up by the groins.

Construction or stabilization of inlets though barrier islands or beaches often includes the construction of groin-like jetties on either side of the channel, to prevent sand from entering and closing the channel. Like groins, these jetties prevent sand transportation by longshore drift, causing beaches to grow on the updrift side of the jetty. Sand that used to replenish the beach on the downdrift side gets blocked, or washed around the jetty into the tidal channel, where it moves into the lagoon to form tidal deltas. The result is that the beaches on the downdrift side of the jetties become sand-starved and thin, eventually disappearing.

See also BEACH.

groundwater All of the water contained within spaces in bedrock, soil, and regolith. The volume of groundwater is 35

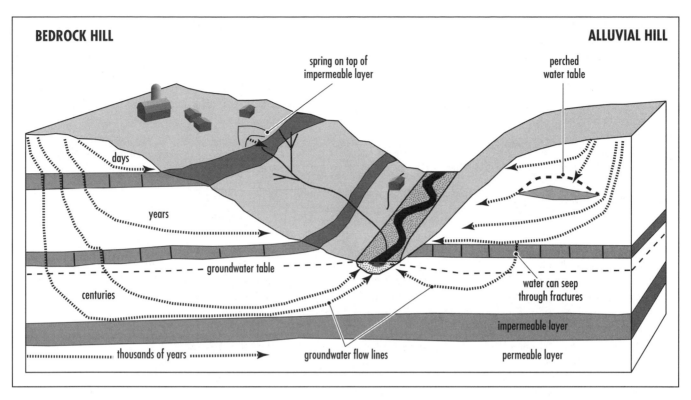

BEDROCK HILL

ALLUVIAL HILL

spring on top of impermeable layer

perched water table

days

years

groundwater table

centuries

thousands of years

groundwater flow lines

water can seep through fractures

impermeable layer

permeable layer

The groundwater system. Water enters the system on hillslopes, follows complex pathways simplified as dotted lines, and emanates lower on hills as springs and in effluent stream.

times the volume of freshwater in lakes and streams, but overall it accounts for less than 1 percent of the planet's water. America and other nations have come to realize that groundwater is a vital resource for their survival and only recently began to appreciate that much of the world's groundwater resources have become contaminated by natural and human-aided processes. Approximately 40 percent of drinking water in the United States comes from groundwater reservoirs; about 80 billion gallons of groundwater are pumped out of these reservoirs every day in the United States.

Groundwater comes from rainfall and surface flow, where it seeps into the ground and slowly makes its way downhill toward the sea. There is water everywhere beneath the ground surface, and most of this occurs within 2,500 feet (750 m) of the surface. The volume of groundwater is estimated to be equivalent to a layer 180 feet (55 m) thick spread evenly over the Earth's land surface.

The distribution of water in the ground can be divided into the unsaturated and the saturated zones. The top of the water table is defined as the upper surface of the saturated zone; below this surface, all openings are filled with water.

Movement of Groundwater

Most of the water under the ground is constantly in motion, although rates are typically only centimeters per day. The rates of movement are controlled by the amount of open space in the bedrock or regolith, and how the spaces are connected.

Porosity is the percentage of total volume of a body that consists of open spaces. Sand and gravel typically have about 20 percent open spaces, while clay has about 50 percent. The sizes and shapes of grains determine the porosity, which is also influenced by how much they are compacted, cemented together, or deformed.

In contrast, permeability is a body's capacity to transmit fluids or to allow the fluids to move through its open pore spaces. Permeability is not directly related to porosity. All the pore spaces in a body may be isolated (high porosity), but the water may be trapped and unable to move through the body (low permeability). Permeability is also affected by molecular attraction, the force that makes thin films of water stick to objects, instead of being forced to the ground by gravity. If the pore spaces in a material are very small, as in a clay, then the force of molecular attraction is strong enough to stop the water from flowing through the body. When the pores are large, the water in the center of the pores is free to move.

After a rainfall, much of the water stays near the surface, because clay in the near-surface horizons of the soil retains much water due to molecular attraction. This forms a layer of soil moisture in many regions and is able to sustain seasonal plant growth.

Some of this near-surface water evaporates and is used by plants. Other water runs directly off into streams. The remaining water seeps into the saturated zone, or into the water table. Once in the saturated zone it moves slowly by percolation, from high areas to low areas, under the influence of gravity. These lowest areas are usually lakes or streams. Many streams form where the water table intersects the surface of the land.

Once in the water table, the paths that individual particles follow varies. The transit time from surface to stream may vary from days to thousands of years along a single hillside. Water can flow upward because of high pressure at depth and low pressure in streams.

The Groundwater System

Groundwater is best thought of as a system of many different parts. Some of these act as conduits and reservoirs, and others as off-ramps and on-ramps for the groundwater system.

Recharge areas are where water enters the groundwater system, and discharge areas are where water leaves the groundwater system. In humid climates, recharge areas encompass nearly the land's entire surface (except for streams and floodplains), whereas in desert climates, recharge areas consist mostly of the mountains and alluvial fans. Discharge areas consist mainly of streams and lakes.

The level of the water table changes with different amounts of precipitation. In humid regions it reflects the topographic variation, whereas in dry times or places it tends to flatten out to the level of the streams and lakes. Water flows faster when the slope is greatest, so groundwater flows faster during wet times. The fastest rate of groundwater flow observed in the United States is 800 feet per year (250 m/yr).

Aquifers include any body of permeable rock or regolith saturated with water through which groundwater moves. Gravel and sandstone make good aquifers, as do fractured rock bodies. Clay is so impermeable that it makes bad aquifers, or even aquicludes that stop the movement of water.

Springs are places where groundwater flows out at the ground surface. They can form where the ground surface intersects the water table, or at a vertical or horizontal change in permeability, such as where water in gravels on a hillslope overlie a clay unit and the water flows out on the hill along the gravel/clay boundary.

Most wells fill with water simply because they intersect the water table. However, the rocks below the surface are not always homogeneous, which can result in a complex type of water table know as a perched water table. Perched water tables result from discontinuous bodies in the subsurface that create bodies of water at elevations higher than the main water table.

In many regions, a permeable layer, typically a sandstone, is confined between two impermeable beds, creating a confined aquifer. In these systems, water only enters the system in a small recharge area—if this is in the mountains, then the aquifer may be under considerable pressure. This is known as an artesian system. Water that escapes the system from the fracture or well reflects the pressure difference between the elevation of the source area and the discharge area (hydraulic gradient) and rises above the aquifer as an artesian spring, or artesian well. Some of these wells have made fountains that have spewed water 200 feet (60 m) high.

One example of an artesian system is in Florida, where water enters in the recharge area and is released near Miami about 19,000 years later.

Groundwater Dissolution

Groundwater also reacts chemically with the surrounding rocks; it may deposit minerals and cement together grains, causing a reduction in porosity and permeability, or form features like stalagtites and stalagmites in caves. In other cases, particularly when acidic water moves through limestone, it may dissolve the rock, forming caves and underground tunnels. Where these dissolution cavities intersect the surface of the Earth, they form sinkholes.

Groundwater Contamination

Natural groundwater is typically rich in dissolved elements and compounds derived from the soil, regolith, and bedrock that the water has migrated through. Some of these dissolved elements and compounds are poisonous, whereas others are tolerable in small concentrations but harmful in high concentrations. Groundwater is also increasingly becoming contaminated by human and industrial waste, and the overuse of groundwater resources has caused groundwater levels to drop and has led to other problems, especially along coastlines. Seawater may move in to replace depleted freshwater, and the ground surface may subside when the water is removed from the pore spaces in aquifers.

The U.S. Public Health Service has established limits on the concentrations of dissolved substances (called total dissolved solids, or t.d.s.) in natural waters that are used for domestic and other uses. The table below lists these standards for the United States. It should be emphasized that many other countries, particularly those with chronic water shortages, have much more lenient standards. Sweet water is preferred for domestic use and has less than 500 milligrams (mg) of total dissolved solids per liter (l) of water. Fresh and slightly saline water, with t.d.s. of 1,000–3,000 mg/l, is suitable for use by livestock and for irrigation. Water with higher concentrations of t.d.s. is unfit for humans or livestock. Irrigation of fields using waters with high concentrations of t.d.s. is also not recommended, as the water will evaporate but leave the dissolved salts and minerals behind, degrading and eventually destroying the productivity of the land.

The quality of groundwater can be reduced or considered contaminated by either a high amount of total dissolved solids

Drinking Water Standards for the United States

Water Classification	Total Dissolved Solids (T.D.S.)
Sweet	500 mg/l
Fresh	500–1,000 mg/l
Slightly saline	1,000–3,000 mg/l
Moderately saline	3,000–10,000 mg/l
Very Saline	10,000–35,000 mg/l
Brine	>35,000 mg/l

or by the introduction of a specific toxic element. Most of the total dissolved solids in groundwater are salts that have been derived from dissolution of the local bedrock or soils derived from the bedrock. Salts may also seep into groundwater supplies from the sea along coastlines, particularly if the water is being pumped out for use. In these cases, seawater often moves in to replace the depleted freshwater. This process is known as seawater intrusion, or seawater incursion.

Dissolved salts in groundwater commonly include the bicarbonate (HCO_3) and sulfate (SO_4) ions, often attached to other ions. Dissolved calcium (Ca) and magnesium (Mg) ions can cause the water to become "hard." Hard water is defined as containing more than 120 parts per million dissolved calcium and magnesium. Hard water makes it difficult to lather with soap and forms a crusty mineralization that builds up on faucets and pipes. Adding sodium (Na) in a water softener can soften hard water, but people with heart problems or those who are on a low-salt diet should not do this. Hard water is common in areas where the groundwater has moved through limestone or dolostone rocks, which contain high concentrations of Ca- and Mg-rich rocks that are easily dissolved by groundwater.

Groundwater may have many other contaminants, some natural and others that are the result of human activity, including animal and human waste, pesticides, industrial solvents, petroleum products, and other chemicals. Groundwater contamination, whether natural or human induced, is a serious problem because of the importance of the limited water supply. Pollutants in the groundwater system do not simply wash away with the next rain, as many dissolved toxins in the surface water system do. Groundwater pollutants typically have a residence time (average length of time that they remain in the system) of hundreds or thousands of years. Many groundwater systems are capable of cleaning themselves of natural biological contaminants in a shorter amount of time using bacteria, but other chemical contaminants have longer residence times.

See also HYDROLOGIC CYCLE.

Further Reading

Alley, William M., Thomas E. Reilly, and O. L. Franke. *Sustainability of Ground-Water Resources.* U.S. Geological Survey Circular 1186, 1999.

Keller, Edward A. *Environmental Geology,* 8th ed. Engelwood Cliffs, N.J.: Prentice Hall, 2000.

Skinner, Brian J., and Stephen C. Porter. *The Dynamic Earth, an Introduction to Physical Geology.* New York: John Wiley and Sons, 1989.

guyot *See* SEAMOUNT.

H

Hadean The first of the four major eons of geological time: the Hadean, Archean, Proterozoic, and Phanerozoic. Some time classification schemes use an alternative division of early time, in which the Hadean is considered the earliest part of the Archean. As the earliest phase of Earth's evolution, ranging from accretion to approximately the age of first rocks (4.55 to 4.0 Ga [Ga = giga annee, or 10^9 years]), it is the least known interval of geologic time. Only a few mineral grains and rocks have been recognized from this eon, so most of what we think we know about the Hadean is based on indirect geochemical evidence, meteorites, and models.

Although the universe formed about 14 billion years ago, it was not until 5 Ga ago that the Earth started forming in a solar nebula, consisting of hot solid and gaseous matter spinning around a central protosun. As the solar nebula spun and slowly cooled, the protoearth swept up enough matter by its gravitational attraction to have formed a small protoplanet by 4.6 Ga. Materials accreted to the protoearth as they sequentially solidified out of the cooling solar nebula, with the high-temperature elements solidifying and accreting first. The early materials to accrete to the protoplanet were rich in iron (which forms solids at high temperatures), whereas the later materials to accrete were rich in H, He, and Na (which form solids at lower temperatures and would not accrete until the solar nebula cooled). Heat released by gravitational condensation, the impact of late large asteroids, and the decay of short-lived radioactive isotopes caused the interior of the Earth to melt, perhaps even forming a magma ocean to a depth of 310 miles (500 km). This melting allowed dense iron and nickel that was accreted during condensation of the solar nebula to begin to sink to the core of the planet, releasing much more heat in the process, and causing more widespread melting. This early differ-

entiation of the Earth happened by 4.5 or 4.4 Ga and caused the initial division of the Earth into layers, including the dense iron-nickel rich core and the silicate rich mantle. The outer layer of the Earth was probably a solid crust for most of this time since it would have cooled by conductive heat transfer to the atmosphere and space. The composition of this crust is unknown and controversial, since none of it is known to be preserved. Some models would have a dense ultramafic crust (komatiite), whereas others suggest a lighter anorthositic (made of essentially all plagioclase feldspar) crust. Still other models suggest that the early crust resembled modern oceanic crust. In any case, the crust was a conductively cooled rigid layer capping a hot, convecting magma ocean. By analogy with magma lakes that form in calderas such as Hawaii, this crust was probably also moving with currents in the underlying molten magma and showing early plate tectonic behavior. On magma lakes and on the early Earth, the outer crust moves apart at divergent boundaries, where molten material from below wells up to fill the open space, then cools to become part of the surface crust. This crust is broken into numerous rigid plates that slide past each other along transform faults and converge at several types of convergent boundaries. Subduction zones form where the crust of one plate slides below another, and collision zones form where the two crustal plates deform each other's edges. A third type of convergent boundary not known on the present Earth has been recognized on magma oceans. In these cases, the crust of both converging plates sinks back into the magma ocean, forming a deep V-shaped depression on the surface where they join and sink together. This early form of plate tectonics would begin to mature as the magma ocean crystallized, and crustal slabs began to partially melt yielding buoyant magmas of silicic composition that would rise, crystallize, and form protocontinents.

Formation of the Earth and Solar System

Understanding the origin of the Earth, planets, Sun, and other bodies in the solar system is a fundamental yet complex problem that has intrigued scientists and philosophers for centuries. Most of the records from the earliest history of the Earth have been lost to tectonic reworking and erosion, so most of what we know about the formation of the Earth and solar system comes from the study of meteorites, the Earth's moon, and observations of the other planets and interstellar gas clouds.

The solar system displays many general trends with increasing distance from the Sun, and systematic changes like these imply that the planets were not captured gravitationally by the Sun but rather formed from a single event that occurred about 4.6 billion years ago. The nebular theory for the origin of the solar system suggests that a large spinning cloud of dust and gas formed and began to collapse under its own gravitational attraction. As it collapsed, it began to spin faster to conserve angular momentum (much as ice skaters spin faster when they pull their arms in to their chests), and eventually formed a disk. Collisions between particles in the disk formed protoplanets and a protosun, which then had larger gravitational fields than surrounding particles, and began to sweep up and accrete loose particles.

The condensation theory states that particles of interstellar dust (many of which formed in older supernova) act as condensation nuclei that grow through accretion of other particles to form small planetesimals that then have a greater gravitational field that attracts and accretes other planetesimals and dust. Some collisions cause accretion, other collisions are hard and cause fragmentation and breaking up of the colliding bodies. The Jovan planets became so large that their gravitational fields were able to attract and accrete even free hydrogen and helium in the solar nebula.

The main differences among the planets with distance from the Sun are explained by this condensation theory, since the temperature of the solar nebula would have decreased away from the center where the Sun formed. The temperature determines which materials condense out of the nebula, so the composition of the planets was determined by the temperature at their position of formation in the nebula. The inner terrestrial planets are made of rocky and metallic material because high temperatures near the center of the nebula only allowed the rocky and metallic material to condense from the nebula. Farther out, water and ammonia ices also condensed out of the nebula, because temperatures were cooler at greater distances from the early Sun.

Early in the evolution of the solar system, the Sun was in a T-Tauri stage and possessed a strong solar wind that blew away most gases from the solar nebula, including the early atmospheres of the inner planets. Gravitational dynamics caused many of the early planetesimals to orbit in the Oort Cloud, where most comets and many meteorites are found. Some of these bodies have eccentric orbits that occasionally bring them into the inner solar system, and it is thought that collisions with comets and smaller molecules brought the present atmospheres and oceans to Earth and the other terrestrial planets. Thus air and water, some of the basic building blocks of life, were added to the planet after it formed, being thrown in from the deep space of the Oort Cloud.

These protocontinents would gradually coalesce, the magma ocean would solidify as the mantle but continue to convect, and the continental crust would rapidly grow. This style of plate tectonics led into the Archean.

Between 4.55 Ga and 3.8 Ga, the Earth was bombarded by meteorites, some large enough to severely disrupt the surface, vaporize the atmosphere and ocean, and even melt parts of the mantle. By about 4.5 Ga, it appears as if a giant impactor, about the size of Mars, hit the protoearth. This impact ejected a huge amount of material into orbit around the protoearth, and some undoubtedly escaped. The impact probably also formed a new magma ocean, vaporized the early atmosphere and ocean (if present), and changed the angular momentum of the Earth as it spins and orbits the Sun. The material in orbit coalesced to form the Moon, and the Earth-Moon system was born. Although not certain, this impact model for the origin of the Moon is the most widely accepted hypothesis, and it explains many divergent observations. First, the Moon orbits at 5.1° from the ecliptic plane, whereas the Earth orbits at 23.4° from the ecliptic, suggesting that some force, such as a collision, disrupted the angular momentum and rotational parameters of the Earth-Moon system. The Moon is retreating from the Earth, resulting in a lengthening of the day by 15 seconds per year, but the Moon has not been closer to the Earth than 149,129 miles (240,000 km). The Moon is significantly less dense than the Earth and other terrestrial planets, being depleted in iron and enriched in aluminum, titanium, and other related elements. These relationships suggest that the Moon did not form by accretion from the solar nebula at its present location in the solar system. The oxygen isotopes of igneous rocks from the Moon are the same as from the Earth's mantle, suggesting a common origin. The age of the Moon rocks shows that it formed at 4.5 Ga, with some magmatism continuing until 3.1 Ga, consistent with the impactor hypothesis.

The atmosphere and oceans of the Earth probably formed from early degassing of the interior by volcanism within the first 50 million years of Earth history. It is likely that our present atmosphere is secondary, in that the first or primary atmosphere would have been vaporized by the late great impact that formed the Moon, if it survived being blown away by an intense solar wind when the Sun was in a T-Tauri stage of evolution. The primary atmosphere would have been composed of gases left over from accretion, including primarily hydrogen, helium, methane, and ammonia, along with nitrogen, argon, and neon. However, since the atmosphere has much less than the expected amount of these elements, and is quite depleted in these volatile elements relative to the Sun, it is thought the primary atmosphere has been lost to space.

Gases are presently escaping from the Earth during volcanic eruptions and are also being released by weathering of surface rocks. The secondary atmosphere was most likely produced from degassing of the mantle by volcanic eruptions, and perhaps also by cometary impact. Gases released from volcanic eruptions include N, S, CO_2, and H_2O, closely matching the suite of volatiles that comprise the present atmosphere and oceans. However, there was no or little free oxygen in the early atmosphere, as oxygen was not produced until later, by photosynthetic life.

The early atmosphere was dense, with H_2O, CO_2, S, N, HCl. The mixture of gases in the early atmosphere would have made greenhouse conditions similar to that presently existing on Venus. However, since the early Sun during the Hadean Era was approximately 25 percent less luminous than today, the atmospheric greenhouse served to keep temperatures close to their present range, where water is stable, and life can form and exist. As the Earth cooled, water

vapor condensed to make rain that chemically weathered igneous crust, making sediments. Gases dissolved in the rain made acids, including carbonic acid (H_2CO_3), nitric acid (HNO_3), sulfuric acid (H_2SO_4), and hydrochloric acid (HCl). These acids were neutralized by minerals (which are bases) that became sediments, and chemical cycling began. These waters plus dissolved components became the early hydrosphere, and chemical reactions gradually began changing the composition of atmosphere, getting close to the dawn of life.

It is of great intellectual interest to speculate on the origin of life. In the context of the Hadean, when life most likely arose, we are forced to consider different options for the initial trigger of life. It is quite possible that life came to Earth on late accreting planetesimals (comets) as complex organic compounds, or perhaps it came from interplanetary dust. If true, this would show how life got to Earth, but not how, when, where, or why it originated. Life may also have originated on Earth, in the deep sea near a hydrothermal vent, or in shallow pools with the right chemical mixture. To start, life probably needed an energy source, such as light-

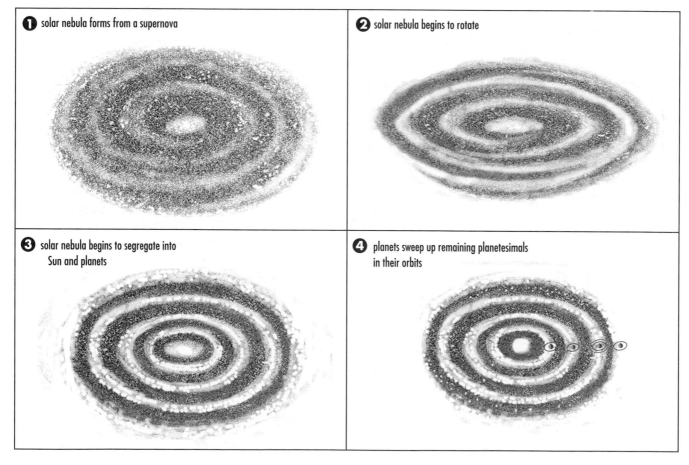

❶ solar nebula forms from a supernova

❷ solar nebula begins to rotate

❸ solar nebula begins to segregate into Sun and planets

❹ planets sweep up remaining planetesimals in their orbits

Formation of the solar system from condensation and collapse of a solar nebula

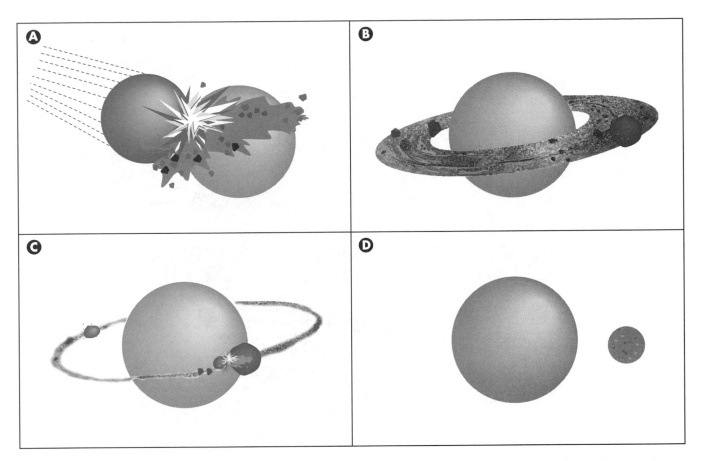

The late great impact hypothesis for the origin of the Moon: A Mars-sized planetesimal (A) colliding with the protoearth, resulting in a gigantic explosion and the jetting outward of both planetesimal and protoearth material into orbit around the planet (B). A protomoon begins to form from a prelunar disk (C), and matter accretes to form the Moon (D).

ning, or perhaps submarine hydrothermal vents, to convert simple organic compounds into building blocks of life (RNA- ribonucleic acid) and amino acids.

See also LIFE'S ORIGINS AND EARLY EVOLUTION; METEOR; PLATE TECTONICS.

Further Reading
Cloud, Preston. *Oasis in Space.* New York: W. W. Norton, 1988.
Condie, Kent C., and Robert E. Sloan. *Origin and Evolution of Earth, Principles of Historical Geology.* Upper Saddle River, N.J.: New Jersey Prentice Hall, 1997.
Schopf, J. William. *Cradle of Life, The Discovery of Earth's Earliest Fossils.* Princeton: Princeton University Press, 1999.

Hadley cells   The globe-encircling belts of air that rise along the equator, dropping moisture as they rise in the Tropics. As the air moves away from the equator at high elevations, it cools, becomes drier, and then descends at 15–30°N and S latitude where it either returns to the equator or moves toward the poles. The locations of the Hadley cells move north and south annually in response to the changing apparent seasonal movement of the Sun. High-pressure systems form where the air descends, characterized by stable clear skies and intense evaporation, because the air is so dry. Another pair of major global circulation belts is formed as air cools at the poles and spreads toward the equator. Cold polar fronts form where the polar air mass meets the warmer air that has circulated around the Hadley Cell from the Tropics. In the belts between the polar front and the Hadley cells, strong westerly winds develop. The position of the polar front and extent of the west-moving wind is controlled by the position of the polar jet stream (formed in the upper troposphere), which is partly fixed in place in the Northern Hemisphere by the high Tibetan Plateau and the Rocky Mountains. Dips and bends in the jet stream path are known as Rossby Waves, and these partly determine the location of high- and low-pressure systems. These Rossby Waves tend to be semi-stable in different seasons and have predictable patterns for summer and winter. If the pattern of Rossby Waves in the jet stream changes significantly for a season or longer, it may cause storm systems to track to different locations than normal, causing local

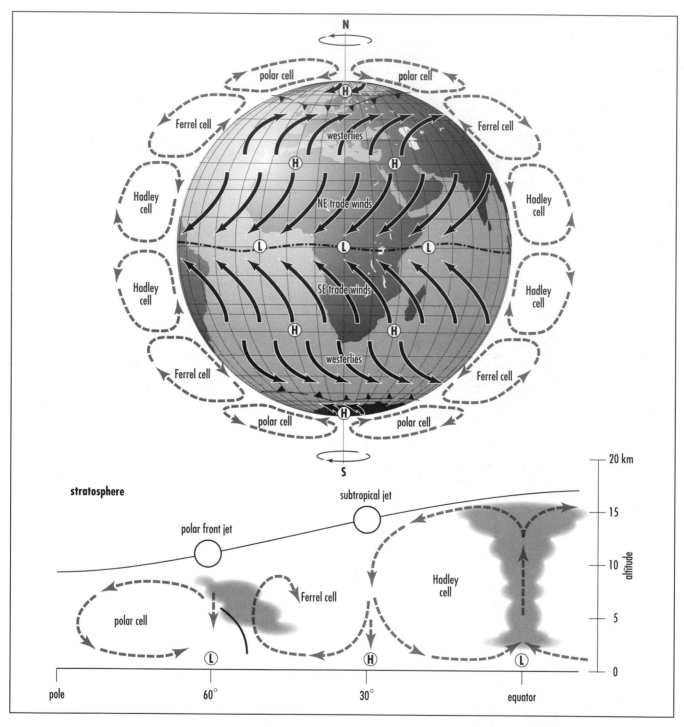

The Earth showing dominant wind patterns and the Hadley cells, Ferrel cells, and polar cells that circulate air around the globe

droughts or floods. Changes to this global circulation may also change the locations of regional downwelling of cold dry air. This can cause long-term drought and desertification. Such changes may persist for periods of several weeks, months, or years and may explain several of the severe droughts that have affected Asia, Africa, North America, and elsewhere.

See also ATMOSPHERE.

hail   *See* PRECIPITATION.

half-life *See* GEOCHRONOLOGY.

halide minerals Minerals that have a halogen such as fluorine, chlorine, iodine, or bromine as the anion. Two common halides are fluorite (CaF_2) and halite (NaCl). Both minerals are cubic. Fluorite has perfect octahedral cleavage and vitreous luster and commonly occurs as purple, violet, green, yellow, or brightly colored crystals. It occurs as a late-crystallizing hydrothermal mineral in some granites, syenites, and other plutonic rocks and is most often found in pegmatites. It also occurs in some hydrothermal veins along with barite, sphalerite, galena, calcite, and quartz. Halite, including the form commonly known as rock salt, is normally colorless but may be red, yellow, or blue because of iron and other impurities. It can be distinguished by its perfect cubic cleavage, saline taste, softness, and solubility in water. Halite forms primarily in sedimentary rocks where it has been formed by evaporation from seawater, or saline lakes or sabkhas. Halite begins to precipitate when the volume of seawater has been reduced to about 10 percent of its original volume by evaporation. It may also form as a volcanic sublimate and as a surface efflorescence in arid regions.

Halite is a very weak rock, and it often deforms into salt domes when it gets buried and is subject to large overburden pressures. It may also act as a decollement horizon in mountain and extensional belts, accommodating large contractional or extensional strains because it is typically one of the weakest, most deformable horizons in a sedimentary/volcanic sequence of rocks.

See also MINERALOGY; DIAPIR.

hanging valley When glaciers carve out a mountainous landscape some small, high-elevation tributary glaciers will merge with larger valley or trunk glaciers near the upper parts of the larger glacier. These two glaciers then flow together as one, with the upper glacier carving out a high-elevation glacial valley, and the main valley or trunk glacier carving out a deeper valley at a lower elevation. When the glaciers recede from the landscape, the valley from the higher glacier will merge with the lower valley high on the steep slopes of the lower valley, forming a hanging valley. Streams that flow in the upper valley typically form cascades and waterfalls where the hanging valley meets the main valley, a famous example being Yosemite Falls in Yosemite Valley, California.

See also GLACIER.

hanging wall *See* FAULT.

Harker, Alfred (1859–1939) British *Petrologist* Alfred Harker was principally known for physics, but in 1884 he was appointed university demonstrator in geology at the Sedgwick Museum at Cambridge and was soon recognized as one of the most outstanding figures among British petrologists. Most of his research was conducted in north Wales and the English Lake District. He then combined his university work along with his fieldwork in Scotland for the Geological Survey. His original research concerned five topics: (1) slaty cleavage, (2) the igneous rocks associated with Ordovician sedimentary rocks in Caernarvonshire, (3) plutonic and associated rocks in the English Lake District, (4) the islands of the Inner Hebrides and the studies of the Tertiary igneous activities on the Isle of Skye and smaller islands, and (5) his general works where he looked at philosophical results of his research and thought in his work *The Natural History of Igneous Rocks*.

Further Reading

Harker, Alfred. *The West Highlands and the Hebrides; a Geologist's Guide for Amateurs, xxiii*. Cambridge: University Press, 1941.
———. "Igneous Rock Series and Mixed Igneous Rocks." *Journal of Geology* (1900): 389–399.
———. "Thermometamorphism in Igneous Rocks." *Geological Society of America Bulletin* (1892): 16–22.

harzburgite An ultramafic plutonic rock composed of olivine, orthopyroxene, and minor clinopyroxene. Harzburgite is a common mantle rock found in the mantle sections of ophiolites and deeply exhumed parts of the seafloor, such as regions near large normal faults and near transform faults and fracture zones. The harzburgite section of ophiolites and oceanic mantle is typically strongly deformed into a harzburgite tectonite, indicating very high shear strains formed during flow of the mantle. This flow is thought to be upward beneath the oceanic ridges and generally away from the ridges closer to the surface and reflects mantle convection and seafloor spreading. Oceanic harzburgites are also chemically depleted, having had a basaltic melt extracted from them during their rise beneath the oceanic ridges, and this

Hanging valley in the Chugach Mountains of southern Alaska. Note the large U-shaped valley carved by a glacier, entering the larger valley in the foreground at a high elevation above the main valley floor (not visible in photograph). The hanging valley has lateral moraines formed by a glacier and is dissected by a younger V-shaped gulley carved by a stream. *(Photo by Timothy Kusky)*

melt rose to form the oceanic crust above. The rocks are named after occurrences in the Harz Mountains near Harzburg, Germany.

See also OPHIOLITES.

Hawaii The 50th state to join the United States, in 1959, Hawaii consists of a group of eight major and about 130 smaller islands in the central Pacific Ocean. The islands are volcanic in origin, having formed over a magmatically active hot spot that has melted magmatic channels through the Pacific plate as it moves over the hot spot, forming a chain of southeastward younging volcanoes over the hotspot. Kilauea volcano on the big island of Hawaii is the world's most active volcano, and it often has a lava lake with an actively convecting crust developed in its caldera. The volcanoes are made of low-viscosity basalt and form broad shield types of cones that rise from the seafloor. Only the tops are exposed above sea level, but if the entire height of the volcanoes above the seafloor is taken into account, the Hawaiian Islands form the tallest mountain range on Earth.

The Hawaiian Islands are part of the Hawaiian-Emperor seamount chain that extends all the way to the Aleutian-Kamchatka trench in the northwest Pacific, showing both the great distance the Pacific plate has moved, and the longevity of the hot spot magmatic source. From east to west (and youngest to oldest) the Hawaiian Islands include Hawaii, Maui and Kahoolawe, Molokai and Lanai, Oahu, Kauai, and Niihau.

The volcanic islands are fringed by coral reefs and have beaches with white coral sands, black basaltic sands, and green olivine sands. The climate on the islands is generally mild, and numerous species of plants lend a paradise-like atmosphere to the islands, with tropical fern forests and many species of birds. The islands have few native mammals (and no snakes), but many have been introduced. Some of the islands such as Niihau and Molokai have drier climates, and Kahoolawe is arid.

See also HOT SPOT.

heat flow In geology, crustal heat flow is a measure of the amount of heat energy leaving the Earth, measured in calories per square centimeters per second. Typical heat flow values are about 1.5 microcalories per centimeter squared per second, commonly stated as 1.5 heat flow units. Most crustal heat flow is due to heat production in the crust by radioactive decay of uranium, thorium, and potassium. Heat flow shows a linear relationship with heat production in granitic rocks. Some crustal heat flow, however, comes from deeper in the Earth, beneath the crust.

The Earth shows a huge variation in temperature, from several thousand degrees in the core to essentially zero degrees Celsius at the surface. The Earth's heat was acquired by several mechanisms, including: (1) heat from accretion as potential energy of falling meteorites was converted to heat

energy; (2) heat released during core formation, with gravitational potential energy converted to heat as heavy metallic iron and other elements segregated and sank to form the core soon after accretion; (3) heat production by decay of radioactive elements; and (4) heat added by late impacting meteorites and asteroids, some of which were extremely large in early Earth history. Heat produced by these various mechanisms gradually flows to the surface by conduction, convection, or advection, and accounts for the component of crustal heat flow that comes from deeper than the crust.

Heat flow by conduction involves thermal energy flowing from warm to cooler regions, with the heat flux being proportional to the temperature difference, and a proportionality constant (k), known as thermal conductivity, related to the material properties. The thermal conductivity of most rocks is low, about one-hundreth of that of copper wire.

Advection involves the transfer of heat by the motion of material, such as transport or heat in a magma, in hot water through fractures or pore spaces, and more important on a global scale, by the large-scale rising of heated, relatively low-density buoyant material and the complementary sinking of cooled, relatively high-density material in the mantle. The large-scale motion of the mantle, with hot material rising in some places and colder material sinking in other places, is known as convection, which is an advective heat transfer mechanism. For convection to occur in the mantle, the buoyancy forces of the heated material must be strong enough to overcome the rock's resistance to flow, known as viscosity. Additionally, the buoyancy forces must be able to overcome the tendency of the rock to lose heat by conduction, since this would cool the rock and decrease its buoyancy. The balance between all of these forces is measured by a quantity called the Raleigh number. Convection in Earth materials occurs above a critical value of the Raleigh number, but below this critical value heat transfer will be dominated by conductive processes. Well-developed convection cells in the mantle are very efficient at transporting heat from depth to the surface and are the main driving force for plate tectonics.

Heat transfer in the mantle is dominated by convection (advective heat transfer), except in the lower mantle near the boundary with the inner core (the D" region), and along the top of the mantle and in the crust (in the lithosphere), where conductive and hydrothermal (also advective) processes dominate. The zones where the heat transfer is dominated by conduction are known as conductive boundary layers, and the lithosphere may be thought of as a convecting, conductively cooling boundary layer.

See also CONVECTION AND THE EARTH'S MANTLE; GEOTHERMAL ENERGY; PLATE TECTONICS; RADIOACTIVE DECAY.

heavy minerals Detrital sedimentary grains that have a higher specific gravity than the standard mineral in a sandstone, or other detrital sedimentary rock. Most heavy minerals have a density above 2.9 grams per cubic centimeter and form a minor component of the sediment or sedimentary rock, with common minerals including magnetite, ilmenite, chromite, sphene, zircon, rutile, kyanite, garnet, tourmaline, apatite, olivine, hornblende, and biotite. Less commonly, ore minerals such as gold may occur as detrital heavy mineral concentrates in some deposits.

Heavy minerals are generally eroded from igneous and metamorphic rock terrains and resist weathering more than feldspar, eventually becoming concentrated as detrital grains in sands and other clastic deposits. Certain types of heavy minerals generally form only in specific types of rocks; for instance garnet forms in metamorphic rocks, and chromite forms in ultramafic rocks. Using such relationships, it is often possible to learn much about the source terrane of the sediment or sedimentary rock by examining the heavy mineral concentrates from that deposit.

See also SEDIMENTARY ROCKS.

heliotropism The tendency of a plant to grow toward sunlight is a type of phototropism in which the Sun is the orienting stimulus. Some plants, like sunflowers, orient their flowers toward the sunlight and track the movement of the

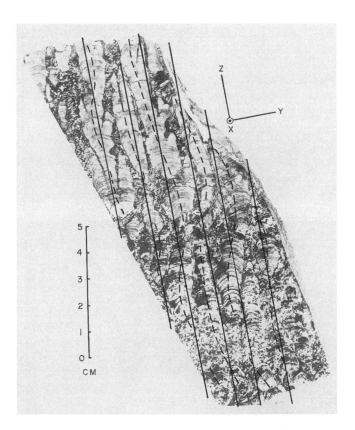

Columnar stromatolites showing sinusoidal growth patterns. The solid lines show the vertical directions, and the dashed lines show the sinusoidal growth directions. *(Photo by J. Vanyo)*

Sun across the sky during the day. Several early scientists, including Charles Darwin in 1880, published observations of heliotropism in plants and related their movement to photosynthesis. Some organisms, such as columnar stromatolites, may also grow toward sunlight, forming columns that are inclined toward the average direction of the Sun for each season. For stromatolites that grow throughout the year, sinusoidal columns may grow upward, reflecting inclination toward the average direction of the Sun during different seasons.

See also STROMATOLITE.

Further Reading

Darwin, Charles. *The Power of Movement in Plants.* New York: P. Appleton and Co., 1880.

Kusky, Tim, and Jim Vanyo. "Plate Reconstructions Using Stromatolite Heliotropism: Principles and Applications." *Journal of Geology* 99 (1991): 321–335.

Hess, Harry Hammond (1906–1969) American *Geologist* Harry Hess was an American geologist who was born in New York and studied at Yale University as an electrical engineer but later changed to study geology. He spent two years in Zimbabwe (then Rhodesia) studying as an exploration geologist before he continued his studies at Princeton University. As a graduate student, Hess worked with a number of students in a submarine gravity study of the West Indies under F. A. Vening Meinesz. Later he extended these studies into the Lesser Antilles using naval submarines. In December 1941, as a U.S. Navy reserve officer, Hess was called to active duty in New York City and was assigned responsibility for detecting enemy submarine operation patterns in the North Atlantic, resulting in the virtual elimination of the submarine threat within two years. Hess arranged a transfer to the decoy vessel U.S.S. *Big Horn* to test the effectiveness of the submarine detection program; he then remained on sea duty for the rest of the war. As commanding officer of the transport vessel U.S.S. *Cape Johnson,* Hess carefully chose his travel routes to Pacific Ocean landings on the Marianas, Philippines, and Iwo Jima, continuously using his ship's echo sounder. This unplanned wartime scientific surveying enabled Hess to collect ocean floor profiles across the North Pacific Ocean, resulting in the discovery of flat-topped submarine volcanoes, which he termed "guyots" after the Princeton Geology Building. Hess contributed greatly to the fields of science, including geophysics, geodesy, tectonophysics, and mineralogy. Because of his outstanding achievements, the American Geophysical Union set up the "Harry H. Hess Medal for outstanding achievements in research in the constitution and evolution of Earth and sister planets."

Further Reading

Hess, Harry H. "Comments on the Pacific Basin." *Geological Survey of Canada* Special Paper (1966): 311–316.

——. "The Oceanic Crust." *Journal of Marine Research* 14 (1955): 423–439.

——. "Geological Hypotheses and the Earth's Crust under the Oceans." *Proceedings of the Royal Society of London, Series A: Mathematical and Physical Sciences* 222 (1954): 341–348.

Himalaya Mountains   The world's tallest mountains, as well as those exhibiting the greatest vertical relief over short distances, form the Himalaya range that developed in the continent-continent collision zone between India and Asia. The range extends for more than 1,800 miles (3,000 km) from the Karakorum near Kabul (Afghanistan), past Lhasa in Tibet, to Arunachal Pradesh in the remote Assam province of India. Ten of the world's 14 peaks that rise to more than 26,000 feet (8,000 m) are located in the Himalayas, including Mount Everest, 29,035 feet (8,850 m); Nanga Parbat, 26,650 feet (8,123 m); and Namche Barwa, 25,440 feet (7,754 m). The rivers that drain the Himalayas feature some with the highest sediment outputs in the world, including the Indus, Ganges, and Brahmaputra. The Indo-Gangetic plain on the southern side of the Himalayas represents a foreland basin filled by sediments eroded from the mountains and deposited on Precambrian and Gondwanan rocks of peninsular India. The northern margin of the Himalayas is marked by the world's highest and largest uplifted plateau, the Tibetan plateau.

The Himalayas is one of the youngest mountain ranges in the world but has a long and complicated history. This history is best understood in the context of five main structural and tectonic units within the ranges. The Subhimalaya includes the Neogene Siwalik molasse, bounded on the south by the Main Frontal Thrust that places the Siwalik molasse over the Indo-Gangetic plain. The Lower Subhimalaya is thrust over the Subhimalaya along the Main Boundary Thrust, and it consists mainly of deformed thrust sheets derived from the northern margin of the Indian shield. The High Himalaya is a large area of crystalline basement rocks, thrust over the Subhimalaya along the Main Central Thrust. Further north, the High Himalaya sedimentary series or Tibetan Himalaya consists of sedimentary rocks deposited on the crystalline basement of the High Himalaya. Finally, the Indus-Tsangpo suture represents the suture between the Himalaya and the Tibetan Plateau to the north.

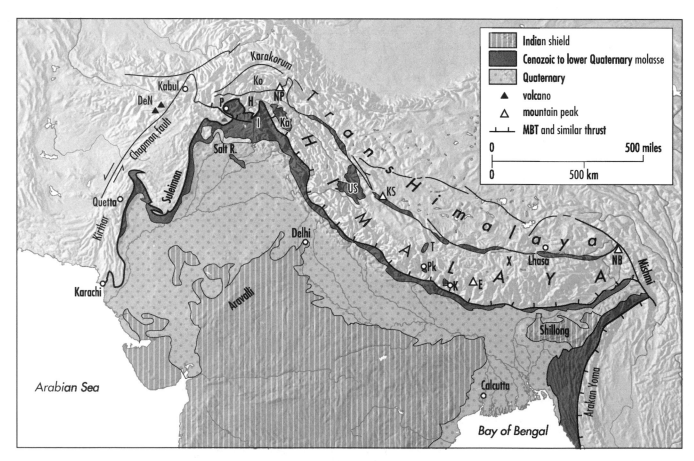

Simplified map of the Himalaya Mountains and surrounding regions showing the main tectonic elements and ages of important strata. Abbreviations as follows: DeN: Dacht-e-Newar; E: Everest; H: Hazara; I: Islamabad; K: Kathmandu; Ka: Kashmir; Ko: Kohistan; KS: Kalais; NB: Namche Barwa; NP: Nanga Parbat; P: Peshawar; Pk: Pokhra; T: Thakkhola; US: Upper Sutlej; X: Xiagaze

Sedimentary rocks in the Himalayas record events on the Indian subcontinent, including a thick Cambrian-Ordovician through Late Carboniferous/Early Permian Gondwanan sequence, followed by rocks deposited during rifting and subsidence events on the margins of the Tethys and Neotethys Oceans. Collision of India with Asia was in progress by the Early Eocene. This collision exposed the diverse rocks in the Himalayas, revealing a rich geologic history that extends back to the Precambrian, where shield rocks of the Aravalli Delhi cratons are intruded by 500-million-year-old granites. Subduction of Tethyan oceanic crust along the southern margin of Tibet formed an Andean-style arc represented by the Transhimalaya batholith that extends west into the Kohistan island arc sequence, in a manner similar to the Alaskan range-Aleutians of western North America. The obduction of ophiolites and high-pressure (blueschist facies) metamorphism dated to have occurred around 100 million years ago is believed to be related to this subduction. Thrust stacks began stacking up on the Indian subcontinent, and by the Miocene, deep attempted intracrustal subduction of the Indian plate beneath Tibet along the Main Central Thrust formed high-grade metamorphism and generated a suite of granitic rocks in the Himalayas. After 15–10 million years ago, movements were transferred to the south to the Main Frontal Thrust, which is still active.

Further Reading

Molnar, Peter. "The Geologic History and Structure of the Himalaya." *American Scientist* 74 (1986).

hinterland   The interior or region bordering the internal zone of a mountain belt is sometimes referred to as the hinterland. It contrasts to the foreland, the area in front of the mountain belt and toward which most folds are overturned and thrusts have moved, and the backland, the area behind an orogenic belt. Hinterlands tend to have exposed high-grade metamorphic rocks, show strong deformation, and may be highly elevated if the mountain belt is young.

See also CONVERGENT PLATE MARGIN PROCESSES; PLATE TECTONICS.

Hoffman, Paul Felix (1941–) Canadian *Geologist, Tectonicist*   Paul F. Hoffman was born on March 21, 1941, in Toronto, Canada, and obtained his bachelor's degree from McMaster University in 1964. From there, he moved to Johns Hopkins University where he received his master's (1965) and Ph.D. (1970) degrees. He has had professional appointments as a lecturer at Franklin and Marshall College from 1968 to 1969, and then built a remarkable career as a research scientist at the Geological Survey of Canada from 1969 to 1992. After that he moved to a position as professor at the University of Victoria in British Columbia, and since 1994 he has been the Sturgis Hooper Professor of Geology at Harvard University.

Paul Hoffman started his career as a sedimentologist, studying Precambrian sedimentary rocks of the Great Slave Lake area of Canada, and the use of stromatolites as paleoenvironmental indicators. From there he moved to the Great Bear Lake area and was one of the pioneers in the field of demonstrating how the plate tectonics operated in the Proterozoic, through his detailed analysis of the 1.9-billion-year-old Wopmay Orogen. Soon after this, Hoffman began several years of tectonic syntheses of the Precambrian geology of North America, pioneering the synthesis of several billion years of Earth history from the perspective of North America. By the late 1980s, Hoffman was synthesizing the Proterozoic tectonics and processes on a global scale, and he was one of the first scientists to correlate different rock packages on different continents and propose reconstructions and histories for several Proterozoic supercontinents, including Rodinia and Gondwana. By the early 1990s, Hoffman's interests and work began to shift focus to the interactions between tectonics and climate, testing specifically the Neoproterozoic Snowball Earth hypothesis.

Paul Hoffman has received a number of awards for his important contributions to the fields of Precambrian geology, tectonics, and climate. In 1989 Ohio State University awarded him the Bownocker Medal, while in 1992 the Geological Association of Canada awarded Hoffman the prestigious Logan Medal. Hoffman received the Miller Medal from the Royal Society of Canada in 1997, the Henno Martin Medal from the Geological Society of Namibia in 2000, and the Alfred Wegener Medal from the European Union of Geosciences in 2001.

hogback   Steeply inclined strata may form sharp ridges with approximately equal dips on each side, known as hogbacks because of their resemblance to the back of the swine. Dips of strata in hogbacks are typically greater than 20°, with one slope being a bedding plane surface and the other being an eroded face cutting through the strata. Cuestas are similar hills where the bedding planes dip less than 20°.

Holocene   An epoch in the Quaternary period, starting at the end of the Pleistocene epoch about 10,000 years ago, until the present time. It is sometimes referred to as the Recent. The term also includes the rocks and unconsolidated deposits formed and deposited during this interval. The Holocene was defined by Charles Lyell in 1833 as the time period in which the Earth was inhabited by humans. This time approximately corresponds archaeologically to the beginning of the Neolithic or New Stone Age at about 8,000 years ago, when humans first moved into structured villages and began agricultural and animal domestication practices. These events heralded the remarkable explosion of population of humans and their artifacts across the globe. By about 5,000 years ago, the Bronze Age began in the Near East, and

Hogback-shaped mountain, Oman. Steeply inclined bedding planes form the slopes on the right-hand side of the mountain, and a steep erosional cliff forms the slopes on the left. *(Photo by Timothy Kusky)*

people used metals widely, migrated extensively, and expanded in numbers to a global population in the tens of millions.

The beginning of the Holocene is now defined as 10,000 years ago, a time when the great ice sheets were retreating and the planet experienced dramatic warming. With the melting of the glaciers, global sea level has been rising, defining the ongoing Flandrian transgression during which sea levels have risen about 558 feet (170 m).

See also NEOGENE; PLEISTOCENE; QUATERNARY; TERTIARY.

homocline   A uniformly dipping set of strata, such as that found on a fold limb, monocline, or a fault block. A homocline should ideally not contain bedding parallel faults or hidden isoclinal folds, as the name suggests a simple stratigraphic succession from older rocks to younger rocks.

See also STRUCTURAL GEOLOGY.

horse latitudes   The mid-latitude regions at about 30–35°N/S over the oceans characterized by weak winds. In the past, sailing ships frequently would become stranded in these regions because of the lack of driving winds, and horses on board had

to be eaten or thrown overboard. The stable air masses in these latitudes form because air that moves poleward from the tropics in Hadley cells cools by radiation, and converges in midlatitudes, since the circumference of a line of latitude is smaller toward the pole than near the equator. This convergence increases the mass of the air, and hence the pressure at the surface, forming subtropical high-pressure systems. This converging, relatively dry air descends, and warms by compression, producing clear skies and warm surface conditions, and resulting in the formation of many of the world's deserts. The weak pressure gradients in these systems produce only very weak winds over the oceans, resulting in the former loss of many horses at sea.

See also ATMOSPHERE; DESERT; HADLEY CELLS.

horst   An elongate, relatively uplifted fault block bounded by faults on its long sides. Horsts are commonly found in extensional tectonic settings, alternating with down-dropped fault blocks known as grabens. Many are tilted on rotational normal faults, whereas others may essentially stay at one elevation while the surrounding terrane sinks in horsts. Some well-known large horsts include the Danakil horst in the Afar region of Ethiopia and Eritrea, and many of the fault blocks in the basin and range province of the western United States.

See also STRUCTURAL GEOLOGY.

hot spot A center of volcanic and plutonic activity that is not associated with an arc and generally not associated with an extensional boundary. Most hot spots are 60–125 miles (100–200 km) across and are located in plate interiors. A few, such as Iceland, are found on oceanic ridges and are identified on the basis of unusually large amounts of volcanism on the ridge. Approximately 200 hot spots are known, and many others have been proposed but their origin is uncertain.

Hot spots are thought to be the surface expression of mantle plumes that rise from deep in the Earth's mantle, perhaps as deep as the core/mantle boundary. As the plumes rise to the base of the lithosphere they expand into huge, even thousand-kilometer-wide plume heads, parts of which partially melt the base of the lithosphere and rise as magmas in hot spots in plate interiors.

See also CONVECTION AND THE EARTH'S MANTLE; MANTLE PLUMES.

hot spring Thermal springs in which the temperature is greater than that of the human body are known as hot springs. Hot springs are found in places where porous structures such as faults, fractures, or karst terrains are able to channel meteoric water (derived from rain or snow) deep into the ground where it can get heated, and also where it can escape upward fast enough to prevent it from cooling by conduction to the surrounding rocks. Most hot springs, especially those with temperatures above 140°F (60°C), are associated with regions of active volcanism or deep magmatic activity, although some hot springs are associated with regions of tectonic extension without known magmatism. Active faulting is favored for the development of hot springs since the fluid pathways tend to become mineralized and closed by minerals that precipitate out of the hot waters, and the faulting is able to repeatedly break and reopen these closed passageways.

When cold descending water gets heated in a hot spring thermal system it expands, and the density of the water decreases, giving it buoyancy. Typical geothermal gradients increase about 155°F–170°F per mile (25°C–30°C per km) in the Earth, so for surface hot springs to attain temperatures of greater than 60°C it is usually necessary for the water to circulate to at least two or three kilometers' depth. This depth may be less in volcanically active areas where hot magmas may exist at very shallow crustal levels, even reaching several hundred degrees at two or three kilometers' depth. Boiling of hot springs may occur when the temperatures of the waters are greater than 212°F (100°C), and if the rate of upward flow is fast enough to allow decompression. In these cases, boiling water and steam may be released at the surface, sometimes forming geysers.

Hot springs are often associated with a variety of mineral precipitates and deposits, depending on the composition of the waters that come from the springs. This composition is typically determined by the type of rocks the water circulates through and is able to leach minerals from, with typical deposits including mounds of travertine, a calcium carbonate precipitate, siliceous sinters, and hydrogen sulfides.

Hot springs are common on the seafloor, especially around the oceanic ridge system where magma is located at shallow levels. The great pressure of the overlying water column on the seafloor elevates the boiling temperature of water at these depths, so that vent temperatures may be above 572°F (300°C). Submarine hot springs often form several-meter or taller towers of sulfide minerals with black clouds of fine metallic mineral precipitates emanating from the hot springs. These systems, known as black smoker chimneys, are also host to some of the most primitive known life-forms on Earth, some of which do not require sunlight but derive their energy from the sulfur and other minerals that come out of the hot springs.

See also BLACK SMOKER CHIMNEYS; GEOTHERMAL ENERGY; GEYSER.

Hudson Bay, Canada A 475,000-square-mile (1,230,250-km²) inland sea in eastern Canada. James Bay is a southern embayment of Hudson Bay that extends along the Ontario–Quebec border, and the bay extends northward to the Foxe Channel, across which lie the Foxe basin and Baffin Island. Hudson Strait connects the bay with the Atlantic Ocean, whereas the Foxe Channel leads to the Arctic Ocean. The bay is shallow and has numerous island groups such as

Numerous raised beach terraces on the shore of Hudson Bay, Canada. The beaches form at sea level, then as the crust is uplifted in response to glacial rebound, new beaches form at the new relatively lower sea level. *(Photo by Timothy Kusky)*

the Belcher Islands, and it is surrounded by a gently dipping glacially leveled shield. Numerous raised beach terraces surround the bay reflecting a rise in the land surface caused by release of stresses from the weight of the glaciers, which only geologically recently retreated from the region. Hudson Bay is located on the sparsely populated Canadian shield, and the tree line, north of which no significant stands of trees can grow, cuts diagonally across the southwest side of the bay.

Hudson Bay used to be ice-free from middle July through October. However, a series of dams has been built in Canada along rivers that flow into Hudson Bay, and these dams are used to generate clean hydroelectric energy. The problem that has arisen is that these dammed rivers have annual spring floods, which before the dams were built would flush the pack ice out of Hudson Bay. Since the dams have been built, the annual spring floods are diminished, resulting in the pack ice remaining on the bay through the short summer. This has drastically changed the summer season on the Ungava peninsula. As the warm summer winds blow across the ice they pick up cool moist air, and cold fogs now blow across the Ungava all summer. This has drastically changed the local climate and has hindered growth and development of the region.

Hudson Bay was named by Henry Hudson in 1610 during his explorations of the area in his search for a northwest passage from the Atlantic to the Pacific. The region was inhabited by French fur trappers, but France gave up its claims to the region in 1713. Hudson's Bay company was founded in 1670 as a group of trading posts at river mouths, and it survives to this day.

Huronian   A division of the Proterozoic of the Canadian Shield, originally based on Sir William Logan's 1845 field examination of the region north of Lake Huron, in the Lake Timiskaming area. He named the Huronian after a group of conglomerates and slates that were found to rest unconformably over older Laurentian granite, now known to be Archean in age. The Huronian was later extended to the Lake Superior region by Alexander Murray during 1847–1858.

See also CONTINENTAL CRUST; PROTEROZOIC.

hurricane   Intense tropical storms with sustained winds of more than 74 miles per hour (119 km/hr) are known as hurricanes if they form in the northern Atlantic or eastern Pacific Oceans, cyclones if they form in the Indian Ocean or near Australia, and typhoons if they form in the western North Pacific Ocean. Most large hurricanes have a central eye with calm or light winds and clear skies or broken clouds, surrounded by an eye wall, a ring of very tall and intense thunderstorms that spin around the eye, with some of the most intense winds and rain of the entire storm system. The eye is surrounded by spiral rain bands that spin counterclockwise

Galveston Island Hurricane, 1900

The deadliest natural disaster to affect the United States was when a category 4 hurricane hit Galveston Island, Texas, on September 8, 1900. Galveston is a low-lying barrier island located south of Houston and in 1900 served as a wealthy port city. Residents of coastal Texas received early warning of an approaching hurricane from a Cuban meteorologist, but most chose to ignore this advice. Later, perhaps too late, U.S. forecasters warned of an approaching hurricane, and many people then evacuated the island to move to relative safety inland. However, many others remained on the island. In the late afternoon the hurricane moved in to Galveston, and the storm surge hit at high tide covering the entire island with water. Even the highest point on the island was covered with one foot of water. Winds of 120 miles per hour (190 km/hr) destroyed wooden buildings, as well as many of the stronger brick buildings. Debris from destroyed buildings crashed into other structures, demolishing them and creating a moving mangled mess for residents trapped on the island. The storm continued through the night, battering the island and city with 30-foot (9-m) high waves. In the morning, residents who found shelter emerged to see half of the city totally destroyed, and the other half severely damaged. But worst of all, thousands of bodies were strewn everywhere, 6,000 on Galveston Island, and another 1,500 on the mainland. There was no way off the island as all boats and bridges were destroyed, so survivors were in additional danger of disease from the decaying bodies. When help arrived from the mainland, the survivors needed to dispose of the bodies before cholera set in, so they put the decaying corpses on barges, and dumped them at sea. However, the tides and waves soon brought the bodies back, and they eventually had to be burned in giant funeral pyres built from wood from the destroyed city. Galveston was rebuilt, and a seawall built from stones was supposed to protect the city; however, in 1915, another hurricane struck Galveston, claiming 275 additional lives.

The Galveston seawall has since been reconstructed and is higher and stronger, although some forecasters believe that even this seawall will not be able to protect the city from a category 5 hurricane. The possibility of a surprise storm hitting Galveston again is not so remote, as demonstrated by the surprise tropical storm of early June 2001. Weather forecasters were not successful in predicting the rapid strengthening and movement of this storm, which dumped 23–48 inches (58–122 cm) of rain on different parts of the Galveston-Houston area and attacked the seawall and coastal structures with huge waves and 30-mile-per-hour (48-km/hr) winds. Twenty-two people died in the area from the surprise storm, showing that even modern weather forecasting cannot always adequately predict tropical storms. It is best to heed early warnings and prepare for rapidly changing conditions when hurricanes and tropical storms are approaching vulnerable areas.

NASA satellite image of Southern Hemisphere hurricane Manoun, which hit the island of Madagascar in June 2003. Note that the spiral rain bands have a clockwise inward spin, opposite to that of hurricanes that form in the Northern Hemisphere. *(Photo by NASA)*

in the Northern Hemisphere (clockwise in the Southern Hemisphere) in toward the eye wall, moving faster and generating huge waves as they approach the center. Wind speeds increase toward the center of the storm and the atmospheric pressure decreases to a low in the eye, uplifting the sea surface in the storm center. Surface air flows in toward the eye of the hurricane, then moves upward, often above nine miles (15 km), along the eye wall. From there it moves outward in a large outflow, until it descends outside the spiral rain bands. Air in the rain bands is ascending, whereas between the rain bands, belts of descending air counter this flow. Air in the very center of the eye descends to the surface. Hurricanes drop enormous amounts of precipitation, typically spawn numerous tornadoes, and cause intense coastal damage from winds, waves, and storm surges, where the sea surface may be elevated many meters above its normal level.

Most hurricanes form in the summer and early fall over warm tropical waters when the winds are light and the

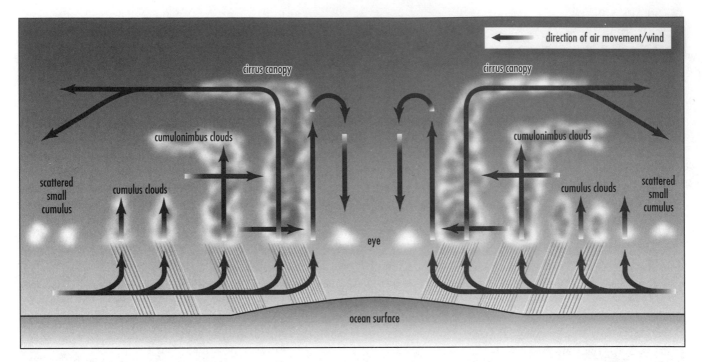

cirrus canopy   cirrus canopy

cumulonimbus clouds   cumulonimbus clouds

scattered small cumulus   cumulus clouds   cumulus clouds   scattered small cumulus

eye

ocean surface

Cross section of a typical hurricane showing eye, eye wall, and circulating spiral bands of cumulonimbus clouds

humidity is high. In the North Atlantic, hurricane season generally runs from June through November, when the tropical surface waters are warmer than 26.5°C (80°F). They typically begin when a group of unorganized thunderstorms are acted on by some trigger that causes the air to begin converging and spinning. These triggers are found in the intertropical convergence zone that separates the northeast trade winds in the Northern Hemisphere from the southeast trade winds in the Southern Hemisphere. Most hurricanes form within this zone, between 5° and 20° latitude. When a low-pressure system develops in this zone in hurricane season, the isolated thunderstorms can develop into an organized convective system that strengthens to form a hurricane. Many Atlantic hurricanes form in a zone of weak convergence on the eastern side of tropical waves that form over North Africa, then move westward where they intensify over warm tropical waters.

In order for hurricanes to develop, high level winds must be mild, otherwise they would disperse the tops of the growing thunderclouds. In addition, high level winds must not be descending, since this would also inhibit the upward growth of the thunderstorms. Once the mass of thunderstorms is organized, hurricanes gain energy by evaporating water from the warm tropical oceans. When the water vapor condenses inside the thunderclouds, this heat energy is then converted to wind energy. The upper level clouds then move outward, causing the storm to grow stronger, and decreasing the pressure in the center of the storm. The low pressure in the storm's center draws the outlying thunderstorms in toward the surface low, and these rain bands then spiral inward because of the Coriolis force. The clouds spin pro-

gressively faster as they move inward, because of the conservation of angular momentum.

The strength of hurricanes is measured using the Saffir-Simpson scale, which measures the damage potential of a storm, considering such factors as the central barometric pressure, maximum sustained wind speeds, and the potential height of the storm surge. Category 1 hurricanes have central pressures of greater than 980 millibars, sustained winds of 74–95 miles per hour (119–153 km/hr), and a likely 4–5 foot (1–1.5 m) storm surge. Damage potential is minimal, with likely effects including downed power lines, ruined crops, and minor damage to weak parts of buildings. Category 2 hurricanes have central barometric pressures at 979–965 millibars, maximum sustained winds of 96–110 miles per hour (155–177 km/hr), and 6–8 foot (1.8–2.4 m) storm surges. Damage is typically moderate, including roof and chimney damage, beached and splintered boats, and destroyed crops, road signs, and traffic lights. Category 3 hurricanes have central barometric pressures falling of 964 and 945 millibars, sustained winds of 111–130 miles per hour (179–209 km/hr), and storm surges of 9–12 feet (2.7–3.6 m). Category 3 hurricanes are major storms capable of extensive property damage, including uprooting large trees, and the destruction of mobile homes and poorly constructed coastal houses. Category 4 storms can be devastating, with central barometric pressures falling to 940–920 millibars, sustained winds of 131–155 miles per hour (211–249 km/hr), and storm surges of 13–18 feet (4–5.5 m). These storms typically rip the roofs off homes and businesses, destroy sea piers, and throw boats well inland. Waves may breach seawalls causing large-scale coastal flooding. Category 5 storms are truly massive with central barometric pressures dropping below 920 millibars, maximum sustained winds above 155 miles per hour (249 km/hr), and storm surges over 18 feet (5.5 m). Storms with this power rarely hit land, but when they do they are capable of leveling entire towns, moving large amounts of coastal sediments, and causing large death tolls.

Hurricanes inflict some of the most rapid and severe damage and destruction to coastal regions and sometimes cause large numbers of deaths. The number of deaths from hurricanes has been reduced dramatically in recent years with our increased ability to forecast the strength and landfall of hurricanes, and our ability to monitor their progress with satellites. However, the costs of hurricanes in terms of property damage have greatly increased, as more and more people build expensive homes along the coast. The greatest number of deaths from hurricanes has been from effects of the storm surge. Storm surges typically come ashore as a wall of water that rushes onto land at the forward velocity of the hurricane, as the storm waves on top of the surge are pounding the coastal area with additional energy. For instance, when Hurricane Camille hit Mississippi in 1969 with 200-mile-per-hour winds (322 km/hr), a 24-foot (7.3-m) high storm surge moved into coastal areas, killing most of the 256 people that perished in this storm. Winds and tornadoes account for more deaths. Heavy rains from hurricanes also cause considerable damage. Flooding and severe erosion is often accompanied by massive mudflows and debris avalanches, such as those caused by Hurricane Mitch in Central America in 1998. In a period of several days, Mitch dropped 25–75 inches (63.5–190.5 cm) of rain on Nicaragua and Honduras, initiating many mudslides that were the main cause of the more than 11,000 deaths from this single storm. One of the worst events was the filling and collapse of a caldera on Casitas volcano—when the caldera could hold no more water, it gave way sending mudflows (lahars) cascading down on several villages, killing 2,000 people.

Many cyclones are spawned in the Indian Ocean. Bangladesh is a densely populated low-lying country, sitting mostly at or near sea level, between India and Myanmar. It is a delta environment, built where the Ganges and Brahmaputra rivers drop their sediment eroded from the Himalaya Mountains. It sits directly in the path of many Bay of Bengal tropical cyclones and has been hit by seven of the nine most deadly hurricane disasters in the history of the world. On November 12 and 13 of 1970, a category 5 typhoon hit Bangladesh with 155-mile-per-hour (249.5 km/hr) winds, and a 23-foot (7-m) high storm surge that struck at the astronomically high tides of a full moon. The result was devastating, with 400,000 human deaths and half a million farm animals perishing. Again in 1990, another cyclone hit the same area, this time with a 20-foot (6-m) storm surge and 145-mile-per-hour (233 km/hr) winds, killing another 140,000 people and another half-million farm animals.

Hurricane Andrew was the most destructive hurricane in U.S. history, causing more than $30 billion in damage in August of 1992. Andrew began to form over North Africa and grew in strength as it was driven across the Atlantic by the trade winds. On August 22 Andrew had grown to hurricane strength and moved across the Bahamas with 150-mile-per-hour (241 km/hr) winds, killing four people. On August 24 Andrew smashed into southern Florida with a nearly 17-foot (5.2-m) high storm surge, steady winds of 145 miles per hour (233 km/hr), and gusts to 200 miles per hour (322 km/hr). Andrew's path took it across a part of south Florida that had hundreds of thousands of poorly constructed homes and trailer parks, and hurricane winds caused intense and widespread destruction. Andrew destroyed 80,000 buildings, severely damaged another 55,000, and demolished thousands of cars, signs, and trees. In southern Florida, 33 people died. By August 26 Andrew had traveled across Florida, losing much of its strength, but had moved back into the warm waters of the Gulf of Mexico and regained much of that strength. On August 26 Andrew made landfall again, this time in Louisiana with 120-mile-per-hour (193 km/hr) winds, where it killed another 15 people. Andrew's winds stirred up the fish-rich marshes of southern Louisiana, where the muddied waters were agitated so much that the decaying organic material overwhelmed the oxygen-rich surface layers, suffocating millions of fish. Andrew then continued to lose strength but dumped flooding rains over much of Mississippi.

See also AIR PRESSURE; BEACH; CLOUDS; EXTRATROPICAL CYCLONES; POLAR LOW; STORM SURGES; THUNDERSTORMS; TRADE WINDS.

Further Reading

Ahrens, C. Donald. *Meteorology Today, An Introduction to Weather, Climate, and the Environment,* 7th ed. Pacific Grove, Calif.: Thomson/Brookscolc, 2003.

Longshore, David. *Encyclopedia of Hurricanes, Typhoons, and Cyclones.* New York: Facts On File, 1998.

Hutton, James (1726–1797) Scottish *Geologist* James Hutton's most important contribution to science was his book entitled *Theory of the Earth.* The theory was simple and yet contained such fundamental ideas that he was later known as the founder of modern geology. In the 30-year period when he was developing and writing his ideas, other areas of earth sciences had been explored, however geology had not really been recognized as an important science. Hutton's "Theory of the Earth" consisted of three main ideas: (1) the amount of time the Earth had existed as a "habitable world;" (2) the changes it had undergone in the past; and (3) whether any end to the present state of affairs could be foreseen. He also talked about how the rocks were good at indicating the different periods and how they could tell us roughly when the Earth was formed. Initially his ideas were not well received by his fellow scientists. His ideas contradicted the natural conservatism of many geologists, including a reluctance to abandon belief in the biblical account of creation, and the widespread catastrophism. By the 1830s even though geologists were still conservative, they were better equipped to assess the value of Hutton's theory.

Further Reading

Hutton, James. "Theory of the Earth with Proofs and Illustrations." 2 volumes. London: Messrs Cadell, Junior, Davies, 1795.

hydrocarbon   Gaseous, liquid, or solid organic compounds consisting of hydrogen and carbon. Petroleum is a mixture of different types of hydrocarbons (fossil fuels) derived from the decomposed remains of plants and animals that are trapped in sediment and can be used as fuel. When plants and animals are alive, they absorb energy from the Sun (directly through photosynthesis in plants, and indirectly through consumption in animals) to make complex organic molecules which, after they die, decay to produce hydrocarbons and other fossil fuels. If organic matter is buried before it is completely decomposed, some "solar energy" may become stored in the rocks as fossil fuels (less than 1 percent of total organic matter gets buried). In most industrial nations the chief source of energy is fossil fuels.

The type of organic matter that gets buried in sediment plays an important role in the type of fossil fuel that forms. Oceanic organisms (such as bacteria and phytoplankton) are buried by shales and muds and form proteins, lipids, and carbohydrates that form oil and natural gas when heated. Terrigenous plants (such as trees and bushes) form resins, waxes, lignins, and cellulose, which form coals. Incompletely broken down organics in shale form kerogens, or oil shales, which require additional heat to convert to oil.

The first people on the planet to use oil were the ancient Iraqis, 6,000 years ago. Oil is fluid and is lighter than water, which strongly influences where it is found. An oil "pool" is an underground accumulation of oil and gas occurring in the pore spaces of rock. An oil field is a group of oil pools of similar type.

Once oil forms by the decay of organic material, it migrates upward until it seeps out at the surface or it encounters a trap. The migration of oil is like the movement of groundwater. Migration is slow, and since petroleum is lighter than water, water forces it upward to the tops of the traps. Because most oil eventually finds its way to the surface, it is not surprising that most oil is found in relatively young rocks.

The formation of oil requires that the source has been through a critical range of pressure and temperature conditions, known as the oil window. If the geothermal gradient is too low or too high, oil will not form. Oil and gas can only accumulate if five basic requirements are met. First, an appropriate source rock is needed to provide the oil. Second, a permeable reservoir is needed, and third it must have an impermeable roof rock. Fourth, a trap (stratigraphic or structural) is needed to hold the oil, and finally, the formation of the trap must have occurred before the oil has escaped from the system. Thus, it is extremely lucky if all five criteria are met and a petroleum deposit is formed.

Geologists know the location of approximately 600 billion barrels of oil; estimates for total global oil reserves are 1,500–3,000 billion barrels. We have used up about 500 billion barrels. Many of the unknown reserves include small deposits, but not tars, tar sands, and oil shales, which must be heated and extensively processed to make them useful, and thus are very expensive. We now use 30 million barrels of oil a year, therefore, known reserves will last 20–50–100 years at the most. Oil is running out and is becoming an increasingly powerful political weapon. The oil-rich nations can effectively hold the rest of the world hostage, being that we have become so dependent on oil. Future energy sources may include nuclear fuels, solar energy, hydroelectric power, geothermal energy, biomass, wind, gas hydrates, and tidal energy.

See also COAL; GAS HYDRATES.

hydroelectric   Energy may be generated by the movement of water. In places where water is moving rapidly, such as over a waterfall, the water may be redirected to move through a turbine that moves and powers a generator, producing hydroelectric power. Hydroelectric power is often heralded as clean energy, since it does not burn hydrocarbons or other fuels, and hydroelectric plants can be constructed with minimal environmental impact.

hydrofracturing   In water-gas reservoirs, the common practice of pumping water at high pressure into wells, so that the rocks in the reservoir fracture, generating new porosity that will aid in the oil extraction. The pressure needed to fracture the rocks is equal to the minimum principal stress. The fractures that form will be parallel to the maximum principal stress, enabling greater fluid flow through the reservoir.

See also STRUCTURAL GEOLOGY.

hydrologic cycle   The hydrologic cycle describes changes, both long and short term, in the Earth's hydrosphere. It is powered by heat from the Sun, which causes evaporation and transpiration. This water then moves in the atmosphere, and precipitates as rain or snow, which then drains off in streams, evaporates, or moves as groundwater, eventually to begin the cycle over and over again. The time required for individual molecules of water to complete the cycle varies greatly and may range from a few weeks for some molecules to many thousands of years for others.

hydrology   The study of water, in liquid, solid, and vapor form, on local to global scales. Hydrology includes analysis of the properties of water, its circulation, and distribution on and below the surface in reservoirs, streams, lakes, oceans, and the groundwater system. Many hydrologists assess the movement of water through different parts of the hydrologic system and evaluate the influence of human activities on the system in attempts to maximize the benefits to society. Hydrology may also involve environmental and economic aspects of water use.

hydrosphere One of the external layers of the Earth, consisting of the oceans, lakes, streams, and the atmosphere. The hydrosphere is a dynamic mass of liquid, continuously on the move. It includes all the water in oceans, lakes, streams, glaciers, and groundwater, although most water is in the oceans.

See also HYDROLOGIC CYCLE; HYDROLOGY.

hydrothermal Hot waters in the Earth's crust are known as hydrothermal fluids. Most heated subsurface waters contain dissolved minerals or other substances and are then known as hydrothermal solutions. Hydrothermal solutions are very important because they are able to dissolve, transport, and redistribute many elements in the Earth's crust and are responsible for the concentration and deposition of many ore deposits, including many gold, copper, silver, zinc, tin, and sulfide deposits. These mineral deposits are known as hydrothermal deposits.

Hydrothermal solutions are typically derived from one or more sources, including fresh or saline groundwater, water trapped in rocks as they are deposited, water released during metamorphic reactions, or water released from magmatic systems. The minerals, metals, and other compounds dissolved in hydrothermal solutions are typically derived from the dissolution of the rocks that the fluids migrate through, or released from magmatic systems. Hydrothermal solutions are commonly formed during the late stages of crystallization of a magma, and these fluids will contain many of the chemical elements that do not readily fit into the atomic structures of the minerals crystallizing from the magma. These fluids tend to be enriched in lead, copper, zinc, gold, silver tin, tungsten, and molybdenum. Many hydrothermal fluids are also saline, with the salts derived from leaching of country rocks. Saline solutions are much more effective at carrying dissolved metals than nonsaline solutions, so these hydrothermal solutions tend to be enriched in dissolved metals.

As hydrothermal solutions move up through the crust they cool from as high as 1,112°F (600°C), and with lower temperatures the solutions can hold less dissolved material. Therefore, as the fluids cool, hydrothermal veins and ore deposits form, with different minerals precipitating out of the fluid at different temperatures. Some minerals may also pre-

cipitate out when the fluids come into contact with rocks of a certain composition, with a fluid-wall rock reaction.

hypsometric curve The distribution of elevation of continents and oceans can be portrayed on a curve showing percentage of land at a certain elevation versus elevation, known as the hypsometric curve, or the hypsographic curve. The curve is a cumulative frequency profile representing the statistical distribution of areas of the Earth's solid surface above or below mean sea level. The hypsometric curve is strongly bimodal, reflecting the two-tier distribution of land in continents close to sea level, and on ocean floor abyssal plains 1.9–2.5 miles (3–4 km) below sea level. Relatively little land surface is found in high mountains or in deep-sea trenches.

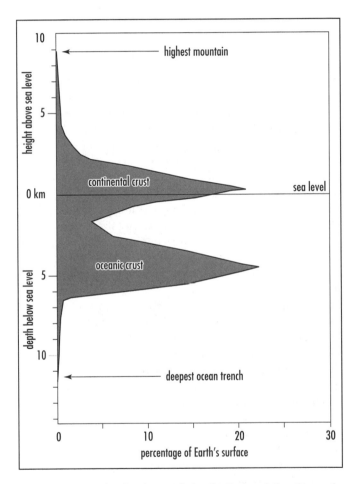

Hypsometric curve showing the cumulative distribution of elevations and depths of the solid surface of the Earth. Continental crust has an average elevation of a few hundred meters above sea level, whereas oceanic crust has an average depth of nearly five kilometers below sea level.

I

ice age Times when the global climate was colder, and large masses of ice covered many continents are referred to as ice ages. At several times in Earth's history, large portions of the Earth's surface have been covered with huge ice sheets. About 10,000 years ago, all of Canada, much of the northern United States, and most of Europe were covered with ice sheets, as was about 30 percent of the world's landmass. These ice sheets lowered sea level by about 320 feet (100 m), exposing the continental shelves and leaving cities including New York, Washington, and Boston 100 miles from the sea. In the last 2.5 billion years, several periods of ice ages have been identified, separated by periods of mild climate similar to that of today. Ice ages seem to form through a combination of several different factors. One of the variables is the amount of incoming solar radiation, and this changes in response to several astronomical effects. Another variable is the amount of heat that is retained by the atmosphere and ocean, or the balance between the incoming and outgoing heat. A third variable is the distribution of landmasses on the planet. Shifting continents can influence the patterns of ocean circulation and heat distribution, and a large continent on one of the poles can cause ice to build up on that continent, increasing the amount of heat reflected back to space, and lowering global temperatures in a positive feedback mechanism.

Glaciations have happened frequently in the past 55 million years and could occur again at almost any time. In the late 1700s and early 1800s, Europe experienced a "little ice age" when many glaciers advanced out of the Alps and destroyed many small villages. Ice ages have occurred at several other times in the ancient geologic past, including in Late Paleozoic (about 350–250 million years ago), Silurian (435 million years ago), and Late Proterozoic (about 800–600 million years ago). During parts of the Late Proterozoic glacia-

tion, it is possible that the entire Earth surface temperature was below freezing and covered by ice.

In the Late Proterozoic, the Earth experienced one of the most profound ice ages in the history of the planet. Isotopic records and geologic evidence suggest that the entire Earth's surface was frozen, though some workers dispute the evidence and claim that there would be no way for the Earth to recover from such a frozen state. In any case it is clear that in the Late Proterozoic, during the formation of the supercontinent of Gondwana, the Earth experienced one of the most intense glaciations ever, with the lowest average global temperatures in known Earth history.

One of the longest lasting glacial periods was the Late Paleozoic ice age that lasted about 100 million years, indicating a long-term underlying cause of global cooling. Of the variables that operate on these long time scales, it appears that the distribution and orientation of continents seems to have caused the Late Paleozoic glaciation. The Late Paleozoic saw the amalgamation of the planet's landmasses into the supercontinent of Pangea. The southern part of Pangea, known as Gondwana, consisted of present-day Africa, South America, Antarctica, India, and Australia. During the drift of the continents in the Late Paleozoic, Gondwana slowly moved across the South Pole, and huge ice caps formed on these southern continents during their passage over the pole. The global climate was overall much colder, with the subtropical belts becoming very condensed and the polar and subpolar belts expanding to low latitudes.

It seems that during all major glaciations there was a continent situated over one of the poles. We now have Antarctica over the South Pole, and this continent has huge ice sheets on it. When continents rest over a polar region they accumulate huge amounts of snow that gets converted into several-kilometer-thick ice sheets, which reflect more solar radiation back to space and lower global seawater temperatures and sea levels.

Another factor that helps initiate glaciations is to have continents distributed in a roughly north-south orientation across equatorial regions. Equatorial waters receive more solar heating than polar waters. Continents block and modify the simple east to west circulation of the oceans induced by the spinning of the planet. When continents are present on or near the equator, they divert warm water currents to high latitudes, bringing warm water to higher latitudes. Since warm water evaporates much more effectively than cold water, having warm water move to high latitudes promotes evaporation, cloud formation, and precipitation. In cold high-latitude regions the precipitation falls as snow, which persists and builds up glacial ice.

The Late Paleozoic glaciation ended when the supercontinent of Pangea began breaking apart, suggesting a further link between tectonics and climate. It may be that the smaller landmasses could not divert the warm water to the poles any more, or perhaps enhanced volcanism associated with the breakup caused additional greenhouse gases to build up in the atmosphere, raising global temperatures.

The planet began to enter a new glacial period about 55 million years ago, following a 10-million-year-long period of globally elevated temperatures and expansion of the warm subtropical belts into the subarctic. This Late Paleocene global hothouse saw the oceans and atmosphere holding more heat than at any other time in Earth history, but temperatures at the equator were not particularly elevated. Instead, the heat was distributed more evenly around the planet such that there were probably fewer violent storms (with a small temperature gradient between low and high latitudes), and overall more moisture in the atmosphere. It is thought that the planet was so abnormally warm during this time because of several factors, including a distribution of continents that saw the equatorial region free of continents. This allowed the oceans to heat up more efficiently, raising global temperatures. The oceans warmed so much that the deep ocean circulation changed, and the deep currents that are normally cold became warm. These melted frozen gases (known as methane hydrates) accumulated on the seafloor, releasing huge amounts of methane to the atmosphere. Methane is a greenhouse gas, and its increased abundance in the atmosphere trapped solar radiation in the atmosphere, contributing to global warming. In addition, this time saw vast outpourings of mafic lavas in the North Atlantic Ocean realm, and these volcanic eruptions were probably accompanied by the release of large amounts of CO_2, which would have increased the greenhouse gases in the atmosphere and further warmed the planet. The global warming during the Late Paleocene was so extreme that about 50 percent of all the single-celled organisms living in the deep ocean became extinct.

After the Late Paleocene hothouse, the Earth entered a long-term cooling trend which we are still currently in, despite the present warming of the past century. This current ice age was marked by the growth of Antarctic glaciers, starting about 36 million years ago, until about 14 million years ago, when the Antarctic ice sheet covered most of the continent with several miles of ice. At this time global temperatures had cooled so much that many of the mountains in the Northern Hemisphere were covered with mountain and piedmont glaciers, similar to those in southern Alaska today. The ice age continued to intensify until 3 million years ago, when extensive ice sheets covered the Northern Hemisphere. North America was covered with an ice sheet that extended from northern Canada to the Rocky Mountains, across the Dakotas, Wisconsin, Pennsylvania, and New York, and on the continental shelf. At the peak of the glaciation (18,000–20,000 years ago), about 27 percent of the land surface was covered with ice. Midlatitude storm systems were displaced to the south, and desert basins of the southwest United States, Africa, and the Mediterranean received abundant rainfall and hosted many lakes. Sea level was lowered by 425 feet (130 m) to make the ice that covered the continents, so most of the world's continental shelves were exposed and eroded.

The causes of the Late Cenozoic glaciation are not well known but seem related to Antarctica coming to rest over the south pole and other plate tectonic motions that have continued to separate the once contiguous landmasses of Gondwana, changing global circulation patterns in the process. Two of the important events seems to be the closing of the Mediterranean Ocean around 23 million years ago and the formation of the Panama isthmus at 3 million years ago. These tectonic movements restricted the east-to-west flow of equatorial waters, causing the warm water to move to higher latitudes where evaporation promotes snowfall. An additional effect seems to be related to uplift of some high mountain ranges, including the Tibetan Plateau, which has changed the pattern of the air circulation associated with the Indian monsoon.

The closure of the Panama isthmus is closely correlated with the advance of Northern Hemisphere ice sheets, suggesting a causal link. This thin strip of land has drastically altered the global ocean circulation such that there is no longer an effective communication between Pacific and Atlantic Ocean waters, and it diverts warm currents to near-polar latitudes in the North Atlantic, enhancing snowfall and Northern Hemisphere glaciation. Since 3 million years ago, the ice sheets in the Northern Hemisphere have alternately advanced and retreated, apparently in response to variations in the Earth's orbit around the Sun and other astronomical effects. These variations change the amount of incoming solar radiation on timescales of thousands to hundreds of thousands of years (Milankovitch Cycles). Together with the other longer-term effects of shifting continents, changing global circulation patterns, and abundance of greenhouse gases in the atmosphere, most variations in global climate can be approximately explained. This knowledge may help predict where the climate is heading in the future and may help model and mitigate the effects of human-induced changes to the atmospheric greenhouse gases. If we are heading into another warm phase and the existing ice on the planet melts, sea level will quickly rise by 210 feet (64 m), inundating many of the world's cities and farmlands. Alternately, if we enter a new ice sheet stage, sea levels will be lowered, and the planet's climate zones will be displaced to more equatorial regions.

See also ATMOSPHERE; GLACIER; GREENHOUSE EFFECT.

iceberg Calving refers to a process in which large pieces of ice break off from the fronts of tidewater glaciers, ice shelves, or sea ice. Typically, the glacier will crack with a loud noise that sounds like an explosion, and then a large chunk of ice will splash into the water, detaching from the glacier. Glaciers may retreat rapidly by calving. Ice that has broken off an ice cap or polar sea, or calved off a glacier and is floating in open water is known as sea ice or, more commonly, as icebergs. Icebergs present a serious hazard to ocean traffic and shipping lanes and have sunk numerous vessels, including the famous sinking of the *Titanic* in 1912, killing 1,503 people. Icebergs from sea ice float on the surface, but between 81 and 89 percent of the ice will be submerged. The exact level that sea ice floats in the water depends on the exact density of the ice, as determined by the total amount of air bubbles trapped in the ice, and how much salt got trapped in the ice during freezing.

There are several main categories of sea ice that may break up to form many icebergs. The first comes from ice that formed on polar seas in the Arctic Ocean and around Antarctica. The ice that forms in these regions is typically about 10–15 feet (3–4 m) thick. Antarctica becomes completely surrounded by this sea ice every winter, and the Arctic Ocean is typically about 70 percent covered in the winter. During summer, many passages open up in this sea ice, but during the winter they re-close, forming pressure ridges of ice that may be up to tens of meters high. Recent observations suggest that the sea ice in the Arctic Ocean is thinning dramatically and rapidly and may soon disappear altogether. The icecap over the Arctic Ocean rotates clockwise, in response to the spinning of the Earth. This spinning is analogous to putting an ice cube in a glass and slowly turning the glass. The ice cube will rotate more slowly than the glass, because it is decoupled from the edge of the glass. About one-third of the ice is removed every year by the East Greenland current. This ice then moves south and becomes icebergs, forming a hazard to shipping in the North Atlantic.

A second type of sea ice forms as pack ice in the Gulf of St. Lawrence, along the southeast coast of Canada, in the Bering, Beaufort, and Baltic Seas, in the Seas of Japan and Okhotsk, and around Antarctica. Pack ice builds up especially along the western sides of ocean basins, where cold currents are more common on the west sides of the oceans. Occasionally, during cold summers, pack ice may persist throughout the summer.

Pack ice presents hazards when it gets so extensive that it effectively blocks shipping lanes, or when leads (channels) into the ice open and close, forming pressure ridges that become too thick to penetrate with icebreakers. Ships attempting to navigate through pack ice have become crushed when leads close, and the ships are trapped. Pack ice has terminated or resulted in disaster for many expeditions to polar seas, most notably Franklin's expedition in the Canadian Arctic and Scott's expedition to Antarctica. Pack ice also breaks up, forming many small icebergs, but because these are not as thick as icebergs of other origins they do not present as significant a hazard to shipping.

Pack ice also presents hazards when it drifts to shore, usually during spring breakup. With significant winds, pack ice can pile up on flat shorelines and accumulate in stacks up to 50 feet (15 m) high. The force of the ice is tremendous and is enough to crush shoreline wharves, docks, buildings, and boats. Pack ice that is blown ashore also commonly pushes up high piles of gravel and boulders that may be 35 feet (10.5 m) high in places. These ridges are common around many of the Canadian Arctic islands and mainland. Ice that forms initially attached to the shore presents another type of hazard. If it breaks free and moves away from shore, it may carry with it significant quantities of shore sediment, causing rapid erosion of beaches and shore environments.

Pack ice also forms on many high-latitude lakes, and the freeze-thaw cycle causes cracking of the lake ice. When lakewater rises to fill the cracks, the ice cover on the lake expands and pushes over the shoreline, resulting in damage to any structures built along the shore. This is a common problem on many lakes in northern climates and leads to widespread damage to docks and other lakeside structures.

Icebergs derived from glaciers present the greatest danger to shipping. In the Northern Hemisphere most icebergs calve off glaciers in Greenland or Baffin Island, then they move south through the Davis Strait into shipping lanes in the North Atlantic off Newfoundland. Some icebergs calve off glaciers adjacent to the Barents Sea, and others come from glaciers in Alaska and British Columbia. In the Southern Hemisphere, most icebergs come from Antarctica, though some come from Patagonia.

Once in the ocean icebergs drift with ocean currents, but because of the Coriolis force they are deflected to the right in the Northern Hemisphere, and to the left in the Southern Hemisphere. Most icebergs are about 100 feet to 300 feet (30.5–91.5 m) high, and up to about 2,000 feet (609.5 m) in length. However, in March of 2000 a huge iceberg broke off the Ross Ice Shelf in Antarctica, and this berg was roughly the size of the state of Delaware. It had an area of 4,500 square miles (11,655 km²) and stuck 205 feet (62.5 m) out of the water. Icebergs in the Northern Hemisphere pose a greater threat to shipping, as those from Antarctica are too remote and rarely enter shipping lanes. Ship collisions with icebergs have resulted in numerous maritime disasters, especially in the North Atlantic on the rich fishing grounds of the Grand Banks off the coast of Newfoundland.

Icebergs are now tracked by satellite, and ships are updated with their positions so they can avoid any collisions that could prove fatal for the ships. Radio transmitters are placed on larger icebergs to more closely monitor their locations, and many ships now carry more sophisticated radar and navigational equipment that help track the positions of large icebergs and the ship, so that they avoid collision.

Icebergs also pose a serious threat to oil drilling platforms and seafloor pipelines in high-latitude seas. Some precautions have been taken, such as building seawalls around nearshore platforms, but not enough planning has gone into preventing an iceberg colliding with and damaging an oil platform, or from one being dragged across the seafloor and rupturing a pipeline.

See also GLACIER.

ice cap Glaciers are any permanent body of ice (recrystallized snow) that shows evidence of gravitational movement. Ice caps form dome-shaped bodies of ice and snow over mountains and flow radially outward. They cover high peaks of some mountain ranges, such as parts of the Kenai and Chugach Mountains in Alaska, the Andes, and many in the Alpine-Himalayan system. Ice caps are relatively small, less than 20,000 miles (50,000 km²), whereas ice sheets are similar but larger. Ice sheets are huge, continent-sized masses of ice that presently cover Greenland and Antarctica. Ice sheets contain about 95 percent of all the glacier ice on the planet. If global warming were to continue to melt the ice sheets, sea level would rise by 230 feet (70 m). A polar ice sheet covers Antarctica, consisting of two parts that meet along the Transantarctic Mountains. It shows ice shelves, which are thick glacial ice that floats on the sea. By calving these form many icebergs, which move northward into shipping lanes of the Southern Hemisphere.

Global sea levels are currently rising, partly as a result of the melting of the Greenland and Antarctica ice sheets. We are presently in an interglacial stage of an ice age, and sea levels have risen nearly 400 feet (120 m) since the last glacial maximum 20,000 years ago, and about six inches (15 cm) in the past 100 years. The rate of sea-level rise seems to be accelerating and may presently be as much as a centimeter every five years. If all the ice on both ice sheets were to melt, global sea levels would rise by another 230 feet (70 m), inundating most of the world's major cities, and submerging large parts of the continents under shallow seas. The coastal regions of the world are densely populated and are experienc-

ing rapid population growth. Approximately 100 million people presently live within 3.2 feet (1 m) of the present-day sea level. If the sea level were to rise rapidly and significantly, the world would experience an economic and social disaster on a magnitude not yet experienced by the civilized world. Many areas would become permanently flooded or subject to inundation by storms, beach erosion would be accelerated, and water tables would rise.

The Greenland and Antarctic ice sheets have some significant differences that cause them to respond differently to changes in air and water temperatures. The Antarctic ice sheet is about 10 times as large as the Greenland ice sheet, and since it sits on the south pole, Antarctica dominates its own climate. The surrounding ocean is cold even during summer, and much of Antarctica is a cold desert with low precipitation rates and high evaporation potential. Most meltwater in Antarctica seeps into underlying snow and simply refreezes, with little running off into the sea. Antarctica hosts several large ice shelves fed by glaciers moving at rates of up to a thousand feet per year. Most ice loss in Antarctica is accomplished through calving and basal melting of the ice shelves, at rates of 10–15 inches (25–38 cm) per year.

In contrast, Greenland's climate is influenced by warm North Atlantic currents, and by its proximity to other landmasses. Climate data measured from ice cores taken from the top of the Greenland ice cap show that temperatures have varied significantly in cycles of years to decades. Greenland also experiences significant summer melting, abundant snowfall, has few ice shelves, and its glaciers move quickly at rates of up to miles per year. These fast-moving glaciers are able to drain a large amount of ice from Greenland in relatively short amounts of time.

The Greenland ice sheet is thinning rapidly along its edges, losing an average of 15–20 feet (4.5–6 m) in the past decade. In addition, tidewater glaciers and the small ice shelves in Greenland are melting at an order of magnitude faster than the Antarctic ice sheets, with rates of melting around 25–65 feet (7–20 m) per year. About half of the ice lost from Greenland is through surface melting that runs off into the sea. The other half of ice loss is through calving of outlet glaciers and melting along the tidewater glaciers and ice shelf bases.

These differences between the Greenland and Antarctic ice sheets lead them to play different roles in global sea-level rise. Greenland contributes more to the rapid short-term fluctuations in sea level, responding to short-term changes in climate. In contrast, most of the world's water available for raising sea level is locked up in the slowly changing Antarctic ice sheet. Antarctica contributes more to the gradual, long-term sea-level rise.

See also GLACIER.

Iceland The Mid-Atlantic Ridge rises above sea level on the North Atlantic island of Iceland, lying 178 miles off the

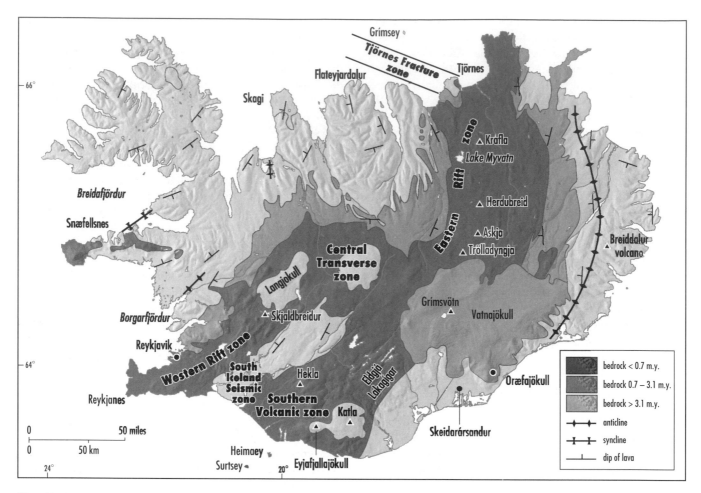

Map of Iceland showing the Western, Central, Southern, and Eastern Rift zones, where active seafloor spreading is exposed on the surface

coast of Greenland and 495 miles from the coast of Scotland. Iceland has an average elevation of more than 1,600 feet (500 m) and owes its elevation to a hot spot that is interacting with the mid-ocean ridge system beneath the island. The Mid-Atlantic Ridge crosses the island from southwest to northeast and has a spreading rate of 1.2 inches per year (3 cm/yr) with the mean extension oriented toward an azimuth of 103°. The oceanic Reykjanes ridge and sinistral transform south of the island rises to the surface and continues as the Western Rift zone. Active spreading is transferred to the Southern Volcanic zone across a transform fault called the South Iceland Seismic zone, then continues north through the Eastern Rift zone. Spreading is offset from the oceanic Kolbeinsey ridge by the dextral Tjörnes fracture zone off the island's northern coast.

During the past 6 million years the Iceland hot spot has drifted toward the southeast relative to the North Atlantic, and the oceanic ridge system has made a succession of small jumps so that active spreading has remained coincident with the plume of hottest weakest mantle material. These ridge jumps have caused the active spreading to propagate into regions of

older crust that have been remelted, forming alkalic and even silicic volcanic rocks that are deposited unconformably over older tholeiitic basalts. Active spreading occurs along a series of 5–60-mile (10–100-km) long zones of fissures, graben, and dike swarms, with basaltic and rhyolitic volcanoes rising from central parts of fissures. Hydrothermal activity is intense along the fracture zones with diffuse faulting and volcanic activity merging into a narrow zone within a few kilometers depth beneath the surface. Detailed geophysical studies have shown that magma episodically rises from depth into magma chambers located a few kilometers below the surface, then dikes intrude the overlying crust and flow horizontally for tens of miles to accommodate crustal extension of several feet to several tens of feet over several hundred years.

Many Holocene volcanic events are known from Iceland, including 17 eruptions of Hekla from the Southern Volcanic zone. Iceland has an extensive system of glaciers and has experienced a number of eruptions beneath the glaciers that cause water to infiltrate the fracture zones. The mixture of water and magma induces explosive events, including Plinian eruption clouds, phreo-magmatic, tephra-producing eruptions, and sudden floods known as jokulhlaups induced when the glacier experiences rapid melting from contact with magma. Many Icelanders have learned to use the high geothermal gradients to extract geothermal energy for heating, and to enjoy the many hot springs on the island.

ice sheet *See* ICE CAP.

ichnofossils Fossilized structures such as tracks, burrows, trails, footprints, tubes, borings, or tunnels produced by the activities of an animal. The activities may include walking, crawling, climbing, boring, tunneling, or resting on or in soft sediment. Ichnofossils, also known as trace fossils, do not preserve any parts of the animal that made the structures, but only their traces. They are typically marked as a raised line or depression in a sedimentary rock.

igneous rocks Rocks that have crystallized from a melt or partially molten material (known as magma). Magma is a molten rock within the Earth; if it makes its way to the surface, it is known as lava. Different types of magma form in different tectonic settings, and many processes act on the magma as it crystallizes to produce a wide variety of igneous rocks.

Most magma solidifies below the surface, forming igneous rocks (*ignis* is Latin for fire). Igneous rocks that form below the surface are called intrusive (or plutonic) rocks, whereas those that crystallize on the surface are called extrusive (or volcanic) rocks. Rocks that crystallize at a very shallow depth are called hypabyssal rocks. Intrusive igneous rocks crystallize

slowly, giving crystals an extended time to grow, thus forming rocks with large mineral grains that are clearly distinguishable with the naked eye. These rocks are called phanerites. In contrast, magma that cools rapidly forms fine-grained rocks. Aphanites are igneous rocks in which the component grains can not be distinguished readily without a microscope and are formed when magma from a volcano falls or flows across the surface and cools quickly. Some igneous rocks, known as porphyries, have two populations of grain size—a very large group of crystals (called phenocrysts) mixed with a uniform groundmass (or matrix), filling the space between the large crystals. This indicates two stages of cooling, as when magma has resided for a long time beneath a volcano, growing big crystals. When the volcano erupts, it spews out a mixture of the large crystals and liquid magma that then cools quickly.

Once magmas are formed from melting rocks in the Earth, they intrude the crust, and may take several forms. A pluton is a general name for a large cooled igneous intrusive body in the Earth. The specific type of pluton is based on its geometry, size, and relations to the older rocks surrounding the pluton, known as country rock. Concordant plutons have boundaries parallel to layering in the country rock, whereas discordant plutons have boundaries that cut across layering in the country rock. Dikes are tabular but discordant intrusions, and sills are tabular and concordant intrusives. Volcanic necks are conduits connecting a volcano with its underlying magma chamber (a famous example of a volcanic neck is Devils Tower in Wyoming). Some plutons are so large that they have special names. Batholiths, for example, have a surface area of more than 60 square miles (100 km²).

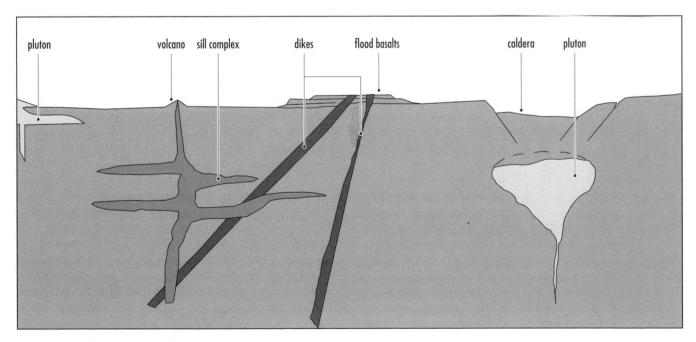

Forms of plutons and intrusive rocks showing names given to various types of intrusions

Geologists have long speculated on how such large volumes of magma intrude the crust, and what relationships these magmas have to the style of volcanic eruption. One mechanism that may operate is assimilation, where the hot magma melts surrounding rocks as it rises, causing them to become part of the magma. In doing this, the magma becomes cooler, and its composition changes to reflect the added melted country rock. It is widely thought that magmas may rise only a very limited distance by the process of assimilation. Some magmas may forcefully push their way into the crust if there are high pressures in the magma. One variation of this forceful emplacement style is diapirism, where the weight of surrounding rocks pushes down on the melt layer, which then squeezes its way up through cracks that can expand and extend, forming volcanic vents at the surface. Stoping is a mechanism whereby big blocks get thermally shattered, drop off the top of the magma chamber, and fall into the chamber, much like a glass ceiling breaking and falling into the space below.

Names of Igneous Rocks

Determining whether an igneous rock is phaneritic or aphanitic is just the first stage in giving it a name. The second stage is determining its mineral constituents. The chemical composition of a magma is closely related to how explosive and hazardous a volcanic eruption will be. The variation in the amount of silica (SiO_2) in igneous rocks is used to describe the variation in composition of igneous rocks and the magmas that formed them. Rocks with low amounts of silica (basalt, gabbro) are known as mafic rocks, whereas rocks with high concentrations of silica (rhyolite, granite) are known as silicic or felsic rocks.

Some of the variation in the nature of different types of volcanic eruptions can be understood by examining what causes magmas to have such a wide range in composition.

The Origin of Magma

Magmas come from deep within the Earth, but what conditions lead to the generation of melts in the interior of the Earth? The geothermal gradient is a measure of how temperature increases with depth in the Earth, and it provides information about the depths at which melting occurs and the depths at which magmas form. The differences in the composition of the oceanic and continental crusts lead to differing abilities to conduct the heat from the interior of the Earth, and thus different geothermal gradients. The geothermal gradients show that temperatures within the Earth quickly exceed 1,832°F (1,000°C) with increasing depth, so why are these rocks not molten? The answer is that pressures are very high, and pressure influences the ability of a rock to melt. As the pressure rises, the temperature at which the rock melts also rises. This effect of pressure on melting is modified greatly by the presence of water, because wet minerals melt at lower temperatures than dry minerals. As the pressure rises, the amount of water that can be dissolved in a melt increases. Therefore, increasing the pressure on a wet mineral has the opposite effect to increasing the pressure on a dry mineral: it decreases the melting temperature

Partial Melting

If a rock melts completely, the magma has the same composition as the rock. However, rocks are made of many different minerals, all of which melt at different temperatures. So if a rock is slowly heated, the resulting melt or magma will first have the composition of the first mineral that melts, and then the first plus the second minerals that melts, and so on. If the rock continues to melt, the magma will eventually end up with the same composition as the starting rock, but this does not always happen. What often happens is that the rock only partially melts, so that the minerals with low melting temperatures contribute to the magma, whereas the minerals with high melting temperatures do not melt and are left as a residue (or restite). In this way, the end magma can have a composition different from the rock it came from.

The phrase "magmatic differentiation by partial melting" refers to the process of forming magmas with differing compositions through the incomplete melting of rocks. For magmas formed in this way, the composition of the magma depends on both the composition of the parent rock and the percentage of melt.

Basaltic Magma

Partial melting in the mantle leads to the production of basaltic magma, which forms most of the oceanic crust. By looking at the mineralogy of the oceanic crust, which is dominated by olivine, pyroxene, and feldspar, we conclude that very little water is involved in the production of the oceanic crust. These minerals are all anhydrous, that is without water in their structure. Thus dry partial melting of the upper mantle may lead to the formation of oceanic crust. By collecting samples of the mantle that have been erupted through volcanoes, we know that it has a composition of garnet peridotite (olivine + garnet + orthopyroxene). By taking samples of this back to the laboratory and raising its temperature and pressure so that it is equal to 62 miles (100 km) depth, we find

Relationships of Igneous Rocks

MAGMA TYPES

	SiO₂%	volcanic rock	plutonic rock
mafic	45–52%	basalt	gabbro
intermediate	53–65%	andesite	diorite
felsic	65%	rhyolite	granite

that 10 percent to 15 percent partial melt of this garnet peridotite yields a basaltic magma.

Magma that forms at 50 miles (80 km) depth is less dense than the surrounding solid rock, so it rises, sometimes quite rapidly (at rates of half a mile per day measured by earthquakes under Hawaii). In fact, it may rise so fast that it does not cool off appreciably, erupting at the surface at more than 1,832°F (1,000°C); this is where basalt comes from.

Granitic Magma

Granitic magmas are very different from basaltic magmas. They have about 20 percent more silica, and the minerals in granite (mica, amphibole) have a lot of water in their crystal structures. Also, granitic magmas are found almost exclusively in regions of continental crust. From these observations we infer that the source of granitic magmas is within the continental crust. Laboratory experiments suggest that when rocks with the composition of continental crust start to melt at temperature and pressure conditions found in the lower crust, a granitic liquid is formed, with 30 percent partial melting. These rocks can begin to melt by either the addition of a heat source, such as basalt intruding the lower continental crust, or by burying water bearing minerals and rocks to these depths.

These granitic magmas rise slowly (because of their high SiO$_2$ and high viscosities), until they reach the level in the crust where the temperature and pressure conditions are consistent with freezing or solidification of a magma with this composition. This is about 3–6 miles (5–10 km) beneath the surface, which explains why large portions of the continental crust are not molten lava lakes. There are many regions with crust above large magma bodies (called batholiths) that are heated by the cooling magma. An example is the Yellowstone National Park, where there are hot springs, geysers, and many features indicating that there is a large hot magma body at depth. Much of Yellowstone Park is a giant valley called a caldera, formed when an ancient volcanic eruption emptied an older batholith of its magma, and the overlying crust collapsed into the empty hole formed by the eruption.

Andesitic Magma

The average composition of the continental crust is andesitic, or somewhere between the composition of basalt and rhyolite. Laboratory experiments show that partial melting of wet oceanic crust yields an andesitic magma. Remember that oceanic crust is dry, but after it forms it interacts with seawater, which fills cracks to several kilometers depth. Also, the sediments on top of the oceanic crust are full of water, but these are for the most part non-subductable.

Solidification of Magma

Just as rocks partially melt to form different liquid compositions, magmas may solidify to different minerals at different times to form different solids (rocks). This process also

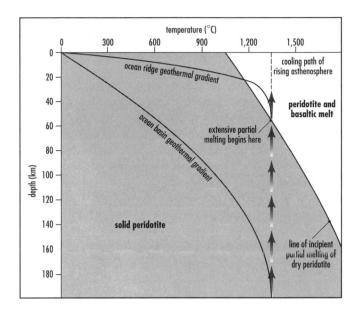

Typical geothermal gradient beneath the oceans, showing also the line of incipient partial melting of a dry peridotite

results in the continuous change in the composition of the magma—if one mineral is removed, the resulting composition is different. Removal of solidified crystals results in a new magma composition.

The removal of crystals from the melt system may occur by several processes, including the squeezing of melt away from the crystals or by sinking of dense crystals to the bottom of a magma chamber. These processes lead to magmatic differentiation by fractional crystallization, as first described by Norman L. Bowen. Bowen systematically documented how crystallization of the first minerals changes the composition of the magma and results in the formation of progressively more silicic rocks with decreasing temperature.

See also MINERALOGY; OPHIOLITES; PLATE TECTONICS; VOLCANO.

impact crater   The collision of meteorites with Earth produces impact craters, which are generally circular bowl-shaped depressions. There are more than 200 known impact structures on Earth, although processes of weathering, erosion, volcanism, and tectonics have undoubtedly erased many thousands more. The Moon and other planets show much greater densities of impact craters, and since the Earth has a greater gravitational pull than the Moon, it should have been hit by many more impacts.

Meteorite impact craters have a variety of forms but are of two basic types. Simple craters are circular bowl-shaped craters with overturned rocks around their edges and are generally less than 3 miles (5 km) in diameter. They are thought to have been produced by impact with objects less than 100 feet (30 m) in diameter. Examples of simple craters include the Barringer Meteor Crater in Arizona and Roter Kamm in Namibia. Complex craters are larger, generally greater than 2 miles (3 km) in diameter. They have an uplifted peak in the center of the crater and have a series of concentric rings around the excavated core of the crater. Examples of complex craters include Manicougan, Clearwater Lakes, and Sudbury in Canada, Chicxulub in Mexico, and Gosses Bluff in Australia.

The style of impact crater depends on the size of the impacting meteorite, the speed it has as it strikes the surface, and to a lesser extent the underlying geology and the angle at which the meteor strikes the Earth. Most meteorites hit the Earth with a velocity between 2.5 and 25 miles per second (4–40 km/s), releasing tremendous energy when they hit. Meteor Crater in Arizona was produced about 50,000 years ago by a meteorite 100 feet (30 m) in diameter that released the equivalent of four megatons of TNT. The meteorite body and a large section of the ground at the site were suddenly melted by shock waves from the impact, which released about twice as much energy as the eruption of Mount Saint Helens. Most impacts generate so much heat and shock pressure that the entire meteorite and a large amount of the rock it hits are melted and vaporized. Temperatures may exceed thousands of degrees in a fraction of a second as pressures increase a million times atmospheric pressure during passage of the shock wave. These conditions cause the rock at the site of the impact to accelerate downward and outward, and then the ground rebounds and tons of material are shot outward and upward into the atmosphere.

Impact cratering is a complex process. When the meteorite strikes it explodes, vaporizes, and sends shock waves through the underlying rock, compressing the rock, crushing it into breccia, and ejecting material (conveniently known as ejecta) back up into the atmosphere, from where it falls out as an ejecta blanket around the impact crater. Large impact events may melt the underlying rock forming an impact melt and may form distinctive minerals that only form at exceedingly high pressures.

After the initial stages of the impact crater forming process, the rocks surrounding the excavated crater slide and fall into the deep hole, enlarging the diameter of the crater, typically making it much wider than it is deep. Many of the rocks that slide into the crater are brecciated or otherwise affected by the passage of the shock wave and may preserve these effects as brecciated rocks, high-pressure mineral phases, shatter cones, or other deformation features.

Impact cratering was probably a much more important process in the early history of the Earth than it is at present. The flux of meteorites from most parts of the solar system was much greater in early times, and it is likely that impacts totally disrupted the surface in the early Precambrian. At present the meteorite flux is about 100 tons per day (somewhere between 10^7 and 10^9 kg/yr), but most of this material burns up as it enters the atmosphere. Meteorites that are about a tenth of an inch to several feet (.25–60 cm) in diameter make a flash of light (a shooting star) as they burn up in the atmosphere, and the remains fall to Earth as a tiny glassy sphere of rock. Smaller particles, known as cosmic dust, escape the effects of friction and slowly fall to Earth as a slow rain of extraterrestrial dust.

Meteorites must be greater than 3.2 ft (1 m) in diameter to make it through the atmosphere without burning up from friction. The Earth's surface is currently hit by about one small meteorite per year. Larger impact events occur much less frequently, with meteorites 328 feet (100 m) in diameter hitting once every 10,000 years, 3,280 feet (1,000 m) in diameter hitting the Earth once every million years, and 6.2 miles (10 km) in diameter hitting every 100 million years. Meteorites of only hundreds of meters in diameter could create craters about 0.6–1.2 miles (1–2 km) in diameter, or if they hit in the ocean, they would generate tsunami more than 16.5 feet (5 m) tall over wide regions. The statistics of meteorite impact show that the larger events are the least frequent.

index fossil   A fossil that precisely identifies the age or range of strata in which it is found. Index fossils must be morphologically distinctive, relatively common, and have populated a broad geographic range, but be limited in stratigraphic or time range. The most useful index fossils are those that floated or swam in marine environments and evolved or changed morphologically in short time intervals. Graptolites, ammonites, and radiolarians have proven to be excellent and widely used index fossils.

India's Eastern and Western Ghats   Two converging mountain ranges that run along the east and west coasts of southern India and form the east and west boundaries of the Deccan plateau. The two ranges are joined by the Nilgiri Hills in the south, and the highest point in the ranges is Anai Mudi with an elevation of 8,841 feet (2,697 m). The eastern Ghats have an average elevation of 2,000 feet (600 m) and generally lie at 50 to 150 miles (80–240 km) from the Coromandel coastline but locally reach the coast forming steep coastal cliffs. The Eastern Ghats are crossed by the Godavari, Krishna, and Kaveri Rivers and are covered by many hardwood trees. The Western Ghats extend along the Malabar coast from the Tapi River to Cape Comorin at the southern tip of India and are generally very close to the coastline. Elevations in the northern part of the Western Ghats reach 4,000 feet (1,200 m) and 8,652 feet (2,637 m) at Doda Beta in the south. The western side of the Western Ghats receive heavy monsoonal rainfalls, but the eastern side of the Western Ghats is generally dry.

Geologically, the Western Ghats extend from the Deccan flood basalt plateau in the north to the Precambrian basement shield, including the Dharwar craton, in the south. Isotopic ages of gneisses and greenstone belts in the Dharwar craton range from 2.6 billion to 3.4 billion years old. The Dharwar craton is well known for gold deposits associated with greenstone belts and banded iron formations, with the most well-known greenstone belt being the Chitradurga. The Dharwar craton is divided into eastern and western parts by the elongate north-northwest-striking 2.6-billion-year-old Closepet granite, probably of Andean arc affinity. Late Proterozoic metamorphism, locally to granulite grade and including large areas of charnockites, affects much of the southern part of peninsular India. Rocks of the Dharwar craton are overlain by Paleozoic sedimentary deposits of Gondwanan affinity. The Deccan flood basalts erupted at the end of the Cretaceous and overlie Gondwana and continental margin sequences that began developing with the breakup of Gondwana. The Eastern Ghats are entirely within the Precambrian basement rocks of the Indian subcontinent and the Aravalli craton in the north. The Aravalli craton is somewhat younger than the Dharwar craton, with isotopic ages falling in the range of 3.0 billion years to 450 million years.

See also CRATONS; FLOOD BASALT; GONDWANA; GREENSTONE BELTS.

Further Reading

Naqvi, S. M., S. Mahmood, and John J. W. Rogers. *Precambrian Geology of India*. Oxford: Oxford University Press, 1983.
Rogers, John J. W. "The Dharwar Craton and the Assembly of Peninsular India." *Journal of Geology* 94 (1986).

Indo-Gangetic Plain The Indo-Gangetic Plain is the active foreland basin of the India-Asia collision, with sediments derived from erosion of the Himalaya Mountains and carried by numerous rivers that feed into the Indus and Ganges Rivers. Alluvial deposits of the Indo-Gangetic Plain stretch from the Indus River in Pakistan to the Punjab Plain in India and Pakistan, to the Haryana Plain and Ganges delta in Bangladesh. Sediments in the foreland basin are up to 24,500 feet (7,500 m) thick over the basement rocks of the Indian Shield, thinning toward the southern boundary of the basin plain. The plain has very little relief, with only occasional bluffs and terraces related to changes in river levels.

The northern boundary of the plain is marked by two narrow belts known as Terai, containing small hills formed by coarse remnant gravel deposits emerging from mountain streams. Many springs emanate from these gravel deposits forming large swampy areas along the major rivers. In most places the Indo-Gangetic Plain is about 250 miles (400 km) wide. The southern boundary of the plain is marked by the front of the Great Indian Desert in Rajasthan, then continues eastward to the Bay of Bengal along the hills of the Central Highlands.

It is possible to divide the Indo-Gangetic Plain into three geographically and hydrologically distinct sections. The Indus Valley in the west is fed by the Indus River that flows out of Kashmir, the Hindu Kush, and the Karakoram Range. The Punjab and Haryana plains are fed by runoff from the Siwaliks and Himalaya Mountains into the Ganges River, and the Lower Ganga and Brahmaputra drainage systems in the east. The lower Ganga plains and Assam Valley are lush and heavily vegetated, and the waters flow into the deltaic regions of Bangladesh.

Clastic sediments of the foreland basin deposits under the Indo-Gangetic Plain include Eocene-Oligocene (about 50–30-million-year-old) deposits, grading up to the Miocene to Pleistocene Siwalik clastic rocks eroded from the Siwalik and Himalayan Ranges. The basement of the Indian Shield dips about 15° beneath the Great Boundary and other faults marking the deformation front at the toe of the Himalayas.

See also CONVERGENT PLATE MARGIN PROCESSES; FORELAND BASIN; HIMALAYA MOUNTAINS.

infrared The part of the electromagnetic spectrum with wavelengths between 0.7 microns and about one millimeter. Infrared radiation is used in earth sciences for remote sensing of surfaces, since it has some properties that optical and other wavelength sensors do not possess. Some types of shortwave infrared radiation (from 2.1–2.4 microns) lead to molecular interactions with minerals in rocks, particularly those minerals with Al-OH, Mg-OH, and C-O bonds. When viewed in the infrared, rocks with minerals with these bonds will show a number of narrow absorption features whose wavelengths depend on the content of minerals with these bonds. Infrared radiation can therefore be used to detect and differentiate between minerals such as micas, clays, magnesium silicates, and carbonates. At longer infrared wavelengths, the signature is dominated by the temperature of the surface, in reflected sunlight. Since many minerals have very different thermal emission spectra, thermal imagery also provides a potentially powerful way to differentiate between rock and mineral types.

See also REMOTE SENSING.

inselberg Steep-sided mountains or ridges that rise abruptly out of adjacent monotonously flat plains in deserts. Ayres Rock in central Australia is perhaps the world's best-known inselberg. These are produced by differential erosion, leaving behind as a mountain rocks that for some reason are more resistant to erosion. Many inselbergs are produced by a process known as circumdenudation, where surrounding rocks are removed from all directions, by ice or rivers, leaving only the prominent isolated knob or mountain behind.

intrusive rocks *See* PLUTON.

inversion Under normal conditions, the temperature of the atmosphere decreases with increasing height. Occasionally, conditions are set up such that the temperature actually increases with height, a condition known as a temperature inversion, or simply an inversion. A radiation temperature inversion may occur on clear, cold, generally windless nights. On these nights the ground may undergo strong radiation cooling, cooling the dense surface layer of air. At the same time the slightly higher layers of air that are not in contact with the ground do not cool as much, forming a radiation temperature or a nocturnal inversion.

Atmospheric inversions often cause increased air pollution. The temperature profile of the atmosphere helps determine the atmospheric stability—when the temperature increases with height, the air is very stable. Any air parcel that tries to rise up into the inversion layer will at some point be cooler and heavier than the air around it in the inversion and will not be able to rise any further. This, and warm air from smokestacks, cars, and other pollutants that normally rises and is removed from the local area and dispersed, gets trapped below the inversion and stays close to the source. This can form particularly acute conditions in places where cities are located in valleys and the air has no way to escape, increasing levels of toxic chemicals.

ion microprobe   An ion microprobe allows the study and analysis of complex isotopic composition of most of the elements in the periodic table. The ion microprobe is used widely in geology to decipher the geochronological dating of samples and in the study of the formation and creation of mineralizing systems. The ion microprobe is used mainly in the study of stable isotope geochemistry that examines the changes in isotopic composition of elements that are caused by either physical or chemical processes. The use of the ion microprobe is increasing in the study of isotopic composition of materials because modern technology has increased the accuracy and made the use of the ion probe easier and less expensive.

ionosphere   An electrified region in the upper atmosphere where a large number of ions and free electrons exist. The ionosphere is above approximately 37–50 miles (60–80 km) and extends to the outer reaches of the atmosphere. It absorbs much of the harmful ultraviolet radiation and high-energy particles from the Sun, protecting life on Earth from their ill effects. Some of the ultraviolet radiation that passes through this region of the atmosphere is absorbed by the ozone layer. The ionosphere forms where high-energy particles from the Sun begin colliding with molecules in the Earth's atmosphere, splitting some electrons off their molecules, forming ions. The ionosphere has both positively charged ions that have lost electrons and negatively charged ions that have absorbed electrons.

The auroras are produced in the ionosphere when atmospheric ions charged by interaction with solar particles decay to their former state, emitting light in the process. The ionosphere also plays a major role in radio communications. The lower part of the ionosphere reflects AM radio waves, while absorption in the daytime weakens the signal. At night, the radio waves penetrate more deeply and are not strongly absorbed, so at night AM radio signals are able to travel much farther than in the day. FM radio waves have a shorter wavelength than AM waves and are able to penetrate the ionosphere without reflection.   *See also* ATMOSPHERE; AURORA.

island arc   *See* CONVERGENT PLATE MARGIN PROCESSES; JAPAN'S PAIRED METAMORPHIC BELT.

isograd   *See* METAMORPHISM.

isostasy   The principle of hydrostatic equilibrium applied to the Earth, referring to the position of the lithosphere essentially floating on the asthenosphere, much like the level at which ice floats on water. Isostatic forces are of major importance in controlling the topography of the Earth's surface. There are several different models for how topography is supported, referred to as isostatic models. The simplest models have blocks of crust that are essentially floating as isolated blocks in a fluid substrate (the asthenosphere) and are free to migrate vertically and do not interact with neighboring blocks. There are two main variations of these simple isostatic models. In the Pratt model, crustal blocks of different density are assumed to extend to a constant depth known as the depth of compensation, and the height of the topography varies inversely with the density of each block. Thus, high-density oceanic crust resides at a lower level than lower-density continental crust. In the Airy model, the level of isostatic compensation varies for each block, but the crustal layer is assumed to have a constant density. This way, thick blocks have high topography and a thick root to compensate the topography, whereas thin crustal blocks have subdued topography. Both of these models are simplistic descriptions of a complex lithosphere, and they were both derived in the 1700s before an appreciation of plate tectonics. The Airy model is generally more applicable than the Pratt model, but the Airy model does not accommodate variations in crustal density that are known to exist, such as that between continents and oceans.

Isostatic anomalies are variations in measured gravity values from those expected using an assumed isostatic model and depth of compensation. The anomaly indicates that the model used or the compensation depth assumed needs to be adjusted in the model.
See also GRAVITY ANOMALY.

isostatic anomaly   *See* ISOSTACY.

isotope   Atoms are the smallest unit of matter that can take part in chemical reactions, and they cannot be broken down chemically into smaller parts. The simple structure of an atom consists of a nucleus, containing the heavy particles including the protons and neutrons, orbited by shells of negatively charged electrons. Protons are positively charged, whereas neutrons have no charge. Every element has its own characteristic nucleus with the same number of protons, known as the atomic number. For an atom to be electrically neutral it must have the same number of protons and electrons, and if the charge is unbalanced the particle is called an ion. The atomic weight refers to the number of protons plus the numbers of neutrons. When atoms of the same element (same number of protons) have different numbers of neutrons, they are referred to as isotopes of the same element. Some isotopes are stable, whereas others are unstable and decay emitting radioactivity. Despite this, all isotopes of the same element exhibit identical chemical properties.

See also GEOCHEMISTRY.

J

jade   A very hard and compact gemstone with uneven colors, ranging from dark green to greenish white. It has been used for thousands of years as an ornamental stone for jewelry, notably by native Central Americans, and western North and South American tribes. The source of much of the jade used by the Aztecs has long been a mystery, but sources were recently discovered in remote parts of Guatemala. Jade is made of either the pyroxene mineral jadeite, or the amphibole mineral nephrite, although many other minerals have been falsely marketed as jade. Some of the minerals that have been used as substitutes for jade include serpentine, vesuvianite, sillimanite, pectolite, garnet, and sausserite.

Japan's paired metamorphic belt   In a classic 1961 work, Akiho Miyashiro recognized that many circum-Pacific orogenic belts can be divided into two component parts. On the oceanward side of these orogens, Miyashiro recognized that the rocks had experienced low-temperature high-pressure metamorphism, whereas on the continentward side, he recognized suites of high-temperature low-pressure metamorphism. He named these orogens paired metamorphic belts and used his native Japan as the prime example of such a paired metamorphic belt.

Japan is an arc-trench system that rests as a sliver of the North American plate above the Pacific plate outboard of the Eurasian plate. This tectonic scenario has existed since the mid-Tertiary. From the beginning of the Phanerozoic to the Tertiary, Japan was part of the Eurasian continental margin and was involved in interactions between Eurasia and the Tethys and Panthalassian Oceans. A few occurrences of middle Paleozoic rocks are known from Japan, but the vast majority of strata are younger than middle Paleozoic. Most of the rocks are aligned in strongly deformed structural belts that parallel the coast for 1,850 miles (3,000 km) and include fossiliferous marine strata, weakly to strongly metamorphosed pelitic to psammitic rocks, and granitic intrusions. Since most rocks are strongly deformed in fold-thrust belt structures, it is inferred that the more strongly metamorphosed units have been uplifted from deeper in the arc-accretionary wedge system, or metamorphosed near the plutons. The complexly deformed zones are overlain by little-deformed Mesozoic-Cenozoic non-marine to shallow-marine basin deposits. In addition, abundant Tertiary volcanic and volcaniclastic deposits are present along the western side of the islands, and in the Fossa Magna in central Honshu island.

In Japan the Sanbagawa belt represents a high-pressure, low-temperature metamorphic belt, and the adjacent Ryoke-Abukuma belt represents a high-temperature, low-pressure metamorphic belt. Together, these two contrasted metamorphic belts represent Japan's paired metamorphic belt. In Japan Miyashiro and others have deduced that these adjacent belts with contrasted metamorphic histories formed during subduction of the oceanic plates beneath Japan. The low-temperature metamorphic series forms in the trench and immediately above the subduction zone, where the cold subducting slab insulates overlying sediments from high mantle temperatures as they are brought down to locally deep high-pressure conditions. These rocks then get accreted to the overriding plate and may become exhumed and exposed at the surface in the high-pressure low-temperature belt. It is not uncommon to find blueschist facies rocks containing the diagnostic mineral galucophane, formed with pressures greater than four kilobars and temperatures of 390–840°F (200–450°C).

The adjacent Ryoke-Abukuma belt contains low-pressure high-temperature, as well as medium-pressure, high-temperature metamorphic rocks. These metamorphic rocks form near the axis of the arc in close association with subduction-derived magmas. Since the rocks in this belt form over a considerable crustal thickness, and are associated with many high-temperature magmas, they were metamorphosed at a range of pressures and generally high temperatures.

A number of other paired metamorphic belts have been recognized throughout the world. In the western United States, the Franciscan complex contains rocks metamorphosed at high pressures and low temperatures, whereas rocks in the Sierra Nevada and Klamath Mountains contain high-temperature low-pressure metamorphic facies. Other paired metamorphic belts are recognized in Alaska, New Zealand, Indonesia, Chile, Jamaica, and the Alps.

See also CONVERGENT PLATE MARGIN PROCESSES; METAMORPHISM.

Further Reading

Miyashiro, Akiho. *Metamorphism and Metamorphic Belts*. London: Allen and Unwin, 1973.

jet streams   High-level, narrow, fast-moving currents of air that are typically thousands of kilometers long, hundreds of kilometers wide, and several kilometers deep. Jet streams typically form near the tropopause, 6–9 miles (10–15 km) above the surface, and may reach speeds of 115–230 miles per hour (100–200 knots). Rapidly moving cirrus clouds often reveal the westerly jet streams moving air from west to east. Several jet streams are common—the subtropical jet stream forms about 8 miles (13 km) above the surface, at the pole-ward limit of the tropical Hadley Cell, where a tropospheric gap develops between the circulating Hadley Cells. The polar jet stream forms at about a 6–mile (10–km) height, at the tropospheric gap between the cold polar cell and the mid-latitude Ferrel cell. The polar jet stream is often associated with polar front depressions. The jet streams, especially the subtropical jet, are fairly stable and drive many of the planet's weather systems. The polar jet stream tends to meander and develop loops more than the subtropical jet. A third common jet stream often develops as an easterly flow, especially over the Indian subcontinent during the summer monsoon.

Joides Resolution   The *Joides Resolution* is a naval research ship named after the famous H.M.S. *Resolution* that was commanded by Captain James Cook and served as his main vessel as he explored the Pacific Ocean and Antarctic area. The ship is equipped with a drill that enables it to obtain core samples and sediment samples from the seafloor more than three miles (5 km) below the sea surface. Several coordinated thrusters enable the ship to remain stationary in violent seas to obtain accurate and precise drilling samples. The samples enable scientists to study such things as earthquakes, volcanoes, seafloor spreading, and natural resources.

The *Joides Resolution* is a privately owned ship that is leased to the Ocean Drilling Program (ODP) that is funded by the U.S. National Science Foundation and several other partner countries.

joints   *See* FRACTURES.

Jupiter   The fifth planet from the Sun, Jupiter is the gaseous giant of the solar system, named after the most powerful Roman god of the Pantheon. It has more than twice the mass of all the other planets combined, estimated at 1.9×10^{27} kilograms, or 318 Earth masses, and a radius of 71,400 kilometers, or 11.2 Earth radii. Volumetrically, it would take 1,400 Earths to fill the space occupied by Jupiter. Jupiter is the third brightest object in the night sky, following the Moon and Venus. Four of its many moons are visible from the Earth. It orbits the Sun at a distance of 483 million miles (778 million km, at its semi-major axis) and takes 11.9 Earth years to complete each orbit.

Visual observations of the surface of Jupiter indicate that the gaseous surface has a rapid differential rotation rate, with the equatorial zones rotating with a period of 9 hours and 50 minutes, and higher latitudes rotating with a period of 9 hours and 56 minutes. The interior of the planet is thought to be rotating with a period of 9 hours 56 minutes, since the

Jupiter, showing swirling clouds, including the Great Red Spot. This true-color simulated view of Jupiter is composed of four images taken by NASA's *Cassini* spacecraft on December 7, 2000. To illustrate what Jupiter would have looked like if the cameras had a field-of-view large enough to capture the entire planet, the cylindrical map was projected onto a globe. The resolution is about 89 miles (144 km) per pixel. Jupiter's moon Europa is casting the shadow on the planet. *Cassini* is a cooperative mission of NASA, the European Space Agency, and the Italian Space Agency. The Jet Propulsion Laboratory (JPL), a division of the California Institute of Technology in Pasadena, manages *Cassini* for NASA's Office of Space Science, Washington, D.C. *(Image courtesy of NASA/JPL/University of Arizona)*

magnetic field rotates at this rapid rate. The rapid rotation has distorted the planet so that the equatorial radius (44,365 miles, or 71,400 km) is 6.5 percent greater than the polar radius (41,500 miles, or 66,800 km).

The outer layer of Jupiter is made of gases, with temperatures at the top of the cloud layers estimated to be 185°F) and a pressure of 10 bars. This is underlain by a layer of molecular hydrogen extending to 12,425 miles (20,000 km) below the surface. The temperature at this depth is estimated to be 19,340°F (11,000°K), with 4 Megabars pressure. Below this a layer of metallic hydrogen extends to 37,280 miles (60,000 km), with basal temperatures of 44,540°F (25,000°K), and 12 Megabars pressure. An internal rocky core extends another 6,215 miles (10,000 km).

Jupiter's surface and atmosphere are visibly dominated by constantly changing colorful bands extending parallel to the equator, and a great red spot, which is a huge hurricane-like storm. The bands include yellows, blues, browns, tans, and reds, thought to be caused by chemical compounds at different levels of the atmosphere. The most abundant gas in the atmosphere is molecular hydrogen (86.1 percent), followed by helium (13.8 percent). Other chemical elements such as carbon, nitrogen, and oxygen are chemically mixed with helium. Hydrogen is so abundant on Jupiter because the gravitational attraction of the planet is so large that it can retain hydrogen, and most of the planet's original atmosphere has been retained.

Since Jupiter has no solid surface layer, the top of the troposphere is conventionally designated as the surface. A haze layer lies above the troposphere, then grades up into the stratosphere. The colorful bands on Jupiter are thought to reflect views deep into different layers of the atmosphere. Several to tens of kilometers of white wispy clouds of ammonia ice are underlain by a layer of red ammonium hydrosulfide ice, then blue water ice extending to about 62 miles (100 km) below the troposphere. The cloud layers are constantly changing, reflecting different weather and convective systems in the atmosphere, creating the bands, and the Great Red Spot. The leading hypothesis about the origin of the bands is that the light or bright colored bands represent regions where the atmosphere is warm and upwelling, whereas the darker bands represent places where the atmosphere is downwelling back to deeper levels. The rapid rotation of Jupiter causes these convective bands to be wrapped around the planet in elongate bands, unlike on Earth where they tend to form isolated convective cells. The rotation of the planet causes a strong zonal flow, with most wind belts moving the atmosphere to the east at tens to several hundreds of kilometers per hour. Several belts are moving westward however, with the largest and fastest being the 124-mile-per-hour (200 km/hr) westward-moving belt associated with the northern edge of the Great Red Spot. The southern edge of the Great Red Spot is in an eastward flowing zone (also about 200

km/hr), and the Great Red Spot rotates with the planet, caught between these two powerful belts. Many smaller oval-shaped vortices spin off the edges of the Great Red Spot and are thought to be smaller storms that may persist for several or several tens of years. Many similar features are found elsewhere on the planet.

Jupiter has many moons, with the Galilean satellites resembling a miniature solar system. The four largest moons include Io (1.22 Earth/Moon masses), Europa (0.65 Earth/Moon masses), Ganymede (2.02 Earth/Moon masses), and Callisto (1.47 Earth/Moon masses). Each moon is distinct and fascinating, showing different effects of the gravitational attraction of nearby Jupiter. Io and Europa are rocky planet-like bodies, with Io exhibiting active sulfur-rich volcanism and very young surface material. The energy for the volcanism is thought to be the gravitational attraction of Jupiter, and the sulfur particles emitted from the volcanoes get entrained as charged ions in Jupiter's magnetosphere, forming a plasma torus ring around the planet. Europa has an icy surface with a rocky interior, crisscrossed by cracks on the surface that may be analogous to pressure ridges on terrestrial ice flows. The surface is not heavily cratered and must be relatively young. Ganymede and Callisto are both thought to be icy planets with low densities, and Ganymede is heavily cratered, reflecting that it has an old surface. Callisto also has many craters, including two huge ones with multiple rings, reflecting cataclysmic impacts in its history.

Jurassic The second of three periods of the Mesozoic, lasting from 206 to 144 million years ago. The name Jurassic comes from studies of the Jura Mountains of Switzerland by Alexander von Humboldt, published in 1795, and compared with strata of England and Germany by Leopold von Buch in 1839. Rocks of the Jurassic are well preserved throughout western Europe, and many of the subdivisions of Jurassic time are based on studies in that region. Ammonites evolved rapidly in this time and are widely used as index fossils for the Jurassic, which includes more than 70 ammonite zones.

Pangea was breaking up and dispersing in the Jurassic and had already broken into the northern Laurentian and southern Gondwanan landmasses, separated by the growing Tethys Ocean. North America separated from Africa, forming the incipient Atlantic Ocean. Rifting of Antarctica, India, and Australia also contributed to the pattern of global breakup and separation, contributing to global sea-level rise. At the same time, convergent margin tectonism was affecting the western Cordilleran margins of North and South America. Massive volcanism associated with plate breakup and eruption of the Karoo flood basalt volcanism contributed to climate warming.

The Jurassic is famous as a time when dinosaurs and pterosaurs roamed freely on the land, with many herbivores and giant bipedal carnivores. Gymnosperms including conifers and ginkgos, along with ferns and horsetails, covered much of the land, creating great forests. Marine life also flourished in the Jurassic, with coral reefs, algae, sponges, and bryozoans common, as were free-moving benthic forms including bivalves, gastropods, belemnites, and foraminifera. Fish were generally large with enameled scales, and several types of sharks roaming the seas. Marine vertebrates included the long-necked plesiosaurs, the large ichthyosaurs, giant crocodiles, turtles, and other reptiles.

See also MESOZOIC; SUPERCONTINENT CYCLE.

K

Kaapvaal craton, South Africa The Archean Kaapvaal craton of southern Africa contains some of the world's oldest and most intensely studied Archean rocks, yet nearly 86 percent of the craton is covered by younger rocks. The craton covers approximately 363,000 square miles (585,000 km²) near the southern tip of the African continent. The craton is bordered on the north by the high-grade Limpopo mobile belt, initially formed when the Kaapvaal and Zimbabwe cratons collided at 2.6 billion years ago. On its southern and western margins the craton is bordered by the Namaqua-Natal Proterozoic orogens, and it is overlapped on the east by the Lebombo sequence of Jurassic rocks recording the breakup of Gondwana.

Most of the rocks comprising the Archean basement of the Kaapvaal craton are granitoids and gneisses, along with less than 10 percent greenstone belts known locally as the Swaziland Supergroup. The oldest rocks are found in the Ancient Gneiss complex of Swaziland, where a 3.65–3.5-billion-year-old bimodal gneiss suite consisting of interlayered tonalite-trondhjemite-granite and amphibolite are complexly folded together with migmatitic gneiss, biotite-hornblende tonalitic gneiss, and lenses of 3.3–3.0-billion-year-old quartz monzonite. Several folding and deformation events are recognized from the Ancient Gneiss complex, whose history spans an interval of 700 million years, longer than the entire Phanerozoic.

There are six main greenstone belts in the Kaapvaal craton, the most famous of which is the Barberton greenstone belt. Although many studies have attempted to group all of the greenstone sequences of the Kaapvaal craton into the term *Swaziland Supergroup*, there is little solid geochronologic or other evidence that any of these complexly deformed belts are contemporaneous or related to each other, so this usage is not recommended. Other greenstone belts include the Murchison,

Sutherland, Amalia, Muldersdrif, and Pietersburg belts. U-Pb isotopic ages from these belts span the interval of 3.5–3.0 billion years ago, a period of 500 million years. The greenstone belts include structurally repeated and complexly folded and metamorphosed sequences of tholeiitic basalts, komatiites, picrites, cherts (or metamorphosed felsic mylonite), felsic lava, clastic sediments, pelites, and carbonates. Possible partial ophiolite sequences have been recognized in some of these greenstone belts, particularly in the Jamestown section of the Barberton belt.

One of the long-held myths about the structure of greenstone belts in the Kaapvaal craton is that they represent steep synclinal keels of supracrustal rocks squeezed between diapiric granitoids. However, detailed structural studies of the Murchison greenstone belt have established that there is a complete lack of continuity of strata from either side of the supposed syncline of the Murchison belt, and that the structure is much more complex than the pinched-synform model predicts. Downward-facing structures and fault-bounded panels of rocks with opposing younging directions have been documented, emphasizing that the "stratigraphy" of this and other belts cannot be reconstructed until the geometry of deformation is better understood; early assumptions of a simple synclinal succession are invalid.

Detailed mapping in a number of greenstone belts in the Kaapvaal craton has revealed early thrust faults and associated recumbent nappe-style folds. Most do not have any associated regional metamorphic fabric or axial planar cleavage, making their identification difficult without very detailed structural mapping. Furthermore, in some cases, late intrusive rocks have utilized the zone of structural weakness provided by the early thrusts for their intrusion.

Elucidation of the structure of South African greenstone belts has undergone a recent revolution. Several different tectonic events are responsible for the present structural geometry of the greenstone belts. Early regional recumbent folds, thrust faults, inverted stratigraphy, juxtaposition of deep and shallow water facies, nappes, and precursory olistostromes related to the northward tectonic emplacement of the circa 3.5 Ga Barberton greenstone collage on gneissic basement have been documented. The thrusts may have been zones of high fluid pressure resulting from hydrothermal circulation systems surrounding igneous intrusions, and are locally intruded by syn-tectonic 3.43–3.44-billion-year-old felsic igneous rocks. Confirmation of thrust-style age relationships comes from recent U-Pb zircon work, which has shown that older (circa 3.482 ± 5 Ga) Komatii Formation rocks lie on top of younger (circa 3.453 ± 6 Ga) Theespruit Formation.

The Pietersburg greenstone belt is located north of the Barberton and Murchison belts, near the high-grade Limpopo belt. Greenschist to amphibolite facies, oceanic-affinity pillow basalts, gabbros, peridotites, tuffs, metasedimentary rocks, and banded iron formation are overlain unconformably by a terrestrial clastic sequence deposited during D_2 northward-directed thrusting between 2.98 billion and 2.69 billion years ago. Coarse clastic rocks deposited in intermontaine basins are imbricated with the oceanic affinity rocks and were carried piggyback on the moving allochthon. Syn-thrusting depositional troughs became tightened into synclinal structures during the evolution of the thrust belt, and within the coarse-clastic section it is possible to find thrusts that cut local unconformities, and unconformities that cut thrusts.

The granite-greenstone terrane is overlain unconformably by the 3.1-billion-year-old Pongola Supergroup, which has been proposed to be the oldest well-preserved continental rift sequence in the world. Deposition of these shallow water tidally influenced sediments was followed by a widespread granite intrusion episode at 3.0 billion years ago. The next major events recorded include the formation of the

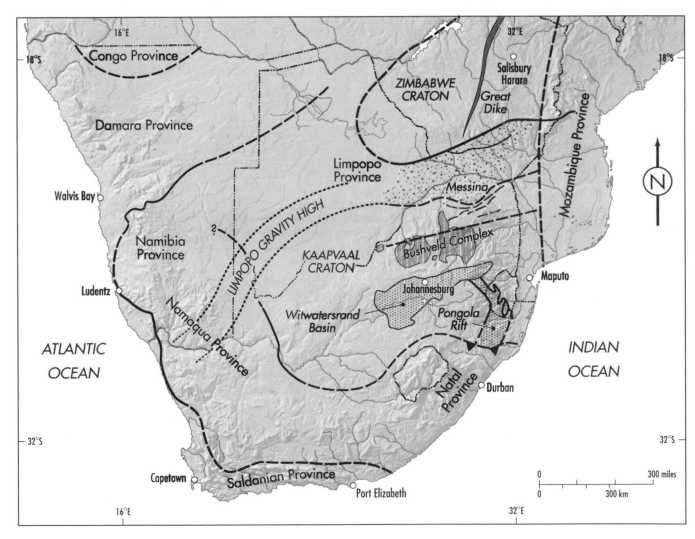

Map of southern Africa showing the main features of the Kaapvaal craton and their relationships with surrounding tectonic elements

West Rand Group of the Witwatersrand basin on the cratonward side of an Andean arc around 2.8 billion years ago, then further deposition of the extremely auriferous sands of the Central Rand Group in a collisional foreland basin formed when the Zimbabwe and Kaapvaal cratons collided. This collision led to the formation of a continental extensional rift province in which the Ventersdorp Supergroup was deposited at 2.64 billion years ago, with the extension occurring at a high angle to the collision. The latest Archean through Early Proterozoic is marked by deposition of the 2.6–2.1-billion-year-old Transvaal Supergroup in a shallow sea, perhaps related to slow thermal subsidence following Ventersdorp rifting. The center of the Witwatersrand basin is marked by a large circular structure called the Vredefort dome. This structure, several tens of kilometers wide, is associated with shock metamorphic structures, melts, and extremely high-pressure phases of silica suggesting that it represents a meteorite impact structure.

The Bushveld complex is the world's largest layered mafic-ultramafic intrusion, located near the northern margin of the Kaapvaal craton. The complex occupies an area of 40,000 square miles (65,000 km²) and intrudes Late Archean–Early Proterozoic rocks of the Transvaal Supergroup. Isotopic studies using a variety of methods have yielded age estimates of 2.0–2.1 billion years, with some nearby intrusions yielding ages as young as 1.6 billion years. The complex consists of several lobes with a cone-like form and contains numerous repeating cycles of mafic, ultramafic, and lesser felsic rocks. Several types of ores are mined from the complex, including chromite, platinum group metals, cobalt, nickel, copper, and vanadiferous iron ores. Nearly 70 percent of the world's chrome reserves are located in the Bushveld complex. The mafic phases of the complex include dunite, pyroxenite, harzburgite, norite, anorthosite, gabbro, and diorite. The center of the complex includes felsic rocks, including granophyres and granite.

Much of the Kaapvaal craton is covered by rocks of the Karoo basin, including fluvial-deltaic deposits and carbonaceous deposits including coal. The top of the Karoo Sequence includes mafic and felsic lavas that were erupted soon before the breakup of Gondwana 200 million years ago.

See also ARCHEAN; CRATONS; GREENSTONE BELTS; PRECAMBRIAN; WITWATERSRAND BASIN.

Further Reading

Burke, Kevin, William S. F. Kidd, and Timothy M. Kusky. "Archean Foreland Basin Tectonics in the Witwatersrand, South Africa." *Tectonics* 5 (1986).
———. "Is the Ventersdorp Rift System of Southern Africa Related to a Continental Collision between the Kaapvaal and Zimbabwe Cratons at 2.64 Ga ago?" *Tectonophysics* 11 (1985).
———. "The Pongola Structure of Southeastern Africa: the World's Oldest Recognized Well-Preserved Rift?" *Journal of Geodynamics* 2 (1985).
de Wit, Maarten J. "Gliding and Overthrust Nappe Tectonics in the Barberton Greenstone Belt." *Journal of Structural Geology* 4 (1982).
de Wit, Maarten J., Chris Roering, Roger J. Hart, Richard A. Armstrong, Charles E. J. de Ronde, R. W. E. Green, Marian Tredoux, E. Pederdy, and R. A. Hart. "Formation of an Archean Continent." *Nature* 357 (1992).
Kusky, Timothy M., and Julian Vearncombe. "Structure of Archean Greenstone Belts." In *Tectonic Evolution of Greenstone Belts*, edited by Maarten J. de Wit and Lewis D. Ashwal. Oxford: Oxford Monograph on Geology and Geophysics, 1997.
Tankard, Anthony J., M. P. A. Jackson, Ken A. Eriksson, David K. Hobday, D. R. Hunter, and W. E. L. Minter. *Crustal Evolution of Southern Africa: 3.8 Billion Years of Earth History.* New York: Springer-Verlag, 1982.
Vearncombe, Julian R., Peter E. Cheshire, J. H. de Beer, A. M. Killick, W. S. Mallinson, Steve McCourt, and E. H. Stettler. "Structures Related to the Antimony Line, Murchison Schist Belt, South Africa." *Tectonophysics* 154 (1988).

kame A glacial landform such as a low mound, knob, or ridge composed of stratified fluvial/glacial sand and gravel, deposited by a glacial stream. Some kames form as deltas or fans along the margins of deltas, others form in low spots beneath glaciers, and still others form in ponds on the glacial surface or on stagnant ice left by retreating glaciers. Kames may occur in groups, forming kame fields, or as a long line of kames along the front of a stagnant glacier, leaving a kame moraine behind when the glacier retreats. Kame terraces consist of stratified fluvial/glacial sand and gravel deposited between a glacier and a valley wall, commonly left standing as a ridge on and parallel to the slope of a valley wall separating unglaciated topography above from the glaciated valley floor below.

See also GLACIER.

kaolinite A common clay mineral belonging to the kaolin group, with the chemical formula $Al_2Si_2O_5(OH)_4$. Consisting of sheets of tetrahedrally coordinated silicon joined by oxygen shared with octahedrally coordinated aluminum, kaolinite does not expand with the addition of water and does not exchange magnesium or iron. A high-alumina clay mineral and one of the most common of the clay minerals, it often occurs with quartz, iron oxides, pyrite, and siderite and is formed by the weathering of feldspars and other silicate minerals.

karst Areas that are affected by groundwater dissolution, cave complexes, and sinkhole development are known as karst terrains. Globally, several regions are known for spectacular karst systems, including the cave systems of the Caucasus, southern Arabia including Oman and Yemen, Borneo, and the mature highly eroded karst terrain of southern China's Kwangsi Province.

The formation of karst topography begins with a process of dissolution. Rainwater that filters through soil and rock may work its way into natural fractures or breaks in the rock, and chemical reactions that remove ions from the limestone slowly dissolve and carry away in solution parts of the limestone. Fractures are gradually enlarged, and new passageways are created by groundwater flowing in underground stream networks through the rock. Dissolution of rocks is most effective if the rocks are limestone, and if the water is slightly acidic (acid rain greatly helps cave formation). Carbonic acid (H_2CO_3) in rainwater reacts with the limestone, rapidly (at typical rates of a few millimeters per thousand years) creating open spaces, cave and tunnel systems, and interconnected underground stream networks.

When the initial openings become wider, they are known as caves. Many caves are small pockets along enlarged or widened cavities, whereas others are huge open underground spaces. In many parts of the world, the formation of underground cave systems has led to parts of the surface collapsing into the caverns and tunnels, forming a distinctive type of topography known as karst topography. Karst is named after the Kars Limestone plateau region in Serbia (the northwest part of the former Yugoslavia) where it is especially well developed. Karst topography may take on many forms in different stages of landscape evolution but typically begins with the formation of circular pits on the surface known as sinkholes. These form when the roof of an underground cave or chamber suddenly collapses, bringing everything on the surface suddenly down into the depths of the cave. Striking examples of sinkhole formation surprised residents of the Orlando region in Florida in 1981, when series of sinkholes swallowed many businesses and homes with little warning. In this and many other examples, sinkhole formation is initiated after a prolonged drought, or drop in the groundwater levels. This drains the water out of underground cave networks, leaving the roofs of chambers unsupported, and making them prone to collapse.

The sudden formation of sinkholes in the Orlando area is best illustrated by the formation of the Winter Park sinkhole on May 8, 1981. The first sign that trouble was brewing was provided by the unusual spectacle of a tree suddenly disappearing into the ground at 7:00 P.M. as if being sucked in by some unseen force. Residents were worried, and rightfully so. Within 10 hours a huge sinkhole nearly 100 feet (30 m) across and more than 100 feet deep had formed. It continued to grow, swallowing six commercial buildings, a home, two streets, six Porsches, and the municipal swimming pool, causing more than $2 million in damage. The sinkhole has since been converted into a municipal park and lake. More than one thousand sinkholes have formed in part of southern Florida in recent years.

Sinkhole topography is found in many parts of the world, including Florida, Indiana, Missouri, Pennsylvania, and Tennessee in the United States, the Karst region of Serbia, the Salalah region of Arabia, southern China, and many other places where the ground is underlain by limestone.

Sinkholes have many different forms. Some are funnel-shaped, with boulders and unconsolidated sediment along their bottoms; others are steep-walled pipe-like features that have dry or water-filled bottoms. Some sinkholes in southern Oman are up to 900-feet (247-m) deep pipes with caves at their bottoms, where residents would get their drinking water until recently, when wells were drilled. Villagers, mostly women, would have to climb down precarious vertical walls and then back out carrying vessels of water. The bottoms of some of these sinkholes are littered with bones, some dating back thousands of years, of water carriers who slipped on their route. Some of the caves are decorated with prehistoric cave art, showing that these sinkholes were used as water sources for thousands or tens of thousands of years.

Sinkhole formation is intricately linked to the lowering of the water table, as exemplified by the Winter Park example. When water fills the underground caves and passages, it slowly dissolves the walls, floor, and roof of the chambers, carrying the limestone away in solution. When the water table is lowered by drought, by overpumping of groundwater by people, or by other mechanisms, the roofs of the caves may no longer be supported, and they may catastrophically collapse into the chambers forming a sinkhole on the surface. In Florida many of the sinkholes formed because officials lowered the water table level to drain parts of the Everglades, to make more land available for development. This ill-fated decision was rethought and attempts have been made to restore the water table, but in many cases it was too late and the damage was done.

Many sinkholes form suddenly and catastrophically, with the roof of an underground void suddenly collapsing, dropping all of the surface material into the hole. Other sinkholes form more gradually, with the slow movement of loose unconsolidated material into the underground stream network, eventually leading to the formation of a surface depression that may continue to grow into a sinkhole.

The pattern of surface subsidence resulting from sinkhole collapse depends on the initial size of the cave that collapses, the depth of the cavity, and the strength of the overlying rock. Big caves that collapse can cause a greater surface effect. In order for a collapse structure at depth to propagate to the surface, blocks must fall off the roof and into the cavern. The blocks fall by breaking along fractures and falling by the force of gravity. If the overlying material is weak, the fractures will propagate outward, forming a cone-shaped depression with its apex in the original collapse structure. In contrast, if the overlying material is strong, the fractures will propagate vertically upward, resulting in a pipe-like collapse structure.

When the roof material collapses into the cavern, blocks of wall rock accumulate on the cavern floor. There is abundant pore space between these blocks, so the collapsed blocks take up a larger volume than they did when they were

Large sinkhole in Shelby County, Alabama, measuring 425 feet long, 350 feet wide, and 150 feet deep *(Photo courtesy of the U.S. Geological Survey)*

attached to the walls. In this way, the underground collapsed cavern may become completely filled with blocks of the roof and walls before any effect migrates to the surface. If enough pore space is created, almost no subsidence may occur along the surface. In contrast, if the cavity collapses near the surface, a collapse pit will eventually form on the surface.

It may take years or decades for a deep-collapse structure to migrate from the depth where it initiates to the surface. The first signs of a collapse structure migrating to the surface may be tensional cracks in the soil, bedrock, or building foundations, formed as material pulls away from unaffected areas as it subsides. Circular areas of tensional cracks may enclose an area of contractional buckling in the center of

The Teyq sinkhole in southern Oman is the largest sinkhole in the world, measuring nearly 5,000 feet across and hundreds of feet deep. The bright dry streambeds in the bottom of the sinkhole merge, then drain into a cave known as a sparrow hole in the bottom of the sinkhole. *(Photo by Timothy Kusky)*

the incipient collapse structure, as bending in the center of the collapsing zone forces material together.

After sinkholes form, they may take on several different morphological characteristics. Solution sinkholes are saucer-shaped depressions formed by the dissolution of surface limestone, and have a thin cover of soil or loose sediment. These grow slowly and present few hazards, since they are forming on the surface and are not connected to underground stream or collapse structures. Cover-subsidence sinkholes form where the loose surface sediments move slowly downward to fill a growing solution sinkhole. Cover-collapse sinkholes form where a thick section of sediment overlies a large solution cavity at depth, and the cavity is capped by an impermeable layer such as clay or shale. A perched water table develops over the aquiclude. Eventually, the collapse cavity becomes so large that the shale or clay aquiclude unit collapses into the cavern, and the remaining overburden rapidly sinks into the cavern, much like sand sinking in an hourglass. These are some of the most dangerous sinkholes, since they form rapidly and may be quite large. Collapse sinkholes are simpler but still dangerous. They form where the strong layers on the surface collapse directly into the cavity, forming steep walled sinkholes.

Sinkhole topography may continue to mature into a situation where many of the sinkholes have merged into elongate valleys, and the former surface is found as flat areas on surrounding hills. Even this mature landscape may continue to evolve, until tall steep-walled karst towers reach to the former land surface, and a new surface has formed at the level of the former cave floor. The Cantonese region of southern China's Kwangsi Province best shows this type of karst tower terrain.

See also CAVES.

Further Reading
White, William B. *Geomorphology and Hydrology of Karst Terrains.* Oxford: Oxford University Press, 1988.

katabatic winds   A local wind that moves down a slope, usually as a result of cooling at night. Katabatic winds are also commonly known as mountain or valley breezes, foehns, and glacier winds, but the term *katabatic wind* is usually reserved for particularly strong downslope winds. Strong, even hurricane-force winds form particularly well on elevated plateaus surrounded by mountains, with an opening that slopes steeply downhill. When these plateaus get covered in snow in the winter, the cold air on top of them produces a small high-pressure system that flows down the hills and is particularly strong in the valleys or gaps that lead off the plateau. The katabatic winds can become extremely strong when low-pressure systems approach, such as when snow-covered plateaus are formed along the coast. Such conditions are found along the coast of parts of southern Alaska, along the Greenland and Antarctic ice sheets, along the Columbia

River gorge, through Yosemite Valley coming off the Sierras, in the Rhone Valley of France, and coming off the Russian Plateau into the northern Adriatic Sea.

See also CHINOOK WINDS; SANTA ANA WINDS.

kimberlite Large volcanic pipes that come from deep in the Earth and explode their way to the surface with such force that they may blow holes through the stratosphere. Some kimberlite pipes carry diamonds from hundreds of kilometers in the Earth and seem to form in places that were abnormally rich in fluids. Kimberlites and related diatremes represent rare types of continental volcanic rock types, produced by generally explosive volcanism with an origin deep within the mantle. They form pipe-like bodies extending vertically downward and are the source of many of the world's diamonds. Kimberlites were first discovered in South Africa during diamond exploration and mining in 1869, when the source of many alluvial diamonds on the Vaal, Orange, and Riet Rivers was found to be circular mud "pans," later appreciated to be kimberlite pipes. In 1871 two very diamond-rich kimberlite pipes were discovered on the Vooruitzigt Farm, owned by Nicolas de Beer. These discoveries led to the establishment of several large mines and one of the most influential mining companies in history.

Kimberlites are very complicated volcanic rocks, with mixtures of material derived from the upper mantle and complex water-rich magma of several different varieties. A range of volcanic intrusive styles, including some extremely explosive events, characterizes kimberlites. True volcanic lavas are only rarely associated with kimberlites, so volcanic styles of typical volcanoes are not typical of kimberlites. Most near-surface kimberlite rocks are pyroclastic deposits formed by explosive volcanism filling vertical pipes, and surrounded by rings of volcanic tuff and related features. The pipes are typically a couple of hundred yards wide, with the tuff ring extending another hundred yards or so beyond the pipes. The top part of many kimberlite pipes includes reworked pyroclastic rocks, deposited in lakes that filled the kimberlite pipes after the explosive volcanism blasted much of the kimberlite material out of the hole. Geologic studies of kimberlites reveal that they intrude the crust suddenly and behave differently from typical volcanoes. Kimberlites intrude violently and catastrophically, with the initial formation of a pipe filled with brecciated material from the mantle, reflecting the sudden and explosive character of the eruption. As the eruption wanes, a series of tuffs falls out of the eruption column and deposits the tuff ring around the pipes. Unlike most volcanoes, kimberlite eruptions are not followed by the intrusion of magma into the pipe. The pipes simply get eroded by near-surface processes, lakes form in the pipes, and nature tries to hide the very occurrence of the explosive event.

Below these upward-expanding craters are deep vertical pipes known as diatremes that extend down into the mantle source region of the kimberlites. Many diatremes have features that suggest the brecciated mantle and crustal rocks were emplaced at low temperature, nonviolently, presenting a great puzzle to geologists. How can a deep source of broken mantle rocks passively move up a vertical pipe to the surface, suddenly explode violently, and then disappear beneath a newly formed lake?

Early ideas for the intrusion and surface explosion of kimberlites suggested that they rose explosively and catastrophically from their origin in the mantle. Subsequent studies revealed that the early deep parts of their ascent did not seem to be explosive. It is likely that kimberlite magma rises from deep in the upper mantle along a series of cracks and fissures until it gets to shallow levels where it mixes with water and becomes extremely explosive. Other diatremes may be more explosive from greater depths and may move as gas-filled bodies rising from the upper mantle. As the gases move into lower pressure areas they would expand, resulting in the kimberlite moving faster until it explodes at the surface. Still other ideas for the emplacement of kimberlites and diatremes invoke hydrovolcanism, or the interaction of the deep magma with near-surface water. Magma may rise slowly from depth until it encounters groundwater in fractures or other voids then explodes when the water mixes with the magma. The resulting explosion could produce the volcanic features and upward-expanding pipe found in many kimberlites.

It is likely that some or all of these processes play a role in the intrusion of kimberlites and diatremes, the important consequence being a sudden, explosive volcanic eruption at the surface, far from typical locations of volcanism, and the relatively rapid removal of signs of this volcanism. The initial explosions are likely to be so explosive that they may blast material to the stratosphere, though other kimberlite eruptions may only form small eruptions and ash clouds.

klippe An erosional remnant or outlier of the hanging wall of a nappe that is completely surrounded by rocks of the footwall. Klippen were originally part of a larger thrust or nappe complex, but erosion removed the parts of the nappe between the klippe and the main part of the nappe.

See also NAPPE; STRUCTURAL GEOLOGY.

Kola Peninsula The Kola Peninsula occupies 50,000 square miles (129,500 km²) in northwestern Russia as an eastern extension of the Scandinavian peninsula, on the shores of the Barents Sea, east of Finland and north of the White Sea. Most of the peninsula lies north of the Arctic Circle. The peninsula is characterized by tundra in the northeast and taiga forest in the southwest. Winters are atypically warm and snowy for such a northern latitude because of nearby warm Atlantic Ocean waters, and warm summers are filled with long daylight hours.

The Kola Peninsula is part of the Archean Baltic shield, containing medium to high-grade mafic and granitic gneisses including diorite, tonalite, trondhjemite, granodiorite, and granite. Metasedimentary schist, metapelitic gneiss, quartzite, and banded iron formation known as the Keivy assemblage form linear outcrop belts in the eastern part of the Kola Peninsula. Mafic/ultramafic greenstone belts and several generations of intrusions are found on the peninsula, and these may correlate with ophiolitic rocks of the North Karelian greenstone belts further south in the Baltic shield. Metamorphism is mostly at amphibolite facies but locally reaches granulite facies, and deformation is complex with abundant fold interference patterns and early isoclinal folds possibly associated with early thrust faults. The Kola schist belts are intruded by several generations of mafic to granitic intrusions.

See also BALTIC SHIELD.

komatiite A high-magnesium ultramafic lava exhibiting spinifex (quench) textures as shown by bladed olivine or pyroxene crystals. The composition of komatiite may range from peridotite, with 30 percent MgO and 44 percent SiO_2, to basalt, with 8 percent MgO and 52 percent SiO_2. The name is from the type section where these rocks were first identified on the Komati River in Barberton, South Africa. Komatiites are very rare in Phanerozoic orogenic belts and have been recovered from few places such as fracture zones on the modern seafloor. They are more abundant but still rare in Archean greenstone belts. Early work on komatiites suggested that they reflected high degrees of partial melting of a high-temperature mantle, with mantle melting temperatures estimated to be as high as 2,912°F–3,272°F (1,600°C–1,800°C). Since these temperatures are much higher than those in the melting region of the mantle today, and komatiites are more abundant in Archean greenstone belts than in younger orogenic belts, some workers used komatiites as evidence that the Archean mantle was much hotter than the mantle is today. However, more recent petrological work has shown that the earlier estimates were based on dry melting experiments, and komatiites have water in their structure. Therefore, by adding a small percent of water to the melting calculations, new estimates of komatiite source region melting temperatures fall in the range of 2,192°F–2,552°F (1,200°C–1,400°C), much more similar to present-day mantle temperatures.

See also ARCHEAN; CRATONS; GREENSTONE BELTS.

Kuwait Kuwait is located in the northwest corner of the Arabian Gulf between 28°30' and 30° north latitude and 46°30' to 48°30' east longitude. It is approximately 10,700 square miles (17,818 km²) in area; the extreme north-south distance is 120 miles (200 km), the east-west distance is 100 miles (170 km). To the south it shares a border with Saudi Arabia; to the west and north, it shares a border with Iraq. The climate of Kuwait is semiarid and is characterized by two seasons: a long, hot and humid summer, and a short cold winter. Summer temperatures range at 84.2°F–113°F (29°C–45°C)

with relatively high humidity. The prevailing shamal winds from the northwest bring severe dust storms and sandstorms from June to early August with gusts up to 60 miles per hour (100 km/hr). Winter temperatures range at 46.4°F–64.4°F (8°C–18°C). Occasionally, there are simoan winds from the southwest during the month of November. Annual precipitation averages 4.5 inches (115 mm) and infiltrates the sandy soil very rapidly, leaving no surface water except in very few depressions. Most of the limited rainfall occurs in sudden squalls during the winter season.

Most of Kuwait is a flat, sandy desert. There is a gradual decrease in elevation from an extreme of 980 feet (300 m) in the southwest near Shigaya to sea level. The southeast is generally lower than the northwest. There are no mountains or rivers. The country can be divided into roughly two parts, including a hard, flat stone desert in the north with shallow depressions and low hills running northeast to southwest. The principal hills in the north are Jal al-Zor (475 feet; 145 m) and the Liyah ridge. Jal al-Zor runs parallel to the northern coast of Kuwait Bay for a distance of 35 miles (60 km). The southern region is a treeless plain covered by sand. The Ahmadi Hills (400 feet; 125 m) are the sole exception to the flat terrain. Along the western border with Iraq lies Wadi Al-Batin, one of the few valleys in Kuwait. The only other valley of note is Ash Shaqq, a portion of which lies within the southern reaches of the country. Small playas, or enclosed basins, are covered intermittently with water. During the rainy season they may be covered with dense vegetation; during the dry season they are often devoid of all vegetation. Most playas range 650–985 feet (200–300 m) in length, with depths at 16–50 feet (5–15 m).

There are few sand dunes in Kuwait, occurring mainly near Umm Al-Neqqa and Al-Huwaimiliyah. The dunes at Umm Al-Neqqa are crescent-shaped or barchan dunes with an average width of 550 feet (170 m) and average height of 25 feet (8 m). Those near Al-Huwaimiliyah are smaller, averaging 65 feet (20 m) wide and 7 feet (2 m) in height, and are clustered into longitudinal dune belts. Both mobile and stable sand sheets occur in Kuwait. A major mobile sand belt crosses Kuwait in a northwest to southeast direction, following the prevailing wind pattern. Smaller sheets occur in the Al-Huwaimiliyah area, in the Al-Qashaniyah in the northeast, and in much of the southern region.

During the past few years, overgrazing and an increase of motor vehicles in the desert have caused great destruction to the desert vegetation. Stabilized vegetated sheets have changed to mobile sheets as the protective vegetation is destroyed. The largest stable sheet occurs at Shugat Al-Huwaimiliyah. Recently, smaller sheets have begun to develop at Umm Al-Neqqa and Burgan oil field due to an increase in desert vegetation resulting from a prohibition of traffic.

Kuwait Bay is a 25-mile (40-km) long indentation of marshes and lagoons. The coast is mostly sand interspersed with sabkhas and gravel. Sabkhas are flat, coastal areas of clay, silt and sand that are often encrusted with salt. The northern portion of the bay is very shallow, averaging less than 15 feet (5 m) in depth. This part of the shore consists mostly of mudflats and sandy beaches. The more southern portion is relatively deep with a bed of sand and silicic deposits. Most of the ports are situated in the southern area.

Kuwaiti territory includes 10 islands. Most are covered by scrub, and a few serve as breeding grounds for birds. Bubiyan, the largest island measuring approximately 600 square miles (1,000 km²) is a low, level bare piece of land with mudflats along much of its north and west coasts that are covered during high tides. It is connected to the mainland by a concrete causeway. To its north lies Warba, another low-lying island covered with rough grass and reeds. East of Kuwait Bay and on the mudflats extending from Bubiyan lie three islands: Failaka, Miskan, and Doha. Failaka is the only inhabited island belonging to Kuwait. A small village, located near an ancient shrine, is set on a 30-foot (9-m) hill at the northwest point of the island. The rest of the island is flat with little vegetation. There are a few trees in the center of the island, and date trees are grown in the village. West of Kuwait City in the Bay are two islets: Al-Qurain and Umm Al-Naml. On the south side lie three more small islands: Qaruh, Kubbar, and Umm Al-Maradim. The last two are surrounded by reefs on three sides.

Kuwait occupies one of the most petroleum-rich areas in the world, situated in a structurally simple region on the Arabian platform in the actively subsiding foreland of the Zagros Mountains to the north and east. Principal structural features of Kuwait include two subsurface arches (Kuwait arch and Dibdibba arch) and the fault-bounded Wadi Al-Batin. Faults defining Wadi Al-Batin are related to Tertiary extension in the region. The Kuwait and Dibdibba arches have no geomorphic expression, whereas the younger Bahra anticline and Ahmadi ridge have a surface expression and are structurally superimposed on the Kuwait arch. Major hydrocarbon accumulations are associated with the Kuwait arch. The stratigraphy of Kuwait includes a nearly continuous section of Arabian platform sediments ranging in age from Cambrian through Holocene, although the pre-Permian rocks are poorly known. The Permian through Miocene section is 3.5–4 miles (6–7 km) thick in Kuwait but thickens toward the northeast. These rocks include continental and shallow marine carbonates, evaporites, sandstones, siltstones, and shales, with less common gravels and cherts. Plio-Pleistocene sand and gravel deposits of the Dibdibba Formation outcrop in northwestern Kuwait, and Miocene sands, clay and nodular limestone of the Fars and Ghar Formation outcrop in the southeast. A small area of Eocene limestone and chert (Dammam Formation) outcrops south of Kuwait City on the Ahmadi Ridge.

The structural arches in Kuwait are part of a regional set of north-trending arches known as the Arabian folds, along which many of the most important oil fields in the Arabian Gulf are located. These arches are at least mid-Cretaceous in age. The orientation of the Arabian folds has been interpreted to be inherited from older structures in the Precambrian basement, with possible amplification from salt diapirism. The north-south trends may continue northward beneath the Mesopotamian basin and the Zagros fold belt.

The northwest-trending anticlinal structures of the Ahmadi ridge and Bahra anticline are younger than the Arabian folds and related to the Zagros collision, initiated in post-Eocene times. These younger folds seem to have a second-order control on the distribution of hydrocarbon reservoirs in Kuwait, as oil wells (and since the first Gulf War in 1990, oil lakes) are concentrated in northwest trending belts across the north-striking Kuwait arch. The Kuwait arch has a maximum structural relief in the region between Burgan and Bahra, with closed structural contours around the Wafra, Burgan, Magwa, and Bahra areas, and a partial closure indicating a domal structure beneath Kuwait City and Kuwait Bay. The superposition of the Kuwait arch and the shallow anticlinal structure of the Ahmadi ridge form a total structural relief of at least one mile (1.6 km).

The northwest-trending Dibdibba arch represents another subsurface anticline in western Kuwait. The ridge is approximately 45 miles (75 km) long and is an isolated domal structure, but it has not as yet yielded any significant hydrocarbon reservoirs.

Wadi Al-Batin is a large valley, 4–6 miles (7–10 km) wide and with relief of up to 185 feet (57 m). In the upper valley of the wadi, the valley sides are steep, but in southwestern Kuwait few ravines have steep walls greater than 15 feet (5 m) in height. The wadi has a length of more than 45 miles (75 km) in Kuwait and extends 420 miles (700 km) southwestward into Saudi Arabia, where it is referred to as Wadi Ar-Rimah. The ephemeral drainage in the wadi drains from the southwest and has transported Quaternary and Tertiary gravels consisting of igneous and metamorphic rock fragments from the Saudi Arabian and Syrian deserts during Pleistocene pluvial episodes. The wadi widens toward the northeast, and it becomes indistinguishable from its surroundings northwest of Kuwait City. Ridges made of Dibdibba gravel define paleodrainage patterns of a delta system draining Wadi Al-Batin, and many of these gravel ridges stand out as prominent lineaments. Some of these gravel ridges are marked by faults on at least one side, suggesting a structural control on the drainage pattern.

Numerous small and several relatively large faults are revealed on seismic reflection lines across the wadi, and hydrological pumping tests show a break in the drawdown slope at the faults. The steep Miocene–late Eocene faults parallel to the wadi have displaced the block in the center of the wadi upward by 15–20 feet (25–35 m) relative to the strata outside the wadi, and the displacements die out toward the northeast.

Further Reading

Al-Sarawi, A. Mohamed. "Tertiary Faulting beneath Wadi Al-Batin (Kuwait)." *Geological Society of America Bulletin,* v. 91 (1980).

Directorate of General Geologic Surveys and Mineral Investigation. Tectonic map of Iraq, scale 1: 1,000,000, 1984.

Koop, Walter J., and R. Stonely. "Subsidence History of the Middle East Zagros Basin, Permian to Recent." *Philosophical Transactions of the Royal Society of London* A305 (1982).

Kuwait Oil Company. Geological Map of the State of Kuwait, Scale 1:250,000, 1981.

Milton, D. I. "Geology of the Arabian Peninsula, Kuwait." *U.S. Geological Society* Professional Paper 560-F (1967).

Murris, Roelef J. "Middle East: Stratigraphic Evolution of Oil Habitat." *American Association of Petroleum Geologists Bulletin* 64 (1980).

Stocklin, Jovan, and M. H. Nabavi. Tectonic map of Iran, Geologic Survey of Iran, scale 1:2,500,000, 1973.

Warsi, Waris E. K. "Gravity Field of Kuwait and Its Relevance to Major Geological Structures." *American Association of Petroleum Geologists Bulletin* 74 (1990).

L

lagoon A shallow body of water that is connected to a larger body of water, typically the ocean or sea, by a narrow passage. Most lagoons are elongate, parallel to the coastline, and separated from the main ocean or sea by a narrow elongate barrier island or reef complex. Many lagoons have shallow stagnant water, and the salinity of lagoon waters is highly variable, depending on the amount of runoff, mixing with ocean water, and evaporation rates. Lagoons are broadly similar to estuaries, which are more typically elongate perpendicular to the shoreline and are fed by river systems from the mainland, whereas river input to lagoons is generally negligible.

Most lagoons are fed and flushed by tidal inlets, with waters exchanged with the change of tides. Wind and wave formation may strongly affect the turbidity of waters in shallow lagoons, but in deeper lagoons the turbidity is controlled by the strength of tidal currents.

Lagoons commonly form on coastlines that are subsiding, or where sea level is rising. The sedimentary evolution of the lagoon depends on the relative rates of migration of the landward and seaward coastlines, which may migrate at different rates, widening or shortening the lagoon. The rate at which the lagoon fills depends on the amount of sediment flux from the mainland, the amount of material that moves in through the tidal inlets, the amount of overwash from the barrier or reef, and the amount of organic and muddy sediment produced within the lagoon.

See also BEACH.

lahar Volcanic mudflows are known as lahars, some of which have killed tens of thousands of people in single events. When pyroclastic flows and nuée ardentes move into large rivers, they quickly cool and mix with water, becoming fast-moving mudflows. Lahars may also result from the extremely rapid melting of icecaps on volcanoes. A type of lahar in which ash, blocks of rock, trees, and other material is chaotically mixed together is known as a debris flow. Such lahars and mudflows were responsible for much of the burial of buildings and deaths from the trapping of automobiles during the 1980 eruption of Mount Saint Helens. Big river valleys were also filled by lahars and mudflows during the 1991 eruption of Mount Pinatubo in the Philippines, resulting in extensive property damage and loss of life. One of the greatest volcanic disasters of the 20th century resulted from the generation of a huge lahar when the icecap on the volcano Nevado del Ruiz in Colombia catastrophically melted during the 1985 eruption. Nevado del Ruiz entered an active phase in November 1984 and began to show harmonic tremors on November 10, 1985. At 9:37 P.M. that night, a large Plinian eruption sent an ash cloud several miles into the atmosphere, and this ash settled onto the ice cap on top of the mountain. This ash, together with volcanic steam, quickly melted large amounts of the ice, which mixed with the ash and formed giant lahars down the east side of the mountain into the village of Chinchina, killing 1,800 people. The eruption continued and melted more ice that mixed with more ash and sent additional larger lahars westward. Some of these lahars moved nearly 30 miles (48 km) at nearly 30 miles per hour (48 km/hr) and buried the town of Armero under 26 feet (8 m) of mud, killing 23,000 people in Armero and surrounding communities.

lake An inland body of water that occupies a sizable depression in the Earth's surface and is too deep to allow terrestrial vegetation to grow across the entire body. Lakes have many sizes and shapes, ranging from just bigger than a pond to virtual inland seas thousands of square kilometers in area. Lakes may form in depressions left by erosion, melting ice, in wide parts in streams, behind natural and manmade dams, and in fault-controlled depressions. Many lakes are located in tectonic depressions, including extremely deep rift basins such as Lake Baikal in Russia, Lakes Malawi, Tanganyika, Edward, Albert, and Turkana in the East African rift. Lake waters have a large variation in chemistry, including fresh and saline varieties, as well as those with variable contributions from hot springs and other deep crustal fluids. Other chemical variations depend in part on the chemistry of inflowing streams, evaporation, and biological activity.

Being water-filled depressions, lakes tend to become filled with sediments over time. There is a huge variety in the types and styles of sedimentation in lakes, depending on the types of sediment available in the source area, the physiography of the lake and surrounding area, the climate, fauna and flora present, and the manner in which the sediments settle in the basin. Many lakes exhibit a density stratification, with warm, oxygenated, well-mixed waters near the surface in the epilimnion layer, and cold, dense, anoxic waters in the hypolimnion layer near the bottom. Oligotrophic lakes are those where winter surface cooling leads to overturning of the lake waters and mixing of the waters. In low latitudes where there is no appreciable winter cooling, the surface layer does not cool and the lake waters do not overturn. These eutrophic lakes have bottom waters that remain anoxic.

See also EUTROPHICATION; FACIES; SEDIMENTARY ROCKS.

Lake Eyre, Australia

A shallow, frequently dry salt lake in central Australia that occupies the lowest point on the continent at 39 feet (12 m) below sea level. The lake occupies 3,430 square miles (8,884 km²), but the drainage basin is one of the world's largest internally draining river systems covering 1.2 million square miles (1.93 million km²), with no outlet to the sea. All water that enters the Lake Eyre basin flows into the lake and eventually evaporates leaving salts behind. Lake Eyre is located in the driest part of Australia, where the evaporation potential is 8.175 feet (2.5 m) but the annual precipitation is only half an inch (1.25 centimeters). However, flows in the river system are highly variable and unpredictable, since rare rainfall events may cause flash flooding. All rivers in the system are ephemeral, typically with no water in the system. Aridity increases downstream toward the lake, and the basin is characterized by huge braided stream networks, floodplains, and waterholes. The stream systems leading into Lake Eyre are one of the largest unregulated river systems in the world.

See also DESERT.

Lake Titicaca

Sitting at 12,500 feet (3,815 m) above sea level along the border between Peru and Bolivia in the South American Andes, Lake Titicaca is the world's highest lake navigable to large vessels. The lake basin is situated between Andean ranges on the Altiplano plateau and is bordered to the northeast by some of the highest peaks in the Andes in the Cordillera Real, where several mountains rise to more than 21,000 feet (6,400 m). Covering 3,200 square miles, Lake Titicaca is the largest freshwater lake in South America, although it is divided into two parts by the Strait of Tiquina. The body of water north of the Strait is called Chucuito in Bolivia and Lake Grande in Peru, and south of the strait the smaller body of water is called Lake Huinaymarca in Bolivia and Lake Pequeno in Peru. Most of the lake is 460–600 feet (140–180 m) deep, but it reaches 920 feet (280 m) deep near the northeast corner of the lake. The lake is fed by many short tributaries from surrounding mountains and is drained by Desaguadero River that flows into Lake Poopo. However, only 5 percent of water loss is through this single outlet—the remainder is lost by evaporation in the hot dry air of the Altiplano. Lake levels fluctuate on seasonal and several longer time cycles, and the water retains a relatively constant temperature of 56°F (14°C) at the surface but cools to 52°F (11°C) below a thermocline at 66 feet (20 m). Salinity ranges from 5.2 to 5.5 parts per thousand.

Lake Titicaca, translated variously as Rock of the Puma or Craig of Lead, has been the center of culture since pre-Inca times (600 years before present [b.p.]), and its shoreline is presently covered by Indian villages and terraced rice fields. Some of the oldest civilizations are preserved in ruins around Lake Titicaca, including those at Tiahuanaco, on the southern end of the lake, and others on the many islands in the lake. Ruins of a temple on Titicaca Island mark the spot where Inca legends claim that Manco Capac and Mama Ocllo, the founders of the Inca dynasty, were sent to Earth by the Sun.

Lake Victoria

The largest lake in Africa, and the second largest freshwater lake in the world, Lake Victoria, also known as Victoria Nyanza, covers 26,830 square miles (69,490 km²) along the Kenya-Uganda-Tanzania border. The lake is shallow (75 feet; 22.5 m), occupying a shallow depression on the Equatorial plateau between the Eastern and Western Arms of the East African Rift System. The shoreline of the lake is irregular and typically marked by transitional marshes and also by numerous small islands. As the main headwater reservoir for the Nile, the lake is fed by many small streams and drains out to the north into the Victoria Nile.

The lake basin is densely populated and the lake is used for fishing and transportation. Many species of fish are found in the lake, although some nonindigenous species introduced to the lake have been decimating native populations and adapting into different environmental niches. The first European to see the lake was the British explorer John Speke in 1858 who called the lake Ukerewe, and the lake was later explored by Henry Stanley in 1875.

landslide *See* MASS WASTING.

Langmuir Circulation   Wind-driven circulation of the water, in which the wind forms cells with their long axes parallel to the wind direction. Adjacent cells rotate in opposite directions and show elliptical or screw-like rotation patterns known as left-handed and right-handed helical vortices. Langmuir Cells can often be recognized, as they leave regularly spaced elongate streaks of foamy water or debris on the surface in the zones of convergence, elongated parallel to the wind direction. They typically form when wind speeds exceed a few kilometers per hour and increase in intensity when the wind speed increases.

See also EKMAN SPIRALS; THERMOHALINE CIRCULATION; WAVES.

La Niña   *See* EL NIÑO AND THE SOUTHERN OSCILLATION.

laterite   A highly weathered red soil rich in iron oxides, aluminum, and kaolin, and nearly devoid of primary silicates except for quartz. It is a common residual product of extreme tropical weathering, developing on many different types of bedrock. Many third-world populations cut and dry bricks from laterites, using these for building.

See also SOILS.

Laurasia   The northern continents of North America and Eurasia, that existed north of the Tethys Ocean in Alex du Toit's alternate reconstruction of Alfred Wegener's Pangea. Du Toit argued that Pangea was not a single simple supercontinent but for much of the Paleozoic consisted of the northern Laurasian landmass, joined in the west, and separated by the Tethys Ocean from the southern continents of Gondwana.

See also GONDWANA.

lava   Molten rock or magma that flows on the surface of the Earth. Lavas have a wide range in composition, texture, temperature, viscosity, and other physical properties, based on the composition of the melt and the amount of volatiles present. They tend to be very viscous (sticky, resistant to flow) when they are rich in silica, and they form slow-moving and steep-sided flows. If a large amount of volatiles are added to silicic magma it may erupt explosively. Mafic, or low-silica, lavas are less viscous and tend to flow more easily, forming planar flows with gently sloping surfaces. Some basaltic flood lavas have flowed over hundreds or thousands of square kilometers, forming flat-lying layers of crystallized lava. Other mafic and intermediate lavas form shield volcanoes such as the Hawaiian Islands, with gently sloping sides built by numerous eruptions. If mafic lavas are rich in volatiles they tend to become abundant in empty gas bubbles known as vesicles, forming pumice. Mafic lavas that flow on the surface often form ropey lava flows known as pahoehoes, or blocky flows known as aa lavas.

See also AA LAVA; IGNEOUS ROCKS; VOLCANO.

Lawson, Andrew Cowper   (1861–1952) *Scottish Geologist* Andrew Lawson is known for his work in isostasy. He studied in public schools in Hamilton, Ontario, and later received his bachelor's degree in 1883 from the University of Toronto. He then received his master's degree and his doctorate from the Johns Hopkins University in 1888. Lawson then spent seven years with the Geological Survey of Canada until he moved to the University of California, Berkeley, where he remained for 60 years. During this time, his studies of isostasy continued over three decades, and these studies brought special attention to the importance of isostasy in diastrophism. He also developed the logical consequences of isostatic adjustment as an important factor in orogenesis. In some cases this was seen as the determinative agent in the elevation of mountains and the depression of deep troughs and basins. His research areas included the Sierra Nevada, the Great Valley of California, the Mississippi delta, the Cordillera, and the Canadian shield. Lawson also spent some 40 years studying the northwest region of Lake Superior where he provided new information and revised the correlation of the pre-Cambrian rocks over a large part of North America. His earlier studies of Lake of the Woods and Rainy Lake showed that the Laurentian granites were intrusive into metamorphosed volcanic and sedimentary rocks that he called the Keewatin Series. Under this series he found another sedimentary series which he called Coutchiching and saw these as the oldest rocks in the series. Later research on this area showed that two periods of batholithic invasion were followed by a great period of peneplaination. These surfaces, the Laurentian peneplain and the Eparchean peneplain, were used as references for the correlation of the invaded formations and for the subsequently deposited sedimentary beds over the areas. Lawson also worked as a consultant in a number of construction engineering projects including the Golden Gate Bridge.

levee   A natural or man-made topographically high embankment on the side of a river, stream, or body of water. Natural levees are built by stream systems during floods, where the water leaves the fast-moving channel and loses velocity as it expands onto the floodplain. When the water loses velocity, it has a lower capacity to hold material in suspension and drops much of its load on the side of the channel, forming the levee. Levees are also commonly built or enhanced along riverbanks to protect towns and farmlands from river floods. These levees are usually successful at the job they were intended to do, but they also cause some other collateral effects. First, the levees do not allow waters to spill onto the floodplains, so the floodplains do not receive the annual fertilization by thin layers of silt, and they may begin to deflate and slowly degrade as a result of this loss of nourishment by the river. The ancient Egyptians relied on such yearly floods to maintain their fields' productivity, which has declined since the Nile has been dammed and altered in recent times. Another effect of levees is that they constrict the river to a narrow channel, so that floodwaters that once spread slowly over a large region are now focused into a narrow space. This causes floods to rise faster, reach greater heights, have a greater velocity, and reach downstream areas faster than rivers without levees. The extra speed of the river is in many cases enough to erode the levees and return the river to its natural state.

One of the less appreciated effects of building levees on the sides of rivers is that they sometimes cause the river to slowly rise above the height of the floodplain. Many rivers naturally aggrade or accumulate sediment along their bottoms. In a natural system without levees, this aggradation is accompanied by lateral or sideways migration of the channel so that the river stays at the same height with time. However, if a levee is constructed and maintained, the river is forced to stay in the same location as it builds up its bottom. As the bottom rises, the river naturally adds to the height of the levee, and people will build up the height of the levee as well as the river rises, to prevent further flooding. The net result is that the river may gradually rise above the floodplain, until some catastrophic flood causes the levee to break, and the river establishes a new course.

The process of breaking through a levee is known as avulsion. Avulsion has occurred seven times in the last 6,000 years along the lower Mississippi River. Each time, the river has broken through a levee a few hundred miles from the mouth of the river and has found a new shorter route to the Gulf of Mexico. The old river channel and delta are then abandoned, and the delta subsides below sea level, as the river no longer replenishes it. A new channel is established and this gradually builds up a new delta until it too is abandoned in favor of a younger shorter channel to the Gulf.

Some of the most tragic examples of the effects of rivers rising above and breaking through their levees are from the Yellow River in China. The Yellow River flows out of the Kunlun Mountains across much of China into the wide lowland basin between Beijing and Shanghai. The river has switched courses in its lower reaches at least 10 times in the last 2,500 years. It currently flows into Chihli (Bohai) Bay and then into the Yellow Sea. The Chinese have attempted to control and modify the course of the Yellow River since dredging operations in 2356 B.C.E., and the construction of levees in 602 B.C.E. One of the worst modern floods along the Yellow River was in 1887 when the river rose over the top of the 75-foot (22-m) high levees and covered the lowlands with water. More than one million people died from the floods and subsequent famine, along with widespread destruction of crops and livestock.

The Yellow River was also the site of an unnatural disaster in 1938. As part of the war effort, in 1938 Japan attacked and bombed the levees along the Yellow River. The river escaped and took another million lives. The Yellow River is continuing its natural process of building up its bottom, and

the people along the river continue to raise the level of the levees in an attempt to keep the river's floods out of their fields. Today, the river bottom rests an astounding 65 feet (20 m) above the surrounding floodplain, a testament to the attempts of the river to find a new lower channel and to abandon its current channel in the process of avulsion.

See also FLOOD; RIVER SYSTEM.

life's origins and early evolution   The origin of life and its early evolution from simple single-celled organisms to more complex forms has intrigued scientists, philosophers, theologians, and people from all parts of the world for much of recorded history. The question of the origin of life relates to where we came from as humans, why we are here, and what the future holds for our species. It is one of the most interdisciplinary of sciences, encompassing cosmology, chemistry, astrophysics, biology, geology, and math.

Several ideas about the location of the origin of life have received the most support from the scientific community. Some scientists believe that life originated by chemical reactions in a warm little pond, whereas others suggest that it may have started at surface hot springs. Another model holds that the energy for life was first derived from deep within the Earth, at a hydrothermal vent on the seafloor. Still others hold that life may have fallen to Earth from outer reaches of the solar system, though this does not answer the question of where and how it began.

Life on the early Earth would have to have been compatible with conditions very different from what they are on the present-day Earth. The Earth's early atmosphere had no or very little oxygen, so the partial pressure of oxygen was lower and the partial pressure of CO_2 was higher on the early Archean. The Sun's luminosity was about 25 percent less than that of today, but since the early atmosphere was rich in CO_2, CH_4 (methane), NH_3 (ammonia), and N_2O (Nitrous oxide), an early greenhouse effect warmed the surface of the planet. CO_2 was present at about 100 times its present abundance, so the surface was probably even hotter than today, despite the Sun's decreased luminosity. Evidence suggests that the surface temperature was about 140°F (60°C), so thermophyllic bacteria (heat adaptive) were favored to evolve over organisms that could not tolerate such high temperatures. The lack of free oxygen and radiation-shielding ozone (O_3) in the early atmosphere led to a 30 percent higher ultraviolet flux from the Sun, which would have been deadly to most early life. The impact rate from meteorites was higher, and heat flow from the interior of Earth was about three times higher than at present. Early life would have to have been compatible with these conditions, so it would have to have been thermophyllic and chemosynthetic. The best place for life under these extreme conditions would be deep in the ocean. The surface would have been downright unpleasant.

Since the Earth is cool today, some process must have removed CO_2 from the atmosphere, otherwise it would have had a runaway greenhouse effect, similar to Venus. Processes that remove CO_2 from atmosphere include deposition of limestone ($CaCO_3$), and burial of organic matter (CH_2O). These processes are aided by chemical weathering of silicates (e.g., $CaSiO_3$) by CO_2 rich rainwater, that produces dissolved Ca^{2+}, SiO_2, and bicarbonate (HCO_3^-), which is then deposited as limestone and silica. Life evolved in the early Precambrian and began to deposit organic carbon, removing CO_2 from the atmosphere. Limestones formed as a result of organic processes acted as big CO_2 sinks and served to lower global temperatures.

The present-day levels of CO_2 in the atmosphere are balanced by processes that remove CO_2 from the atmosphere and processes that return CO_2 to the atmosphere. Today 78,000 billion tons of carbon are stored in sedimentary rocks. It would take a few hundred million years to accumulate this from the atmosphere. The return part of the carbon cycle is dominated by a few processes. The decomposition of organic matter releases CO_2. Limestone deposited on continental margins is eventually subducted, or metamorphosed into calc-silicate ($CaSiO_3$) rocks, both processes that release CO_2. This system of CO_2 cycling regulates atmospheric CO_2, and thus global temperature, on long time scales. Changes in the rates of carbon cycling are intimately associated with changes in rates of plate tectonics, showing that tectonics, atmospheric composition and temperature, and the development of life are closely linked in many different ways.

It is often difficult to recognize signs of life in very old deformed rocks. Signs of life in rocks are often detected by searching for geochemical isotope fractionation. Metabolism produces distinctive isotopic signatures in C—there is about a 5 percent difference between $^{13}C/^{12}C$ of organic v. inorganic carbon. So, the presence of isotopically light carbon in old rocks suggests the influence of life. However, life may have been diverse, photosynthesizing, methanogenic, and methylothropic bacteria by 3.5 billion or even 3.85 billion years ago. Early life, in a pre–oxygen-rich atmosphere, had to be adapted to the reducing environment.

Early life consisted of 3.8 Ga primitive bacteria (prokaryotes means before nuclei), that may have been photosynthetic. These bacteria made food from CO_2, water, and energy from the Sun, but did not release O_2. Many of the bacteria oxidized sulfur from S_2^- to sulfate SO_4^{2-}. Oxygen is toxic to these bacteria, so they must have lived in environments with no oxygen, reducing the sulfate ion for their energy. By 3.5 Ga cyanobacteria, or blue-green algae, used CO_2 and emitted oxygen to the atmosphere. These organisms began to form the protective ozone (O_3) layer, blocking UV from the Sun and making the surface habitable for other organisms.

Recently, 2.5-billion-year-old Archean ophiolites with black smoker types of hydrothermal vents and evidence for primitive life-forms have been discovered in northern China. The physical conditions at these and even older mid-ocean ridges permit the inorganic synthesis of amino acids and other prebiotic organic molecules, and this environment would have been sheltered from early high ultraviolet radiation and many effects of the impacts. In this environment, the locus of precipitation and synthesis for life might have been in small iron-sulfide globules emitted by hydrothermal vents on the seafloor. Right now, this day, we have very primitive bacteria on the East Pacific Rise on black smoker chimneys, living at 230°F (110°C), the highest temperature in which life exists on Earth. Life at the black smokers draws energy from the internal energy of the Earth (not the Sun), via oxidation in a reducing environment.

Life apparently remained relatively simple for more than a billion years. However, sometime roughly around 2.5 billion years ago, life changed from primitive prokaryotes to eukaryotes, with cell nuclei. Molecular biology yields some clues about life at 2.5 Ga. Molecular phylogenies compare genetic sequences and show that all living species cluster into three groups, Archea, Bacteria, and Eukarya (plants and animals). They all have a common ancestor that is thermophyllic, or heat-loving. So, the deepest branches of the "Universal Tree of Life" are dominated by heat-loving species. This amazing fact suggests that hydrothermal systems may have been the location for the development of early life. Furthermore, the oldest thermophiles are all chemosynthetic organisms that use H and S in their metabolism. H and S are readily available at the black smoker chimneys, adding further support to the idea that submarine hydrothermal vents may have been the site of the development of Earth's earliest life.

Late Archean (2.5 Ga) BIF (banded iron formation) associated with the Dongwanzi, Zunhua, and Wutai Shan ophiolites in North China have black smoker chimneys associated with them, and some of these bear signs of early life. This is a time when the Earth surface environment began a dramatic shift from reducing environments to highly oxidizing conditions. This may be when photosynthesis (i.e., the metabolic strategy common today) developed in sulfur bacteria. Oxygenic photosynthesis first developed in cyanobacteria and later transferred to plants (Eukaryotes) through an endosymbiotic association.

Life continued to have a major role in controlling atmospheric composition and temperature for the next couple of billion years. The first well-documented ice age is at the Archean/Proterozoic boundary, although some evidence points to other ice ages in the Archean. The Archean/Proterozoic ice age may have been related to decreasing tectonic activity and to less CO_2 in the atmosphere. Decreasing plate tectonic activity results in less CO_2 released by metamorphism and volcanism. These trends resulted in global levels of atmospheric CO_2 falling, and this in turn caused a less effective greenhouse, enhancing cooling, and leading to the ice age.

Prolonged cool periods in Earth history are called ice-houses, and most result from decreased tectonic activity and the formation of supercontinents. Intervening warm periods are called hothouses, or greenhouses. In hothouse periods, higher temperatures cause more water vapor to be evaporated and stored in the atmosphere, so more rain falls during hothouses than in normal times, increasing the rates of chemical weathering, especially of calcium silicates. These free Ca and Si ions in the ocean combine with atmospheric CO_2 and O_2, to form limestone and silica that gets deposited in the oceans. This in turn causes increased removal of CO_2 from the atmosphere, which cools the planet in a self-regulating mechanism. The cooling reduces the rate of chemical weathering. Eventually, previously deposited calc-silicates are buried, metamorphosed, and release CO_2, which counters the cooling from a runaway ice age effect, warming the planet in another self-regulating step.

The earliest bacteria appear to have been sulfate-reducing thermophilic organisms that dissolved sulfate by reduction to produce sulfide. In this process the bacteria oxidize organic matter, to produce CO_2 and H_2O, and deposit FeS_2 (pyrite).

Banded iron formations (BIFs) are rocks rich in iron and silica and are very common in 2.2 to 1.6 Ga old rock sequences. They are the source of 90 percent of the world's iron ore. BIFs were probably deposited during a hothouse interval and require low oxygen in the atmosphere/hydrosphere system to form. It is hypothesized that water with high concentrations of Fe^{2+} was derived from weathering of crust.

Eukaryotes with membrane-bound cell nuclei emerged at about 2 Ga. Aerobic photosynthetic cells evolved and very effectively generated oxygen. These organisms rapidly built up atmospheric oxygen, to high levels by 1.6 Ga. The eukaryotes evolved into plants and animals. For the next billion years, oxygen increased and CO_2 fell in the atmosphere until the late Proterozoic, when the explosion of invertebrate metazoans (jellyfish) marked the emergence of complex Phanerozoic styles of life on Earth. This transition occurred during the formation and breakup of the supercontinent of Gondwana, with associated climate changes from a 700 Ma global icehouse (supercontinent), with worldwide glaciations, to equatorial regions. This was followed by warmer climates and rapid diversification of life.

See also ARCHEAN; BANDED IRON FORMATION; CARBON CYCLE; HADEAN.

Further Reading

Farmer, Jack. "Hydrothermal Systems: Doorways to Early Biosphere Evolution." *GSA Today* 10, no. 7 (2000).

Mojzsis, Stephen, and Mark Harrison. "Vestiges of a Beginning: Clues to the Emergent Biosphere Recorded in the Oldest Known Sedimentary Rocks." *GSA Today* 10, no. 4 (2000).

Rasmussen, Birger. "Life at 3.25 Ga in Western Australia." *Nature* 405 (2000): 676.

limestone   *See* CHEMICAL SEDIMENT; SEDIMENTARY ROCKS.

limonite   A group of brown or yellowish-brown amorphous hydrous ferric oxides that occur naturally, usually as a weathering product of other iron-bearing minerals. Limonite also occurs as an organic or inorganic precipitate in lakes, ponds, bogs, springs, and in marine settings, and in some places coats surfaces such as twigs, or around earthen masses. The exact composition of limonite is unknown, but it is probably a mixture of hematite, goethite, and lepidocrocite. Occurring in a variety of forms, including botryoids, fibrous or stalactitic columns, and reniform and mammillary masses, limonite also goes by the names bog iron ore and brown ochre.

lithification   The process of changing an unconsolidated sediment into a solid coherent rock. It may include a variety of processes that may act separately or together. For instance, cementing of grains together by minerals, typically quartz or calcite, that flow in the pore spaces between grains may change a loose sediment into a hard rock. Compaction of sediment may force grains together, causing one to impinge upon another, and causing them to be tightly joined together. Compaction may also drive the water out of a group of sediments, forcing grains closer together and aiding lithification.

See also SEDIMENTARY ROCKS.

lithosphere   The top of the mantle and the crust is a relatively cold and rigid boundary layer called the lithosphere, about 60 miles (100 km) thick. Heat escapes through the lithosphere largely by conduction, transport of heat in igneous melts, and in convection cells of water through mid-ocean ridges. The lithosphere is about 75 miles (125 km) thick under most parts of continents, and 45 miles (75 km) thick under oceans, whereas the asthenosphere extends to about a 155-mile (250-km) depth. Lithospheric roots, also known as the tectosphere, extend to about 155 miles beneath many Archean cratons.

The base of the crust, known as the Mohorovicic discontinuity (the Moho), is defined seismically and reflects the rapid increase in seismic velocities from basalt to peridotite at 5 miles per second (8 km/s). However, some petrologists distinguish between the seismic Moho, as defined above, and the petrologic Moho, reflecting the difference between the crustal cumulate ultramafics and the depleted mantle rocks that the crustal rocks were extracted from. This petrological Moho boundary is not recognizable seismically. In contrast, the base of the lithosphere is defined rheologically as where the same rock type on either side begins to melt, and it corresponds roughly to the 2,425°F (1,330°C) isotherm.

Since the lithosphere is rigid, it cannot convect, and it loses its heat by conduction and has a high temperature contrast (and geothermal gradient) across it compared with the upper mantle, which has a more uniform temperature profile. The lithosphere thus forms a rigid, conductively cooling thermal boundary layer riding on mantle convection cells, becoming convectively recycled back into the mantle at convergent boundaries.

The elastic lithosphere is that part of the outer shell of the Earth that deforms elastically, and the thickness of the elastic lithosphere increases significantly with the time from the last heating and tectonic event. This thickening of the elastic lithosphere is most pronounced under the oceans, where the elastic thickness of the lithosphere is essentially zero to a few kilometers at the ocean ridges. This thickness increases proportionally to the square root of age to about a 35-mile (60-km) thickness at an age of 160 million years.

The thickness of the lithosphere may also be measured by the wavelength and amplitude of the flexural response to an induced load. The lithosphere behaves in some ways like a thin beam or ruler on the edge of a table that bends and forms a flexural bulge inward from the main load. The wavelength is proportional to, and the amplitude is inversely proportional to, the thickness of the flexural lithosphere under an applied load, providing a framework to interpret the thickness of the lithosphere. Natural loads include volcanoes, sedimentary prisms, thrust belts, and nappes. Typically, the thermal, seismic, elastic, and flexural thicknesses of the lithosphere are different because each method is measuring a different physical property, and also because elastic and other models of lithospheric behavior are overly simplistic.

See also ASTHENOSPHERE; CONTINENTAL CRUST; CRATONS; OPHIOLITES; PLATE TECTONICS.

littoral The nearshore intertidal environment between the high-tide and the low-tide marks, and the organisms that live in this zone. The bottom sediments in this zone are strongly affected by wave action, constantly being stirred and mixed, so only very hardy benthic organisms live in this zone. Because waves often approach the shoreline at an oblique angle and return perpendicular to the shore, the sediments on the bottom in the littoral zone are subject to a slow, steady displacement parallel to the shoreline. This slow movement of sand grains is known as littoral drift or longshore drift. The greatest amount of transport occurs beneath the breaker zone where individual sand grains may move as much as hundreds of meters per day. Littoral drift of sand is strongly affected by changes to the beach, including the construction of groins and jetties. There are many examples where groins were constructed, and beaches down-drift from the groins disappear soon after. Likewise, there are many examples of where jetties were constructed at the ends of inlets, disrupting the flow of sand to the down-drift beaches, which gradually disappear. The sand that used to replenish these beaches either moves into the channel and forms deltas in the lagoon or is carried to deep water by rip currents set up by the jetties.

See also BEACH; GROIN; WAVES.

load The total amount of material that is transported by various mechanisms by a stream, glacier, the wind, waves, the tides, or other transporting agent is referred to as the load. Streams carry a variety of materials as they make their way to the sea. These materials range from minute dissolved particles and pollutants to giant boulders moved only during the most massive floods. The bed load consists of the coarse particles that move along or close to the bottom of the streambed. Particles move more slowly than the stream, by rolling or sliding. Saltation is the movement of a particle by short intermittent jumps caused by the current lifting the particles. Bed load typically constitutes 5–50 percent of the total load carried by the stream, with a greater proportion carried during high-discharge floods. The suspended load consists of the fine particles suspended in the stream. This makes many streams muddy, and it consists of silt and clay that moves at the same velocity as the stream. The suspended load generally accounts for 50–90 percent of the total load carried by the stream. The dissolved load of a stream consists of dissolved chemicals, such as bicarbonate, calcium, sulfate, chloride, sodium, magnesium, and potassium. The dissolved load tends to be high in streams fed by groundwater. Pollutants such as fertilizers and pesticides from agriculture and industrial chemicals also tend to be carried as dissolved load in streams.

There is a wide range in the sizes and amounts of material that can be transported by a stream. The competence of a stream is the size of particles a stream can transport under a given set of hydraulic conditions, measured in the diameter of the largest bed load. A stream's capacity is the potential load it can carry, measured in the amount (volume) of sediment passing a given point in a set amount of time. The amount of material carried by streams depends on a number of factors. Climate studies show erosion rates are greatest in climates between a true desert and grasslands. Topography affects stream load, as rugged topography contributes more detritus, and some rocks are more erodable. Human activity, such as farming, deforestation, and urbanization, all strongly affect erosion rates and stream transport. Deforestation and farming greatly increase erosion rates and supply more sediment to streams, increasing their loads. Urbanization has complex effects, including decreased infiltration and decreased times between rainfall events and floods, as discussed in detail below.

The amount of sediment load available to the stream is also independent of the stream's discharge, so different types of stream channels develop in response to different amounts of sediment load availability. If the sediment load is low, streams tend to have simple channels, whereas braided stream channels develop where the sediment load is greater than the stream's capacity to carry that load. If a large amount of sediment is dumped into a stream, the stream will respond by straightening, thus increasing the gradient and stream velocity and increasing the stream's ability to remove the added sediment.

When streams enter lakes or reservoirs along their path to the sea, the velocity of the stream will suddenly decrease. This causes the sediment load of the stream or river to be dropped as a delta on the lake bottom, and the stream attempts in this way to fill the entire lake with sediment. The stream is effectively attempting to regain its gradient by filling the lake, then eroding the dam or ridge that created the lake in the first place. When the water of the stream flows over the dam, it does so without its sediment load and therefore has greater erosive power and can erode the dam more effectively.

Glaciers transport enormous amounts of rock debris, including some large boulders, gravel, sand, and fine silt in their load. The glacier may carry this at its base, on its surface, or internally. Glacial deposits are characteristically poorly sorted or non-sorted, with large boulders next to fine silt. Most of a glacier's load is concentrated along its base and sides, because in these places plucking and abrasion are most effective. In contrast, wind can only transport relatively small particles in its load, and windblown deposits tend to be

Windblown loess in Shanxii Province, China. (A) Thick loess deposit forming fertile layer used for agriculture; (B) Loess that is blown out of the Gobi Desert covers many fields and mountainsides, forming thick deposits. *(Photos by Timothy Kusky)*

well sorted reflecting the strength of the wind. Currents along the seashore and on the beach environment move huge amounts of sediment in littoral drift, and in seasonal adjustments to the beach profile.

See also GLACIER; RIVER SYSTEM; SAND DUNES.

loess A name for silt and clay deposited by wind. It forms a uniform blanket that covers hills and valleys at many altitudes, which distinguishes it from deposits of streams. Strong winds that blow across desert regions sometimes pick up dust made of silt and clay particles and transport them thousands of kilometers from their source. For instance, dust from China is found in Hawaii, and the Sahara commonly drops dust in Europe. This dust is a nuisance, has a significant influence on global climate, and has, at times, as in the dust bowl days of the 1930s, been known to nearly block out the sun.

Recently, it has been recognized that windblown dust contributes significantly to global climate. Dust storms that come out of the Sahara can be carried around the world and can partially block out some of the Sun's radiation. The dust particles may also act as small nuclei for raindrops to form around, perhaps acting as a natural cloud-seeding phenomenon. One interesting point to ponder is that as global warming increases global temperatures, the amount and intensity of storms increase, and some of the world's deserts expand. Dust storms may serve to reduce global temperatures and increase precipitation. In this way, dust storms may represent some kind of self-regulating mechanism, whereby the Earth moderates its own climate.

See also DESERT.

longshore drift *See* LITTORAL.

Long Valley The Long Valley caldera is a 10-mile by 20-mile (15–30 km) oval-shaped depression located 12 miles (20 km) south of Mono Lake along the eastern side of the Sierra Nevada in eastern California, at the boundary with the Basin and Range Province. This region is one of the most dangerous volcanically active areas in North America, having produced numerous volcanic eruptions over the past 3 million years. A massive caldera-forming event occurred 760,000 years ago, and the most recent eruptions occurred 250 and 600 years ago. A period of new unrest including swarms of seismic activity, changes in thermal springs and gas emissions, and doming of the southern part of the caldera indicates renewed magmatic activity and the potential of an eruption. The activity is being monitored closely by the U.S. Geological Survey, but warnings of volcanic danger are resisted by local businesses because of fear of decreased tourism.

The Long Valley caldera is one of the largest Quaternary rhyolitic volcanic centers in North America. The caldera floor has an elevation of 6,500–8,500 feet (2,000–2,600 m), and the walls reach heights of 9,800–11,500 feet (3,000–3,500 m). The main eruptive phase from the caldera began 3.6 million years ago with the eruption of basaltic-andesitic lavas that covered 1,500 square miles (4,000 km²), followed by rhyo-dacite flows and domes. The caldera produced a catastrophic eruption 730,000 years ago when the roof of the caldera collapsed along with the expulsion of 150 cubic miles (600 km³) of rhyolitic magma in Plinian ash clouds, more than 10–20 times as much as is typically produced from the largest convergent margin stratovolcano eruptions such as Pinatubo or Tambora. This eruption formed the Bishop Tuff that covers large parts of southern California, Nevada, Mexico, all of Arizona, Utah, Colorado, most of New Mexico, Wyoming, and parts of Texas, Oklahoma, Kansas, Nebraska, South Dakota, and Idaho. Small eruptions continued after the main catastrophic eruption, and a resurgent dome formed within 100,000 years after the main eruption. New eruptions produced rhyolitic domes at 500,000, 300,000, and 100,000 years ago.

A dozen of the domes that were extruded from the southwest margin of the caldera from 200,000 to 50,000 years ago have merged to produce the 11,050-feet (3,370-m) high Mammoth Mountain, a popular ski resort. The Mono Lake–Inyo Craters area just to the north of Long Valley is part of the same volcanic province, and the eruption history of these craters overlaps that of Long Valley. Mafic eruptions from the Mono Lake caldera began 300,000–200,000 years ago, and the youngest mafic flow is 13,300 years old. Volcanism migrated to the north around 35,000 years ago, and the Mono Craters chain northwest of the caldera consists of about 30 coalesced domes, flows, and craters, the youngest of which is 600 years old. On the northwest rim of the caldera, the Inyo Craters consists of domes, flows, and craters that range in age from 6,000 to 500 years old, with abundant evidence for explosive phreatic eruptions. There is an approximate 500-year interval between major eruptions in this area. Recent increased activity in the area suggests that a new eruptive phase could be imminent in the Long Valley–Mono Lake–Inyo Craters region.

Further Reading
Miller, C. D., Donal R. Mullineaux, Dwight R. Crandell, and R. C. Bailey. "Potential Hazards of Future Volcanic Eruptions in the Long Valley-Mono Lake Area, East Central California and Southwest Nevada, United States." *Geological Survey Circular* 877 (1982).

Lyell, Charles (1779–1875) British *Geologist, Evolutionary Biologist* Charles Lyell was born in Scotland, and he attended Oxford University where his main interests at the time were mathematics, classics, law, and geology. He continued his career in law but later turned to geology and became one of the most well-renowned geologists of our time. His zoology skills helped him in his extensive studies and observations around the world. Lyell is well known for his argument that stated that geological processes that were easy to observe were adequate enough to explain the geological history of the Earth, and that the rain, oceans, volcanoes, and earthquakes explained the geological history of the Earth in ancient times. He put these ideas in his books entitled *The Geological Evidence of the Antiquity of Man* and *Principles of Geology*. Lyell was against the geological theories of his time. He believed that it was necessary to create a timescale that depicted the Earth's history. His book *Principles of Geology* was divided into several volumes. The first three were particularly important because they introduced ideas pertaining to igneous, metamorphic, and sedimentary rocks. His third volume also dealt with paleontology and stratigraphy. According to Charles Darwin, "The greatest merit of the *Principles* was that it altered the whole tone of one's mind, and therefore that, when seeing a thing never seen by Lyell, one yet saw it through his eyes."

Further Reading
Lyell, Charles. "On the Upright Fossil Trees Found at Different Levels in the Coal Strata of Cumberland, N. S." *Quarterly Journal of the Geological Society of London* (1843): 176–178.
———. *Principles of Geology*. New Haven, Conn.: Hezekiah Howe and Co., 184?.

M

Madagascar The world's fourth largest island, covering 388,740 square miles (627,000 km²) in the western Indian Ocean off the coast of southeast Africa. Madagascar is in a great many ways one of the most distinctive locations in the world. It is recognized as biologically distinguished because of its singular community of fauna and flora and its unique role in the evolution and sustenance of life as evidenced in the geologic record. Madagascar is a valuable natural laboratory for the study of many phenomena ranging from biological systems, cultural adaptations, and fundamental geologic and geomorphic structures and material that form the foundation of the island and the platform on which all other systems have been situated and developed. Although, much effort has been focused in recent years on many aspects of Madagascar, from a geologic and geomorphic material perspective the island is not sufficiently understood in detail. Unlocking the secrets to the geology of this region could reveal information about the formation, breakup, and dispersal of several supercontinents including Rodinia, Gondwana, and Pangea. French colonial geologists pioneered by Bernard Moine mapped and described the principal geologic elements of Madagascar.

The island consists of a highland plateau fringed by a lowland coastal strip on the east, with a very steep escarpment dropping thousands of feet from the plateau to the coast over a distance of only 50 to 100 miles. The highest points in Madagascar are Mount Maromokotro in the north, which rises to 9,450 feet (2,882 m), and the Ankaratra Mountains in the center of the plateau, which rise to 8,670 feet (2,645 m). The plateau dips gently to the western coast of the island toward the Mozambique Channel, where wide beaches are located. There are several islands around the main island, including Isle St. Marie in the northeast and Nosy-Be in the north. Most of the high plateau of Madagas-

car was once heavily forested, but intense logging over the last century has left most of the plateau a barren, rapidly eroding soil and bedrock-covered terrain. Red soil eroded from the plateau has filled many of the river estuaries along the coast.

Precambrian rocks underlie the eastern two-thirds of Madagascar, and the western third of the island is underlain by sedimentary and minor volcanic rocks that preserve a near-complete record of sedimentation from the Devonian to Recent. The Ranotsara fault zone divides the Precambrian bedrock of Madagascar into two geologically different parts. The northern part is underlain by Middle and Late Archean orthogneisses, variably reworked in the Early and Late Neoproterozoic, whereas the southern part, known as the Bekily block, consists dominantly of graphite-bearing paragneisses, bounded by north-south trending shear zones that separate belts with prominent fold-interference patterns. All rocks south of the Ranotsara fault zone have been strongly reworked and metamorphosed to granulite conditions in the latest Neoproterozoic. Because the Ranotsara and other sinistral fault zones in Madagascar are subvertical, their intersections with Madagascar's continental margin provide ideal piercing points to match with neighboring continents in the East African Orogen. Thus, the Ranotsara fault zone is considered an extension of the Surma fault zone or the Ashwa fault zone in East Africa, or the Achankovil or Palghat-Cauvery fault zones in India. The Palghat-Cauvery fault zone changes strike to a north-south direction near the India-Madagascar border and continues across north Madagascar. The Precambrian rocks of northern Madagascar can be divided into three north-south trending tectonic belts defined, in part, by the regional metamorphic grade. These belts include the Bemarivo block, the Antongil block, and the Antananarivo block.

The Bemarivo block of northernmost Madagascar is underlain by calc-alkaline intrusive igneous rocks (Andriba Group) with geochemical compositions suggestive of rapid derivation from depleted mantle sources. These rocks are strikingly similar in age, chemistry, and isotopic characteristics to the granitoids of the Seychelles and Rajasthan (India). The Andriba granitoids are overlain by the Daraina-Milanoa Group (~750–714 million years old) in the north, and juxtaposed against the Sambirano Group in the south. A probable collision zone separates the Sambirano Group from the Andriba Group. The Daraina-Milanoa Group consists of two parts: a lower, largely clastic metasedimentary sequence and an upper volcanic sequence dominated by andesite with lesser basalt and rhyolite. Like the Andriba Group, volcanic rocks of the Daraina-Milanoa Group are calc-alkaline in chemistry and have Nd isotope signatures indicating a juvenile parentage. Cu and Au mineralization occurs throughout the belt. The Sambirano Group consists of pelite schist, and lesser quartzite and marble, which are variably metamorphosed to greenschist grade (in the northeast) and amphibolite grade (in the southwest). In its central part, the Sambirano Group is invaded by major massifs of migmatite gneiss and charnockite. The depositional age and provenance of the Sambirano Group is unknown.

The Antongil block, surrounding the Bay of Antongil and Isle St. Marie, consists of late Archean biotite granite and granodiorite, migmatite, and tonalitic and amphibolitic gneiss, bound on the west by a belt of Middle Archean metasedimentary gneiss and migmatite. The tonalitic gneisses of this region are the oldest rocks known on the island of Madagascar, dated (U-Pb on zircon) at $3{,}187{\pm}2$ million years. The older gneisses and migmatites are intruded by circa $2{,}522{\pm}2$ million-year-old epidote-bearing granite and granodiorite. Late Archean gneisses and migmatites near the coast in the Ambositra area may be equivalent to those near the Bay of Antongil, although geochronological studies are sparse and have not yet identified middle Archean rocks in this area. Rocks of the Antongil block have greenschist to lower-amphibolite metamorphic assemblages, in contrast to gneisses in the Antananarivo block, which tend to be metamorphosed to granulite facies. This suggests that the Antongil block may have escaped high-grade Neoproterozoic events that affected most of the rest of the island. Gneisses in this block are broadly similar in age and lithology to the peninsular gneisses of southern India. High-grade psammites of the Sambirano Group unconformably overlie the northern part of the Antongil block and become increasingly deformed toward the north in the Tsarantana thrust zone, a Neoproterozoic or Cambrian collision zone between the Bemarivo block and central Madagascar. The western margin of the Antongil block is demarcated by a 30-mile (50-km) wide belt of pelitic metasediments with tectonic blocks of gabbro, harzburgite, and chromitites, with nickel and emerald deposits. This belt,

named the Betsimisiraka suture, may mark the location of the closure of a strand of the Mozambique Ocean that separated the Antongil block (and southern India?) from the Antananarivo block.

The Antananarivo block is the largest Precambrian unit in Madagascar, consisting mainly of 2,550–2,490-million-year-old granitoid gneisses, migmatites, and schist intruded by 1,000–640-million-year-old calc-alkaline granites, gabbro, and syenite. Rocks of the Antananarivo block were strongly reworked by high-grade Neoproterozoic tectonism between 750 million and 500 million years ago and metamorphosed to granulite facies. Large, sheet-like granitoids of the stratoid series intruded the region, perhaps during a phase of extensional tectonism. Rocks of the Antananarivo block were thrust to the east on the Betsimisiraka suture over the Antongil block around 630–515 million years ago, then intruded by post-collisional granites (such as the 537–527-million-year-old Carion granite, and the Filarivo and Tomy granites) 570–520 million years ago.

The Sèries Quartzo-Schisto-Calcaire or QSC (also known as the Itremo Group) consists of a thick sequence of Mesoproterozoic stratified rocks comprising, from presumed bottom to top, quartzite, pelite, and marble. Although strongly deformed in latest Neoproterozoic time (~570–540 million years ago), the QSC is presumed to rest unconformably on the Archean gneisses of central Madagascar because both the QSC and its basement are intruded by Early Neoproterozoic (~800-million-year-old) granitoids, and no intervening period of tectonism is recognized. The minimum depositional age of the QSC is ~800 million years ago—and its maximum age of ~1,850 million years ago—is defined by U-Pb detrital zircon geochronology. The QSC has been variably metamorphosed (~570–540 million years ago; greenschist grade in the east; amphibolite grade in the west) and repeatedly folded and faulted, but original sedimentary structures and facing-directions are well preserved. Quartzite displays features indicative of shallow subaqueous deposition, such as flat lamination, wave ripples, current ripple cross lamination, and dune cross-bedding, and carbonate rocks have preserved domal and pseudo-columnar stromatolites. To the west of the Itremo Group, rocks of the Amboropotsy and Malakialana Groups have been metamorphosed to higher grade, but include pelites, carbonates, and gabbro that may be deeper water equivalents of the Itremo Group. A few areas of gabbro/amphibolite-facies pillow lava/marble may represent strongly metamorphosed and dismembered ophiolite complexes.

Several large greenstone belts crop out in the northern part of the Antananarivo block. These include the Maevatana, Andriamena, and Beforana-Alaotra greenstone belts, collectively called the Tsarantana sheet. Rocks in these belts include metamorphosed gabbro, mafic gneiss, tonalites, norite, and chromitites, along with pelites and minor magnetite-iron formation. Some early intrusions in these belts

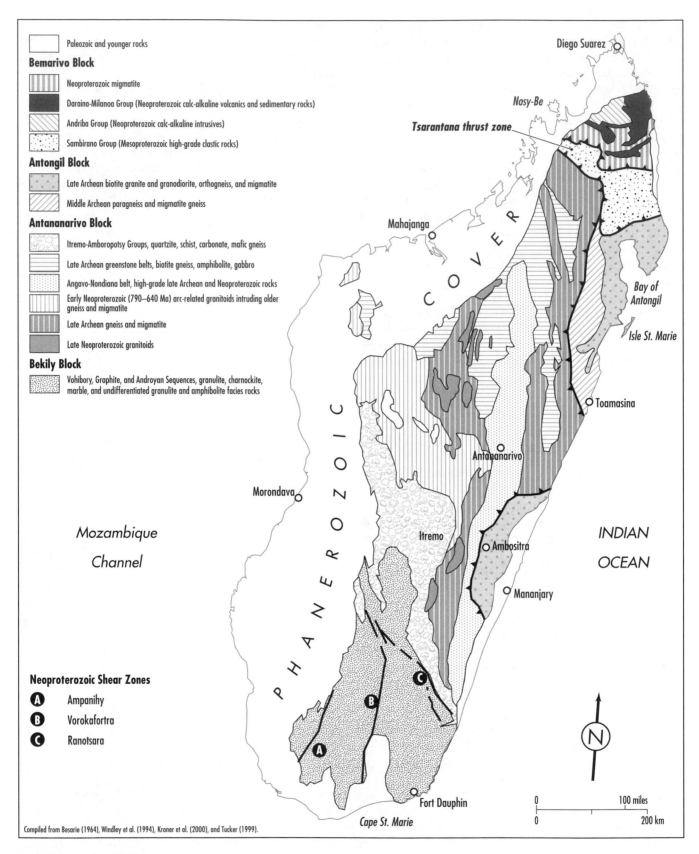

Simplified geologic map of Madagascar

Paleozoic and younger rocks

Bemarivo Block

Neoproterozoic migmatite

Daraina-Milanoa Group (Neoproterozoic calc-alkaline volcanics and sedimentary rocks)

Andriba Group (Neoproterozoic calc-alkaline intrusives)

Sambirano Group (Mesoproterozoic high-grade clastic rocks)

Antongil Block

Late Archean biotite granite and granodiorite, orthogneiss, and migmatite

Middle Archean paragneiss and migmatite gneiss

Antananarivo Block

Itremo-Amboropotsy Groups, quartzite, schist, carbonate, mafic gneiss

Late Archean greenstone belts, biotite gneiss, amphibolite, gabbro

Angavo-Nondiana belt, high-grade late Archean and Neoproterozoic rocks

Early Neoproterozoic (790–640 Ma) arc-related granitoids intruding older gneiss and migmatite

Late Archean gneiss and migmatite

Late Neoproterozoic granitoids

Bekily Block

Vohibory, Graphite, and Androyan Sequences, granulite, charnockite, marble, and undifferentiated granulite and amphibolite facies rocks

Neoproterozoic Shear Zones

A Ampanihy

B Vorokafortra

C Ranotsara

Diego Suarez

Nosy-Be

Tsarantana thrust zone

Mahajanga

C O V E R

Bay of Antongil

Isle St. Marie

Toamasina

Antananarivo

Morondava

Mozambique Channel

P H A N E R O Z O I C

Itremo

Ambositra

Mananjary

INDIAN OCEAN

Fort Dauphin

Cape St. Marie

N

0   100 miles

0   200 km

Compiled from Besarie (1964), Windley et al. (1994), Kroner et al. (2000), and Tucker (1999).

have been dated by Robert Tucker to be between 2.75 billion and 2.49 billion years old, with some 3.26-billion-year-old zircon xenocrysts and Middle Archean Nd isotopic signatures. The chemistry, age, and nature of chromite mineralization all suggest an arc setting for the mafic rocks of the Tsarantana sheet, which is in thrust contact with underlying gneisses of the Antananarivo block. The thrust zone is not yet well documented, but limited studies indicate east-directed thrusting. The 800–770-million-year-old gabbro cut early fabrics but are deformed into east-vergent asymmetric folds cut by east-directed thrust faults.

The effects of Neoproterozoic orogenic processes are widespread throughout the Antananarivo block. Archean gneisses and Mesoproterozoic stratified rocks are interpreted as the crystalline basement and platformal sedimentary cover, respectively, of a continental fragment of undetermined tectonic affinity (East or West Gondwanan, or neither). This continental fragment (both basement and cover) was extensively invaded by subduction-related plutons in the period from about 1,000 million to ~720 million years ago, that were emplaced prior to the onset of regional metamorphism and deformation. Continental collision related to Gondwana's amalgamation began after ~720 million years ago and before ~570 million years ago and continued throughout the Neoproterozoic with thermal effects that lasted until about 520 million years ago. The oldest structures produced during this collision are kilometer-scale fold and thrust-nappes with east or southeast-directed vergence (present-day direction). They resulted in the inversion and repetition of Archean and Proterozoic rocks throughout the region. During this early phase of convergence warm rocks were thrust over cool rocks thereby producing the present distribution of regional metamorphic isograds. The vergence of the nappes and the distribution of metamorphic rocks are consistent with their formation within a zone of west- or northwest-dipping continental convergence (present-day direction). Later upright folding of the nappes (and related folds and thrusts) produced km-scale interference fold patterns. The geometry and orientation of these younger upright folds is consistent with east-west horizontal shortening (present-day direction) within a sinistral transpressive regime. This final phase of deformation may be related to motion along the Ranotsara and related shear zones of south Madagascar, and to the initial phases of lower crustal exhumation and extensional tectonics within greater Gondwana.

South Madagascar, known as the Bekily block, consists of upper amphibolite and higher-grade paragneiss bounded by north-south–striking shear zones that separate belts with prominent fold interference patterns. Archean rocks south of the Ranotsara shear zone have not been positively identified, but certain orthogneisses have Archean ages (~2,905 million years) that may represent continental basement to the paragneisses of the region. All rocks south of the Ranotsara shear

zone have been strongly reworked and metamorphosed in the latest Neoproterozoic. The finite strain pattern of refolded folds results from the superimposition of at least two late Neoproterozoic deformation events characterized by early subhorizontal foliations and a later network of kilometer-scale vertical shear zones bounding intensely folded domains. These latest upright shears are clearly related to late Neoproterozoic horizontal shortening in a transpressive regime under granulite facies conditions.

The western third of Madagascar is covered by Upper Carboniferous (300 million years old) to Mid-Jurassic (180 million years old) basinal deposits that are equivalents to the Karoo and other Gondwanan sequences of Africa and India. There are three main basins, including from south to north, the Morondova, Majunga (or Mahanjanga), and Diego (or Ambilobe) basins. Each has a similar three-fold stratigraphic division including the Sakoa, Sakamena, and Isalo Groups, with mainly sandstones, limestones, and basalts, overlain by unconsolidated sands in the south and along the western coast. These basins formed during rifting of Madagascar from Africa and have conjugate margins along the east coast of southern and central Africa. The base of the Morondova basin, the oldest of the three, has spectacular glacial deposits including diamictites, tillites, and glacial outwash gravels. These are overlain by coals and arkoses, along with plant fossil– (Glossopteris-) rich mudstones thought to represent meandering stream deposits. Marine limestones cap the Sakoa Group. Fossiliferous deltaic and lake deposits of the Sakamena Group prograde (from the East) over the Sakoa Group. The uppermost Isalo Group is 0.6–3.7 miles (1–6 km) thick, consisting of large-scale cross-bedded sandstones, overlain by red beds and fluvial deposits reflecting arid conditions. Mid-Jurassic limestones (Ankara and Kelifely Formations) mark a change to subaqueous conditions throughout the region.

See also GONDWANA; SUPERCONTINENT CYCLE.

Further Reading
Ashwal, Lewis D., Robert D. Tucker, and Ernst K. Zinner. "Slow Cooling of Deep Crustal Granulites and Pb-Loss in Zircon." *Geochimica et Cosmochimica Acta* 63 (1999).
Besairie, Henri. Carte géologique de Madagascar: Antananarivo, Service Géologique de Madagascar, scale: 1:1,000,000, color (3 sheets), 1964.
Collins, Alan S., Ian C. W. Fitzsimons, Bregje Hulscher, and Theodore Razakamanana. "Structure of the Eastern East African Orogen in Central Madagascar." In *Evolution of the East African and Related Orogens, and the Assembly of Gondwana,* edited by Timothy M. Kusky, Mohamed Abdelsalam, Robert Tucker, and Robert Stern, 111–134. Precambrian Research, 2003.
Cox, Ronadh, Richard A. Armstrong, and Lewis D. Ashwal. "Sedimentology, Geochronology and Provenance of the Proterozoic Itremo Group, Central Madagascar, and Implications for Pre-Gondwana Palaeogeography." *Journal of the Geological Society, London* 155 (1998).

de Wit, Maarten J. "Madagascar: Heads It's a Continent, Tails It's an Island." *Annual Reviews of Earth and Planetary Sciences* 31 (2003): 213–248.

Fernandez, Alan., Guido Schreurs, I. M. Villa, S. Huber, and Michel Rakotondrazafy. "Tectonic Evolution of the Itremo Region (Central Madagascar) and Implications for Gondwana Assembly." In *Evolution of the East African and Related Orogens, and the Assembly of Gondwana,* edited by Timothy M. Kusky, Mohamed Abdelsalam, Robert Tucker, and Robert Stern. Precambrian Research, 2003.

Goncalves, Phillipe, Christian Nicollet, and Jean M. Lardeaux. "Finite Strain Pattern in Andriamena Unit (North-Central Madagascar): Evidence for Late Neoproterozoic-Cambrian Thrusting during Continental Convergence." In *Evolution of the East African and Related Orogens, and the Assembly of Gondwana,* edited by Timothy M. Kusky, Mohamed Abdelsalam, Robert Tucker, and Robert Stern. Precambrian Research, 2003.

Handke, Michael J., Robert D. Tucker, and Lewis D. Ashwal. "Neoproterozoic Continental Arc Magmatism in West-Central Madagascar." *Geology* 27 (1999).

Kröner, Alfred, E. Hegner, Alan S. Collins, Brian F. Windley, Tim S. Brewer, Theodore Razakamanana, and Robert T. Pidgeon. "Age and Magmatic History of the Antananarivo Block, Central Madagascar, as Derived from Zircon Geochronology and Nd Isotopic Systematics." *American Journal of Science* 300 (2000).

Kröner, Alfred, Brian F. Windley, P. Jaeckel, Tim S. Brewer, Theodore Razakamanana. "New Zircon Ages and Regional Significance for the Evolution of the Pan-African Orogen in Madagascar." *Journal of the Geological Society, London* 156 (1999).

Martelat, Jean E., Jean M. Lardeaux, Christian Nicollet, and Raymond Rakotondrazafy. "Strain Pattern and Late Precambrian Deformation History in Southern Madagascar." *Precambrian Research* 102 (2000): 1–20.

Moine, Bernard. "Caractères de sédimentation et de métamorphisme des séries précambriennes épizonales à catazonales du centre de Madagascar (Région d'Ambatofinandrahana)." *Sciences de la terre, Mémoire* 31, 1974.

———. Carte du massif schisto-quartzo-dolomitique, region d'Ambatofinandrahana, centre-ouest du socle cristallin précambrien de Madagascar. Centre de l'Institut Géographique National à Tananarive (Imprimeur), *Sciences de la terre,* Nancy (Editeur), scale: 1:200,000, 1968.

———. "Relations stratigraphiques entre la série <schisto-quartzo-calcaire> et les gneiss environants (centre-ouest de Madagascar): Données d'une première étude géochimique." *Comptes Rendus de la Semaine Géologique de Madagascar,* 1967.

———. "Grand traits structuraux du massif schisto-quartzo-calcaire (centre-ouest de Madagascar." *Comptes Rendus de la Semaine Géologique de Madagascar,* 1966.

———. "Étude du massif schisto-quartzo-calcaire du centre-ouest de l'Ile." *Rapport Annuel du Service Géologique* (1965).

———. "Contribution a l'étude géologique du massif schisto-quartzo-calcaire du centre-ouest de Madagascar." *Comptes Rendus de la Semaine Géologique de Madagascar,* 1965.

Moine, Bernard, Betrondro Ambatondradama Miandrivazo. Carte Géologique, 1:100,000. Service Géologique, Tananarive, 1964.

Nédélec, Ann, W. E. Stephens, and Anthony E. Fallick. "The Panafrican Stratoid Granites of Madagascar; Alkaline Magmatism in a Post-Collisional Extensional Setting." *Journal of Petrology* 36 (1995).

Tucker, Robert T., Timothy M. Kusky, Robert Buchwaldt, and Michael Handke. "Neoproterozoic Nappes and Superimposed Folding of the Itremo Group, West-Central Madagascar." *Precambrian Research* (2004).

Tucker, Robert D., Lewis D. Ashwal, Michael J. Handke, M. A. Hamilton, M. LeGrange, and R. A. Rambeloson. "U-Pb Geochronology and Isotope Geochemistry of the Archean and Neoproterozoic Rocks of Madagascar." *Journal of Geology* 107 (1999).

Windley, Brian F., Adriantefison Razafiniparany, Theodore Razakamanana, D. Ackemand. "Tectonic Framework of the Precambrian of Madagascar and its Gondwanan Connections: A Review and Reappraisal." *Geologische Rundschau* 83 (1994).

mafic *See* IGNEOUS ROCKS.

magma *See* IGNEOUS ROCKS.

magnetic anomalies *See* PALEOMAGNETISM; PLATE TECTONICS.

magnetic compass The magnetic compass is an instrument that indicates the whole circle bearing from the magnetic meridian to a particular line of sight. It consists of a needle that aligns itself with the Earth's magnetic flux, and with some type of index that allows for a numeric value for the calculation of bearing. A compass can be used for many things. The most common application is for navigation. People are able to navigate throughout the world by simply using a compass and map. The accuracy of a compass is dependent on other local magnetic influences such as man-made objects or natural abnormalities such as local geology. The compass needle does not really point true north but is attracted and oriented by magnetic force lines that vary in different parts of the world and are constantly changing. For example, when you read north on a compass you are reading the direction toward the magnetic north pole. To offset this phenomenon we use calculated declination values to convert the compass reading to a usable map reading. Since the magnetic flux changes through time it is necessary to replace older maps with newer maps to insure accurate and precise up-to-date declination values.

See also GEOMAGNETISM; PALEOMAGNETISM.

magnetic field *See* GEOMAGNETISM; PALEOMAGNETISM.

magnetosphere The limits of the Earth's magnetic field, as confined by the interaction of the solar wind with the planet's internal magnetic field. The natural undisturbed state of the Earth's magnetic field is broadly similar to a bar magnet, with magnetic flux lines (of equal magnetic intensity and direction) coming out of the south polar region and returning back into the north magnetic pole. The solar wind, consisting of supersonic H^+ and $^4He^{2+}$ ions expanding away from the Sun, deforms this ideal state into a teardrop-shaped configuration known as the magnetosphere. The magnetosphere has a rounded compressed side with about 6–10 Earth radii facing the sun, and a long tail (magnetotail) on the opposite side that stretches past the orbit of the moon. It is likely that the magnetotail is open, meaning that the magnetic flux lines probably never close but instead merge with the interplanetary magnetic field. The magnetosphere shields the Earth from many of the charged particles from the Sun by deflecting them around the edge of the magnetosphere, causing them to flow harmlessly into the outer solar system.

The Sun periodically experiences periods of high activity when many solar flares and sunspots form. During these periods the solar wind is emitted with increased intensity, and the solar plasma is emitted with greater velocity, in greater density, and with more energy than in its normal state. During these periods of high solar activity, the extra pressure of the solar wind distorts the magnetosphere and causes it to move around and also causes increased auroral activity.

See also AURORA.

Further Reading

Merrill, Ronald T., and Michael W. McElhinny. *The Earth's Magnetic Field; Its History, Origin and Planetary Perspective.* International Geophysics Series, vol. 32. London: Academic Press, 1983.

manganese nodule   Small black to brown concretionary nodules made mostly of manganese salts and manganese oxides are common on the floors of many parts of the world's oceans and great lakes, especially in areas of slow pelagic sedimentation. They generally contain about 15–30 percent manganese (Mn) and so represent a significant manganese resource, though technically difficult to mine because of their depth beneath the ocean. They also contain smaller amounts of copper, cobalt, nickel, and other minerals. Manganese nodules were first discovered during the 1872–76 voyage of the research vessel H.M.S. *Challenger.*

The nodules consist of a number of different minerals, including compounds of manganese and iron. The nodules are typically concentrically layered like an onion and have a nucleus of foreign material where growth started. The nuclei are highly varied between different nodules, with examples including radiolaria, diatoms, clays, basalts, and even shark's teeth. Growth rates are thought to be about a few millimeters per million years. Manganese occurs primarily as the oxide MnO_2, but also as the mineral todorokite and birnessite. Iron occurs as goethite and as ferric hydroxide. The cobalt, copper, and nickel typically occur in the manganese minerals and absorbed on the surface of manganese dioxide.

The mechanism of growth of manganese nodules and the origin of the Mn and other metals has remained elusive. Manganese may be derived from erosion of the continents or from seafloor volcanism. Once in the oceans, the metals may accumulate in the sedimentary pile that forms on the seafloor and gradually move upward through the sediments where it becomes oxidized and accumulates in nodules on the seafloor. There has been speculation that microorganisms might be involved in the process, but no evidence of biologic processes has yet been documented.

mantle   Volumetrically, the mantle forms about 80 percent of the Earth, occupying the region between the crust and upper core, between about 20 and 1,800 miles (35–2,900 km) depth. It is divided into two regions, the upper and lower mantle, and is thought to be composed predominantly of silicate minerals in closely packed high-pressure crystal structures. Some of these are high-pressure forms of more common silicates, formed under high temperature and pressure conditions found in the deep Earth. The upper mantle extends to a depth of about 146 miles (670 km), and the lower mantle extends from there to the core-mantle boundary near 1,800 miles (2,900 km) depth.

Most of our knowledge about the mantle comes from seismological and experimental data, as well as rare samples of upper mantle material that has made its way to the surface in kimberlite pipes, volcanoes, and in some ophiolites and some other exhumed subcrustal rocks. Most of the mantle rock samples that have made their way to the surface are composed of peridotite, with a mixture of the minerals olivine, pyroxene, and some garnet. The velocities of seismic waves depend on the physical properties of the rocks they travel through. Seismic experiments show that S-waves (shear waves) propagate through the mantle, so it is considered to be a solid rocky layer since S-waves do not propagate through liquids. The temperature does not increase dramatically from the top to the bottom of the mantle, indicating that an effective heat transfer mechanism is operating in this region. Heat transfer by convection, in which the material of the mantle is flowing in large-scale rotating cells, is a very efficient mechanism that effectively keeps this region at nearly the same temperature throughout. In contrast, the lithosphere (occupying the top of the mantle and crust) is not convecting but cools by conduction, so it shows a dramatic temperature increase from top to bottom.

There are several discontinuities in the mantle, where seismic velocities change across a discrete layer or zone. Between about 62 and 155 miles (100–250 km) depth, both P and S-waves (compressional and shear waves) decrease in velocity, indicating that this zone probably contains a few percent partial melt. This low-velocity zone is equated with the upper part of the asthenosphere, upon which the plates of the lithosphere move. The low-velocity zone extends around most of the planet; however, it has not been detected beneath many Archean cratons that have thick roots, leading to uncertainty about the role that the low-velocity zone plays in allowing these cratons to move with the tectonic plates. The asthenosphere extends to about 415–435 miles (670–700 km) depth. At 250 miles (400 km), seismic velocities increase rather abruptly, associated with an isochemical phase change of the mineral olivine ($[Mg,Fe]_2SiO_4$) to a high-pressure mineral known as wadsleyite (or beta-phase), then at 325 miles (520 km) this converts to a high-pressure spinel known as ringwoodite. The base of the asthenosphere at 415 miles (670 km) is also associated with a phase change from spinel to the minerals perovskite ($[Mg,Fe]SiO_3$) and magnesiowustite ($[Mg,Fe]O$), stable through the lower mantle or mesosphere that extends to 1,800 miles (2,900 km).

The nature of the seismic discontinuities in the mantle has been debated for decades. A. E. Ringwood proposed a model in which the composition of the mantle started off essentially homogeneous, consisting of the hypothetical composition pyrolite (82 percent harzburgite and 18 percent basalt). Extraction of basalt from the upper mantle has led to its depletion relative to the lower mantle. An alternative model, proposed by Don Anderson, poses that the mantle is compositionally layered, with the 670-kilometer discontinuity representing a chemical as well as a phase boundary, with more silica-rich rocks at depth. As such, this second model requires that convection in the mantle be of a two-layer type, with little or no mixing between the upper and lower mantle to maintain the integrity of the chemical boundary. Recent variations on these themes include a two-layered convecting mantle with subducting slab penetration downward through the 670-kilometer discontinuity, and mantle plumes that move up from the core mantle boundary through the 670-kilometer discontinuity. In this model, the unusual region at the base of the mantle known as D" may be a place where many subducted slabs have accumulated.

The basal, D" region of the mantle is unusual and represents one of the most significant boundaries in the Earth. The viscosity contrast across the boundary is huge, being several times that of the rock/air interface. D" is a boundary layer, so temperatures increase rapidly through the layer, and there is a huge seismic discontinuity at the boundary. P-waves drop in velocity from about 8.5 to 4 miles per second (14 to 8 km/s), and S-waves do not propagate across the boundary since the outer core is a liquid. Research into the nature of the D" layer is active, and several ideas have emerged as possibilities for the nature of this region. It may be a slab graveyard where subducted slabs temporarily accumulate, or it could be a chemical reaction zone between the lower mantle and outer core. It could be a remnant of chemical layering formed during the early accretion and differentiation of the Earth, or it could be material that crystallized from the core and floated to accumulate at the core/mantle boundary. Whatever the case, it has been speculated that there may be an analogue to plate tectonics operating in this region, since it is a viscosity and thermal boundary layer, subjected to basal traction forces by the rapidly convecting outer core.

See also CONVECTION AND THE EARTH'S MANTLE; KIMBERLITE; LITHOSPHERE; PLATE TECTONICS.

Further Reading

Anderson, Don L. *Theory of the Earth*. Oxford: Blackwell Scientific Publications, 1989.
Ringwood, A. E. *Composition and Petrology of the Earth's Mantle*. New York: McGraw-Hill, 1975.

mantle plumes   The mantle of the Earth convects with large cells that generally upwell beneath the oceanic ridges and downwell with subduction zones. These convection cells are the main way that the mantle loses heat. In addition to these large cells, a number of columnar plumes of hot material upwell from deep within the mantle, perhaps even from the core-mantle boundary. Heat and material in these plumes move at high velocities relative to the main mantle convection cells, and therefore they burn their way through the moving mantle and reach the surface forming thick sequences of generally basaltic lava. These lavas are chemically distinct from mid-ocean ridge and island arc basalts, and they form either as continental flood basalts, oceanic flood basalts (on oceanic plateaux), or shield volcanoes.

Mantle plumes were first postulated to be upper mantle hot spots that were relatively stationary with respect to the moving plates, because a number of long linear chains of islands in the oceans were found to be parallel, and all old at one end and younger at the other end. In the 1960s when plate tectonics was first recognized, it was suggested that these hot spot tracks were formed when the plates moved over hot, partially molten spots in the upper mantle that burned their way, like a blow torch, through the lithosphere, and erupted basalts at the surface. As the plates moved, the hot spots remained stationary, so the plates had a series or chain of volcanic centers erupted through them, with the youngest volcano sitting above the active hot spot. The Hawaiian-Emperor island chain is one of the most exemplary of these hot spot tracks. They are about 70 million years old in the northwest near the Aleutian arc, show a sharp bend in the middle of the chain where the volcanoes are 43 million years old, and then are progressively younger to essentially zero age beneath the island of Hawaii. The bend in the chain is thought to represent a change in the plate motion direction and is reflected in a similar change in direction of many other hot spot tracks in the Pacific Ocean.

More recently, geochemical data and seismic tomography has shown that the hot spots are produced by plumes of deep mantle material that probably rise from the D" layer at the core-mantle boundary. These plumes may rise as a mechanism to release heat from the core, or as a response to greater heat loss than is accommodated by convection. If heat is transferred from the core to D", parts of this layer may become heated, become more buoyant, and rise as thin narrow plumes

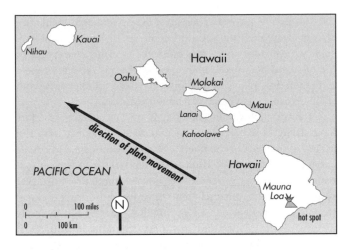

Map of Hawaii-Emperor chain formed by a hotspot track as the Pacific plate moved over a mantle plume. The youngest volcanoes are found on the island of Hawaii (Mauna Loa), and they get progressively older toward the northwest.

that rise buoyantly through the mantle. As they approach the base of the lithosphere, the plumes expand outward, forming a mushroom-like plume head that may expand to more than 600 miles (1,000 km) in diameter. Flood basalts may rise from these plume heads, and large areas of uplift, doming, and volcanism may be located above many plume heads.

There are thought to be several plumes located beneath the African plate, such as beneath the Afar region, which has experienced uplift, rifting, and flood basalt volcanism. This region exemplifies a process whereby several (typically three) rifts may propagate off of a dome formed above a plume head, and several of these may link up with rifts that propagated off other plumes formed over a large stationary plate. When several rifts link together, they may form a continental rift system that could become successful and expand into a young ocean basin, similar to the Red Sea. The linking of plume-related rifts has been suggested to be a mechanism to split supercontinents that have come to rest (in a geoid low) above a number of plumes. The heat from these plumes must eventually escape by burning through the lithosphere, forming linked rift systems that eventually rip apart the supercontinent.

Some areas of anomalous young volcanism may also be formed above mantle plume heads. For instance, the Yellowstone area has active volcanism and geothermal activity and is thought to rest above the Yellowstone hot spot, which has left a track extending northwest back across the flood basalts of the Snake River plain. Other flood basalt provinces probably also formed in a similar way. For instance, the 65-million-year-old Deccan flood basalts of India formed when this region was over the Reunion hot spot that is presently in the Indian Ocean, and these may be related to a mantle plume.

Mantle plumes may also interact with mid-ocean ridge volcanism. For instance, the island of Iceland is located on the Reykjanes Ridge, part of the Mid-Atlantic Ridge system, but the height of the island is related to unusually thick oceanic crust produced in this region because a hot spot (plume) has risen directly beneath the ridge. Other examples of mantle plumes located directly beneath ridges are found in the south Atlantic Ocean, where the Walvis and Rio Grande Ridges both point back to an anomalously thick region on the present-day ridge where the plume head is located. As the South Atlantic opened, the thick crust produced at the ridge on the plume head was split, half being accreted to the African plate, and half being accreted to the South American plate.

See also AFAR; CONVECTION AND THE EARTH'S MANTLE; FLOOD BASALT; HOT SPOT; MANTLE.

marble A metamorphic rock derived from recrystallized limestone or dolostone. It typically has a granoblastic texture with interlocking crystals and may contain colorful swirly patterns that represent deformed primary layers such as beds or veins and other secondary features. Marble is commonly used as an ornamental building stone, and rocks that qualify to be called marble by the building industry include true marbles, as well as any limestone or dolostone that can take a polish. Pure marbles may be pure white, and easily sculpted, and have thus been favored for ornamental buildings and statues for centuries.

Marianas trench Located 210 miles (338 km) southwest of Guam, the Marianas trench is an elongated depression on the floor of the Pacific Ocean. The trench is formed by the bending and subduction of the Pacific plate beneath the Marianas island arc, located south of Japan and east of the Philippines. The deepest part of the trench, the Challenger Deep, reaches 36,198 feet (11,040 m) below sea level and is the deepest known place on Earth. This deep is named after the research exploration vessel H.M.S. *Challenger II.* The trench was first detected by echo-soundings by Soviet scientists, and its bottom was first reached by two U.S. Navy divers in a bathyscaph in 1960.

The Marianas Islands are located just to the west of the trench and include volcanoes formed above the subduction zone marked by the Marianas trench to the east. Some of the islands are surrounded by limestones and reefs, deposited on a volcanic base.

See also CONVERGENT PLATE MARGIN PROCESSES; SUBDUCTION ZONE.

Mars The fourth planet from the Sun, Mars is only 11 percent of the mass of Earth. It has an average density of 3.9 grams per cubic centimeter and a diameter of 4,222 miles (6,794 km). Mars orbits the Sun every 687 days at a distance of 142 million miles (228 million km) and has a period of rotation about its axis of 24 hours, 37 minutes, and 23 seconds.

When viewed from Earth, Mars shows several striking surface features, including bright polar caps that consist mostly of frozen carbon dioxide (dry ice) that change in size with the seasons, almost disappearing in the Martian summer. Some spectacular canyons are also visible, including the 2,485-mile (4,000-km) long Valles Marineris. This canyon probably formed as a giant crack or fracture on the surface of the expanding bulge in the Tharsis region. Its surface may have later been modified by running water. Mars is prone to strong surface winds that kick up a lot of dust and generate dust storms that occasionally obscure the surface for long periods of time, an observation that led early observers to suggest that the planet may host vegetation and other life-forms.

Mars shows evidence for widespread volcanism in its past, and its surface is covered with basaltic volcanic rocks, flows, and cones and hosts several large shield volcanoes in the Tharsis and Elysium regions. The Tharsis region is a huge, North America-sized bulge on the planet that rises on average 6.2 miles (10 km) above the elevation of surrounding regions. These volcanoes are huge compared to shield volcanoes on Earth. The largest volcano, Olympus Mons, is 435 miles (700 km) across, and several others are 220 to 250 miles (350–400 km) across and rise 12.5 miles (20 km) over the surrounding terrain. The northern hemisphere of Mars is made of rolling volcanic plains, similar to lunar maria, but formed by much larger volcanic flows than on Earth or the Moon. In contrast, the southern hemisphere consists of heavily cratered highlands, with a mean elevation several kilometers higher than the volcanic plains to the north. It is estimated that the average age of the highland plains is 4 billion years, whereas the age of most of the volcanic plains in the north may be 3 billion years, with some volcanoes as young as 1 billion years old.

Recent high-resolution images of the surface of Mars have strengthened earlier views that water may once have run across the surface of the planet. Outflow channels and runoff channels are common. Runoff channels form extensive systems in the southern hemisphere and resemble dried-up river systems. Outflow channels may have formed during a catastrophic flooding episode about 3 billion years ago in the early history of the planet and are most common in equatorial regions. Flow rates are estimated to have been at least 100 times that of the Amazon River. Since there is no longer any water, or visible water ice on the surface, it has been suggested that much of this water is frozen beneath the surface in a permafrost layer, reflecting a severe global cooling since 4 billion years ago.

The atmospheric pressure on Mars is only 1/150th that of Earth, and carbon dioxide makes up most of the gas in the atmosphere, with a few percent nitrogen, argon, oxygen, carbon monoxide, and less than 0.1 percent water vapor. The temperature rises from about –244°F at 62 miles (120°K at 100 km) above the surface, to about –10°F (250°K) at the

Is There Life on Mars?

For centuries people have gazed at Mars, the red planet, and wondered if life could exist on this distant world. In the early 1900s, Percival Lowell saw long, linear features on Mars and suggested that these marks were canals or ditches carved into the surface by an advanced or perhaps dying civilization, and the imagination of the world was captured. Science-fiction novels raised people's hopes and expectations that our nearest outward neighbor might have advanced life-forms, or once have hosted superior civilizations. As technology and understanding of planetary evolution advanced, hopes of finding advanced life-forms on Mars diminished and were replaced by hopes of finding even simple life, answering the question "Are we alone?" once and for all. As exploration of the solar system continued, remnants of ice and traces of once-flowing water were discovered on Mars, increasing the chances for life. Excitement was generated twice in the late 20th century, first when meteorites thought to be from Mars had organic remains identified in them, and then again when samples taken from Mars yielded what appeared to be fossil organisms. Unfortunately, both claims were proven false, but hopes of finding life continue, and a new generation of Mars orbiters, landers, and rovers are currently exploring the planet with the hopes of finding signs of life.

Like Earth, Mars experiences cyclic climate fluctuations in response to orbital fluctuations analogous to Milankovitch cycles on Earth. However, Mars experiences much more severe fluctuations in temperature in response to the huge variations in orbital tilt of the planet. The orientation of the orbital axis of Mars with respect to the ecliptic plane of the solar system has fluctuated from 15° to 35° at least 50 times in the last 5 million years, and has swung from 0° to 60° less frequently. These dramatic changes in tilt cause huge variations in solar insolation and correspondingly large variations in temperature on the planet. When the tilt axis is most inclined, the polar regions get the most solar radiation, causing ice to vaporize and crystallize in equatorial regions, where evidence of glacial cycles have recently been documented. When Mars has a low obliquity (not tilted much), the equatorial regions get the most sunlight, and the ice in low latitudes vaporizes and condenses at the poles. Most life-forms can not tolerate such large variations in environmental conditions, making it much less likely that life will be found on the red planet, unless the life-forms have found some mechanism to minimize or tolerate the astronomically large variations in climate. However, fleeting traces of ice and flowing water offer such hope and potential for hosting life that scientists have not given up hope of finding some adaptable forms that may hide in the nearly constant temperatures of ice, deep in the soil, or elsewhere. Some models for the origin of life on Earth suggest that primitive organic compounds may have been delivered to Earth by cometary or asteroid impacts, and if this is true, it is likely that the same compounds were delivered to Mars as well. If so, we may not be alone.

2003 Mars Closest Approach

August 26, 2003
23:00 UT

August 27, 2003
10:00 UT

Hubble Space Telescope • WFPC2

NASA, J. Bell (Cornell University) and M. Wolff (Space Science Institute)
STScI-PRC03-22a

Mars. Frosty, white water ice clouds and swirling orange dust storms above a vivid, rusty landscape reveal Mars as a dynamic planet in this sharpest view ever obtained by an Earth-based telescope. NASA's Earth-orbiting Hubble Space Telescope took these pictures on August 26 and 27, 2003, when Mars was approximately 43 million miles (68 million km) from Earth—the closest Mars has been to Earth since 1988. Hubble can see details as small as 10 miles (16 km) across. Especially striking is the large amount of seasonal dust storm activity seen in this image. One large storm system is churning high above the northern polar cap (top of image), and a smaller dust storm cloud can be seen nearby. Another large dust storm is spilling out of the giant Hellas impact basin in the southern hemisphere (lower right). Hubble has observed Mars before but never in such detail. The biennial close approaches of Mars and Earth are not all the same. Mars's orbit around the Sun is markedly elliptical; the close approaches to Earth can range from 35 million to 63 million miles. Astronomers are interested in studying the changeable surface and weather conditions on Mars, in part, to help a pair of NASA missions that landed rovers on the planet's surface in 2004. The Mars opposition of 2001 preceded that of 2003 when Mars and Earth came within 35 million miles of each other, the closest since 1924 and not to be matched until 2287. *(Photo by NASA and the Hubble Heritage Team. Acknowledgment: J. Bell [Cornell University], and M. Wolff [Space Science Institute].)*

surface, with surface temperatures on average about 370°F (50°K) cooler than on Earth.

Many early speculations centered on the possibility of life on Mars, and several spectacular claims of evidence for life have been later found to be invalid. To date, no evidence for life, either present or ancient, has been found on Mars.

mass extinctions   Most species are present on Earth for about 4 million years. Many species come and go during a typically low rate of background level extinctions and evolution of new species from old, but the majority of changes occur during distinct mass-death and repopulations of the environment. The Earth's biosphere has experienced five major and numerous less-significant mass extinctions in the past 500 million years (in the Phanerozoic Era). These events occurred at the end of the Ordovician, in the Late Devonian, at the Permian-Triassic boundary, the Triassic-Jurassic boundary, and at the Cretaceous-Tertiary (K-T) boundary.

The Early Paleozoic saw many new life-forms emerge in new environments for the first time. The Cambrian explosion led to the development of trilobites, brachiopods, conodonts, mollusks, echinoderms, and ostracods. Bryozoans, crinoids, and rugose corals joined the biosphere in the Ordovician, and reef-building stromatoporoids flourished in shallow seas. The end-Ordovician extinction is one of the greatest of all Phanerozoic time. About half of all species of brachiopods and bryozoans died off, and more than 100 other families of marine organisms disappeared forever.

The cause of the mass extinction at the end of the Ordovician appears to have been largely tectonic. The major landmass of Gondwana had been resting in equatorial regions for much of the Middle Ordovician but migrated toward the South Pole at the end of the Ordovician. This caused global cooling and glaciation, lowering sea levels from the high stand where they had been resting for most of the Cambrian and Ordovician. The combination of cold climates with lower sea levels, leading to a loss of shallow shelf environments for habitation, probably were enough to cause the mass extinction at the end of the Ordovician.

The largest mass extinction in Earth history occurred at the Permian-Triassic boundary, over a period of about 5 million years. The Permian world included abundant corals, crinoids, bryozoans, and bivalves in the oceans, and on land, amphibians wandered about amid lush plant life. Of all oceanic species, 90 percent were to become extinct, and 70 percent of land vertebrates died off at the end of the Permian. This greatest catastrophe of Earth history did not have a single cause but reflects the combination of various elements.

Before the extinction event began, plate tectonics was again bringing many of the planet's landmasses together in a supercontinent (this time, Pangea), causing greater competition for fewer environmental niches by Permian life-forms. Drastically reduced were the rich continental shelf areas. As the continents collided mountains were pushed up, reducing the effective volume of the continents available to displace the sea, so sea levels fell, putting additional stress on life by further limiting the availability of favorable environmental niches. The global climate became dry and dusty, and the supercontinent formation led to widespread glaciation. This lowered sea level even more, lowered global temperatures, and put many life-forms on the planet in a very uncomfortable position, and many perished.

In the final million years of the Permian, the Northern Siberian plains let loose a final devastating blow. The Siberian flood basalts began erupting at 250 million years ago, becoming the largest known outpouring of continental flood basalts ever. Carbon dioxide was released in hitherto unknown abundance, warming the atmosphere and melting the glaciers. Other gases were also released, perhaps also including methane, as the basalts probably melted permafrost and vaporized thick accumulations of organic matter that accumulate in high latitudes like that at which Siberia was located 250 million years ago.

The global biosphere collapsed, and evidence suggests that the final collapse happened in less than 200,000 years,

and perhaps in less than 30,000 years. Entirely internal processes may have caused the end-Permian extinction, although some scientists now argue that an impact may have dealt the final death blow. After it was over, new life-forms populated the seas and land, and these Mesozoic organisms tended to be more mobile and adept than their Paleozoic counterparts. The great Permian extinction created opportunities for new life-forms to occupy now empty niches, and the most adaptable and efficient organisms took control. The toughest of the marine organisms survived, and a new class of land animals grew to new proportions and occupied the land and skies. The Mesozoic, time of the great dinosaurs, had begun.

The Triassic-Jurassic extinction is not as significant as the Permian-Triassic extinction. Mollusks were abundant in the Triassic shallow marine realm, with fewer brachiopods, and ammonoids recovered from near total extinction at the Permian-Triassic boundary. Sea urchins became abundant, and new groups of hexacorals replaced the rugose corals. Many land plants survived the end-Permian extinction, including the ferns and seed ferns that became abundant in the Jurassic. Small mammals that survived the end-Permian extinction re-diversified in the Triassic, many only to become extinct at the close of the Triassic. Dinosaurs evolved quickly in the late Triassic, starting off small, and attaining sizes approaching 20 feet (6 m) by the end of the Triassic. The giant pterosaurs were the first known flying vertebrate, appearing late in the Triassic. Crocodiles, frogs, and turtles lived along with the dinosaurs. The end of the Triassic is marked by a major extinction in the marine realm, including total extinction of the conodonts, and a mass extinction of the mammal-like reptiles known as therapsids, and the placodont marine reptiles. Although the causes of this major extinction event are poorly understood, the timing is coincident with the breakup of Pangea and the formation of major evaporite and salt deposits. It is likely that this was a tectonic-induced extinction, with supercontinent breakup initiating new oceanic circulation patterns, and new temperature and salinity distributions.

After the Triassic-Jurassic extinction, dinosaurs became extremely diverse and many quite large. Birds first appeared at the end of the Jurassic. The Jurassic was the time of the giant dinosaurs, which experienced a partial extinction affecting the largest varieties of Stegosauroids, Sauropods, and the marine Ichthyosaurs and Plesiosaurs. This major extinction is also poorly explained but may be related to global cooling. The other abundant varieties of dinosaurs continued to thrive through the Cretaceous.

The Cretaceous-Tertiary (K-T) extinction is perhaps the most famous of mass extinctions because the dinosaurs perished during this event. The Cretaceous land surface of North America was occupied by bountiful species, including herds of dinosaurs both large and small, some herbivores, and other carnivores. Other vertebrates included crocodiles, turtles, frogs, and several types of small mammals. The sky had flying dinosaurs including the vulture-like pterosaurs, and insects including giant dragonflies. The dinosaurs had dense vegetation to feed on, including the flowing angiosperm trees, tall grasses, and many other types of trees and flowers. Life in the ocean had evolved to include abundant bivalves including clams and oysters, ammonoids, and corals that built large reef complexes.

Near the end of the Cretaceous, though the dinosaurs and other life-forms did not know it, things were about to change. High sea levels produced by mid-Cretaceous rapid seafloor spreading were falling, decreasing environmental diversity, cooling global climates, and creating environmental stress. Massive volcanic outpourings in the Deccan traps and the Seychelles formed as the Indian Ocean rifted apart and magma rose from an underlying mantle plume. Massive amounts of greenhouse gases were released, raising temperatures and stressing the environment. Many marine species were going extinct, and others became severely stressed. Then, one bright day, a visitor from space about six miles (10 km) across slammed into the Yucatán Peninsula of Mexico, instantly forming a fireball 1,200 miles (1,931 km) across, followed by giant tsunamis perhaps thousands of feet tall. The dust from the fireball plunged the world into a dusty fiery darkness, months or years of freezing temperatures, followed by an intense global warming. Few species handled the environmental stress well, and more than a quarter of all the plant and animal kingdom families, including 65 percent of all species on the planet, became extinct forever. Gone were dinosaurs, mighty rulers of the Triassic, Jurassic, and Cretaceous. Oceanic reptiles and ammonoids died off, and 60 percent of marine planktonic organisms went extinct. The great K-T deaths affected not only the numbers of species but also the living biomass—the death of so many marine plankton alone amounted to 40 percent of all living matter on Earth at the time. Similar punches to land-based organisms decreased the overall living biomass on the planet to a small fraction of what it was before the K-T 1–2–3 knockout blows.

Some evidence suggests that the planet is undergoing the first stages of a new mass extinction. In the past 100,000 years, the ice ages have led to glacial advances and retreats, sea-level rises and falls, the appearance and rapid explosion of human (*Homo sapiens*) populations, and the mass extinction of many large mammals. In Australia 86 percent of large (greater than 100 pounds) animals have become extinct in the past 100,000 years, and in South America, North America, and Africa the extinction is an alarming 79 percent, 73 percent, and 14 percent. This ongoing mass extinction appears to be the result of cold climates and, more important, predation and environmental destruction by humans. The loss of large-bodied species in many cases has immediately followed the arrival of humans in the region, with the clearest examples being found in Australia, Madagascar, and New Zealand. Similar loss of races through disease and famine has accompanied many invasions and explorations of new lands by humans, suggesting we are causing a new mass extinction.

mass spectrometer   The mass spectrometer is a tool used by scientists to identify and analyze unknown compounds, to analyze and quantify known compounds, and to provide information and analysis of structural and chemical properties. The application and use of the mass spectrometer in society is known as mass spectrometry and is useful in many scientific fields. The mass spectrometer can be used to analyze almost any substance known to man. The mass spectrometer consists of an analyzer that contains a vacuum by which ions are removed from the sample by an electromagnetic field. The captured ions are then sorted according to size, velocity, and charge.

See also GEOCHRONOLOGY.

mass wasting   Movement of soil, rock and other earth materials (together called regolith) downslope by gravity without the direct aid of a transporting medium such as ice, water, or wind. It is estimated that more than 2 million mass movements occur each year in the United States alone. Mass movements occur at various rates, from a few inches per year to sudden catastrophic rock falls and avalanches that can bury entire towns under tons of rock and debris. In general, the faster the mass movement the more hazardous it is to humans, although even slow movements of soil down hill slopes can be extremely destructive to buildings, pipelines, and other societal constructions. In the United States alone, mass movements kill tens of people and cost more than $1.5 billion a year. Other mass movement events overseas have killed tens to hundreds of thousands of people in a matter of seconds. Mass wasting occurs under a wide variety of environmental conditions and forms a continuum with weathering, as periods of intense rain reduce friction between regolith and bedrock, making movement easier. Mass movements also occur underwater, such as the giant submarine landslides associated with the 1964 Alaskan earthquake.

Mass movements are a serious concern and problem in hilly or mountainous terrain, especially for buildings, roadways and other features engineered into hillsides. Mass movements are also a problem along riverbanks and in places with large submarine escarpments, such as along deltas (like the Mississippi Delta in Louisiana). The problems are further compounded in areas prone to seismic shaking or severe storm-related flooding. Imagine building a million-dollar mansion on a scenic hillside, only to find it tilting and sliding down the hill at a few inches per year. Less spectacular but common effects of slow downhill mass movements are the slow tilting of telephone poles along hillsides, and the slumping of soil from oversteepened embankments onto roadways during storms.

Mass wasting is becoming more of a problem as the population moves from the overpopulated flat land to new

Lahars of Nevado del Ruiz, Colombia, 1985

Lahars are volcanic mudflows that form on the slopes of many active volcanoes and can rush downhill at up to 40 miles per hour (64 km/hr) causing widespread death and destruction. Lahars may form after explosive eruptions deposit thick accumulations of volcanic ash and debris on a volcano's slopes. When the ash mixes with heavy rain, or water derived from melted ice and snow, the ash and water mixture forms a material with a consistency of wet concrete that begins to flow downhill. The mixture of ash and water is typically about 40 percent ash and rock fragments, and 60 percent water. This mixture can flow with speeds greater than natural streams, up to 40 miles per hour (64 km/hr), and travel large distances, typically about 30–60 miles (50–100 km). Since these flows are denser and faster than water, they are much more destructive than normal floods. Additionally, when the lahar event is over it leaves behind a thick mud and debris layer that hardens and entombs all that was unfortunate enough to be caught in the flow.

Lahars have formed on many volcanoes. Some of the most devastating lahars in recent history have been associated with the Nevado del Ruiz volcano from the Andes. The Nevado del Ruiz volcano in Colombia entered an active phase in November 1984 and began to show harmonic earthquake tremors on November 10, 1985. At 9:37 P.M. that night, a large eruption sent an ash cloud several miles into the atmosphere, and this ash settled onto the ice cap on top of the mountain. The warm ash, together with volcanic steam, quickly melted large amounts of the ice, which mixed with the ash and formed giant lahars (mudflows) crashing down the east side of the mountain into the village of Chinchina, killing 1,800 people. The eruption continued and melted more ice that mixed with more ash and sent additional larger lahars westward. Some of these lahars moved nearly 30 miles at almost 30 miles per hour (48 km at 48 km/hr), and under a thunderous roar buried the town of Armero under 26 feet (8 m) of mud. Twenty-three thousand people died in Armero that night. Other volcanoes also have produced disastrous lahars. For instance, since the 1991 eruption of Mount Pinatubo, the homes of more than 100,000 people have been destroyed by lahars. Lahar deposits filled the channels of the Toutle River in Washington State after the 1980 eruption of Mount Saint Helens, causing much of the death and destruction from that eruption.

These lahars could have been predicted, and the huge loss of life prevented, if a program had been installed to map the deposits of previous lahars, and map the topography showing where future lahars might flow. With such warnings perhaps the towns of Armero and Chinchina could have been moved to slightly higher ground, saving tens of thousands of lives. However, even with such warnings, many communities would choose to live with the risks instead of accepting the costs and inconvenience of moving to a higher location.

developments in hilly terrain. In the past, small landslides in the mountains, hills, and canyons were not a serious threat to people, but now, with large numbers of people living in landslide-prone areas, landslide hazards and damage are rapidly increasing.

Driving Forces of Mass Wasting

Gravity is the main driving force behind mass wasting processes, as it is constantly adding material and attempting to force it downhill. On a slope, gravity can be resolved into two components, one perpendicular to the slope, and one parallel to the slope. The steeper the angle of the slope, the greater the influence of gravity. The effect of gravity reaches a maximum along vertical or overhanging cliffs.

The tangential component of gravity tends to pull material downhill and results in mass wasting. When the tangential component of gravity is great enough to overcome the force of friction at the base of the loose mass, it falls downhill. The friction is really a measure of the resistance to gravity—the greater the friction, the greater the resistance to gravity's pull. Friction can be greatly reduced by lubrication of surfaces in contact, allowing the two materials to slide past one another more easily. Water is a common lubricating agent, so mass wasting events tend to occur more frequently during times of heavy or prolonged rain. For a mass wasting event or a mass movement to occur, the lubricating forces must be strong enough to overcome the resisting forces that tend to hold the boulder in place, against the wishes of gravity. Lubricating forces include the cohesion between similar particles (like one clay molecule to another) and the adhesion between different or unlike particles (like the boulder to the clay beneath it). When the resisting forces are greater than the driving force (tangential component of gravity) the slope is steady, and the boulder stays in place. When lubricating components reduce the resisting forces so much that the driving forces are greater than the resisting forces, slope failure occurs.

The process of the movement of regolith downslope (or under water) may occur rapidly, as in this case, or it may proceed slowly. In any case, slopes on mountainsides typically evolve toward steady-state angles, known as the angle of repose, balanced by material moving in from upslope, and out from downslope. This angle of repose is also a function of the grain size of the regolith.

Driving forces for mass wasting can also be increased by human activity. Excavation for buildings, roads, or other cultural features along the lower portions of slopes may actually remove parts of the slopes, causing them to become steeper than they were before construction and to exceed the angle of repose. This will cause the slopes to be unstable (or metastable), and susceptible to collapse. Building structures on the tops of slopes will also make them unstable, as the extra weight of the building adds extra stresses to the slope that may be enough to initiate the collapse of the slope.

Physical Conditions That Control Mass Wasting

Whether or not mass wasting occurs—and the type of resulting mass wasting—is controlled by many factors. These include characteristics of the regolith and bedrock, and the presence or absence of water, overburden, angle of the slope, and the way that the particles are packed together.

Mass wasting in solid bedrock terrain is strongly influenced by preexisting weaknesses in the rock that make movement along them easier than if the weaknesses were not present. For instance, bedding planes, joints, and fractures, if favorably oriented, may act as planes of weakness along which giant slabs of rock may slide downslope. If the rock or regolith has many pores, or open spaces between grains, it will be weaker than a rock without pores. This is because there is no material in the pores, whereas if the open spaces were filled the material in the pore space could hold the rock together. Furthermore, pore spaces allow fluids to pass through the rock or regolith, and the fluids may further dissolve the rock creating more pore space and further weakening the material. Water in open pore space may also exert pressure on the surrounding rocks, pushing individual grains apart and making the rock weaker.

Water may act to either enhance or inhibit movement of regolith and rock downhill. Water inhibits downslope movement when the pore spaces are only partly filled with water, and the surface tension (bonding of water molecules along the surface) acts as an additional force holding grains together. This surface tension is able to bond water grains to each other, water grains to rock particles, and rock particles to each other. An everyday example of how effective surface tension may be at holding particles together is found in sand castles at the beach—when the sand is wet, tall towers can be constructed, but when the sand is dry, only simple piles of sand can be made.

Water more typically acts to reduce the adhesion between grains, promoting downslope movements. When the pore spaces are filled, the water acts as a lubricant and may actually exert forces that push individual grains apart. The weight of the water in pore spaces also exerts additional pressure on underlying rocks and soils, known as loading. The loading from water in pore spaces is in many cases enough so that the strength of the underlying rocks and soil is exceeded, and the slope fails, resulting in a downslope movement.

Another important effect of water in pore spaces occurs when the water freezes; freezing causes the water to expand by a few percent, and this expansion exerts enormous pressures on surrounding rocks, in many cases pushing them apart. The freeze-thaw cycles found in many climates are responsible for many of the downslope movements.

Steep slopes are less stable than shallow slopes. Loose unconsolidated material tends to form slopes at specific angles that range about 33°–37°, depending on the specific characteristics of the material. The way that the particles are arranged or packed in the slope is also a factor; the denser the packing, the more stable the slope.

Processes of Mass Wasting

Mass movements are of three basic types, distinguished from each other by the way that the rock, soil, water, and debris move. Slides move over and in contact with the underlying surface, while flows include movements of regolith, rock, water, and air in which the moving mass breaks into many pieces that flow in a chaotic mass movement. Falls move freely through the air, and land at the base of the slope or escarpment. There is a continuum between different processes of mass wasting, but many differ in terms of the velocity of downslope movement and also in the relative concentrations of sediment, water, and air. A landslide is a general name for any downslope movement of a mass of bedrock, regolith, or a mixture of rock and soil, and it is used to indicate any mass wasting process.

Slumps

A slump is a type of sliding slope failure in which a downward and outward rotational movement of rock or regolith occurs along a concave up slip surface. This produces either a singular or a series of rotated blocks, each with the original ground surface tilted in the same direction. Slumps are especially common after heavy rainfalls and earthquakes and are common along roadsides and other slopes that have been artificially steepened to make room for buildings or other structures. Slump blocks may continue to move after the initial sliding event, and in some cases this added slippage is

Roadside slump in Wutai Shan, China. A rain-soaked oversteepened slope failed, causing soil and rock to slump downhill, blocking the road. Note the tilted telephone poles on the slumped area. *(Photo by Timothy Kusky)*

enhanced by rainwater that falls on the back-tilted surfaces, infiltrates along the fault, and acts as a lubricant for added fault slippage.

A translational slide is a variation of a slump in which the sliding mass moves not on a curved surface but downslope on a preexisting plane, such as a weak bedding plane or a joint. Translational slides may remain relatively coherent or break into small blocks forming a debris slide.

Sediment Flows

When mixtures of rock debris, water, and air begin to move under the force of gravity, they are said to flow. This is a type of deformation that is continuous and irreversible. The way in which this mixture flows depends on the relative amounts of solid, liquid, and air, the grain size distribution of the solid fraction, and the physical and chemical properties of the sediment. Mass wasting processes that involve flow are transitional within themselves, and to stream-type flows in the amounts of sediment/water and in velocity. There are many names for the different types of sediment flows, including slurry flows, mudflows, debris flows, debris avalanches, earthflows, and loess flows. Many mass movements begin as one type of flow and evolve into another during the course of the mass wasting event. For instance, it is common for flows to begin as rock falls or debris avalanches and evolve into debris flows or mudflows along its length as the flow picks up water and debris and flows over differing slopes.

Creep

Creep is the imperceptible slow downslope flowing movement of regolith. It involves the very slow plastic deformation of the regolith, as well as repeated microfracturing of bedrock at nearly imperceptible rates. Creep occurs throughout the upper parts of the regolith, and there is no single surface along which slip has occurred. Creep rates range from a fraction of an inch per year up to about two inches per year on steep slopes. Creep accounts for leaning telephone poles, fences, and many of the cracks in sidewalks and roads. Although creep is slow and not very spectacular, it is one of the most important mechanisms of mass wasting and it accounts for the greatest total volume of material moved downhill in any given year. One of the most common creep mechanisms is through frost heaving. Creep through frost heaving is extremely effective at moving rocks, soil, and regolith downhill. The ground freezes and ice crystals form and grow, pushing rocks upward perpendicular to the surface. As the ice melts in the freeze-thaw cycle, gravity takes over and the pebble or rock moves vertically downward, ending up a fraction of an inch downhill from where it started. Creep can also be initiated by other mechanisms of surface expansion and contraction, such as warming and cooling, or the expansion and contraction of clay minerals with changes in moisture levels. In a related phenomenon, the freeze-thaw

cycle can push rocks upward through the soil profile, as revealed by farmers' fields in New England and other northern climates, where the fields seem to grow boulders. The fields are cleared of rocks, and years later, the same fields are filled with numerous boulders at the surface. In these cases, the freezing forms ice crystals below the boulders that push them upward, and, during the thaw cycle, the ice around the edges of the boulder melt first and mud and soil seep down into the crack, finding their way beneath the boulder. This process, repeated over years, is able to lift boulders to the surface, keeping the northern farmer busy.

The operation of the freeze-thaw cycle makes rates of creep faster on steep slopes than on gentle slopes, with more water, and greater numbers of freeze-thaw cycles. Rates of creep of up to half an inch per year are common.

Solifluction

Solifluction is the slow viscous downslope movement of waterlogged soil and debris. Solifluction is most common in polar latitudes where the top layer of permafrost melts, resulting in a water-saturated mixture resting on a frozen base. It is also common in very wet climates, as found in the Tropics. Rates of movement are typically an inch or two per year, which is slightly faster than downslope flow by creep. Solifluction results in distinctive surface features, such as lobes and sheets, carrying the overlying vegetation; sometimes the lobes override each other, forming complex structures. Solifluction lobes are relatively common sights on mountainous slopes in wet climates, especially in areas with permafrost. The frozen layer beneath the soil prevents drainage of water deep into the soil or into the bedrock, so the uppermost layers in permafrost terrains tend to be saturated with water, aiding solifluction.

Slurry Flows

A slurry flow is a moving mass of sediment saturated in water that is transported with the flowing mass. The mixture, however, is so dense that it can suspend large boulders or roll them along the base. When slurry flows stop moving, the resulting deposit therefore consists of a non-sorted mass of mud, boulders, and finer sediment.

Debris Flows

Debris flows involve the downslope movement of unconsolidated regolith, most of which is coarser than sand. Some debris flows begin as slumps but then continue to flow downhill as debris flows. They typically fan out and come to rest when they emerge out of steeply sloping mountain valleys onto lower-sloping plains. Rates of movement in debris flows vary from several feet per year to several hundred miles per hour. Debris flows are commonly shaped like a tongue with numerous ridges and depressions. Many form after heavy rainfalls in mountainous areas, and the number of debris flows is increasing with greater deforestation of mountain and hilly areas. This is particularly obvious on the island of Madagascar, where deforestation in places has taken place at an alarming rate, removing most of the island's trees. What was once a tropical rain forest is now a barren (but geologically spectacular) landscape, carved by numerous landslides and debris flows that bring the terra rossa soil to rivers, making them run red to the sea.

Most debris flows that begin as rock falls or avalanches move outward in relatively flat terrain less than twice the distance they fell. Internal friction (between particles in the flow) and external friction (especially along the base of the flow) slow them. However, some of the largest debris flows that originated as avalanches or debris falls travel exceptionally large distances at high velocities—these are debris avalanches.

Mudflows

Mudflows resemble debris flows, except that they have higher concentrations of water (up to 30 percent), making them more fluid, with a consistency ranging from soup to wet concrete. Mudflows often start as a muddy stream in a dry mountain canyon, which, as it moves, picks up more and more mud and sand, until eventually the front of the stream is a wall of moving mud and rock. When this comes out of the canyon, the wall commonly breaks open, spilling the water behind it in a gushing flood, which moves the mud around on the valley floor. These types of deposits form many of the gentle slopes at the bases of mountains in the southwest United States.

Mudflows have also become a hazard in highly urbanized areas such as Los Angeles, where most of the dry riverbeds have been paved over, and development has moved into the mountains surrounding the basin. The rare rainfall events in these areas then have no place to infiltrate, and rush rapidly into the city picking up all kinds of street mud and debris, and forming walls of moving mud that cover streets and low-lying homes in debris. Unfortunately, after the storm rains and water recedes, the mud remains and hardens in place. Mudflows are also common with the first heavy rains after prolonged droughts or fires, as residents of many California and other western states know. After the drought and fires of 1989 in Santa Barbara, California, heavy rains brought mudflows down out of the mountains filling the riverbeds and inundating homes with many feet of mud. Similar mudflows followed the heavy rains in Malibu in 1994, which remobilized barren soil exposed by the fires of 1993. Three to four feet of mud filled many homes and covered parts of the Pacific Coast highway. Mudflows are part of the natural geologic cycle in mountainous areas, and they serve to maintain equilibrium between the rate of uplift of the mountains, and their erosion. Mudflows are only catastrophic when people have built homes, highways, and businesses in places that mudflows must go.

Volcanoes too can produce mudflows: layers of ash and volcanic debris, sometimes mixed with snow and ice, are easily remobilized by rain or by an eruption and may travel many tens of kilometers. Volcanic mudflows are known as lahars.

Mudflows have killed tens of thousands of people in single events and have been some of the most destructive of mass movements.

Granular Flows and Earthflows

Granular flows are unlike slurry flows, in that in granular flows the full weight of the flowing sediment is supported by grain-to-grain contact between individual grains. Earthflows are relatively fast granular flows with velocities ranging from three feet per day to 1,180 feet per hour (1 m/day–360 m/hr).

Rockfalls and Debris Falls

Rockfalls are the free falling of detached bodies of bedrock from a cliff or steep slope. They are common in areas of very steep slopes, where rockfall deposits may form huge deposits of boulders at the base of the cliff. Rockfalls can involve a single boulder or the entire face of a cliff. Debris falls are similar to rockfalls but consist of a mixture of rock and weathered debris and regolith.

Rockfalls have been responsible for the destruction of parts of many villages in the Alps and other steep mountain ranges, and rock fall deposits have dammed many a river valley, creating lakes behind the newly fallen mass. Some of these natural dams have been extended and heightened by engineers to make reservoirs, with examples including Lake Bonneville on the Columbia River and the Cheakamus Dam in British Columbia. Smaller examples abound in many mountainous terrains.

Rockslides and Debris Slides

Rockslides is the term given to the sudden downslope movement of newly detached masses of bedrock (or debris slides, if the rocks are mixed with other material or regolith). These are common in glaciated mountains with steep slopes and also in places where there are planes of weakness, such as bedding planes, or fracture planes that dip in the direction of the slope. Like rockfalls, rockslides may form fields of huge boulders coming off mountain slopes. The movement to this talus slope is by falling, rolling, and sliding, and the steepest angle at which the debris remains stable is known as the angle of repose. The angle of repose is typically 33°–37° for most rocks.

Debris Avalanches

Debris avalanches are granular flows moving at very high velocity and covering large distances. These rare, destructive (but spectacular) events have ruined entire towns, killing tens of thousands of people in them without warning. Some have been known to move as fast as 250 miles per hour (400 km/hr). These avalanches thus can move so fast that they move down one slope, then thunder right up and over the next slope and into the next valley. One theory of why these avalanches move so fast is that when the rocks first fall, they trap a cushion of air, and then travel on top of it like a hovercraft. Two of the worst debris avalanches in recent history originated from the same mountain, Nevado Huascarán, the highest peak in the Peruvian Andes. More than 25,000 people died in these two debris avalanches.

Subaqueous Mass Wasting

Mass wasting is not confined to land. Submarine mass movements are common and widespread on the continental shelves, slopes, and rise, and also in lakes. Mass movements under water however typically form turbidity currents, which leave large deposits of graded sand and shale. Under water, these slope failures can begin with very gentle slopes, even of less than 1°. Other submarine slope failures are similar to slope failures on land.

Slides, slumps, and debris flows are also common in the submarine realm. Submarine deltas, deep-sea trenches, and continental slopes are common sites of submarine slumps, slides, and debris flows. Some of these are huge, covering hundreds of square miles. Many of the mass wasting events that produced these deposits must have produced large tsunami. The continental slopes are cut by many canyons, produced by submarine mass wasting events, which carried material eroded from the continents into the deep ocean basins.

See also SOILS; WEATHERING.

Further Reading

Armstrong, B. R., and K. Williams. *The Avalanche Book*. Armstrong, Colo.: Fulcrum Publishing, 1992.

Brabb, Earl E. "Landslides: Extent and Economic Significance." In *Proceedings of the 28th International Geological Congress: Symposium on Landslides, Washington D.C., July 17, 1989*, edited by Earl E. Brabb and Betty L. Harrod, 25–50. Rotterdam, Netherlands: A. A. Balkema, 1989.

Coates, Donald R., ed., "Landslides." *Geological Society of America Reviews in Engineering Geology* 3 (1977): 278.

Hsu, K. J. "Catastrophic Debris Streams (Sturzstroms) Generated by Rockfalls." *Geological Society of America Bulletin* 86 (1989): 129–140.

Norris, Robert M. "Sea Cliff Erosion." *Geotimes* 35 (1990): 16–17.

Pinter, Nicholas, and Mark Brandon. "How Erosion Builds Mountains." *Scientific American, Earth from the Inside Out* (2000): 24–29.

Plafker, George, and George E. Ericksen. "Nevados Huascaran Avalanches, Peru." In *Rockslides and Avalanches*, edited by B. Voight. Amsterdam: Elsevier, 1978.

Schultz, Arthur P., and C. Scott Southworth, eds., *Landslides in Eastern North America*. U.S. Geological Survey Circular 1008, 1987.

Schuster, R. L., and R. W. Fleming. "Economic Losses and Fatalities due to Landslides." *Bulletin of the Association of Engineering Geologists* 23 (1986): 11–28.

Shaefer, S. J., and S. N. Williams. "Landslide Hazards." *Geotimes* 36 (1991): 20–22.

Varnes, David J. "Slope Movement Types and Processes." In *Landslides, Analysis and Control*, edited by R. L. Schuster and Raymond J. Krizek. Washington, D.C.: National Academy of Sciences, 1978.

meander Most streams move through a series of bends known as meanders. Meanders are always migrating across the floodplain, by the process of the deposition of the point bar deposits, and the erosion of the bank on the opposite side of the stream with the fastest flow. The erosion typically occurs through slumping of the stream bank. Meanders typically migrate back and forth, and also down-valley at a slow rate. If the downstream portion of a meander encounters a slowly erodable rock, the upstream part may catch up and cut off the meander. This forms an oxbow lake, which is an elongate and curved lake formed from the former stream channel.

meandering stream   *See* MEANDER.

mechanical weathering   *See* WEATHERING.

mélange   Complex, typically chaotic tectonic mixtures of sedimentary, volcanic, and other types of rocks, typically in a highly sheared sedimentary or serpentinitic matrix, are known as mélanges. Mélanges must be mappable units, and most show inclusions of material of widely diverse origins at many different scales, suggesting that mélanges are fractal systems. Some mélanges may be sedimentary in origin, formed by the slumping of sedimentary sequences down marine escarpments. These mélanges are more aptly termed olistostromes. Tectonic mélanges are formed by the structural mixing between widely different units, typically in subduction zone settings.

Tectonic mélanges are one of the hallmarks of convergent margins, yet understanding their genesis and relationships of specific structures to plate kinematic parameters has proved elusive because of the complex and seemingly chaotic nature of these units. Many field-workers regard mélanges as too deformed to yield useful information and simply map the distribution of mélange-type rocks without further investigation. Other workers map clasts and matrix types, search for fossils or metamorphic index minerals in the mélange, and assess the origin and original nature of the highly disturbed rocks. Recent studies have made progress in being able to relate some of the structural features in mélanges to the kinematics of the shearing and plate motion directions responsible for the deformation at plate boundaries.

One of the most persistent questions raised in mélange studies relates to the relative roles of soft-sediment versus tectonic processes of disruption and mixing. Many mélanges have been interpreted as deformed olistostromes, whereas other models attribute disruption entirely to tectonic or diapiric processes. Detailed structural studies have the potential to differentiate between these three end-member models, in that soft-sedimentary and some diapiric processes will produce clasts, which may then be subjected to later strains, whereas purely tectonic disruption will have a strain history beginning with continuous or semi-continuous layers which become extended parallel to initial layering. Detailed field, kinematic, and metamorphic studies may be able to further differentiate between mélanges of accretionary tectonic versus diapiric origin. Structural observations aimed at these questions should be completed at regional, outcrop, and hand-sample scales.

Analysis of deformational fabrics in tectonic mélange may also yield information about the kinematics of past plate interactions. Asymmetric fabrics generated during early stages of the mélange-forming process may relate to plate kinematic parameters such as the slip vector directions within an accretionary wedge setting. This information is useful for reconstructing the kinematic history of plate interactions along ancient plate boundaries, or how convergence was partitioned into belts of head-on and margin-parallel slip during oblique subduction.

See also CONVERGENT PLATE MARGIN PROCESSES.

Further Reading

Kusky, Timothy M., and Dwight C. Bradley. "Kinematics of Mélange Fabrics: Examples and Applications from the McHugh Complex, Kenai Peninsula, Alaska." *Journal of Structural Geology* 21, no. 12 (1999): 1,773–1,796.

Raymond, Loren, ed., *Mélanges: Their Nature, Origin, and Significance.* Boulder, Colo.: Geological Society of America Special Paper 198, 1984.

Mercury

Mercury The closest planet to the Sun, Mercury, is a midget. It has a mass of only 5.5 percent of the Earth, with a diameter of 3,031 miles (4,878 kilometers). It has an average density of 5.4 grams per cubic centimeter, and it rotates once on its axis every 59 Earth days. It orbits the Sun once every 88 days at a distance of 36 million miles (58 million km). Since it is so close to the Sun, it is only visible to the naked eye when the sun is blotted out, such as just before dawn, after sunset, or during total solar eclipses.

Mercury has such a weak gravitational field that it lacks an atmosphere, although bombardment by the solar wind releases some sodium and potassium atoms from surface rocks, and these may rest temporarily near the planet's surface. It has a very weak magnetic field, approximately 1/100th as strong as Earth's. The surface of Mercury is heavily cratered and looks much like the Earth's moon. It shows no evidence for ever having sustained water, dust storms, ice, plate tectonics, or life. The surface of Mercury is less densely cratered than the Moon, however, and some planetary geologists suggest that the oldest craters may be filled in by volcanic deposits. It also has some surface scarps, estimated to

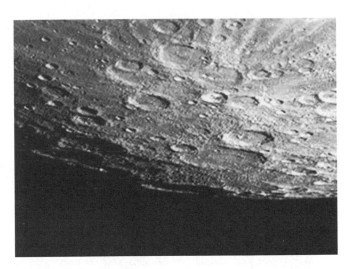

Mosaic of Mercury. After passing on the dark side of the planet, *Mariner 10* photographed the other, somewhat more illuminated hemispheres of Mercury. Note the rays of ejecta coming from some of the impact craters in the image. The *Mariner 10* spacecraft was launched in 1974. The spacecraft took images of Venus in February 1974 on the way to three encounters with Mercury in March and September 1974 and March 1975. The spacecraft took more than 7,000 images of Mercury, Venus, the Earth, and the Moon during its mission. The *Mariner 10* mission was managed by the Jet Propulsion Laboratory for NASA's Office of Space Science in Washington, D.C. Image Note: Davies, M. E., S. E. Dwornik, D. E. Gault, and R. G. Strom, Atlas of Mercury, NASA SP-423 (1978). *(Photo by NASA)*

be more than 4 billion years old. These are thought to represent contraction of the surface associated with the core formation and shrinking of the planet in the first half-billion years of its history.

The density of Mercury and the presence of a weak magnetic field suggest that the planet has a differentiated iron-rich core with a radius of approximately 1,118 miles (1,800 km), but it is not known whether this is solid or liquid. A mantle probably exists between the crust and core, extending to 311–373 miles (500–600 km) depth. The small size of Mercury means it did not have enough internal energy to sustain plate tectonics or volcanism for long in its history, so the planet has been essentially dead for the past 4 billion years.

mesosphere In solid Earth geophysics, the mesosphere refers to the solid lower mantle beneath the asthenosphere. The term is not widely used, however, and this region is more commonly referred to simply as the lower mantle. The same term is used by atmospheric scientists for the region that lies between the stratosphere and the thermosphere, at an average distance of 31–50 miles (50–80 km) above the surface of the Earth. It accounts for only less than 0.1 percent of the total mass of the atmosphere.

There is not much significant weather or heating in the mesosphere. Temperatures average about 32°F (0°C) at the base of the mesosphere, with slight heating in the summer and cooling in the winter. A small amount of heat is transferred upward from the stratosphere, which becomes heated by absorption of ultraviolet radiation. The temperature at the top of the mesosphere is about −230°F (−110°C) in summer, when the mesosphere is the coldest part of the atmosphere, and −140°F (−60°C) in winter.

Sometimes very thin clouds called noctilucent clouds form in the mesosphere, but these are so thin that they can only be seen at sunset or sunrise, when the lower atmosphere is in shadow and the mesosphere is still receiving direct sun rays.

See also ASTHENOSPHERE; ATMOSPHERE; MANTLE.

Mesozoic   The fourth of five main geological eras, falling between the Paleozoic and Cenozoic, and the erathem of rocks deposited in this era. It includes the Triassic, Jurassic, and Cretaceous Periods. The era begins at 248 million years ago at the end of the Permian-Triassic extinction event and continues to 66.4 million years ago at the Cretaceous-Tertiary (K-T) extinction event. Named by Charles Lyell in 1830, the term means middle life, recognizing the major differences in the fossil record between the preceding Paleozoic era, and the succeeding Cenozoic era. Mesozoic life saw the development of reptiles and dinosaurs, mammals, birds, many invertebrate species that are still flourishing, and saw flowering plants and conifers inhabit the land. The era is commonly referred to as the age of reptiles, since they dominated the terrestrial, marine, and aerial environments.

Pangea continued to grow in the Early Mesozoic, with numerous collisions in eastern Asia, but as the supercontinent grew in some areas, it was breaking apart in others. As the fragments drifted apart especially in the later part of the Mesozoic, continental fragments became isolated and life-forms began to evolve separately in different places, allowing independent forms to develop, such as the marsupials of Australia. The Atlantic Ocean began opening as arcs and oceanic terranes collided with western North America. In the later part of the Mesozoic in the Cretaceous, sea levels were high and shallow seas covered much of western North America and central Eurasia, depositing extensive shallow marine carbonates. Many marine organisms such as plankton rapidly diversified and bloomed, and thick organic rich deposits formed source rocks for numerous coal and oil fields. With the high Cretaceous sea levels, Cretaceous rocks are abundant on many continents and are the most represented of the Mesozoic strata. These strata are rich in fossils that show both ancient and modern features, including dinosaurs, ammonoids, plus newly developed bony fishes, and flowering plants. The dinosaurs and reptiles continued to rule the land, sea, and air until the devastating series of events, culminating with the collision of an asteroid or comet with Earth at the end of the Cretaceous, eliminating the dinosaurs, and causing the extinction of 45 percent of marine genera including the ammonites, belemnites, inoceramid clams, and large marine reptiles.

See also CRETACEOUS; JURASSIC; MASS EXTINCTIONS; TRIASSIC.

metamorphism   Metamorphism, a term derived from Greek, means change of form or shape. Geologists use the term to describe changes in the minerals, chemistry, and texture within a rock. Metamorphism is typically induced by increases in pressure and temperature from burial, regional tectonics, or nearby igneous intrusions.

Any previously formed rocks may be deeply buried by sedimentary cover, affected by regional plate-boundary processes, or be heated close to an igneous intrusion, changing the temperature and pressure conditions from when and where they were formed. Early changes that occur to rocks, generally less than 390°F (200°C), are referred to as diagenesis. However, when temperatures rise above 200°C the changes become more profound and are referred to as metamorphism.

When sedimentary rocks are deposited they contain many open spaces filled with water-rich fluids. When these rocks are deeply buried and subjected to very high temperatures and pressures, these fluids react with the mineral grains in the rock and play a vital role in the metamorphic changes that occur. These fluids act as a hot, reactive juice that transports chemical elements from mineral to fluids to new minerals. This is confirmed by observations of rocks heated to the same temperature and pressure without fluids, and these hardly change at all.

When rocks are heated, certain minerals become unstable and others become stable. Chemical reactions transform one assemblage of minerals into a new assemblage. Most temperature changes are accompanied by pressure changes, and it is the combined P-T fluid composition that determines how the rock will change.

In liquids pressures are equal in all directions, but in rocks pressures may be greater or lesser in one direction, and we refer to them as stresses. Textures in metamorphic rocks often reflect stresses that are greater in one direction than in another. Sheets of planar minerals become oriented with their flat surfaces perpendicular to the strongest or maximum stress. This planar arrangement of platy minerals is known as foliation.

Time is also an important factor in metamorphism. In general, the longer the reaction time the larger the mineral grains, and the more complete the metamorphic changes.

Grades of Metamorphism
Low-grade metamorphism refers to changes that occur at low temperatures and pressures whereas high-grade metamorphism refers to changes that occur at high temperatures and pressures. At progressively higher grades of metamorphism the high temperature drives the water out of the pore spaces and eventually out of the hydrous mineral structures, so that at very high grades of metamorphism, the rocks contain fewer hydrous minerals (e.g., micas). Prograde metamorphism refers

to changes that occur while the temperature and pressure are rising and pore fluids are abundant, whereas retrograde metamorphism refers to changes that occur when temperature and pressure are falling. At this stage, most fluids have already been expelled and the retrograde changes are less pronounced. If this were not so, then all metamorphic rocks would revert back to clays stable at the surface.

Metamorphic Changes

FOLIATION If you were to take a piece of paper and compress it, you would find that the flat dimensions would orient themselves perpendicular to the direction that you compressed it from. Likewise, when a metamorphic rock is compressed or stressed, the platy minerals, such as chlorite and micas, orient themselves so that their long dimensions are perpendicular to the maximum compressive stress. The planar fabric that results from this process is known as a foliation.

Slaty cleavage is a specific type of foliation, in which the parallel arrangement of microscopic platy minerals causes the rock to break in parallel plate-like planes. Schistosity forms at higher metamorphic grades and is a foliation defined by a wavy or distorted plane containing large visible oriented minerals such as quartz, mica, and feldspar.

MINERAL ASSEMBLAGES As rocks are progressively heated and put under more pressure during metamorphism, different mineral assemblages are stable. Even though the rock may retain a stable overall composition, the minerals will become progressively recrystallized and the mineral assemblages (or parageneses) will change under different P-T conditions.

Kinds of Metamorphic Rocks

The names of metamorphic rocks are derived from their original rock type, their texture, and mineral assemblages.

METAMORPHISM OF SHALE AND MUDSTONE Shales and mudstones have an initial mineral assemblage of quartz, clays, calcite, and feldspar. Slate is the low-grade metamorphic equivalent of shale and, with recrystallization, is made of quartz and micas. At intermediate grades of metamorphism, the mica grains grow larger so that individual grains are now visible to the naked eye and the rock is called a phyllite. At high grades of metamorphism, the rock (ex-shale) now becomes a schist, which is coarse grained and the foliation becomes a bit irregular. Still higher grades of metamorphism separate the quartz and the mica into different layers; this rock is called a gneiss. For both schists and gneisses, a prefix is commonly added to the names to denote some of the minerals present in the rock. For instance, if garnet grows in a biotite schist, it could be named a garnet-biotite schist.

METAMORPHISM OF BASALT Fresh basalts contain olivine, pyroxene, and plagioclase, none of which contains abundant water. When metamorphosed, however, water typically enters the rock from outside the system. At low grades of metamorphism, the basalt is turned into a greenstone or greenschist, which has a distinctive color because of its mineral assemblage of chlorite (green)+albite (clear)+epidote (green)+calcite (clear).

At higher metamorphic grades, the greenschist mineral assemblage is replaced by one stable at higher temperature and pressure, typically plagioclase and amphibole, and the rock is known as amphibolite. Amphiboles have a chain structure, which gives them an elongated shape. When they crystallize in a different stress field like that found in a metamorphic rock, the new minerals tend to align themselves so that their long axes are parallel to the least compressive stress, forming a lineation. At even higher metamorphic grades the amphiboles are replaced by pyroxenes and the rock is called a granulite.

METAMORPHISM OF LIMESTONE When limestone is metamorphosed, it is converted to marble, which consists of a network of coarsely crystalline interlocking calcite grains. Most primary features, such as bedding, are destroyed during metamorphism and a new sugary texture appears.

METAMORPHISM OF SANDSTONE When sandstone is metamorphosed, the silica is remobilized and fills in the pore spaces between the grains, making a very hard rock called a quartzite. Primary sedimentary structures may still be seen through the new mineral grains.

Kinds of Metamorphism

Metamorphism is a combination of chemical reactions induced by changing pressure and temperature conditions and mechanical deformation caused by differential stresses. The relative importance of physical and chemical processes changes with metamorphism in different tectonic settings.

THERMAL OR CONTACT METAMORPHISM Near large plutons or hot igneous intrusions, rocks are heated to high temperatures without extensive mechanical deformation. Therefore, rocks next to plutons typically show growth of new minerals but lack strong foliations formed during metamorphism. Rocks adjacent to these large plutons develop a contact metamorphic aureole of rocks, altered by heat from the intrusion. Large intrusions carry a lot of heat and typically have large contact aureoles, several kilometers wide.

The contact metamorphic aureole is made of several concentric zones each with different mineral groups related to higher temperatures closer to the pluton. A hornfels is a hard, fine-grained rock composed of uniform interlocking grains, typically from metamorphosed and suddenly heated shale.

BURIAL METAMORPHISM When rocks are buried by the weight of overlying sedimentary rocks, they undergo small

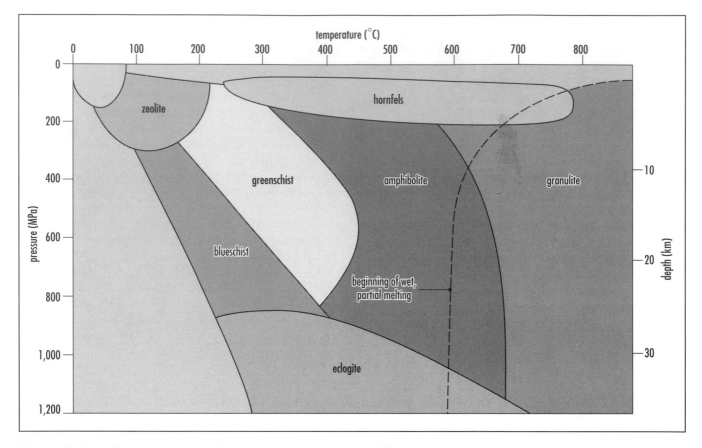

Metamorphic facies showing the relationship between pressure, temperature, and different grades of metamorphism

changes called diagenesis, until they reach 390°F (200°C). At about 570°F (300°C), some recrystallization may begin, particularly the formation of a group of water-rich minerals known as zeolites.

REGIONAL METAMORPHISM The most common types of metamorphic rocks are the regional metamorphic rocks. Regional metamorphism involves a combination of chemical and mechanical effects and so these rocks tend to have a pronounced foliation (slate, schist). Most regional metamorphic rocks are found in mountain belts or old eroded mountain belts, which were formed by the collision of two tectonic plates. In regional metamorphic conditions, the rocks are compressed horizontally, resulting in large folds and faults, which place some rocks on top of other ones, burying them quickly and elevating their pressure and temperature conditions. In this type of environment there is a wide range of pressure/temperature conditions over which the rocks were metamorphosed, and geologists have defined a series of different metamorphic zones which reflect these conditions. These metamorphic zones are each defined by the appearance of a new metamorphic index mineral, which include, in progressively higher grade order (for shale), chlorite-biotite-

garnet-staurolite-kyanite-sillimanite. In the field, the geologist examines the rocks and looks for the first appearance of these different minerals and plots them on a map. By mapping out the distribution of the first appearance of these minerals on a regional scale, the geologist then defines isograds, which are lines on a map marking the first appearance of a given index mineral on the map. The regions between isograds are known as metamorphic zones.

Metamorphic Facies

When rocks are metamorphosed their bulk chemistry remains about the same, except for water and CO_2, which are fairly mobile. The mineral assemblages constantly change but the chemistry remains the same. Thus, the temperature and pressure of metamorphism control the mineral assemblages in metamorphic rocks. In 1915 Pentti Eskola presented the concept of metamorphic facies. Simply put, this concept states that different assemblages of metamorphic minerals that reach equilibrium during metamorphism within a specific range of physical conditions belong to the same metamorphic facies.

Eskola studied rocks of basaltic composition, so he named his facies according to the metamorphic names for basaltic rocks. His classification, shown here, stands to this day.

Metamorphism and Tectonics

Regional metamorphism is a response to tectonic activity and different metamorphic facies are found in different tectonic environments.

The figure on page 274 shows the distribution of metamorphic facies in relationship to the structure of a subduction zone. Burial metamorphism occurs in the lower portion of the thick sedimentary piles that fill the trench, whereas deeper down the trench blue schist facies metamorphism reflect the high pressures and low temperatures where magmas come up off the subduction slab and form an island arc; metamorphism is of greenschist to amphibolite facies. Closer to the plutons of the arc, the temperatures are high, but the pressures are low, so contact metamorphic rocks are found in this region.

See also MINERALOGY; PETROLOGY; STRUCTURAL GEOLOGY.

metasomatic   The metamorphic process of changing a rock's composition or mineralogy by the gradual replacement of one component by another through the movement and reaction of fluids and gases in the pore spaces of a rock is called metasomatism, and these processes are metasomatic. Metasomatic processes are thought to be responsible for the formation of many ore deposits, which have extraordinarily concentrated abundances of some elements. They may also play a role in the replacement of some limestones by silica, and the formation of dolostones.

Many metasomatic rocks and ore deposits are formed in hydrothermal circulation systems that are set up around igneous intrusions. When magmas, particularly large batholiths, intrude country rocks they set up a large thermal gradient between the hot magma and the cool country rock. Any water above the pluton gets heated and rises toward the surface, and water from the sides of the pluton moves in to replace that water. A hydrothermal circulation system is thus set up, and the continuous movement of hot waters in such systems often leaches elements from some rocks and from the pluton and deposits them in other places, in metasomatic processes.

See also METAMORPHISM.

Metazoa   Complex multicellular animals in which the cells are arranged in two layers in the embryonic gastrula stage. The Metazoa are extremely diverse and include 29 phyla, most of which are invertebrates. The phylum Chordata are an exception.

The Metazoa appeared about 620 million years ago and experienced a rapid explosion around the Precambrian-Cambrian boundary, probably associated with the formation and breakup of the supercontinents Rodinia and Gondwana and the rapidly changing environments associated with the supercontinent cycle. They probably evolved from the eukaryotes, single-celled organisms with a nucleus, that appeared around 1,600 million years ago. Prokaryotes are older, probably extending back past 3,800 million years ago.

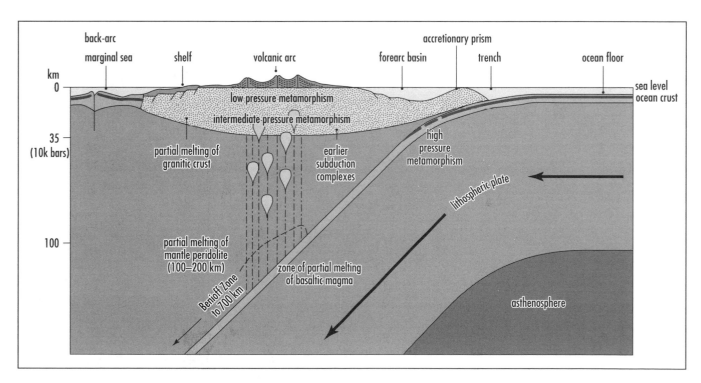

Relationship between metamorphism and tectonics in a convergent margin setting

Some of the oldest soft-bodied Metazoa are remarkably well preserved in the Ediacarian fauna from southeast Australia and other locations around the world. These fauna include a remarkable group of very unusual shallow marine forms, including some giants up to a meter in length. This explosion from simple small single-celled organisms that existed on Earth for the previous 3 billion years (or more) is truly remarkable. The Ediacarian fauna (and related fauna, collectively called the Vendoza fauna) died off after the period between 620 million and 550 million years ago, as these organisms show no affinity with modern invertebrates.

After the Ediacarian and Vendoza fauna died off, other marine invertebrates saw a remarkable explosion through the Cambrian. These organisms in the Cambrian included shelly fossils, trilobites, brachiopods, mollusks, archeocyathids, and echinoderms, and eventually in the Ordovician were joined by crinoids and bryozoans. Modern Metazoa include corals, gastropods, bivalves, and echinoids.

See also SUPERCONTINENT CYCLE; VENDIAN.

meteor   Rocky objects from space that strike the Earth. When meteorites pass through Earth's atmosphere, they get heated and their surfaces become ionized, causing them to glow brightly and form a streak moving across the atmosphere known as a shooting star or fireball.

At certain times of the year, the Earth passes through parts of our solar system that are rich in meteorites, and the night skies become filled with shooting stars and fireballs, sometimes as frequently as several per minute. These times of high-frequency meteorite encounters are known as meteor showers and include the Perseid showers that appear around August 11 and the Leonid showers that appear about November 14 (both occur annually).

There is ample evidence that many small meteorites have hit the Earth frequently throughout time. Eyewitness accounts describe many events, and fragments of meteorites are regularly recovered from places like the Antarctic ice sheets, where rocky objects on the surface have no place to come from but space. Although meteorites may appear as flaming objects moving across the night skies, they are generally cold icy bodies when they land on Earth, and only their outermost layers get heated from the deep freeze of space during their short transit though the atmosphere.

Meteorites consist of several different main types. Stony meteorites include chondrites, which are very primitive and ancient meteorites made of silicate minerals, like those common in the Earth's crust and mantle, but chondrites contain small spherical objects known as chondrules. These chondrules contain frozen droplets of material that is thought to be remnants of the early solar nebula from which the Earth and other planets initially condensed. Achondrites are similar to chondrites in mineralogy, except that they do not contain the chondritic spheres. Iron meteorites are made of an iron-nickel alloy with textures that suggest they formed from slow crystallization inside a large asteroid or small planet that has since been broken into billions of small pieces, probably by an impact with another object. Stony-irons are meteorites that contain mixtures of stony and iron components and probably formed near the core-mantle boundary of the broken planet or asteroid. Almost all meteorites found on Earth are stony varieties.

Origin of Meteorites and Other Earth Orbit-Crossing Objects
Most meteorites originate in the asteroid belt, situated between the orbits of Mars and Jupiter. There are at least one million asteroids in this belt with diameters greater than 0.6 miles (1 km), 1,000 with diameters greater than 18 miles (30 km), and 200 with diameters greater than 60 miles (100 km). Asteroids and meteorites are distinguished only by their size, asteroids being greater than 328 feet (100 m) in diameter. Meteorites are referred to as meteors only after they enter the Earth's atmosphere. These are thought to be either remnants of a small planet that was destroyed by a large impact event, or perhaps fragments of rocky material that failed to coalesce into a planet, probably due to the gravitational effects of the nearby massive planet of Jupiter. Most scientists favor the second hypothesis but recognize that collisions between asteroids have fragmented a large body to expose a planet-like core and mantle now preserved in the asteroid belt.

Collisions between asteroids can alter their orbits and cause them to head into an Earth orbit-crossing path. At this point, the asteroid becomes hazardous to life on Earth and is known as an Apollo object. Presently, about 150 Apollo objects with diameters of greater than 0.6 miles (1 km) are known, but there are bound to be many more. In 1996 an asteroid about one-quarter-mile across barely missed hitting the Earth, speeding past at a distance about equal to the distance to the Moon. A similar near-miss event was recorded in 2001; Asteroid 2001 YB5 passed Earth at a distance of twice that to the Moon. Had YB5 hit Earth, it would have released energy equivalent to 350,000 times the energy released during the nuclear bomb blast in Hiroshima.

The objects that are in an Earth orbit-crossing path could not have been in this path for very long, because gravitational influences of the Earth, Mars, and Venus would cause them to hit one of the planets or be ejected from the solar system within about 100 million years. The abundance of asteroids in an Earth orbit-crossing path demonstrates that ongoing collisions in the asteroid belt are replenishing the source of potential impacts on Earth. A few rare meteorites found on Earth have chemical signatures that suggest they originated on Mars and on the Moon, probably being ejected toward the Earth from giant impacts on those bodies.

Other objects from space (such as comets) may collide with Earth. Comets are masses of ice and carbonaceous material mixed with silicate minerals that are thought to originate in the outer parts of the solar system, in a region called the Oort Cloud. Other comets have a closer origin, in the Kuiper Belt just beyond the orbit of Neptune. There is considerable debate about whether small icy Pluto, long considered the small outermost planet, should actually be classified as a large Kuiper Belt object. Comets may be less common near Earth than meteorites, but they still may hit the Earth with severe consequences. There are estimated to be more than a trillion comets in our solar system. Since they are lighter than asteroids, and have water-rich and carbon-rich compositions, many scientists have speculated that cometary impact may have brought water, the atmosphere, and even life to Earth.

See also ASTEROID; COMET.

Further Reading

Alvarez, W. *T Rex and the Crater of Doom.* Princeton: Princeton University Press, 1997.

Chapman, C. R., and D. Morrison. "Impacts on the Earth by Asteroids and Comets: Assessing the Hazard." *Nature* 367 (1994): 33–39.

Martin, P. S., and R. G. Klein, eds., *Quaternary Extinctions.* Tucson: University of Arizona Press, 1989.

Melosh, H. Jay. *Impact Cratering: A Geologic Process.* New York: Oxford University Press, 1988.

Poag, C. Wylie. *Chesapeake Invader, Discovering America's Giant Meteorite Crater.* Princeton: Princeton University Press, 1999.

Sharpton, Virgil L., and P. D. Ward. "Global Catastrophes in Earth History." *Geological Society of America Special Paper* 247, 1990.

Stanley, S. M. *Extinction.* New York: Scientific American Library, 1987.

meteoric Water that has recently come from the Earth's atmosphere. The term is usually used in studies of groundwater, to distinguished water that has resided in ground for extended periods of time versus water that has recently infiltrated the system from rain, snow melt, or stream infiltration. Measurements of oxygen isotopes and other elements are typically used to aid this differentiation, as water from different source shows different isotopic compositions.

See also GROUNDWATER.

meteorite *See* METEOR.

meteorology The study of the Earth's atmosphere, along with its movements, energy, interactions with other systems, and weather forecasting. Different aspects of meteorology include the study of the structure of the atmosphere, such as its compositional and thermal layers, and how energy is distributed within these layers. It includes analysis of the composition of the atmosphere and how the relative and absolute abundance of elements has changed with time, and how different interactions of the atmosphere, biosphere, and lithosphere contribute to the atmosphere's chemical stability. A fundamental aspect of meteorology is relating how different factors, including energy from the Sun, contribute to cloud formation, movement of air masses, and weather patterns at specific locations. Meteorologists are responsible for being able to interpret these complex energy changes and moisture changes and for using this knowledge to predict the weather. Increasingly, meteorologists are able to use data collected from orbiting satellites to aid their interpretation of these complex phenomena. Satellites have immensely increased the ability to monitor and predict the strength and paths of severe storms such as hurricanes, as well as monitor many aspects of the atmosphere, including moisture content, pollution, and wind patterns.

See also ATMOSPHERE; CLOUDS.

mica A group of platy minerals with the general formula $(K,Na,Ca) (Mg,Fe,Li,Al)_{2-3} (Al,Si)_4 O_{10}(OH,F)_2$, that have perfect basal cleavage and exhibit a tendency to split into thin elastic sheets. The basal cleavage reflects the layered atomic structure of complex phyllosilicates that form repeating sheets and typically show one type of mica interlayered with another. Some common micas include muscovite $(K_2Al_4Si_6Al_2)$, Paragonite $(Na_2Al_4Si_6Al_2)$, phlogopite $(K_2[Mg,Fe^{+2}]_6Si_6Al_2$, and biotite $(K_2[Mg,Fe,Al]_6Si_{6-5}Al_{2-3})$.

Micas are common constituents of a number of different types of igneous, metamorphic, and sedimentary rocks. Muscovite is found in many granites, pegmatites, phyllites, schists, gneisses, and as a detrital sedimentary mineral. Phlogopite is found in ultramafic peridotites and metamorphosed carbonates. Biotite is a common component in gabbros, diorites, granites, pegmatites, in schists, phyllites, and gneisses. Paragonite is found in schists, gneisses, and as detrital and authigenic sediments.

See also IGNEOUS ROCKS; METAMORPHISM; MINERALOGY.

mid-latitude cyclone *See* EXTRATROPICAL CYCLONES.

mid-ocean ridge system The mountain ranges on the seafloor that mark the mid-ocean ridge system form the longest linear feature on the Earth's surface, with a total length of approximately 40,000 miles (65,000 km). The ridge system ranges from about 600 to 2,500 miles (1,000 to 4,000 km) wide, and rises an average of 1–2 miles (2–3 km) above the surrounding seafloor, or about 1.5–2 miles (2.5 km) below the sea surface. The ridges are broken into segments by transform faults that accommodate the differential motion caused by spreading on offset ridge segments. The amount of offset on the transforms ranges from a few miles to hundreds of miles. The sense of motion on the transform faults is opposite to that of what would be expected if the faults had offset a previously continuous ridge. This relationship shows that the transforms accommodate geometric consequences of spreading on a sphere, and the ridge segments were always arranged in an offset manner since they formed.

The mid-ocean ridge system is divided into several main branches located in each of the world's main oceans. The Mid-Atlantic Ridge bisects the North and South Atlantic Oceans and connects in the south with the Antarctic ridge that surrounds the Antarctic continent. The northern extension of the Mid-Atlantic Ridge strikes through Iceland (where it is known as the Reykjanes Ridge) and then continues to connect with the Arctic Ocean ridge. The East Pacific Rise branches off the Antarctic ridge between Australia and South America and in various places separates the Antarctic, Pacific, Nazca Cocos, and North American plates. The ridge disappears beneath North America in the Gulf of California where plate boundary motions are taken up by the San Andreas Fault system. Remnants of the once-larger East Pacific rise are located along the North American–Pacific plate boundary, such as off the coast of British Columbia where the Juan de Fuca Ridge separates the Pacific and Juan de Fuca plates. The Indian Ocean ridge branches off the Antarctic ridge and extends into the Gulf of Aden, where it continues as the immature Red Sea rift and branches into the Afar triple junction where incipient spreading is occurring in the East African rift system.

Morphological studies of these ridge systems has led to the division of mid-ocean ridges into slow-spreading or Atlantic-type ridges, and fast-spreading or Pacific-type ridges. Atlantic-type ridges are characterized by a broad, 900–2,000-mile (1,500–3,000-km) wide swell in which the seafloor rises 0.5–2 miles (1–3 km) from abyssal plains at 2.5 miles (4.0 km) below sea level to about 1.7 miles (2.8 km) below sea level along the ridge axis. Slopes on the ridge are generally less than 1°. Slow-spreading ridges have a median rift, typically about 19 miles (30 km) wide at the top to 0.5–2.5 miles (1–4 km) wide at the bottom of the 0.5-mile (1-km) deep medial rift. Many constructional volcanoes are located along the base and inner wall of the medial rift. Rugged topography and many faults forming a strongly block-faulted slope characterize the central part of Atlantic-type ridges.

Fast-spreading or Pacific-type ridges are generally 1,250–2,500 miles (2,000–4,000 km) wide, and rise 1–2 miles (2–3 km) above the abyssal plains, with 0.1° slopes. Pacific-type ridges have no median valley but have many shallow earthquakes, high heat flow, and low gravity in the center of the ridge, suggesting that magma may be present at shallow levels beneath the surface. Pacific-type ridges have much smoother flanks than Atlantic-type ridges.

See also DIVERGENT OR EXTENSIONAL BOUNDARIES.

Further Reading

Nicolas, Adolphe. *The Mid-Ocean Ridges: Mountains Below Sea Level*. Berlin: Springer-Verlag, 1995.

Milankovitch, Milutin M. (1879–1958) Serbian *Mathematician, Physicist*

Milutin Milankovitch was born and educated in Serbia and was appointed to a chair in the University of Belgrade in 1909, where he taught courses in mathematics, physics, mechanics, and celestial mechanics. He is well known for his research on the relationship between celestial mechanics and climate on the Earth, and he is responsible for developing the idea that rotational wobbles and orbital deviations combine in cyclic ways to produce the climatic changes on the Earth. He determined how the amount of incoming solar radiation changes in response to several astronomical effects such as orbital tilt, eccentricity, and wobble. These changes in the amount of incoming solar radiation in response to changes in orbital variations occur with different frequencies, and they produce cyclical variations known as Milankovitch cycles. Milankovitch's main scientific work was published by the Royal Academy of Serbia in 1941, during World War II in Europe. He was able to calculate that the effects of orbital eccentricity, wobble, and tilt combine every 40,000 years to change the amount of incoming solar radiation, lowering temperatures and causing increased snowfall at high latitudes. His results have been widely used to interpret the climatic variations especially in the Pleistocene record of ice ages, and also in the older rock record.

See also CLIMATE CHANGE; MILANKOVITCH CYCLES.

Milankovitch cycles

Systematic changes in the amount of incoming solar radiation, caused by variations in Earth's orbital parameters around the Sun. These changes can affect many Earth systems, causing glaciations, global warming, and changes in the patterns of climate and sedimentation.

Astronomical effects influence the amount of incoming solar radiation; minor variations in the path of the Earth in its orbit around the Sun, and the inclination or tilt of its axis cause variations in the amount of solar energy reaching the top of the atmosphere. These variations are thought to be responsible for the advance and retreat of the Northern and Southern Hemisphere ice sheets in the past few million years. In the past 2 million years alone, the Earth has seen the ice sheets advance and retreat approximately 20 times. The climate record, as deduced from ice-core records from Greenland and isotopic tracer studies from deep ocean, lake, and cave sediments, suggests that the ice builds up gradually over periods of about 100,000 years, then retreats rapidly over a period of decades to a few thousand years. These patterns result from the cumulative effects of different astronomical phenomena.

Several movements are involved in changing the amount of incoming solar radiation. The Earth rotates around the Sun following an elliptical orbit, and the shape of this elliptical orbit is known as its eccentricity. The eccentricity changes cyclically with time with a period of 100,000 years, alternately bringing the Earth closer to and farther from the Sun in summer and winter. This 100,000-year cycle is about the same as the general pattern of glaciers advancing and retreating every

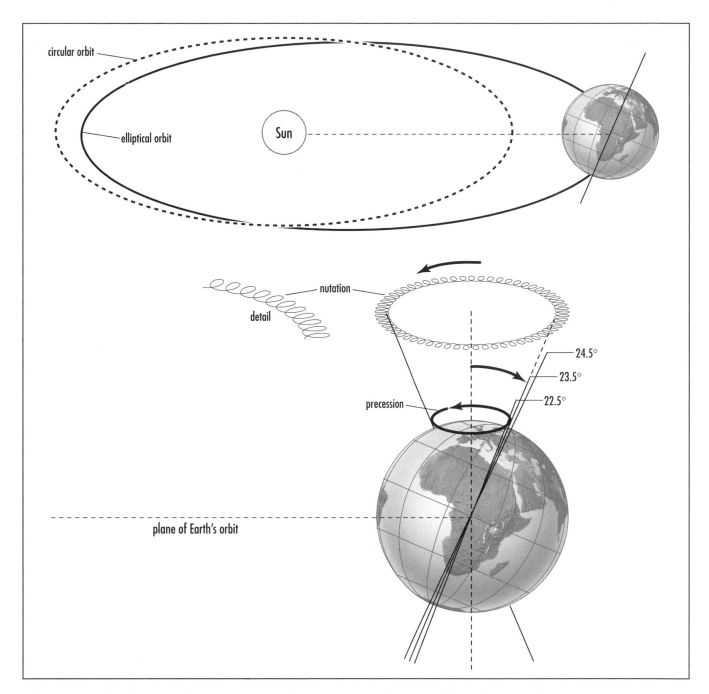

circular orbit

elliptical orbit

Sun

nutation

detail

24.5°

23.5°

22.5°

precession

plane of Earth's orbit

Orbital variations of the Earth cause changes in the amount of incoming solar radiation known as Milankovitch cycles. Shown here are changes in the eccentricity of the orbit, the tilt of the spin axis (nutation), and precession of the equinoxes.

100,000 years in the past 2 million years, suggesting that this is the main cause of variations within the present-day ice age.

The Earth's axis is presently tilting by 23.5°N/S away from the orbital plane, and the tilt varies between 21.5°N/S and 24.5°N/S. The tilt changes by plus or minus 1.5°N/S from a tilt of 23°N/S every 41,000 years. When the tilt is greater, there is greater seasonal variation in temperature.

Wobble of the rotation axis describes a motion much like a top rapidly spinning and rotating with a wobbling motion, such that the direction of tilt toward or away from the Sun changes, even though the tilt amount stays the same. This wobbling phenomenon is known as precession of the equinoxes, and it has the effect of placing different hemispheres closest to the Sun in different seasons. Presently the precession of the equinoxes is such that the Earth is closest to the Sun during the Northern Hemisphere winter. This precession changes with a double cycle, with periodicities of 23,000 years and 19,000 years.

Because each of these astronomical factors act on different timescales, they interact in a complicated way, known as Milankovitch cycles, after a Yugoslavian (Milutin Milankovitch) who first analyzed them in the 1920s. Using the power of understanding these cycles, we can make predictions of where the Earth's climate is heading, whether we are heading into a warming or cooling period, and whether we need to plan for sea-level rise, desertification, glaciation, sea-level drops, floods, or droughts.

Milankovitch cycles have been invoked to explain the rhythmic repetitions of layers in some sedimentary rock sequences. The cyclical orbital variations cause cyclical climate variations, which in turn are reflected in the cyclical deposition of specific types of sedimentary layers in sensitive environments. There are numerous examples of sedimentary sequences where stratigraphic and age control are sufficient to be able to detect cyclical variation in the timescales of Milankovitch cycles, and studies of these layers have proven consistent with a control of sedimentation by the planet's orbital variations. Some examples of Milankovitch-forced sedimentation have been documented from the Dolomite Mountains of Italy, the Proterozoic Rocknest Formation of northern Canada, and from numerous coral reef environments.

See also STRATIGRAPHY; SEQUENCE STRATIGRAPHY.

Further Reading

Allen, P. A., and J. R. Allen. *Basin Analysis, Principles and Applications.* Oxford: Blackwell Scientific Publications, 1990.

Goldhammer, Robert K., Paul A. Dunn, and Lawrence A. Hardie. "High-Frequency Glacial-Eustatic Sea Level Oscillations with Milankovitch Characteristics Recorded in Middle Triassic Platform Carbonates in Northern Italy." *American Journal of Science* 287 (1987): 853–892.

Grotzinger, John P. "Upward Shallowing Platform Cycles: A Response to 2.2 Billion Years of Low-Amplitude, High-Frequency (Milankovitch Band) Sea Level Oscillations." *Paleoceanography* 1 (1986): 403–416.

Hayes, James D., John Imbrie, and Nicholas J. Shakelton. "Variations in the Earth's Orbit: Pacemaker of the Ice Ages." *Science* 194 (1976): 2,212–2,232.

Imbrie, John. "Astronomical Theory of the Pleistocene Ice Ages: A Brief Historical Review." *Icarus* 50 (1982): 408–422.

mineralogy The branch of geology that deals with the classification and properties of minerals. It is closely related to petrology, the branch of geology that deals with the occurrence, origin, and history of rocks. Minerals are the basic building blocks of rocks, soil, and sand. Most beaches are made of the mineral quartz, which is very resistant to weathering and erosion by the waves. Most minerals, like quartz or mica, are abundant and common, although some minerals like diamonds, rubies, sapphires, gold, and silver are rare and very valuable. Minerals contain information about the chemical and physical conditions in the regions of the Earth that they formed in. They can often help discriminate which tectonic environment a given rock formed in, and they can tell us information about the inaccessible portions of Earth. For example, mineral equilibrium studies on small inclusions in diamonds show that they must form below a depth of 90 miles (145 km). Economies of whole nations are based on exploitation of mineral wealth; for instance, South Africa is such a rich nation because of its abundant gold and diamond mineral resources.

The two most important characteristics of minerals are their composition and structure. The composition of minerals describes the kinds of chemical elements present and their proportions, whereas the structure of minerals describes the way in which the atoms of the chemical elements are packed together.

We know of 3,000 minerals, most made out of the eight most common mineral-forming elements. These eight elements make up greater than 98 percent of the mass of the continental crust. Most of the other 133 scarce elements do not occur by themselves, but occur with other elements in compounds by ionic substitution. For example, olivine may contain trace amounts of Cu, Ni, Co, Mn, and other elements.

The two elements oxygen and silicon make up more than 75 percent of the crust, with oxygen alone forming nearly half of the mass of the continental crust. Oxygen forms a simple anion (O^{-2}), and silicon forms a simple cation (Si^{+4}). Silicon and oxygen combine together to form a very stable complex anion that is the most important building block for minerals—the silicate anion $(SiO_4)^{-4}$. Minerals that contain this anion are known as the silicate minerals, and they are the most common naturally occurring inorganic compounds in the solar system. The other, less common, building blocks of minerals (anions) are oxides (O^{-2}), sulfides (S^{-2}), chlorides (Cl^{-1}), carbonates $(CO_3)^{-2}$, sulfates $(SO_4)^{-2}$, and phosphates $(PO_4)^{-3}$.

The Rock-Forming Minerals

Approximately 20 minerals are so common that they account for greater than 95 percent of all the minerals in the continental and oceanic crust; these are called the rock-forming minerals. Most rock-forming minerals are silicates, and they have some common features in the way their atoms are arranged.

The Silicate Tetrahedron

The silicate anion is made of four large oxygen atoms and one small silicon atom that pack themselves together to occupy the smallest possible space. This shape, with big oxygen atoms at four corners of the structure, and the silicon atom at the center, is known as the silicate tetrahedron. Each silicate tetrahedron has four unsatisfied negative charges (Si has a charge of +4, whereas each oxygen has a charge of −2). To make a stable compound the silicate tetrahedron must therefore combine to neutralize this extra charge, which can happen in one of two ways:

1. Oxygen can form bonds with cations (positively charged ions). For instance, Mg^{+2} has a charge of +2, and by combining with Mg^{+2}, the silicate tetrahedron makes a mineral called olivine $(Mg_2)SiO_4$.
2. Two adjacent tetrahedra can share an oxygen atom, making a complex anion with the formula $(Si_2O_7)^{-6}$. This process commonly forms long chains, so that the charge is balanced except at the ends of the structure. This process of linking silicate tetrahedra into large anion groups is called polymerization. It is the most common way to build minerals, but in making the various possible combinations of tetrahedra, one rule must be followed, that is, tetrahedra can only be linked at their apices.

Olivine is one of the most important minerals on Earth, forming much of the oceanic crust and upper mantle. It has the formula $(Mg,Fe)_2SiO_4$ and forms the gem peridot.

Garnet is made of isolated silicate tetrahedra packed together without polymerizing with other tetrahedra. There are many different kinds of garnets, with almandine being one of the more common, deep red varieties that forms a common gemstone. Ionic substitution is common, with garnet having the chemical formula $A_3B_2(SiO_4)^3$, where:

$$A = Mg^{+2}$$
$$Fe^{+2}$$
$$Ca^{+2}$$
$$Mn^{+2}$$

$$B = Al^{+3}$$
$$Fe^{+3}$$

Pyroxene and amphibole both contain continuous chains of silicate tetrahedra. Pyroxenes are built from a polymerized chain of single tetrahedra, whereas amphiboles are built in double chains or linked rings. In both of these structures, the chains are bound together by cations such as Ca, Mg, and Fe, which satisfy the negative charges of the polymerized tetrahedra. Pyroxenes are common minerals in the oceanic crust and mantle, and they also occur in the continental crust. Amphiboles are common in metamorphic rocks. They have a complicated chemical formula and can hold large varieties of cations in their crystal structure.

Clays, micas, and chlorites are all closely related to sheet silicates, made of polymerized sheets of tetrahedra. By sharing three oxygens with adjacent tetrahedra, there is only a single unbalanced oxygen in each tetrahedra on which is typically balanced by Al^{+3} cations, which occupy spaces between

The 8 Most Common Mineral-Forming Elements

Element	Abbreviation	Percentage of Continental Crust Mass
Oxygen	O	46.6
Silicon	Si	27.7
Aluminum	Al	8.1
Iron	Fe	5.0
Calcium	Ca	3.6
Sodium	Na	2.8
Potassium	K	2.6
Magnesium	Mg	2.1

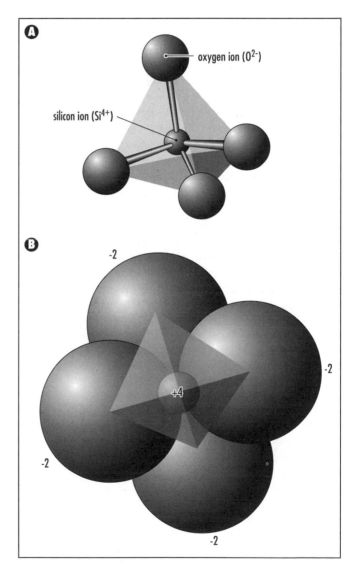

Diagram showing silicate tetrahedra

the sheets. The sheet structure is why micas are easy to peel apart on cellophane-like surfaces.

Quartz, one of the most common minerals, also has one of the most common polymerizations. It has all of its charges satisfied by sharing all of its oxygen in a three-dimensional network. Quartz typically has six-sided crystals and has many other different forms and colors.

Feldspars are the most common minerals in the Earth's crust. They account for 60 percent of all minerals in the continental crust, and 75 percent of the volume. Feldspars are also common in the oceanic crust. Like quartz, feldspars also have a structure formed by polymerization of all the oxygen atoms, and some of the silicon atoms are replaced by Al^{+3}. There are many different kinds of feldspar minerals, formed by different cations added to the structure. For instance, potassium feldspar has the formula $K(Si_3Al)O_8$, albite has the formula $Na(Si_3Al)O_8$, and anorthite has the formula $Ca(Si_2Al_2)O_8$. There is a complete range of chemical compositions of feldspars possible between the albite and anorthite varieties. These feldspar minerals are known as the plagioclase feldspars.

Silicates are the most abundant rock-forming minerals, but other types do occur in sufficient quantities to call them rock-forming minerals. Oxides use the oxygen anion and include ore minerals such as chromium, uranium, tin, and magnetite (FeO_4). Sulfides are minerals such as copper, lead, zinc, cobalt, mercury, and silver that use the sulfur anion. For instance, FeS_2 is the formula for pyrite, commonly known as fool's gold. The carbonates calcite, aragonite, dolomite are formed with the complex carbonate anion $(CO_3)^{-2}$. Phosphates are formed using the complex anion $(PO_4)^{-3}$. An example is the mineral apatite, used for fertilizers, and the same substance as our teeth and bones are made from. Sulfate minerals are formed using the complex sulfate ion $(SO_4)^{-2}$. Gypsum and anhydrite are sulfate minerals formed by evaporation of saltwater, commonly used to make plaster.

The Properties of Minerals

Minerals have specific properties determined by their chemistry and crystal structure. Certain properties are characteristic of certain minerals, and we can identify minerals by learning these properties. The most common properties are crystal form, color, hardness, luster, cleavage, specific gravity, and taste.

When a mineral grows freely, it forms a characteristic geometric solid bounded by geometrically arranged plane surfaces (this is the crystal form). This symmetry is an external expression of the symmetric internal arrangement of atoms, such as in repeating tetrahedron arrays. Individual crystals of the same mineral may look somewhat different because the relative sizes of individual faces may vary, but the angle between faces is constant and diagnostic for each mineral.

Every mineral has a characteristic crystal form. Some minerals have such distinctive forms that they can be readily identified without measuring angles between crystal faces. For instance, pyrite is recognized as interlocking growth of cubes, whereas asbestos forms long silky fibers. These distinctive characteristics are known as growth habit.

Cleavage is the tendency of a mineral to break in preferred directions along bright reflective planar surfaces. The planar surface along which cleavage occurs is deterred by external structure; cleavage occurs along planes where the bands between the atoms are relatively weak.

Luster is the quality and intensity of light reflected from a mineral. Typical lusters include metallic (like a polished metal), vitreous (like a polished glass), resinous (like resin), pearly (like a pearl), and greasy (oily).

Color is not reliable for identification of minerals, since it is typically determined by ionic substitution. For instance, sapphires and rubies are both varieties of the mineral corundum, with different types of ionic substitution. However, the color of the streak a mineral leaves on a porcelain plate is often diagnostic for opaque minerals with metallic lusters.

The density of a mineral is a measure of mass per unit volume (g/cm^3). Density describes "how heavy the mineral feels." Specific gravity is an indirect measure of density; it is the ratio of the weight of a substance to the weight of an equal volume of water (specific gravity has no units because it is a ratio).

Hardness is a measure of the mineral's relative resistance to scratching. Hardness is governed by the strength of bonds between atoms and is very distinctive and useful for mineral identification. A mineral's hardness can be determined by the ease with which one mineral can scratch another. For instance, talc (used for talcum powder) is the softest mineral, whereas diamond is the hardest mineral. Hardness is commonly measured using Moh's Hardness Scale.

See also PETROLOGY.

Further Reading

Skinner, Brian J., and Stephen C. Porter. *The Dynamic Earth, an Introduction to Physical Geology.* New York: John Wiley and Sons, 1989.

Moh's Hardness Scale

10	Diamond
9	Corundum-(ruby, sapphire)
8	Topaz
7	Quartz
6	Potassium feldspar-(pocketknife, glass)
5	Apatite-(teeth, bones)
4	Fluoride
3	Calcite-(copper penny)
2	Gypsum-(fingernail)
1	Talc

Mississippian   *See* CARBONIFEROUS.

Mississippi River   The largest river system in the United States, the Mississippi stretches 2,350 miles (3,780 km) from northern Minnesota to the end of the Mississippi delta in Louisiana. The Missouri River is longer than the Mississippi but carries less water, and the two systems merge at St. Louis, Missouri, continuing as a larger Mississippi System. Together, the two rivers have a total length (from the Missouri headwaters to the mouth of the Mississippi) of 3,740 miles (6,020 km) and drain 1,231,000 square miles (3,188,290 km²) including parts of 31 states and two Canadian provinces. The area covered by the basin is about 40 percent of the United States, or about 13 percent of North America. This combined system ranks as the world's third longest river system, after the Nile and the Amazon. A sediment-free passage is maintained in the river for navigation all the way from the South Pass area in the delta to St. Anthony Falls in Minneapolis, Minnesota. This passage in the river is extensively used for shipping, making the Mississippi a major economic waterway. The Mississippi is connected with the Intercoastal Waterway in the south, and the Great Lakes–St. Lawrence Seaway in the north through the Illinois Waterway, allowing commerce to move from Canada to the Gulf of Mexico. The name Mississippi is derived from an Ojibwa (Chippewa) Indian word meaning "great river of gathering of waters."

The source of the Mississippi is in small streams that feed into Lake Itasca in northern Minnesota. From there it flows south until it meets the Missouri north of St. Louis, causing the river to expand to a width of 3,500 feet (1,070 m), and then expands to 4,500 feet (1,375 m) at Cairo, Illinois, where the Ohio River joins the flow. South of Cape Girardeau, Missouri, the river begins to meander in large loops as the gradient decreases on a broad alluvial plain, which continues through to the delta section that begins in Mississippi and continues through Louisiana and out into the Gulf of Mexico. The delta of the Mississippi is a bird's foot type of delta, characterized by channels that shift every few hundred years causing a new lobe to be deposited while the once-active lobe generally subsides below sea level. Presently, the active delta lobe that extends past New Orleans to Venice is overextended, and the river has attempted to change courses to follow the distributary Atchafalaya River that provides a much shorter route to the Gulf. The U.S. Army Corps of Engineers has done everything possible to prevent such a change, as it would be disastrous for the economies of New Orleans (which would subside further below sea level) and the entire United States.

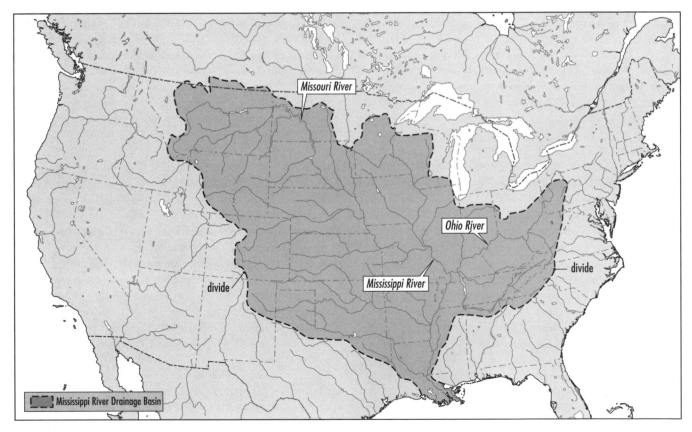

Map showing the Mississippi River drainage basin and the main tributaries and river systems in the United States

Hernando de Soto of Spain is thought to have been the first European to discover and explore the Mississippi in 1541, although in 1682 Sieur de La Salle traveled down the length of the river and claimed the entire territory for France. After the territory changed hands several times, the United States acquired the territory as part of the Louisiana Purchase in 1803. The river soon became the main trade and commerce route for the new territories in the United States, with the first steamboat moving up the river in 1811. During the Civil War the river became a major invasion route from north to south and was the scene of many major battles, including the capture of New Orleans by Union forces in 1863. With the construction of railroads, much of the river's commerce was shifted to overland routes. However, the river continues to be used for shipping bulky freight such as petroleum, sand and gravel, coal, chemical products, and limestone.

The Mississippi River is divided into three main segments including the headwaters, the Upper Mississippi River, and Lower Mississippi River. The headwaters stretch from the source in Lake Itasca to St. Anthony Falls in Minneapolis and are characterized by the steepest gradients, including a drop in elevation from 1,443 feet (440 m) at Lake Itasca to 669 feet (204 m) at St. Anthony Falls, along a distance of 492 miles (794 km). The headwaters flow through thick spruce forests, swamps, wild rice beds, cattail marshes, natural lakes, rapids, and glacial deposits. Water in the headwaters is stained a reddish brown color by organic acids leached from decaying bog vegetation.

The Upper Mississippi River flows 906 miles (1,462 km) from St. Anthony Falls to the mouth of the Ohio River at Cairo, Illinois. The river follows a glacial valley scoured out during melting of the Wisconsin glaciation starting about 15,000 years ago, and these glaciers and meltwater streams followed older tectonic features. Water that spilled out of glacial Lake Agassiz and Lake Duluth (now Lake Superior) for 2,700 years provided huge amounts of water to the Upper Mississippi through the Minnesota River Valley, and this meltwater carved out a river valley to the bedrock that was 300 feet (90 m) deep. By the time the glaciers had retreated into Canada by 9,200 years ago, the torrents of meltwater had scoured the Upper Mississippi River basin to depths of 820 feet (250 m) deep, attesting to the strength of the meltwater floods. Since then the Upper Mississippi basin has been filling in with glacial outwash, a process that is continuing to the present day.

The Lower Mississippi River begins at Cairo, Illinois, where the river enters the gently sloping lowlands of the Lower Mississippi alluvial valley that continues all the way to the Gulf of Mexico. Meandering channels on flat to gently sloping floodplains with low river terraces characterize this section of the Mississippi. The river is silt laden, as are many of its main tributaries including the Black River, Tensas River, Yazoo, Big Sunflower, White, and St. Francis Rivers. There are many oxbow lakes and abandoned meander channels, and tributaries that are trapped by former Mississippi River embayments (Yazoo type rivers) and flow parallel to the main river for many miles before joining the Mississippi. Hardwood forests and swamps cover much of the area that has not been cleared for agriculture. Near New Orleans the river narrows to about 3,300 feet (1 km) and is about 200 feet (60 m) deep. Beyond New Orleans the river flows on the modern Mississippi bird's foot delta, and the river breaks into numerous, levee-bounded distributary channels that are only slightly above sea level.

See also DELTAS; FLOOD.

Moho *See* MOHOROVICIC DISCONTINUITY.

Mohorovicic discontinuity The transition from crust to mantle is generally marked by a dramatic increase in the velocity of compressional seismic waves, from typical values of less than 4.7 miles per second to values greater than 4.85 miles per second (7.6–7.8 km/s). This boundary was first noted by the Yugoslavian seismologist Andrija Mohorovicic in his study of a Balkan earthquake in 1909. Although the Mohorovicic discontinuity is present in most places in continents at depths of several tens of kilometers, there are some regions beneath continents where there is not a significant variation between the seismic velocities of the lower continental crust and the upper mantle. In these regions the Moho is either absent or difficult to detect. The Moho generally lies at depths of 3–6 miles (5–10 km) beneath most regions in the oceans. The Moho probably represents the change from basaltic or gabbroic material above to peridotitic or dunitic material below.

See also CONTINENTAL CRUST; LITHOSPHERE; OCEANIC CRUST.

Moine thrust *See* SCOTTISH HIGHLANDS.

molasse Thick sequences of coarse-grained post-orogenic sandstones, conglomerates, shales, and marls that form in response to the erosion of orogenic mountain ranges are known as molasse. The name is derived from the classic Miocene-Oligocene-Pliocene molasse of the European foreland, deposited across much of France, Switzerland, and Germany and overlying the Alpine flysch sequence. These sediments are up to four miles (7 km) thick on the Swiss Plateau and represent rapid erosion of the Alps. Lower parts of the molasse include shallow marine and tidally influenced sediments, overlain by alluvial fan deltas, alluvial fan complexes, and overbank deposits.

See also FLYSCH; PLATE TECTONICS.

monsoons A wind system that changes direction with the seasons is known as a monsoon, after the Arabic term *mausim,* meaning seasons. The Arabian Sea is characterized by monsoons, with the wind blowing from the northeast for six months, then from the southeast for the other half of the year. Seasonal reversal of winds is probably best known from India and southern Asia, where monsoons bring seasonal rains and floods.

The Asian and Indian monsoon originates from differential heating of the air over the continent and ocean with the seasons. In the winter monsoon, the air over the continents becomes much cooler than the air over the ocean, and a large, shallow high-pressure system develops over Siberia. This produces a clockwise movement of air that rotates over the South China Sea and Indian Ocean, producing northeasterly winds and fair weather with clear skies over eastern and southern Asia. In contrast, in the summer monsoon the air pattern reverses itself as the air over the continents becomes warmer than the air over the oceans. This produces a shallow low-pressure system over the Indian subcontinent, within which the air rises. Air from the Indian Ocean and Arabian Sea rotates counterclockwise into the low-pressure area, bringing moisture-laden winds into the subcontinent. As the air rises due to convergence and orographic effects, it cools below its saturation point, resulting in heavy rains and thunderstorms that characterize the summer monsoons of India from June through September. Some regions of India, especially the Cherrapunji area in the Khasi Hills of northeastern India, receive more than 40 inches (1,000 cm) of rain during a summer monsoon. A similar pattern develops over Southeast Asia. Other less intense monsoons are known from Australia, South America, Africa, and parts of the desert southwest, Pacific coast, and Mississippi Valley of the United States.

The strength of the Indian monsoon is related to the El Niño–Southern Oscillation. During the El Niño events, surface water near the equator in the central and eastern Pacific is warmer than normal, forming excessive rising air, thunderstorms, and rains in this region. This pattern causes air to sink over eastern Asia and India, leading to a summer monsoon with much lower than normal rainfall totals.

See also EL NIÑO.

moraine A landform made of unstratified, unsorted glacial drift, produced by dumping, lodging, or pushing by a glacier. It has a constructional topography that is independent of the surface that it rests on. There are many different types of moraines, produced in a variety of ways by direct action of glacial ice. An advancing or stationary glacier may have hundreds of meters of glacial till wedged in front of it, and this till will fall, making an elongate terminal moraine when the glacier begins to retreat. As a glacier retreats, it may leave a rather level gravel plain, outwash plain, or ground moraine if the retreat is water-rich. Recessional moraines are similar to terminal moraines, except that they mark places where the glacier temporarily paused during its retreat. Knob and kettle

This medial moraine formed where two glaciers came out of separate valleys. The material scraped from the sides of the valley formed lateral moraines along the sides, and the glaciers merged, forming a single large medial moraine. McCarthy Glacier, Kenai Fiords National Park, Alaska. *(Photo by Timothy Kusky)*

or hummocky topography is produced during jerky, on-again, off-again styles of glacial retreat, where many large blocks of ice may get left behind, forming large depressions known as kettle holes in the outwash plain as the glacier retreats. These typically become filled by water to form kettle lakes after the retreat.

There are many types of moraines formed in glaciated terranes that may or may not be experiencing retreat. Lateral moraines form on the sides of mountain and valley glaciers in response to plucking of material off the mountainsides by the ice as it passes, and by material that avalanches off from oversteepened slopes. Medial moraines form where two glaciers join together, and their lateral moraines merge in the center of the new larger glacier.

Glacial moraines form extensive deposits of gravel that are easily mined, since they rest on the surface. A wide variety of grain sizes of gravel and sand may be found in an individual moraine, with steeper surface slopes reflecting coarse-grained material, since larger grains and clasts have a steeper angle of repose than finer grains.

See also GLACIER.

mountain breeze *See* KATABATIC WINDS.

Mount Everest The highest mountain in the world is Mount Everest, reaching 29,035 feet (8,853.5 m) on the border between Tibet and Nepal in the central Himalaya ranges. The summit was first reached by Sir Edmund Hillary and Tenzing Norgay on May 28, 1953. The local Tibetan name for the mountain is Sagarmatha, meaning goddess of the sky, but the name Everest was given to the peak after Sir George Everest in 1865, the British surveyor general of India who first determined the height and location of the mountain known then as Peak 15. Mount Everest continues to rise at a few millimeters per year because of the continued convergence between India and Asia, uplifting the Himalaya and Tibetan plateau.

The base of Mount Everest consists of a series of gneisses and leucocratic granite sheets and dikes. A normal fault cuts across the south face of the mountain and juxtaposes unmetamorphosed Paleozoic marine sediments against the underlying crystalline basement. The normal fault is thought to represent the early stages of orogenic collapse, caused by gravitational stresses leading to collapse and spreading of mountain ranges that reach elevations too high to be held up by the strength of underlying rocks.

See also HIMALAYA MOUNTAINS.

Mount Fuji   Fuji-san, also known as Mount Fuji (and as Fujiyama in some literature) is a large stratovolcano located in central Honshu, Japan, about 60 miles (100 km) west of Tokyo. Fuji rises 12,385 feet (3,776 m) above its base and has a basal diameter of nearly 20 miles (30 km) and is the highest mountain in Japan. The volume of volcanic material added to Japan from Mount Fuji is approximately 540 cubic miles (870 km³), with the last major eruption in 1707, which covered the present site of Tokyo with several inches of volcanic debris. In the past 800 years the volcano has experienced at least 16 major eruptions, averaging one every 50 years. Thus, with the last major eruption in 1707, it appears that activity on the mountain is at least temporarily slowing down. The mountain has a basaltic to andesitic composition. The Japanese revere Mount Fuji as the most beautiful mountain in the world, and its image appears on many works of art and currency and is a major object in Japanese culture and tradition. More than 200,000 people climb Mount Fuji each year.

See also CONVERGENT PLATE MARGIN PROCESSES; JAPAN'S PAIRED METAMORPHIC BELT.

Mount Logan   The highest mountain in Canada and second highest mountain in North America, Mount Logan rises to 19,850 feet (6,054 m) above a huge tableland in the Wrangell–St. Elias Mountains of the southwest Yukon Territories, just east of Alaska. The mountain is composed predominantly of granodiorite and is being actively uplifted. The mountain is located in the most heavily glaciated region of North America, with many mountain and valley glaciers feeding tidewater and large trunk glaciers. Logan is situated in Kluane National Park Reserve, west of Whitehorse and less than 60 miles (100 km) from the coast. The mountain is covered by thick snow and ice that in places is about 1,000 feet (300 m) deep and at least several thousand years old. The climate has been cold and extreme on Mount Logan since well before the last Pleistocene glaciation, and it is possible that some of the ice is even several millions of years old.

Mount Logan was named in 1890 by I. C. Russell of the U.S. Geographical Survey after Sir William Edmond Logan (1798–1875), who founded the Geological Survey of Canada in 1842. However, in October of 2000, Pierre Chretien, the prime minister of Canada, proposed renaming the mountain Mount Trudeau, after former prime minister Pierre Trudeau.

Mount McKinley   The tallest mountain in North America, Mount McKinley soars to 20,320 feet (6,194 m) in central Alaska's Alaska Range. The mountain also boasts one of the steepest vertical rises in the world, with its base nearly at sea level, and the south slope rises an amazing 17,000 feet (5,180 m) in 12 miles (19 km). The mountain was named in 1897 by W. A. Dickey, after President William McKinley. However, the local Athabaskan Indians called the mountain Denali, meaning the high one, and the national park that surrounds the mountain is also called Denali. In 1913, Hudson Stuck, Harry Karstens, Robert Tatum, and Walter Harper became the first people to successfully climb Mount McKinley. Five major and many smaller glaciers and ice falls emanate from

Geological fieldwork in the Chugach mountains; Mount Logan is in the background. *(Photo by Timothy Kusky)*

Massive Mount McKinley, looming over Lake Minchumina, more than 50 miles (100 km) north of the mountain *(Photo by Timothy Kusky)*

the mountain and flow into lower elevations. The mountain is located between branches of the active Denali fault and is actively rising and affected by numerous earthquakes.

Mount Washington The tallest peak in the northeastern United States is Mount Washington, rising to 6,288 feet (1,886 m) in the Presidential Range of New Hampshire's White Mountains. The mountain was formerly known as Agiocochook by the local Indians, and the first known ascent was by Darby Field in 1642. By the middle of the 1800s a summit house was built, soon followed by the Cog Railroad and Carriage Road, making the top of New England's tallest mountain accessible to many visitors. Many structures were built, but a large fire in 1908 destroyed most buildings on the summit. The Mount Washington Observatory was founded in 1932 and has recorded daily weather conditions ever since. Mount Washington is now known to have some of the most extreme weather because of its location and height at the junction of two jet streams. The highest wind gust ever recorded on Earth was 231 miles per hour (371.75 km/hr) on top of Mount Washington, on April 12, 1934.

Mount Whitney, California The tallest mountain in the lower 48 states is Mount Whitney, California, reaching 14,491 feet (4,417 m) in the Sierra Nevada mountain ranges. The mountain is located at the eastern border of Sequoia National Park and is surrounded by many peaks reaching over 12,000 feet (3,650 m). Mount Whitney is thought to be the most frequently climbed peak in the United States.

The Sierra Nevada are part of a large continental margin plutonic belt including granites, granodiorites, diorites, and other rocks, intruded from Late Jurassic through Cretaceous above a long-lasting convergent margin and subduction zone. This tectonism and plutonism probably cause some uplift of the Sierra Nevada. However, much of the topography is a result of relatively recent uplift in the past 3–4 million years, and the eastern margin of the Sierra Nevada is bounded by a series of normal faults that mark the transition into the Basin and Range Province. The western margin of the Sierra is more gently inclined and slopes into the Great Valley of California. The precise cause for the uplift of the Sierra Nevada is not certain, but many geologists relate the initial uplift to flat-slab subduction during the latest Cretaceous to Eocene Laramide Orogeny. Continued uplift could be related to strike-slip motions, isostatic compensation of crust with a thick light root, or thermal effects of lithospheric thinning.

See also CONVERGENT PLATE MARGIN PROCESSES.

muskeg A high-latitude bog that is found in many areas of permafrost and boreal forest. Muskegs typically have tussocks of grass separated by patches of water and may have tamarack and black spruce trees rimming the edges of the bog. Muskegs may eventually be filled in by organic material and convert to boreal forest.

mylonite A fine-grained, typically banded rock produced by the grain size reduction and recrystallization produced during ductile deformation of other rocks by shearing and flattening. Mylonites usually form in shear zones that may be a few millimeters to many kilometers wide and may be linked to other structures that accommodate regional tectonic deformations. They form at depths that are at or below the brittle ductile transition at 10–15 kilometers, at medium to high metamorphic grades.

Protomylonites represent the early stages of formation of mylonitic rocks and are characterized by flattened and stretched mineral grains, showing signs of internal strain such as undulose extinction, deformation twins, and kink bands. Some types of minerals may be more deformed than other types, depending on their relative strength and the temperature and pressure conditions of deformation. As deformation becomes more intense, the grain size is reduced and a strongly foliated matrix develops, with some isolated larger grains known as porphyroclasts. The matrix foliation may flow around these porphyroclasts and become more closely spaced on sides of the grains that face the principal flattening direction, with pressure shadow zones developed on the opposing two sides. This asymmetric structure is one of the ways to determine the kinematics, or sense of shear within the mylonite zone.

With continued deformation, the mylonite may become extremely fine-grained and very strongly foliated, forming an ultramylonite. If the temperature and pressure are high enough, new metamorphic grains may form, known as porphyroblasts, and these may grow over and rotate with the mylonite fabric forming distinctive kinematic indicators useful for determining the sense of shear of the mylonite zone.

See also DEFORMATION OF ROCKS; STRUCTURAL GEOLOGY.

Further Reading

Passchier, Cees W., and Rudolph A. J. Trouw. *Micro-tectonics.* Berlin: Springer-Verlag, 1996.

Granitic mylonite showing large porphyroclasts (relict feldspar grains) and mylonitic foliation (from 1-billion-year-old shear zone in Parry Sound, southern Canada) *(Photo by Timothy Kusky)*

N

nappe Relatively thin or sheet-like masses of rock that have moved large distances on subhorizontal surfaces. They may move on thin thrust faults, or along zones of recumbent folding and attenuation of fold limbs. Nappes can be very large in scale, some being many tens of kilometers in dimensions. Recumbent fold nappes have one inverted and one upright limb, with bedding planes typically being parallel for many kilometers on either limb.

Large-scale thrust and nappe tectonics were first recognized in the Helvetic Alps in the late 1880s. Nappes have since been recognized in the internal zone of many mountain belts worldwide, and of all ages, stretching back to the Precambrian.

See also DEFORMATION OF ROCKS; STRUCTURAL GEOLOGY.

National Academy of Sciences (NAS) The National Academy of Sciences (http://www.nas.edu/) was established by President Abraham Lincoln on March 3, 1863. It has served to "investigate, examine, experiment, and report upon any subject of science or art" whenever called upon to do so by any department of the government. The NAS has expanded to include the so-called sister organizations, including the National Research Council, the National Academy of Engineering, and the Institute of Medicine. Collectively, these organizations are called the National Academies.

The mission of the National Academies is to advise on matters of science, technology, and medicine. They enlist committees of the nation's top scientists, engineers, and other experts that frequently pervade policy decisions. The results of their deliberations have inspired some of America's most significant and lasting efforts to improve the health, education, and welfare of the population.

Their service to government has become so essential that Congress and the White House have issued legislation and executive orders over the years that reaffirm their unique role.

National Aeronautics and Space Administration (NASA) The National Aeronautics and Space Administration (http://www.nasa.gov) was established on October 1, 1958. Since then NASA has accomplished many great scientific and technological feats in air and space. NASA has become the world's leading force in scientific research of the solar system and beyond and in aerospace exploration, as well as science and technology in general. NASA's main research fields include space science, aerospace technology, biological and physical research, earth science, and human exploration and development of space.

National Association of Geoscience Teachers (NAGT) NAGT (http://www.nagt.org) was established in 1938 in order to: (1) foster improvement in the teaching of earth sciences at all levels of formal and informal instruction, (2) emphasize the cultural significance of the earth sciences, and (3) disseminate knowledge in this field to the general public. NAGT organizes the technical program in geoscience education at every Geological Society of America (GSA) annual meeting and in this context also sponsors numerous technical sessions, workshops, and field trips. The organization also fosters advances in the teaching and learning of geoscience by means of scholarship aid to summer field-camp students, professional workshops for geoscience faculty, its Distinguished Speaker Series, and the *Journal of Geoscience Education,* and through its partnership with the National Association for Black Geologists and Geophysicists.

National Oceanographic and Atmospheric Association (NOAA) This government agency conducts research and gathers data about the global oceans, atmosphere, space, and sun and applies this knowledge to science and service that touch lives of all Americans. NOAA's mission is to describe and predict changes in the Earth's environment and conserve and wisely manage the nation's coastal and marine resources. NOAA's strategy consists of seven interrelated strategic goals for environmental assessment, prediction, and stewardship. These include: (1) advance short-term warnings and forecast services, (2) implement seasonal to interannual climate forecasts, (3) assess and predict decadal to centennial change, (4) promote safe navigation, (5) build sustainable fisheries, (6) recover protected species, and (7) sustain healthy coastal ecosystems. NOAA runs a Web site (http://www.noaa.gov/) that includes links to current satellite images of weather hazards, issues warnings of current coastal hazards and disasters, and has an extensive historical and educational service. The National Hurricane Center (http://www.nhc.noaa.gov/) is a branch of NOAA and posts regular updates of hurricane paths and hazards. NOAA includes the following organizations:

1. NOAA National Weather Service (http://www.nws.noaa.gov/),
2. NOAA Satellites and Information (http://www.nesdis.noaa.gov/),
3. NOAA Fisheries (http://www.nmfs.noaa.gov/),
4. NOAA Ocean Service (http://www.oceanservice.noaa.gov/),
5. NOAA Research (http://www.research.noaa.gov/),
6. NOAA Marine & Aviation Operations (http://www.nmao.noaa.gov/).

National Weather Service (NWS) The National Weather Service (http://www.nws.noaa.gov/) provides weather, hydrologic, and climate forecasts and warnings for the United States, its territories, adjacent waters, and ocean areas. NWS data and products form a national information database and infrastructure, which can be used by other governmental agencies, the private sector, the public, and the global community. "The NWS receives and processes over 1 million observations each day to develop forecasts and warnings for the American people."

natural gas Petroleum that occurs naturally under normal conditions of temperature and pressure in the ground in a gaseous state. Most natural gas is methane, followed by ethane, propane, butane, and pentane, with common impurities of inorganic gases including nitrogen, carbon dioxide, and hydrogen sulfide. Natural gas has an origin similar to other hydrocarbons, being derived from the decomposition of buried organic matter. It is simply the lighter end member of the spectrum of compositions of hydrocarbons, and being a gas, it has only gaseous hydrocarbons (C_1-C_5) with no C_{6+} compounds. All types of organic matter can contribute to the formation of natural gas, when buried and heated to more than 320°F (160°C). Some natural gas is generated during the decomposition of coal and petroleum when they are heated above 320°F, whereas other gas is produced along with the generation of other hydrocarbons. An additional type of natural gas is biogenic methane, produced at shallow levels by the biodegradation of petroleum, and when bacteria reduce carbon dioxide to methane in shallow sediments. Natural gas is abundant in shallow crustal reservoirs and is useful as a fossil fuel. It generally burns much cleaner than petroleum or coal and so is increasingly being sought as an energy source. Reserves of natural gas are huge and may greatly exceed the remaining reserves of petroleum in the world.

See also HYDROCARBON; PETROLEUM.

nekton Animals that move through the water primarily by swimming are nektons, and free-swimming pelagic animals are said to be nektonic. They are distinguished from other pelagic organisms (plankton) that float in the water. The most important nektons in the water today are the fish, whereas in the Paleozoic several other forms were common. The ammonoids of the Devonian were coiled cephalopod mollusks that evolved from earlier nautilids, and these existed with the free-swimming scorpion-like eurypterids. Fish first appeared in the marine record in the Cambrian-Ordovician and included the early bony-skinned fish known as ostracoderms, followed in the Late Silurian by the finned acanthodians. Heavily armored large-jawed fish known as placoderms are found in many Late Devonian deposits, as are lungfish, ray-finned fish, and lobe-finned fish that include the coelacanths, one species of which survives to this day. Lobe-finned fish are the ancestors of all terrestrial vertebrates. Sharks were very common in the marine realm by the Late Paleozoic.

Neogene The second of three periods of the Cenozoic, including the Paleogene, Neogene, and Quaternary, and the second of two subperiods of the Tertiary, younger than Paleogene. Its base is at 23.8 million years ago, and its top is at 1.8 million years ago, followed by the Quaternary period. Subdivision of the Neogene into the Miocene, Pliocene, Pleistocene, and Recent Epochs was proposed by Charles Lyell in 1833, in his book *Principles of Geology*. The division of the Neogene was formally proposed by R. Hornes in 1835 and included only the older parts of Lyell's Neogene.

The Atlantic and Indian Oceans were open in the Neogene, and the plate mosaic looked quite similar to the modern configuration. India had begun colliding with Asia, Australia had already rifted and was moving away from Antarctica, isolating Australia and leading to the development of the cold circumpolar current and the Antarctic ice cap. Subduction and accretion events were active along the Cordilleran margins of North and South America. Basin and Range extension was active, and the Columbia River basalts were erupted in the northwestern United States. The San Andreas Fault developed in California during subduction of the East Pacific Rise.

One of the more unusual events to mark the Neogene is the development of up to 1.2 miles (2 km) of salt deposits between 5.5 million and 5.3 million years ago in the Mediterranean region. This event, known as the Messinian salt crisis, was caused by the isolation of the Mediterranean Sea by collisional tectonics and falling sea levels, that caused the sea to at least partially evaporate several times during the 200,000-year-long crisis. Rising sea level ended the Messinian crisis at 5.3 million years ago, when waters of the Atlantic rose over the natural dam in the Strait of Gibraltar, probably forming a spectacular waterfall.

A meteorite impact event occurred about 15 million years ago, forming the 15-mile (24-km) wide Ries Crater near Nordlingen, Germany. The meteorite that hit the Earth in this event is estimated to have been one kilometer in diameter, releasing the equivalent of a 100,000-megaton explosion. About 55 cubic feet (155 m³) of material were displaced from the crater, some of which formed fields of tektites, unusually shaped melted rock that flew through the air for up to 248.5 miles (400 km) from the crater.

The Neogene saw the spread of grasses and weedy plants across the continents, and the development of modern vertebrates. Snakes, birds, frogs, and rats expanded their niches, whereas the marine invertebrates experienced few changes. Humans evolved from earlier ape-like hominids. Continental glaciations in the Northern Hemisphere began in the Neogene and continue to this day.

See also CENOZOIC; EVAPORITE; TERTIARY.

Neolithic An archaeological term for the last division of the Stone Age, during which time humans developed agriculture and domesticated animals. The transition from hunter-gatherer and nomadic types of existence to the development of farming took place about 10,000–8,000 years ago in the Fertile Crescent, a broad stretch of land that extends from southern Israel through Lebanon, western Syria, Turkey, and through the Tigris-Euphrates Valley of Iraq and Iran. The Neolithic revolution and the development of stable agricultural practices led an unprecedented explosion of the human population that continues to this day. About a million years ago there are estimated to have been a few thousand migratory humans on the Earth, and by about 10,000 years ago this number had increased to only 5 million–10 million. When humans began stable agricultural practices and domesticated some species of animals, the population rate started to increase substantially. The increased standards of living and nutrition caused the population growth to soar to about 20 million by 2,000 years ago, and 100 million by 1,000 years ago. By the 18th century, humans began manipulating their environments more, began public health services, and began

to recognize and seek treatments for diseases that previously took many lives. The average life span began to soar, and world population surpassed 1 billion in the year 1810. A mere 100 years later, world population doubled again to 2 billion and had reached 4 billion by 1974. World population is now close to 7 billion and climbing more rapidly than at any time in history, doubling every 50 years.

Further Reading

Diamond, John. *Guns, Germs, and Steel, The Fates of Human Societies.* New York: W. W. Norton and Co., 1999.

Leonard, Jonathan N. *The First Farmers, The Emergence of Man.* New York: Time-Life Books, 1973.

Neoproterozoic The last of three major eras of the Proterozoic era, the Neoproterozoic stretches from 900 to 544 million years ago. The Neoproterozoic was a time of major global-scale tectonic, biologic, and climatic change. The supercontinent of Rodinia broke up and the continental fragments rearranged themselves in the supercontinent of Gondwana, closing several large ocean basins in the process. The breakup, dispersal, and re-amalgamation of the main continents in the Neoproterozoic caused many global climatic changes, which together with the changing environments on the continental margins provided many environments and stimuli for life to dramatically change in this period. One of the great debates that are currently active in geology is whether and how the dramatic biologic, climatic, and geologic events that mark Earth's transition from the Precambrian into the Cambrian can be linked to the distribution of continents and to the breakup and reassembly of a supercontinent.

See also GONDWANA.

Further Reading

Grotzinger, John P., Samuel A. Bowring, Beverly Z. Saylor, and Alan J. Kaufman. "Biostratigraphic and Geochronologic Constraints on Early Animal Evolution." *Science* 270 (1995): 598–604.

Kaufman, Alan J., Andrew J. Knoll, and Guy M. Narbonne. "Isotopes, Ice Ages, and Terminal Proterozoic Earth History." *National Academy of Sciences Proceedings* 94 (1997): 6,600–6,605.

Kusky, Timothy M., Mohamed Abdelsalam, Robert Tucker, and Robert Stern, eds., *Evolution of the East African and Related Orogens, and the Assembly of Gondwana.* Amsterdam: Elsevier, Precambrian Research, 2003.

Neptune The eighth planet from the center of the solar system, the giant Jovian planet Neptune orbits the Sun at a distance of 2.5 billion miles (4.1 billion km, or 30.1 astronomical units), completing each circuit every 165 years. Rotating about its axis every 16 hours, Neptune has a diameter of 31,400 miles (50,530 km) and has a mass of more than 17.21 times that of Earth. It has a density of 1.7 grams per cubic centimeter, showing that the planet has a dense rocky interior sur-

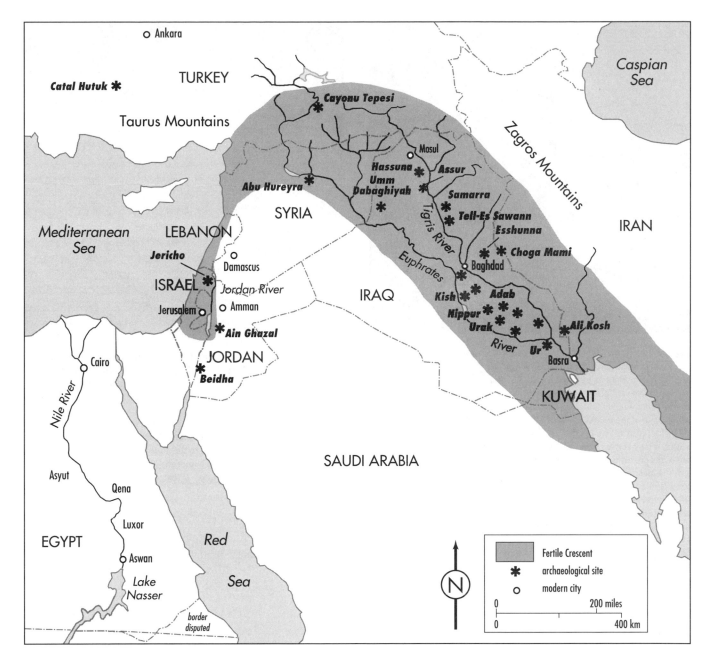

Map of the Fertile Crescent stretching from the Levant (Israel and Lebanon) through rolling hills in parts of Syria, southern Turkey, Iraq, and Iran. Ancient cities and agriculture arose in this area, with many early cities located in the Sumerian region between the Tigris and Euphrates Rivers, in what is now southern Iraq.

rounded by metallic, molecular, and gaseous hydrogen, helium, and methane, giving the planet its blue color.

Neptune is unusual in that it generates its own heat, radiating 2.7 times more heat than it receives from the Sun. The source of this heat is uncertain, but it may be heat trapped from the planet's formation that is only slowly being released by the dense atmosphere. The cloud systems that trap this heat are visible from Earth-based telescopes

and include some large hurricane-like storms such as the former Great Dark Spot, a storm about the size of the Earth, similar in many ways to the Great Red Spot on Jupiter, but that has dissipated.

Neptune has two large moons visible from Earth, Triton and Nereid, and six other smaller moons discovered by the *Voyager 2* spacecraft. Triton has a diameter of 1,740 miles (2,800 km) and orbits Neptune at a distance of 220,000

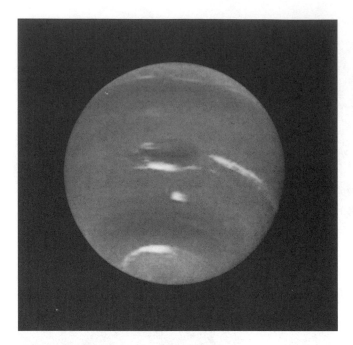

Neptune and its swirling clouds. Neptune's blue-green atmosphere is shown in greater detail than ever before by the *Voyager 2* spacecraft as it rapidly approaches its encounter with the giant planet. This image, produced from a distance of about 9.6 million miles (16 million km), shows several complex and puzzling atmospheric features. The Great Dark Spot (GDS) seen at the center is about 8,080 miles (13,000 km) by 4,100 miles (6,600 km) in size—as large along its longer dimension as the Earth. The bright, wispy "cirrus-type" clouds seen hovering in the vicinity of the GDS are higher in altitude than the dark material of unknown origin that defines its boundaries. A thin veil often fills part of the GDS interior, as seen on the image. The bright cloud at the southern (lower) edge of the GDS measures about 600 miles (1,000 km) in its north-south extent. The small, bright cloud below the GDS, dubbed the "scooter," rotates faster than the GDS, gaining about 30° eastward (toward the right) in longitude every rotation. Bright streaks of cloud at the latitude of the GDS, the small clouds overlying it, and a dimly visible dark protrusion at its western end are examples of dynamic weather patterns on Neptune, which can change significantly on timescales of one rotation (about 18 hours). *(Photo by NASA)*

Composite view showing Neptune on Triton's horizon. Neptune's south pole is to the left; clearly visible in the planet's southern hemisphere is the Great Dark Spot, a large anticyclonic storm system located about 20° south. The foreground is a computer-generated view of Triton's maria as they would appear from a point approximately 30 miles (45 km) above the surface. This three-dimensional view was created from a *Voyager* image, and the relief has been exaggerated roughly 30-fold. Neptune would not appear to be rising or setting; due to the motion of Triton relative to Neptune, it would appear to move laterally along the horizon, eventually rising and setting at high latitudes. *(Photo by NASA)*

miles (354,000 km) from the planet, and it is the only large moon in the solar system that has a retrograde orbit.

New Zealand; Alpine fault and Otago Schist   New Zealand lies along the junction between the Indo-Australian plate and the Pacific plate in the southern Pacific Ocean. The plate boundary is curved and the two plates have various relationships to each other along the boundary. The Pacific plate is being subducted toward the west beneath the North Island, forming the overlying Taupo volcanic zone. On South Island, the Pacific and Indo-Australian plates are moving roughly parallel to each other and have formed a dextral continental transform fault known as the Alpine fault. South of New Zealand the plate margin curves again so there is contraction across the boundary, but here the Indo-Australian

plate is being subducted beneath the Campbell plateau on the Pacific plate. The Alpine fault has a component of contraction across the plate boundary, causing the main part of the South Island to be thrust over the Australian plate, and uplifting the mountains of the Southern Alps by 0.3 inches per year (7 mm/yr).

The South Island of New Zealand can be divided into a number of different belts, cut by the active Alpine fault. Western Fiordland is underlain by high temperature/high pressure crystalline metamorphic rocks, while Southland is underlain by volcanosedimentary rocks of the Hokonui facies. A scattered belt of ultramafic rocks separates the Hokonui facies and high-grade metamorphic rocks from rocks in the east that include non-schistose and schistose rocks of the Torlesse facies. About 12,000 square miles (19,500 km²) of the Southern Alps are underlain by a structurally complex rock unit known as the Otago Schist, part of the Torlesse facies. These rocks form a strongly curved outcrop belt that is deflected and offset by the Alpine fault and occupy rugged terrain with peaks reaching 5,000 feet (1,500 m), the upper part of which is above the tree line. The curved outcrop pattern, strong strain gradient toward the Alpine fault, and well-preserved structures in the schist led to many structural interpretations of the area, elevating the region to a classic area for structural geology. The schists are bounded on the northeast and southwest by Permian-Triassic graywackes, on the west by the Livingstone fault, and on the northwest by the Alpine fault. The schists include a group of quartzofeldspathic graywackes and argillites of a turbidite sequence. They have been metamorphosed to the chlorite grade, with biotite, garnet, and oligoclase zones. The graywacke layers in the schists show a wide variety of fold morphologies that vary according to metamorphic grade, amount of strain, and proximity to the Alpine fault. The regional structure of the belt includes a large stack of recumbent fold nappes that are overturned toward the northeast.

Late Tertiary faulting and uplift on South Island formed many of the visible landforms including valleys and plateaus. The western part of the island was heavily glaciated in the Pleistocene, while parts of the eastern region are covered in loess deposits.

See also STRUCTURAL GEOLOGY; TRANSFORM PLATE MARGIN PROCESSES.

Further Reading

George, Annette D. "Deformation Processes in an Accretionary Prism: A Study from the Torlesse Terrane of New Zealand." *Journal of Structural Geology* 12 (1990).
Turner, F. J. "Structural Petrology of the Schist of Eastern Otago, New Zealand." *American Journal of Science* 238 (1940).
Wood, B. L. "Structure of the Otago Schists." *New Zealand Journal of Geology and Geophysics* 6 (1963).

Nile River   The longest river in the world, the Nile flows 4,184 miles (6,695 km) from its remotest headwater stream, the Luvironza River in central Africa's Burundi, to its delta on the Mediterranean coast. The Nile drains more than 1,100,000 square miles (2,850,000 km²) amounting to 10 percent of Africa, including parts of Egypt, Sudan, Ethiopia, Uganda, Kenya, Rwanda, Burundi, and Zaire. Water from the Nile is used for almost all of the irrigation and agriculture in Egypt and about 20 percent of the irrigation in Sudan. The river is widely used for transportation, hydroelectric power, and as a source of food.

The Nile has two main tributaries, the White Nile and the Blue Nile, that meet in Khartoum, Sudan, then flow together to the north to the sea. Lake Victoria in Uganda is the main headwater reservoir for the White Nile, whereas Lake Tana in Ethiopia is the main headwater reservoir for the Blue Nile. The White Nile flows north out of Lake Victoria over Owen Falls into the Victoria Nile that flows into Lake Albert on the floor of the East African rift system and then is known as the Albert Nile as it flows north out of Uganda. This region receives high rainfall all year round, so the flow of water in the White Nile is close to constant. In southern Sudan the river becomes known as the Mountain Nile, before it makes a bend to the

east and is joined by the Bahr al-Ghazal River and it turns back to the north as the White Nile. This relatively straight stretch continues to Khartoum where the White Nile is joined by the waters of the Blue Nile. The Blue Nile begins its journey on the northern slope of the Ethiopian highlands, formed by uplift and volcanism on the Ethiopian dome around the Afar triple junction. This region receives heavy summer rains, and these are the source of the floodwaters that reach Egypt nearly every September. The Blue Nile and its many tributaries flow over rugged volcanic terrane into Sudan, where a reservoir has been created by damming Lake al-Azraq. The Blue Nile then meanders across southern Sudan until it meets the White Nile at Khartoum. More than half of the Nile's water is contributed by the Blue Nile, and during flood season, most of the silt in the river comes from erosion of the Ethiopian highlands. From Khartoum, the merged White and Blue Niles form a new trunk stream, that flows in a narrow channel with virtually no floodplain around a great bend where the waters flow first northeast, then swing northwest, and then southwest before continuing to the north. Old river channels followed a more direct route through the Bayuda Desert region the river now flows around, so it is evident that the Bayuda Desert region is experiencing active uplift that has deflected the river to the east. From Khartoum to the Egyptian border, the river must flow over six sections containing rapids, known locally as cataracts. As the river flows out of Sudan it enters Lake Nasser in Egypt, created by construction of the Aswan High Dam. The Nile then continues north through Upper Egypt, past Assuit and Cairo, and splits into tributaries in the delta region before flowing into the Mediterranean Sea. The floodplain of the river broadens in Egypt, first to around 10 miles (16 km) south of Cairo, then broadens dramatically to more than 100 (160 km) miles wide on the delta north of Cairo. The delta is the main agricultural region of Egypt, containing more than 60 percent of the cultivated lands in the country.

In Sudan, the region between the White Nile and Blue Nile is known as Al-Jazirah (or Gezira), meaning the island. This region is extremely fertile and serves as the principal agricultural area in Sudan, and it is the only large area outside Egypt where Nile waters are used extensively for irrigation.

The Egyptian government is currently attempting a massive project to form a second arm of the Nile extending out of Lake Nasser and flowing across the Western Desert to eventually reach the sea near Alexandria. This ambitious project starts in the Tushka Canal area, where water is drained from Lake Nasser and steered into a topographic depression that winds its way north through some of the hottest, driest desert landscape on Earth. The government plans to move thousands of farmers and industrialists from the familiar Nile Valley into this national frontier, hoping to alleviate overcrowding. Cairo's population of 13 million is increasing at a rate of nearly one million per year. If successful, this plan could reduce the water demands on the limited resources of the river. However, there are many obstacles with this plan. Will people stay in a desert where temperatures regularly exceed 120°F (49°C)? Will the water make it to Alexandria, having to flow through unsaturated sands, and through a region where the evaporation rate is 200 times greater than the precipitation rate? How will drifting sands and blowing dust affect plans for agriculture in the Western Desert? Much of the downriver part of the Nile is suffering from lower water and silt levels than are needed to sustain agriculture and even the current land surface. So much water is used, diverted, or dammed upstream that parts of the Nile Delta have actually started to subside (sink) beneath sea level. These regions desperately need to receive the annual silt layer from the flooding Nile to rebuild the land surface and keep it from disappearing beneath the sea.

There are also political problems with establishing the New River through the Western Desert. Ethiopia contributes about 85 percent of the water to the Nile, yet it is experiencing severe drought and famine in the eastern part of the country. There is no infrastructure to get the water from the Nile to the thirsty lands and people to the east. Sudan and Egypt have long-standing disputes over water allotments, and Sudan is not happy that Egypt plans to establish a new river that will further their use of the water. Water is currently flowing out of Lake Nasser, filling up several small lake depressions to the west, and evaporating between the sands.

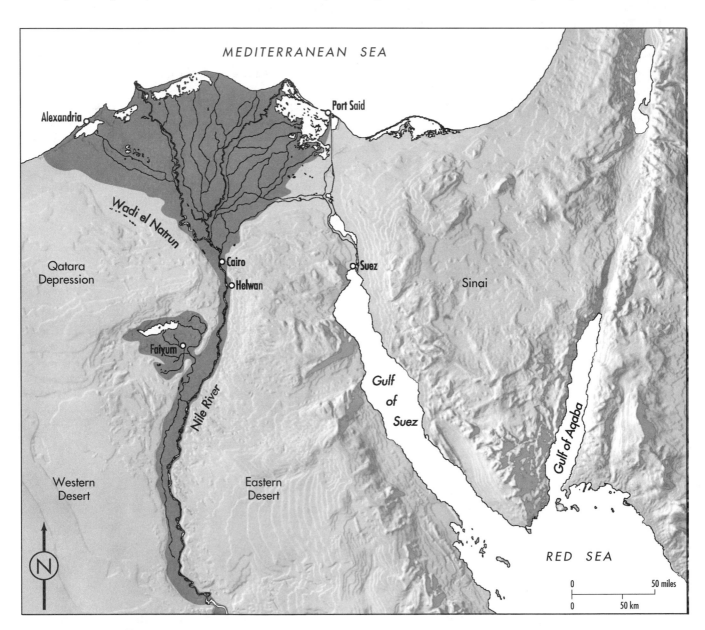

Map of the Nile River in Egypt

The Nile has been used as a water source for irrigation for more than 6,000 years, with ancient Egyptians trapping water and silt from the yearly floods in small basins, then growing crops in the water-soaked soils. The ancient system has been replaced with modern perennial irrigation systems, in which a series of dams (barrages) keep the water at levels to grow several crops per year, and some water is siphoned off to irrigate areas such as the Faiyum depression. Canals have been built to Alexandria and even across the Suez Canal to the northern Sinai.

Ancient people of the Nile Valley did not know the source of the Nile, or why it was prone to annual floods that showed a seven-year cycle in intensity. The cyclicity has since been understood to be a consequence of the El Niño Southern Oscillation cycle, and explorers have entered the central African regions that are the sources of the Nile. The ancient Greek philosopher Ptolemy suggested that the source of the Nile was the Mountains of the Moon, since identified as the Ethiopian highlands around Lake Tana by Scottish explorer James Bruce in 1770. John Speke, a British explorer, identified Lake Victoria as the source of the White Nile in 1861.

See also RIVER SYSTEM.

nonconformity *See* UNCONFORMITY.

normal fault *See* FAULT.

North China craton The North China craton (NCC) occupies about one million square miles (1.7 million km²) in northeastern China, Inner Mongolia, the Yellow Sea, and North Korea. It is bounded by the Qinling–Dabie Shan orogen to the south, the Yinshan-Yanshan orogen to the north, the Longshoushan belt to the west and the Qinglong-Luznxian and Jiao-Liao belts to the east. The North China craton includes a large area of intermittently exposed Archean crust, including circa 3.8–2.5 billion-year-old gneiss, tonalite, trondhjemite, and granodiorite. Other areas include granite, migmatite, amphibolite, ultramafite, mica schist and dolomitic marble, graphitic and sillimanitic gneiss (khondalites), banded iron formation (BIF), and metaarkose. The Archean rocks are overlain by the 1.85–1.40 billion-year-old Mesoproterozoic Changcheng (Great Wall) system. In some areas in the central part of the North China Craton, 2.40–1.90 billion-year-old Paleoproterozoic sequences deposited in cratonic graben are preserved.

The North China craton is divided into two major blocks separated by the Neoarchean Central orogenic belt in which virtually all U-Pb zircon ages fall between 2.55 billion and 2.50 billion years old. The Western block, also known as the Ordos block, is a stable craton with a thick mantle root, no earthquakes, low heat flow, and a lack of internal deformation since the Precambrian. In contrast, the Eastern block is atypical for a craton in that it has numerous earthquakes, high heat flow, and a thin lithosphere reflecting the lack of a thick mantle root. The North China craton is one of the world's most unusual cratons in that it had a thick tectosphere (subcontinental lithospheric mantle) developed in the Archean, which was present through the Ordovician as shown by deep xenoliths preserved in Ordovician kimberlites. However, the eastern half of the root appears to have delaminated or otherwise disappeared during Paleozoic, Mesozoic, or Cenozoic tectonism. This is demonstrated by Tertiary basalts that bring up mantle xenoliths of normal "Tertiary mantle" with no evidence of a thick root. The processes responsible for the loss of this root are enigmatic but are probably related to the present-day high-heat flow, Phanerozoic basin dynamics, and orogenic evolution.

The Central orogenic belt includes belts of tonalite-trondhjemite-granodiorite, granite, and supracrustal sequences metamorphosed from granulite to greenschist facies. It can be traced for about 1,000 miles (1,600 km) from west Liaoning to west Henan. Widespread high-grade regional metamorphism including migmatization occurred throughout the Central orogenic belt between 2.6 billion and 2.5 billion years ago, with final uplift of the metamorphic terrain at 1.9–1.8 billion years ago associated with extensional tectonism or a collision on the northern margin of the craton. Amphibolite to greenschist-grade metamorphism predominates in the southeastern part of the Central orogenic belt, but the northwestern part of the orogen is dominated by granulite facies to amphibolite facies rocks, including some high-pressure assemblages (10–13 kilobars at 850±50°C). The high-pressure assemblages can be traced for more than 400 miles (700 km) along a linear belt trending east-northeast. Internal (western) parts of the orogen are characterized by thrust-related horizontal foliations, flat-dipping shear zones, recumbent folds, and tectonically interleaved high-pressure granulite migmatite and metasediments. It is widely overlain by sediments deposited in graben and continental shelf environments and intruded by several dike swarms (2.4–2.5 billion and 1.8–1.9 billion years ago). Several large anorogenic granites with ages of 2.2–2.0 billion years are identified within the belt. Recently, two linear units have been documented within the belt, including a high-pressure granulite belt in the west and a foreland-thrust fold belt in the east. The high-pressure granulite belt is separated by normal-sense shear zones from the Western block, which is overlain by thick metasedimentary sequences (khondalite, younger than 2.4 billion years, and metamorphosed at 1.86 billion years).

The Hengshan high-pressure granulite belt is about 400 miles (700 km) long, consisting of several metamorphic terrains, including the Hengshan, Huaian, Chengde, and west Liaoning complexes. The high-pressure assemblages commonly occur as inclusions within intensely sheared tonalite-trondhjemite-granodiorite (2.6–2.5 billion years) and granitic gneiss (2.5 billion years) and are widely intruded by K-granite (2.2–1.9 billion years) and mafic dike swarms (2.40–2.45 Ga,

1.77 billion years). Locally, khondalite and turbiditic slices are interleaved with the high-pressure granulite rocks, suggesting thrusting. The main rock type is garnet-bearing mafic granulite with characteristic plagioclase-orthopyroxene corona around the garnet, which show rapid exhumation-related decompression. An isothermal decompressive pressure-temperature-time path can be documented within the rocks, and the peak pressures and temperatures are in the range of 1.2–1.0 GPa, at 1,290°F–1,470°F (700°C–800°C). At least three types of geochemical patterns are shown by mafic rocks of the high-pressure granulites, indicating a tectonic setting of active continental margin or island arc. The high-pressure granulites were formed through subduction-collision, followed by rapid rebound-extension, recorded by 2.5–2.4 billion-year-old mafic dike swarms and graben-related sedimentary sequences in the Wutai Mountain–Taihang Mountain areas.

The Qinglong foreland basin and fold-thrust belt is north to northeast-trending and is now preserved as several relict folded sequences (Qinglong, Fuping, Hutuo, and Dengfeng). Its general sequence from bottom to top can be further divided into three subgroups of quartzite-mudstone-marble, turbidite, and molasse, respectively. The lower subgroup of quartzite-mudstone-marble is well preserved in central sections of the Qinglong foreland basin (Taihang Mountain), with flat-dipping structures, interpreted as a passive margin developed prior to 2.5 Ga on the Eastern block. It is overlain by lower grade turbidite and molasse type sediments. The western margin of Qinglong foreland basin is intensely reworked by thrusting and folding and is overthrust by the overlying orogenic complex (including the tonalitic-trondhjemitic-granodiorotic gneiss, ophiolites, accretionary sediments). To the east its deformation becomes weaker in intensity. The Qinglong foreland basin is intruded by a gabbroic dike complex consisting of 2.4 billion-year-old diorite and is overlain by graben-related sediments and flood basalts. In the Wutai and North Taihang basins, many ophiolitic blocks are recognized along the western margin of the foreland thrust-fold belt. These consist of pillow lava, gabbroic cumulates, and harzburgite. The largest ophiolitic thrust complex imbricated with foreland basin sedimentary rocks is up to five miles (10 km) long, preserved in the Wutai-Taihang Mountains.

Several dismembered Archean ophiolites have been identified in the Central orogenic belt, including some in Liaoning Province, at Dongwanzi, north of Zunhua, and at Wutai Mountain. The best studied of these are the Dongwanzi and Zunhua ophiolitic terranes. The Zunhua structural belt of the Eastern Hebei Province preserves a cross section through most of the northeastern part of the Central orogenic belt. This belt is characterized by highly strained gneiss, banded iron formation, 2.6–2.5 billion-year-old greenstone belts and mafic to ultramafic complexes in a high-grade ophiolitic melange. The belt is intruded by widespread 2.6–2.5 billion-year-old tonalite-trondhjemite gneiss, 2.5 billion-year-old granites, and is cut by ductile shear zones. The Neoarchean high-pressure granulite belt (Chengde-Hengshan HPG) strikes through the northwest part of the belt. The Zunhua structural belt is thrust over the Neoarchean Qianxi-Taipingzhai granulite-facies terrane, consisting of enderbitic to charnockitic gneiss forming several small dome-like structures southeast of the Zunhua belt. The Zunhua structural belt clearly cuts across the dome—like Qian'an-Qianxi structural patterns to the east. The Qian'an granulite-gneiss dome (3.8–2.5 billion years old) forms a large circular dome in the southern part of the area and is composed of tonalitic-trondhjemitic gneiss and biotite-granite. Mesoarchean (2.8–3.0 billion years old) and Paleoarchean (3.50–3.85 billion years old) supracrustal sequences outcrop in the eastern part of the region. The Qinglong Neoarchean amphibolite to greenschist facies supracrustal sequence strikes through the center of the area and is interpreted to be a foreland fold-thrust belt, intruded by large volumes of 2.4 billion-year-old diorite in the east. The entire North China craton is widely cut by at least two Paleoproterozoic mafic dike swarms (2.5–2.4 and 1.8–1.7 billion years old), associated with regional extension. Mesozoic-Cenozoic granite, diorite, gabbro, and ultramafic plugs occur throughout the NCC and form small intrusions in some of the belts.

The largest well-preserved sections of the Dongwanzi ophiolite are located approximately 120 miles (200 km) northeast of Beijing in the northeastern part of the Zunhua structural belt, near the villages of Shangyin and Dongwanzi. The belt consists of prominent amphibolite-facies mafic-ultramafic complexes in the northeast sector of Zunhua structural belt. The southern end of the Dongwanzi ophiolite belt near Shangyin is complexly faulted against granulite-facies gneiss, with both thrust faults and younger normal faults present. The main section of the ophiolite dips steeply northwest, is approximately 30 miles (50 km) long, and is 3–6 miles (5–10 km) wide. A U/Pb-zircon age of 2.505 billion years for two gabbro samples from the Dongwanzi ophiolite shows that this is the oldest, relatively complete ophiolite known in the world. However, parts of the central belt are intruded by a mafic/ultramafic Mesozoic pluton with related dikes.

The base of the ophiolite is strongly deformed and is intruded by the 2.391 billion-year-old Cuizhangzi diorite-tonalite complex. The Dongwanzi ophiolite is associated with a number of other amphibolite-facies belts of mafic plutonic and extrusive igneous rocks in the Zunhua structural belt. These mafic to ultramafic slices and blocks can be traced regionally over a large area from Zunhua to West Liaoning (about 120 miles or 200 km). Much of the Zunhua structural belt is interpreted as a high-grade ophiolitic mélange, with numerous tectonic blocks of pillow lava, BIF, dike complex, gabbro, dunite, serpentinized harzburgite, and podiform chromitite in a biotite-gneiss matrix, intruded extensively by tonalite and granodiorite. Cross-cutting granite has yielded an age of 2.4 billion years. Blocks in the mélange correlate with the Dongwanzi and other ophiolitic fragments in the Zunhua structural belt. This correlation is supported by Rhenium-Osmium (Re-Os) age determinations on several of these blocks, revealing that they are 2.54 Ga old.

The Eastern and Western blocks of the North China craton collided at 2.5 billion years ago during an arc/continent collision, forming a foreland basin on the Eastern block, a granulite facies belt on the Western block, and a wide orogen between the two blocks. This collision was followed rapidly by post-orogenic extension and rifting that formed mafic dike swarms and extensional basins along the Central orogenic belt and led to the development of a major ocean along the north margin of the craton. An arc terrane developed in this ocean and collided with the north margin of the craton by 2.3 Ga, forming an 850-mile (1,400-km) long orogen known as the Inner Mongolia–Northern Hebei orogen. A 1,000-mile (1,600-km) long granulite-facies terrain formed on the southern margin of this orogen, representing a 120-mile (200-km) wide uplifted plateau formed by crustal thickening. The orogen was converted to an Andean-style convergent margin between 2.20 billion and 1.85 billion years ago, recorded by belts of plutonic rocks, accreted metasedimentary rocks, and a possible back arc basin. A pulse of convergent deformation is recorded at 1.9–1.85 billion years across the northern margin of the craton, perhaps related to a collision outboard of the Inner Mongolia–Northern Hebei orogen and closure of the back arc basin. This event caused widespread deposition of conglomerate and sandstone of the basal Changcheng Series in a foreland basin along the north margin of the craton. At 1.85 billion years the tectonics of the North China craton became extensional, and a series of aulacogens and rifts propagated across the craton, along with the intrusion of mafic dike swarms. The northern granulite facies belt underwent retrograde metamorphism and was uplifted during extensional faulting. High-pressure granulites are now found in the areas where rocks were metamorphosed to granulite facies and exhumed two times, at 2.5 billion and 1.8 billion years ago, exposing rocks that were once at lower crustal levels. Rifting led to the development of a major ocean along the southwest margin of the craton, where oceanic records continue until 1.5 billion years ago.

See also ARCHEAN; OPHIOLITES.

Further Reading
Kusky, Timothy M., and Jianghai Li. "Paleoproterozoic Tectonic Evolution of the North China Craton." *Journal of Asian Earth Sciences* 22 (2004).
Kusky, Timothy M., Jianghai Li, and Robert T. Tucker. "The Archean Dongwanzi Ophiolite Complex, North China Craton: 2.505 Billion Year Old Oceanic Crust and Mantle." *Science* 292 (2001).
Kusky, Timothy M., Jianghai Li, Adam Glass, and Alan Huang. "Archean Ophiolites and Ophiolite Fragments of the North China Craton." *Earth Science Reviews* (2004).
Li, Jianghai., Timothy M. Kusky, and Alan Huang. "Neoarchean Podiform Chromitites and Harzburgite Tectonite in Ophiolitic Melange, North China Craton, Remnants of Archean Oceanic Mantle." *GSA Today* (2002).

North Slope of Alaska and ANWR The Brooks Range of northern Alaska gradually diminishes northward, sloping toward the Beaufort Sea along the North Slope coastal plain. The region referred to as the North Slope stretches from the Yukon border in the British and Romanzof Mountains, stretches west past Prudhoe Bay and Deadhorse, past Barrow, and to Point Lay and Cape Lisburne on the Chukchi Sea. This arctic coastal plain and mountainous area is one of the last large relatively untouched wilderness areas in the United States, and constant battles are fought between environmentalists who would like to preserve the area's pristine state and proponents of the petroleum industry who would like to see additional petroleum exploration, especially in the Arctic National Wildlife Refuge (ANWR) in the eastern part of the plain. This region is continuous with the petroleum-rich Prudhoe Bay area and may contain large reserves of oil. The region is the summer home and breeding grounds for herds of caribou, bears, and many other species of animals. In 2005 Congress approved legislation allowing exploration for petroleum in ANWR.

Paleozoic rocks of the Brooks Range are separated from Mesozoic rocks of the foothills by a major imbricate fault zone. Along this zone, many east-west striking faults dip south under the Brooks Range, and folds are overturned toward the north (and dip south by 20°–40°). Many minor thrust faults are present in the contact zone between the Brooks Range and foothills, with the thrust faults becoming steeper toward the Brooks Range in the south. Displacements on the faults range up to several thousand feet (hundreds of meters).

The southern foothills are located just north of the Brooks Range and include mainly Mesozoic rocks that are folded but rarely faulted, forming elongate rolling hills. Many of the folds have steep northern limbs and shallow-dipping southern limbs, indicating they formed by northward tectonic movements. Thick units exhibit long-wavelength parallel folds, while thin units show small tight folds. Some of the folds are cored and cut by minor thrust faults. The southern foothills grade into the northern foothills, characterized by laterally persistent anticlines, some of which bifurcate and separate canoe-shaped synclines. There is a nearly constant distance of 7 miles (11 km) between the crests of adjacent anticlines. Dips on the limbs of the folds range between 6° and 20°, and there is a gradual northward decrease in the amplitude of the folds, and in the number of faults cutting the section. The region is cut by several northnortheast striking faults that show strike-slip displacements and are probably tear faults related to northward thrusting and folding, with differential movement between blocks on either side of the faults. These faults control the courses of several of the major rivers in the area.

The Arctic Foothills province grades north into the Arctic Coastal Plain province. North of about 70° latitude, the beds become largely flat-lying, and the surface is covered by permafrost and tundra. The eastern part of this region was designated as the Arctic National Wildlife Refuge in 1960 by the secretary of the interior, including about 13,000 square miles (21,000 km²) inhabited by grizzlies, caribou, dall sheep, and millions of migratory birds. The Arctic National Wildlife Refuge is the largest wildlife refuge in the United States, and the only refuge in the Arctic. The mission of the refuge is to preserve the flora and fauna of the Arctic in its natural state. It will be a daunting task for oil companies to meet this mission while exploring for and extracting oil to meet America's energy needs.

See also BROOKS RANGE; TAIGA; TUNDRA.

Further Reading

Oldow, John S., and Hans G. Ave Lallemant, eds., "Architecture of the Central Brooks Range Fold and Thrust Belt, Arctic Alaska." *Geological Society of America Special Paper* 324 (1998).

nuée ardente   Volcano-generated hot glowing clouds of dense gas and ash that may reach temperatures of nearly 1,850°F (1,000°C), rush down volcanic flanks at 450 miles per hour (700 km/hour), and travel more than 60 miles (100 km) from the volcanic vent. They are one of the most dangerous and devastating types of volcanic flows known. Nuée ardentes have been the nemesis of many a volcanologist and curious observer, as well as thousands upon thousands of unsuspecting or trusting villagers. Nuée ardentes are but one type of pyroclastic flow, which include a variety of mixtures of volcanic blocks, ash, gas, and lapilli that produce volcanic rocks called ignimbrites. Nuée ardentes were responsible for the destruction and burial of Pompeii by an eruption from Vesuvius in 79 C.E., preserving people, animals, and artifacts in a dense layer of ash. In 1902 an eruption of Mount Pelée buried the city of St. Pierre on Martinique killing approximately 29,000 people.

See also CONVERGENT PLATE MARGIN PROCESSES; VOLCANO.

O

ocean basin The surface of the Earth is divided into two fundamentally different types of crust, including relatively light quartz and plagioclase-rich sial, forming the continental regions, and relatively dense olivine and pyroxene-rich sima underlying the ocean basins. The ocean basins may be defined as submarine topographic depressions underlain by oceanic (simatic) crust. Ocean basins are quite diverse in size, shape, depth, characteristics of the underlying seafloor topography, and types of sediments deposited on the oceanic crust. The largest ocean basins include the Pacific, Atlantic, Indian, and Arctic Oceans, and the Mediterranean Sea, whereas dozens of smaller ocean basins are located around the globe.

The ocean basins depths were first extensively explored by the H.M.S. *Challenger* in the 1800s, using depth reading from a weight attached to a several-kilometer-long cable that was dropped to the ocean floor. Results from these studies suggested that the oceans were generally about 3–4 miles (5–6 km) in depth. Later, with the development of echo-sounding technologies and war-induced mapping efforts, the variety of seafloor topography became appreciated. Giant submarine mountain chains were recognized where the depth rises to 1.7 miles (2.7 km), and these were later understood to be oceanic ridges where new oceanic crust is created. Deep-sea trenches with depths exceeding 5 miles (8 km) were delineated and later recognized to be subduction zones where oceanic crust is sinking back into the mantle. Other anomalous regions of thick oceanic crust (and reduced depths) were recognized, including large oceanic plateaux where excessive volcanism produced thick crust over large regions, and smaller seamounts (or guyots) where smaller, off-ridge volcanism produced isolated submarine mountains. Some of these rose above sea level, were eroded by waves, and grew thick reef complexes as they subsided with the cooling of the oceanic crust. Such guyots and coral atolls were made famous by Charles Darwin, in his study of coral reefs of the Pacific Ocean basin.

Pelagic sediments are deposited in the ocean basins, and generally form a blanket of sediments draping over preexisting topography. Carbonate rocks produced mainly by the tests of foraminifera and nannofossils may be deposited on the ocean ridges and guyots that are above the CCD (carbonate compensation depth), above which the seawater is saturated with $CaCO_3$, and below which it dissolves in the water. Below this, sediments comprise red clays and radiolarian and diatomaceous ooze. Manganese nodules are scattered about on some parts of the ocean floor.

The abyssal plains are relatively flat parts of the ocean basins where the deep parts of the seafloor topography have been filled in with sediments, forming flat plains. Some of these abyssal plains are quite large, such as the 386,100-square-mile (1 million-km²) Angolan abyssal plain in the South Atlantic, and the 1,428,578-square-mile (3.7 million-km²) abyssal plain in the Antarctic Ocean basin. Other abyssal plains are much smaller, such as the 1,003-square-mile (2,600-km²) Alboran Sea in the Mediterranean. Abyssal plains may also be characterized and distinguished on the basis of their sediment composition, their geometry, depth, and volume and thickness of the sediments they contain.

See also ABYSSAL PLAINS; CORALS; OCEANIC CRUST; PLATE TECTONICS; REEF.

ocean currents Like the atmosphere, the ocean is constantly in motion. Ocean currents are defined by the movement paths of water in regular courses, driven by the wind and thermohaline forces across the ocean basins. Shallow currents are driven primarily by the wind but are systematically deflected by the Coriolis force to the right of the atmospheric wind directions in the Northern Hemisphere, and to

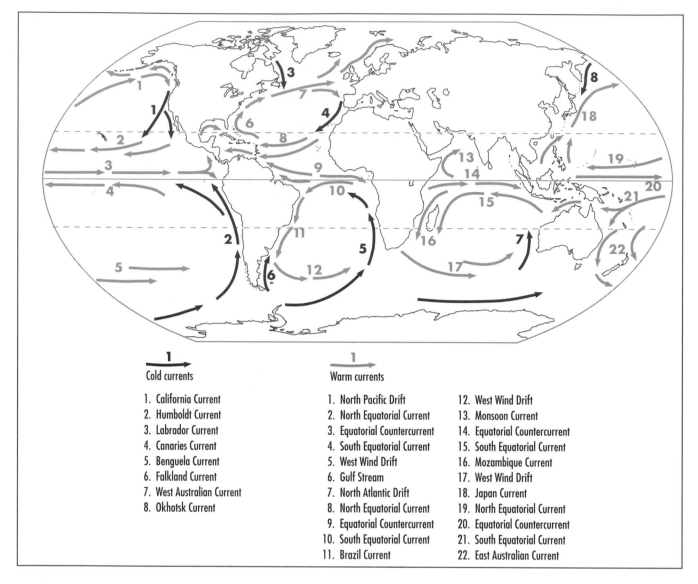

Cold currents

1. California Current
2. Humboldt Current
3. Labrador Current
4. Canaries Current
5. Benguela Current
6. Falkland Current
7. West Australian Current
8. Okhotsk Current

Warm currents

1. North Pacific Drift
2. North Equatorial Current
3. Equatorial Countercurrent
4. South Equatorial Current
5. West Wind Drift
6. Gulf Stream
7. North Atlantic Drift
8. North Equatorial Current
9. Equatorial Countercurrent
10. South Equatorial Current
11. Brazil Current

12. West Wind Drift
13. Monsoon Current
14. Equatorial Countercurrent
15. South Equatorial Current
16. Mozambique Current
17. West Wind Drift
18. Japan Current
19. North Equatorial Current
20. Equatorial Countercurrent
21. South Equatorial Current
22. East Australian Current

Map showing major cold and warm ocean currents of the world

the left of the prevailing winds in the Southern Hemisphere. Therefore, shallow water currents tend to be oriented about 45° from the predominant wind directions.

Deepwater currents, however, are driven primarily by thermohaline effects, that is the movement of water driven by differences in temperature and salinity. The Atlantic and Pacific Ocean basins both show a general clockwise rotation in the Northern Hemisphere, and a counterclockwise spin in the Southern Hemisphere, with the strongest currents in the midlatitude sectors. The pattern in the Indian Ocean is broadly similar but seasonally different and more complex because of the effects of the monsoon. Antarctica is bound on all sides by deep water and has a major clockwise current surrounding it known as the Antarctic Circumpolar Current,

lying between 40° and 60° south. This is a strong current, moving at 1.6–5 feet per second (0.5–1.5 m/s), and has a couple of major gyres in it at the Ross Ice Shelf and near the Antarctic Peninsula. The Arctic Ocean has a complex pattern, because it is sometimes ice covered and is nearly completely surrounded by land with only one major entry and escape route east of Greenland, called Fram Strait. Circulation patterns in the Arctic Ocean are dominated by a slow, 0.4–1.6-inch per second (1–4 cm/s) transpolar drift from Siberia to the Fram Strait, and by a thermohaline-induced anticyclonic spin known as the Beaufort Gyre that causes ice to pile up on the Greenland and Canadian coasts. Together the two effects in the Arctic Ocean bring numerous icebergs into North Atlantic shipping lanes and send much of the cold deep water around Greenland into the North Atlantic ocean basin.

See also EKMAN SPIRALS; GEOSTROPHIC CURRENTS; THERMOHALINE CIRCULATION.

oceanic crust   The distinctive type of crust that is produced at mid-ocean ridges and underlies the ocean basins is known as oceanic crust. About 70 percent of the Earth's surface is covered by oceanic crust. It is rich in iron and magnesium minerals such as olivine and pyroxene and is also known as sima or simatic crust. Most oceanic crust is 3–6 miles (5–10 km) thick, thickening considerably by cooling from where it is created by magmatic upwelling at the mid-ocean ridges as part of the seafloor spreading process. Oceanic crust has an average density of 3.0 grams per cubic centimeter, and an average seismic P-wave velocity of 3.85 miles per second (6.2 km/s).

Since most oceanic crust is produced by broadly similar processes, it has a characteristic general layered structure that can be recognized on the basis of seismic refraction studies. Geologists have compared the seismic velocities of rock samples from ophiolites and oceanic crust and concluded that many are broadly similar. This observation has been used to suggest that ophiolites are pieces of oceanic crust tectonically emplaced onto the continents and also to use the geology of ophiolites to infer details about the deep parts of oceanic crust. Layer 1 of the oceanic crust consists of deep-sea sediments, underlain by layer 2A, consisting of up to several kilometers of pillow basalts and mafic dikes. Layer 2B, where present, is interpreted to represent a sheeted dike complex that may also be several kilometers thick, underlain by Layer 3, consisting of gabbro and gabbro cumulate rocks. These layers are underlain by a layer marking a sharp increase in seismic velocity known as the seismic Moho, interpreted from ophiolite studies to represent the transition from cumulate gabbros to denser cumulate ultramafic rocks. Since there is a sharp seismic boundary at this transition, but the processes of crystal accumulation from a melt are similar across the boundary, a separate and deeper petrological Moho is distinguished. This is marked at the boundary from crystal cumulate rocks to mantle rocks from which the melts that formed the overlying crust were extracted. This petrological Moho is not determinable seismically but only from ophiolite or deep drilling studies.

See also DIVERGENT OR EXTENSIONAL BOUNDARIES; OCEAN BASIN; OPHIOLITES; PLATE TECTONICS.

oceanic plateau   Regions of anomalously thick oceanic crust and topographically high seafloor are known as oceanic plateaus. Many have oceanic crust that is 12.5–25 miles (20–40 km) thick and rise thousands of meters above surrounding oceanic crust of normal thickness. The Caribbean ocean floor represents one of the best examples of an oceanic plateau, with other major examples including the Ontong-

Java Plateau, Manihiki Plateau, Hess Rise, Shatsky Rise, and Mid-Pacific Mountains. All of these oceanic plateaus contain thick piles of volcanic and subvolcanic rocks representing huge outpourings of lava, most erupted in a few million years. They typically do not show the magnetic stripes that characterize normal oceanic crust produced at oceanic ridges and are thought to have formed when mantle plume heads reached the base of the lithosphere, releasing huge amounts of magma. Some oceanic plateaus have such large volumes of magma that the total magmatic flux in the plateaus would have been similar to or larger than all of the magma erupted at the mid-ocean ridges during the same interval.

The Caribbean seafloor preserves 5–7.5-mile (8–21-km) thick oceanic crust formed before about 85 million years ago in the eastern Pacific Ocean. This unusually thick ocean floor was transported eastward by plate tectonics, where pieces of the seafloor collided with South America as it passed into the Atlantic Ocean. Pieces of the Caribbean oceanic crust are now preserved in Colombia, Ecuador, Panama, Hispaniola, and Cuba, and some scientists estimate that the Caribbean oceanic plateau may have once been twice its present size. In either case, it represents a vast outpouring of lava that would have been associated with significant outgassing with possible consequences for global climate and evolution.

The western Pacific ocean basin contains several large oceanic plateaus, including the 20-mile (32-km) thick crust of the Alaskan-sized Ontong-Java Plateau, which is the largest outpouring of volcanic rocks on the planet. It apparently formed in two intervals, at 122 million and 90 million years ago, entirely within the ocean, and represents magma that rose in a plume from deep in the mantle and erupted on the seafloor. It is estimated that the volume of magma erupted in the first event was equivalent to that of all the magma being erupted at mid-ocean ridges at the present time. Sea levels rose by more than 10 meters in response to this volcanic outpour-

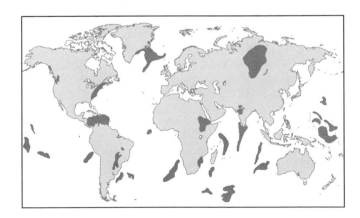

Map of the world showing major flood basalt regions, including oceanic plateaus

ing. The gases released during these eruptions are estimated to have raised average global temperatures by 23°F (13°C).

See also FLOOD BASALT.

oceanography   The study of the physical, chemical, biological, and geological aspects of the ocean basins is called oceanography. The oceans are increasingly being studied with an Earth system science approach, with the appreciation that many of the different systems are related, and changes in the biological, chemical, physical, or geological conditions will result in changes in the other systems and also influence other Earth systems such as the atmosphere and climate. The oceans contain important geological systems, since the ocean basins are the places where new oceanic crust is both created at mid-ocean ridges and destroyed at deep-sea trenches. Being topographic depressions, they are repositories for many of the sediments eroded from the continents and carried by rivers and the wind to be deposited in submarine settings. Seawater is the host of much of the life on Earth and also holds huge quantities of dissolved gases and chemicals that buffer the atmosphere, keeping global temperatures and climate hospitable for humans. Energy is transferred around the planet in ocean currents and waves, which interact with land, eroding or depositing shoreline environments. Being host to some of the planet's largest and most diverse biota, the oceans may hold the key to feeding the planet. Mineral resources are also

History of Ocean Exploration

The earliest human exploration of the oceans is poorly known, but pictures of boats on early cave drawings in Norway illustrate Viking-style ocean vessels known to be used by the Vikings centuries later. Other rock drawings around the world show dugout canoes, boats made of reeds, bark, and animal hides. Early migrations of humans must have utilized boats to move from place to place. For instance, analysis of languages and genetics shows that the Polynesians moved south from China into southeast Asia and Polynesia, then somehow made it, by sea, all the way to Madagascar off the east coast of Africa. Other oceanic migrations include the colonization of Europe by Africans about 10,000 years ago, explorations and trade around and out of the Mediterranean by the Phoenicians about 3,000 years ago, and the colonization of North America by the Siberians and Vikings. Ming Dynasty ocean explorations in the early 1400s were massive, involving tens of thousands of sailors on 317 ships. The Chinese ships were huge, including as many as nine masts, over 444 feet (135 m) in length and 180 feet (55 m) in width. The Chinese mounted these expeditions to promote Chinese culture, society, and technology, but they did not contribute significantly to understanding the oceans.

The first European to reach North America was probably Leif Eriksson, who, in the year 1000, landed at L'Anse-aux Meadows in the Long Range Peninsula of Newfoundland, after becoming lost on his way from Greenland to Norway. The Vikings established a temporary settlement in Newfoundland, and there are some speculations of further explorations by the Vikings to places as far south as New England and Narragansett Bay in Rhode Island. Before this, Ptolemy (in the year 140) published maps of Europe's coastline that were largely inaccurate, and took many years of ocean exploration to correct. The Greek and Islamic explorers had made great strides in understanding the geography of the world centered on the Mediterranean Sea and Arabian Peninsula, and the records of these explorations eventually made it into European hands where this knowledge was used for further explorations. The Portuguese, most notably Prince Henry the Navigator (1392–1460), were the most avid explorers of the Atlantic, exploring northwest Africa and the Azores in the early 1400s. In the late 1400s, Vasco de Gama (1460–1524) made it to southern Africa and eventually around the Cape of Good Hope, past Madagascar, and all the way to India in 1498. These efforts initialized economically important trade routes between Portugal and India, building the powerful Portuguese Empire. The timing was perfect for establishment of ocean trade routes, as the long-used overland Silk Roads had become untenable and dangerous with the collapse of the Mongol Empire and the Turk conquest of Constantinople (Istanbul) in 1453.

In the late 15th and early 16th centuries, many ocean exploration expeditions were mounted, as a precursor to more widespread use of the oceans for transportation. In 1492 Christopher Columbus sailed for Spain to the east coast of North America, and from the late 1400s to 1521 Ferdinand Magellan sailed around the world, including a crossing of the Pacific Ocean, followed by Sir Francis Drake of England. Later, Henry Hudson explored North American waters, including attempts to find a northwest passage between the Atlantic and Pacific. During the 1700s, Captain James Cook made several voyages in the Pacific and coastal waters of western North America, improving maps of coastal and island regions.

The early explorations of the oceans were largely concerned with navigation and determining the positions of trade routes, coastlines, and islands. Later seagoing expeditions aimed at understanding the physical, chemical, biological, and geological conditions in the ocean were mounted. In the late 1800s the British Royal Society sponsored the world's most ambitious scientific exploration of the oceans ever, the voyage of the H.M.S. *Challenger*. The voyage of the *Challenger* in 1872–76 established for the first time many of the basic properties of the oceans and set the standard for the many later expeditions. Ocean exploration today is led by American teams based at several universities and the Woods Hole Oceanographic Institute, where the deep submersible *Alvin* is based, and from where many oceanographic cruises are coordinated. The Ocean Drilling Program (formerly the Deep-Sea Drilling Project) has amassed huge quantities of data on the sediments and volcanic rocks deposited on the ocean floor, as well as information about biology, climate, chemistry, and ocean circulation. Many other nations, including Japan, China, France, and Russia, have mounted ocean exploration campaigns, with a trend toward international cooperation in understanding the evolution of the ocean basins.

abundant on the seafloor, many formed at the interface between hot volcanic fluids and cold seawater, forming potentially economically important reserves of many minerals.

The oceans cover two-thirds of the Earth's surface, yet we have explored less of the ocean's depths and mysteries than the surfaces of several nearby planets. The oceans have hindered migration of peoples and biota between distant continents, yet paradoxically now serve as a principal means of transportation. Oceans provide us with incredible mineral wealth and renewable food and energy sources, yet they also breed devastating hurricanes. Life may have begun on Earth in environments around hot volcanic events on the seafloor, and we are just beginning to explore the diverse and unique fauna that can still be found living in deep dark waters around similar vents today.

Ocean basins have continually opened and closed on Earth, and the continents have alternately been swept into large single supercontinents and then broken apart by the formation of new ocean basins. The appearance, evolution, and extinction of different life-forms is inextricably linked to the opening and closing of ocean basins, partly through the changing environmental conditions associated with the changing distribution of oceans and continents.

Early explorers were slowly able to learn about ocean currents and routes to distant lands, and some dredging operations discovered huge deposits of metals on the seafloor. Tremendous leaps in our understanding of the structure of the ocean basin seafloor were acquired during surveying for the navigation of submarines and detection of enemy submarines during World War II. Magnetometers towed behind ships and accurate depth measurements provided data that led to the formulation of the hypothesis of seafloor spreading, which added the oceanic counterpart to the idea of continental drift. Together, these two theories became united as the plate tectonic paradigm.

Ocean circulation is responsible for much of the world's climate. For instance, mild foggy winters in London are caused by warm waters from the Gulf of Mexico flowing across the Atlantic in the Gulf Stream to the coast of the British Isles. Large variations in ocean and atmospheric circulation patterns in the Pacific lead to alternating wet and dry climate conditions known as El Niño, and La Niña. These variations affect Pacific regions most strongly but are felt throughout the world. Other movements of water are more dramatic, including the sometimes devastating tsunami that may be initiated by earthquakes, volcanic eruptions, and giant submarine landslides. Two of the most tragic tsunami in recent history were generated by the eruption of the Indonesian volcano Krakatau in 1883, and by an earthquake in Indonesia in 2004. When Krakatau erupted, it blasted out a large part of the center of the volcano, and seawater rushed in to fill the hole. This seawater was immediately heated and it exploded outward in a steam eruption and a huge wave of hot water. The tsunami generated by this eruption reached more than 120 feet (36.5 m) in height and killed an estimated 36,500 people in nearby coastal regions. In 1998 a catastrophic 50-foot (15-m) high wave unexpectedly struck Papua New Guinea, killing more than 2,000 people and leaving more than 10,000 homeless. The December 2004 Indian Ocean tsunami generated by a magnitude 9.0–9.2 earthquake in Sumatra was more destructive, killing an estimated 300,000 people.

The oceans are full of rich mineral deposits, including oil and gas on the continental shelves and slopes, and metalliferous deposits formed near mid-ocean ridge vents. Much of the world's wealth of manganese, copper, and gold may lie on the seafloor. The oceans also yield rich harvests of fish, and care must be taken that we do not deplete this source. Sea vegetables are growing in popularity and their use may help alleviate the growing demand for space in fertile farmland. The oceans may offer the world a solution to growing energy and food demands in the face of a growing world population. New life-forms are constantly being discovered in the depth of the oceans and we need to take precautions to understand these creatures, before any changes we make to their environment causes them to perish forever.

See also OCEAN BASIN; OCEAN CURRENTS; OCEANIC CRUST; PLATE TECTONICS.

Further Reading

Erickson, Jon. *Marine Geology; Exploring the New Frontiers of the Ocean, Revised Edition.* New York: Facts On File, 2003.

Oman Mountains The Oman or Hajar Mountains in northern Oman and the United Arab Emirates are located on the northeastern margin of the Arabian plate, 60–120 miles (100–200 km) from the active deformation front in the Gulf of Oman between Arabia and the Makran accretionary wedge of Asia. They are made up of five major structural units ranging in age from Precambrian to Miocene. These include the pre-Permian basement, Hajar Unit, Hawasina nappes, Semail ophiolite and metamorphic sole, and post-nappe structural units.

The Hajar Mountains are up to 1.8 miles (3 km) high, displaying many juvenile topographic features such as straight mountain fronts and deep steep-walled canyons that may reflect active tectonism causing uplift of these mountains. The present height and ruggedness of the Hajar mountainous area is a product of Cretaceous ophiolite obduction, Tertiary extension, and rejuvenated uplift and erosion that initiated at the end of the Oligocene and continues to the present. The Sayq Plateau southwest of Muscat is 1.2–1.8 miles (2–3 km) in elevation. Jabal Shams on the margin of the Sayq Plateau is the highest point in Arabia, rising more than 1.8 miles (3 km) in the central Hajar Mountains. The heights decrease gradually northward reaching 1.2 miles (2 km) on the Musandam peninsula. There the mountain slopes drop directly into the sea.

Pre-Permian basement is exposed mainly in the Jabal Akhdar, Saih Hatat, and Jabal J'Alain areas. The oldest structural unit includes a Late Proterozoic basement gneiss correlative with the Arabian-Nubian Shield, overlain by a Late Proterozoic/Ordovician volcano-sedimentary sequence. The later is divided into the Late Proterozoic/Cambrian Huqf Group, and the Ordovician Haima Group. The Huqf Group is mainly composed of diamictites, siltstone, graywacke, dolostone, and intercalated mafic volcanics. The Ordovician Haima Group consists of a series of sandstones, siltstones, quartzites, skolithos-bearing sandstones, and shales, interpreted as subtidal to intertidal deposits.

The Hajar Unit represents the main part of the Permian/Cretaceous Arabian platform sequence that formed on the southern margin of the Neo-Tethys Ocean. These carbonates form most of the rugged peaks of Jabal Akhdar, form a rim around the southwestern parts of Saih Hatat, and continue in several thrust sheets in the Western Hajar region. They are well-exposed on the Musandam peninsula. The Hajar Unit contains the Akhdar, Sahtan, Kahmmah, and Waisa Groups of mainly carbonate lithologies, overlain by the Muti Formation in the eastern Hajar, and the equivalent Ruus al Jibal, Elphinstone, Musandam and Thamama Groups on the Musandam peninsula.

The Hawasina nappes consist of a series of Late Permian/Cretaceous sedimentary and volcanic rocks deposited in the Hawasina basin, between the Arabian continental margin and the open Neo-Tethys Ocean. The Hawasina nappes include the Hamrat Duru, Al Aridh, Kawr, Umar Groups. Chaotic deposits of the Baid Formation are interpreted as a foundered carbonate platform. The Hamrat Duru Group includes radiolarian chert, gabbro, basaltic and andesitic pillow lava, carbonate breccia, shale, limestone and sandstone turbidites. The Al Aridh Group contains an assemblage of basaltic andesite, hyaloclastite and pillow lavas, micrites, pelagic carbonates, carbonate breccias, chert, and turbidites. This is overlain by the Kawr Group, which includes basalts, andesites, and shallow marine carbonates. The Umar Group contains basaltic and andesitic pillow lavas, cherts, carbonate breccias, and micrites.

The Semail nappe forms the largest ophiolitic sheet in the world, and it is divided into numerous blocks in the northern Oman Mountains. The Semail ophiolite contains a complete classic ophiolite stratigraphy, although parts of it are unusual in that it contains two magmatic sequences, including upper and lower units. The upper magmatic unit grades downward from radiolarian cherts and umber of the Suhaylah Formation, to basaltic and andesitic pillow lavas locally intruded by trondhjemites, through a sheeted diabase dike unit, and into massive and layered gabbros, and finally into cumulate gabbro, wehrlite, dunite, and clinopyroxenite. This upper magmatic sequence grades down from basaltic pillow lavas into a sheeted dike complex, through isotropic then layered gabbros, then into cumulate gabbro and dunite. The Mohorovicic discontinuity is well exposed throughout the northern Oman Mountains, separating the crustal and the mantle sequences. The mantle sequence consists of tectonized harzburgite, dunite, and lherzolite, cut by pyroxenite dikes and local chromite pods.

The metamorphic sole or dynamothermal aureole of the Semail ophiolite formed through metamorphism of rocks immediately under the basal thrust, heated and deformed during emplacement of the hot allochthonous sheets. In most places it consists of two units, including a lower metasedimentary horizon and an upper unit of banded amphibolites. The metamorphic grade increases upward through the unit to upper amphibolite facies near the contact with the Semail nappe.

Post-nappe units consist of Late Cretaceous and Tertiary rocks. The Cretaceous Aruma Group consists of a lower unit of Turonian-Santonian polymict conglomerate, sandstone, and shale of the Qahlah Formation, and an upper unit of Campanian-Maastrictian marly limestone and polymict breccia (Thaqab Formation). The Tertiary Hadhramaut Group comprises Paleocene to Eocene limestones, marly limestone, dolostone, conglomerate, and sandstones that outcrop along the southern edge of the Batinah coastal plain at the border with the northeast flank of the Hajar Mountains.

Several levels of Quaternary fluvial terraces are preserved along the flanks of the Hajar Mountains in Oman. These can be divided in most places into an older lower cemented terrace and an upper younger uncemented terrace group. The lower cemented terrace is one of the youngest geological units that we have been able to use as a time marker to place constraints on the ages of structures. The terraces are younger than and unconformably overlie most faults and folds, but in several places faults and fracture intensification zones demonstrably cut through the Quaternary terraces, providing some of the best evidence for the young age of some of the faults described here. These terraces grade both northward and southward into coalesced alluvial fans forming bajada flanking the margins of the mountains. The northern alluvial plains grade into a narrow coastal plain along the Gulf of Oman.

The Oman (Hajar) Mountains are situated at the northeastern margin of the Arabian plate. This plate is bounded to the south and southwest by the active spreading axes of the Gulf of Aden and Red Sea. On the east and west its border is marked by transcurrent fault zones of the Owen Fracture Zone and the Dead Sea Transform. The northern margin of the plate is marked by a complex continent-continent to continent-oceanic collision boundary along the Zagros and Makran fold and thrust belts.

Rocks of the Hajar Supergroup preserve a history of Permian through Cretaceous subsidence of the Arabian Platform on the margin of the Neo-Tethys Ocean. Formations that now comprise the Hawasina nappes have biostratigraphic ages of 260–95 Ma, interpreted to have been deposited on the continental slope and in abyssal environments of the Neo-Tethys Ocean. By about 100 Ma, spreading in the Neo-Tethys generated the oceanic crust of the Semail ophiolite, which was detached in the oceanic realm and thrust over adjacent oceanic crust soon after its formation. Metamorphic ages for the initiation of thrusting range from 105 Ma to 89 Ma. The ophiolitic nappes moved toward the Arabian margin, forming the high-grade metamorphic sole during transport and progressively scraping off layers of the Hawasina sediments and incorporating them as thrust nappes to the base of the ophiolite. The ophiolite reached the Arabian continental margin and was thrust over it before 85–75 Ma as indicated by greenschist facies metamorphism in the metamorphic sole and by deformation of the Arabian margin sedi-

ments. Initial uplift of the dome-shaped basement cored antiforms of Jabal Akhdar and Saih Hatat may have been initiated during the late stages of the collision of the ophiolite with the Arabian passive margin and may have been localized by preexisting basement horst and graben structures. The location and geometry of these massive uplifts is probably controlled by basement ramps. Uplift of these domes was pronounced during the Oligocene/Miocene, as shown by tilting of Late Cretaceous/Tertiary formations on the flanks of

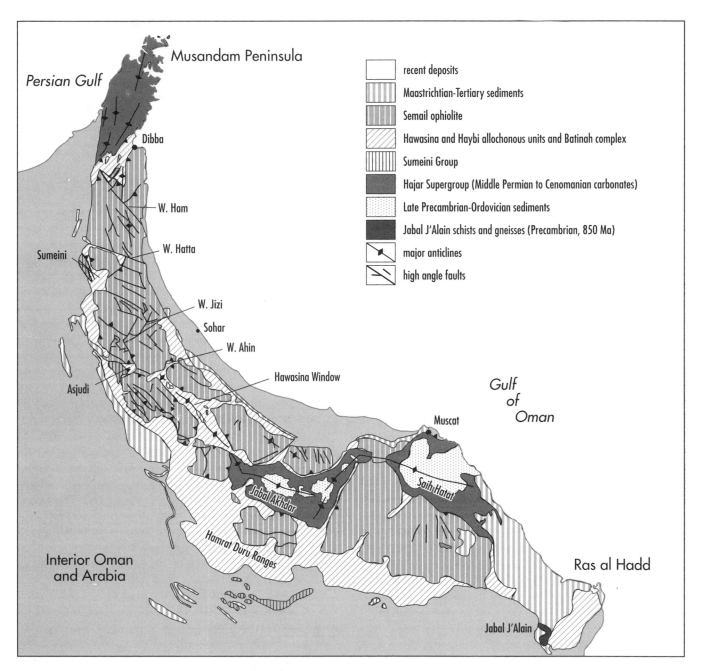

Geological map of the northern Oman (Hajar) Mountains, showing the main ophiolitic nappes, thrust-emplaced continental margin sediments, and major structures

the domes. Uplift of the domes may have begun in the Oligocene, resulting from the propagation of a fault beneath the southern limbs of the folds. The uplift of the domes includes a complex history, involving several different events. Some uplift of the domes continues at present, whereas much of the Batinah coastal plain is subsiding.

In most of the Hajar Mountains, the Hawasina nappes structurally overlie the Hajar Supergroup and form a belt of north- or northeastward-dipping thrust slices. However, on the southern margins of Jabal Akhdar, Saih Hatat, and other domes, the Hawasina form south-dipping thrust slices. Major valleys typically occupy the contact between the Hajar Supergroup and the Hawasina nappes, because of the many, easily erodable shale units within the Hawasina nappes. Several very large (~10 km scale) allochthonous limestone blocks known as the "Oman Exotics" are also incorporated into mélange zones within the Hawasina nappes. These form light-colored, erosionally resistant cuestas including Jabal Kawr and several smaller mountains south of Al Hamra.

South and southwest of the belt of ophiolite blocks, sediments of the Hamrat Duru Group are complexly folded and faulted in a regional foreland-fold-thrust belt and then grade into the Suneinah foreland basin. The Hamrat Duru rocks include radiolarian cherts, micritic limestones, turbiditic sandstones, shales, and calcarenite, all complexly folded and thrust faulted in a 18.5-mile (30-km) wide fold/thrust belt.

A belt of regional anticlinal uplifts brings up carbonates of the Hajar Supergroup in the central part of the basin, as exposed at Jabal Salakh. These elongate anticlinal domes have gentle to moderate dips on their flanks and are cut by several thrust faults that may be linked to a deeper system. This could be a blind thrust, or the folds could be flower structures developed over deep strike-slip faults. South of the Jabal Salakh fold belt, the surface is generally flat, covered by Miocene/Pliocene conglomerates of the Barzaman Formation, and cut by an extensive network of Quaternary channels of the active alluvial plain.

Tertiary/Quaternary uplift of the northern Oman Mountains may account for the juvenile topography of the area. One of the best pieces of evidence for young uplift of the Northern Oman Mountains comes from a series of uplifted Quaternary marine terraces, best exposed in the Tiwi area 31–62 miles (50–100 km) southeast of Muscat. The uplift is related to the contemporaneous collision between the northeastern margin of the Arabian plate and the Zagros fold belt and the Makran accretionary prism. The Hajar Mountains lie on the active forebulge of this collision, and the fault systems are similar to those found in other active and ancient forebulge environments. The amount of Quaternary uplift, estimated between 300 and 1,600 feet (100–500 m), is also similar to uplift in other forebulge environments developed on continental margins. This Quaternary uplift is superimposed on an older, Cretaceous/Tertiary (Oligocene) topography.

See also CONVERGENT PLATE MARGIN PROCESSES; OPHIOLITES.

Further Reading

Al-Lazki, A. I., Don Seber, and Eric Sandvol. "A Crustal Transect across the Oman Mountains on the Eastern Margin of Arabia." *GeoArabia*, v. 7, no. 1 (2002).

Boote, D. R. D., D. Mou, and R. I. Waite. "Structural Evolution of the Suneinah Foreland, Central Oman Mountains." In *The Geology and Tectonics of the Oman Region,* edited by A. H. F. Robertson, M. P. Searle, and A. C. Reis. *Geological Society of London Special Publication* 49 (1990).

Glennie, Ken W., M. G. A. Boeuf, M. W. Hughes-Clarke, Stuart M. Moody, W. F. H. Pilar, and B. M. Reinhardt. *The Geology of the Oman Mountains.* 3 Vols. Verhandelinge (Transactions), 31. The Netherlands: Royal Dutch Geological and Mining Society (KNGMG), 1974.

Kusky, Timothy M., Cordula Robinson, and Farouk El-Baz. "Tertiary and Quaternary Faulting and Uplift of the Hajar Mountains of Northern Oman and the U.A.E." *Journal of the Geological Society of London* (2005).

Searle, Mike, and J. Cox. "Tectonic Setting, Origin, and Obduction of the Oman Ophiolite." *Geological Society of America Bulletin* 111 (1999).

Ontong-Java plateau, Pacific Ocean The Ontong-Java plateau is the largest igneous province in the world not associated with the oceanic ridge spreading center network, covering an area roughly the size of Alaska (9,300,000 square miles, or 15,000 km²). The plateau is located northeast of Papua New Guinea and the Solomon Islands in the southwest Pacific Ocean, centered on the equator at 160°E longitude. Most of the plateau formed about 120 million years ago in the Cretaceous period, probably as a result of a mantle plume rising to the surface and causing massive amounts of volcanism over a geologically short interval likely lasting only about a million years. Smaller amounts of volcanic material were erupted later, at about 90 million years ago. Together, these events formed a lava plateau that is 20 miles (32 km) thick. The amount of volcanic material produced to form the plateau is estimated to be approximately the same as that erupted from the entire global ocean ridge spreading center system in the same period. Such massive amounts of volcanism cause worldwide changes in climate and ocean temperatures and typically have great impacts on the biosphere. Sea levels rose by more than 30 feet (9 m) in response to this volcanic outpouring. The gases released during these eruptions are estimated to have raised average global temperatures by 23°F (13°C). Perhaps more remarkably, the Ontong-Java plateau is but one of many Cretaceous oceanic plateaus in the Pacific, suggesting that the Cretaceous was characterized by long-standing eruption of massive amounts of deeply derived magma. Some geologists have suggested that events like this may be related to major mantle overturn events, when heat loss from the Earth is dominated by plumes instead of oceanic ridge spreading as occurs in the present plate mosaic.

The plateau is thought to be composed largely of basalt, based on limited sampling, deep-sea drilling, and seismic velocities. Great difficulties are encountered trying to sample the plateau because it is covered by a thick veneer of sediments exceeding thousands of feet (a kilometer or more) in most places. The plateau is colliding with the Solomon trench, but thick oceanic plateaus like the Ontong-Java are generally unsubductable. When oceanic plateaus confront subduction, they typically get accreted to the continents, leading to continental growth.

See also FLOOD BASALT.

Further Reading

Cloos, Mark. "Lithospheric Buoyancy and Collisional Orogenesis: Subduction of Oceanic Plateaus, Continental Margins, Island Arcs, Spreading Ridges, and Seamounts." *Geological Society of America Bulletin* 105 (1993): 715–737.

Kusky, Timothy M., and Ali Polat. "Growth of Granite-Greenstone Terranes at Convergent Margins and Stabilization of Archean Cratons." *Tectonophysics* 305 (1999): 43–73.

Stein, Mordechai, and A. W. Hofmann. "Mantle Plumes and Episodic Crustal Growth." *Nature* 372 (1994): 63–68.

ophiolites A distinctive group of rocks that include basalt, diabase, gabbro, and peridotite. They may also be associated with chert, metalliferous sediments (umbers), trondhjemite, diorite, and serpentinite. Many ophiolites are altered to serpentinite, chlorite, albite, and epidote rich rocks, possibly by hydrothermal seafloor metamorphism. The term was introduced by G. Steinman in 1905 for a tripartite assemblage of rocks, including basalt, chert, and serpentinite, that he recognized as a common rock association in the Alps. Most ophiolites are interpreted as having formed in an ocean floor environment, including at mid-ocean ridges, in back arc basins, in extensional fore arcs, or within arcs. Ophiolites are detached from the oceanic mantle and have been thrust upon continental margins during the closure of ocean basins. Lines of ophiolites decorate many sutures around the world marking places where oceans have closed. In the 1960s and 1970s much research was aimed at defining a type of ophiolite succession, which became known as the Penrose-type of ophiolite. More recent research has revealed that the variations between individual ophiolites are as significant as any broad similarities between them.

A classic Penrose-type of ophiolite is typically three to nine miles (5–15 km) thick, and if complete, consists of the following sequence from base to top, with a fault marking the base of the ophiolite. The lowest unit in some ophiolites is an ultramafic rock called lherzolite, consisting of olivine + clinopyroxene + orthopyroxene, generally interpreted to be fertile, undepleted mantle. The base of most ophiolites con-

The World's Oldest Ophiolite

Ophiolites are a distinctive association of allochthonous rocks interpreted to form at oceanic spreading centers in back arc basins, forearcs, arcs, and in major oceans. A complete ophiolite grades downward from pelagic sediments into a mafic volcanic complex comprised mostly of pillow basalts, underlain by a sheeted dike complex. These are underlain by gabbros exhibiting cumulus textures, then tectonized peridotite, resting above a thrust fault that marks the contact with underlying rock sequences. The term *ophiolite* refers to this distinctive rock association, although many workers interpret the term to mean allochthonous oceanic lithosphere rocks formed exclusively at mid-ocean ridges.

Prior to 2001, no complete Phanerozoic-like ophiolite sequences had been recognized in Archean greenstone belts, leading some workers to the conclusion that no Archean ophiolites or oceanic crustal fragments are preserved. These ideas were recently challenged by the discovery of a complete 2.5 billion-year-old ophiolite sequence in the North China craton. This remarkable rock sequence includes chert and pillow lava, a sheeted dike complex, gabbro and layered gabbro, cumulate ultramafic rocks, and a suite of strongly deformed mantle harzburgite tectonites. The mantle rocks include a distinctive type of intrusion with metallic chrome nodules called a podiform chromite deposit, known to form only in oceanic crust.

Well-preserved black smoker chimney structures in metallic sulfide deposits have also been discovered in some sections of the Dongwanzi ophiolite belt, and these ancient seafloor hydrothermal vents are among the oldest known. Deep-sea hydrothermal vents host the most primitive thermophyllic, chemosynthetic, sulfate-reducing organisms known, believed to be the closest relatives of the oldest life on Earth, with similar vents having possibly provided nutrients and protected environments for the first organisms. These vents are associated with some unusual microscale textures that may be remnants of early life forms, most likely bacteria. These ancient fossils provide tantalizing suggestions that early life may have developed and remained sheltered in deep-sea hydrothermal vents until surface conditions became favorable for organisms to inhabit the land.

Archean oceanic crust was possibly thicker than Proterozoic and Phanerozoic counterparts, resulting in accretion predominantly of the upper basaltic section of oceanic crust. The crustal thickness of Archean oceanic crust may in fact have resembled modern oceanic plateaux. If this were the case, complete Phanerozoic-like ophiolite sequences would have been very unlikely to be accreted or obducted during Archean orogenies. In contrast, only the upper, pillow-lava–dominated sections would likely be accreted. Remarkably, Archean greenstone belts contain an abundance of tectonic slivers of pillow lavas, gabbros, and associated deep-water sedimentary rocks. The observation that Archean greenstone belts have such an abundance of accreted ophiolitic fragments compared with Phanerozoic orogens suggests that thick, relatively buoyant, young Archean oceanic lithosphere may have had a rheological structure favoring delamination of the uppermost parts during subduction and collisional events.

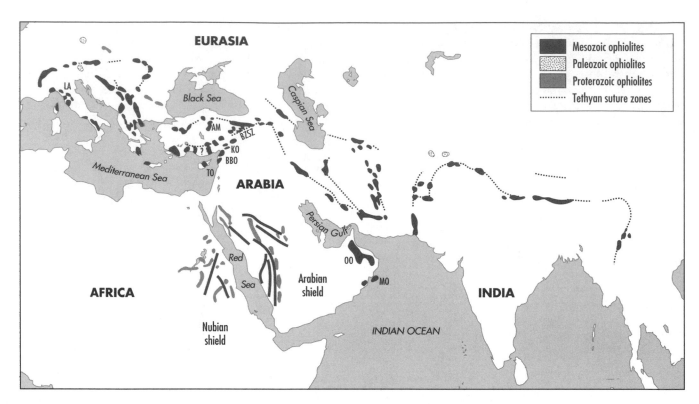

Distribution of ophiolites in the Tethyan orogenic belt, showing the location of Proterozoic, Paleozoic, and Mesozoic ophiolites

sists of an ultramafic rock known as harzburgite, consisting of olivine + orthopyroxene (± chromite), often forming strongly deformed or transposed compositional layering, forming a distinctive rock known as harzburgite tectonite. In some ophiolites, harzburgite overlies lherzolite. The harzburgite is generally interpreted to be the depleted mantle from which overlying mafic rocks were derived, and the deformation is related to the overlying lithospheric sequence flowing away from the ridge along a shear zone within the harzburgite. The harzburgite sequence may be six miles (10 km) or more thick in some ophiolites, such as the Semail ophiolite in Oman and the Bay of Island ophiolite in Newfoundland.

Resting above the harzburgite is a group of rocks that were crystallized from a magma derived by partial melting of the harzburgite. The lowest unit of these crustal rocks includes crystal cumulates of pyroxene and olivine, forming distinctive layers of pyroxenite, dunite, and other olivine + clinopyroxene + orthopyroxene peridotites including wehrlite, websterite, and pods of chromite + olivine. The boundary between these rocks (derived by partial melting and crystal fractionation) and those below from which melts were extracted is one of the most fundamental boundaries in the crust, known as the Moho, or base of the crust. It is named after Andrija Mohorovicic, a Yugoslavian geophysicist who noted a fundamental seismic boundary beneath the continental crust. In this case, the Moho is a chemical boundary, without a sharp seismic discontinuity. A seismic

discontinuity occurs about half a kilometer higher than the chemical Moho in ophiolites.

The layered ultramafic cumulates grade upward into a transition zone of interlayered pyroxenite and plagioclase-rich cumulates, then into an approximately half-mile (1-km) thick unit of strongly layered gabbro. Individual layers within this thin unit may include gabbro, pyroxenite, and anorthosite. The layered gabbro is succeeded upward by one to three miles (2–5 km) of isotropic gabbro, which is generally structureless but may have a faint layering. The layers within the isotropic gabbro in some ophiolites define a curving trajectory, interpreted to represent crystallization along the walls of a paleomagma chamber. The upper part of the gabbro may contain many xenoliths of diabase, pods of trondhjemite (plagioclase plus quartz), and may be cut by diabase dikes.

The next highest unit in a complete, Penrose-style ophiolite is typically a sheeted dike complex, consisting of a 0.3–1.25-mile (0.5–2-km) thick complex of diabasic, gabbroic, to silicic dikes that show mutually intrusive relationships with the underlying gabbro. In ideal cases, each diabase dike intrudes into the center of the previously intruded dike, forming a sequence of dikes that have chilled margins developed only on one side. These dikes are said to exhibit one-way chilling. In most real ophiolites, examples of one-way chilling may be found, but statistically the one-way chilling may only show directional preference in 50–60 percent of cases.

The sheeted dikes represent magma conduits that fed basaltic flows at the surface. These flows are typically pillowed, with lobes and tubes of basalt forming bulbous shapes distinctive of underwater basaltic volcanism. The pillow basalt section is typically 0.3–0.6-mile (0.5–1-km) thick. Interstices between the pillows may be filled with chert, and sulfide minerals are common.

Many ophiolites are overlain by deep-sea sediments, including chert, red clay, in some cases carbonates, or sulfide layers. Many variations are possible, depending on tectonic setting (e.g., conglomerates may form in some settings) and age (e.g., siliceous biogenic oozes and limestones would not form in Archean ophiolites, before the life-forms that contribute their bodies developed).

Processes of Ophiolite and Oceanic Crust Formation

The sequence of rock types described above are a product of a specific set of processes that occurred along the oceanic spreading centers that the ophiolites formed along. As the mantle convects and the asthenosphere upwells beneath mid-ocean ridges, the mantle harzburgites undergo partial melting of 10–15 percent in response to the decreasing pressure. The melts derived from the harzburgites rise to form a magma chamber beneath the ridge, forming the crustal section of the oceanic crust. As the magma crystallizes, the densest crystals gravitationally settle to the bottom of the magma chamber, forming layers of ultramafic and higher mafic cumulate rocks. Above the cumulate a gabbroic fossil magma chamber forms, typically with layers defined by

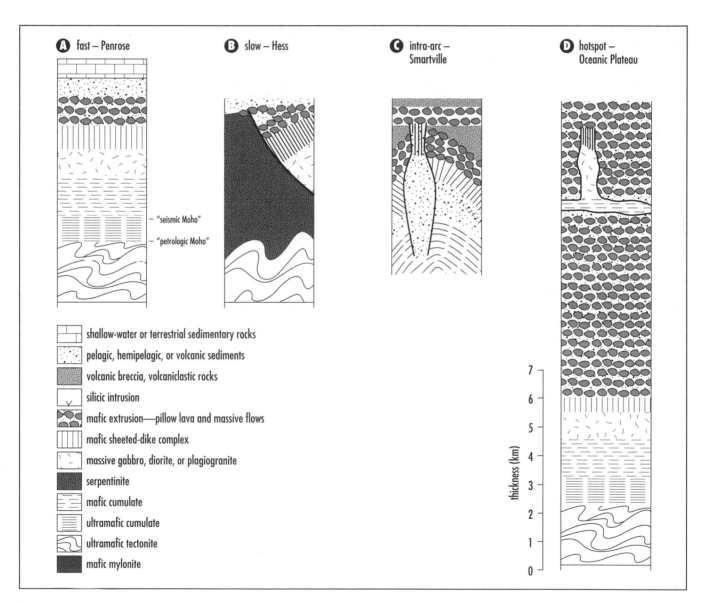

Cross sections through typical ophiolites, including different types of ophiolites produced at slow, intra-arc, and hotspot types of tectonic settings

varying amounts of pyroxene and feldspar crystals. In many examples the layering in ophiolites has been shown to be parallel to the fossil margins of the magma chamber. An interesting aspect of the magma chamber is that periodically, new magma is injected into the chamber, changing the chemical and physical dynamics. These new magmas are injected during extension of the crust so the magma chamber may effectively expand infinitely if the magma supply is continuous, as in fast-spreading ridges. In slow-spreading ridges the magma chamber may completely crystallize before new batches of melt are injected.

As extension occurs in the oceanic crust, dikes of magma shoot out of the gabbroic magma chamber, forming a diabasic (fine-grained rapidly cooled magma with the same composition as gabbro) sheeted dike complex. The dikes have a tendency to intrude along the weakest, least crystallized part of the previous dike, which is usually in the center of the last dike to intrude. In this way each dike intrudes the center of the previous dikes, forming a sheeted dike complex characterized by dike that have only one chill margin, most of which face in the same direction.

Many of the dikes reach the surface of the seafloor, where they feed basaltic lava flows. Basaltic lava flows on the seafloor are typically in the form of bulbous pillows that stretch out of magma tubes, forming the distinctive pillow-lava section of ophiolites. The top of the pillow-lava section is typically quite altered by seafloor metamorphism including having deposits of black smoker-type hydrothermal vents. The pillow lavas are overlain by sediments deposited on the seafloor. If the oceanic crust forms above the calcium carbonate compensation depth, the lowermost sediments may be calcareous. These would be succeeded by siliceous oozes, pelagic shales, and other sediments as the seafloor cools, subsides, and moves away from the mid-ocean ridge. A third sequence of sediments may be found on the ophiolites. These would include sediments shed during detachment of the ophiolite from the seafloor basement, and its thrusting (obduction) onto the continental margin.

The type of sediments deposited on ophiolites may have been very different in some of the oldest ophiolites that formed in the Precambrian. For instance, in the Proterozoic and especially the Archean, organisms that produce the carbonate and siliceous oozes would not be present, as the organisms that produced these sediments had not yet evolved.

There is considerable variation in the classical ophiolite sequence described above, as first formally defined by the participants of a Penrose conference on ophiolites in 1972. First, most ophiolite sequences are deformed and metamorphosed so it is difficult to recognize many of the primary magmatic units, especially sheeted dikes. Deformation associated with emplacement typically causes some or several sections of the complete sequence to be omitted, and others to be repeated along thrust faults. Therefore the adjectives *metamorphosed, partial,* and *dismembered* are often added as prefixes to descriptions of individual ophiolites. There is also considerable variation in the thickness of individual units, some may be totally absent, and different units may be present in specific examples. Similar variations are noted from the modern seafloor and island arc systems, likely settings for the formation of ophiolites. Most ophiolites are interpreted to be fragments of the ocean floor generated at mid-ocean ridges, but the thickness of the modern oceanic crustal section is about 4 miles (7 km), whereas the equivalent units in ophiolites average about 1.8–3.1 miles (3–5 km).

Some of the variations may be related to the variety of tectonic environments that ophiolites form in. The Ocean Drilling Program, in which the oceanic crust has been drilled in a number of locations, has resulted in the recognition that differences in spreading rate and magma supply, among other factors, may determine which units in what thickness are present in different sections of oceanic crust. Fast-spreading centers such as the East Pacific Rise typically show the complete ophiolite sequence, whereas slow-spreading centers such as the Mid-Atlantic Ridge may be incomplete, in some cases entirely lacking the magmatic section. Other ophiolites may form at or near transform faults, in island arcs, back arc basins, forearcs, or above plumes.

See also CONVERGENT PLATE MARGIN PROCESSES; DIVERGENT OR EXTENSIONAL BOUNDARIES.

Further Reading

Anonymous. "Ophiolites." *Geotimes* 17 (1972): 24–25.

Dewey, John F., and John M. Bird. "Origin and Emplacement of the Ophiolite Suite: Appalachian Ophiolites in Newfoundland, in Plate Tectonics." *Journal of Geophysical Research* 76 (1971): 3,179–3,206.

Kusky, Timothy M., Jianghai Li, and Robert T. Tucker. "The Archean Dongwanzi Ophiolite Complex, North China Craton: 2.505 Billion Year Old Oceanic Crust and Mantle." *Science* 292 (2001): 1,142–1,145.

Moores, Eldridge M. "Origin and Emplacement of Ophiolites." *Review Geophysics* 20 (1982): 735–750.

Ordovician The second period of the Paleozoic era, and the corresponding rock series, falling between the Cambrian and the Silurian. It is commonly referred to as the age of marine invertebrates. The base of the Ordovician is defined on the Geological Society of America timescale (1999) as 490 million years ago, and the top or end of the Ordovician is defined at 444 million years ago. The period was named by Charles Lapworth in 1879 after the Ordovices, a Celtic tribe that inhabited the Arenig-Bala area of northern Wales, where rocks of this series are well exposed.

By the Early Ordovician, North America had broken away from the supercontinent of Gondwana that amalgamated during the latest Precambrian and early Cambrian period.

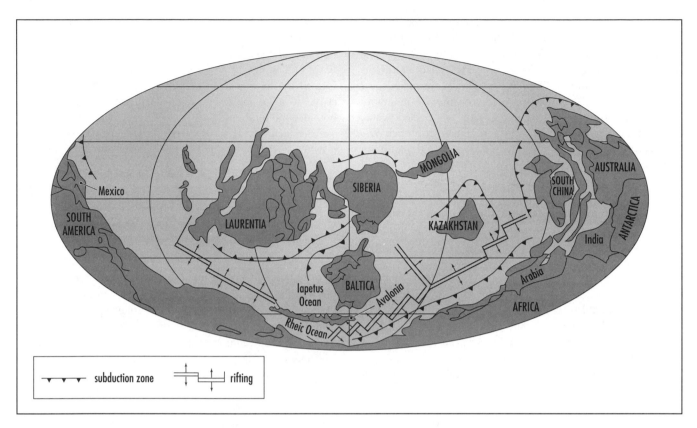

Ordovician paleogeography showing the distribution of continents approximately 500 million years ago

It was therefore surrounded by shallow water passive margins, and being at equatorial latitudes, these shallow seas were well suited for the proliferation of marine life-forms. The Iapetus Ocean separated what is now the east coast of North America from the African and South American segments of the remaining parts of Gondwana. By the Middle Ordovician, convergent tectonics brought an island arc system to the North American margin, initiating the Taconic Orogeny as an arc/continent collision. This was followed by a sideways sweep of parts of Gondwana past the North American margin, leaving fragments of Gondwana attached to the modified eastern margin of North America.

During much of the Ordovician, carbonate sediments produced by intense organic productivity covered shallow epeiric seas in the tropical regions, including most of North America. This dramatic increase in carbonate sedimentation may reflect a combination of tectonic activity that brought many low-lying continental fragments into the Tropics, high sea-level stands related to the breakup of Gondwana, and a sudden increase in the number of different organisms that started to use calcium carbonate to build their skeleton or shell structures.

Marine life included diverse forms of articulate brachiopods, communities of echinoderms such as the crinoids or sea lilies, and reef-building stromatoporoids, rugose, and tabulate corals. Trilobites roamed the shallow seafloors, and many forms emerged. The Ordovician saw rapid diversification and wide distribution of several planktonic and pelagic faunas, especially the graptolites and conodonts which form useful index fossils for this period. Nautiloids floated across the oceans, and some attained remarkably large sizes, reaching up to several meters across. Fish fossils are not common from Ordovician deposits but there may have been some primitive armored types present. The end of the Ordovician is marked by a marine extinction event, apparently caused by rapid cooling of the shallow seas, perhaps related to continental glaciation.

See also APPALACHIANS; CONODONT; CRINOID; PALEOZOIC.

Further Reading
Condie, Kent C., and Robert E. Sloan. *Origin and Evolution of Earth: Principles of Historical Geology.* Upper Saddle River, N.J.: Prentice Hall, 1997.

origin of universe *See* BIG BANG THEORY.

orogeny The building of mountains. Many early philosophers, theologians, and scientists going back at least as far as Francis Bacon were formulating theories about the forces involved in uplifting and deforming mountains for the past several or tens of centuries. By the middle 1800s the processes involved in the formation of mountains became known as orogeny. Early ideas suggested that mountains were deformed and uplifted by magmatic intrusions, or reflected an overall contraction of the Earth with mountain belts representing cooling wrinkles as on a shriveled prune. In a classical work in 1875, Eduard Suess published *Die Entstehung der Alpen* in which he argued that mountain belts on the planet did not follow any regular pattern that would indicate global contraction, and he suggested that the mountain belts represented contraction between rigid blocks (now called cratons) and surrounding rocks on the margins of these massifs. However, he still believed that the main driving force was global contraction induced by cooling of the Earth. In Suess's last volume of *Das Antlitz der Erde* (*The Face of the Earth*) (1909), he admitted that the amount of shortening observed in mountain belts was greater than could be explained by global cooling and contraction, and he suggested that perhaps other translations have occurred in response to tidal forces and the rotation of the planet.

In 1915 and 1929 Alfred Wegener published *Die Entstehung der Kontinente* (*The Origin of the Continents*) and *Die Entstehung der Kontinente und Ozeane* (*The Origin of the Continents and Oceans*). Wegener argued strongly for large horizontal motions between cratons made of sial, and using such data as the match of restored coastlines and paleontological data, he founded the theory of continental drift. Several geologists, including Alex du Toit, Reginald Daly, and Arthur Holmes, documented geological ties between different continents supporting the idea of continental drift. In the 1940s–1960s, geophysical exploration of the seafloor led to the recognition of seafloor spreading and provided the data that J. Tuzo Wilson needed to propose the modern theory of plate tectonics in 1965.

With the development of the ideas of plate tectonics, geologists now recognize that mountain belts are of three basic types, including fold and thrust belts, volcanic mountain ranges, and fault block ranges. Fold and thrust belts are contractional mountain belts, formed where two tectonic plates collided, forming great thrust faults, folds, metamorphic rocks, and volcanic rocks. Detailed mapping of the structure in the belt can enable geologists to reconstruct their history and essentially pull them apart. It is found that many of the rocks in fold and thrust belt types of mountain ranges were deposited on the bottom of the ocean, continental rises, slope, shelves, or on ocean margin deltas. When the two plates collide, many of the sediments get scraped off and deformed, forming the mountain belts. Thus, fold and thrust mountain belts mark places where oceans have closed. Volcanic mountain ranges include places such as Japan's Mount Fuji, and Mount St. Helens in the Cascades of the western

United States. These mountain ranges are not formed primarily by deformation, but by volcanism associated with subduction and plate tectonics. Fault-block mountains, such as the Basin and Range Province of the western United States, are formed by the extension or pulling apart of the continental crust, forming elongate valleys separated by tilted fault-bounded mountain ranges.

See also CONTINENTAL DRIFT AND PLATE TECTONICS: CONVERGENT PLATE MARGIN PROCESSES; PLATE TECTONICS.

Further Reading

Miyashiro, Akiho, Keiti Aki, and A. M. Celal Sengör. *Orogeny.* Chichester: John Wiley and Sons, 1982.

Wilson, J. Tuzo. "A New Class of Faults and their Bearing on Continental Drift." *Nature* 207 (1965): 343–347.

oxbow lake   An elongate or crescent-shaped body of standing water formed in a former channel of a meandering river. Oxbow lakes form when a typically strongly curved channel is abandoned when its neck is cut off, usually during a high-flow interval in the stream, and the ends of the bend are silted up. Meandering rivers may develop many strongly curved oxbows, where exaggerated U-shaped bends isolate teardrop-shaped parcels of land only connected to the riverbank by narrow necks. These necks may slowly or rapidly erode until the river takes a newer, shorter course and the old curved segment is abandoned. The oxbow lakes then evolve from channels with the former depth of the river but gradually get filled in by silt during flood stages of the main river. Some river systems have thousands of oxbows and oxbow lakes, and the entire floodplain may be marked by variably sedimented oxbow lakes and former oxbow lakes, perhaps only marked by curved patterns of differently aged vegetation.

See also RIVER SYSTEM.

View from a helicopter of the Kuskokwim River plain in central Alaska, showing numerous oxbow lakes representing abandoned channels of the river *(Photo by Timothy Kusky)*

oxide minerals Mineral compounds that are formed by the bonding of oxygen with one or more metallic elements. Common examples include periclase (MgO), cassiterite (SnO_2), corundum (Al_2O_3), ruby (a corundum with minor amounts of chromium ions), hematite (Fe_2O_3), ilmenite ($FeTiO_3$), rutile (TiO_2), anatase (TiO_2), and perovskite ($[Ca,Na,Fe^{+2},Ce][Ti,Nb]O_3$). The spinel group consists of a number of oxide minerals with the general formula AB_2O_4. Included are spinel ($MgAl_2O_4$), hercynite ($Fe^{+2}Al_2O_4$), gahnite ($ZnAl_2O_4$), galaxite ($MnAl_2O_4$), magnesioferrite ($MgFe^{+3}_2O_4$), magnetite ($Fe^{+2}Fe^{+3}_2O_4$), ulvospinel ($Fe^{+2}Ti_2O_4$), franklinite ($ZnFe^{+3}_2O_4$), jacobsite ($MnFe^{+3}_2O_4$), trevorite ($NiFe^{+3}_2O_4$), magnesiochromite ($MgCr_2O_4$), and chromite ($Fe^{+2}Cr_2O_4$). Oxide minerals are of great economic importance since they include ores of most metals that are necessary for industrial, technological, and manufacturing applications. Most of the oxide minerals exhibit ionic bonds, with the size of lattice structures dependent on the size of the metallic cations. Some minerals in the spinel group are used as gemstones.

ozone hole Ozone is a poisonous gas (O_3) that is present in trace amounts in much of the atmosphere but reaches a maximum concentration in a stratospheric layer at 9–25 miles (15–40 km) above the Earth, with a peak at 15.5 miles (25 km). The presence of ozone in the stratosphere is essential for most life on Earth, since it absorbs the most carcinogenic part of the solar spectrum with wavelengths between 0.000011 and 0.0000124 inches (280 and 315 nm). If these ultraviolet rays were to reach the Earth they would cause many skin cancers and possibly depress the human immune system. These harmful rays would greatly reduce photosynthesis in plants and reduce plant growth to such an extent that the global ecosystems would crash.

Ozone naturally changes its concentration in the stratosphere and is also strongly affected by human or anthropogenically produced chemicals that make their way into the atmosphere and stratosphere. Ozone is produced by photochemical reactions above 25 kilometers mostly near the equator and moves toward the poles where it is most abundant, and where it is gradually destroyed. The concentration of ozone does not vary greatly in equatorial regions, but at the poles it tends to be the greatest in the winter and early spring. Stratospheric circulation in the winter over Antarctica is characterized by the formation of a strong vortex that isolates the stratospheric air over the pole during the night in the Antarctic winter.

Atmospheric and stratospheric flow dynamics can change the distribution of ozone, solar flare and sunspot activity can enhance ozone, and volcanic eruptions can add sulfates to the stratosphere that destroy ozone. In the 1970s it was realized that some aerosol chemicals and refrigerants, the chlorofluorocarbons (CFCs), could make their way into the stratosphere and be broken down by ultraviolet light to release chlorine, which can destroy ozone. The use of CFCs was subsequently curtailed, but the aerosols and chlorine have very long residence times in the stratosphere, and each chlorine ion is capable of destroying large amounts of ozone. Since the middle 1980s, a large hole marked by large depletions of ozone in the stratosphere has been observed above Antarctica every spring, its growth aided by the polar vortex. The hole has continued to grow, but the relative contributions to the destruction of ozone by CFCs, other chemicals (such as supersonic jet and space shuttle fuel), volcanic gases, and natural fluctuations is uncertain. In 1999 the size of the Antarctic ozone hole was measured at more than 9,650,000 square miles (25 million km²), more than two and half times the size of Europe. However, the appearance of ozone depletion above arctic regions has added credence to models that show the ozone depletion being largely caused by CFCs. Many models suggest that the CFCs may lead to a 5–20 percent reduction in global ozone, with consequent increases in cancers, disease, and loss of crop and plant yield.

See also ATMOSPHERE.

P

pahoehoe lava A term of Hawaiian origin for a basaltic lava flow with a twisted billowy surface that resembles a coiled rope. The term was introduced by Clarence E. Dutton in 1882. It is contrasted with other main types of basaltic flows in Hawaii known as aa, characterized by a blocky, rubbly surface, resembling bulbous pillows. There seem to be several variables that cause one type of flow to form instead of the other. Pahoehoe flows are generally richer in gas, are hotter, and are slightly less viscous than their aa counterparts. Pillow lavas form underwater.

There are several varieties of pahoehoe flows. Massive pahoehoe are 6.5–50 feet (2–15 m) thick and are smooth for large parts of the flow, whereas scaly pahoehoe flows resemble a fish's scales or a shingled roof, where numerous typically 2–12-inch (5–30-cm) thick lobes overlap preceding lobes. Scaly pahoehoe grade into entrail pahoehoe, in which the lobes expand into piled up masses or may grade into pillow-like lobes. Shelly pahoehoe is characterized by numerous burst bubbles on the surface producing a frothy texture, and when the bubbles (vesicles) are very numerous they produce a flow with large holes and thin vesicle walls known as reticulite or thread-lace scoria. Slabby pahoehoe have surfaces consisting of a number of broken, piled-up slabs, probably produced by cracking of the surface when underlying lava tubes drained.

Pahoehoe lava flows may have many other irregular features on their surface, including spatter cones, linear ridges, lava blisters, pressure plateaus, and squeeze-ups, where lava from beneath the surface has pushed its way through the overlying crust. Explosion tubes, or thin pipes known as pipe amygdules, form where lava has overrun water, and the water is converted to steam which explodes and bores a tube to the surface. Where the lava flow has overrun something, such as a house or tree, the object may leave a mold in the lava flow,

especially if it burns after the lava has surrounded it. Lava tubes and tunnels are well developed in many pahoehoe flows, forming where lava on the surface has cooled and lava from the tube has drained by flowing to lower elevation areas, leaving a tube-like cave behind. These tubes typically form a branching network following the branching paths of the lava flows across the surface. These tubes may be small, but may be up to several tens of meters in diameter and many are even tens of kilometers long. Lava tubes have made excellent hiding places for bandits and chased armies at various times and places in history.

See also AA; HAWAII; VOLCANO.

Further Reading
Williams, Howel, and Alexander R. McBirney. *Volcanology.* San Francisco: Freeman, Cooper and Co., 1979.

paleoclimatology The study of past and ancient climates and their distribution and variation in space and time, and of the mechanism of long-term climate variations. Various types of data are used to determine past climates, such as the distribution of certain fauna and flora that are climate sensitive, and the distribution of certain rock types that form in restricted climate conditions. Other types of data are used as paleoclimate indicators, including tree ring studies (dendrochronology), ice core data, cave deposits (speleothems), and lake sediment studies. Increasingly, isotopic data such as ratios between light and heavy oxygen isotopes are used in these studies as paleoclimate indicators, since these ratios are very sensitive to past global climates, glaciations, and elevations at which rainwater fell.

Most paleoclimate studies reveal that there have been major climate shifts on the planet throughout Earth history, with periods of near global glaciation; periods of intense hot and humid, or hot and dry, weather; and more temperate periods, such as the interglacial stage we are currently experiencing. Many factors may play roles in climate change, including orbital and astronomical variations described by Milankovitch cycles, plate tectonics and the distribution of continental landmasses, and volcanic productivity.

See also CLIMATE; CLIMATE CHANGE; DENDROCHRONOLOGY; MILANKOVITCH CYCLES.

Paleogene *See* CENOZOIC; TERTIARY.

Paleolithic The first division of the Stone Age in archaeological time, marked by the first appearance of humans and their associated tools and workings. The time of the Paleolithic corresponds generally with the Pleistocene (from 1.8 million years ago until 100,000 years ago) of the geological time scale, but varies somewhat from place to place.

See also PLEISTOCENE.

paleomagnetism The study of natural remnant magnetism in rocks to understand the intensity and direction of the Earth's magnetic field in the geologic past, and to understand the history of plate motion. The Earth's magnetic field can be divided into two components at any location, including the declination and the inclination. The declination measures the angular difference between the Earth's rotational north pole and the magnetic north pole. The inclination measures the angle at which the magnetic field lines plunge into the Earth. The inclination is 90° at the magnetic poles, and 0° halfway between the poles.

Studies of paleomagnetism in young rocks have revealed that the Earth's magnetic poles may flip suddenly, over a period of thousands or even hundreds of years. The magnetic poles also wander by about 10°–20° around the rotational poles. On average, however, the magnetic poles are coincident with the Earth's rotational poles. This coincidence can be used to estimate the direction to the north and south poles in ancient rocks that have drifted or rotated in response to plate tectonics. Determination of the natural remnant magnetism in rock samples can, under special circumstances, reveal the paleo-inclination and paleo-declination, which can be used to estimate the direction and distance to the pole at the time the rock acquired the magnetism. If these parameters can be determined for a number of rocks of different ages on a tectonic plate, then an apparent polar wander path for that plate can be constructed. These show how the magnetic pole has apparently wandered with respect to (artificially) holding the plate fixed—when the reference frame is switched, and the pole is held fixed, the apparent polar wander curve shows how the plate has drifted on the spherical Earth.

Paleomagnetism played an enormous role in the confirmation of seafloor spreading, through the discovery and understanding of seafloor magnetic anomalies. In the 1960s geophysicists surveyed the magnetic properties of the ocean floor and began to discover some amazing properties. The seafloor has a system of linear magnetic anomalies where one "stripe" has its magnetic minerals all orientated the same way as the present magnetic field, and the alternate stripes have all their magnetic minerals orientated in completely the opposite sense. These stripes are orientated parallel to the mid-ocean ridge system; where the ridges are "offset" by transform faults, the anomalies are also "offset." The anomalies are symmetric on either side of the ridge, and the same symmetry is found across ridges worldwide.

Understanding the origin of seafloor magnetic stripes was paramount in acceptance of the plate tectonic paradigm. The magnetic stripes form in the following way. As oceanic crust is continuously formed as on a conveyor belt, all the magnetic minerals tend to align with the present magnetic field when the new crust forms. The oceanic crust thus contains a record of when and for how long the Earth's magnetic field has been in the "normal" position, and when and for how long it has been "reversed." Similar reversals of the Earth's magnetic field are known from rock sequences on land, and many of these have been dated. Using these data geologists have now established a magnetic polarity reversal timescale. The last reversal was about 700,000 years ago, and the one before that, about 2.2 million years ago. Oceanic crust is as old as Jurassic, and documentation of the age of seafloor magnetic stripes has led to the construction of the magnetic polarity timescale back to 170 million years ago.

See also PLATE TECTONICS.

paleontology The study of past life based on fossil evidence, the lines of descent of organisms, and the relationships between life and other geological phenomena. Information from fossil distributions is used to understand ancient environments and climates and to constrain former ties between now separated landmasses. Many paleontologists are concerned with mechanisms of extinction and the appearance of new organisms, as well as the mode of life of organisms and their evolution. In the past, paleontology was mostly a descriptive science describing the morphology of fossils, but in recent years it has become much closer to biology, with a new science of geobiology emerging. In this approach, many biological methods are applied to the study of fossils, including cladistic methods, functional morphology, and even paleogenetic studies.

Paleozoic The era of geological time that includes the interval between 544 million and 250 million years ago, and the erathem of rocks deposited in this interval. It includes seven geological periods and systems of rocks deposited in those periods, including the Cambrian, Ordovician, Silurian, Devonian, Carboniferous (Mississippian and Pennsylvanian), and Permian. The Paleozoic was named by Adam Sedgwick in 1838 for the deformed rocks underlying the Old Red Sand-

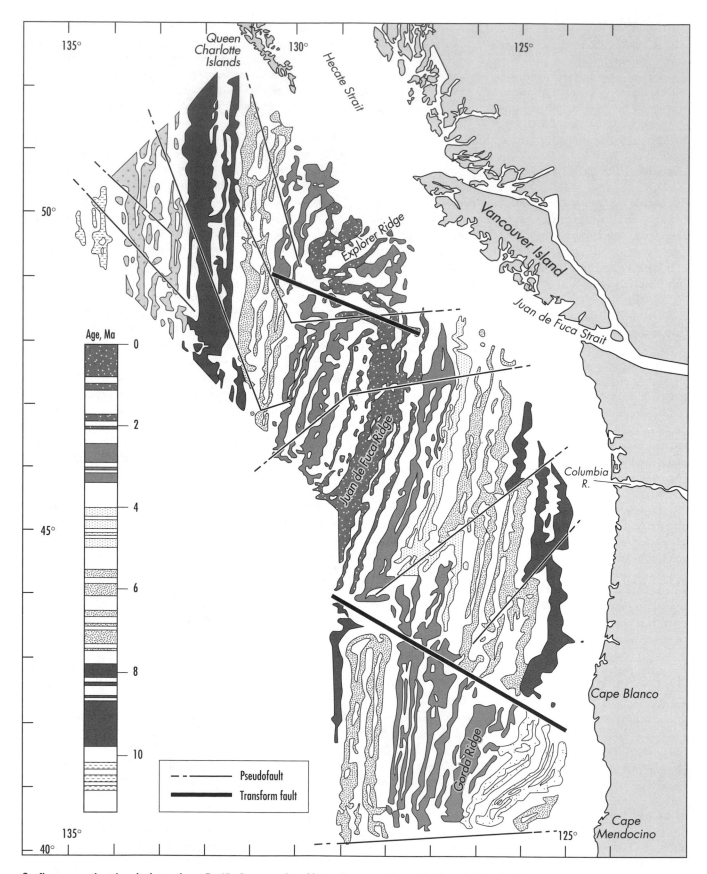

Seafloor magnetic stripes in the northeast Pacific Ocean produced by seafloor spreading on the Juan de Fuca, Gorda, and Explorer ridges

stone in Wales, and the name means ancient life. The base of the Paleozoic is defined as the base of the Cambrian period, conventionally taken as the lowest occurrence of trilobites. Recently, however, with the recognition of the advanced Vendian and Ediacaran fauna, the base of the Cambrian was reexamined and has been defined using fossiliferous sections in eastern Newfoundland and Siberia to be the base of an ash bed dated at 544 million years ago.

At the beginning of the Paleozoic, the recently formed supercontinent of Gondwana was breaking apart, but by the Carboniferous it regrouped as Pangea. This supercontinent included the southern continents in the Gondwanan landmass, and the northern continents in Laurasia, separated by the Pleionic and Tethys Oceans, and surrounded by the Panthalassa Ocean. With the breakup of the late Precambrian supercontinent, climates changed from icehouse to hothouse conditions. The cause of this dramatic change was the volume of CO_2 emitted to the atmosphere by the mid-ocean ridge system. During supercontinent periods, the length of the ridge system is small, and relatively small amounts of CO_2 are emitted to the atmosphere. During supercontinent breakup, however, much more CO_2 is released during enhanced volcanism associated with the formation of new ridge systems. Since CO_2 is a greenhouse gas, supercontinent breakup is associated with increasing temperatures and the establishment of hothouse conditions. The Pangean supercontinent then experienced continental climates ranging from hot and dry to icehouse conditions, with huge continental ice sheets covering large parts of the southern continents. There were many collisional and rifting events, especially along the active margins of Pangea, and huge tracts of oceanic crust must have been subducted to accommodate these collisional events.

The dramatic changes in continental configurations, the arrangement of ecological niches, and the huge climatic fluctuations at the base of the Paleozoic are also associated with the most dramatic explosion of life in the history of the planet. Hard-shelled organisms first appeared in the lower Cambrian and are abundant in the fossil record by the Mid-Cambrian. Fish first appeared in the Ordovician. All of the modern animal phyla and most of the plant kingdom are represented in the Paleozoic record, with fauna and flora inhabiting land, shallow seas, and deep-sea environments. There are several mass extinction events in the Paleozoic, in which large numbers of species suddenly died off and were replaced by new species in similar ecological niches.

In addition to the development of hard-shelled organisms and skeletons, the Paleozoic saw the dramatic habitation of the terrestrial environment. Bacteria and algae crept into different environments such as soils before the Paleozoic, with land plants appearing in the Silurian. Dense terrestrial flora expanded by the Devonian and culminated in the dense forests of the Carboniferous. This profoundly changed the weathering, erosion, and sedimentation patterns from those of the Precambrian and also significantly affected the atmo-

sphere-ocean composition. Terrestrial fauna rapidly followed the plants onto land, with tetrapods roaming the continents by Middle or Late Devonian. By the Devonian, invertebrates including spiders, scorpions, and cockroaches had invaded the land. Fish became abundant in the oceans.

In the Carboniferous much organic carbon got buried, ending the reduction of atmospheric CO_2, and ending the hothouse conditions. With the formation of Pangea in the Carboniferous and Permian, global ridge lengths were reduced, and less CO_2 was released to the atmosphere. Together with the burial of organic carbon, new icehouse conditions were established, stressing the global fauna and flora. The largest mass extinction in geological history marks the end of the Paleozoic (end Permian mass extinction), and the start of the Mesozoic. The causes of this dramatic event seem to be multifold. Conditions on the planet included the formation of a supercontinent (Pangea), sea-level regression, evaporite formation, and rapidly fluctuating climatic conditions. At the boundary between the Paleozoic and Mesozoic Periods (245 million years ago), 96 percent of all species became extinct, including the rugose corals, trilobites, many types of brachiopods, and marine organisms including many foraminifera species.

The Siberian flood basalts were erupted over a period of less than one million years 250 million years ago, at the end of the Permian at the Permian-Triassic boundary. They are remarkably coincident in time with the major Permian-Triassic extinction, implying a causal link. They cover a large area of the Central Siberian Plateau northwest of Lake Baikal and are more than half a mile thick over an area of 210,000 square miles (544,000 km²) but have been significantly eroded from an estimated volume of 1,240,000 cubic miles (5,168,500 km³). It has been postulated that the rapid volcanism and degassing released enough sulfur dioxide to cause a rapid global cooling, inducing a short ice age with associated rapid fall of sea level. Soon after the ice age took hold the effects of the carbon dioxide took over and the atmosphere heated, resulting in a global warming. The rapidly fluctuating climate postulated to have been caused by the volcanic gases is thought to have killed off many organisms, which were simply unable to cope with the wildly fluctuating climate extremes.

It has also been postulated that the end Permian extinction was aided by the impact of a meteorite or asteroid with the Earth, adding environmental stresses to an already extremely stressed ecosystem. If additional research proves this to be correct, it will be shown that a 1–2–3 punch, including changes in plate configurations and environmental niches, dramatic climate changes, and extraterrestrial impacts together caused history's greatest calamity.

See also CAMBRIAN; CARBONIFEROUS; DEVONIAN; GONDWANA; MASS EXTINCTIONS; ORDOVICIAN; PANGEA; PERMIAN; SILURIAN; VENDIAN.

Panama isthmus A narrow body of land that connects Central and South America. Most of the isthmus is occupied by the country of Panama, with Costa Rica occupying the western part of the isthmus, and Colombia located to the west. The isthmus has a wet tropical climate and has rugged volcanic mountains in the west, reaching 11,410 feet (3,478 m) on Chiriqui. The center of the isthmus is characterized by low hills, and there is a low mountain range in the east. The most famous feature of the isthmus is the Panama Canal. The canal was built by the United States in 1904–14 to connect the Pacific Ocean to the Caribbean and Atlantic Oceans, and to avoid the need to ship materials along the long difficult route around southern South America. The canal is 40 miles (64 km) long from coast to coast, but 51 miles (82 km) long between channel entrances. Ships must pass through a series of locks and sail across Lake Gatun to cross the continental divide, and emerge on the other side of the canal seven–eight hours later. Approximately 240 million cubic yards (184 million m³) were excavated from the canal by the Americans, at a cost of $337 million. The canal and canal zone bordering the canal reverted from U.S. control to Panamanian control in 2000.

The formation of the Isthmus of Panama had profound effects on global ocean circulation models and climate. Prior to its formation in the Pliocene (about 5 million years ago), Caribbean Ocean and Pacific Ocean waters were able to flow through the open passageway between North and South America, and there was greater communication between waters and organisms in the equatorial oceans. When the isthmus formed with the movement of the Caribbean plate to the east, and the Central American subduction-related arc forming on the western side of the Caribbean plate, this passage was blocked. Warm waters that formed in the shallow Caribbean Sea were deflected into the north-flowing Gulf Stream and moved into the North Atlantic. This dramatic change in ocean circulation produced dramatic effects on global climate and may even have indirectly triggered the Pleistocene ice ages in the Northern Hemisphere. Warm water in the North Atlantic increased humidity and snowfall at high latitudes, which then increased albedo (surface reflectance of solar energy), leading to temperature decreases. The formation of continental glaciers reduced sea levels, reduced the size of the North Atlantic basin, and forced the Gulf Stream closer to Europe.

Many species of plants and animals were also strongly affected by the formation of the isthmus. The isthmus served as a new land bridge that allowed mammals to migrate between North and South America. South America was inhabited by a diverse population of marsupial mammals that developed in isolation from other landmasses after the breakup of Pangea in the Late Mesozoic, so the mammal population was unique and different from those of Australia, Antarctica, and Africa. Some of the marsupials included giant land sloths and armadillos, as well as opossums. Some of these marsupials migrated north, including the armadillos, porcupines, opossums, anteaters, and monkeys. However, most of the species that migrated across the isthmus moved in the opposite direction, invading the south from the north. Among these animals were the pig, deer, horse, elephant, camel, rhinoceros, squirrel, raccoon, rabbit, rat, bear, dog, and cat.

Further Reading
Stanley, Steven. *Earth and Life Through Time*. New York: W.H. Freeman and Company, 1986.

Pangea   The supercontinent that formed in the Late Paleozoic, lasting from about 300–200 million years ago, and included most of the planet's continental masses. The former existence of Pangea, meaning all land, was first postulated by Alfred Wegener in 1924, when he added the Australian and Antarctic landmasses to an 1885 supercontinent reconstruction of Gondwana by Eduard Suess that included Africa, India, Madagascar, and South America. He used the fit of the shapes of the coastlines of the now dispersed continental fragments, together with features such as mineral belts, faunal and floral belts, mountain ranges, and paleoclimate zones that matched across his reconstructed Pangean landmass to support the hypothesis that the continents were formerly together. Wegener proposed that the supercontinent broke up first into two large fragments including Laurasia in the north, and Gondwana in the south, and then continued breaking up, leading to the present distribution of continents and oceans. Wegener's ideas were not generally accepted at first but since the discovery of seafloor magnetic anomalies and the plate tectonic revolution, the general framework of his Pangea model has become recognized as generally valid.

The Pangean supercontinent began amalgamating from different continental fragments with the collision of Gondwana and Laurentia and Baltica in the Middle Carboniferous, resulting in the Alleghenian, Mauritanide, and Variscan orogenies. Final assembly of Pangea involved the collision of the South China and Cimmerian blocks with the Paleo-Tethyan margin, resulting in the early Yenshanian and Indonesian orogenies in the Middle to Late Triassic.

The formation of Pangea is associated with global climate change and rapid biological evolution. The numerous collisions caused an overall thickening of the continental crust that decreased continental land area and resulted in a lowering of sea level. The uplift and rapid erosion of many carbonate rocks that had been deposited on trailing or passive margins caused a decrease in the carbonate $^{87}Sr/^{86}Sr$ ratios in the ocean. During the final stages of the coalescence of Pangea, drainage systems were largely internal, erosion rates were high, and the climate, with large parts of the supercontinent lying between 15° and 30° latitude, became arid, with widespread red-bed deposition. Soon, however, the effects of the erosion and burial of large amounts of carbonate and the associated drawdown of atmospheric CO_2 caused climates to rapidly cool, resulting in high-latitude glaciations.

The main glaciations of Pangea started in the late Devonian and early Carboniferous, began escalating in intensity

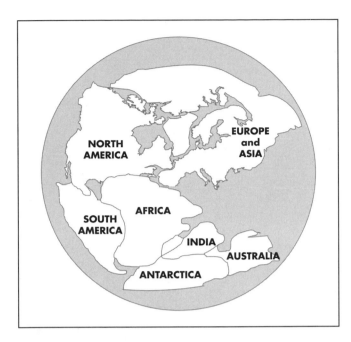

Map showing the distribution of landmasses in the supercontinent of Pangea

by 333 million years ago, peaked in the Late Carboniferous by 292 million years ago, and ended in the early Permian by 272 million years ago. These glaciations resulted in major global regressions as the continental ice sheets used much of the water on the planet. Wegener, and many geologists since, used the distribution of Pangean glacial deposits as one of the main lines of reasoning to support the idea of continental drift. If the glacial deposits of similar age are plotted on a map of the present distribution of the continents, the ice flow patterns indicate that the oceans too must have been covered. However, there is not enough water on the planet to make ice sheets so large that they can cover the entire area required if the continents have not moved. If the glacial deposits are plotted on a map of Pangea however, they cover a much smaller area, the ice flow directions are seen to be radially outward from depocenters, and the total volume of ice is able to be accommodated by the amount of water on Earth.

Pangea began rifting and breaking apart about 230 million years ago, with numerous continental rifts, flood basalts, and mafic dikes intruding into the continental crust. Major breakup and seafloor spreading began about 175 million years ago in the central Atlantic, when North and South America broke away from Pangea, 165 million years ago off Somalia, and 160 million years ago off the coast of northwest Australia. Sea levels began to rise with breakup because of the increase in volume of the mid-ocean ridges that displaced seawater onto the continents, forming marine transgressions. Episodic transgressions and evaporation of seawater from restricted basins led to the deposition of thick salts in parts of the Atlantic, with some salt deposits reaching 1.2 miles (2

km) thick off the east coast of North America, Spain, and northwest Africa. Several rifts along these margins have several to 10 kilometers of non-marine sandstones, shales, redbeds, and volcanics, associated with breakup of Pangea. Many are very fossiliferous, including plants, mudcracks, and even dinosaur footprints attesting to the shallow water and subaerial nature of these deposits.

Breakup of the supercontinent was also associated with a dramatic climate change and high sea levels. The increased volcanism at the oceanic ridges released many gases to the atmosphere, inducing global warming, leading to a global greenhouse, ideal for carbonate production on passive or trailing continental margins.

See also PALEOZOIC; SUPERCONTINENT CYCLE.

Panthalassa The large ocean that surrounded Pangea in the Late Paleozoic and Early Mesozoic. This ocean was huge, even by Pacific standards, since virtually all of the planet's landmasses were grouped together in a single supercontinent. Sea levels were low during Pangean glaciations and rose dramatically during the breakup of the supercontinent in the Triassic and Jurassic.

See also PANGEA.

passive margin Continental margins that are attached to the adjacent oceanic crust and do not have a plate boundary along the margin are known as passive or trailing margins. Most parts of the east coasts of North and South America, the west coasts of Europe and Africa, and most of the coastlines of India, Antarctica, and Australia are passive margins, many of which are characterized by thick accumulations of marine carbonates, shales, and sandstones. Conditions for the formation, accumulation, and preservation of hydrocarbons are met along many passive margins, so contemporaneous and ancient passive margin sequences are the focus of intense petroleum exploration.

Tilted fault block defining graben along the Gulf of Suez, on the Red Sea of Sinai, Ras Sharatibi area. Note fences and trenches from military actions in the area. *(Photo by Timothy Kusky)*

Carbonate and shale deposited on the Ordovician passive margin of eastern North America Mohawk Valley, New York *(Photo by Timothy Kusky)*

Trailing or passive margins typically develop from a continental rift and first form an immature passive margin, with the Red Sea being the main example extant on the planet at this time. Rifting along the Red Sea began in earnest by 30 million years ago, separating Arabia from Africa. The Red Sea is characterized by uplifted rift shoulders that slope generally away from the interior of the sea but have narrow down-dropped coastal plains where steep mountain fronts are drained by wadis (dry streambeds) with alluvial fans that form a typically narrow coastal plain. These have formed over stretched continental crust, forming many rotated fault blocks and grabens, intruded by mafic dike swarms that in some places feed extensive young volcanic fields, especially in Saudi Arabia. Since much of the Red Sea is located in tropical to subtropical latitudes, it has developed thick carbonate platforms along the stretched continental crust. As rifting and associated stretching of the continental crust proceeded, areas that were once above sea level subsided below sea level, but different parts of the Red Sea basin reached this point at different times. Together with global rises and falls of sea level, this led to episodic spilling of salty seawater into restricted basins which would then evaporate, leaving thick deposits of salt behind. As rifting continued these salts became buried beneath the carbonates and shales and sandstones, but when salt gets buried deeply it rises buoyantly, forming salt domes that pierce overlying sediments and form exceptionally good petroleum traps. Parts of the passive margins along the Red Sea are several kilometers thick and currently exhibit some of the world's best coral reefs. The center of the Red Sea is marked by steep slopes off the carbonate platforms, leading to the embryonic spreading center that is only present in southern parts of the sea. Abundant volcanism, hot black smoker vents, and active metalliferous and brine mineralization on the seafloor characterize this spreading center.

As spreading continues on passive margins, the embryonic or Red Sea stage gradually evolves into a young oceanic or mature passive margin stage where the topographic relief on the margins decreases and the ocean to passive margin transition becomes very flat, forming wide coastal plains such as those along the east coast of North America. This transition is an important point in the evolution of passive margins, as it marks the change from rifting and heating of the lithosphere to drifting and cooling of the lithosphere. The cooling of the lithosphere beneath the passive margin leads to gradual subsidence typically without the dramatic faulting that characterized the rifting and Red Sea stages of the margin's evolution. Volcanism wanes, and sedimentation on the margins evolves to exclude evaporites, favoring carbonates, mudstones, sandstones, and deltaic deposits. The overall thickness of passive margin sedimentary sequences can grow to nine or even 12.5 miles (15–20 km), making passive margin deposits among the thickest found on Earth.

See also CONTINENTAL MARGIN; DIAPIR; DIVERGENT OR EXTENSIONAL BOUNDARIES; HYDROCARBON; PLATE TECTONICS.

Patagonia A vast, gently east-sloping windswept, semiarid plateau located primarily in southern Argentina. It is located south of the Rio Negro east of the Andes and includes a 300,000-square-mile (776,996-km²) area that also includes parts of southeast Chile and northern Tierra del Fuego. It is located east of the Andes Mountain range and terminates in cliffs on the Atlantic Ocean. The region is known for sheep raising and tourism in its lake region but also has mineral wealth including petroleum production and deposits of coal and iron ore. Geologically, Patagonia is probably most famous for its rich paleontological record and has been explored for its diverse fossils since the region was visited by Charles Darwin. It is home to the legendary Patagonian Giants (the Tehuelches), later found to be not so giant by Spanish conquerors.

Early explorers of Patagonia include Amerigo Vespucci in 1501 and Ferdinand Magellan in 1520. In the late 1800s an Argentine general named Julio A. Roca led campaigns against the local Indians, leading the way for Argentine ranchers to move into the area. The area is also inhabited by guanaco, rhea, puma, and deer, including the pudu, the world's smallest deer species. Bird life is also common among Patagonia's forests and waters and ranges from the Andean condor, the world's largest land bird, to hummingbirds and parakeets. Some of the world's oldest trees stand in Parque Nacional Los Alerces.

Patagonia has several large rivers that flow through it to the Atlantic, and one river, the Manso, forms a natural pathway of water cutting across the barrier of the Andes to the Pacific Ocean. Along its shores, animals and travelers formed paths allowing for cultures to spread and cities to develop. The Manso headwaters lie high in the Cordillera Patagonica in Argentina and descend southeast into several lakes. The river turns and then cuts through a gorge to the west, passing through Lago Tagua-Tagua, then reaches the Pacific Ocean.

Charles Darwin described many details of the landscape of Patagonia. He noted that the surface is quite level (a peneplain) and is composed of a well-rounded mix of gravel and white soil, where scattered tufts of brown wiry grass have established a shaky foothold. Standing in the middle of the desert plains Darwin noted another, second and higher escarpment marking a higher plain, equally level and desolate.

The western side of Patagonia is marked by the Andean Mountain range, where subduction to the west has produced a series of volcanic-intrusive magmatic pulses. The southern Andes and Patagonia are host to a number of precious base metal deposits related to this magmatism. A lucrative belt of epithermal gold deposits and subvolcanic porphyry copper-gold deposits is found in southern Chile and Argentina. Deposits in Patagonia are associated with the Somuncura and Deseado volcanic massifs and include Anglo Gold's Cerro Vanguardia deposit (9.1 million tons of open-pit mineable ore, with 9.7 grams per ton gold and 113 grams per ton silver), and Meridian's Esquel deposit (3.8 million ounces gold at 8 grams per ton and 6.4 million ounces of silver).

pediment Desert surfaces that slope away from the base of a highland, generally with slopes of 11° or less. They have longitudinal profiles that are flat or slightly concave upward, and they range in size from less than 0.4 square miles (1 km²) to more than hundreds of square kilometers. Many are covered by a thin or discontinuous layer of alluvium and rock fragments. Pediments are erosional features, formed by running water, and are typically cut by shallow channels and dotted by erosional rock remnants known as inselbergs that stand above the pediment surface. Pediments grow as mountains are eroded, and they seem to represent a delicate balance between climate, erosion, time, deposition of alluvial fan material, and the ability of fluvial and other processes to transport material across the pediment surface. The boundary between deposited alluvial material and the pediment surface represents an equilibrium line between the erosional and depositional processes

Pediment surface, southern Sinai Peninsula, Egypt. Flat erosional surface on the dark-toned rocks in foreground slopes away from mountains in background. *(Photo by Timothy Kusky)*

and can move in response to changes in the system. Pediments were first described from the Henry Mountains of Utah by Grove Karl Gilbert (see biography) as "hills of planation."

See also GEOMORPHOLOGY.

pegmatite A very coarse-grained igneous rock that typically forms dikes, irregular lenses, or veins along the margins of batholiths and plutons. Most crystals in pegmatites are larger than 0.4 inches (1 cm) in diameter and form an interlocking network of grains. The majority of pegmatites have approximately granitic compositions, containing quartz, feldspars, and mica, but many pegmatites also contain rare minerals including gems and semiprecious minerals such as emeralds, rubies, sapphires, and tourmalines. They may also be enriched in rare elements including rare earth elements, lithium, uranium, boron, fluorine, niobium, tantalum, and beryllium. Pegmatites form in the late stages of crystallization of granitic plutons when the residual fluids are enriched in volatiles and there is significant interaction between the fluids from the pluton and hydrothermal fluids from the country rock.

See also GRANITE; PLUTON.

Pennsylvanian *See* CARBONIFEROUS.

Penobscottian orogeny The northern Appalachians have been divided into a number of different terranes or tectonostratigraphic zones, reflecting different origins and accretionary histories of different parts of the orogen. The Laurentian craton and autochthonous sedimentary sequences form the Humber zone, whereas fragments of peri-Gondwanan continents are preserved in the Avalon zone. The Dunnage terrane includes material accreted to the Laurentian and Gondwanan continents during closure of the Paleozoic Iapetus Ocean. Gander terrane rocks were initially deposited adjacent to the Avalonian margin of Gondwana. A piece of northwest Africa left behind during Atlantic rifting is preserved in the Meguma terrane. For many years a pre-Middle Ordovician (Taconic age) deformation event has been recognized from parts of central Maine, New Brunswick, and a few other scattered parts of the Appalachian orogen. This Late Cambrian–Early Ordovician event probably occurred between two exotic terranes in the Iapetus Ocean, before they collided with North America. Understanding the nature of this event, known as the Penobscottian orogeny, has become one of the more enigmatic features of the northern Appalachians. The overprinting of the original deformational fabrics by the subsequent Taconic orogeny, despite the outboard nature of the Penobscot, has led to numerous interpretations of this particular event.

Because the Penobscot orogenic deformation is less extensive than other orogenies in the Appalachians, interpretations have been based on information in limited areas, and thus a comprehensive model is difficult to propose. The deformation is confined to the Gander terrane in the Northern Appalachians and the Piedmont in the Southern Appalachians. The timing of this orogenic event can be best constrained in western Maine where an exposure of sialic basement, an ophiolite, and associated accretionary complexes crop out. It has also been suggested that the Brunswick subduction complex in New Brunswick, Canada, represents the remains of the Penobscottian orogeny. In western Maine, the major units include the Chain Lakes massif, interpreted to be Grenville age; the Boil Mountain ophiolite, interpreted to be a relict of oceanic crust obducted during accretion of the Boundary Mountains terrane to the previously existing Gander terrane; and the Hurricane Mountain mélange, expressing the flysch deposits of the accretionary prism during amalgamation. These rock units together record a portion of the complex deformational history of the closing of the Iapetus Ocean. The Boil Mountain ophiolite occurs in a structurally complex belt of ophiolitic slivers, exotic microterranes (e.g., Chain Lakes massif), and mélange along the boundary between the Dunnage and Gander terranes in central Maine. Its geological evolution is critical to understanding the Late Cambrian–Early Ordovician Penobscottian orogeny.

The Penobscot orogeny in the Northern Appalachians was initiated sometime during the Late Cambrian. Isotopic ages of the Boil Mountain ophiolite range from 500 million to 477 million years old. Fossil evidence further supports this Middle Cambrian to Early Ordovician age range. Sponges of this age have been found in the black shales of the Hurricane Mountain mélange constraining the maximum age for the mélange.

The Boil Mountain ophiolite and the associated Chain Lakes massif are instrumental in the interpretation of the Penobscottian orogeny. In order to better comprehend their relationship it is important to understand their lithologies and regional context. The Chain Lakes massif is characterized by (meta-) diamictite that is composed of mostly metasandstone, minor amphibolite, granofels, and gneiss. The structure has been interpreted to be an elongate dome, and an exposed thickness of 9,840 feet (3,000 m) is estimated. The unit is bounded by faults that strike northwest and comes in contact with several intrusive bodies, including the Attean batholith to the northeast and the Chain of Ponds pluton to the southwest. Along its eastern and western margin, Silurian and Devonian strata overlie the massif. The Boil Mountain ophiolite is in fault-bounded contact with the Chain Lakes massif on its southern and southeastern margin. Seismic reflection profiling has shown that the Chain Lakes massif is floored by a decollement dipping toward the southeast. This was interpreted to represent a thrust that originated to the southeast and occurred either during the Acadian or the Taconic orogenies.

The massif may be divided into eight facies based on structural aspects and lithology of the complex. The structurally lowest sequence is first divided into three facies: (1) the Twin Bridges semipelitic gneiss; (2) the Appleton epidior-

ite; and (3) the Barrett Brook polycyclic epidiorite breccia. The next four facies represent the principal diamictite sequence and include (1) the McKenney Pond chaotic rheomorphic granofels; (2) the Coburn Gore semipelitic gneissic granofels; (3) the Kibby Mountain flecky gneiss; and (4) the Sarampus Falls massive to layered granofels. The structurally highest facies is the Bag Pond Mountain bimodal metavolcanic section and feldspathic meta-arenite. The highest facies are interpreted to represent the return to a passive margin sequence after metamorphism and deformation. However, the structural and metamorphic history of the area is complicated, and it is difficult to be certain about stratigraphic relationships across structural boundaries.

The metamorphic history of the Chain Lakes massif is one of repeated deformation over a long span of time. The metamorphism of this complex spans a period of 800 million years and has been interpreted to record a pressure-temperature time path of prograde and retrograde events that end with the emplacement of the Late Ordovician batholith. U/Pb, Rb/Sr, and Nd/Sm ages of the Chain Lakes massif range from approximately 1,500 Ma to 684 Ma. The wide variance in the ages may be caused by deposition of material in the massif sometime between these ages as the deformational history was proceeding, or the zircons that produced the Precambrian dates must have been derived from a preexisting Precambrian unit that was eroding into the basin. The diamictite may have been deposited as a fanglomerate along the rifted margin of the Iapetus and was subsequently overridden by the Boil Mountain ophiolite complex.

The Boil Mountain ophiolite lies in fault-bounded contact with the Chain Lakes massif along its southern boundary. This complex extends about 20 miles (30 km) along strike and has a maximum exposure across strike of about four miles (6 km). Units typical of ophiolites such as serpentinite, pyroxenite, metagabbro, mafic volcanics, and sediments consisting of metaquartzwacke, metapelite, slate, and metaconglomerate characterize the Boil Mountain ophiolite. The stratigraphic units consist of, from bottom to top: an ultramafic unit, a gabbroic unit, a tonalitic unit, mafic and felsic volcanics, and metasedimentary units. Thus, the ophiolite has all the components of an ophiolite sequence except the tectonite ultramafic unit and the sheeted dike unit. The basal contact is difficult to differentiate in the large scale, but any contacts that are present suggest that ductile faulting accompanied their emplacement. However, some of the units, including the serpentinite, are in sharp structural contact with the Chain Lakes massif along its southern boundary.

The complex may be divided based on chemistry into several units, including ultramafics, gabbros, two mafic volcanic units, and felsic volcanics. A felsic unit separates the two mafic volcanic units, termed upper and lower. Trace element and rare earth element (REE) patterns reveal two distinct crystallization trends within the complex. The first magma to be erupted was the lower mafic volcanic unit. The trace element and REE trends suggest that the crystallization of olivine, clinopyroxene, and plagioclase resulted in this magma composition. This is accomplished by the formation of the ultramafics and gabbroic unit. The upper mafic unit has geochemical affinities to the Island Arc Tholeiites (IAT) zone while the upper mafic unit is similar to Mid-Ocean Ridge Basalts (MORB). Since there is such a large volume of felsic material associated with the ophiolite, it is likely that there were two phases of extrusion for the mafic volcanic unit. Because the presence of tonalites and other felsic volcanic rocks imply that there must be hydrous fluids, the first phase must include a subduction zone. The lower mafic unit and the felsics represent this phase. The upper mafic unit represents the volcanism at a marginal basin because the mantle sources for these units were not affected by the subducting slab.

The presence of tonalites in the Boil Mountain ophiolite, with small amounts of trondhjemite, is an unusual feature not found in "normal" ophiolite sequences. Other ophiolites that include a unit of tonalite are the Semail ophiolite in Oman and the Canyon Mountain ophiolite in California. Tonalites suggest the presence of fluids during crystallization and are instrumental in the interpretation of the ophiolites. The placement of these ophiolites with respect to the rest of the sequence is also of interest. The tonalite of the Boil Mountain complex appears in the sequence above the gabbros and below the volcanogenic units. It has the form of a sill, with abundant intrusive contacts evident in float and rare, poorly exposed outcrops in the northeast part of ophiolite. The tonalite was probably derived from partial melting of the lower mafic volcanics and gabbros, which then intruded as a sill.

The Boil Mountain ophiolite tonalite has yielded a U/Pb zircon age of 477±1 million years. The age of 477 million years places a minimum plutonic age for the tonalites of the Boil Mountain ophiolite, which is significantly less than any previously determined age associated with the ophiolite. A Late Cambrian to Early Ordovician age for the Boil Mountain ophiolite has been previously suggested based on several pieces of information. Felsic volcanics in the upper part of the ophiolite give a U-Pb zircon age of 500 Ma±10 million years. The age of 477 million years for the Boil Mountain tonalites is interpreted as a late-stage intrusive event, possibly related to partial melting of hydrated oceanic crust and the intrusion of a tonalitic extract.

A comparison of the age of the Boil Mountain ophiolite with nearby ophiolitic sequences in the Taconic allochthons of Quebec shows that ophiolite obduction was occurring on the Humber zone of the Appalachian margin of Laurentia, at similar times as the Boil Mountain ophiolite was being emplaced over the Chain Lakes massif, interpreted as a piece of the Gander margin of Gondwana. Hornblendes from the metamorphic sole to the Thetford Mines ophiolite have yielded $^{40}Ar/^{39}Ar$ ages of 477±5 million years, with an initial detachment age of 479±3 million years for the ophiolitic crust. Detachment of the circa 479-million-year-old sheet began at a ridge segment in a fore-arc environment, in contrast to an older, circa 491±11-million-year-old ($^{40}Ar/^{39}Ar$ from amphibole) oceanic slab preserved as the Pennington sheet in the Flintkote mine that is interpreted as a piece of oceanic crust originally attached to Laurentia. Thus, there may be a protracted history of Taconian ophiolite obduction in the Quebec Appalachians. In Gaspe a $^{40}Ar/^{39}Ar$ analysis on basal amphibolite tectonite gave an emplacement age of 456±3 million years for the Mount Albert ophiolite.

The age and origin of the tonalites of the Boil Mountain ophiolite, and their relationships to the Chain Lakes massif, have considerable bearing on the Penobscottian orogeny. First, since tonalitic intrusives are confined to the

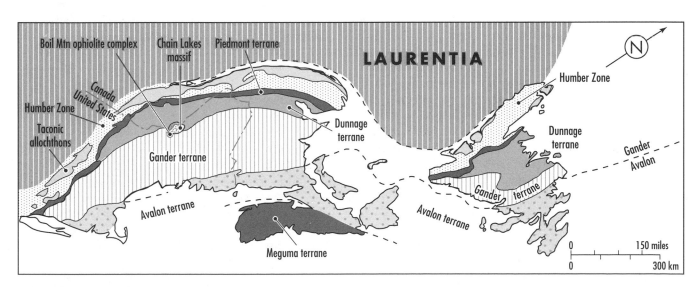

Simplified map of the Northern Appalachians showing the location of the Chain Lakes massif, Boil Mountain ophiolite complex, Dunnage and Gander terranes, and other important features

Pillow lavas of the Boil Mountain ophiolite, Maine *(Photo by Timothy Kusky)*

setting is compatible with the IAT characteristics of the lower volcanic unit and the MORB characteristics of the upper mafic volcanic unit of the ophiolite. Sedimentary rocks intercalated with the upper volcanic unit include iron formation, graywacke, phyllite, and chert, consistent with deposition in a back arc basin setting. The tonalites formed by hydrous melting of mafic crust of the lower volcanic unit, during initial stages of back arc basin evolution.

The Boil Mountain ophiolite and Chain Lakes massif thus reveal critical insights about the Penobscottian orogeny. The lower volcanic group of the Boil Mountain ophiolite are island arc tholeiites formed within an immature arc setting 500 million years ago. This phase of volcanism is similar to other Exploits subzone (Dunnage zone) ophiolites, such as the 493-million-year-old Pipestone Pond complex, the 489-million-year-old Coy Pond complex, the South Lake ophiolite of Newfoundland, and 493-million-year-old rhyolites associated with serpentinites of the Annidale area of southern New Brunswick. The Pipestone Pond and Coy Pond complexes both occur as allochthons overlying domal metamorphic cores of Gander zone rocks (Mount Cormack and Meelpaeg Inliers), in a structural arrangement reminiscent of the Boil Mountain complex resting on the margin of the Chain Lakes massif. These ophiolites, including the Boil Mountain ophiolite, probably represent part of the Penobscot arc, which developed in the forearc over a west-dipping subduction zone, near the Gander margin of Gondwana.

The 513-million-year-old Tally Pond volcanics of the Lake Ambrose volcanic belt of south-central Newfoundland may also be time-correlative with the lower volcanics of the Boil Mountain ophiolite; both preserve a mixed mafic-felsic volcanic section with arc related geochemical affinities, and both are interpreted as parts of the Penobscot-Exploit's arc accreted to Gondwana in the Early Ordovician. Thus, the Boil Mountain, Coy Pond, and Pipestone Pond ophiolites formed the fore-arc to the Penobscot-Exploit's arc, and the Lake Ambrose volcanic belt represents volcanism in more central parts of the arc. Obduction of Exploit's subzone forearc ophiolites over Gander zone rocks in Newfoundland and New Brunswick occurred in Tremadocian-Arenigian (490–475 million years ago), was followed by arc reversal in the Arenigian (475–465 million years ago), which formed a new arc (Popelogan arc in northern New Brunswick) and a back arc basin (Fournier Group ophiolites at 464 Ma) that widened rapidly to accommodate the collision of the Popelogan arc with the Notre Dame (Taconic) arc by 445 million years ago. This rifting event may have detached a fragment of Avalonian basement and Gander Zone sediments, with overlying allochthonous ophiolitic slabs, now preserved as the Chain Lakes massif and the lower sections of the Boil Mountain ophiolite.

A second episode of magma generation in the Boil Mountain ophiolite is represented by the upper tholeiitic volcanic unit and by the 477-million-year-old tonalite sill, generated by partial melting of the lower volcanic unit. This phase

allochthonous Boil Mountain ophiolite, it can be inferred that the ophiolite was not structurally emplaced in its final position over the Chain Lakes massif until after 477 million years ago (Arenigian). This does not, however, preclude earlier emplacement (previous to tonalite intrusion) of the ophiolite in a different structural position. Dates for the Taconic orogeny range from 491 million to 456 million years ago. Early ideas for the Penobscottian orogeny suggested that it took place prior to the Taconic orogeny. However, it seems more likely that these two orogenies were taking place at the same time, although not necessarily on the same margin of Iapetus. Geochemical data suggest a two-stage evolution for the Boil Mountain ophiolite, including an early phase of arc or forearc volcanism, followed by a tholeiitic phase of spreading in an intra-arc or back arc basin. The first phase includes felsic volcanics dated at 500±10 million years, and the second phase, associated with partial melting of the older oceanic crust and the formation of the upper volcanic sequence, occurred at 477±1 million years ago. This tectonic of magma generation is associated with development of the Tetagouche back arc basin behind the Popelogan arc during Late Arenigian-Llanvirnian. The timing of this event appears to be similar in southern Newfoundland, where several Ordovician granites, including the 477.6±1.8-million-year-old Baggs Hill granite, and the 474±3-million-year-old Partridgeberry Hills granite, intrude ophiolites obducted onto the Gondwanan margin during the early Ordovician.

Since the Boil Mountain tonalites are allochthonous and are not known to occur in the rocks beneath the ophiolite, final obduction of the Boil Mountain complex probably occurred after collision of the Popelogan and Notre Dame (Taconic) arcs, during post-450-million-year-old collision of the Gander margin of Avalon with the active margin of Laurentia. This event could be represented by the 445-million-year-old Attean pluton.

As an alternative model to that presented above, the entire Boil Mountain ophiolite complex may have formed in the fore-arc of the Popelogan arc, which formed after the Penobscot arc collided with the Avalonian margin of Gondwana. In this model, the Boil Mountain ophiolite would be correlative with other 480–475-million-year-old ophiolites of the Robert's Arm–Annieopsquotsch belt, that occur along the main Iapetus suture between the Notre Dame (Taconic) arc accreted to Laurentia, and the Penobscot-Exploit's arcs accreted to Gondwana. This model would help explain why no evidence of pre-477-million-year-old obduction-related fabrics have been documented from the Boil Mountain complex–Chain Lakes massif contact, but it does not adequately account for the Cambrian ages of the lower volcanic unit of the Boil Mountain ophiolite.

The Penobscottian orogeny has presented difficulties to geologists for quite some time. However, recent studies of the exposure in Maine, including the Chain Lakes unit and the Boil Mountain ophiolite, have led to new models for the tectonic evolution of this complex terrane. New isotopic dates and geochemical evidence show that the Taconian and Penobscottian orogenies were most likely simultaneous events. The models presented in this entry take into account several factors, including the formation of the Chain Lakes unit, the timing of emplacement of the ophiolite, and the timing of the intrusion of tonalites that are thought to represent the suture of the Gander to the Boundary Mountains terrane. However, further work needs to be done to truly constrain the timing and sequence of events that created the rocks of the Penobscottian orogeny.

See also APPALACHIANS; CALEDONIDES.

Further Reading

Boone, Gary M., David T. Doty, and Matt T. Heizler. "Hurricane Mountain Formation Mélange: Description and Tectonic Significance of a Penobscottian Accretionary Complex." In *Maine Geological Survey: Studies of Maine Geology.* Vol. 2, *Structure and Stratigraphy,* edited by Robert D. Tucker and Robert G. Marvinney, 33–83. Maine Geological Survey, 1989.

Boudette, Eugene L. "Ophiolite Assemblage of Early Paleozoic Age in Central Western Maine." In "Major Structural Zones and Faults of the Northern Appalachians," edited by Pierre St-Julien and J. Beland. *Geological Association of Canada* Special Paper 24 (1982): 209–230.

Dunning, Greg R., H. Scott Swinden, Baxter F. Kean, D. T. W. Evans, and G. A. Jenner. "A Cambrian Island Arc in Iapetus: Geochronology and Geochemistry of the Lake Ambrose Volcanic Belt, Newfoundland Appalachians." *Geological Magazine* 1 (1991): 1–17.

Kusky, Timothy M., James S. Chow, and Samuel A. Bowring. "Age and Origin of the Boil Mountain Ophiolite and Chain Lakes Massif, Maine: Implications for the Penobscottian Orogeny." *Canadian Journal of Earth Sciences* 34 (1997): 646–654.

Trzcienski, W. E., Jr., John Rodgers, and Charles V. Guidotti. "Alternative Hypothesis for the Chain Lakes 'Massif,' Maine and Quebec." *American Journal of Science* 292 (1992): 508–532.

Tucker, Robert D., S. J. O'Brien, and B. H. O'Brien. "Age and Implications of Early Ordovician (Arenig) Plutonism in the Type Area of the Bay du Nord Group, Dunnage Zone, Southern Newfoundland Appalachians." *Canadian Journal of Earth Sciences* 31 (1994): 351–357.

van Staal, Cees R., and Leslie R. Fyffe. "Dunnage Zone-New Brunswick." In "Geology of the Appalachian-Caledonian Orogen in Canada and Greenland," edited by H. Williams. *Geological Society of America,* The Geology of North America F-1 (1995): 166–178.

peridotite A coarse-grained ultramafic igneous plutonic rock characterized by abundant olivine, plus lesser quantities of orthopyroxene, clinopyroxene, amphibole, and micas, with little or no feldspar. The term *peridotite* is a general term for many narrowly defined ultramafic rock compositions including harzburgite, lherzolite, websterite, wehrlite, dunite, and pyroxenite. Peridotites are not common in the continental crust but are common in the lower cumulate section of ophiolites, in the mantle, and in continental layered intrusions and ultramafic dikes. Peridotites have unstable compositions under shallow crustal metamorphic conditions, and in the presence of shallow surface hydrating weathering conditions, and commonly become altered to serpentinites through the addition of water to the mineral structures.

See also IGNEOUS ROCKS.

permafrost Any soil, gravel, porous bedrock, or any part of the regolith that has remained below freezing temperatures for long periods of time, generally ranging from 2,000 to more than 10,000 years. It underlies about one-fifth of the world's continental land area and is found in arctic, subarctic, and alpine environments. Much of Canada, Siberia, Alaska, the Tibetan Plateau, and even submerged areas around the Arctic Ocean are underlain by permafrost. Most permafrost extends to depths between 12 inches and 3,281 feet (30 cm–1,000 m) and may be laterally continuous or discontinuous.

The upper surface of permafrost is known as the permafrost table and is typically overlain by an active layer consisting of up to a few meters of material that freezes

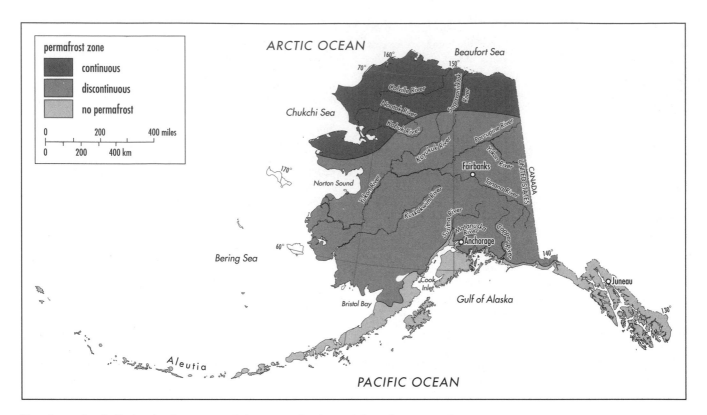

Map of permafrost in Alaska, showing permanently frozen ground and zone of discontinuous permafrost

and thaws on an annual basis. The thickness of the active layer is determined by soil type, latitude, and the presence of standing bodies of water that tend to limit the layer's thickness. The active layer in permafrost layers tends to become saturated with water since the meltwaters cannot percolate downward through the permafrost. This results in the layer becoming very weak, enhancing the formation of landslides, solifluction lobes, and other slip features on slopes.

The presence of permafrost in an area can make it extremely difficult to construct roads, pipelines, buildings, and other features that can change the temperature of the ground, melting the permafrost layer and resulting in the sinking or destruction of the artifact. This problem has been growing in recent decades with increasing populations moving into subarctic and arctic environments, and petroleum exploration in places including northern Alaska. Houses must be built on pillars that extend to bedrock, roads often cannot be paved, and pipelines must be built either above ground, or insulated and with special expansion/contraction joints. Many recent studies have focused on the influence of climate change and global warming on the distribution of permafrost and have noted that permafrost regions contain large amounts of trapped organic carbon that, if released to the atmosphere, could increase CO_2 levels, escalating global warming.

See also SOILS.

permeability A body's capacity to transmit fluids or to allow the fluids to move through its open pore spaces. Permeability is not directly related to porosity, because all the pore spaces in a body can be isolated (high porosity), but the water may be trapped and unable to move through the body (low permeability). Permeability is also affected by molecular attraction (the force that causes a thin film of water to stick to objects), instead of being forced to the ground by gravity. If the pore spaces in a material are very small, as in a clay, then the force of molecular attraction is strong enough to stop the water from flowing through the body. When the pores are large, the water in the center of the pores is free to move. The customary unit of measure of permeability is the millidarcy (.001 Darcy). One Darcy represents the passage through a porous medium of one cubic centimeter of fluid with one centipoise viscosity per second, flowing under one atmosphere of pressure, through a cross-sectional area of one square centimeter and a length of centimeter.

See also AQUIFER; POROSITY.

Permian The last period in the Paleozoic era, lasting from 290 million to 248 million years ago, and the corresponding system of rocks. Sir Roderick Murchison named it in 1841 after the Perm region of northern Russia where rocks of this age were first studied in detail. The supercontinent of Pangea included most of the planet's landmasses during the Permian, and this continental landmass extended from the South Pole, across the equator to high northern latitudes, with a wide Tethys Sea forming an open wedge of water near the equator. The Siberian continental block collided with Laurasia in the Permian, forming the Ural Mountains. Most of Pangea was influenced by hot and dry climate conditions and saw the formation of continental red-bed deposits and large-scale cross-bedded sandstones such as the Coconino sandstone of the southwestern United States, and the New Red sandstone of the United Kingdom. Ice sheets covered the south-polar region, amplifying already low sea levels so they fell below the continental shelves, causing widespread mass extinctions. The glaciations continued to grow in intensity through the Permian and, together with weathering of continental calc-silicates, were able to draw enough CO_2 out of the atmosphere to drastically lower global temperatures. This dramatic climate change enhanced the already widespread extinctions, killing off many species of corals, brachiopods, ammonoids, and foraminifers in one of history's greatest mass extinctions, in which about 70–90 percent of all marine invertebrate species perished, as did large numbers of land mammals.

See also MASS EXTINCTIONS; PANGEA; STRATIGRAPHY; SUPERCONTINENT CYCLE.

petrogenesis *See* PETROLOGY.

petrography The description and systematic classification of rocks. Identification of minerals and textures is typically done using a polarizing microscope. Petrography is largely a descriptive science, and it is different from petrology, which uses petrographic and other data to deduce the origin and history of rocks.

See also PETROLOGY.

petroleum A mixture of different types of hydrocarbons (fossil fuels) derived from the decomposed remains of plants and animals that are trapped in sediment and can be used as fuel. The petroleum group of hydrocarbons includes oil, natural gas, and gas condensate. When plants and animals are alive, they absorb energy from the Sun (directly through photosynthesis in plants, and indirectly through consumption in animals) to make complex organic molecules. After these plants and other organisms die, they may decay to produce hydrocarbons and other fossil fuels if the plant remains are buried before they completely decay.

Crude oil and natural gas may become concentrated in some regions and become mineable for use under some special conditions. First, for oil and gas to form, more organic matter must be produced than is destroyed by scavengers and organic decay, conditions that are met in relatively few places. One of the best places for oil and gas to form is on offshore continental shelves, passive margins, or carbonate platforms, where organic productivity is high, and the oxygen contents of bottom waters is low so organic decay is low and inadequate to destroy the amount of organic material produced. Organic material may also be buried before it decays in sufficient quantities to make petroleum in some deltaic and continental rise environments.

Once the organic material is buried it must reach a narrow window of specific pressure and temperature conditions to make petroleum. If these temperatures and pressures are not met, or are exceeded, petroleum will not form or will be destroyed. When organic rich rocks are in this petroleum window of specific temperature and pressure, organic rich beds known as source rocks become compacted and the organic material undergoes chemical reactions to form hydrocarbons including oil and gas. These fluids and gases have a lower density than surrounding rocks and a lower density than water, so they tend to migrate upward until they escape at the surface, or are trapped between impermeable layers where they may form a petroleum reservoir.

Oil traps are of many varieties, divided into mainly structural and stratigraphic types. Structural traps include anticlines, where the beds of rocks are folded into an upward arching dome. In these types of traps, petroleum in a permeable layer that is confined between impermeable layers (such as a sandstone bed between shale layers) may migrate up to the top of the anticlinal dome, where it becomes trapped. If a fault cuts across beds, it may form a barrier or it may act as a conduit for oil to escape along, depending on the physical properties of the rock in the fault zone. In many cases faults juxtapose an oil-bearing permeable unit against an impermeable horizon, forming a structural trap. Salt domes in many places form diapirs that pierce through oil-bearing stratigraphic horizons. They typically cause an upwarping of the rock beds around the dome, forming a sort of anticlinal trap that in many regions has yielded large volumes of petroleum. Stratigraphic traps are found mainly where two impermeable layers such as shales are found above and below a lens-shaped sandstone unit that pinches out laterally, forming a wedge-shaped trap. These conditions are commonly met along passive margins, where transgressions and regressions of the sea cause sand and mud facies to migrate laterally. When combined with continuous subsidence, passive margin sequences typically develop many sandstone wedges caught between shale layers. River systems and sandstone channels in muddy overbank delta deposits also form good trap and reservoir systems, since the porous sandstone channels are trapped between impermeable shales.

Most of the world's industrialized nations get the majority of their energy needs from petroleum and other fossil fuels, so exploration for and exploitation of petroleum is a major national and industrial endeavor. Huge resources are spent in petroleum exploration, and thousands of geologists are employed in the oil industry. In the early days of exploration the oil industry gained a reputation of being environmentally degrading, but increased regulations and awareness by these companies has greatly alleviated these problems, and most petroleum is now explored for and extracted with minimal environmental consequences. The burning of fossil fuels, however, continues to release huge amounts of carbon dioxide and other chemicals into the atmosphere, contributing to global warming.

See also HYDROCARBON; PASSIVE MARGIN.

petrology The branch of geology that attempts to describe and understand the origin, occurrence, structure, and evolution of rocks. It is similar to but an extension of petrography that describes the minerals and textures in rock bodies. The petrologic classification of rocks recognizes three main categories with different modes of origin and histories. Igneous rocks crystallized from magma, and include plutonic and volcanic varieties that cooled below and at the surface, respectively. Metamorphic rocks are those that have been changed in some way, such as the growth of new minerals or structures, during heating and pressure from being subjected to tectonic forces. Sedimentary rocks include clastic varieties that represent the broken down, transported, deposited, and cemented fragments of older rocks, as well as chemical and biochemical varieties that represent chemicals that precipitated from a solution.

See also IGNEOUS ROCKS; PETROGRAPHY; SEDIMENTARY ROCKS.

Pettijohn, Francis John (1904–1999) *American Sedimentologist, Field Geologist* Francis Pettijohn was born in Waterford, Wisconsin, on June 20, 1904, and is widely known as the "Father of Modern Sedimentology." He graduated from high school in Indianapolis in 1921, then entered the University of Minnesota where he received his B.A. in 1924 and his M.A. in 1925. In 1927 he entered graduate school at the University of California at Berkeley and then transferred to the University of Minnesota, where he received his Ph.D. in 1930.

Pettijohn is most famous for his studies of the sedimentology of the rocks in the Appalachian Mountains, and for the 24 books that he authored or coauthored. Perhaps his most famous book is *Sedimentary Rocks,* which has been reprinted many times since its first publication in 1949 and remained a standard in the field for more than 50 years. Francis Pettijohn published his own autobiography in 1984, a humorous and anecdotal work entitled *Memoirs of an Unrepentant Field Geologist.*

Pettijohn received numerous awards for his work, including the Sorby Medal of the International Association of Sedimentologists in 1983, the Twenhofel Medal from the Society of Economic Paleontologists and Mineralogists, the Wollaston Medal from the Geological Society, the Penrose Medal from the Geological Society of America, the Francis J. Pettijohn Medal from the Society for Sedimentary Geology, and an Honorary Doctorate of Science from the University of Minnesota. He was active in professional life, serving as president of the Society of Economic Paleontologists and Mineralogists

and councilor of the Geological Society of America. Pettijohn was elected a member of the National Academy of Sciences and a Fellow of the American Academy of Arts and Sciences.

Further Reading
Pettijohn, Francis J. "In Defense of Outdoor Geology." *Bulletin of the American Association of Petroleum Geologists* 40 (1956): 1,455–1,461.
———. *Sedimentary Rocks.* New York: Harper and Brothers, 1949.
Pettijohn, Francis J., and Paul E. Potter. *Atlas and Glossary of Primary Sedimentary Structures.* New York: Springer-Verlag, 1964.

Phanerozoic The eon of geological time since the base of the Cambrian at 544 million years ago and extending to the present. Introduced by George H. Chadwick in 1930, the eon is characterized by the appearance of abundant visible life in the geological record, in contrast to the earlier eon referred to by Chadwick as the Cryptozoic but now generally referred to as the Precambrian. Although paleontologists now recognize that many forms of life existed prior to the Phanerozoic, the first appearance of shelly fossils corresponds to the base of the Phanerozoic.

Phanerozoic time is divided mainly on the basis of fossil correlations and absolute ages, for the divisions are continuously being revised based on new geochronological studies of important fossil-bearing units. It is divided into three main fundamental time divisions—eras—the Paleozoic, Mesozoic, and Cenozoic. These eras are in turn divided into smaller divisions known as periods, epochs, and ages.

See also CENOZOIC; MESOZOIC; PALEOZOIC.

photosynthesis The process in green plants of trapping solar energy and using this energy to drive a series of chemical reactions that results in the production of carbohydrates such as sugar or glucose. These carbohydrates then form the basic food for the plant, and for insects and animals that eat these plants. Photosynthesis is therefore one of the most important processes for life on Earth. In order for photosynthesis to occur, the plant must contain chloroplasts with the green pigment chlorophyll and have supplies of carbon dioxide and water. One of the products of photosynthesis is oxygen, also necessary for most life to exist on the planet. Before simple single-celled organisms developed photosynthesis in the Precambrian, there was probably very little oxygen in the atmosphere. Therefore, the process has also been responsible for changing the conditions on the planet's surface to become more hospitable for life to develop into forms that are more complex, and for conditions to evolve so that they are suitable for humans.

pillow lavas A morphological type of lava that forms under water, typically in mafic lava flows. They have a wide variety of shapes and sizes, ranging from simple pillow shapes that drape over underlying pillows, to long complex tubes that branch and splay. Overlying pillows tend to fill in

2.7-billion-year-old pillow lavas from the Yellowknife greenstone belt, Nunavut (northern Canada) *(Photo by Timothy Kusky)*

any depressions that develop on or between underlying pillows, so typically have an apical region that points downward, and an upward-pointing convex surface. Most pillows are several inches to three–six feet (a few tens of cm to 1–2 m) in cross section and may be similar sized or larger along their long axes. A thin fine-grained chill margin that is glassy in young pillows develops as the lava next to the seawater cools quickly, with coarser grained parts forming in the pillow center where the magma cooled more slowly. Pillow lavas form in many tectonic settings, the most abundant of which is in the upper volcanic layer of oceanic crust. Piles of pillows in oceanic crust and in pieces of oceanic crust thrust onto land (ophiolites) may be up to a few miles (2–3 km) thick. Pillow lavas are also common in submarine sections of island arc systems, hot spot volcanoes such as Hawaii, and in other submarine volcanic settings.

placer A surficial mineral deposit formed by the mechanical concentration of heavy minerals eroded from weathered rock masses by currents in streams, along beaches, or by wind. Many types of valuable minerals including gold, platinum, cassiterite, ilmenite, zircon, rutile, sapphire, ruby, and diamond are considerably denser than the average sand or sediment in an area. When the air or water current moves these sediments, the heavy minerals tend to be concentrated by several processes. Denser grains tend to become trapped in riffles, cracks, and in areas of low flow velocity. Less dense material also may be winnowed away and removed by the current, concentrating the heavy minerals. For the placer minerals to be concentrated, they must also be resistant to chemical weathering, mechanical abrasion, and fragmentation during transport. Famous alluvial placer deposits include the California and Klondike alluvial gold deposits of California, Alaska, and the Yukon, the sites of the famous gold rushes of the mid and late 1800s. The largest gold placer deposit in the world is an ancient, 2.6–2.7-billion-year-old paleoplacer system found in the Witwatersrand basin of South Africa.

It has accounted for nearly half of the world's production of gold. Beach placers include the diamonds found in ancient offshore beach deposits off southwest Africa, gold in beach ridges near Nome, Alaska, and Ti-placers that have concentrated ilmenite along the southeast shores of Madagascar. An additional class of placers includes colluvial deposits, in which weathered material accumulates on a slope but less dense, more weatherable material is preferentially removed downslope, so the soil profile tends to become concentrated in the heavy weather-resistant minerals.

Placer mining is one of the oldest forms of mining and involves the further concentration of economical minerals by using running water to remove the less dense material. It has been responsible for the production of much of the world's gold, tin, titanium, platinum, diamonds, rubies, emeralds, and sapphires.

Further Reading

Antrobus, Edmund S. A., ed., *Witwatersrand Gold–100 Years, A Review of the Discovery and Development of the Witwatersrand Goldfield as Seen from the Geological Viewpoint*. Johannesburg: Geological Society of South Africa, 1986.
Jensen, Mead L., and Alan M. Bateman. *Economic Mineral Deposits*, 3rd ed. New York: John Wiley, 1981.

plankton Aquatic organisms that float, drift freely, or swim lightly. These include a large variety of species in the marine realm, including bacteria, phytoplankton (one-celled plants), and zooplankton that are animals such as jellyfish, invertebrates, as well as numerous non-marine aquatic species. Planktonic species are contrasted with nektonic organisms, which are strong swimmers, and benthic organisms that are bottom-dwellers.

Planktonic species tend to be small and lack strong skeletons, and they utilize the density of surrounding water to support their dominantly water-filled bodies. Many types sink or float to specific depths or levels where light and salinity characteristics meet their needs. They move vertically by changing the amount of air in their bodies, thus getting the nutrients they require, and avoiding becoming food for predators. Other plankton utilize their transparency to avoid being eaten or live in large schools of similar organisms.

Phytoplankton are dominantly microscopic one-celled floating plants that form an extremely important part of the biomass and food chain. Diatoms are the most abundant, secreting walls of silica, and dinoflagellates exhibit characteristics of both plants and animals. Coccolithopores are one-celled floating plants covered with an armor of small calcareous plates, whereas silicoflagellates are similar but have plates made of silica.

Zooplankton have a wide range of temperature and salinity tolerances and include a huge variety of species. Holoplankton are those that remain free-floating throughout their life stages, whereas meroplankton include the larval stages of dominantly benthic organisms. Holoplankton include the extremely important foraminifera and radiolaria that live throughout the oceans, producing calcium carbonate and silica tests, respectively. The Crustacean (insects of the sea) include arthropods with stiff chitinous outer shells. They include cocopods and euphausiids (krill) that form the dominant food for many fish and whale species. Meroplankton are common in coastal waters and include the eggs and sperm of many benthic animals and fish larvae and tend to be free-floating for a few weeks.

Gelatinous plankton such as jellyfish include the siphonophores that paralyze prey with stinging cells made of barbs attached to poison sacs. The siphonophores are colonies of animals that live together but function as a single animal. Ctenophores resemble jellyfish and have trailing tentacles, used to trap prey. They are carnivorous and may occur in large swarms, greatly reducing local populations of crustaceans and small or young fish. Tunicates are primitive planktonic creatures with backbones inside a barrel-shaped gelatinous structure.

plate tectonics   The study of the large-scale evolution of the lithosphere of the Earth. In the 1960s the Earth Sciences experienced a scientific revolution, when the paradigm of plate tectonics was formulated from a number of previous hypotheses that attempted to explain different aspects about the evolution of continents, oceans, and mountain belts. New plate material is created at mid-ocean ridges and destroyed when it sinks back into the mantle in deep-sea trenches. Scientists had known for some time that the Earth is divided into many layers defined mostly by chemical characteristics, including the inner core, outer core, mantle, and crust. The plate tectonic paradigm led to the understanding that the Earth is also divided mechanically and includes a rigid outer layer, called the lithosphere, sitting upon a very weak layer containing a small amount of partial melt of peridotite, termed the asthenosphere. The lithosphere is about 78 miles (125 km) thick under continents, and 47 miles (75 km) thick under oceans, whereas the asthenosphere extends to about 155 miles (250 km) depth.

Plate tectonics has been a unifying science, bringing together diverse fields such as structural geology, geophysics, sedimentology and stratigraphy, paleontology, geochronology, and geomorphology, especially with respect to active tectonics (also known as neotectonics). Plate motion almost always involves the melting of rocks, so other fields are also important, including igneous petrology, metamorphic petrology, and geochemistry (including isotope geochemistry).

The base of the crust, known as the Mohorovicic discontinuity, is defined seismically and reflects the difference

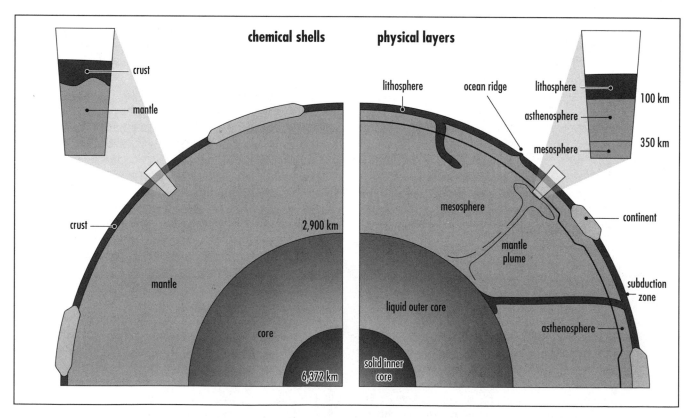

Cross sections of the Earth showing chemical shells (crust, mantle, and core) and physical layers (lithosphere, asthenosphere, mesosphere, outer core, inner core)

in seismic velocities of basalt and peridotite. However, the base of the lithosphere is defined rheologically as where the same rock type on either side begins to melt, and it corresponds roughly to the 2,425°F (1,330°C) isotherm. The main rock types of interest to tectonics include granodiorite, basalt, and peridotite. The average continental crustal composition is equivalent to granodiorite (the density of granodiorite is 2.6 g/cm^3; its mineralogy includes quartz, plagioclase, biotite, and some potassium feldspar). The average oceanic crustal composition is equivalent to that of basalt (the density of basalt is 3.0 g/cm^3; its mineralogy includes plagioclase, clinopyroxene, and olivine). The average upper mantle composition is equivalent to peridotite (the density of peridotite is 3.3 g/cm^3; its mineralogy includes olivine, clinopyroxene, and orthopyroxene). Considering the densities of these rock types, the crust can be thought of as floating on the mantle; rheologically, the lithosphere floats on the asthenosphere.

The plate tectonic paradigm states that the Earth's outer shell, or lithosphere, is broken into 12 large and about 20

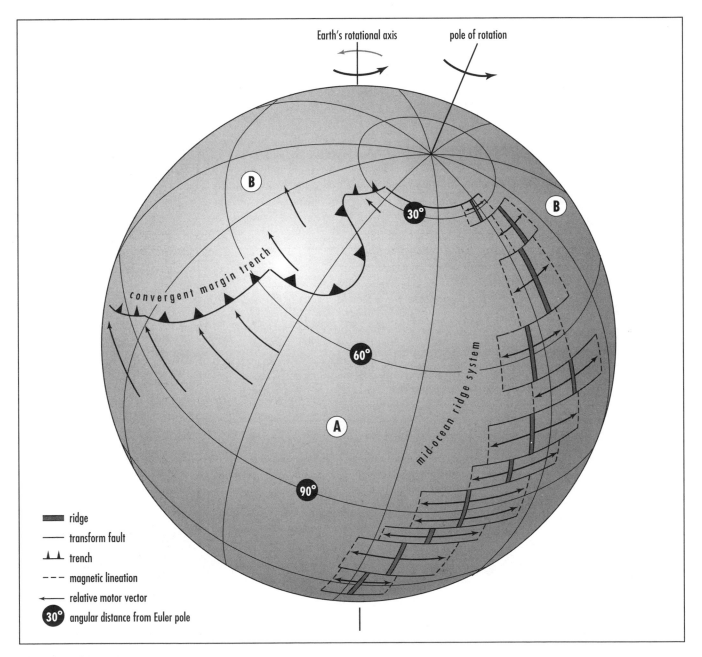

Pole of rotation on a sphere. Plate A rotates away from Plate B, with ridge axes falling on great circles intersecting at the pole of rotation, and oceanic transform faults falling along small circles that are concentric about the pole of rotation. The angular velocity of the plates increases with increasing distance from the pole of rotation.

smaller rigid blocks, called plates, that are all moving with respect to each other. The plates are rigid and do not deform internally when they move, but only deform along their edges. The edges of plates are therefore where most mountain ranges are located and are where most of the world's earthquakes occur and active volcanoes are located. The plates are moving as a response to heating of the mantle by radioactive decay, and are somewhat analogous to lumps floating on a pot of boiling stew.

The movement of plates on the spherical Earth can be described by a rotation about a pole of rotation, using a theorem first described by Euler in 1776. Euler's theorem states that any movement of a spherical plate over a spherical surface can be described by a rotation about an axis that passes through the center of the sphere. The place where the axis of rotation passes through the surface of the Earth is referred to as the pole of rotation. The pole of rotation can be thought of analogous to a pair of scissors opening and closing. The motions of one side of the scissors can be described as a rotation of the other side about the pin in a pair of scissors, either opening or closing the blades of the scissors. The motion of plates about a pole of rotation is described using an angular velocity. As the plates rotate, locations near the pole of rotation experience low angular velocities, whereas points on the same plates that are far from the pole of rotation experience much greater angular velocities. Therefore, oceanic spreading rates or convergence rates along subduction zones may vary greatly along a single plate boundary. This type of relationship is similar to a marching band going around a corner. The musicians near the corner have to march in place and pivot (acting as a pole of rotation) while the musicians on the outside of the corner need to march quickly to keep the lines in the band formation straight as they go around the corner.

Rotations of plates on the Earth lead to some interesting geometrical consequences for plate tectonics. We find that mid-ocean ridges are oriented so that the ridge axes all point toward the pole of rotation and are aligned on great circles about the pole of rotation. Transform faults lie on small circles that are concentric about the pole of rotation. In contrast, convergent boundaries may lie at any angle with respect to poles of rotation.

Since plates do not deform internally, all the action happens along their edges, and we can define three fundamental types of plate boundaries. Divergent boundaries are where two plates move apart, creating a void that typically becomes filled by new oceanic crust that wells up to fill the progressively opening hole. Convergent boundaries are where two plates

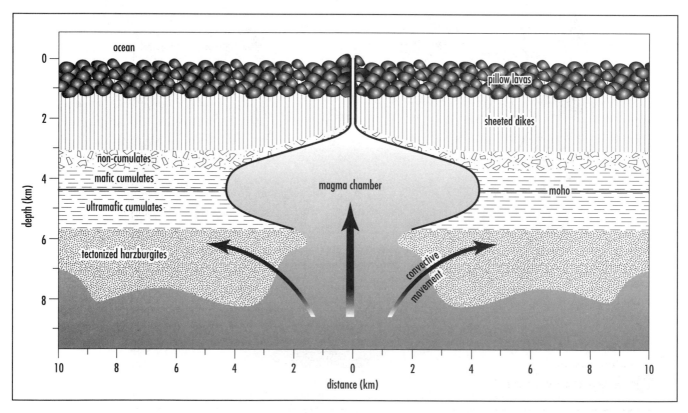

Formation of oceanic crust at mid-ocean ridges. Magma forms by partial melting in the asthenosphere and upwells to form a magma chamber beneath the ridge axis. As the plates move apart, dikes intrude upward from the magma chamber and feed lava flows on the surface. Heavy crystals settle out of the magma chamber and form layers of crystal cumulates on the magma chamber floor.

move toward each other, resulting in one plate sliding beneath the other (when a dense oceanic plate is involved), or collision and deformation (when continental plates are involved). Transform boundaries form where two plates slide past each other, such as along the San Andreas Fault in California.

Since all plates are moving with respect to each other, the surface of the Earth is made up of a mosaic of various plate boundaries, and the geologist has an amazing diversity of different geological environments to study. Every time one plate moves, the others must move to accommodate this motion, creating a never-ending saga of different plate configurations.

Divergent Plate Boundaries and the Creation of Oceanic Crust

Where plates diverge, new oceanic crust is produced by seafloor spreading. As the plates move apart, the pressure on deep underlying rocks is lowered, which causes them to rise and partially melt by 15–25 percent. Basaltic magma is produced by a partial melting of the peridotitic mantle, leaving a "residue" type of rock in the mantle known as harzburgite. The magma produced in this way upwells from deep within the mantle to fill the gap opened by the diverging plates. This magma forms a chamber of molten or partially molten rock that slowly crystallizes to form a coarse-grained igneous rock known as gabbro, which has the same composition as basalt. Before crystallization, some of the magma moves up to the surface through a series of dikes and forms the crustal sheeted dike complex, and basaltic flows. Many of the basaltic flows have distinctive forms with the magma forming bulbous lobes known as pillow lavas. Lava tubes are also common, as are fragmented pillows formed by the implosion of the lava tubes and pillows. Back in the magma chamber, other crystals grow in the gabbroic magma, including olivine and pyroxene, which are heavier than the magma and sink to the bottom of the chamber. These crystals form layers of dense minerals known as cumulates. Beneath the cumulates, the mantle material from which the magma was derived gets progressively more deformed as the plates diverge and form a highly deformed ultramafic rock known as a harzburgite or mantle tectonite. This process can be seen on the surface in Iceland along the Reykjanes Ridge.

TRANSFORM PLATE BOUNDARIES AND TRANSFORM FAULTS

In many places in the oceanic basins, the mid-ocean ridges are apparently offset along great escarpments or faults, which fragment the oceanic crust into many different segments. In 1965 J. Tuzo Wilson correctly interpreted these not as offsets but as a new class of faults, known as transform faults. The actual sense of displacement on these faults is opposite to the apparent "offset," so the offset is apparent, not real. It is a primary feature of Wilson's model, proven correct by earthquake studies.

These transform faults are steps in the plate boundary where one plate is sliding past the other plate. Transform faults are also found on some continents, with the most famous examples being the San Andreas Fault, the Dead Sea Transform, North Anatolian Fault and Alpine Fault of New Zealand. All of these are large strike-slip faults with horizontal displacements and separate two different plates.

Convergent Plate Boundaries

Oceanic lithosphere is being destroyed by sinking back into the mantle at the deep ocean trenches in a process called subduction. As the oceanic slabs sink downward, they experience higher temperatures that cause water and other volatiles to be released, causing melts to be generated in the mantle wedge overlying the subducting slab. These melts then move upward to intrude the overlying plate, where the magma may become contaminated by melting through and incorporating minerals and elements from the overlying crust. Since subduction zones are long narrow zones where large plates are being subducted into the mantle, the melting produces a long line of volcanoes above the down-going plate. These volcanoes form a volcanic arc, either on a continent or over an oceanic plate, dependent on which type of crust composes the overlying plate.

Island arcs are extremely important for understanding the origin of the continental crust because the magmas and sediments produced here have the same composition as the average continental crust. A simple model for the origin of the continental crust is that it represents a bunch of island arcs which formed at different times and which collided during plate collisions.

Since the plates are in constant motion, island arcs, continents, and other terranes often collide with each other. Mountain belts or orogens typically mark the places where lithospheric plates have collided, and the zone that they collided along is referred to as a suture. Suture zones are very complex and include folded and faulted sequences of rocks that form on the two colliding terranes and in any intervening ocean basin. Often, slices of the old ocean floor are caught in these collision zones (we call these ophiolites), and the process by which they are emplaced over the continents is called obduction (opposite of subduction).

In some cases, subduction brings two continental plates together, which collide forming huge mountain belts like the Himalaya Mountain chain. In continent-continent collisions, deformation may be very diffuse and extend beyond the normal limit of plate boundary deformation that characterizes other types of plate interactions. For instance, the India-Asia collision has formed the huge uplifted Tibetan Plateau, a series of mountain ranges to the north including the Tien Shan and Karakoram, and deformation of the continents extends as far into Asia as Lake Baikal.

Historical Development of the Plate Tectonic Paradigm

The plate tectonic paradigm was developed from a number of different models, ideas, and observations that were advanced over the prior century by a number of scientists on different

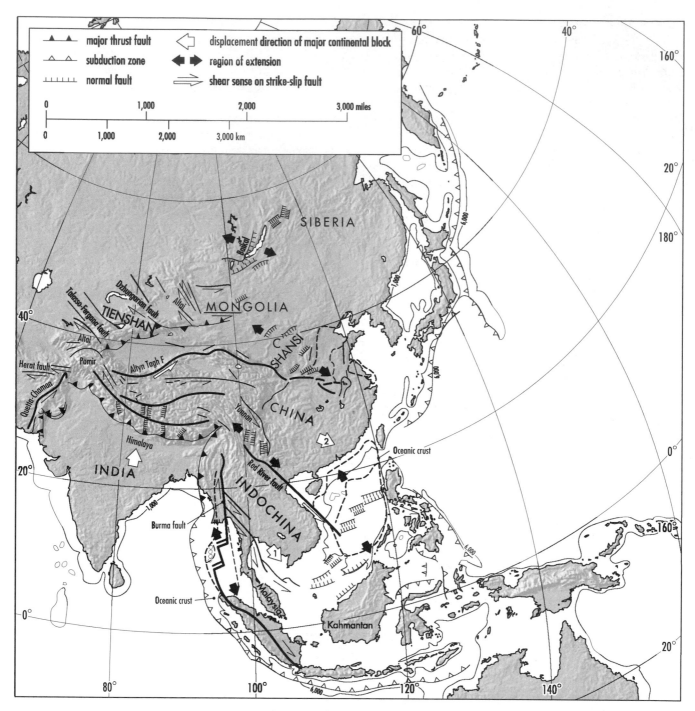

Map of central and eastern Asia collision showing the wide area affected by the collision of India with Asia

continents. Between 1912 and 1925, Alfred Wegener, a meteorologist, published a series of papers and books outlining his ideas for the evolution of continents and oceans. Wegener was an early proponent of continental drift. He looked for a driving mechanism to move continents through the mantle, and invoked an imaginary force (which he called *Pohlfluicht*) that he proposed caused the plates to drift toward the equator because of the rotation of the Earth. Geophysicists were able to show that this force was unrealistic, and since Wegener's idea of continental drift lacked a driving mechanism, it was largely disregarded.

In 1929, Arthur Holmes proposed that the Earth produces heat by radioactive decay and that there are not enough volcanoes to remove all this heat. He proposed that a combination of volcanic heat loss and mantle convection can lose the heat, and that the mantle convection drives continental drift. Holmes wrote a widely used textbook on this subject, which became widely respected. Holmes proposed that the upwelling convection cells were in the ocean basins, and that downwelling areas could be found under Andean-type volcano chains.

Alex du Toit was a South African geologist who worked on Gondwana stratigraphy and published a series of important papers between 1920 and 1940. Du Toit compared stratigraphic sections on the various landmasses that he thought were once connected to form the supercontinent of Gondwana (Africa, South America, Australia, India, Antarctica, Arabia). He showed that the stratigraphic columns of these places were very similar for the periods he proposed the continents were linked, supporting his ideas of an older large linked landmass. Du Toit also was able to show that the floral distributions had belts that matched when the continents were reconstructed but appeared disjointed in the continents' present distribution.

In the 1950s paleomagnetism began developing as a science. The Earth has a dipolar magnetic field, with magnetic field lines plunging into and out of the Earth at the north and south magnetic poles. Field or flux lines are parallel or inclined to the surface at intermediate locations, and the magnetic field can be defined by the inclination of the field lines

and their deviation from true north (declination) at any location. When igneous rocks solidify, they pass through a temperature at which any magnetic minerals will preserve the ambient magnetic field at that time. In this way, some rocks acquire a magnetism when they solidify. The best rocks for preserving the ambient magnetic field are basalts, which contain 1–2 percent magnetite; they acquire a remnant inclination and declination as they crystallize. Other rocks, including sedimentary "red beds" with iron oxide and hematite cements, shales, limestones, and plutonic rocks also may preserve the magnetic field, but they are plagued with their own sets of problems for interpretation.

In the late 1950s Stanley K. Runcorn and Earl Irving first worked out the paleomagnetism of European rocks and discovered a phenomenon they called apparent polar wandering (APW). The Tertiary rocks showed very little deviation from the present poles, but rocks older than Tertiary showed a progressive deviation from the expected results. They initially interpreted this to mean that the magnetic poles were wandering around the planet, and the paleomagnetic rock record was reflecting this wandering. Runcorn and Irving made an APW path for Europe by plotting apparent position of the poles, while holding Europe stationary. However, they found that they could also interpret their results to mean that

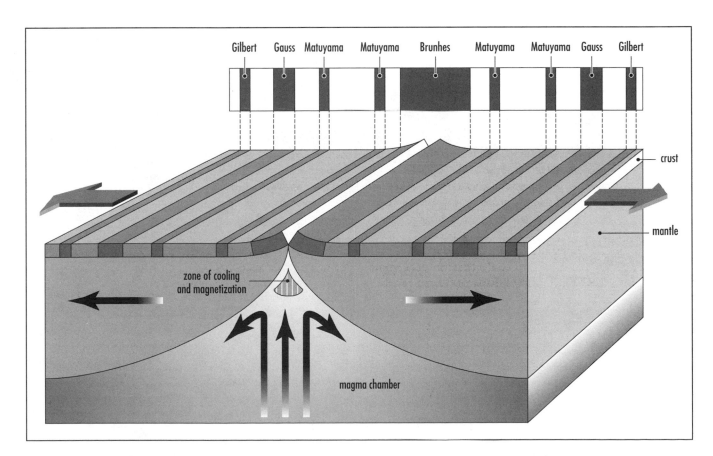

Symmetric magnetic anomalies produced by conveyor-belt style production of oceanic crust in an alternating magnetic field

the poles were stationary and the continents were drifting around the globe. Additionally, they found that their results agreed with some previously hard to interpret paleolatitude indicators from the stratigraphy.

Next, Runcorn and Irving determined the APW curve for North America. They found it similar to Europe from the late Paleozoic to the Cretaceous, implying that the two continents were connected for that time period, and moved together, and later (in the Cretaceous) separated, as the APW curves diverged. This remarkable data set converted Runcorn from a strong disbeliever of continental drift into a drifter.

In 1954 Hugo Benioff, a seismologist, studied worldwide, deep-focus earthquakes, to about 435-mile (700-km) depth. He plotted earthquakes on cross sections of island arcs and found earthquake foci were concentrated in a narrow zone to about 700 km depth. He noted that volcanoes of the island arc systems were located about 62 miles (100 km) above this zone. He also noted compression in island arc geology and proposed that island arcs are overthrusting oceanic crust. Geologists now recognize that this narrow zone of seismicity is the plate boundary between the subducting oceanic crust and the overriding island arc and have named this area the Benioff Zone.

Development of technologies associated with World War II led to remarkable advancements in understanding some basic properties of the ocean basins. In the 1950s the ocean basin bathymetry, gravity, and magnetic fields were mapped for the U.S. Navy submarine fleet. After this, research vessels from oceanographic institutes such as Scripps, Woods Hole, and Lamont-Doherty Geological Observatory studied the immense sets of oceanographic data. In the 1950s raw data was acquired, and the extent of the mid-ocean ridge system was recognized and documented by geologists including Bruce Heezen, Maurice Ewing, and Harry Hess. They also documented the thickness of the sedimentary cover overlying igneous basement and showed that the sedimentary veneer is thin along the ridge system and thickens away from the ridges. Walter Pitman happened to cross the oceanic ridge in the South Pacific perpendicular to the ridge and noticed the symmetry of the magnetic anomalies on either side of the ridge. In 1962 Harry Hess from Princeton proposed that the mid-ocean ridges were the site of seafloor spreading and the creation of new oceanic crust, and that Benioff Zones were sites where oceanic crust was returned to the mantle. In 1963 Frederick John Vine and D. H. Matthews combined Hess's idea and magnetic anomaly symmetry with the concept of

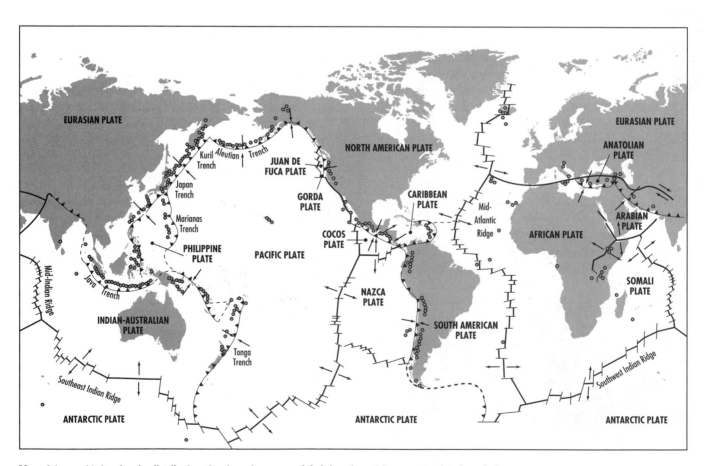

Map of the world showing the distribution of major volcanoes and their location with respect to plate boundaries

geomagnetic reversals. They suggested that the symmetry of the magnetic field on either side of the ridge could be explained by conveyor-belt style formation of oceanic crust forming and crystallizing in an alternating magnetic field, such that the basalts of similar ages on either side of the ridge would preserve the same magnetic field properties. Their model was based on earlier discoveries by a Japanese scientist, Matuyama, who in 1910 discovered recent basalts in Japan that were magnetized in a reversed field and proposed that the magnetic field of the Earth experiences reversals. Allan Cox (Stanford University) had constructed a geomagnetic reversal timescale in 1962, so it was possible to correlate the reversals with specific time periods and deduce the rate of seafloor spreading.

With additional mapping of the seafloor and the mid-ocean ridge system, the abundance of fracture zones on the seafloor became apparent with mapping of magnetic anomalies. In 1965 J. Tuzo Wilson wrote a classic paper, "A New Class of Faults and Their Bearing on Continental Drift," published in *Nature*. This paper connected previous ideas, noted the real sense of offset of transform faults, and represented the final piece in the first basic understanding of the kinematics or motions of the plates. Wilson's model was proved about one year later by Lynn Sykes and other seismologists, who used earthquake studies of the mid-ocean ridges. They noted that the ridge system divided the Earth into areas of few earthquakes, and that 95 percent of the earthquakes occur in narrow belts. They interpreted these belts of earthquakes to define the edges of the plates. They showed that about 12 major plates are all in relative motion to each other. Sykes and others confirmed Wilson's model, and showed that transform faults are a necessary consequence of spreading and subduction on a sphere.

See also CONVERGENT PLATE MARGIN PROCESSES; DIVERGENT OR EXTENSIONAL BOUNDARIES; TRANSFORM PLATE MARGIN PROCESSES.

Further Reading

Moores, Eldridge M., and Robert Twiss. *Tectonics*. New York: W.H. Freeman, 1995.

platform *See* CARBONATE; PASSIVE MARGIN; REEF.

playa Dry lake beds in low-lying flat areas of desert environments that may only have water in them once every few years are known as playas or hardpans in the United States, sabkhas in Africa and the Middle East, and salinas in South America. When there is water in these basins, they are known as playa lakes. Playas may form 5–10 percent of desert basin areas in many mountainous desert regions of the world, where water may drain from mountains and occasionally fill intermountain basins with water that quickly evaporates, leaving salts, silts, and evaporite minerals behind. Playas typically have deposits of white salts mixed with lake bottom silt and clay that form when water from storms flows into the dry lake and then evaporates, leaving the lakes dry. Coarser grained material is deposited between the mountains and the playa in large alluvial fan systems in which the grain size of clasts increases toward the mountains. Sedimentation rates vary from a few tenths of an inch to three feet (2–3 cm–1 m) or more per year, depending on rainfall, erosion rates, and tectonic subsidence rates. Playas tend to become more level with time since water and sediment flows into the center of the basin and preferentially fills in the lowest parts of the basin as the water evaporates, erasing topographic variations.

There are more than 100 playas in the American southwest, including Lake Bonneville, which formed during the last ice age and now covers parts of Utah, Nevada, and Idaho. Large playas are also visible in parts of Death Valley, California, and some of these are fed by springs that keep water in them most of the time. Playas form the flattest geomorphological surfaces on Earth and make excellent racetracks and runways. The U.S. Space Shuttles commonly land on Rogers Lake playa at Edwards Air Force Base in California.

See also ALLUVIAL FANS; DESERT; EVAPORITE.

Pleistocene The Pleistocene is the older of two epochs of the Quaternary period, lasting from 1.8 million years ago until 10,000 years ago, at the beginning of the Holocene epoch. Charles Lyell formally proposed the name Pleistocene in 1839, after earlier informal proposals, based on the appearance of species of North Sea mollusks in Mediterranean strata. The Pleistocene is recognized as an epoch of widespread glaciation, with glaciers advancing through much of Europe and North America, and across the southern continents. Glaciers covered about 30 percent of the northern continents, most as huge ice sheets that advanced across Canada, the northern United States, and Eurasia, and smaller alpine glaciers that dissected the mountain ranges forming the glacial landforms visible today including horns, aretes, U-shaped valleys, and giant eskers and moraines. Some of the ice sheets were up to three kilometers thick, forming huge bulldozers that removed much of the soil from Canada and scraped the bedrock clean, depositing giant outwash plains in lower latitudes.

The continental ice sheets are known to have advanced and retreated several times during the Pleistocene, based on correlations of moraines, sea surface temperatures deduced from oxygen isotope analysis of deep-sea cores, and magnetic stratigraphy. Eighteen major glacial expansions and retreats are now recognized from the past 2.4 million years, including four major glacial stages in North America. The Nebraskian glacial maximum peaked at 700,000 years ago, followed by the Kansan, Illinoisan, and the Wisconsin maximums. Ice from the Wisconsin glacial maximum only retreated from northern United States and Canada 11,000 years ago and may return in a short amount of geological time.

Homo sapiens sapiens and Neandertal Migration and Relations in the Ice Ages

A race of ancient hominids known as Neandertals (also spelled Neanderthals) inhabited much of central Asia, the Middle East, Near East, western Siberia, and Europe during the last 200,000 years during the Pleistocene ice ages. Neandertals were heavily built and had large brains and were probably adapted to the cold climate conditions in the periglacial environments they inhabited (i.e., they were probably hairy and fat). These premodern humans were few in number, dwindled in numbers around 40,000 years ago, and disappeared into extinction around 27,000 years ago. The first remains of modern humans (*Homo sapiens sapiens*) are dated at around 30,000 to 40,000 years old and inhabited some of the same areas as the Neandertals. The overlapping time and space ranges of Neandertals and modern humans raises some interesting questions about the origins of modern humans, and relationships between the two groups of hominids. Did the Neandertals go extinct because they were hunted and killed by the modern humans, or did climate conditions become unbearable to the Neandertals? Did modern humans and Neandertals live in peace side by side, possibly interbreeding, or did they fight? Could modern humans have evolved from Neandertals, or are the two races of hominids unrelated? A possible transitional form between Neandertals and *Homo sapiens sapiens* has been described from Mount Carmel in Israel, although most genetic evidence so far suggests that modern humans did not evolve from Neandertals. Modern humans seem to have descended from a single African female that lived 200,000 years ago, although some theories suggest that humans evolved in different parts of the world at virtually the same time.

Understanding of the evolution of humans is a controversial and constantly changing field. Genetic studies show that humans and chimpanzees had a common ancestor that lived about 5–10 million years ago. The earliest known human ancestor is australopithecines, from 3.9–4.4-million-year-old hominids found in Ethiopia and Kenya. It is thought that australopithecines evolved into *Homo habilis* by 2 million years ago. *Homo habilis* was larger than australopithecines, walked upright, and was the first hominid to use stone tools. By 1.7 million years ago, *Homo erectus* appeared in Africa, probably evolving from *Homo habilis*. *Homo erectus* had prominent brow ridges, a flattened cranium, a rounded jawbone, and was the first hominid to use fire and migrate out of Africa as far as China, Europe, and the British Isles. Modern humans (*Homo sapiens sapiens*) and Neandertals (*Homo sapiens neanderthalensis*) are both probably descendants of *Homo erectus*.

Neandertals were hunters and gatherers who roamed the plains, forests, and mountains of Europe and Eurasia. They left many stone tools, clothing, and possibly some art including cave drawings and sculptures. Around 30,000–40,000 years ago, the Neandertals found their environments increasingly inhabited by modern humans, who had smaller, less-robust skeletons, smaller brains, and lacked many of the primitive traits that characterized earlier humans. Some early modern humans had some Neandertal traits, but there is considerable debate in the anthropological community about whether this indicates an evolutionary trend, or more likely that the two races interbred producing mixed offspring. Whether the interbreeding was peaceful or a consequence of war and raids, there is no evidence that the mixed offspring were successful at producing a separate mixed race. The genetic and most archaeological evidence suggests that modern humans evolved separately, from a single African female, whose descendants came out of Africa and inhabited the Near East, Europe, and Asia.

These debates in the scientific community highlight two competing hypotheses for the origin of modern humans. The Out-of-Africa theory follows the genetic evidence that modern humans arose about 200,000 years ago in Africa and spread outward, replacing older indigenous populations of Neandertals and other hominids by 27,000 years ago. An opposing theory, called the multiregional evolution theory, argues that all modern humans are not descended from a single 200,000-year-old African ancestor. This model supposes that modern humans have older ancestors such as *Homo erectus* that spread out to Europe and Asia by 1 or 2 million years ago, then evolved into separate races of *Homo sapiens sapiens* independently in different parts of the world. There are many arguments against this theory of multiregional parallel evolution, the strongest of which notes the unlikelihood of the same evolutionary path being followed independently in several different places at the same time. However, the multiregionalists argue that the evolutionary advances were driven by similar technological and lifestyle advances, and that many adjacent groups may have been interbreeding and thereby exchanging genetic material. The multiregionalists need to allow enough genetic exchange between regional groups to form an early worldwide-web dating or genetic exchange system, whereby enough traits are transmitted between groups to keep *Homo sapiens sapiens* the same species globally, but to keep enough isolation so that individual groups maintain certain distinctive traits.

The multiregionalists see a common ancient ancestor 1–2 million years ago, with different groups evolving to different degrees toward what we call modern humans. In contrast, the Out-of-Africa theorists see different branches from *Homo erectus* 200,000 years ago, with Neandertals first moving into Eurasia and the Middle East, to be later replaced by migrating early modern humans that followed the migrating climate zones north with the retreat of the Pleistocene glaciers.

Many species became extinct or otherwise changed in response to the rapid climate changes in the Pleistocene. Many species lived in the climate zone close to the glacier front, including the woolly mammoth, giant versions of mammals now living in the arctic, rhinoceros, and caribou. Further from the ice, giant deer, mastodons, dogs and cats, ground sloths, and other mammals were common. Both humans (*Homo sapiens*) and Neandertals roamed through Eurasia, but the nature of the interaction between these two hominid species is unknown. Many of the giant mammals became extinct in the latter part of the Pleistocene, especially between 18,000 and 10,000 years ago, and there is currently considerable debate about the relative roles of climate change and predation by hominids in these extinctions.

See also HOLOCENE; NEOGENE; QUATERNARY; TERTIARY.

Pluto The ninth, and perhaps the most distant planet from the Sun, Pluto has a variable orbital distance of about 30–40 astronomical units (between 2.7 billion miles, or 4.4–6 billion km) from the Sun, circling once every 249 Earth years, and has a retrograde rotation period of 6.4 Earth days. It is a small planet with a mass of 0.003 Earth masses, and a diameter of 1,400 miles (2,250 km; only 20 percent that of Earth), and a density of 2.3 grams/cubic centimeter. It has one known moon, Charon. Some scientists debate whether Pluto should be considered a planet or a captured asteroid, supported by the large, 17.2° inclination of its orbital plane with respect to the ecliptic plane.

The physical properties of the Pluton-Charon system suggest that it is an icy planet system similar to some of the Jovian moons, being most similar to Neptune's moon Triton. Models for the origin of Pluto range from it being a captured icy asteroid, an escaped moon, to being a remnant of material left over from the formation of the solar system. The great distance and small size of the system make it difficult to observe, and certainly as deep planetary probes explore the

Image of Pluto and its moon Charon. This is the clearest view yet of the distant planet and its moon, as revealed by NASA's Hubble Space Telescope (HST). The image was taken by the European Space Agency's Faint Object Camera on February 21, 1994, when the planet was 2.6 billion miles (4.4 billion km) from Earth; or nearly 30 times the separation between Earth and the Sun. Hubble's corrected optics show the two objects as clearly separate and sharp disks. Though Pluto was discovered in 1930, Charon was not detected until 1978. That is because the moon is so close to Pluto that the two worlds are typically blurred together when viewed through ground-based telescopes. The two worlds are 12,200 miles apart (19,640 km). Hubble's ability to distinguish Pluto's disk at a distance of 2.6 billion miles (4.4 billion km) is equivalent to seeing a baseball at a distance of 40 miles (64 km). Pluto typically is called the double planet because Charon is half the diameter of Pluto (our Moon is one-quarter the diameter of Earth). This image and other images and data received from the HST are posted on the World Wide Web on the Space Telescope Science Institute home page at http://oposite.stsci.edu/pubinfo. *(Photo by Dr. R. Albrecht, ESA/ESO Space Telescope European Coordinating Facility, NASA. Courtesy of NASA.)*

outer reaches of the solar system, new theories and models for the origin and evolution of this system will emerge. In 2005 scientists announced that they may have discovered a tenth planet orbiting the sun, beyond the orbit of Pluto.

pluton A general name for a large cooled igneous intrusive body in the Earth. The classification of the specific type of pluton is based on its geometry, size, and relations to the older rocks surrounding the pluton, known as country rock. Concordant plutons have boundaries parallel to layering in the country rock, whereas discordant plutons have boundaries that cut across layering in the country rock. Dikes are tabular but discordant intrusions, and sills are tabular and concordant intrusives. Volcanic necks are conduits connecting a volcano with its underlying magma chamber. A famous example of a volcanic neck is Devils Tower in Wyoming. Some plutons are so large that they have special names. Batholiths have a surface area of more than 60 square miles (100 km²).

Geologists have long speculated on how such large volumes of magma intrude the crust, and what relationships these magmas have to the style of volcanic eruption. One mechanism that may operate is assimilation, where the hot magma melts surrounding rocks as it rises, causing it to become part of the magma. In doing so, the magma becomes cooler, and its composition changes to reflect the added melted country rock. It is widely thought that magmas may rise only a very limited distance by the process of assimilation. Some magmas may forcefully push their way into the crust if there are high pressures in the magma. One variation of this forceful emplacement style is diapirism, where the weight of surrounding rocks pushes down on the melt layer, which squeezes its way up through cracks that can expand and extend, forming volcanic vents at the surface. Stoping is a mechanism whereby big blocks get thermally shattered and drop off the top of the magma chamber and fall into the chamber, much like a glass ceiling breaking and falling into the space below. Many if not most plutons seem to be emplaced into structures such as faults, utilizing the weakness provided by the structure for the emplacement of the magma. Some plutons are emplaced into active faults, intruding into spaces created by gaps that open up between misaligned segments of the moving fault zone.

See also PETROLOGY; STRUCTURAL GEOLOGY.

plutonic *See* PLUTON.

plutonism *See* PLUTON.

polarizing microscope The polarizing microscope is an optical measuring device and is also used as an instrument for the detailed examination of specimens. In addition to standard microscope optics, the microscope contains two polarizing lenses. These lenses polarize the incoming light source to enable the measurements to be made. There is one polarizer in the condenser and another on the slider above the objective, both are adjustable and can be set to examine different specimens. The specimen is illuminated with the polarized light, and its rotation or twinning of the light can be determined. The polarizing microscope is particularly useful in the study of rocks and minerals with particular characteristics. Scientists in the fields of chemical microscopy and optical mineralogy mostly use the polarizing microscope for its ability to polarize the light onto a specimen.

See also PETROGRAPHY; PETROLOGY.

polar low Hurricane or gale-strength storms that form over water behind (poleward) the main polar front are known as polar lows. They may form over either the Northern or Southern Hemisphere oceans but are a larger menace to the more-populated regions around the North Atlantic, North Sea, and Pacific Oceans, as well as the Arctic Ocean. Most polar lows are much smaller than tropical and midlatitude cyclones, with diameters typically less than 620 miles (1,000 km). Like hurricanes, many polar lows have spiral bands of precipitation (snow in this case) that circle a central warmer low-pressure eye, whereas other polar lows develop a comma-shaped system.

Most polar lows develop during winter months. In the Northern Hemisphere they form along an arctic front, where frigid air blows off landmasses and encounters relatively warm current-fed ocean water, resulting in a rising column of warm air and sinking columns of cold air. This situation sets up an instability that induces condensation of water vapor in the rising air, along with the associated release of latent heat that then warms the atmosphere. The warming lowers the surface pressure adding convective updrafts to the system and starting the classical spiral cloud band formation. Polar lows may attain central barometric pressures comparable to hurricanes (28.9 inches or 980 mbars) but tend to dissipate more quickly when they move over the cold polar landmasses.

See also EXTRATROPICAL CYCLONES.

polynyas Pools or large, nonlinear openings in sea ice or pack ice. Some polynyas are persistently open, others are transient features that open and then disappear. They are smaller in size than large ice-free regions called open-air areas. Many form offshore from the mouths of large rivers. In the Southern Hemisphere winter (June–September) large polynyas may form in the Antarctic ice pack, and they are thought to be of two main origins. Coastal polynyas are typically 30–60 miles (50–100 km) across and are formed by winds blowing the ice away from the shore. Once these areas open up, new sea ice forms in them, and it is thought that a large percent of the sea ice in the Antarctic ice shelf may form

in this way. So much ice forms in Antarctic polynyas that the amount of salt released (it is not incorporated into the ice) is estimated to be 2.5 million cubic miles (10 million m³/s), roughly equivalent to all of the salt released into the oceans by river systems. This salt mixes with the cold Antarctic waters, becomes very dense, and sinks to the bottom where the salt is isolated from the atmosphere for long periods of time. Open ocean polynyas form in the middle of sea ice, and may recur in the same places year after year. They are probably controlled by convective and other water currents that are influenced by ocean bottom topography, influencing their persistent location.

See also SEA ICE.

porosity The percentage of total volume of a body that consists of open spaces. Sands and gravels typically have about 20 percent open spaces, while clays have about 50 percent. The sizes and shapes of grains determine the porosity, which is also influenced by how much they are compacted, cemented together, or deformed.

Porosity may be of several different types. Primary porosity is the open space that forms during the deposition of sedimentary rocks, such as the open spaces between individual sand grains in a sandstone. Secondary porosity is the amount of open space that forms in a rock by secondary processes, such as dissolution of limestone. Fracture porosity measures the amount of open space created by fracturing, jointing, and faulting of a rock. Fracture porosity may be enhanced by dissolution, and thus, fracture and secondary porosity typically act together to increase the primary porosity of a rock.

Minerals that precipitate between grains, cementing them together, may reduce the porosity of a rock. It may also be reduced if high pressure and temperatures during compaction or tectonic forces deform the grains.

See also AQUIFER; PERMEABILITY.

Powell, John Wesley (1834–1902) *American Geologist, Explorer* John Wesley Powell is probably most famous for his early explorations and geological descriptions of the Grand Canyon and Colorado River in 1869. Powell led a group of 10 men on four rowboats down the Colorado, and his geological observations led to the understanding that the canyon was formed by the river gradually cutting down through the rocks of the region as the plateau was slowly uplifted. Between 1874 and 1879, Powell was the director of the U.S. Geological and Geographical Survey of the Territories, and he led explorations into the Rocky Mountains of the southwestern regions. During these field excursions Powell became convinced of the limits on development posed by the paucity of water in the desert southwest, and he completed many surveys of the region's water resources. In 1881 Powell became director of the U.S. Geological Survey, but he resigned in 1894 to pursue studies of the native peoples of the land.

Further Reading

Powell, John W. *The Exploration of the Colorado River.* New York: Doubleday and Company, 1961.

———. "The Laws of Hydraulic Degradation." *Science* (1888): 229–233.

———. "Remarks on the Structural Geology of the Valley of the Colorado of the West." *Bulletin—Philosophical Society of Washington* (1874): 48–51.

Precambrian Comprising nearly 90 percent of geological time, the Precambrian eon includes the time interval in which all rocks older than 544 million years old formed. The Precambrian is preceded by the Hadean eon, representing the period during which the Earth and other planets were accreting and no rocks are preserved and is succeeded by the Cambrian, the dawn of advanced life on Earth. It is divided into two eras including the Archean, ranging in age from the oldest known rocks at about 4.0 billion years old, to 2.5 billion years ago, and the Proterozoic, ranging from 2.5 billion years ago until 540 million years ago. The Archean is further divided into the Early (4.0 Ga–3.0 Ga) and Late (3.0 Ga–2.5 Ga), and the Proterozoic is divided into the Early or Paleoproterozoic (2.5 Ga–1.6 Ga), Middle or Mesoproterozoic (1.6 Ga–1.3 Ga), and Late or Neoproterozoic (1.3 Ga–0.54 Ga).

Most Precambrian rocks are found in cratons, areas of generally thick crust that have been stable since the Precambrian, and exhibit low heat flow, subdued topography, and few earthquakes, and many also preserve a thick lithospheric keel known as the tectosphere. Exposed parts of Precambrian cratons are known as shields. Many of the rocks in cratons are preserved in granite-greenstone terrains, fewer are preserved as linear high-grade gneiss complexes, and still fewer form relatively undeformed platformal or basinal volcanosedimentary sequences resting on older Precambrian rocks. Platformal sequences form a thin veneer over many older Precambrian terrains, so geological maps of cratons and continents show many essentially flat-lying platformal units, but these are volumetrically less significant than the underlying sections of the crust. Many other areas of Precambrian rocks are found as linear tectonic blocks within younger orogenic belts. These probably represent fragments of older cratons that have been rifted, dispersed, and accreted to younger orogens by plate tectonic processes, some traveling huge dis-

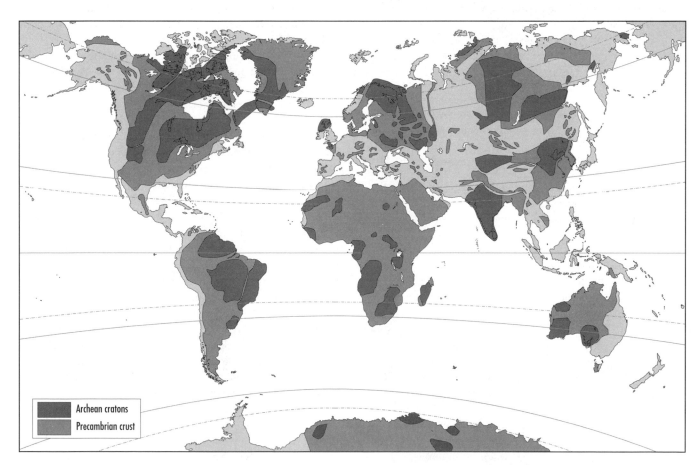

Archean cratons
Precambrian crust

Map of the world showing the distribution of Precambrian rocks in Archean cratons and other undifferentiated areas of Precambrian crust

tances from the rocks adjacent to them which they initially formed in their primary tectonic settings.

The Precambrian is the most dramatic of all geological eons. It marks the transition from the accretion of the Earth to a planet that has plate tectonics, a stable atmosphere-ocean system, and a temperature range all delicately balanced in a way that allows advanced life to develop and persist on the planet. The planet has been cooling steadily since accretion and was producing more heat by radioactive decay in the Precambrian than it has been since. However, it is uncertain if this greater amount of heat significantly heated the mantle and crust, or if this additional heat was simply lost faster by the present style of plate tectonics. It is likely that more rapid seafloor spreading or a greater total length of oceanic ridges with active volcanism was able to accommodate this higher heat flow, keeping mantle and crustal temperatures close to what they have been in the Phanerozoic.

Understanding of the development of life in the Precambrian has been undergoing rapid advancement, and the close links between life, atmospheric chemistry, plate tectonics, and global heat loss are only recently being explored. Many mysteries remain about the events that led to the initial creation of life, its evolution to more complex forms, and the eventual development of multicelled complex organisms at the end of the Precambrian.

See also ARCHEAN; CONTINENTAL CRUST; CRATONS; GREENSTONE BELTS; HADEAN; LIFE'S ORIGINS AND EARLY EVOLUTION; PROTEROZOIC; TECTOSPHERE.

Further Reading

Condie, Kent C., and Robert Sloan. *Origin and Evolution of Earth, Principles of Historical Geology.* Upper Saddle River, N.J.: Prentice Hall, 1997.

Goodwin, Alan M. *Precambrian Geology.* London: Academic Press, 1991.

Kusky, Timothy M. *Precambrian Ophiolites and Related Rocks.* Amsterdam: Elsevier, 2004.

Windley, Brian F. *The Evolving Continents,* 3rd ed. Chinchester, U.K.: John Wiley, 1995.

precipitation Water that falls to the surface from the atmosphere in liquid or fluid form. Whether it falls as rain, drizzle, fog, snow, sleet, freezing rain, or hail, it is measured as a liquid-water equivalent. The types and amounts of precipitation in different parts of the world vary greatly, from places that have never had any measurable precipitation to places that regularly receive hundreds of centimeters of rain per year. Precipitation is strongly seasonal is some places, with dry and wet seasons, and distributed more regularly in other climates.

Rain is liquid precipitation with droplets greater than 0.02 inches (0.5 mm) in diameter, whereas drizzle has droplets of 0.008–0.02 inch (0.2–0.5 mm) in diameter. Fog is a cloud whose base is at the surface, and it has smaller parti-

cles that only truly become precipitation when wind drives them against surfaces or the ground. Freezing rain and drizzle both fall in liquid form but freeze upon hitting cold surfaces on the ground, creating a frozen coating known as glaze. Sleet consists of frozen ice pellets less than 0.2 inch (5 mm) in diameter, and hail consists of larger transparent to opaque particles that typically have diameters of 0.2–0.8 inch (5 mm–2 cm), but sometimes as large as golf balls or rarely even grapefruits. Snow is frozen precipitation consisting of complex hexagonal ice crystals that fall to the ground.

In tropical regions and temperate climates in warmer parts of the year, most precipitation falls as rain and drizzle. Heavy rain is defined as more than 0.16 inch (4 mm) of precipitation per hour, moderate rain falls at 0.16–0.02 inch (4 mm–0.5 mm) per hour, and light rain (commonly called drizzle) is less than 0.02 inch (0.5 mm) per hour. Frequent and steady rains characterize some regions; others are characterized by infrequent but intense downpours, including thunderstorms that may shed hailstones. At high elevation, high latitudes, and in midlatitudes in colder months, most precipitation falls as frozen solid particles. Most frozen precipitation falls as snow that typically has a water equivalent of one-tenth the amount of snow that falls (i.e., 10 cm of snow equals 10 mm of rain). More freezing rain and sleet than snow characterize some regions, particularly coastal regions influenced by warm ocean currents.

Uplift within clouds or larger scale systems are generally necessary to initiate the formation of water droplets that become precipitation. Convection cells in thunderheads, air forced over mountains, zones of convergence along fronts, and cyclonic systems can all produce dramatic uplift and induce precipitation. One of the biggest obstacles that must be overcome for precipitation to form is that very small water (or ice droplets) that are separated by very wide spaces must coalesce to form particles large enough to fall as precipitation. Additionally, the particles must overcome the forces of evaporation as they rise or fall through unsaturated air in order to make it to the ground. Rapid lateral and vertical motions in clouds, leading to collisions between particles, aid the coalescence of particles, and gravity then accelerates particles to the ground with larger particles initially falling faster than smaller ones since they are less affected by updrafts. Large particles therefore tend to collide with and incorporate smaller particles. Frozen particles form in the upper levels of vertically extensive cloud systems and may alternately fall into and rise out of lower levels, where they partially melt, grow, and rise on updrafts. Such cycling can produce relatively large particles that may fall as precipitation.

See also CLIMATE; CLOUDS; HURRICANE.

Proterozoic The younger of the two Precambrian eras, and the erathem of rocks deposited in this era. The Proterozoic is divided into several intervals, including the Early or Paleoproterozoic (2.5 Ga–1.6 Ga), Middle or Mesoproterozoic (1.6 Ga–1.3 Ga), and Late or Neoproterozoic (1.3 Ga–0.54 Ga). Proterozoic rocks are widespread on many continents, with large areas preserved especially well in North America, Africa and Saudi Arabia, South America, China, and Antarctica.

Like the Archean, Proterozoic terrains are of three basic types, including rocks preserved in cratonic associations, orogens (often called mobile belts in Proterozoic literature), and cratonic cover associations. Many Proterozoic terrains are cut by wide shear zones, extensive mafic dike swarms, and layered mafic-ultramafic intrusions. Proterozoic orogens have long linear belts of arc-like associations, metasedimentary belts, and widespread, well-developed ophiolites. Many geologists have suggested that clear records of plate tectonics first appeared in the Proterozoic, although many others have challenged this view, placing the operation of plate tectonics earlier, in the Archean. This later view is supported by the recent recognition of Archean ophiolites (including the Dongwanzi ophiolite) in northern China.

The Proterozoic saw the development of many continental-scale orogenic belts, many of which have been recently recognized to be parts of global-scale systems that reflect the formation, breakup, and reassembly of several supercontinents. Paleoproterozoic orogens include the Wopmay in northern Canada, interpreted to be a continental margin arc that rifted from North America, then collided soon afterward, closing the young back arc basin. There are many 1.9–1.6 Ga orogens in many parts of the world, including the Cheyenne belt in the western United States, interpreted as a suture that marks the accretion of the Proterozoic arc terrains of the southwestern United States with the Archean Wyoming Province.

The supercontinent Rodinia formed in Mesoproterozoic times by the amalgamation of Laurentia, Siberia, Baltica, Australia, India, Antarctica, and the Congo, Kalahari, West Africa, and Amazonia cratons between 1.1 Ga and 1.0 Ga. The joining of these cratons resulted in the terminal collisional events at convergent margins on many of these cratons, including the circa 1.1–1.0 Ga Ottawan and Rigolet orogenies in the Grenville Province of Laurentia's southern margin. Globally, these events have become known as the Grenville orogenic period, named after the Grenville orogen of eastern North America. Grenville-age orogens are preserved along eastern North America, as the Rodonia-Sunsas belt in Amazonia, the Irumide and Kibaran belts of the Congo craton, the Namaqua-Natal and Lurian belts of the Kalahari craton, the Eastern Ghats of India, and the Albany-Fraser belt of Australia. Many of these belts now preserve deep-crustal metamorphic rocks (granulites) that were tectonically buried to 19–25-mile (30–40-km) depth, then the overlying crust was removed by erosion, forcing the deeply buried rocks to the surface. Since 30–40 kilometers of crust still underlie these regions, it has been surmised that during the peak of metamorphism, they had double crustal thicknesses. Such

thick crust is today produced in regions of continent-continent collision, and locally in Andean arc settings. Since the Grenville-aged orogens are so linear and widely distributed, they are generally interpreted to mark the sites of continent-continent collisions where the various cratonic components of Rodinia collided between 1.1 Ga and 1.0 Ga.

The Neoproterozoic breakup of Rodinia and the formation of Gondwana at the end of the Precambrian and the dawn of the Phanerozoic represents one of the most fundamental problems being studied in earth sciences today. There have been numerous and rapid changes in our understanding of events related to the assembly of Gondwana. One of the most fundamental and most poorly understood aspects of the formation of Gondwana is the timing and geometry of closure of the oceanic basins which separated the continental fragments that amassed to form the Late Neoproterozoic supercontinent. It appears that the final collision between East and West Gondwana most likely followed the closure of the Mozambique Ocean, forming the East African Orogen. The East African Orogen encompasses the Arabian-Nubian Shield in the north and the Mozambique Belt in the south. These and several other orogenic belts are commonly referred to as Pan-African belts, recognizing that many distinct belts in Africa and other continents experienced deformation, metamorphism, and magmatic activity spanning the period of 800–450 Ma. Pan-African tectonothermal activity in the Mozambique Belt was broadly contemporaneous with magmatism, metamorphism, and deformation in the Arabian-Nubian Shield. The difference in lithology and metamorphic grade between the two belts has been attributed to the difference in the level of exposure, with the Mozambican rocks interpreted as lower crustal equivalents of the juvenile rocks in the Arabian-Nubian Shield. Recent geochronologic data indicate the presence of two major "Pan-African" tectonic events in East Africa. The East African Orogeny (800–650 Ma) represents a distinct series of events within the Pan-African of central Gondwana, responsible for the assembly of greater Gondwana. Collectively, paleomagnetic and age data indicate that another later event at 550 Ma (Kuunga orogeny) may represent the final suturing of the Australian and Antarctic segments of the Gondwana continent. The Arabian-Nubian shield in the northern part of the East African orogen preserves many complete ophiolite complexes, making it one of the oldest orogens with abundant Penrose-style ophiolites with crustal thicknesses similar to those of Phanerozoic orogens.

The Proterozoic record preserves several continental scale rift systems. Rift systems with associated mafic dike swarms cut across the North China craton at 2.4 billion and 1.8 billion years, as well as in many other cratons. One of the best-known of Proterozoic rifts is the 1.2–1.0 Ga Keweenawan rift, a 932-mile (1,500-km) long, 93-mile (150-km) wide trough that stretches from Lake Superior to Kansas in North America. This trough, like many Proterozoic rifts, is filled with a

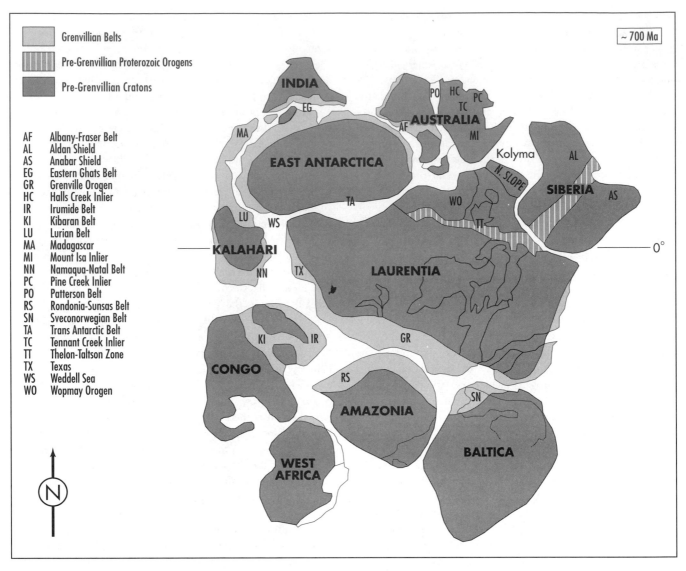

Plate reconstruction of the continents at 700 million years ago showing the supercontinent of Rodinia, with North America (Laurentia) situated in the center

mixture of basalts, rhyolites, arkose, conglomerate, and other, locally red, immature sedimentary rocks, all intruded by granite and syenite. Some of the basalt flows in the Keweenawan rift are 1–4 miles (2–7 km) thick.

Massive Proterozoic diabase dike swarms cut straight across many continents and may be related to some of the Proterozoic rift systems, or to mantle plume activity. Some of the dike swarms are more than 1,865 miles (3,000 km) long, hundreds of kilometers wide, and made of thousands of individual dikes ranging from less than three feet to more than 1,600 feet (1–500 m). Some dike swarms, such as the 1.267 Ga Mackenzie swarm of North America, show radial patterns and point to a source near the Coppermine River basalts in northern Canada. Other dike swarms are more lin-

ear and may parallel failed or successful rift arms. Magma flow directions in the dikes is generally parallel to the surface, except in the central 300–650 miles (500–1,000 km) of the swarms, suggesting that magma may have fed upward from a plume that initiated a triple-armed rift system, then the magma flowed away from the plume head. In some cases, such as the Mackenzie swarm, one of the rift arms may have become successful, forming an ocean basin.

Cratonic cover sequences are well-preserved from the Proterozoic in many parts of the world. In China the Mesoproterozoic Changcheng Series consists of several-kilometer-thick accumulations of quartzite, conglomerate, carbonate, and shale. In North America the Paleoproterozoic Huronian Supergroup of southern Canada consists of up to 7.5 miles (12 km) of coarse clastic rocks dominated by clean beach and fluvial sandstones, interbedded with carbonates and shales. Thick sequences of continentally derived clastic rocks interbedded with marine carbonates and shales represent deposition on passive continental margins, rifted margins of back arc basins, and as thin cratonic cover sequences from epicontinental seas. Similar cratonic cover sequences are known from many parts of the world, showing that continents were stable by the Proterozoic, that they were at a similar height with respect to sea level (freeboard), and that the volume of continental crust at the beginning of the Proterozoic was at least 60 percent of the present volume of continental crust.

One of the more unusual rock associations from the Proterozoic record is the 1.75–1.00 Ga granite-anorthosite association. The anorthosites (rocks consisting essentially of plagioclase) have chemical characteristics indicating that they were derived as cumulate rocks from fractional crystallization of a basaltic magma extracted from the mantle, whereas the granites were produced by partial melting of lower crustal rocks. The origin of these rocks is not clearly understood— some geologists suggest they were produced on the continental side of a convergent margin, others suggest an extensional origin, still others suggest an anorogenic association.

Proterozoic life began with very simple organisms similar to those of the Archean, and by 2.0 Ga planktonic algae and stromatolitic mounds with prokaryotic filaments and spherical forms are well preserved in many cherts and carbonates. The stromatolites are formed by cyanobacteria and form a wide variety of morphological forms, including columns, branching columns, mounds, cones, and cauliflower type forms. In the 1960s many geologists, particularly from the Russian academies, attempted to correlate different Precambrian strata based on the morphology of the contained stromatolites, but this line of research proved futile as all forms are found in rocks of all ages. The diversity and abundance of stromatolites peaked about 750 million years ago and declined rapidly after that time period. The decline is probably related to the sudden appearance of grazing multicellular metazoans such as worms at this same time. Eukaryotic cells (with a membrane-lined cell wall) are preserved from at least as old as 1.8 Ga, reflecting increased oxygen in the atmosphere and ocean. The Acritarchs are single-celled spherical fossils that are interpreted as photosynthetic marine plankton and are found in a wide variety of rock types. Around 750 million years ago the prokaryotes experienced a sudden decline, and their niches were replaced by eukaryotic forms. This dramatic change is not understood, but its timing coincident with the breakup of Rodinia and the formation of Gondwana is notable. It could be that tectonic changes induced atmospheric and environmental changes, favoring one type of organism over the other.

A wide range of metazoans, complex multicellular organisms, are recognized from the geological record by 1.0 Ga and probably evolved along several different lines before the record was well established. A few metazoans up to 1.7 Ga old have been recognized from North China, but the fossil record from this interval is poorly preserved since most animals were soft-bodied. The transition from the Proterozoic fauna to the Paleozoic is marked by a remarkable group of fossils known as the Ediacaran fauna, first described from the Ediacara Hills in the Flinders Ranges of southern Australia. These 550–540-million-year-old fauna show an extremely diverse group of multicelled complex metazoa including jellyfish-like forms, flatworm-like forms, soft-bodied arthropods, echinoderms, and many other species. The ages of these fauna overlap slightly with the sudden appearance and explosion of shelly fauna in Cambrian strata at 540 million years ago, showing the remarkable change in life coincident with the formation of Gondwana at the end of the Proterozoic.

See also ARCHEAN; CHINA'S DONGWANZI OPHIOLITE; SUPERCONTINENT CYCLE.

Further Reading

Dalziel, Ian W. D. "Neoproterozoic-Paleozoic Geography and Tectonics: Review, Hypothesis, Environmental Speculation." *Geological Society of America* 109 (1997): 16–42.

Hoffman, Paul F. "Did the Breakout of Laurentia Turn Gondwanaland Inside-Out?" *Science* 252 (1991): 1,409–1,411.

Kusky, Timothy M., and Jianghai H. Li. "Paleoproterozoic Tectonic Evolution of the North China Craton." *Journal of Asian Earth Sciences*, 2003.

Kusky, Timothy M., Mohamed Abdelsalam, Robert Tucker, and Robert Stern. *East African and Related Orogens, and the Assembly of Gondwana*. Amsterdam: Elsevier, Precambrian Research, 2003.

Moores, E. M. "Southwest United States—East Antarctic (SWEAT) Connection: A Hypothesis." Geology 19 (1991): 425–428.

Stern, Robert J. "Arc Assembly and Continental Collision in the Neoproterozoic East African Orogen: Implications for Consolidation of Gondwanaland." *Annual Review of Earth and Planetary Science* 22 (1994): 319–351.

Protozoa   The diverse phylum of single-celled organisms that includes the foraminifera, radiolaria, thecamoebians, silicoflagellates, and tintinnids, and their fossil records. Most protozoa are soft-bodied and do not leave a fossil record. Modern zoological classifications regard the protozoa as a subkingdom of the Animalia Kingdom, excluding the algae, whereas paleontologists include these and other botanical groups with the protozoa. Geologists and paleontologists include botanical unicellular organisms such as acritarchs, calcareous nannofossils, chitinozoans, diatoms, and dinoflagellates in this phylum and refer to them by the terms *protozoans* or *protistids*. The protistids are descendants of the early and simple prokaryotes from the Archean, but since the early forms were soft-bodied, they rarely leave fossil records.

See also ARCHEAN; LIFE'S ORIGINS AND EARLY EVOLUTION.

pull-apart basin An elongate depression that develops along extensional steps on strike-slip faults. Pull-apart basins are features that develop in transtensional regions, in which the principal stresses are compressional, but some areas within in the region are under extension due to the obliquity of the major stress direction with respect to plane of failure. This results in extension of the crust along releasing bends, leading to a break in the crust and the formation of basins. Some pull-apart basins show several progressive stages in their formation. Others initiate along a fracture, progress into lazy Z or S shapes, and finally progress into a basin that ranges in length to width ratio from 2:1 to 10:1. These types of basins are characterized by steep sides on major fault boundaries with normal faults developing on their shorter sides. Continuous movement along the major faults tend to offset deposits from their source inlet to the basin. These basins are characterized by rapid deposition and rapid facies changes along or across the width of the basin and gradual facies changes along the longest axis of the basin. Pull-apart basin deposits are typically made mostly of coarse fanglomerate, conglomerate, sandstone, shales, and shallow water limestones and evaporites. Bimodal volcanics and volcanic sediments are also found interbedded within the basin deposits. These bimodal volcanics are typical of those found in rift settings, but here they are in a transtensional regime. Transcurrent faults can penetrate down deep into the crust reaching the upper mantle and providing a conduit for magma.

See also PLATE TECTONICS; TRANSFORM PLATE MARGIN PROCESSES.

Further Reading

Mann, Paul, Mark R. Hempton, Dwight C. Bradley, and Kevin Burke. "Development of Pull-Apart Basins." *Journal of Geology* 91 (1983): 529–554.

Reading, Harold G. "Characteristics and Recognition of Strike-Slip Faults Systems." In *Sedimentation in Oblique-Slip Mobile Zones*, edited by Peter F. Balance and Harold G. Reading, 7–26. International Association of Sedimentology Special Publication 4 (1980).

pycnocline A steep density gradient in a body of water, or the layer that a strong vertical density gradient occurs within. Changes in salinity and/or temperature cause the changes in density and are referred to as the halocline or thermocline, respectively. In the oceans the pycnocline is about a kilometer thick with its top approximately corresponding to the 50°F (10°C) temperature contour, and its base corresponding to the 39.2°F (4°C) contour. The pycnocline is found below the surface layer or mixed zone and overlies the deep zone where changes in temperature, salinity, and density are less dramatic. The pycnocline is a very stable region in the oceans, rarely affected by movements and seasonal changes in the surface zone. Waters below the pycnocline in the deep zone primarily move along the density gradients, and do not penetrate the pycnocline either. Internal waves may form along the pycnocline surface, since it is a surface with a density contrast, similar to the ocean/air surface. These waves are induced by currents flowing along the base of the pycnocline in the deep zone.

See also OCEANOGRAPHY.

pyroclastic Fragments of rocks and lava that are ejected from volcanoes, and the rocks that are made up from these fragments. The term excludes lava flows but includes most other material that is thrown out of volcanoes during eruptions. The term is closely related to volcaniclastic sediments and rocks, another general term for fragmental rocks or material that includes some volcanic fragments. Volcaniclastic material may be reworked and redeposited away from the volcano, whereas pyroclastic rocks are those deposited directly by the eruption. Volcanic blocks are solid materials thrown out of the volcano, whereas volcanic bombs are hardened pieces of lava that were ejected while still in the molten state.

Pyroclastic deposits typically contain a mixture of material that ranges in size from fine ash to blocks and bombs that may be as large as houses. Most of the very large blocks and bombs are deposited very close to the volcanic vent and may form a deposit type called an agglomerate, in which large volcanic clasts are mixed with fine-grained ash. Smaller particles are deposited farther away, and ash may travel hundreds or thousands of kilometers and spread around the globe if it is ejected into the upper atmosphere. In some cases, hot volcanic ash clouds known as nuée ardentes rush down the sides of volcanoes, destroying all in their path. Rocks formed from hardened ash are known as tuffs, or welded tuffs if they are hardened by the heat of the ash, and ignimbrites if they contain flattened pieces of ash known as fiamme. Some volcanic eruptions generate rainstorms, and when the rain drops fall through the ash cloud they grow concentric rings of ash particles known as lapilli that fall to the ground as mud balls. When preserved, these mud ball rich ash deposits are called lapilli tuffs.

See also NUÉE ARDENTE; VOLCANIC BOMB; VOLCANO.

Further Reading

Fisher, Richard V., and Hans U. Schminke. *Pyroclastic Rocks*. Berlin: Springer-Verlag, 1984.

Quaternary The last 1.8 million years of Earth history are known as the Quaternary period, divided into the older Pleistocene and the younger Holocene. The rocks and unconsolidated deposits formed during this period were first recognized by Jules Desnoyers to be different from older deposits by their characteristic boulder clays and other units deposited by glaciers in Europe. It was soon recognized that these deposits reflected globally cool climates for the first part of the Quaternary, since glacier deposits were found in many parts of both the Northern and Southern Hemispheres. Global climate zones were condensed toward the equator, with ice sheets covering about one-third of the continental surfaces and desert regions converted to moist grasslands. Grasses, plants, and mammals experienced a rapid expansion.

Another major important discovery from this period was the recognition of the first human fossils, which became the basis for dividing the Quaternary into the older Pleistocene, and the younger Holocene, in which the human fossils appear abundantly about 10,000 years ago. Older primate fossils including early hominids are found in the older record going back several million years, and the record of human habitation in the Western Hemisphere may extend back to 13,000 or 14,000 years ago. Genetic evidence suggests that humans are all descendants of a single female ancestor that lived somewhere in East Africa about 100,000 to 300,000 years ago, with the first hominids appearing about 4 million years ago.

See also HOLOCENE; NEOGENE; PLEISTOCENE; TERTIARY.

Further Reading
Charlesworth, J. Kaye. *The Quaternary Era*. London: Arnold, 1957.

R

radiation   The emission of radiant energy as particles or waves such as heat, light, alpha particles, and beta particles. Heat transfer by infrared rays is also known as radiation or radiative heat transfer. Infrared radiation travels at the speed of light and can travel through a vacuum, gets reflected and refracted, and does not affect the medium that it passes through.

The electromagnetic spectrum divides types of radiation according to wavelength, with the shortest wavelengths being cosmic rays, and in increasing wavelength, gamma rays, X rays, ultraviolet rays, visible rays, infrared rays, microwave rays, radio waves, and television waves. The environment contains a low level of background radiation that is always present, most being given off from radioactive decay of minerals and radioactive gases such as radon and thoron. Some background radiation also comes from space and is known as cosmic radiation. The Sun emits solar radiation consisting of visible light, ultraviolet radiation, and infrared waves spanning the entire spectrum of electromagnetic wavelengths from radio waves to X rays. The sun also emits high-energy particles such as electrons, especially from solar flares. Short-wavelength high-frequency electromagnetic waves from 0.0000157 to almost 0.0 inch (400–4 nanometers) are known as ultraviolet radiation, which is powerful and needed by humans but harmful in strong doses. X rays, because of their very wavelength, are able to penetrate soft tissue, some sands and soils, and reflect off internal denser material such as bones or rocks. This property has made X rays useful for medical practices and geologic mapping of subsurface materials. Visible radiation includes all that humans see with their eyes, including the wide range of colors of the rainbow.

radioactive decay   The nuclei of unstable radioactive elements may spontaneously break down to become more sta-

ble, emitting radiation as alpha particles, beta particles, or gamma rays. When these particles and rays are emitted by radioactive decay they move through matter and knock electrons out of surrounding atoms, ionizing these atoms. Alpha decay of an atomic nucleus produces alpha particles that are the most ionizing form of radioactive radiation, consisting of two neutrons and two protons with two positive charges, and are heavy, slow-moving particles. Beta decay of a nucleus converts a neutron into a proton, emitting a high-speed electron and an electron-antineutrino, increasing the atomic number by one and leaving the mass number the same. The beta particles that are emitted are high-speed electrons and are moderately ionizing, being more penetrating than alpha particles. They can travel several meters in the air and are easily deflected by electromagnetic fields. Gamma rays carry no charge, are weakly ionizing, and consist of a very high-frequency type of electromagnetic radiation emitted by the nuclei of radioactive elements during decay, typically as part of alpha or beta decay. Gamma rays may also form from the interaction of high-energy electrons with matter. Gamma rays are deeply penetrating and are not deflected by electromagnetic fields, and they may be used to kill bacteria or to sterilize surfaces. Space-based observatories have detected cosmic gamma ray radiation coming from distant pulsars, quasars, and radio galaxies, but this cosmic gamma ray radiation cannot penetrate the Earth's atmosphere.

When radioactive elements or radioactive isotopes of stable isotopes decay to more stable elements, the atomic mass number of the element is changed, transmuting the parent element into a different element known as a daughter isotope and emitting atomic radiation. For each radioactive element or isotope, decay occurs at a constant rate known as the half-life, determined by the time taken for half of any mass of that isotope to decay from the parent isotope to the daughter isotope. Radioactive decay is an exponential process, with half of the original starting material decaying in the first step, half of the remaining material (25 percent of the original material) decaying after the second step, half of the remaining material (12.5 percent of the original material) decaying after the third step, and so on. The final product of all decay schemes is a stable element.

Radioactive decay may occur in one step or, more commonly, in a series of steps known as a decay series. In some decay series the intermediate steps may be moderately or very short-lived, and the daughter isotope may be more or less radioactive than the parent isotope. There is a very wide range in half-lives for different radioactive isotopes, ranging from 4.4×10^{-22} sec for lithium-5, through 4.551×10^9 years for ^{238}U, to 1.5×10^{24} years for tellurium-128.

See also GEOCHRONOLOGY.

radiosonde A radiosonde is a balloon that contains an instrument package of meteorological equipment that is capable of transmitting real-time data back to Earth. Once launched, the radiosonde transmits valuable meteorological data back to Earth until the instrument is destroyed. A transmitter within the instrument package can send measurements of air temperature, pressure, humidity, and wind speed down to Earth. The United States has been releasing radiosondes into the atmosphere since 1936 and continues to release them twice a day, once at 0000 UTC and once again at 1200 UTC. Radiosondes typically reach a height of 20 miles (30 km) above the surface of the Earth, and only a small fraction are ever found and sent back to the National Weather Service to be used again.

radon A poisonous gas that is produced as a product of radioactive decay product of the uranium decay series. Radon is a heavy gas, and it presents a serious indoor hazard in every part of the country. It tends to accumulate in poorly ventilated basements and well-insulated homes that are built on specific types of soil or bedrock rich in uranium minerals. Radon is known to cause lung cancer, and since it is an odorless, colorless gas, it can go unnoticed in homes for years. However, the hazard of radon is easily mitigated, and homes can be made safe once the hazard is identified.

Uranium is a radioactive mineral that spontaneously decays to lighter "daughter" elements by losing high-energy particles at a predictable rate, known as a half-life. The half-life specifically measures how long it takes for half of the original or parent element to decay to the daughter element. Uranium decays to radium through a long series of steps with a cumulative half-life of 4.4 billion years. During these steps, intermediate daughter products are produced, and high-energy particles including alpha particles, consisting of two protons and two neutrons, are released, which produces heat. The daughter mineral radium is itself radioactive, and it

decays with a half-life of 1,620 years by losing an alpha particle, forming the heavy gas radon. Since radon is a gas, it escapes out of the minerals and ground and makes its way to the atmosphere where it is dispersed, unless it gets trapped in people's homes. If it gets trapped, it can be inhaled and do damage. Radon is a radioactive gas, and it decays with a half-life of 3.8 days, producing daughter products of polonium, bismuth, and lead. If this decay occurs while the gas is in someone's lungs, then the solid daughter products become lodged in their lungs, which is how the damage from radon is initiated. Most of the health risks from radon are associated with the daughter product polonium, which is easily lodged in lung tissue. Polonium is radioactive, and its decay and emission of high-energy particles in the lungs can damage lung tissue, eventually causing lung cancer.

There is a huge variation in the concentration of radon between geographic regions, and in specific places in those regions. There is also a great variation in the concentration of the gas at different levels in the soil, home, and atmosphere. This variation is related to the concentration and type of radioactive elements present at a location. Radioactivity is measured in a unit known as a *picocurie* (pCi), which is approximately equal to the amount of radiation produced by the decay of two atoms per minute.

Soils have gases trapped between the individual grains that make up the soil, and these soil gases have typical radon levels of 20 pCi per liter, to 100,000 pCi per liter, with most soils in the United States falling in the range of 200–2,000 pCi/L. Radon can also be dissolved in groundwater, with typical levels falling between 100–2 million pCi/Liter. Outdoor air typically has 0.1–20 pCi/Liter, and radon inside people's homes ranges 1–3,000 pCi/Liter, with 0.2 pCi/Liter being typical.

There is a large natural variation in radon levels in different parts of the country and world. One of the main variables controlling the concentration of radon at any site is the initial concentration of the parent element uranium in the underlying bedrock and soil. If the underlying materials have high concentrations of uranium, it is more likely that homes built in the area may have high concentrations of radon. Most natural geologic materials contain a small amount of uranium, typically about one to three parts per million (ppm). The concentration of uranium is typically about the same in soils derived from a rock as in the original source rock. However, some rock (and soil) types have much higher initial concentrations of uranium, ranging up to and above 100 ppm. Some of the rocks that have the highest uranium contents include some granites, some types of volcanic rocks (especially rhyolites), phosphate-bearing sedimentary rocks, and the metamorphosed equivalents of all of these rocks.

As the uranium in the soil gradually decays, it leaves its daughter product, radium, in concentrations proportional to the initial concentration of uranium. The radium then decays

Is Your Home Safe from Radon?

Radon becomes hazardous when it enters homes and becomes trapped in poorly ventilated or well-insulated areas. Radon moves up through the soil and moves toward places with greater permeability. Home foundations are often built with a very porous and permeable gravel envelope surrounding the foundation, to allow for water drainage. This also has the effect of focusing radon movement, bringing it close to the foundation, where the radon may enter through small cracks in the concrete, seams, spaces around pipes, sumps, and other openings, as well as through the concrete which may be moderately porous. Most modern homes intake less than one percent of their air from the soil. Some homes, however, particularly older homes with cracked or poorly sealed foundations, low air pressure, and other entry points for radon, may intake as much as 20 percent of their internal air from the soil. These homes tend to have the highest concentrations of radon.

Radon can also enter the home and body through the groundwater. Homes that rely on well water may be taking in water with high concentrations of dissolved radon. This radon can then be ingested, or it can be released from the water by agitation in the home. Radon is released from high-radon water by simple activities such as taking showers, washing dishes, or running faucets. Radon can also come from some municipal water supplies, such as those supplied by small towns that rely on well fields that take the groundwater and distribute it to homes without providing a reservoir for the water to linger in while the radon decays to the atmosphere. Most larger cities, however, rely on reservoirs and surface water supplies, where the radon has had a chance to escape before being used by unsuspecting homeowners.

A greater understanding of the radon hazard risk in an area can be obtained through mapping the potential radon concentrations in an area. This can be done at many scales of observation. Radon concentrations can also be measured locally to learn what kinds of mitigation are necessary to reduce the health risks posed by this poisonous gas.

The broadest sense of risk can be obtained by examining regional geologic maps and determining whether or not an area is located above potential high-uranium content rocks such as granites, shales, and rhyolites. These maps are available through the U.S. Geological Survey and many state geological surveys. The U.S. Department of Energy has flown airplanes with radiation detectors across the country and produced maps that show the measured surface radioactivity on a regional scale. These maps give a very good indication of the amount of background uranium concentration in an area and thus are related to the potential risk for radon gas.

More detailed information is needed by local governments, businesses, and homeowners to assess whether or not they need to invest in radon remediation equipment. Geologists and environmental scientists are able to measure local soil radon gas levels using a variety of techniques, typically involving placing a pipe into the ground and sucking out the soil air for measurement. Other devices may be buried in the soil to more passively measure the formation of the damage produced by alpha particle emission. With such information, the radon concentrations in certain soil types can be established. This information can be integrated with soil characteristic maps produced by the U.S. Department of Agriculture and state and county officials, to make more regional maps of potential radon hazards and risks.

Most homeowners must resort to private measurements of radon concentrations in their homes using commercial devices that detect radon or measure the damage from alpha particle emission. The measurement of radon levels in homes has become a standard part of home sales transactions, so more data and awareness of the problem has risen in the past 10 years. If your home or business does have a radon problem, an engineer or contractor can simply and cheaply (typically less than a thousand dollars for an average home) design and build a ventilation system that can remove the harmful radon gas, making the air safe to breathe.

by forcefully ejecting an alpha particle from its nucleus. This ejection is an important step in the formation of radon, since every action has a reaction. In this case the reaction is the recoil of the nucleus of the newly formed radon. Most radon remains trapped in minerals once it forms. However, if the decay of radium happens near the surface of a mineral, and if the recoil of the new nucleus of radon is away from the center of the grain, the radon gas may escape the bondage of the mineral. It will then be free to move in the intergranular space between minerals, soil, or cracks in the bedrock, or become absorbed in groundwater between the mineral grains. Less than half (10–50 percent) of the radon produced by decay of radium actually escapes the host mineral. The rest is trapped inside, where it eventually decays, leaving the solid daughter products behind as impurities in the mineral.

Once the radon is free in the open or water-filled pore spaces of the soil or bedrock it may move rather quickly. The exact rate of movement is critical to whether or not the radon enters homes, because radon does not stay around for very long with a half-life of only 3.8 days. The rates at which radon moves through a typical soil depend on how much pore space there is in the soil (or rock), how connected these pore spaces are, and the exact geometry and size of the openings. Radon moves quickly through very porous and permeable soils such as sand and gravel, but moves very slowly through less permeable materials such as clay. Radon moves very quickly through fractured material, whether it is bedrock, clay, or concrete.

The large variation in the concentration of radon from place to place is partly the result of how the rates of radon movement are influenced by the geometry of pore spaces in a soil or bedrock underlying a home, and also how the initial concentration of uranium in the bedrock determines the amount of radon available to move. Homes built on dry permeable soils can accumulate radon quickly because it can migrate through the soil quickly. Conversely, homes built on impermeable soils and bedrock are unlikely to concentrate radon beyond their natural background levels.

See also RADIOACTIVE DECAY.

rain *See* PRECIPITATION.

rainbows Translucent concentric arcs of colored bands that are visible in the air under certain conditions when rain or mist is present in the air and the Sun is at the observer's back. Rainbows form where sunlight enters the rain or water drops in the air, and a small portion of this light is reflected off the back of the raindrops and directed back to the observer. When the sunlight enters the raindrops, it is bent and slows and, as in a prism, violet light is refracted the most and red the least. The amount of light that is reflected off the back of each raindrop is small compared to the amount that enters each drop, and only the rays that hit the back of the drop at angles greater than the critical angle are reflected. Since the Sun's rays are refracted and split by color when they enter the water drops, each color hits the back of the raindrop at a slightly different angle, and the reflected light emerges from the raindrop at different angles for each color. Red light emerges at 42° from the incoming beam, whereas violet light emerges at 40°. An observer only sees one color from each drop, but with millions of drops in the sky an observer is able to see a range of colors formed from different raindrops with light reflected at slightly different angles to the observer. Rainbows appear to move as an observer moves, since each ray of light is entering the observer's eyes from a single raindrop, and as the observer moves, light from different drops enters the observer's eyes.

Recent *See* HOLOCENE.

reef Framework-supported carbonate mounds built by carbonate secreting organisms, or in some usages any shallow

Double rainbow over northern Madagascar *(Photo by Timothy Kusky)*

ridge of rock lying near the surface of the water. Reefs contain a plethora of organisms that together build a wave-resistant structure to just below the low-tide level in the ocean waters and provide shelter for fish and other organisms. The spaces between the framework are typically filled by skeletal debris, which together with the framework become cemented together to form a wave-resistant feature that shelters the shelf from high-energy waves. Reef organisms (presently consisting mainly of zooxanthellae) can only survive in the photic zone, so reef growth is restricted to the upper 328 feet (100 m) of the seawater.

Reefs are built by a wide variety of organisms, today including red algae, mollusks, sponges, and cnidarians (including corals). The colonial Scleractinia corals are presently the principal reef builders, producing a calcareous external skeleton characterized by radial partitions known as septa. Inside the skeleton are soft-bodied animals called polyps, containing symbiotic algae that are essential for the life cycle of the coral, and the building of the reef structure. The polyps contain calcium bicarbonate that is broken down into calcium carbonate, carbon dioxide, and water. The calcium carbonate is secreted to the reef building its structure, whereas the algae photosynthesize the carbon dioxide producing food for the polyps.

There are several different types of reefs, classified by their morphology and relationship to nearby landmasses. Fringing reefs grow along and fringe the coast of a landmass, and are often discontinuous. They typically have a steep outer slope, an algal ridge crest, and a flat, sand-filled channel between the reef and the main shoreline. Barrier reefs form at greater distances from the shore than fringing reefs and are generally broader and more continuous than fringing reefs. They are among the largest biological structures on the planet—for instance, the Great Barrier Reef of Australia is 1,430 miles (2,300 km) long. A wide deep lagoon typically separates barrier reefs from the mainland. Atolls or atoll reefs form rings around emergent or submerged volcanic islands, growing progressively upward as the central volcanic island subsides below sea level.

Reefs are rich in organic material and have high primary porosity, so they are a promising target for many hydrocarbon exploration programs. Reefs are well represented in the geological record, with examples including the Permian reefs of west Texas, the Triassic of the European Alps, the Devonian of western Canada, Europe, and Australia, and the Precambrian of Canada and South Africa. Organisms that produced the reefs have changed dramatically with time, but surprisingly, the gross structure of the reefs has remained broadly similar.

See also ATOLL; CARBONATE; CORALS; PASSIVE MARGIN.

regression A retreat or seaward migration of the shoreline caused by either a global sea-level fall, a rise in the land's surface, or a supply of sediment that is greater than the space created for the sediment by subsidence. Global sea-level rises and falls on different timescales depending on the cause. Changes in ridge volume or mantle plume activity cause slow changes to the ocean ridge volumes and slow rises or falls in sea level, whereas changes in the volume of continental glaciers may cause faster changes in the volume of water in the ocean. All of these may be related to the supercontinent cycle and cause sea-level regressions or transgressions. Local tectonic activity may cause the land surface to rise or fall relative to a stable global sea level, causing local regressions or transgressions. For instance, glacial rebound may cause the land to rise rapidly in response to the reduced load when the glaciers melt. The shoreline of a region may retreat seaward if the sediment supplied by a river or other system is so large that the sediment volume fills up and overflows the space available for it to be deposited in. Deltas form at the mouths of rivers where the coastal subsidence can not accommodate the large sediment flux, causing a local regression in these areas. Deltas have a wide variety of forms, resulting from interactions between sediment supply, subsidence, wave, and tidal action. If coastal currents substantially rework sediment supplied by rivers and alluvial systems, then a series of seaward prograding beach ridges may form along a coastline experiencing a regression.

To interpret patterns of global sea-level rise and fall it is necessary to isolate the effects of local tectonic subsidence or uplift, and sediment supply issues, from the global sea-level signature. This can be difficult and requires precise dating and correlation of events along different shorelines, plus a detailed understanding of the local tectonic and sedimentation history. When the local effects are isolated they can be subtracted from the global sea-level curve, and the causes of global sea-level changes investigated.

See also DELTAS; SEA-LEVEL RISE; SUPERCONTINENT CYCLE; TRANSGRESSION.

remote sensing   The acquisition of information about an object by recording devices that are not in physical contact with the object. There are many types of remote sensing, including airborne or spaceborne techniques and sensors that measure different properties of Earth materials, ground-based sensors that measure properties of distant objects, and techniques that penetrate the ground to map subsurface properties. The term *remote sensing* is commonly used to refer only to the airborne and space-based observation systems, with ground-based systems more commonly referred to as geophysical techniques.

Remote sensing grew out of airplane-based photogeologic reconnaissance studies, designed to give geologists a vertically downward-looking regional view of an area of interest, providing information and a perspective not readily appreciated from the ground. Most geological mapping now includes the use of stereo aerial photographs, produced by taking downward-looking photographs at regular intervals along a flight path from an aircraft, with every area on the ground covered by at least two frames. The resolution of typical aerial photographs is such that objects less than 3.2 feet (1 m) across can be easily identified. The camera and lens geometry is set so that the photographs can be viewed with a stereoscope, where each eye looks at one of the overlapping images, producing a visual display of greatly exaggerated topography. This view can be used to pick out details and variations in topography, geology, and surface characteristics that greatly aid geologic mapping. Typically, geologic structures, rock dips, general rock types, and the distribution of these features can be mapped from aerial photographs.

Modern techniques of remote sensing employ a greater range of the electromagnetic spectrum than aerial photographs. Photographs are limited to a narrow range of the electromagnetic spectrum between the visible and infrared wavelengths that are reflected off the land's surface from the Sun's rays. Since the 1960s a wide range of sensors that can detect and measure different parts of the electromagnetic spectrum have been developed, along with a range of different optical-mechanical and digital measuring and recording devices used for measuring the reflected spectrum. In addition many satellite-based systems have been established, providing stable observation platforms and continuous or repeated coverage of most parts of the globe. One technique uses a mirror that rapidly sweeps back and forth across an area measuring the radiation reflected in different wavelengths. Another technique uses line-scanning, where thousands of detectors are arranged to electronically measure the reflected strength of radiation from different wavelengths in equally divided time intervals as the scanner sweeps across the surface, producing a digital image consisting of thousands of lines of small picture elements (pixels) representing each of the measured intervals. The strength of the signal for each pixel is converted to a digital number (dn) for ease of data storage and manipulation to produce a variety of different digital image products. Information from the reflected spectrum is divided into different wavelength bands that correspond to the narrow wavebands measured by the sensors. The digital data encodes this information, and during digital image processing, the strength of the signal from different bands is converted into the strength of the mixture of red, green, and blue, with the mixture producing a colored image of the region. Different bands may be assigned different colors, and bands may even be numerically or digitally combined or ratioed to highlight different geological features.

Optical and infrared imagery are now widely used for regional geological studies, with common satellite platforms including the United States–based Landsat systems, the French SPOT (Système Pour l'Observation de la Terre) satellite, and more recently some multispectral sensors including ASTER (Advanced Spaceborne Thermal Emission and Reflec-

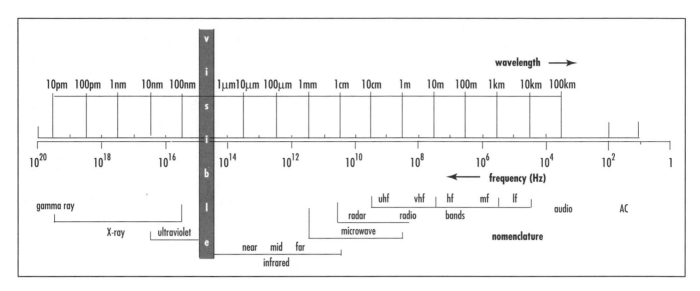

Part of the electromagnetic spectrum, showing relationship between wavelength, frequency, and nomenclature for electromagnetic radiation with different characteristics

tion Radiometer) and AVHRR (Advanced Very High Resolution Radiometer) data. Much optical and infrared imagery is able to detect differences in rock and mineral types because the reflection is sensitive to molecular interactions with solar radiation that highlights differences between Al-OH bonds, C-O bonds, and Mg-OH bonds, effectively discriminating between different minerals such as micas, Mg-silicates, quartz, and carbonates. Bands greater than 2.4 microns are sensitive to the temperature of the surface instead of the reflected light, and studies of surface temperature have proven useful for identifying rock types, moisture content, water and hydrocarbon seeps, and caves.

Microwave remote sensing (wavelengths of less than 0.04 inch, or 1 mm) uses artificial illumination of the surface since natural emissions are too low to be useful. Satellite and aircraft-based radar systems are used to shoot energy of specific wavelength and orientation to the surface, which is then reflected back to the detector. Radar remote sensing is very complex, depending on the geometry and wavelength of the system, and on the nature of the surface. The strength of the received signal is dependent on features such as surface inclination, steepness, orientation, roughness, composition, and water content. Nonetheless, radar remote sensing has proved to be immensely useful for both military and scientific purposes, producing images of topography and surface roughness, and highlighting structural features such as faults, foliations, and other forms that are highlighted by radar reflecting off from sharp edges. Under some special circumstances, radar is able to penetrate the surface of some geological materials (such as dry sand) and produce images of what lies beneath the surface, including buried geologic structures, pipelines, and areas of soil moisture.

Further Reading
Drury, Steven A. *Image Interpretation in Geology.* London: Chapman and Hall, 1993.
Sabins, Floyd F. *Remote Sensing, Principles and Interpretation.* New York: W.H. Freeman and Co., 1997.

resource In economic geology, resources of a particular mineral or commodity include all known reserves of that material, plus all other deposits of the mineral or commodity that may become available over time. This includes deposits that are not yet discovered, plus deposits that are low grade or not economically feasible to recover using present technology. Agencies such as the U.S. Geological Survey routinely make economic resource assessments for small areas, regions, states, countries, and the globe, in order for the government and industry to be able to plan for future availability of a commodity, and for determining prices of the mineral or metal. Such assessments include all known deposits, plus an estimate of how many deposits of a particular grade are likely to be present based on expert knowledge of the geological relationships in the region being studied.

Further Reading
Cox, Dennis P., and Don Singer. *Mineral Deposit Models. United States Geological Survey Bulletin* 1693, 1986.
Singer, Don. "Basic Concepts in Three-Part Quantitative Assessments of Undiscovered Mineral Resources." *Nonrenewable Resources* 2 (1993): 69–81.

rheology *See* VISCOSITY.

Richter scale The Richter scale is a widely used open-ended scale that measures the strength of earthquakes. The scale was devised in 1935 by Dr. Charles Richter, a prominent seismologist who was based at the California Institute of Technology. The Richter scale gives an idea of the amount of energy released during an earthquake and is based on the amplitudes of seismic waves (equal to half the height from wave-base to wave-crest) at a distance of 62 miles (100 km) from the epicenter. The Richter-scale magnitude of an earthquake is calculated using the trace produced on a seismograph by an earthquake, once the epicenter has been located by comparing signals from several different, widely separated seismographs. The Richter scale is logarithmic, where each step corresponds to a tenfold increase in amplitude. This is necessary because the energy of earthquakes changes by factors of more than a hundred million.

The actual energy released in earthquakes changes even more rapidly with each increase in the Richter scale, because the number of high amplitude waves increases with bigger earthquakes and the energy released is according to the square of the amplitude. Thus, it turns out in the end that an increase of one on the Richter scale may correspond to as much as 30 times the energy released at the number below it. The largest earthquakes so far recorded are the 9.2 Alaskan earthquake of 1964, the 9.5 Chilean earthquake of 1960, and the 9.0–9.2 2004 Sumatra earthquake, each of which released the energy equivalent to approximately 1,000–10,000 nuclear bombs the size of the one dropped on Hiroshima.

Before the development of modern inertial seismographs and the Richter scale, earthquake intensity was commonly measured using the modified Mercalli intensity scale. This scale, developed in the late 1800s and named after Father Giuseppe Mercalli, measures the amount of vibration people remember feeling for low-magnitude earthquakes, and the amount of damage to buildings in high-magnitude events. One of the disadvantages of the Mercalli scale is that it is not corrected for distance from the epicenter. Therefore, people near the source of the earthquake may measure the earthquake as a IX or X, whereas people farther from the epicenter might only record a I or II event. However, the modified Mercalli scale has proven very useful for estimating the magnitudes of historical earthquakes that occurred before the development of modern seismographs, since the Mercalli magnitude can be estimated from historical records.

Many seismologists now use a different method of estimating the strength and energy released in an earthquake. The seismic moment accounts better for the low-frequency wave motions produced during an earthquake, but it is more difficult to calculate than the Richter magnitude so it is not commonly used outside the seismological community.

See also EARTHQUAKES.

rift Rift basins are elongate depressions in the Earth's surface where the entire thickness of the lithosphere has ruptured in extension. They are typically bounded by normal faults along their long sides and display rapid lateral variation in sedimentary facies and thicknesses. Rock types deposited in the rift basins include marginal conglomerate, fanglomerate, and alluvial fans, grading basinward into sandstone, shale, and lake evaporite deposits. Volcanic rocks may be intercalated with the sedimentary deposits of rifts and in many cases include a bimodal suite of basalts and rhyolites, some with alkaline chemical characteristics.

Many rifts in continents are associated with incipient breaking apart of the continent to form an oceanic basin. These types of rift system typically form three arms that develop over domed areas above upwelling mantle material, such as is observed in East Africa. Two of the three arms may link with other three-pronged rift systems developed over adjacent domes, forming a linked elongate rift system that then spreads to form an ocean basin. This type of development leaves behind some failed rift arms that will come to reside on the margins of young oceans when the successful rift arms begin to spread. These failed rift arms then become sites of increased sedimentation and subsidence, and they also tend to be low-lying areas and form the tectonic setting where many of the world's major rivers flow (for example, the Nile, Amazon, Mississippi). Other rifts form at high angles to collisional mountain belts, and still others form in regions of widespread continental extension such as the Basin and Range Province of the southwestern United States.

See also DIVERGENT OR EXTENSIONAL BOUNDARIES.

river system Stream and river valleys have been preferred sites for human habitation for millions of years, and they provide routes of easy access through rugged mountainous terrain and also water for drinking, watering animals, and for irrigation. Most of the world's large river valleys are located in structural or tectonic depressions such as rifts, including the Nile, Amazon, Mississippi, Hudson, Niger, Limpopo, Rhine, Indus, Ganges, Yenisei, Yangtze, Amur, and Lena. The soils in river valleys are also some of the most fertile that can be found, as they are replenished by yearly or less frequent floods. The ancient Egyptians appreciated this, as their entire culture developed in the Nile River valley and revolved around the flooding cycles of the river. Rivers now provide easy and relatively cheap transportation on barges, and the river valleys are preferred routes for roads and railways as they are relatively flat and easier to build on than mountains. Many streams and rivers have also become polluted as industry has dumped billions of gallons of chemical waste into our nation's waterways.

However, stream and rivers are dynamic environments. Their banks are prone to erosion, and the rivers periodically flood over their banks. During floods, rivers typically cover their floodplains with several or more feet of water and drop layers of silt and mud. This is part of a river's normal cycle and was relied upon by the ancient civilizations for replenish-

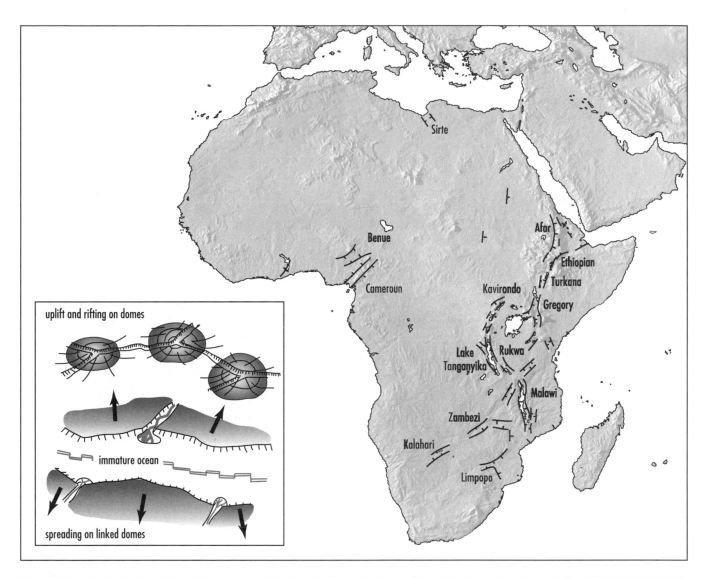

Map of Africa showing the East African Rift system stretching from the Afar to the Limpopo River valley. Inset shows the evolution of RRR (RIFT-RIFT RIFT) domes from uplifts with three linked arms to having two arms spread, leaving one failed rift arm behind. Larger sketch shows how the rift may evolve into an ocean basin, with large rivers draining into the ocean (shown with spreading ridge in the center) along the failed rift arms.

ing and fertilizing their fields. Now, since many floodplains are industrialized or populated by residential neighborhoods, the floods are no longer welcome and natural floods are regarded as disasters. On average, floods kill a couple of hundred people each year in the United States. Dikes and levees have been built around many rivers in attempts to keep the floodwaters out of towns. However, this tends to make the flooding problem worse because it confines the river to a narrow channel, and the waters rise more quickly and cannot seep into the ground of the floodplain.

Streams are important geologic agents that are critical for other earth systems. They carry most of the water from the land to the sea, they transport billions of tons of sediment to the beaches and oceans, and they erode and reshape the

land's surface, forming deep valleys and floodplains and passing through mountains.

Streams and rivers are very dynamic systems and constantly change their patterns, the amount of water (discharge), and sediment being transported in the system. Rivers may transport orders of magnitude more water and sediment in times of spring floods, as compared to low-flow times of winter or drought. Since rivers are dynamic systems, and the amount of water flowing through the channel changes, the channel responds by changing its size and shape to accommodate the extra flow. Five factors control how a river behaves: (1) width and depth of channel, measured in feet (m); (2) gradient, measured in feet per mile (m/km); (3) average velocity, measured in feet per second (m/sec); (4) discharge, measured in cubic feet per second (m³/s); and (5) load, measured as tons per cubic yard (metric tons/m³). All these factors are continually interplaying to determine how the river system behaves. As one factor, such as discharge, changes, so do the others, expressed as:

$$Q = w \times d \times v$$

All factors vary across the stream, so they are expressed as averages. If one term changes then all or one of the others must change too. For example, with increased discharge, the river erodes and widens and deepens its channel. The river may also respond by increasing its sinuosity through the development of meanders, effectively creating more space for the water to flow in and occupy by adding length to the river. The meanders may develop quickly during floods because the increased stream velocity adds more energy to the river system, and this can rapidly erode the cut banks enhancing the meanders.

The amount of sediment load available to the river is also independent of the river's discharge, so different types of river channels develop in response to different amounts of sediment load availability. If the sediment load is low, rivers tend to have simple channels, whereas braided stream and river channels develop where the sediment load is greater than the stream's capacity to carry that load. If a large amount of sediment is dumped into a river, it will respond by straightening, thus increasing the gradient and velocity and increasing its ability to remove the added sediment.

When rivers enter lakes or reservoirs along their path to the sea, the velocity of the water will suddenly decrease. This causes the sediment load of the stream or river to be dropped as a delta on the lake bottom, and the river attempts in this way to fill the entire lake with sediment. The river is effectively attempting to regain its gradient by filling the lake, then eroding the dam or ridge that created the lake in the first place. When the water of the river flows over the dam, it does so without its sediment load and therefore has greater erosive power and can erode the dam more effectively.

Rivers carry a variety of materials as they make their way to the sea. These materials range from minute dissolved particles and pollutants to giant boulders moved only during the most massive floods. The bed load consists of the coarse particles that move along or close to the bottom of the riverbed. Particles move more slowly than the stream, by rolling or sliding. Saltation is the movement of a particle by short intermittent jumps caused by the current lifting the particles. Bed load typically constitutes 5–50 percent of the total load carried by the river, with a greater proportion carried during high-discharge floods. The suspended load consists of the fine particles suspended in the river. These make many rivers muddy, and they consist of silt and clay that moves at the same velocity as the river. The suspended load generally accounts for 50–90 percent of the total load carried by the

river. The dissolved load of a river consists of dissolved chemicals, such as bicarbonate, calcium, sulfate, chloride, sodium, magnesium, and potassium. The dissolved load tends to be high in rivers fed by groundwater. Pollutants such as fertilizers and pesticides from agriculture and industrial chemicals also tend to be carried as dissolved load in rivers.

There is a wide range in the sizes and amounts of material that can be transported by a river. The competence of a stream is the size of particles a river can transport under a given set of hydraulic conditions, measured in diameter of largest bed load. A river's capacity is the potential load it can carry, measured in the amount (volume) of sediment passing a given point in a set amount of time. The amount of material carried by rivers depends on a number of factors. Climate studies show erosion rates are greatest in climates between a true desert and grasslands. Topography affects river load, as rugged topography contributes more detritus, and some rocks are more erodable. Human activity, such as farming, deforestation, and urbanization, all strongly affect erosion rates and river transport. Deforestation and farming greatly increase erosion rates and supply more sediment to rivers, increasing their loads. Urbanization has complex effects, including deceased infiltration and decreased times between rainfall events and floods.

See also BRAIDED STREAM; FLOODPLAIN; FLUVIAL; MEANDER.

Rocky Mountains Extending 3,000 miles (4,800 km) from central New Mexico to northwest Alaska in the easternmost Cordillera, the Rocky Mountains are one of the largest mountain belts in North America. The mountains are situated between the Great Plains on the east and a series of plateaus and broad basins on the west. Mount Elbert in Colorado is the highest mountain in the range, reaching 14,431 feet (4,399 m). The continental divide is located along the rim of the Rockies, separating waters that flow to the Pacific and the Atlantic Oceans. The Rocky Mountains are divided into the Southern, Central, and Northern Rockies in the United States, Canadian Rockies in Canada, and the Brooks Range in Alaska. Several national parks are located in the system, including Rocky Mountain, Yellowstone, Grand Teton, and Glacier Bay National Parks in the United States, and Banff, Glacier, Yoho, Kootenay, and Mount Revelstoke in Canada. The mountains were a major obstacle to traveling west during the expansion of the United States, but western regions opened up when the Oregon trail crossed the ranges through South Pass in Wyoming.

In New Mexico, Colorado, and southern Wyoming the Southern Rockies consist of two north-south ranges of folded mountains that have been eroded to expose Precambrian cores with overlying sequences of layered sedimentary rocks. Three basins are located between these ranges, known as the North, South, and Middle parks. The Southern Rockies are the highest section of the whole range including many peaks more than 14,000 feet (4,250 m).

The Middle Rockies in northeastern Utah and western Wyoming are lower and more discontinuous than the southern Rockies. Most are eroded down to their Precambrian cores, surrounded by Paleozoic-Mesozoic sedimentary rocks. Garnet Peak in the Wind River Range (13,785 feet; 4,202 m) and Grand Teton in the Teton Range (13,766 feet; 4,196 m) are the highest peaks in the Central Rockies.

The Northern Rockies in northeastern Washington, Idaho, and western Wyoming extend from Yellowstone National Park to the Canadian border. This section is dominated by north-south trending ranges separated by narrow valleys, including the Rocky Mountain trench, an especially deep and long valley that extends north from Flathead Lake. The highest peaks in the Northern Rockies include Borah Peak (12,655 feet; 3,857 m) and Leatherman Peak (12,230 feet; 3,728 m) in the Lost River Range.

The Canadian Rockies stretch along the British Columbia–Alberta border and reach their highest point in Canada on Mount Robson (12,972 feet; 3,954 m). The Rocky Mountain trench continues 800 miles (1,290 km) north-northwest from Montana, becomes more pronounced in Canada, and is joined by the Purcell trench in Alberta. In the Northwest Territories (Nunavet) the Rockies expand northeastward in the Mackenzie and Franklin mountains, and near the Beaufort Sea pick up as the Richardson Mountains that gain elevation westward into the Brooks Range of Alaska. Mount Chamberlin (9,020 feet; 2,749 m) is the highest peak in the Brooks Range.

The Rocky Mountains are rich in mineral deposits, including gold, silver, lead, zinc, copper, and molybdenum. Principal mining areas include the Butte-Anaconda district of Montana, Leadville and Cripple Creek in Colorado, Coeur d'Alene in Idaho, and the Kootenay Trail region of British Columbia. Lumbering is an active industry in the mountains but is threatened by growing environmental concerns and tourism in the National Park Systems.

Mesozoic–Early Cenozoic contractional events produced the Rockies during uplift associated with the Cordilleran orogeny. Evidence for older events and uplifts are commonly referred to as belonging to the Ancestral Rocky Mountain System. The Rocky Mountains are part of the larger Cordilleran orogenic belt that stretches from South America through Canada to Alaska, and it is best to understand the evolution of the Rockies through a wider discussion of events in this mountain belt. The Cordillera is presently active and has been active for the past 350 million years, making it one of the longest-lived orogenic belts on Earth. In the Cordillera, many of the structures are not controlled by continent/continent collisions as they are in many other mountain belts, since the Pacific Ocean is still open. In this orogen structures are controlled by the subduction/accretion process, collision of

arcs, islands, oceanic plateaus, and strike-slip motions parallel to the mountain belt. Present-day motions and deformation are controlled by complex plate boundaries between the North American, Pacific, Gorda, Cocos, and some completely subducted plates such as the Farallon. In this active tectonic setting the style, orientation, and intensity of deformation and magmatism depend largely on the relative convergence–strike-slip vectors of motion between different plates.

The geologic history of the North American Cordillera begins with rifting of the present western margin of North America at 750–800 million years ago, which is roughly the same age as rifting along the east coast in the Appalachian orogen. These rifting events reflect the breakup of the supercontinent of Rodinia at the end of the Proterozoic, and they left North America floating freely from the majority of the continental landmass on Earth. Rifting, and the subsequent thermal subsidence of the rifted margin, led to the deposition of Precambrian clastic rocks of the Windemere Supergroup, and carbonates of the Belt and Purcell Supergroups, in belts stretching from Southern California and Mexico to Canada. These are overlain by Cambrian-Devonian carbonates, Carboniferous clastic wedges, and Carboniferous-Permian carbonates, then finally Mesozoic clastic rocks.

The Antler orogeny is a Late Devonian–Early Carboniferous (350–400-million-year-old) tectonic event formed during an arc-continent collision, in which deepwater clastic rocks of the Robert's Mountain allochthon in Nevada were thrust from west to east over the North American carbonate bank, forming a foreland basin that migrated onto the craton. This orogenic event, similar to the Taconic orogeny in the Appalachian mountains, marks the end of passive margin sedimentation in the Cordillera, and the beginning of Cordilleran tectonism.

In the Late Carboniferous (about 300 million years ago), the zone of active deformation shifted to the east with a zone of strike-slip faulting, thrusts, and normal faults near Denver. Belts of deformation formed what is known as the ancestral Rocky Mountains, including the Front Ranges in Colorado and the Uncompahgre uplift of western Colorado, Utah, and New Mexico. These uplifts are only parts of a larger system of strike-slip faults and related structures that cut through the entire North American craton in the Late Carboniferous, probably in response to compressional deformation that was simultaneously going on along three margins of the continent.

The Late Permian–Early Triassic Sonoma orogeny (260–240 million years ago) refers to events that led to the thrusting of deepwater Paleozoic rocks of the Golconda allochthon eastward over autochthonous shallow water sediments just outboard (oceanward) of the Robert's Mountain allochthon. The Golconda allochthon in western Nevada includes deepwater oceanic pelagic rocks, an island arc sequence, and a carbonate shelf sequence, and it is interpreted to represent an arc/continent collision.

In the Late Jurassic (about 150 million years ago) a new, northwest-striking continental margin was established by crosscutting the old northeast striking continental margin. This event is known as the early Mesozoic truncation event and reflects the start of continental margin volcanic and plutonic activity that continues to the present day. There is considerable uncertainty about what happened to the former extension of the old continental margin—it may have rifted and drifted away or may have moved along the margin along large strike-slip faults.

Pacific margin magmatism has been active intermittently from the Late Triassic (220 million years ago) through the Late Cenozoic and in places continues to the present. This magmatism and deformation is a direct result of active subduction and arc magmatism. Since the Late Jurassic, there have been three main periods of especially prolific magmatism including the Late Jurassic/Early Cretaceous Nevadan orogeny (150–130 million years ago), the Late Cretaceous Sevier orogeny (80–70 million years ago), and the Late Cretaceous/Early Cenozoic Laramide orogeny (66–50 million years ago).

Cretaceous events in the Cordillera resulted in the formation of a number of tectonic belts that are still relatively easy to discern. The Sierra Nevada ranges of California and Nevada represent the arc batholith and contain high-temperature, low-pressure metamorphic rocks characteristic of arcs. The Sierra Nevada is separated from the Coast Ranges by flat-lying generally unmetamorphosed sedimentary rocks of the Great Valley, deposited over ophiolitic basement in a fore-arc basin. The Coast Ranges include high-pressure, low-temperature metamorphic rocks, including blueschists in the Franciscan complex. Together, the high-pressure, low-temperature metamorphism in the Franciscan complex, with the high-temperature, low-pressure metamorphism in the Sierra Nevada, represents a paired metamorphic belt, diagnostic of a subduction zone setting.

Several Cretaceous foreland fold-thrust belts are preserved east of the magmatic belt in the Cordillera, stretching from Alaska to Central America. These belts include the Sevier fold-thrust belt in the United States, the Canadian Rockies fold-thrust belt, and the Mexican fold-thrust belt. They are all characterized by imbricate-style thrust faulting, with fault-related folds dominating the topographic expression of deformation.

The Late Cretaceous–Early Tertiary Laramide orogeny (about 70–60 million years ago) is surprisingly poorly understood but generally interpreted as a period of plate reorganization that produced a series of basement uplifts from Montana to Mexico. Some models suggest that the Laramide orogeny resulted from the subduction of a slab of oceanic lithosphere at an unusually shallow angle, perhaps related to its young age and thermal buoyancy.

The late Mesozoic-Cenozoic tectonics of the Cordillera saw prolific strike-slip faulting, with relative northward displacements of terranes along the western margin of North America. The San Andreas Fault system is one of the major transform faults formed in this interval as a consequence of the subduction of the Farallon plate. Previous convergence between the Farallon and North American plates stopped when the Farallon was subducted, and new relative strike-slip motions between the Pacific and North American plates resulted in the formation of the San Andreas system. Remnants of the Farallon plate are still preserved as the Gorda and Cocos plates.

Approximately 15 million years ago the Basin and Range Province and the Colorado Plateau began uplifting and extending through the formation of rifts and normal faults. Much of the Colorado Plateau stands at more than a mile (1.5–2.0 km) above sea level but has a normal crustal thickness. The cause of the uplift is controversial but may be related to heating from below. The extension is related to the height of the mountains being too great for the strength of the rocks at depth to support it, so gravitational forces are able to cause high parts of the crust to extend through the formation of normal faults and rift basins.

See also BROOKS RANGE; CONVERGENT PLATE MARGIN PROCESSES; JAPAN'S PAIRED METAMORPHIC BELT.

Further Reading

Anderson, J. Lawford, ed., *The Nature and Origin of Cordilleran Magmatism.* Geological Society of America Memoir 174, 1990.

Burchfiel, Bert Clark, and George A. Davis. "Nature and Controls of Cordilleran Orogenesis, Western United States: Extensions of an Earlier Synthesis." *American Journal of Science* 275 (1975).

———. "Structural Framework of the Cordilleran Orogen, Western United States." *American Journal of Science* 272 (1972).

Oldow, John S., Albert W. Bally, Hans G. Avé Lallemant, and William P. Leeman. "Phanerozoic Evolution of the North American Cordillera; United States and Canada." In *The Geology of North America; An Overview, Decade of North American Geology,* v. A, edited by Albert W. Bally and Allison R. Palmer. Boulder, Colo.: Geological Society of America, 1989.

Sisson, Virginia B., Sarah M. Roeske, and Terry L. Pavlis. "Geology of a Transpressional Orogen Developed During Ridge-Trench Interaction Along the North Pacific Margin." *Geological Society of America Special Paper* 371 (2003).

rogue waves   Waves usually lose energy and decrease in amplitude as they interact with other waves as they move away from the region in which they were generated. In some cases, however, waves interact with currents in a way that dramatically increases the amplitude of some isolated waves, forming huge towering wave crests capable of capsizing even the largest ships. Some of these rogue waves have been reported to be hundreds of feet high. Regions where large ocean swells meet strong oncoming currents are known for rogue waves. For instance, off the coast of southern Africa, huge ocean swells generated in the southern ocean between Africa and Antarctica move north, and meet the south-flowing

Aghullas current flowing out of the Mozambique Channel. The current causes some of the waves to steepen and become shorter in wavelength, with some becoming so steep they are close to breaking in open waters. Deep holes form in front of these waves, presenting a particular hazard to ships as the holes can often not be seen until the ship is plunging into them, only to be quickly overrun by the towering wave. Huge rogue waves are known to form in other regions such as the North Atlantic, and near Bermuda where large ocean swells interact with the Gulf Stream current. Rogue waves may be the explanation for the large number of ships that have been reported lost without a trace in the Bermuda triangle region.

See also OCEANOGRAPHY; WAVES.

Ross Ice Shelf, Antarctica   Ice shelves form where ice sheets move over ocean waters and form a thick sheet of ice floating on the water and attached to the land on one, two, or three sides. Their seaward sides are typically marked by a steep cliff, up to 1,500 feet (500 m) high, where many glaciers calve off from forming icebergs. Ice shelves are found in Antarctica, Greenland, and along the polar seacoasts of the Canadian Arctic islands. The largest ice shelves are found in Antarctica, which contains 91 percent of the world's glacial ice, around 7 percent of which is contained in ice shelves. These ice shelves cover 50 percent of the coast of Antarctica, forming an area 1/10th the size of the continent. The largest ice shelf in the world is the Ross Ice Shelf (also called the Great Ice Barrier), that fills in the southern half of the Ross

Sea, bisected by longitude 180°W (international date line). The area covered by the Ross Ice Shelf is similar to the area of France. The south and west sides of the ice shelf are bounded by the Trans-Antarctic Mountains in Victoria Land, home to the U.S. scientific research station at McMurdo Bay. The eastern side of the Ross Ice Shelf is occupied by the ice-covered Rockefeller Plateau of Marie Byrd Land. The Ross Ice Shelf is fed by seven major ice streams from the surrounding mountains and plateau, and it breaks up along its northern coast, sending huge icebergs into warmer waters. For many centuries there has been a balance between the amount of ice that has fed the ice shelves of Antarctica and the ice that is lost to icebergs in summer months. More recently, some of the ice shelves on the Antarctic Peninsula have begun to break up, possibly as a result of global warming. The breakup of ice shelves does not change global sea levels because the ice is already floating on and in isostatic equilibrium with seawater, but these early warning signs of significant Antarctic warming and glacial melting could be signaling a start of catastrophic melting of the Antarctic ice sheets. Melting of continental ice would dramatically change global sea levels and could become one of the major environmental catastrophes facing the world in the next few centuries.

See also GLACIER.

Further Reading
Oppenheimer, Michael. "Global Warming and the Stability of the West Antarctic Ice Sheet." *Nature* 393 (1998).

Sahara Desert The world's largest desert, covering 5,400,000 square miles (8,600,000 km²) in northern Africa including Mauritania, Morocco, Algeria, Tunisia, Libya, Egypt, Sudan, Chad, Niger, and Mali. The desert is bordered on the north and northwest by the Mediterranean Sea and Atlas Mountains, on the west by the Atlantic Ocean, and on the east by the Nile River. However, the Sahara is part of a larger arid zone that continues eastward into the Eastern Desert of Egypt and Nubian Desert of Sudan, the Rub' al-Khali of Arabia, and the Lut, Tar, Dasht-e-Kavir, Takla Makan, and Gobi Deserts of Asia. Some classifications include the Eastern and Nubian Deserts as part of the Sahara and call the region of the Sahara west of the Nile the Libyan Desert, whereas other classifications consider them separate entities. The southern border of the Sahara is less well defined but is generally taken as about 16° latitude where the desert grades into transitional climates of the Sahel steppe.

About 70 percent of the Sahara is covered by rocky and stone or gravel-covered denuded plateaus known as hammada, and about 15 percent is covered with sand dunes. The remaining 15 percent is occupied by high mountains, rare oases, and transitional regions. Major mountain ranges in the eastern Sahara include the uplifted margins of the Red Sea that form steep escarpments dropping more than 6,000 feet (2,000 m) from the Arabian Desert into the Red Sea coastal plain. Rocks in these mountains include predominantly Precambrian granitic gneisses, metasediments, and mafic schists of the Arabian Shield and are rich in mineral deposits, including especially gold that has been exploited by the Egyptians since Pharaonic times. The highest point in the Sahara is Emi Koussi in Chad, which rises to 10,860 feet (3,415 m), and the lowest point is the Qattara Depression.

High isolated mountain massifs rise from the plains in the central Sahara, including the massive Ahaggar (Hoggar) in southern Algeria, Tibesti in northern Chad, and Azbine (Air Mountains) in northern Niger. Ahaggar rises to more than 9,000 feet (2,740 m) and includes a variety of Precambrian crystalline rocks of the Ouzzalian Archean craton and surrounding Proterozoic shield. The Air Mountains, rising to more than 6,000 feet (1,830 m), are geologically a southern extension of the Ahaggar to the north, containing metamorphosed Precambrian basement rocks. Tibesti rises to more than 11,000 feet (3,350 m) and also includes a core of Precambrian basement rocks, surrounded by Paleozoic and younger cover. Northeast of Tibesti near the Egypt-Libya-Sudan border, the lower Oweineat (Uwaynat) Mountains form a similar dome, rising to 6,150 feet (1,934 m), and have a core of Precambrian igneous rocks.

The climate in Sahara is among the harshest on the planet, falling in the trade wind belt of dry descending air from Hadley circulation, with strong constant winds blowing from the northeast. These winds have formed elongate linear dunes in specific corridors across parts of the Sahara, with individual sand dunes continuous for hundreds of miles, and virtually no interdune sands. These linear dunes reach heights of more than 1,100 feet (350 m) and may migrate tens of feet or more per year. When viewed on a continental scale (as from space, or on a satellite image) these linear dunes display a curved trace, formed by the Coriolis force deflecting the winds and sand to the right of the movement direction (northeast to southwest). Most parts of the Sahara receive an average of less than five inches (12 cm) of rain a year, and this typically comes in a single downpour every few years, with torrential rains causing flash flooding. Rains of this type run off quickly, and relatively little is captured and returned to the groundwater system for future use. The air is extremely dry, with typical relative humidities ranging from 4 percent to 30 percent. Temperatures can be extremely hot, and the diurnal variation is high. The world's highest recorded temperature is from the Libyan Desert, 136°F (58°C) in the shade, during the fall of 1922. The temperature drop at night can be up to 90°F (30°C), even dropping below freezing after a scorching hot day.

Most of the Sahara is sparsely vegetated, with shrub brushes being common, along with grasses, and trees in the mountains. However, some desert oases and sections along the Nile River are extremely lush, and the Nile Valley has extensive agricultural development. Animal life is diverse, including gazelles, antelopes, jackals, badgers, hyenas, hares, gerbils, sheep, foxes, wild asses, weasels, baboons, mongooses, and hundreds of species of birds.

A variety of minerals are exploited from the Sahara, including major deposits of iron ore from Algeria and Mauritania, Egypt, Tunisia, Morocco, and Niger. Uranium deposits are found throughout the Sahara, with large quantities in Morocco. Manganese is mined in Algeria, and copper is found in Mauritania. Oil is exported from Algeria, Libya, and Egypt.

Much of the Sahara is underlain by vast groundwater reservoirs, both in shallow alluvial aquifers and in fractured bedrock aquifers. The water in these aquifers fell as rain thousands of years ago and reflects a time when the climate over North Africa was much different. In the Pleistocene, much of the Sahara experienced a wet and warm climate, and more than 20 large lakes covered parts of the region. The region experienced several alternations between wet and dry climates in the past couple of hundred thousand years, and active research projects are aimed at correlating these climate shifts with global events such as glacial and interglacial periods, sea surface and current changes (such as the El Niño–Southern Oscillation). The implications for understanding these changes are enormous, with millions of people affected by expansion of the Sahara, and undiscovered groundwater resources that could be used to sustain agriculture and save populations from being decimated. Many of the present drainage and wadi networks in the Sahara follow a drainage network established during the Pleistocene. In the Pliocene the shoreline of the Mediterranean was about 60 miles (40 km) south of its present location, when sea levels were about 300 feet (100 m) higher than today. Sand sheets and dunes, which are currently moving southward, have only been active for the past few thousand years. These are known to form local barriers to wadi channels in the Sahara, Sinai, and Negev Deserts and locally block wadis.

The sand of the Sahara and adjacent Northern Sinai probably originated by fluvial erosion of rocks in the uplands to the south and was transported from south to north by paleo-rivers during wetter climate times, then redistributed by wind. Dry climates such as the present, and low sea levels during glacial maxima, exposed the sediment to wind action that reshaped the fluvial deposits into dunes, whose form

depended on the amount of available sand and prevailing wind directions. This hypothesis was developed to suggest the presence of a drainage network to transport fluvial sediments. Indeed, numerous channels incise into the limestone plateau of the central and northern Sahara, and many lead to elongate areas that have silt deposits. Several of these deposits have freshwater fauna and are interpreted as paleo-lakes and long-standing slack water deposits from floods.

Plio-Pleistocene lakebed sediments have also been identified in many places in the mountains in the Sahara, where erosionally resistant dikes that formed dams in steep-walled bedrock canyons controlled the lakes. The paleo-lake sediments consist of silts and clays interbedded with sands and gravels, cut by channel deposits. These types of lakebeds were formed in a more humid late Plio-Pleistocene climate, based on fossil roots and their continuity with wadi terraces of that age.

The fluvial history of the region reflects earlier periods of greater effective moisture, as evident also from archaeological sites associated with remnants of travertines and playa or lake deposits. An early Holocene pluvial cycle is well documented by archaeological investigations at Neolithic playa sites in Egypt. Late Pleistocene lake deposits with associated early and middle Paleolithic archaeological sites are best known from work in the Bir Tarfawi area of southwest Egypt. Similar associations occur in northwest Sudan and Libya.

An extensive network of sand-buried river and stream channels in the eastern Sahara appears on shuttle imaging radar images. Calcium carbonate associated with some of these buried river channels is thought to have precipitated in the upper zone of saturation during pluvial episodes, when water tables were high. As documented by radiocarbon dating and archaeological investigations, the eastern Sahara experienced a period of greater effective moisture during early and middle Holocene time, about 10,000–5,000 years ago. Uranium-series dating of lacustrine carbonates from several localities indicated that five paleo-lake forming episodes occurred at about 320–250, 240–190, 155–120, 90–65 and 10–5 thousand years ago. These five pluvial episodes may be correlated with major interglacial stages.

These results support the contention that past pluvial episodes in North Africa correspond to the interglacial periods. Isotopic dating results and field relationships suggest that the oldest lake and groundwater-deposited carbonates were more extensive than those of the younger period, and the carbonates of the late wet periods were geographically localized within depressions and buried channels.

This archaeological evidence of previous human habitation, coupled with remains of fauna and flora, suggests the presence of surface water in the past. Indeed, remains of lakes and segments of dry river and stream channels occur throughout the Sahara. Archaeological evidence of human habitation during the early Holocene was recently uncovered in the northeast Sinai Peninsula where an early Middle Paleolithic site shows evidence for habitation at 33,800 years BP.

See also DESERT; FRACTURE ZONE AQUIFERS; NILE RIVER.

Further Reading

Arvidson, Ray E., R. Becker, A. Shanabrook, W. Luo, Neil C. Sturchio, Mohamed Sultan, Z. Lofty, A. M. Mahmood, and Z. El Alfy. "Climatic, Tectonic, and Eustatic Controls on Quaternary Deposits and Landforms, Red Sea Coast, Egypt." *Journal of Geophysical Research* 99 (1994).

Brookes, I. A. "Geomorphology and Quaternary Geology of the Dakhla Oasis Region, Egypt." *Quaternary Science Reviews* 12 (1993).

Burke, Kevin., and G. L. Wells. "Trans-African Drainage System of the Sahara: Was It the Nile?" *Geology* 17 (1989).

Crombie, M. K., Ray E. Arvidson, Neil C. Sturchio, Z. El-Alfy, and K. Abu Zeid. "Age and Isotopic Constraints on Pleistocene Pluvial Episodes in the Western Desert, Egypt." *Paleogeography, Paleoclimatology, and Paleoecology,* v. 130 (1997).

El-Baz, Farouk. "Origin and Evolution of the Desert." *Interdisciplinary Science Reviews* 13 (1988).

El-Baz, Farouk, Timothy M. Kusky, Ibrahim Himida, and Salel Abdel-Mogheeth. "Ground Water Potential of the Sinai Peninsula, Egypt." *Ministry of Agriculture and Land Reclamation, Cairo* (1998).

Guiraud, R. "Mesozoic Rifting and Basin Inversion along the Northern African–Arabian Tethyan Margin: An Overview." In *Petroleum Geology of North Africa,* edited by D. S. MacGregor, R. T. J. Moody, and D. D. Clark-Lowes. *Geological Society of London Special Publication* 133 (1998).

Haynes Jr., C. Vince. "Quaternary Studies, Western Desert, Egypt and Sudan—1979–1983 Field Seasons." *National Geographic Society Research Reports* 16 (1985).

———. "Great Sand Sea and Selima Sand Sheet: Geochronology of Desertification." *Science* 217 (1982).

Haynes Jr., C. Vince, C. H. Eyles, L. A. Pavlish, J. C. Rotchie, and M. Ryback. "Holocene Paleoecology of the Eastern Sahara: Selima Oasis." *Quaternary Science Reviews* 8 (1989).

Henning, D., and H. Flohn. "Climate Aridity Index Map." *United Nations Conference on Desertification,* U.N.E.P., Nairobi, Kenya, 1977.

Klitzsch, E. "Geological Exploration History of the Eastern Sahara." *Geologische Rundschau* 83 (1994).

Kropelin, S. "Paleoclimatic Evidence from Early to Mid-Holocene Playas in the Gilf Kebir, Southwest Egypt." *Paleoecology Africa,* v. 18 (1987).

Kusky, Timothy M., Mohamed A. Yahia, Talaat Ramadan, and Farouk El-Baz. "Notes on the Structural and Neotectonic Evolution of El-Faiyum Depression, Egypt: Relationships to Earthquake Hazards." *Egyptian Journal of Remote Sensing and Space Sciences* 2 (2003).

McCauley, J. F., G. G. Schaber, C. S. Breed, M. J. Grolier, C. Vince Haynes Jr., B. Issawi, C. Elachi, and R. Blom. "Subsurface Valleys and Geoarchaeology of the Eastern Sahara Revealed by Shuttle Radar." *Science* 218 (1982).

Pachur, H. J., and S. Kropelin. "Wadi Howar; Paleoclimate Evidence from an Extinct River System in the Southeastern Sahara." *Science* 237 (1985).

Pachur, H. J., and G. Braun. "The Paleoclimate of the Central Sahara, Libya, and the Libyan Desert." *Paleoecology Africa* 12 (1980).

Pachur, H. J., H. P. Roper, S. Kropelin, and M. Goschin. "Late Quaternary Hydrography of the Eastern Sahara." *Berliner Geowissenschaftliche Abhandlungen, Reihe A: Geologie und Palaeontologie* 75 (2) (1987).

Prell, W. L., and J. E. Kutzbach. "Monsoon Variability Over the Past 150,000 Years." *Journal of Geophysical Research* 92 (1987): 8,411–8,425.

Sestini, G. "Tectonic and Sedimentary History of the NE African Margin (Egypt/Libya)." In *The Geological Evolution of the Eastern Mediterranean,* edited by J. E. Dixon, and A. H. F. Robertson: 161–175. Oxford: Blackwell Scientific Publishers, 1984.

Szabo, B. J., C. Vince Haynes Jr., and Ted A. Maxwell. "Ages of Quaternary Pluvial Episodes Determined by Uranium-Series and Radiocarbon Dating of Lacustrine Deposits of Eastern Sahara." *Paleogeography, Paleoclimatology, and Paleocology* 113 (1995).

Szabo, B. J., W. P. McHugh, G. G. Shaber, C. Vince Haynes Jr., and C. S. Breed. "Uranium-Series Dated Authigenic Carbonates and Acheulian Sites in Southern Egypt." *Science* 243 (1989).

Wendorf, F., A. E. Close, and R. Schild. "Recent Work on the Middle Paleolithic of the Eastern Sahara." *African Archaeology Review* 5 (1987).

Wendorf, F., and R. Schild. *Prehistory of the Eastern Sahara.* New York: Academic Press, 1980.

Sahel, Africa

Sahel, Africa   A large shifting area of sub-Saharan Africa, north of the savannas of southern Africa. The Sahel includes parts of Senegal, Mauritania, Mali, Burkina Faso, Nigeria, Niger, Chad, Sudan, Ethiopia, and Eritrea.

The Sahel region offers one of the world's most tragic examples of how poorly managed agricultural practices, when mixed with long-term drought conditions, can lead to disaster and permanent desertification. Sahel means boundary in Arabic, and the Sahel forms the southern boundary of the world's largest desert, the Sahara. It is home to about 25 million people, most of whom are nomadic herders and subsistence farmers. In the summer months of June and July, heating normally causes air to rise and this is replaced by moist air from the Atlantic, which brings 14–23 inches (35.5–58.5 cm) of rain per year. In the Sahel the normal northward movement of the wet intertropical convergence zone stopped during an El-Niño-Southern Oscillation (ENSO) event in 1968. Further climatic changes in the 1970s led to only about half of the normal rain falling until 1975. The additional lack of moisture brought on complications from the temperature cycles of the northern and southern oceans becoming out of synchronicity at this time, and the region suffered long-term drought and permanent desertification.

As the rains continued to fail to come, and the air masses continued to evaporate surface water, the soil moisture was drastically reduced, which further reduced evaporation and cloud cover. The vegetation soon died off, the soils became dry and hot, and near surface temperatures were further increased. Soon the plants were gone, the soils were exposed to the wind, and the region became plagued with blowing dust and sand. Approximately 200,000 people died, and 12 million head of livestock perished. Parts of the region were altered to desert, with little chance of returning to the previous state.

The desertification of the Sahel was enhanced by the agricultural practices of the people of the region. Nomadic and marginal agriculture was strongly dependent on the monsoon, and when the rains did not come for several years, the natural and planted crops died and many of the remaining plants were used as fuel for fires to offset the cost of fuel. This practice greatly accelerated the desertification process. The Sahara is now thought to be overtaking the Sahel by

migrating southward at approximately three miles per year (4.8 km/yr). Much of Africa including the Sahel region has become increasingly dry and desert-like over the past hundred years or more, and any attempts to restart agriculture and repopulate regions evacuated during previous famines in this region may be fruitless and lead to further loss of life.

See also DESERT; SAHARA DESERT.

salt dome *See* DIAPIR.

saltwater intrusion Encroachment of seawater into drinking and irrigation wells is an increasing problem for many

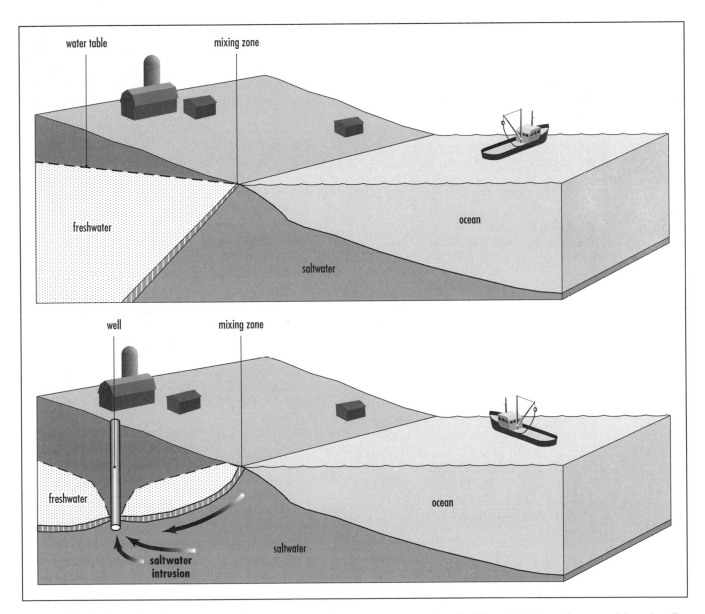

Saltwater intrusion. A wedge of low-density freshwater overlies denser saltwater in the coastal aquifer. When the freshwater is extracted through wells, the salty water moves up to replace the lost water, ruining the aquifer.

coastal communities around the world. Porous soils and rocks beneath the groundwater table in terrestrial environments are generally saturated with freshwater, whereas porous sediment and rock beneath the oceans are saturated with salt water. In coastal environments there must be a boundary between the fresh groundwater and the salty groundwater. In some cases this is a vertical boundary, whereas in other cases the boundary is inclined, with the denser salt water lying beneath the lighter freshwater. In areas where there is complex or layered stratigraphy the boundary may be complex, consisting of many lenses.

In normal equilibrium situations the boundary between the fresh and salty water remains rather stationary. In times of drought the boundary may move landward or upward, and in times of excessive precipitation the boundary may move seaward and downward.

Many coastal communities have been highly developed, with many residential neighborhoods, cities, and agricultural users obtaining their water from groundwater wells. When these wells pump more water out of coastal aquifers than is replenished by new rainfall and other inputs to the aquifer the freshwater lens resting over the saltwater lens is depleted. This causes the salt water to move in to the empty pore spaces to take the place of the freshwater. Eventually as pumping continues the freshwater lens becomes so depleted that the wells begin to draw salt water out of the aquifer, and the well becomes effectively useless. This is called saltwater intrusion or encroachment. In cases of severe drought the process may be natural, but in most cases seawater intrusion is caused by overpumping of coastal aquifers, aided by drought conditions.

Many places in the United States have suffered from seawater intrusion. For instance, many east coast communities have lost use of their wells and had to convert to water piped in from distant reservoirs for domestic use. In a more complicated scenario, western Long Island, New York, experienced severe seawater intrusion into its coastal aquifers because of intense overpumping of its aquifers in the late 1800s and early 1900s. Used water that was once returned to the aquifer by septic systems began to be dumped directly into the sea when sewers were installed in the 1950s, with the result being that the water table dropped more than 20 feet over a period of 20 years. This drop was accompanied by additional seawater intrusion. The water table began to recover in the 1970s when much of the area converted to using water pumped in from reservoirs in the Catskill Mountains to the north of New York City.

See also GROUNDWATER.

San Andreas Fault The transcurrent plate boundary between the North American and Pacific plates in southern California is defined by a number of related faults known as the San Andreas Fault system. The system consists of a number of predominantly right-lateral (dextral) slip faults that accommodate most of the relative motion between North America and the Pacific plate and form a belt about 50–100 miles (80–160 km) wide. Movement along these faults in southern California has caused a number of destructive earthquakes, including the Fort Tejon earthquake of 1857, the 1906 San Francisco earthquake, the 1989 Loma Prieta earthquake, and many others. The fault system has been active for hundreds of millions of years, and as relative plate motions continue between the Pacific and North American plates, it is certain that more large earthquakes will be generated along the system. Since southern California is so densely populated, the San Andreas is one of the most intensely studied and monitored faults in the world.

The San Andreas Fault system extends from an incipient oceanic spreading center in the Gulf of California, past San Diego, Los Angeles, and San Francisco, then joins with the Mendocino transform at Cape Mendocino, where the North American, Pacific, and Gorda plates meet. Individual fault zones along this system range from 0.3 to 0.6 mile (0.5–1 km) wide and extend along strike for hundreds of kilometers. These fault zones are all associated with a number of smaller fault elements, including faults, fault branches, and fault strands.

A number of major elements of the San Andreas Fault system include the main San Andreas Fault, the Hayward and Calaveras Faults (and their extensions) in central California, and the San Jacinto and Elsinore Faults in southern California. North of Los Angeles the San Andreas is met at a high angle by the Garlock Fault, which is linked to the active Owens Valley Fault. Most of the faults of the San Andreas system trend about N35°–40°W, but there are several major bends in the system. The faults bend in a left-stepping manner in a compressional bend north of Los Angeles near Santa Barbara, forming the Transverse Ranges, as relative motion between either side of the fault is compressional, uplifting the mountains in the area. There are many thrust faults and folds associated with the fault in this area. East of San Diego the fault system steps in the opposite direction to the right, forming an extensional bend. The Mojave segment of the San Andreas terminates along the east side of the Salton Sea, and motion is picked up on the Imperial Fault on the west side of the sea. Relative motion across this step is extensional, causing stretching and motion on normal faults, and subsidence in a basin now occupied by the Salton Sea.

The San Andreas Fault system is segmented into a number of different sections with different patterns of behavior, and likelihood of major earthquakes. Some sections are characterized by a slow and steady creep without major earthquake events, whereas other segments move by stick-slip behavior where long periods of quiescence are interrupted by great earthquake events when major slip occurs suddenly. The main segments of the fault, rated from most likely to least likely to slip include the Parkfield segment, the Coachilla Valley, Mojave, San Bernadino Mountains, San

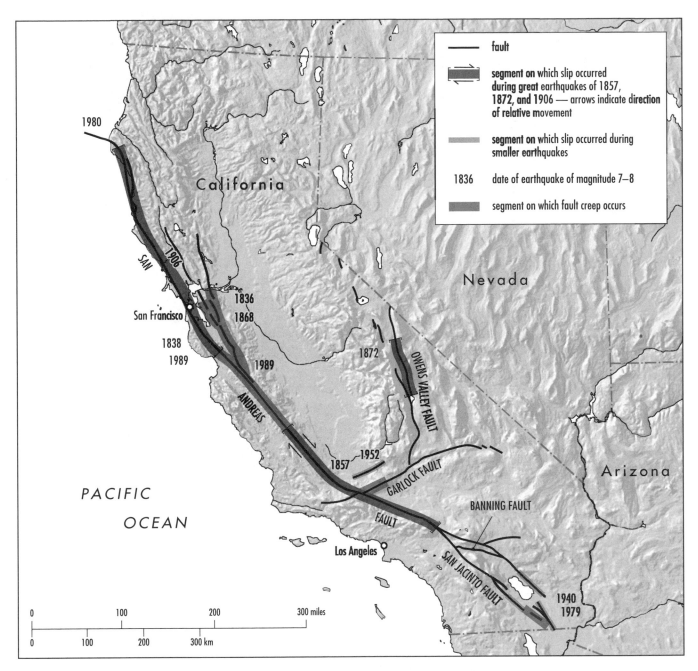

Map of the San Andreas Fault system showing different segments of the fault where different types of behavior (such as stick-slip and creep) occur. The map also shows where major earthquakes have occurred in the past couple of hundred years.

Francisco Peninsula, Carizzo Plain, and Southern Santa Cruz Mountains that slipped in 1989.

See also TRANSFORM PLATE MARGIN PROCESSES.

Further Reading
Atwater, Tanya. "Implications of Plate Tectonics for the Cenozoic Tectonic Evolution of Western North America." *Geological Society of America Bulletin* 81 (1970).

Wallace, Robert E. *The San Andreas Fault System, California.* U.S. Geological Survey Professional Paper No. 1515 (1990).

sand dunes Geometrically regular mounds or ridges of sand that are found in several geological environments, including deserts and beaches. Most people think of deserts as areas covered with numerous big sand dunes and continual swirling winds of dust storms. Really, dunes and dust

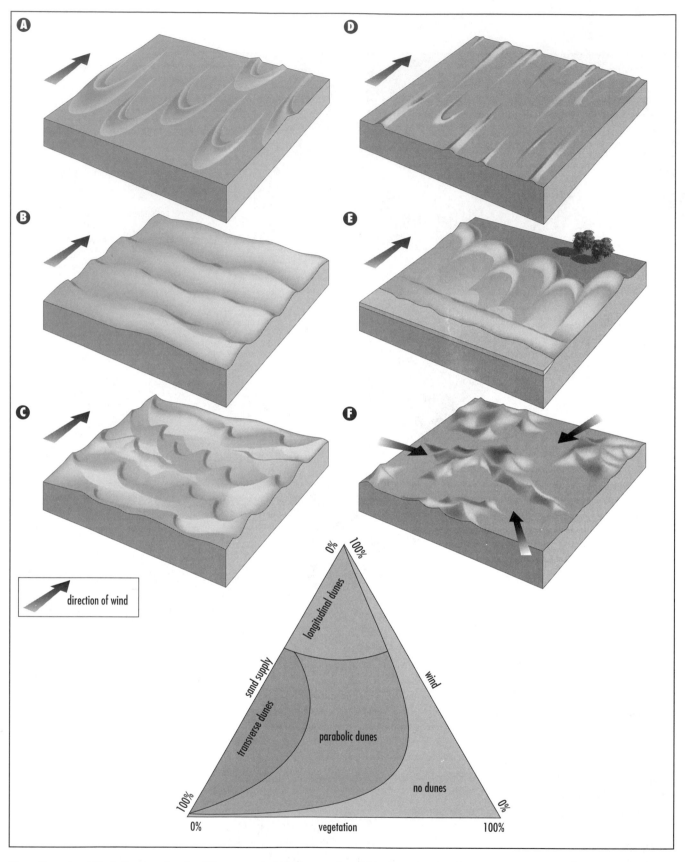

Three-dimensional block diagrams showing different morphological types of sand dunes produced by different combinations of wind strength, sand supply, and density of vegetation

storms are not as common as depicted in popular movies, and rocky deserts are more common than sandy deserts. For instance, only about 20 percent of the Sahara Desert is covered by sand, and the rest is covered by rocky, pebbly, or gravel surfaces. However, sand dunes are locally very important in deserts, and wind is one of the most important processes in shaping deserts worldwide. Shifting sands are one of the most severe geologic hazards of deserts. In many deserts and desert border areas the sands are moving into inhabited areas, covering farmlands, villages, and other useful land with thick accumulations of sand. This is a global problem, as deserts are currently expanding worldwide. The Desert Research Institute in China has recently estimated that in China alone, 950 square miles (2,500 km²) are encroached on by migrating sand dunes from the Gobi Desert each year, costing the country $6.7 billion per year and affecting the lives of 400 million people.

Wind moves sand by saltation—an arching path, in a series of bounces or jumps. You can see this often by looking close to the surface in dunes on beaches or deserts. Wind typically sorts different sizes of sedimentary particles, forming elongate small ridges known as sand ripples, very similar to ripples found in streams. Sand dunes are larger than ripples (up to 1,500 feet high, or almost 0.5 km) and are composed of mounds or ridges of sand deposited by wind. These may form where an obstacle distorts or obstructs the flow of air, or they may move freely across much of a desert surface. Dunes have many different forms, but all are asymmetrical. They have a gentle slope that faces into the wind, and a steep slope that faces away from the wind. Sand particles move by saltation up the windward side, and fall out near the top where the pocket of low-velocity air cannot hold the sand anymore. The sand avalanches, or slips down the leeward slope, known as the slip face. This keeps the slope at 30–34°, the angle of repose. The asymmetry of old dunes is used to tell the directions ancient winds blew.

The steady movement of sand from one side of the dune to the other causes the whole dune to migrate slowly downwind (typically about 80–100 feet per year, or 24–30 m/yr), burying houses, farmlands, temples, and towns. Rates of dune migration of up to 350 feet per year (107 m/yr) have been measured in the Western Desert of Egypt, and the Ningxia Province of China.

A combination of many different factors led to the formation of very different types of dunes, each with a distinctive shape, potential for movement, and hazards. The main variables that determine a dune's shape are the amount of sand that is available for transportation, the strength (and directional uniformity) of the wind, and the amount of vegetation that covers the surface. If there is a lot of vegetation and little wind, no dunes will form. In contrast, if there is very little vegetation, a lot of sand, and moderate wind strength (conditions that you might find on a beach) then a group of dunes

Sand dunes, United Arab Emirates. Burchan dunes have arms and steep slip surfaces that point in the direction of dune movement, in this case toward the Persian Gulf (on the left). Note smaller sand ripples on dune surfaces. *(Photo by Timothy Kusky)*

known as transverse dunes form, with the dune crests aligned at right angles to the dominant wind direction.

Barchan dunes have crescent shapes and have horns pointing downwind, and they form on flat deserts with steady winds and a limited sand supply. Parabolic dunes have a U-shape with the U facing upwind. These form where there is significant vegetation that pins the tails of migrating transverse dunes, with the dune being warped into a wide U-shape. These dunes look broadly similar to barchans, except the tails point in the opposite direction. They can be distinguished because in both cases, the steep side of the dune points away from the dominant wind's direction. Linear dunes are long, straight ridge-shaped dunes elongate parallel to the wind direction. These occur in deserts with little sand supply and strong, slightly variable winds. Star dunes form isolated or irregular hills where the wind directions are irregular.

See also DESERT.

sand sea Deserts covering vast expanses covered by thick sands, including sand dunes of several types and by an absence of other geographic features are known as sand seas, or locally as ergs in the North African Sahara. Interdune areas may be covered by relatively flat tabular sand sheets, or even evaporite basins (sabkhas). Sand seas are abundant in parts of the Sahara of North Africa, the Namib of southern Africa, the Rub' al-Khali (Empty Quarter) of Arabia, the Great Sandy Desert of Australia, the Gobi Desert of Asia, and in the Nebraska Sand Hills of Nebraska.

Sand seas form where the velocity of the transporting wind decreases, dropping its load. The decreased velocity may be caused by a number of factors including their location in topographic lows, or adjacent to topographic barriers such as mountains that cut across the direction of sand transport. A striking example of this process is found in the Wahiba Sand Sea of Oman. Here, the Eastern Hajar Mountains terminate the northward-flowing Wahiba sands, and an intermittent river system at the base of the mountains removes sand that gets close to the mountain front, carrying it to the coast of the Arabian Sea. Longshore transport then carries this sand southward where winds pick it up from beaches and cause it to reenter the Wahiba sand sheet in the south, forming a sort of sand gyre. Sand seas may also form where a large body of water intercepts drifting sand, or where the sand is carried into shifting climate zones where the wind strength decreases.

Surface features in sand seas include bed forms of a variety of scales ranging from several different types of ripples that may be up to an inch (several cm) high, to dunes that are typically up to 300 feet (100 m) tall, to huge bedforms called draa that are giant dunes up to 1,650 feet (500 m) tall, with wavelengths of up to several kilometers. These bedforms are typically superimposed on each other, with dunes migrating over

Dunes, sand ripples, and interdune areas inside Wahiba sand sea *(Photo by Timothy Kusky)*

Landsat thematic mapper image of Wahiba sand sea of northeastern Oman *(Photo by Farouk El-Baz, processed at Boston University Center for Remote Sensing)*

draa, and several different sets of ripples migrating over the dunes. The wind directions inferred from the different sets may also be different, with ripples reflecting the most recent winds, dunes the dominant winds over different seasons, and draa reflecting the very long-term direction of wind in the basin.

See also DESERT.

Further Reading

El-Baz, Farouk. "Origin and Evolution of Sand Seas in the Great Sahara and Implications to Petroleum and Groundwater Exploration." In *Geology of the Arab World,* edited by A. Sadak, 3–17. Cairo: Cairo University Press, 1992.

———. "Origin and Evolution of the Desert." *Interdisciplinary Science Reviews* 13 (1988): 331–347.

McKee, E. D., ed., *A Study of Global Sand Seas.* U.S. Geological Survey Professional Paper 1052 (1979).

sandstone Any medium-grained clastic sedimentary rock composed predominantly of rounded or angular sand-sized particles, set in a finer-grained matrix composed of silt or clay-sized particles, and indurated or cemented by an agent such as carbonate, silica, or iron. It is the consolidated equivalent of sand, and in most cases it is composed of dominantly quartz particles. The names of sandstones that are not composed of 80–90 percent quartz are typically preceded by a prefix describing the main constituents, such as olivine-sandstone, garnet-sandstone, lithic-sandstone, and so on. There is a wide variety of textural and compositional types of sandstones deposited in an equally diverse set of environments ranging from fluvial and aeolian settings to deep marine settings. Various types of sandstones comprise about 10–15 percent of the total amount of sedimentary rocks in the Earth's crust.

Grain size may vary in sandstones between very fine, fine, medium, and coarse-grained sands, and the amount of variation in grain sizes in the sediment is described by the term *sorting,* where well-sorted refers to a fairly uniform grain size, and poorly sorted refers to a wide variation in grain size. The angularity of the edges of individual grains is described by the term *rounding,* and more rounded grains tend to be the ones that have traveled the greatest distance from the source rock. Additionally, windblown grains tend to be more highly rounded and even frosted than grains transported by water.

Numerous studies over several decades have shown that the composition of the major, minor, and heavy-mineral constituents of sandstones may reflect the composition of the source terrane, the weathering processes, the processes that transported the sediment, and the diagenetic and metamorphic effects. Compositionally sandstones are divided into two main groups including the arenites and the wackes. Arenites have less than 15 percent fine-grained matrix material between the sand particles, whereas wackes have more than 15 percent clay and fine-grained material in the matrix. Arkoses are feldspar-rich sandstones that are typically found close to their source in continental settings, since feldspar is unstable and alters to clay minerals with weathering and transportation.

See also CLASTIC ROCKS; SEDIMENTARY ROCKS.

Further Reading

Dickinson, William R., and Chris A. Suczek. "Plate Tectonics and Sandstone Composition." *American Association of Petroleum Geologists Bulletin* V (1978): 2,164–2,182.

Pettijohn, Francis J., D. E. Potter, and Raymond Siever. *Sand and Sandstone.* Berlin: Springer-Verlag, 1975.

Santa Ana winds

Warm, dry winds that blow out of the mountains and desert toward the west or southwest into southern California. They come off the elevated desert plateau of the southwest United States and funnel through the San Bernadino and San Gabriel Mountains, spreading out over Los Angeles, the San Fernando Valley, Santa Barbara, and San Francisco. Similar winds known as the California "northers" can affect northern California. Santa Ana winds develop when high-pressure systems build over the great basin of Utah, Nevada, Arizona, and New Mexico. Clockwise rotation around the high-pressure system forces air downslope into California, and the dry air is compressed, and heated and dried further by compressional heating. Santa Ana winds can be quite dangerous, since they may blow at more than 100 miles per hour (161 km/hr), and they often occur in the autumn or winter, when the brush is extremely dry and prone to flash fires. Some disastrous fires fueled by dry brush and agitated by Santa Ana winds have devastated California communities, causing billions of dollars in losses, including the fires in Bel Air in 1961, and numerous fires in Oakland, Los Angeles, and Santa Barbara in 1991 and 1993.

See also CHINOOK WINDS.

satellite imagery

Satellite imagery forms one of the basic tools for remote sensing, which includes a wide variety of methods to determine properties of a material from a distance. Remote sensing can include subjects as diverse as seismology, satellite image interpretation, and magnetics. The types of satellite images available to the geologist are expanding rapidly, and only the most common in use are discussed here.

The Earth Resources Technology Satellite (ERTS-1) was the first unmanned digital imaging satellite that was launched on July 23, 1972. Four other satellites from the same series, later named Landsat, were launched at intervals of a few years. The Landsat spacecraft carried a Multi-Spectral Scanner (MSS), a Return Beam Vidicon (RBV), and later, a Thematic Mapper (TM) imaging system.

Landsat Multi-Spectral Scanners produce images representing four different bands of the electromagnetic spectrum. The four bands are designated band 4 for the green spectral region (0.5 to 0.6 micron); band 5 for the red spectral region (0.6 to 0.7 micron); band 6 for the near-infrared region (0.7 to 0.8 micron); and band 7 for another near-infrared region (0.8 to 1.1 micron).

Radiation reflectance data from the four scanner channels are converted first into electrical signals, then into digital form for transmission to receiving stations on Earth. The recorded digital data are reformatted into what we know as computer compatible tapes (CCT) and/or converted at special processing laboratories to black-and-white images. These images are recorded on four black-and-white films, from which photographic prints are made in the usual manner.

The black-and-white images of each band provide different sorts of information because each of the four bands records a different range of radiation. For example, the green band (band 4) most clearly shows underwater features, because of the ability of "green" radiation to penetrate shallow water, and is therefore useful in coastal studies. The two near-infrared bands, which measure the reflectance of the Sun's rays outside the sensitivity of the human eye (visible range) are useful in the study of vegetation cover.

When these black-and-white bands are combined, false-color images are produced. For example, in the most popular combination of bands 4, 5 and 7 the red color is assigned to the near-infrared band number 7 (and green and blue to bands 4 and 5 respectively). Vegetation appears red because plant tissue is one of the most highly reflective materials in the infrared portion of the spectrum, and thus, the healthier the vegetation, the redder the color of the image. Also, because water absorbs nearly all infrared rays, clear water appears black on band 7. Therefore, this band cannot be used to study features beneath water even in the very shallow coastal zones. However, it is very useful in delineating the contact between water bodies and land areas.

The Return Beam Vidicon (RBV) was originally flown in the interest of the mapping community. It offered better geometric accuracy and ground resolution (130 feet; 40 m) than was available from the Multi-Spectral Scanner (260 feet/80 m resolution) with which the RBV shared space on Landsats 1, 2, and 3. The RBV system contained three cameras that operated in different spectral bands: blue-green, green-yellow, and red-infrared. Each camera contained an optical lens, a shutter, the RBV sensor, a thermoelectric cooler, deflection and focus coils, erase lamps, and the sensor electronics. The three RBV cameras were aligned in the spacecraft to view the same 70-square mile (185-km^2) ground scene as the MSS of Landsat. Although the RBV is not in operation today, images are available and can be utilized in mapping.

The Thematic Mapper (TM) is a sensor that was carried first on Landsat 4 and 5 with seven spectral bands covering the visible, near infrared, and thermal infrared regions of the spectrum. It was designed to satisfy more demanding performance parameters from experience gained in the operation of the MSS with a ground resolution of 100 feet (30 m).

The seven spectral bands were selected for their band passes and radiometric resolutions. For example, band 1 of the Thematic Mapper coincides with the maximum transmissivity of water and demonstrates coastal water-mapping capabilities superior to those of the MSS. It also has beneficial features for the differentiation of coniferous and deciduous vegetation. Bands 2–4 cover the spectral region that is most significant for the characterization of vegetation. Vegetation and soil moisture may be estimated from band 5 readings, and plant transpiration rates may be estimated from the thermal mapping in band 6. Band 7 is primarily motivated by geological applications, including the identification of rocks altered by percolating fluids during mineralization. The band profiles, which are narrower than those of the MSS, are specified with stringent tolerances, including steep slopes in spectral response and minimal out-of-band sensitivity.

TM band combinations of 7 (2.08–2.35 μm), 4 (0.76–0.90 μm), and 2 (0.50–0.60 μm) are commonly used for geological studies, due to the ability of this combination to discriminate features of interest, such as soil moisture anomalies, lithological variations, and to some extent, mineralogical composition of rocks and sediments. Band 7 is typically assigned to the red channel, band 4 to green, and band 2 to blue. This procedure results in a color composite image; the color of any given pixel represents a combination of brightness values of the three bands. With the full dynamic range of the sensors, there are 16.77×10 possible colors. By convention, this false-color combination is referred to as TM 742 (RGB). In addition to the TM 742 band combination, the thermal band (TM band 6; 10.4–12.5 μm) is sometimes used in geology because it contains useful information potentially relevant to hydrogeology.

The French Système pour l'Observation de la Terre (SPOT) obtains data from a series of satellites in a sun-synchronous 500-mile (830-km) high orbit, with an inclination of 98.7°. The SPOT system was designed by the Centre Nationale d'Etudes Spaciales (CNES) and built by the French industry in association with partners in Belgium and Sweden.

Like the American Landsat it consists of remote sensing satellites and ground receiving stations. The imaging is accomplished by two High-Resolution Visible (HRV) instruments that operate in either a panchromatic (black-and-white) mode for observation over a broad spectrum, or a multispectral (color) mode for sensing in narrow spectral bands. The ground resolutions are 33 and 66 feet (10 and 20 m) respectively. For viewing directly beneath the spacecraft, the two instruments can be pointed to cover adjacent areas. By pointing a mirror that directs ground radiation to the sensors, it is possible to observe any region within 280 miles (450 km) from the nadir, thus allowing the acquisition of stereo photographs for three-dimensional viewing and imaging of scenes as frequently as every four days.

Radar is an active form of remote sensing, where the system provides a source of electromagnetic energy to "illuminate" the terrain. The energy returned from the terrain is detected by the same system and is recorded as images. Radar systems can be operated independently of light conditions and can penetrate cloud cover. A special characteristic of radar is the ability to illuminate the terrain from an optimum position to enhance features of interest.

Airborne radar imaging has been extensively used to reveal land surface features. However, until recently it has not been suitable for use on satellites because: (1) power requirements were excessive; and (2) for real-aperture systems, the azimuth resolution at the long slant ranges of spacecraft would be too poor for imaging purposes. The development of new power systems and radar techniques has overcome the first problem and synthetic-aperture radar systems have remedied the second.

The first flight of the Shuttle Imaging Radar (SIR-A) in November of 1981 acquired images of a variety of features including faults, folds, outcrops, and dunes. Among the revealed features are the sand-buried channels of ancient river and stream courses in the Western Desert of Egypt. The second flight, SIR-B, had a short life; however, the more advanced and higher resolution SIR-C was flown in April 1994 (and was again utilized in August 1994). The SIR-C system acquired data simultaneously at two wavelengths: L band (23.5 cm) and C band (5.8 cm). At each wavelength both horizontal and vertical polarizations are measured. This provides dual frequency and dual polarization data, with a swath width between 18 and 42 miles (30 and 70 km), yielding precise data with large ground coverage.

Different combinations of polarizations are used to produce images showing much more detail about surface geometric structure and subsurface discontinuities than a single-polarization-mode image. Similarly, different wavelengths are used to produce images showing different roughness levels since radar brightness is most strongly influenced by objects comparable in size to the radar wavelength; hence, the shorter wavelength C band increases the perceived roughness.

Interpretation of a radar image is not intuitive. The mechanics of imaging and the measured characteristics of the target are significantly different for microwave wavelengths than the more familiar optical wavelengths. Hence, possible geometric and electromagnetic interactions of the radar waves with anticipated surface types have to be assessed prior to their examination. In decreasing order of effect, these qualities are surface slope, incidence angle, surface roughness, and the dielectric constant of the surface material.

Radar is uniquely able to map the geology at the surface and, in the dry desert environments, up to a maximum 30 feet (10 m) below the surface. Radar images are most useful in mapping structural and morphological features, especially fractures and drainage patterns, as well as the texture of rock types, in addition to revealing sand-covered paleochannels. The information contained in the radar images complements that in the Thematic Mapper (TM) images. It also eliminates the limitations of Landsat when only sporadic measurements can be made; radar sensors have the ability to "see" at night and through thick cloud cover since they are active rather than passive sensors.

Radarsat is an earth observation satellite developed by Canada, designed to support both research on environmental change and research on resource development. It was launched in 1995 on a Delta II rocket with an expected life span of five years. Radarsat operates with an advanced radar sensor called Synthetic Aperture Radar (SAR). The synthetic aperture increases the effective resolution of the imaged area by means of an antenna design in which the spatial resolution of a large antenna is synthesized by multiple sampling from a small antenna. Radarsat's SAR-based technology provides its own microwave illumination, thus can operate day or night, regardless of weather conditions. As such, resulting images are not affected by the presence of clouds, fog, smoke, or darkness. This provides significant advantages in viewing under conditions that preclude observation by optical satellites. Using a single frequency, 5 cm horizontally polarized C band, the Radarsat SAR can shape and steer its radar beam to image swaths between 20 and 300 miles (35 km to 500 km), with resolutions from 33 feet to 330 feet (10 m to 100 m), respectively. Incidence angles can range from less than 20° to more than 50°.

The Space Shuttle orbiters have the capability of reaching various altitudes, which allows the selection of the required photographic coverage. A camera that was specifically designed for mapping the Earth from space using stereo photographs was first flown in October 1984 on the Space Shuttle Challenger Mission 41-G. It used an advanced, specifically designed system to obtain mapping-quality photographs from Earth orbit. This system consisted of the Large Format Camera (LFC) and the supporting Attitude Reference System (ARS). The LFC derives its name from the size of its individual frames, which are 26 inches (66 cm) in length and 9 inches (23 cm) in width. The 992-pound (450-kg) camera has a 12-inch (305-mm) f/6 lens with a 40° × 74° field of view. The film, which is three-fourths of a mile (1,200 m) in length, is driven by a forward motion compensation mechanism as it is exposed on a vacuum plate, which keeps it perfectly flat (Doyle, 1985). The spectral range of the LFC is 400 to 900 nanometers, and its system resolution is 100 lines per millimeter at 1,000:1 contrast and 88 lines per millimeter at 2:1 contrast. This adds up to photo-optical ground resolution of 33–66 feet (10–20 m) from an altitude of 135 miles (225 km) in the 34,200-square-mile (57,000-km²) area that is covered by each photograph. The uniformity of illumination of within 10 percent minimizes vignetting. The framing rate of 5 to 45 seconds allows its operation from various spacecraft altitudes.

The ARS is composed of two cameras with normal axes that take 35-millimeter photographs of star fields at the same instant as the LFC takes a photograph of the Earth's surface. The precisely known positions of the stars allow the calculation of the exact orientation of the Shuttle orbiter, and particularly of the LFC in the Shuttle cargo bay. This accurate orientation data, together with the LFC characteristics, allows the location of each frame with an accuracy of less than half a mile (1 km) and the making of topographic maps of photographed areas at scales of up to 1:50,000.

See also REMOTE SENSING.

Saturn A giant, gaseous planet, Saturn is the sixth planet residing between Jupiter and Uranus, orbiting at 9.54 astronomical units (888 miles, or 1,430 million kilometers) from the Sun, twice the distance from the center of the solar system as Jupiter, and having an orbital period of 29.5 Earth years. The mass of Saturn is 95 times that of Earth, yet it rotates at more than twice the rate of Earth. The average density of this gaseous planet is only 0.7 grams/cm³, less than water. The planet has molecular hydrogen interior with a radius of 37,282 miles (60,000 km), a metallic hydrogen core with a radius of 18,641 miles (30,000 km), and a rocky/icy inner core with a radius of 9,320 miles (15,000 km).

The most striking features of Saturn are its many rings and moons, with the rings circling the planet along its equatorial plane and their appearance from Earth changing with the seasons because of the different tilt of the planet as it orbits the Sun. The rings are more than 124,275 miles (200,000 km) in diameter but are less than 600 feet (200 m) thick. They are composed of numerous small particles, most of which are ice between a few millimeters and a few tens of meters in diameter. The breaks in the rings are a result of gravitational dynamics between the planet and its many moons.

Saturn has a yellowish-tan color produced largely by gaseous methane and ammonia, but the atmosphere consists of 92.4 percent molecular hydrogen, 7.4 percent helium, 0.2 percent methane, and 0.02 percent ammonia. These gases are stratified into three main layers, including a 100–200-kilometer-thick outer layer of ammonia, a 31–62-mile (50–100-km) thick layer of ammonium hydrosulfide ice, and a deeper 31–62-mile (50–100 km) thick layer of water ice. The atmosphere of Saturn is somewhat colder and thicker than that of Jupiter. Atmospheric winds on Saturn reach a maximum eastward-flowing velocity of 930 miles per hour (1,500 km/hr) at the equator and diminish with a few belts of high velocity toward the poles. Like Jupiter, Saturn has atmospheric bands related to these velocity variations, as well as turbulent storms that show as spots, and a few westward-flowing bands.

Many moons circle Saturn, including the large, rocky Titan, possessing a thick nitrogen–argon-rich atmosphere that contains hydrocarbons including methane, similar to the basic building blocks of life on Earth. Other large to midsize moons include, in increasing distance from the planet, Mimas, Enceladus, Tethys, Dione, Rhea, and Iapetus. About a dozen other moons of significant size are known to be circling the planet.

schist A strongly foliated crystalline metamorphic rock characterized by the parallel alignment of more than 50 percent of minerals present, especially platy or elongate prismatic minerals such as micas and hornblende. The strong alignment of platy minerals causes schists to break into slabs or thin flakes. Schists are coarser grained and form at higher metamorphic grades than slates, although the pre-metamorphic protolith of both rock types may be the same muddy sediment. The foliation in schists is called schistosity. There are many different types of schist named on the basis of main and accessory minerals present. Mica schists are the most common, and compound names are typically used to describe schists with prominent accessory minerals, such as

Saturn, showing spectacular ring systems. Saturn puts on a show as the planet and its magnificent ring system nod majestically over the course of its 29-year journey around the Sun. This Hubble Space Telescope image shows Saturn's rings as the planet moves from autumn toward winter in its northern hemisphere. Saturn's equator is tilted relative to its orbit by 27°, very similar to the 23°-tilt of the Earth. As Saturn moves along its orbit, first one hemisphere and then the other is tilted toward the Sun. This cyclical change causes seasons on Saturn, just as the changing orientation of Earth's tilt causes seasons on our planet. Saturn's rings are incredibly thin, with a thickness of only about 30 feet (10 m). The rings are made of dusty water ice, in the form of boulder-sized and smaller chunks that gently collide with each other as they orbit around Saturn. Saturn's gravitational field constantly disrupts these ice chunks, keeping them spread out and preventing them from combining to form a moon. The rings have a slight pale-reddish color due to the presence of organic material mixed with the water ice. Saturn is about 75,000 miles (120,000 km) across and is flattened at the poles because of its rapid rotation. A day is only 10 hours long on Saturn. Strong winds account for the horizontal bands in the atmosphere of this giant gas planet. The delicate color variations in the clouds are due to smog in the upper atmosphere, produced when ultraviolet radiation from the Sun shines on methane gas. Deeper in the atmosphere, the visible clouds and gases merge gradually into hotter and denser gases, with no solid surface for visiting spacecraft to land on. The *Cassini/Huygens* spacecraft, launched from Earth in 1997, reached the Saturn system in 2004 to land a probe on Titan, Saturn's largest moon, and to orbit the planet for four years for a detailed study of the entire Saturn system. This image was taken with the Wide Field Planetary Camera 2 onboard Hubble. *(Photo by NASA and the Hubble Heritage Team [STScI/AURA]: R.G. French [Wellesley College], J. Cuzzi [NASA/Ames], L. Dones [SwRI], and J. Lissauer [NASA/Ames]).*

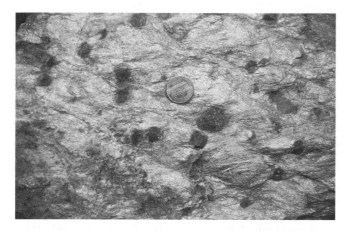

Garnet in mica schist, Fairbanks area, Alaska *(Photo by Timothy Kusky)*

garnet-mica schist, hornblende-biotite schist, or cordierite-mica schists.

See also DEFORMATION OF ROCKS; METAMORPHISM.

Scottish Highlands Located in the northern portion of the United Kingdom, Scotland includes a variety of generally rugged Lower Ordovician to Archean terranes, dissected by numerous northeast trending faults that form deep valleys. The coastline of Scotland is highly irregular and has many narrow to wide indented arms of the sea known respectively as lochs and firths. The Hebrides, Orkney, and Shetland Islands lie off the coast of northern Scotland. The Southern Uplands form a series of high rolling grassy and swampy hills known locally as moors, underlain by a series of strongly folded and faulted Ordovician and Silurian strata. These are separated from the Midland Valley by the Southern Uplands fault, an early Paleozoic strike-slip fault that was later converted to a normal fault. The Midland Valley includes thick deposits of the Devonian-Carboniferous Old Red Sandstone, deposited under continental conditions. The Highland Boundary fault separates the Midlands Valley from the Grampian Highlands, where Precambrian to Early Paleozoic metamorphic and igneous rocks of the Dalradian and Moine Groups are exposed in rugged mountains. The Great Glen fault, a Late Paleozoic left-lateral fault, separates the Grampian Highlands from the Northern Highlands, where Grenvillian age Moine and Archean Lewisian rocks are exposed. The tallest mountain in Scotland, Ben Nevis (4,406 feet; 1,343 meters) is located in the highlands. The Moine thrust forms the northwestern edge of the Caledonian orogen, with Archean and Proterozoic rocks of the Lewisian gneisses forming the basement to the orogen.

The oldest rocks exposed in the Scottish Highlands are the Archean (three billion years old) through Lower Proterozoic Lewisian gneisses formed during the Scourian tectonic cycle, found principally in the Hebrides Islands and the Northern Highlands. The Late Archean gneisses include tonalitic and gabbroic types, with rare ultramafic-mafic plutonic units, probably formed in a volcanic arc setting. Other shallow shelf metasedimentary rocks, including quartzites, limestones, and pelites, were metamorphosed to granulite facies at 2.7 billion years ago.

The Invernian tectonic cycle in the Early Proterozoic deformed large tracts of the Scourian gneisses into steep limbed west-northwest-trending linear structures, accompanied by retrograde amphibolite facies metamorphism. Mafic dike swarms intruded at 2.2 billion and 1.91 billion years ago, and are not metamorphosed. Post-1.9-billion-year-old Laxfordian cycle events include the formation of shear zones and intrusion of granite plutons near 1.72 billion years ago, and a cessation of events by 1.7 billion years ago.

The Moinian Assemblage is a Middle Proterozoic (older than 730 million years, and probably older than 1 billion years) group of pelites and psammites that are complexly folded into fold interference patterns and metamorphosed to amphibolite facies. Late Proterozoic (970–790 million years old) rocks include two sequences of red beds including the Stoer, Sleat, and Torridon Groups. These groups include conglomerates, siltstones, and sandstones that are more than a mile (several kilometers) thick in most places, and up to four miles (6 km) thick in a few places. Most of these Late Proterozoic rocks were probably deposited in fluvial or deltaic environments, perhaps in fault-bounded troughs along a continental margin.

The Dalradian Supergroup is found within the Caledonian orogen south of the Great Glen fault and north of the Highland boundary fault. The Dalradian is more than 12 miles (20 km) thick and is divided into four groups. The lowermost Grampian Group includes shallow to deepwater sandstones and graywackes, overlain by shallow shelf rocks including limestones, shales, and sandstones of the Appin Group. The succeeding Lochaber Group includes sandstones, siltstones, and carbonates deposited in a deltaic environment. The top of the Dalradian consists of the Argyll Group, including a glacial tillite, limestones, and deeper water graywackes, interbedded with Late Proterozoic (595-million-year-old) basalts. The Dalradian rocks were deformed into large nappe structures in the Late Proterozoic Grampian orogeny and metamorphosed to the amphibolite facies. The Paleozoic Era in the Scottish Highlands is marked by a basal transgression of the Durness sequence of shallow-marine skolithos-bearing quartzites and limestones onto the Torridonian and Lewisian gneisses. The basal transgressive sequence is about 1,100 feet (350 m) thick, and is overlain by Lower Cambrian through Ordovician shelf limestones. This sequence is correlated with the basal Cambrian-Ordovician shelf sequence in the Appalachian Mountains, as the Scottish Highlands was linked with Greenland and the Laurentian margin in the Early Paleozoic. However, there is no correlation of rocks of the Scottish Highlands with rocks south of the Highland boundary fault, sup-

Loch Marce, a deep embayment formed by a major fault in Scotland
(Photo by Timothy Kusky)

porting tectonic models that suggest that the southern British Isles were separated from Scotland by a major ocean, known as Iapetus. In latest Cambrian or Early Ordovician times, the region was affected by main phases of the Caledonian orogeny, known as the Athollian orogeny in the Scottish Highlands. Several generations of folds and regional metamorphism are related to the closure of the Iapetus Ocean along the Highland boundary fault, with an oceanic assemblage of cherts, pillow lavas, serpentinites, and Cambro-Ordovician limestones. These events associated with the Early Paleozoic closure of the Iapetus Ocean are correlated with the Taconic and Penobscottian orogenies in the northern Appalachians. In the Southern Uplands, a tectonically complex wedge of imbricated slivers of Ordovician-Silurian deepwater turbidites, shales, and slivers of pillow lavas may represent an oceanic accretionary wedge associated with continued closure of additional segments of the Iapetus Ocean.

Fold interference pattern in Middle Proterozoic Moinian assemblage
(Photo by Timothy Kusky)

The Moine thrust zone in the Northern Highlands formed at the end of the Silurian, and places the Caledonian orogenic wedge over the foreland rocks of the Lewisian and Dalradian sequences to the northwest. The Moine thrust is one of the world's classic zones of imbricate thrust tectonics, clearly displaying a sole thrust and imbricate splays, thrust-related folds, klippen, and windows. These structures formed as a Late-Caledonian effect of convergence and shortening between the formerly separated margins of the Iapetus Ocean and placed the orogen wedge allochthonously over basement rocks of the Laurentian margin.

The Old Red Sandstone is a Silurian-Devonian sequence of conglomerates, sandstones, siltstones, shales, and bituminous limestones that is up to 10,000 feet (3 km) thick. These rocks represent fluvial-delataic to lacustrian deposits eroded from the southeast and are interpreted as a molasse sequence representing denudation of the Caledonides. The Old Red Sandstone is loosely correlated with the Devonian Catskill Mountains deltaic complex in the Appalachians, representing erosion of the Appalachian Mountains after the Devonian Acadian orogeny.

Carboniferous deposits in Scotland include shales, coal measures, basalts, and limestones, deposited in deltaic environments mostly in the Midland Valley. Devonian through Carboniferous sinistral strike-slip faults cut many parts of the Scottish Highlands and are associated with Hercynian tectonic events in Europe and the Acadian-Appalachian orogenies in the Appalachians.

See also APPALACHIANS; CALEDONIDES; PALEOZOIC; STRUCTURAL GEOLOGY.

Further Reading

Craig, G. Y. *Geology of Scotland,* 3rd ed. Bath, U.K.: The Geological Society of London, 1991.

Park, R. G., and John Tarney. *Evolution of the Lewisian and Comparable Precambrian High Grade Terranes.* Oxford: Blackwell Scientific Publications, Geological Society of London Special Publication 27, 1987.

sea breeze A local coastal wind that blows from the ocean over the adjacent land, formed from thermal circulation related to the uneven heating of land and water. Land surfaces heat up more quickly than sea surfaces during the day, creating a shallow thermal low above the land. The air above the water remains cool and develops a shallow thermal high. As the pressure difference builds, a sea breeze begins to blow from the high over the water to the low over the land. The breeze tends to be strongest right at the beach because the pressure gradients are strongest at the boundary between the high- and low-pressure systems. Similar effects may form along large lake shores and are called lake breezes.

During the night the land cools more quickly than the sea, and the opposite pressure situation develops. The air over the land is cooled more than the air over the sea, so a high develops over the land and a low develops over the sea, causing a land breeze to blow from the land over the sea.

Warming of the land in the daylight hours usually causes the sea breeze to start around mid-morning, increasing in strength and distance as it penetrates inland during the afternoon. Strong sea breezes may penetrate 15 miles (24 km) inland. The boundary between the cool ocean air and the warm air over the land is demarcated by the sea breeze front. Passing of the front on the land on a warm summer day may cause temperatures to quickly drop 5°F or more, with an increase in humidity and change in the wind direction. In some cases the ocean air is so moist that water, and other pollutants, condense around salt particles and then the sea breeze moves in as a visible cloud layer, smoke front, or smog front. Small localized storms may also form along the sea breeze front since the air is rising as the front moves inland.

Sea breezes are most common and best-developed year-round in tropical regions, where large temperature differences exist between the sea and the ground. In temperate regions sea breezes are best developed in the spring and fall seasons.

See also ATMOSPHERE; CLOUDS; KATABATIC WINDS.

sea ice Ice that has broken off an ice cap, polar sea, or calved off a glacier and is floating in open water. Sea ice presents a serious hazard to ocean traffic and shipping lanes and has sunk numerous vessels, including the famous sinking of the *Titanic* in 1912, killing 1,503 people.

There are four main categories of sea ice. The first comes from ice that formed on polar seas in the Arctic Ocean and around Antarctica and is typically about 10–13 feet (3–4 m) thick. Antarctica becomes completely surrounded by this sea ice every winter, with the Arctic Ocean typically about 70 percent covered in the winter. During summer, many passages open up in this sea ice, but during the winter they re-close, forming pressure ridges of ice that may be up to tens of meters high. Recent observations suggest that the sea ice in the Arctic Ocean is thinning dramatically and rapidly and may soon disappear altogether. The icecap over the Arctic Ocean rotates clockwise, in response to the spinning of the Earth. This spinning is analogous to putting an ice cube in a glass and slowly turning the glass; the ice cube will rotate more slowly than the glass because it is decoupled from the edge of the glass. About one-third of the Arctic sea ice is removed every year by the East Greenland current. This ice then moves south as icebergs and becomes a hazard to shipping in the North Atlantic.

Recent studies have revealed that the Arctic Ice Cap has thinned dramatically in recent years (by nearly 50 percent), and that the aerial extent of the ice cap is rapidly shrinking. These changes are probably a result of the 10°F average warming of the climate in the Arctic over the past 10 years. It is not certain if this is a human-induced change, but it very likely is a response to increased carbon dioxide, aerosols, and other greenhouse gases put into the atmosphere by human use. Some of the heating may alternatively be the result of natural cycles in global temperature.

Icebergs from sea ice float on the surface, but between 81 and 89 percent of the ice will be submerged. The exact level that sea ice floats in the water depends on the exact density of the ice, as determined by the total amount of air bubbles trapped in the ice, and how much salt got trapped in the ice during freezing.

A second group of sea ice forms as pack ice in the Gulf of St. Lawrence, along the southeast coast of Canada, in the Bering, Beaufort, and Baltic Seas, in the Seas of Japan and Okhotsk, and around Antarctica. Pack ice builds up especially along the western sides of ocean basins where cold currents are more common. Occasionally, during cold summers, pack ice may persist throughout the year.

Several scenarios suggest that new ice ages may begin with pack ice that persists through many summers, gradually growing and extending to lower latitudes. Other models and data show that pack ice varies dramatically with a four or five-year cycle, perhaps related to sunspot activity, and the El-Niño–Southern Oscillation (ENSO).

Pack ice presents hazards when it gets so extensive that it effectively blocks shipping lanes, or when leads (channels) into the ice open and close, forming pressure ridges that become too thick to penetrate with icebreakers. Ships attempting to navigate through pack ice have become crushed when leads close and the ships are trapped. Pack ice has terminated or resulted in disaster for many expeditions to polar seas, most notably Franklin's expedition in the Canadian arctic and Scott's expeditions to Antarctica. Pack ice also breaks up, forming many small icebergs, but because these are not as thick as icebergs of other origins they do not present as significant a hazard to shipping.

Pack ice also presents hazards when it drifts into shore, usually during spring breakup. With significant winds pack ice can pile up on flat shorelines and accumulate in stacks up to 50 feet (15 m) high. The force of the ice is tremendous and is enough to crush shoreline wharves, docks, buildings, and boats. Pack ice that has blown ashore also commonly pushes up high piles of gravel and boulders that may be 35 feet (10 m) high in places. These ridges are common around many of the Canadian Arctic islands and the mainland. Ice that forms initially attached to the shore presents another type of hazard. If it breaks free and moves away from shore, it may carry with it significant quantities of shore sediment, causing rapid erosion of beaches and shore environments.

Pack ice also forms on many high-latitude lakes, and the freeze-thaw cycle causes cracking of the lake ice. When lake water rises to fill the cracks, the ice cover on the lake expands and pushes over the shoreline, resulting in damage to any structures built along the shore. This is a common problem on many lakes in northern climates and leads to widespread damage to docks and other lakeside structures.

Icebergs present the greatest danger to shipping. In the Northern Hemisphere most icebergs calve off glaciers in Greenland or Baffin Island, then move south through the Davis Strait into shipping lanes in the North Atlantic off Newfoundland. Some icebergs calve off glaciers adjacent to the Barents Sea, and others come from glaciers in Alaska and British Columbia. In the Southern Hemisphere, most icebergs come from Antarctica, though some come from Patagonia.

Once in the ocean, icebergs drift with ocean currents, but because of the Coriolis force, they are deflected to the right in the Northern Hemisphere and to the left in the Southern Hemisphere. Most icebergs are approximately 100 feet–300 feet (30.5–90 m) high and up to about 2,000 feet (310 m) in length. However, in March of 2000 a huge iceberg broke off the Ross Ice Shelf in Antarctica that was roughly the size of the state of Delaware. It had an area of 4,500 square miles (11,660 km²) and stuck 205 feet (63 m) out of the water. Icebergs in the Northern Hemisphere pose a greater threat to shipping, as those from Antarctica are too remote and rarely enter shipping lanes. Ship collisions with icebergs have resulted in numerous maritime disasters, especially in the North Atlantic on the rich fishing grounds of the Grand Banks off the coast of Newfoundland.

Icebergs are now tracked by satellite; ships are updated with their positions so they can avoid any collisions that could prove fatal for the ships' occupants. Radio transmitters are placed on larger icebergs to more closely monitor their locations, and many ships now carry more sophisticated radar and navigational equipment that helps track the positions of large icebergs and themselves, so that they avoid collision.

Icebergs also pose a serious threat to oil drilling platforms and seafloor pipelines in high-latitude seas. Some precautions have been taken such as building seawalls around near-shore platforms, but not enough planning has gone into preventing an iceberg colliding with and damaging an oil platform or damage from one being dragged across the seafloor and rupturing a pipeline.

See also GLACIER.

sea-level rise Global sea levels are currently rising as a result of the melting of the Greenland and Antarctica ice sheets and thermal expansion of the world's ocean waters due to global warming. We are presently in an interglacial stage of an ice age. Sea levels have risen nearly 400 feet (122 m) since the last glacial maximum 20,000 years ago and about 6 inches (15 cm) in the past 100 years. The rate of sea-level rise seems to be accelerating and may presently be as much as an inch every eight to 10 years. If all the ice on both the Antarctic and Greenland ice sheets were to melt, global sea levels would rise by 230 feet (70 m), inundating most of the world's major cities and submerging large parts of the continents under shallow seas. The coastal regions of the world are densely populated and are experiencing rapid population growth.

Approximately 100 million people presently live within one meter of the present-day sea level. If sea levels were to rise rapidly and significantly, the world would experience an economic and social disaster of a magnitude not yet experienced by the civilized world. Many areas would become permanently flooded or subject to inundation by storms, beach erosion would be accelerated, and water tables would rise.

The Greenland and Antarctic ice sheets have some significant differences that cause them to respond differently to changes in air and water temperatures. The Antarctic ice sheet is about 10 times as large as the Greenland ice sheet, and since it sits on the South Pole, Antarctica dominates its own climate. The surrounding ocean is cold even during summer, and much of Antarctica is a cold desert with low precipitation rates and high evaporation potential. Most meltwater in Antarctica seeps into underlying snow and simply refreezes, with little running off into the sea. Antarctica hosts several large ice shelves fed by glaciers moving at rates of up to a thousand feet per year. Most ice loss in Antarctica is accomplished through calving and basal melting of the ice shelves at rates of a 10–15 inches per year (25–38 cm/yr).

In contrast, Greenland's climate is influenced by warm North Atlantic currents and its proximity to other landmasses. Climate data measured from ice cores taken from the top of the Greenland ice cap show that temperatures have varied significantly in cycles of years to decades. Greenland also experiences significant summer melting, abundant snowfall, has few ice shelves, and its glaciers move quickly at rates of up to miles per year. These fast-moving glaciers are able to drain a large amount of ice from Greenland in relatively short amounts of time.

The Greenland ice sheet is thinning rapidly along its edges, losing an average of 15–20 feet in different areas (4.5–6 m) over the past decade. In addition, tidewater glaciers and the small ice shelves in Greenland are melting an order of magnitude faster than the Antarctic ice sheets, with rates of melting between 25–65 feet per year (7–20 m/yr). About half of the ice lost from Greenland is through surface melting, forming water that runs off into the sea. The other half of ice loss is through calving of outlet glaciers and melting along the tidewater glaciers and ice shelf bases.

These differences between the Greenland and Antarctic ice sheets lead them to play different roles in global sea-level rise. Greenland contributes more to the rapid short-term fluctuations in sea level, responding to short-term changes in climate. In contrast, most of the world's water available for raising sea level is locked up in the slowly changing Antarctic ice sheet. Antarctica contributes more to the gradual, long-term sea-level rise.

What is causing the rapid melting of the polar ice caps? Most data suggests that the current melting is largely the result of the gradual warming of the planet in the past 100 years through the effects of greenhouse warming. Greenhouse gases have been increasing at a rate of more than 0.2 percent per year, and global temperatures are rising accordingly. The most significant contributor to the greenhouse gas buildup is carbon dioxide, produced mainly by the burning of fossil fuels. Other gases that contribute to greenhouse warming include carbon monoxide, nitrogen oxides, methane (CH_4), ozone (O_3), and chlorofluorocarbons. Methane is produced by gas from grazing animals and termites, whereas nitrogen oxides are increasing because of the increased use of fertilizers and automobiles, and the chlorofluorocarbons are increasing from release of aerosols and refrigerants. Together the greenhouse gases have the effect of allowing short-wavelength incoming solar radiation to penetrate the gas in the upper atmosphere, but trapping the solar radiation after it is reemitted from the Earth in a longer wavelength form. The trapped radiation causes the atmosphere to heat up, leading to greenhouse warming. Other factors also influence greenhouse warming and cooling, including the abundance of volcanic ash in the atmosphere and solar luminosity variations, as evidenced by sunspot variations.

Measuring global (also called *eustatic*) sea-level rise and fall is difficult because many factors influence the relative height of the sea along any coastline. These vertical motions of continents are called epeirogenic movements and may be related to plate tectonics, rebound from being buried by glaciers, or to changes in the amount of heat added to the base of the continent by mantle convection. Continents may rise or sink vertically, causing apparent sea-level change, but these sea-level changes are relatively slow compared to changes induced by global warming and glacial melting. Slow, long-term sea-level changes can also be induced by changes in the amount of seafloor volcanism associated with seafloor spreading. At some periods in Earth's history, seafloor spreading was particularly vigorous, and the increased volume of volcanoes and the mid-ocean ridge system caused global sea levels to rise.

Steady winds and currents can mass water against a particular coastline, causing a local and temporary sea-level rise. Such a phenomena is associated with the ENSO, causing sealevels to rise by 4–8 inches (10–20 cm) in the Australia-Asia region. When the warm water moves east in an ENSO event, sea levels may rise 4–20 inches (10–50 cm) across much of the North and South American coastlines. Other atmospheric phenomena can also change sea level by centimeters to meters locally, on short time scales. Changes in atmospheric pressure, salinity of seawaters, coastal upwelling, onshore winds, and storm surges all cause short-term fluctuations along segments of coastline. Global or local warming of waters can cause them to expand slightly, causing a local sea-level rise. It is even thought that the extraction and use of groundwater and its subsequent release into the sea might be causing sealevel rise of about 0.78 inches per year (1.3 mm/yr). Seasonal changes in river discharge can temporarily change sea levels

along some coastlines, especially where winter cooling locks up large amounts of snow that melt in the spring.

It is clear that attempts to estimate eustatic sea-level changes must be able to average out the numerous local and tectonic effects to arrive at a globally meaningful estimate of sea-level change. Most coastlines seem to be dominated by local fluctuations that are larger in magnitude than any global sea-level rise. Recently, satellite radar technology has been employed to precisely measure sea surface height and to document annual changes in sea level. Radar altimetry is able to map sea surface elevations to the sub-inch scale, and to do this globally, providing an unprecedented level of understanding of sea surface topography. Satellite techniques support the concept that global sea levels are rising at about 0.01 inches per year (0.025 cm/yr).

See also EL NIÑO; GLACIER; PLATE TECTONICS.

Further Reading
Douglas, Bruce C., Michael S. Kearney, and Stephen P. Leatherman. "Sea-Level Rise: History and Consequences." *International Geophysics Series*, vol. 75. San Diego: Academic Press, 2000.
Schneider, D. "The Rising Seas." *Scientific American*, March issue, 1997.

seamount Mountains that rise above the seafloor, typically to a height of 3,000 feet (1,000 meters) or more, are called seamounts. Many parts of the abyssal ocean floor, particularly the western Pacific, are covered by seamounts with different shapes and relationships to each other. Small flat-topped seamounts are called guyots and represent intraplate oceanic volcanoes that grew toward or above the ocean surface and were eroded to have a flat top as they subsided below sea level. Many formed near the oceanic ridge system as off-axis volcanoes, and as the oceanic crust cools and subsides as it ages and moves away from the mid-ocean ridges, the seamounts subside below the surface as well. Once below the surface, waves erode the upper surface making it flat. If the seamount is in tropical or subtropical climates corals may form fringing reefs around the volcanic island. As the island subsides, the reefs keep growing and stay near the surface, forming coral atoll islands.

Other seamounts are much larger and represent voluminous outpourings of mafic lava from hot spots. For example, the Hawaiian-Emperor seamount chain extends for thousands of kilometers across the Pacific Ocean, tracing the movement of the Pacific plate with respect to a fairly stationary hot spot in the underlying mantle. As hot magma upwelled in this hot spot or mantle plume, it acted as a blowtorch against the bottom of the Pacific plate, erupting mafic volcanics as the plate moved overhead. Thus, the youngest volcanic islands are above the surface on the islands of Hawaii and Maui and become older progressively to the northwest. The older islands have cooled and sunk below sea level in northwestern parts of the Hawaiian-Emperor chain.

More voluminous outpourings of lava than on seamount chains are found on thick oceanic plateaus.

There are several other types of seamounts that are less common than the intraplate volcanics. For instance, small submarine intermediate and felsic volcanoes form near some island arcs, creating seamounts along convergent boundaries. Transform faults and aseismic ridges have topographic highs associated with these structures, and some rifted margins have extensional horsts that form seamounts along young passive margins.

See also ATOLL; HOT SPOT; OCEANIC PLATEAU; OCEANOGRAPHY.

seasons Variations in the average weather at different times of the year are known as seasons, controlled by the average amount of solar radiation received at the surface in a specific place for a certain time period. The amount of radiation received at a particular point on the surface is determined by several things, including the angle at which the Sun's rays hit the surface, the length of time the rays warm the surface, and the distance to the Sun. As the Earth orbits the Sun approximately once every 365 days, it follows an elliptical orbit that brings it closest to the Sun in January (91 million miles, or 147 million kilometers) and farthest from the sun in July (94.5 million miles, or 152 million kilometers). Therefore, the Sun's rays are slightly more intense in January than in July but, as any Northern Hemisphere resident can testify, this must not be the main controlling factor on determining seasonal warmth since winters in the Northern Hemisphere are colder than summers. Where the Sun's rays hit a surface directly, at right angles to the surface, they are most effective at warming the surface since they are not being spread out over a larger area on an inclined surface. Also, where the Sun's rays enter the atmosphere directly they travel through the least amount of atmosphere, so are weakened much less than rays that must travel obliquely through the atmosphere, which absorbs some of their energy. The Earth's rotational axis is inclined at 23.5° from perpendicular to the plane it rotates around the sun on (the ecliptic plane), causing different hemispheres of the planet to be tilted toward of away from the sun in different seasons. In the Northern Hemisphere summer, the Northern Hemisphere is tilted toward the Sun so it receives more direct sunlight rays than the Southern Hemisphere, causing more heating in the north than in the south. Also, since the Northern Hemisphere is tilted toward the sun in the summer it receives direct sunlight for longer periods of time than the Southern Hemisphere, enhancing this effect. On the summer solstice on June 21, the Sun's rays are directly hitting 23.5°N latitude (called the tropic of Cancer) at noon. Because of the tilt of the planet, the sun does not set below the horizon for all points north of the arctic circle (66.5°N). Points farther south have progressively shorter days, and points farther north have progressively longer days. At the North Pole, the Sun rises above the horizon on March 20, and does not set again until six months later on September 22. However, since the Sun's rays are so oblique in these northern latitudes, they receive less solar radiation than areas farther south, where the rays hit more directly but for shorter times. As the Earth rotates around the Sun, it finds the Southern Hemisphere tilted at its maximum amount toward the Sun on December 21 and the situation is reversed from the Northern Hemisphere summer, so that the same effects occur in the southern latitudes.

Seasonal variations in temperature and rainfall at specific places are complicated by global atmospheric circulation cells, proximity to large bodies of water and warm or cold ocean currents, and monsoon type effects in some parts of the world. Some seasons in some places are hot and wet, others are hot and dry, cold and wet, or cold and dry.

See also ATMOSPHERE; CLIMATE; HADLEY CELLS.

seawater The oceans cover more than 70 percent or the Earth's surface and extend to an average depth of several kilometers. As part of the hydrologic cycle, each year approximately 1.27×10^{16} cubic feet (3.6×10^{14} m^3) of water evaporates from the oceans with about 90 percent of this returning to the oceans as rainfall. The remaining 10 percent falls as precipitation on the continents where it forms freshwater lakes and streams and seeps into the groundwater system where it is temporarily stored before eventually returning to the sea. During its passage over and in the land, the water erodes huge quantities of rock, soil, and sediment, and dissolves chemical elements such as salts from the continents, carrying these and other sediments as dissolved, suspended, and bed load to the oceans. More than 50 million tons of continental material is transported into the oceans each year. Most of the suspended and bed load materials are deposited as sedimentary layers near passive margins, but the dissolved

Name	Symbol	Concentration in Parts per Thousand	Percentage of Dissolved Material
Chloride	Cl⁻	18.980	55.05
Sodium	Na⁺	10.556	30.61
Sulfate	SO₄²⁻	2.649	7.68
Magnesium	Mg²⁺	1.272	3.69
Calcium	Ca²⁺	0.400	1.16
Potassium	K⁺	0.380	1.10
Bicarbonate	HCO₃⁻	0.140	0.41
Bromide	Br⁻	0.065	0.19
Borate	H₃BO₃⁻	0.026	0.07
Strontium	Sr²⁺	0.008	0.03
Fluoride	F⁻	0.001	0.00
Total		**34.447**	**99.99**

Why Is Seawater Blue?

At some point nearly every child with an inquisitive mind will ask an adult why the sky is blue or why the seawater is blue. In a simple sense, seawater is blue because is it a reflection of the color of the sky above, but then why is the sky blue? The answer lies in phenomena called scattering of light. Sunlight that enters the atmosphere contains the complete visible spectrum of colors, defined by different wavelengths. As this light enters the atmosphere it encounters air molecules of oxygen and nitrogen, each of which is smaller than the wavelength of visible light. These molecules cause the incident light to be scattered when the light hits them, but since the molecules are small they are much more effective at scattering the short wavelengths than the long wavelengths, which tend to pass over the small molecules. This selective scattering is analogous to ocean waves that encounter a buoy in the water. Waves that are small (short wavelength) and about the same size as the buoy will bounce off and be scattered by the buoy, but large waves (long wavelength) will pass right by the buoy and hardly be affected. Similarly, the shorter visible wavelengths of light, including violet, blue, and green, are scattered efficiently by the small air molecules, whereas the longer wavelengths of yellow, orange, and red are scattered very little. The atmosphere scatters blue light about 16 times as much as red light. The result of this scattering is that as we look at the sky, we see blue light coming at us from virtually all directions. In contrast, the yellow, orange, and red light are not scattered effectively and appear to be only coming nearly directly from the direction of the Sun. The same effect causes distant mountains to appear blue, when scattering by small particles is strong near the ground. The presence of larger particles can cause different color sensations. For instance, the presence of larger aerosol pollutants causes a brownish smog color, and the presence of even larger water droplets causes clouds and haze to appear white.

Most of the light and energy from the Sun that strike the sea is absorbed by seawater and converted to heat, but some is reflected. The upper surface of the sea reflects the color of the sky, which is most often blue. However, the presence of suspended particles in seawater can further alter the color of light perceived in the water. For instance, clear ocean waters appear deep blue or violet, whereas coastal waters with large amounts of suspended sediments or dissolved organic substances causes the reflected light to shift to longer wavelength colors such as green. In turbid coastal waters the shift in the wavelength of reflected light is enough to change the color to yellow.

salts and ions derived from the continents play a major role in determining seawater chemistry. The most abundant dissolved salts are chloride and sodium, which together with sulfate, magnesium, calcium, potassium, bicarbonate, bromide, borate, strontium, and fluoride form more than 99.99 percent of the total material dissolved in seawater.

In addition to the elements listed in the table, there are a number of minor and trace elements dissolved in seawater that are important for the life cycle of many organisms. For instance, nitrogen, phosphorous, silicon, zinc, iron, and copper play important roles in the growth of tests and other parts of some marine organisms. Gases, including nitrogen, oxygen, and carbon dioxide are also dissolved in seawater. The amount of oxygen dissolved in the surface layers of seawater is about 34 percent of the total dissolved gases, significantly higher than the 21 percent of total dissolved gases oxygen comprises in the atmosphere. This oxygen is generated through photosynthesis by marine plants, where it is exchanged with the atmosphere across the air–water interface, and also sinks where it is used by deep aerobic organisms. The amount of carbon dioxide dissolved in seawater is about 50 times greater than its concentration in the atmosphere. CO_2 plays an important role in buffering the acidity and alkalinity of seawater where, through a series of chemical reactions, it keeps the pH of seawater between 7.5 and 8.5. Marine organisms make carbonate shells out of the dissolved CO_2, and some is incorporated into marine sediments where it is effectively isolated from the atmosphere. The total amount of CO_2 stored in the ocean is very large, and as a greenhouse gas, if it were to be released to the atmosphere, it would have a profound effect on global climate.

The salinity and temperature of seawater are important in controlling mixing between surface and deep water, and in determining ocean currents. Temperature is controlled largely by latitude, whereas river input, evaporation from restricted basins, and other factors determine the total dissolved salt concentration. Density differences caused by temperature and salinity variations induce ocean currents and thermohaline circulation, distributing heat and nutrients around the globe.

See also HYDROLOGIC CYCLE; OCEAN CURRENTS; OCEANOGRAPHY; THERMOHALINE CIRCULATION.

Further Reading

Allaby, Alisa, and Michael Allaby. *A Dictionary of Earth Sciences,* 2nd ed. Oxford: Oxford University Press, 1999.

Sedgwick, Adam (1785–1873) British *Geologist, Mathematician* Adam Sedgwick was educated as a mathematician at Trinity College, Cambridge, where he was elected to a fellowship in 1810 and ordained in 1817. In 1818 he was appointed a Woodwardian Professor of Geology and was elected as a Fellow of the Royal Society in 1830. Sedgwick made significant contributions to the Permian and Triassic stratigraphy of northeast England but made a bigger mark during his studies of the Lower Paleozoic of Scotland and the Lake District. He was able to demonstrate that the non-marine Old Red Sandstone was laterally equivalent to marine sandstones. The Geological Society of London presented Sedgwick with its highest prize, the Wollaston Medal, in 1851, and the Royal Society awarded him the Copley Medal in 1863, for his work on the Silurian System.

Further Reading

Sedgwick, Adam, and R. I. Murchison. "On the Physical Structure of Devonshire, and on the Subdivisions and Geological Relations of Its Older Stratified Deposits, &c." In *Stratigraphy; Foundations and Concepts*, edited by Barbara M. Conkin and James E. Conkin, 173–189. Stroudsburg, Pa.: Hutchinson Ross Publishing Company, 1984.

sedimentary rocks Those rocks that have consolidated from accumulations of loose sediment, in turn produced by physical, chemical, or biological processes. Common mechanical processes involved in the formation of sediments include the breaking, transportation of fragments, and accumulation of older rocks; chemical processes include the precipitation of minerals by chemical processes or evaporation of water; common biological processes include the accumulation of organic remains.

Soils and other products of weathering of rocks are continuously being removed from their sources and deposited elsewhere as sediments. This process can be observed as gravel in streambeds, on alluvial fans, and in windblown deposits. When these sediments are cemented together, commonly by minerals deposited from water percolating through the ground, they become sedimentary rocks. Other types of sedimentary rocks are purely chemical in origin and were formed by the precipitation of minerals from an aqueous solution.

Clastic sediments (also detritus) are the accumulated particles of broken rocks, some with the remains of dead organisms. The word *clastic* is from the Greek word *klastos,* meaning broken. Most clastic particles have undergone various amounts of chemical change and may have a continuous gradation in size from huge boulders to submicroscopic particles. Size is the main basis for classifying clastic sediments and sedimentary rocks. The texture of the sedimentary rock or individual sedimentary particles forms as additional criteria for classifying sedimentary rocks.

Clastic sediments may be transported by wind, water, ice, or gravity, and each method of transport leaves specific clues as to how it was transported and deposited. For instance, if sediments are transported by gravity in a landslide, the resulting deposit consists of a poorly sorted mixture including everything that was in its path, whereas if the sediment was transported by wind, it has a very uniform grain size and typically forms large dunes. Clastic sediments are deposited when the transporting agent can no longer carry them. For instance, if the wind stops, the dust and sand will fall out, whereas sediments transported by streams are deposited when the river velocity slows down. This happens either where the stream enters a lake or the ocean or when a flood stage lowers and the stream returns to a normal velocity and clears up. Geologists can look at old rocks and tell how fast the water was flowing during deposition and can also use clues such as the types of fossils, or the way the individual particles are arranged, as clues to decipher the ancient environment.

Chemical sediment is sediment formed from the precipitation of minerals from solution in water. They may form from biochemical reactions from activities of plants and animals that live in the water, or they may form from inorganic reactions in the water, induced by things such as hot springs, or simply the evaporation of seawater. This produces a variety of salts, including ordinary table salt. Chemical sedimentary rocks are classified according to their main chemical component, with common types including limestone (made of predominantly calcite), dolostone (consisting of more than 50 percent dolomite), rock salt (composed of NaCl), and chert (whose major component is SiO_2).

Most sedimentary rocks display a variety of internal and surface marking known as sedimentary structures that can be used to interpret the conditions of formation. Stratification results from a layered arrangement of particles in a sediment or sedimentary rock that accumulated at the surface of the earth. The layers are visible and different from adjacent layers because of differences (such as size, shape, or composition) in the particles between different layers, and because of differences in the way the particles are arranged between different layers. Bedding is the layered arrangement of strata in a body of rock. Parallel strata refers to sedimentary layers in which individual layers are parallel. The presence of parallel strata usually means that the sediments were deposited underwater, such as in lakes or in the deep sea. Some sediments with parallel layers have a regular alternation between two or more types of layers, indicating a cycle in the depositional environment. These can be daily, yearly, or some other rhythm influenced by solar cycles. One unusual type of layered rock is a varve, which is a lake sediment that forms a repeating cycle of coarse-grained sediments with spring tides, and fine-clay with winter conditions, when the suspended sediments gradually settle out of the water column. Cross strata are layers that are inclined with respect to larger layers that they occur in. Most cross-laminated deposits are sandy or coarser, and they form as ripples that move along the surface. The direction of inclination of the cross strata is the direction that the water used to flow.

Sorting is a sedimentary structure that refers to the distribution of grain sizes within a sediment or sedimentary rock. Sediments deposited by wind are typically well-sorted, but those deposited by water may show a range of sorting. If the grains have the same size throughout a bed it is called a uniform layer. A gradual transition from coarse to fine-grained, or fine to coarse-grained, is known as a graded bed. Graded beds typically reflect a change in current velocity during deposition. Non-sorted layers represent a mixture of different grain sizes, without order. These are common in rock falls, avalanche deposits, landslides, and from some glaciers. Rounding is a textural term that describes the relative shape or roundness of grains. When sediments first break off from their source area, they tend to be angular, and reflect the shape of joints, or internal mineral forms. With progressive transportation by wind or water, abrasion tends to smooth the grains and make them rounded. The greater the transport distance, in general, the greater the rounding.

Surface features on sedimentary layers also yield clues about the depositional environment. Like ripple marks or footprints on the beach, many features preserved on the surface of strata offer clues about the origin of sedimentary rocks and the environments in which they formed. Ripple marks show the direction of ancient currents, whereas tool marks record places where an object was dragged by a current across a surface. Flute marks are grooves produced by turbulent eddies in the current scouring out small pockets on the paleosurface. Mud cracks reveal that the surface was wet, then desiccated by subaerial exposure. Other types of surface marks may include footprints and animal tracks in shallow water environments, and raindrop impressions in subaerial settings.

Fossils are remains of animals and plants preserved in the rock, that can also tell clues about past environments. For instance, deep marine fossils are not found in lake environments, and dinosaur footprints are not found in deep marine environments.

See also CARBONATE; CLASTIC ROCKS; EVAPORITE.

sedimentation *See* SEDIMENTARY ROCKS.

seiche waves Free oscillations or standing waves that form in enclosed or semi-enclosed bodies of water such as lakes, bays, lagoons, or harbors. The waves may vary in period from a few minutes to hours and may have amplitudes ranging from a few centimeters to several meters. Most seiche waves rock back and forth in the enclosed body of water parallel to the long dimension of the basin, although some propagate in transverse directions. Seiche waves are similar to tsunami, except that they are confined to enclosed bodies of water such as lakes. They are generally generated by similar phenomena as true tsunami and may also be initiated by the rocking motion of the ground associated with large earthquakes. Some seiches are generated by local changes in atmospheric pressure, winds, and tidal currents. Many seiche waves were generated on lakes in southern Alaska during the 1964 m 9.2 earthquake, including some on Kenai Lake that washed away piers and other structures near the shore. Seiche-like waves sometimes resonate in bays and fiords during large earthquakes, but these are not truly seiche waves as they form in bodies of water connected to the sea. Seiche waves were first described from Lake Geneva in Switzerland.

See also EARTHQUAKES; TSUNAMI.

seismograph Seismographs are sensitive instruments that can detect, amplify, and record ground vibrations, especially earthquakes, producing a seismogram. Numerous seismographs have been installed in the ground throughout the world and form a seismograph network, monitoring earthquakes, explosions, and other ground shaking.

The first very crude seismograph was constructed in 1890. While the seismograph could tell that an earthquake was occurring, it was unable to actually record the earthquake. Modern seismographs display Earth movements by means of an ink-filled stylus on a continuously turning roll of graph paper. When the ground shakes, the needle wiggles and leaves a characteristic zigzag line on the paper.

Seismographs are built using a few simple principles. To measure the shaking of the Earth during a quake, the point of reference must be free from shaking, ideally on a hovering platform. To accommodate this need, engineers have designed an instrument known as an inertial seismograph. These make use of the principle of inertia, which is the resistance of a large mass to sudden movement. When a heavy weight is hung from a string or thin spring, the string can be shaken and the big heavy weight will remain stationary. Using an inertial seismograph, the ink-filled stylus is attached to the heavy weight and remains stationary during an earthquake. The continuously turning graph paper is attached to the ground, and moves back and forth during the quake, resulting in the zigzag trace of the record of the earthquake motion on the graph paper.

Seismographs are used in series; some set up as pendulums and some others as springs, to measure ground motion in many directions. Engineers have made seismographs that can record motions as small as one hundred-millionth of an inch, about equivalent to being able to detect the ground motion caused by a car driving by several blocks away. The ground motions recorded by seismographs are very distinctive, and geologists who study them have methods of distinguishing between earthquakes produced along faults, earthquake swarms associated with magma moving into volcanoes, and even between explosions from different types of construction and nuclear blasts. Interpreting seismograph traces has therefore become an important aspect of nuclear test ban treaty verification.

In the late 19th century E. Wiechert introduced a seismograph with a large, damped pendulum used as the sensor, with the damping reducing the magnitude of the pendulum's oscillations. This early seismograph recorded horizontal motions and used a photographic recording device. Wiechert soon introduced a new seismograph with a mechanical recording device, with an inverted pendulum that could

Historical photograph of a turn-of-the-century seismograph. The Wiechert 80 Kg (170 lb) Horizontal Component Seismograph was located in the basement of DuBourg Hall at Saint Louis University, where the seismograph was installed in 1909 and recorded its first earthquake on October 9 of that year. This was one of many seismographs installed at Jesuit universities and colleges in 1909 under the impetus of Fr. Frederick L. Odenbach of St. Ignatius College in Cleveland, Ohio. The Wiechert consisted of an inverted pendulum and recorded the two horizontal components of ground motion on smoked paper, amplifying motions by a factor of 80. The Earthquake Center at St. Louis University continues to archive these old recordings. *(Photo courtesy of the J.B. Macelwane Archives, Saint Louis University)*

vibrate in all horizontal directions. The pendulum was supported by springs that helped stabilize the oscillations and furthered the productivity of the seismograph. Wiechert's assistant, named Schluter, introduced a vertical recording device. He moved the mass horizontally away from the axis of rotation and maintained it there with a vertical spring. In doing so he was able to record vertical displacement, which helped record many of the complex movements associated with earthquakes.

In the 20th century seismographs that recorded movements using a pen on a rotating paper-covered drum were introduced, with alternative devices including those that recorded movements using a light spot on photographic film. More sophisticated seismographs that are able to record movements in three directions (up-down, north-south, and east-west) were introduced, and electronic recording of relative motions became common.

See also EARTHQUAKES; SEISMOLOGY.

seismology The study of the propagation of seismic waves through the Earth, including analysis of earthquake sources, mechanisms, and the determination of the structure of the Earth through variations in the properties of seismic waves.

Measuring the Crust

A Yugoslavian seismologist named Andrija Mohorovičić (1857–1936) noticed slow and fast arrivals of seismic waves in seismic experiments he completed in Europe and proposed that there must be a seismic discontinuity at around 22 miles (35 km) beneath the surface of continents to explain his observations. This discontinuity is now known as the Moho and it is interpreted as the base of the crust. We now use this bounding reflection surface as a measure of the thickness of the crust.

Seismology is also used to determine the detailed structure of shallow levels in the Earth's crust. Seismic reflection techniques are widely used by energy companies in their search for oil and gas trapped in geological structures and formations. Very subtle structures can now be found and mapped in three dimensions using sophisticated seismic surveys and computers. Some techniques involve moving the seismic source (an explosion, sound, thump, etc.) from place to place on the surface and noticing the difference in the receiving functions at different receiving stations. Other techniques involve moving the seismic sources and receivers up and down drilled boreholes and determining the geologic structure and layers between the two boreholes.

Seismologists really are measuring the speed of travel of seismic waves through different rock types. They deduce what the rock types are either by measuring seismic velocities between stations and then correlating this with laboratory studies, or by drilling samples of the area and correlating the samples with places of specific seismic velocity. By putting many observations of this type together, geologists and seismologists are able to obtain a good understanding of the overall structure of the Earth.

Determination of the structure of the deep parts of the Earth can only be achieved by remote geophysical methods such as seismology. There are seismographs stationed all over the world, and by studying the propagation of seismic waves from natural and artificial source earthquake and seismic events, we can calculate changes in the properties of the Earth in different places. If the Earth had a uniform composition, seismic wave velocity would increase smoothly with depth, because increased density is equated with higher seismic velocities. However, by plotting observed arrival time of seismic waves, seismologists have found that the velocity does not increase steadily with depth, but that several dramatic changes occur at discrete boundaries and in transition zones deep within the Earth.

We can calculate the positions and changes across these zones by noting several different properties of seismic waves. Some wave energy is reflected off interfaces, whereas other wave energy is refracted, changing the ray's velocity and path. These reflection and refraction events happen at specific sites in the Earth, and the positions of the boundaries are calculated using wave velocities. The core-mantle boundary at 1,802 miles (2,900 km) depth in the Earth strongly influences both P and S-waves. It refracts P waves, causing a P-wave shadow, and because liquids cannot transmit S waves, none get through causing a huge S-wave shadow. These contrasting properties of P and S-waves can be used to accurately map the position of the core-mantle boundary.

There are several other main properties of the deep Earth illustrated by variations in the propagation of seismic waves. Velocity gradually increases with depth, to about 62 miles (100 km), where the velocity drops a little at 62–124 miles (100–200 km) depth, in a region known as the low velocity zone. The reason for this drop in velocity is thought to be

Seismology and Earth's Internal Structure

How do geologists know that the interior of the Earth is composed of a number of concentric shells of rock with different compositions and physical characteristics? The main way is through studying earthquakes. Geologists have seismographs stationed all around the world, and by studying single earthquake events, changes in the properties of the earth in different places can be determined.

In a very general sense, seismic wave velocity increases smoothly with depth because increased density is equated with higher seismic velocities. Over time, geologists began to note that the velocity of seismic waves does not increase steadily with depth, but that several dramatic changes occur. The positions and degree of changes across these zones can be determined by noting several different properties of seismic waves. Some waves are reflected off interfaces, just as light is reflected off surfaces and other waves are refracted or bent, changing the ray's velocity and path just like a straw appears bent in a glass of water because light rays from it are bent across the water/air surface. The positions of the main boundaries were calculated using observations of where and at what depth these changes occur.

The core-mantle boundary at 2,000 miles (2,900 km) depth in the Earth strongly influences both seismic velocities and properties—it refracts P waves, causing a P-wave shadow in a belt around the globe. Because liquids cannot transmit S waves, none get through causing a huge S-wave shadow on the side of the Earth opposite the earthquake event.

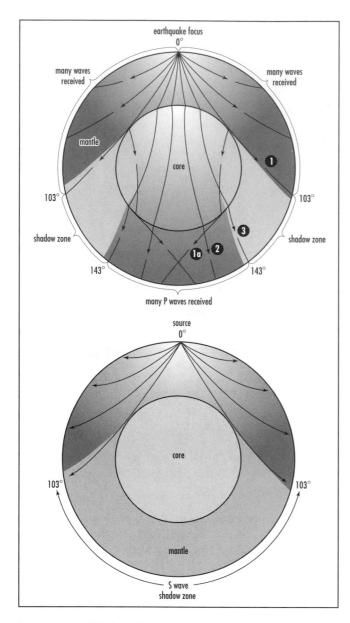

Cross sections of the Earth showing shadow zones that develop in bands around the planet due to refraction of P and S waves at internal boundaries

small amounts of partial melt in the rock, and this corresponds to the asthenosphere, the weak sphere that the plates move on, which is lubricated by partial melts.

There is another seismic discontinuity at 248.5 miles (400 km) depth, where velocity again increases sharply, this time caused by a rearrangement of the atoms of olivine in a polymorphic transition, into spinel structure, corresponding to an approximate 10 percent increase in density.

A major seismic discontinuity at 416 miles (670 km) may be either another polymorphic transition or a compositional change, the topic of many current investigations. Some

models suggest that this boundary separates two fundamentally different types of mantle, circulating in different convection cells, whereas other models suggest that there is more interaction between rocks above and below this discontinuity.

The core-mantle boundary is one of the most fundamental on the planet, with a huge density contrast from 5.5 g/cm^3 above, to 10–11 g/cm^3 below, a contrast greater than that between rocks and air on the surface of the Earth. The outer core is made dominantly of molten iron. An additional discontinuity occurs inside the core at the boundary between the liquid outer core and the solid, iron-nickel inner core.

The properties of seismic waves can also be used to understand the structure of the Earth's crust. Andrija Mohorovicic (a Yugoslavian seismologist) measured slow and fast arrivals from nearby earthquake source events. He proposed that some seismic waves were traveling through the crust, some along the surface, and that some were reflected off a deep seismic discontinuity between seismically slow and fast material at about 18–22 miles (30–35 km) depth. We now recognize this boundary to be the base of the crust and call it the Mohorovicic (or Moho) boundary and use its seismically determined position to measure the thickness of the crust, typically 6–45 miles (10–70 km).

See also CONVECTION AND THE EARTH'S MANTLE; EARTHQUAKES; MANTLE; PLATE TECTONICS; SEISMOGRAPH.

seismometer   *See* SEISMOGRAPH.

sequence stratigraphy   The study of the large-scale three-dimensional arrangement of sedimentary strata, and the factors that influence the geometry of these sedimentary packages. Sequences are defined as groups of strata that are bounded above and below by identifiable surfaces that are at least partly unconformities. Many sequence boundaries show up well in seismic reflection profiles, enabling their identification in deeply buried rock packages. Sequence stratigraphy differs from classical stratigraphy in that it groups together different sedimentary facies and depositional environments that were deposited in the same time interval, whereas classical stratigraphy would separate these units into different formations. By analyzing the three-dimensional shape of time-equivalent packages, the depositional geometry and factors that influenced the deposition are more easily identified. Some of the major factors that control the shape of depositional sequences include global sea-level changes, local tectonic or thermal subsidence or uplift, sediment supply, and differential biologic responses to subsidence in different climate conditions. For instance, carbonate reefs may be expected to keep pace with subsidence in tropic climates, but to be absent in temperate or polar climates. The techniques of sequence stratigraphy are widely used by sedimentologists and tectonicists in the petroleum industry to understand regional controls on sedimentation, and to correlate sequences of similar age worldwide.

See also PASSIVE MARGIN; STRATIGRAPHY.

serpentinite A metamorphic rock consisting almost entirely of minerals from the serpentine group (commonly antigorite, chrysotile, and lizardite), typically with minor talc, chlorite, and magnetite. Most serpentinites are derived through the alteration and hydration of ultramafic and less commonly mafic rocks, through the conversion of olivine, pyroxenes, and amphiboles to serpentine minerals. Serpentinites are typically massive to strongly foliated rocks with a greasy or slippery feel, a concoidal fracture, and may be granular or fibrous. Depending on the dominant serpentine mineral, the color of serpentinites may range from green, yellowish brown, to grayish blue, or a mixture of the three. The serpentine minerals have the general formula $(Mg,Fe)_3Si_2O_5(OH)_4$, with antigorite being the blue-green to colorless variety, lizardite being the green to white variety, and chrysotile being the fibrous white, yellow, or gray variety commonly used as asbestos.

Many serpentinites are associated with tectonically emplaced fragments of oceanic crust and arc rocks known as ophiolites, where the ultramafic rocks have been hydrated and converted to serpentinites during seafloor metamorphism and during regional tectonics associated with emplacement of the ophiolites. Because of their very low strength many serpentinites are strongly deformed and may form serpentinite mélanges containing blocks of ophiolitic and other rock types. These are typically incorporated into the mélange during thrusting and emplacement of the ultramafic body in a convergent margin setting. Some serpentinites have formed diapirs and intruded overlying rocks, most in accretionary wedge settings, where hydration of ultramafic rocks lowered the density of the serpentinized ultramafic rocks. This caused them to become negatively buoyant and to be forced upward as diapirs into the overlying accretionary wedge, typically intruding along faults and other preexisting structures.

See also ASBESTOS; CONVERGENT PLATE MARGIN PROCESSES; MÉLANGE; OPHIOLITES.

shield Large areas of exposed basement rocks in a craton, typically surrounded by flat-lying or gently dipping platformal deposits that either onlap or were eroded from the shield area. Almost all shields expose Precambrian rocks, and most are comprised dominantly of Archean rocks. One of the best known shields is the Canadian Shield, where Archean and Proterozoic cratons and orogenic belts are exposed across a huge expanse of generally low-relief but rugged hills across central, eastern, and northern Canada and Nunavut, and surrounded by gently dipping Phanerozoic sequences of the North American platform. Other well-known shields include the Guinea Shield, the Arabian-Nubian Shield, and the Indian Shield.

See also CRATONS.

Siberian Taiga Forest and Global Carbon Sink The northern third of Asia, stretching from the Ural Mountains in the west to the Pacific coast into the east, is known as Siberia. The southern border of Siberia is generally taken to be the Kazakh steppes in the southwest, the Altai and Sayan Mountains in the south, and the Mongolian steppes in the southeast. This region occupies approximately 3,000,000 square miles (7,500,000 km²). The western third of Siberia is occupied by the Siberian lowland, stretching from the Urals to the Yenisei River. This low marshy area is drained by the Ob River and its tributaries and hosts agriculture, industry, and most of Siberia's population in the wooded steppe. Eastern Siberia stretches from the Yenisei River to a chain of mountains including the Yablonovy, Stanovoy, Verkhoyansk, Kolyma, and Cherskogo Ranges. The eastern half of Siberia is an upland plateau, drained by the Vitim and Aldan Rivers. The Lena runs along the eastern margin of the region, and Lake Baikal, the world's deepest lake, is located in the southeast. Northeasternmost Siberia hosts a smaller plain on the arctic coast between the Lena and Kolyma Rivers, in the Republic of Yakutia (Sakha).

Siberia shows a strong zonation in vegetation zones, including a zone of tundra that extends inland about 200 miles (300 km) from the coast, followed by the taiga forest, a mixed forest belt, and the southern steppes. Siberia's taiga forest accounts for about 20 percent of the world's total forested land, covering about two-thirds of the region. This region accounts for about half of the world's evergreen forest and buffers global warming by acting as a large sink for carbon that otherwise could be released to the atmosphere as carbon dioxide, a greenhouse gas. The forest, and the rich soils derived from the decay of dead trees, represents a very significant sink for global carbon. Much of the taiga forest is currently being logged, at an alarming rate of loss of 12 million hectares per year. Much of this is being done by clearcutting, where 90 percent of the timber is harvested, leading to increased erosion of the soil and runoff into streams. The effects on deforestation could be dramatic for global climate. With so much carbon stored in the taiga forest, both in the trees and in the peat and soils, any logging or development that releases this carbon to the atmosphere will increase global carbon dioxide levels, contributing to global warming. Additional loss of forest is being caused by acid rain and other pollution largely emitted from the coal, nickel, aluminum, and lead smelting plants in the west. Additionally, large tracts of forest are being torn up to explore for and extract oil, natural gas, iron ore, and diamonds.

See also GREENHOUSE EFFECT.

silicate minerals *See* MINERALOGY.

Silurian The third period of Paleozoic time ranging from 443 Ma to 415 Ma, falling between the Ordovician and

Devonian periods, and the corresponding system of rocks. From base to top it is divided into the Llandoverian and Wenlockian ages or series (comprising the Early Silurian), and the Ludlovian and Pridolian ages or series (comprising the Late Silurian). The period is named after a Celtic tribe called the Silures, who inhabited a part of Wales where rocks of the Silurian System are well exposed. The Silurian is also known as the age of fishes.

Rocks of the Silurian System are well exposed on most continents, with carbonates and evaporites covering parts of the mid-west of North America, the Russian platform, and China. Silurian clastic sequences form thick orogenic wedges in eastern and western North America, central Asia, western Europe, China, and Australia. Much of Gondwana was together in the Southern Hemisphere and included the present-day landmasses of South America, Africa, Arabia, India, Antarctica, Australia, and a fragmented China. North America, Baltica, Kazakhstan, and Siberia formed separate landmasses in equatorial and northern latitudes. Much of Gondwana was bordered by convergent margins, and subduction was active beneath the Cordillera of North America. Baltica and Laurentia had collided during early stages of the Acadian-Caledonian orogeny, following an arc-accretion event in the Middle to Late Ordovician, known as the Taconic orogeny in eastern North America.

Land plants first appeared in the Early Silurian and were abundant by the middle of the period. Scorpion-like eurypterids and arthropods inhabited freshwater environments and may have scurried across the land. In the marine realm trilobites, brachiopods, cephalopods, gastropods, bryozoans, crinoids, corals, and echinoderms inhabited shallow waters. Conspicuous reefs were built by stromatoporoids, rugose and tabulate corals, while jawed fish fed on plankton and nekton.

See also PALEOZOIC; PHANEROZOIC.

Sinai Peninsula Egypt's Sinai Peninsula is a 23,554 square mile (61,000 km²) triangular block of land, bounded on the west by the Suez Canal and the Gulf of Suez and on the east by the Araba Rift Valley and the Negev Desert of Israel. The peninsula has three principal physiographic zones: extremely rugged mountains in the southern Pre-Cambrian shield province; a gently northward dipping, highly dissected tableland in the interior; and a gently rolling region of sand dunes in the north. Within these zones, minor subdivisions include a broad coastal plain on the western margin of the southern province, several tectonic depressions on the margins of and within the central plateau area, and a zone of folded and faulted Mesozoic and Tertiary sedimentary strata that forms the northern margin of the interior.

The Sinai Peninsula lies in the east-central portion of the Earth's largest desert system, stretching from the northwest coast of Africa to the eastern coast of the Arabian peninsula.

Over much of this area, precipitation is a rare and unpredictable event and is accompanied by average annual potential evapotranspiration rates greater than 30 times the average precipitation. However, in the Sinai and nearby areas of eastern Egypt, potential evapotranspiration can be as high as 600 times the precipitation (in months where precipitation occurs), and the ratio varies over three orders of magnitude during the year. In the summer, temperatures may reach 45°, and daily temperature excursions are up to 64.4°F (18°C). The typical desert pattern of rainfall is marked by total annual rainfall of less than 15.75 inches per year (400 mm/yr) and large interannual variability. In the case of the Sinai, most of the peninsula except for the extreme northeast corner receives less than 2 inches (50 mm) of rainfall per year. This low amount of rainfall classifies Sinai as an extreme desert, having less than 2.75 inches (70 mm) of precipitation, and the northeast as a semi-desert, characterized as receiving between 6 and 12 inches (150 and 350 mm) of precipitation. Humidity in the Sinai is highest in El Arish and lowest in the stations bordering the Gulf of Suez (El Tor, Abu Suwier, and Ras Gemsa), as well as Fayid and Kabrit (Eastern Desert stations). Humidity is generally higher in the late autumn at El Arish, where it reaches 78 percent, and lowest in the spring (68 percent). This variation is small, however, in comparison to the annual fluctuation at Abu Suweir (Gulf of Suez region), where relative humidity varies between 77 percent in January and 19 percent in May.

Immediately northeast of the Sinai, the desert belt is interrupted by the more temperate climate of the Fertile Crescent, which includes parts of northern Israel, Lebanon, Syria, and Iraq. Except for a slight moderation along the Mediterranean coast, the aridity of the climate in the Sinai, together with the ruggedness of the terrain, has restricted land use to nomadic grazing and small-scale farming. No appreciable moderation is given by the Gulf of Suez on the west, or the Gulf of Aqaba on the east. A slight increase in precipitation (and consequent groundwater recharge) becomes apparent in the northern Negev Desert.

Surface-Water Hydrology

The Sinai Peninsula is drained by an extensive network of channels (wadis) that are generally dry. There are no significant perennial streams flowing except for runoff from the springs at Ain Gederiat and Yaraqa, which infiltrate into the alluvium over a distance of a few hundred meters. No important drainage channels are mappable in the northern sector (except Wadi El Arish and Wadi Hareidan, where dune sands have an infiltration capacity exceeding normal rainfall.

There are five basins to which the wadis empty, including the Mediterranean Sea on the north, the Gulf of Suez to the west, the Gulf of Aqaba to the southeast, the Red Sea at the extreme southern tip, and the Araba (Dead Sea) Rift Valley to the northeast.

The largest catchment basin on the peninsula is that of Wadi El Arish, named for the outlet of the main trunk at the city of El Arish. The area of the basin is 6,757 square miles (17,500 km²) (approximately 28.6 percent of the peninsula). South of El Daiqa Gorge, Wadi El Arish bifurcates into a great number of tributaries, so that, by the time it reaches the Mediterranean, it is a seventh-order stream. North of El Daiqa Gorge, only one significant tributary (Wadi Hareidan) enters Wadi El Arish.

In the Wadi El Arish basin, there are four main divisions including the western, eastern and southern tributaries, and the main trunk and delta. In the western tributaries the floodplain of the principal channel, Wadi El Bruk, averages three miles (5 km) in width. Wadi El Bruk occupies the southern side of the syncline between the Gebel Yelleq anticline and the Gebel Minshera anticline group. The southern tributaries drain the El Egma plateau, forming steep-walled canyons in the limestone. The eastern tributaries drain the northeastern side of the El Egma plateau. The principal channel is Wadi Geraia. The main trunk extends 37 miles (60 km) north of El Daiqa Gorge to the Mediterranean Sea. The other drainage basins of note are the basins of interior drainage (El Maghara and El Hasana) and the Dead Sea drainage basin, which drains a large area in the El Kuntilla depression.

Drainage basins of the Sinai may be divided into two categories including basins of exterior drainage and basins of interior drainage. These correspond to open and closed drainage, where open (exterior) drainage discharges to a base level outside the basin, and closed (interior) drainage has no apparent discharge to a base level or another basin.

There are three major lakes on the peninsula: Lake Bardawil, the Great Bitter Lake, and the Little Bitter Lake. Lake Bardawil lies on the Mediterranean Coast, from which it is separated by a narrow band of sand dunes (barrier islands). Breaches in the barrier island chain provide outlets to the Mediterranean. Lake Bardawil is approximately 19 miles (30 km) long and 14 miles (22 km) wide and is located 12.5 miles (20 km) west of El Arish. An extensive system of lakes existed during past humid climatic phases. Paleoshore terraces are evident along the margins of the flood plains of the larger wadis in central Sinai. Thick deposits of clay underlie the areas behind the natural dams.

Physiography
Igneous and metamorphic rocks of Pre-Cambrian age are exposed only in the Southern Mountainous Region. These rocks belong to the Arabian-Nubian shield, which was assembled during the Pan-African orogeny (1,200–500 million years ago) as island arcs coalesced and extend as basement underlying the sedimentary cover rocks. The basement is deeply buried (7,312 feet; 2,174 m) at Nakhl, < 10,000 feet (3,280 m) at El Khabra, as the thickness of the cover increases to the north.

The southern mountains attain a maximum elevation of 8,665 feet (2,641 m) at St. Katerina, and several other peaks exceed 6,560 feet (2,000 m). The eastern margin of the southern province borders the Gulf of Aqaba and is characterized by a poorly developed coastal plain, consisting of scattered small deltas. This is due to the great depth of the narrow Gulf of Aqaba (3,280 feet; 1,000 m), which prevents the building of a coastal plain by coalescence of deltas. In contrast, on the western margin, the older and relatively shallow Gulf of Suez (984 feet; 300 m) has allowed the formation of a broad coastal plain, as sediments eroding from the shield have been deposited in alluvial fans and deltas.

The southern province displays several sets of prominent lineaments that are clearly visible in satellite images. The lineaments interpreted as faults are the result of tectonism that both pre- and post-dates the deposition of the sedimentary cover. The lineaments are marked by intermontaine valleys that carry the weathering products to the gulfs on either side of the peninsula. Basaltic dikes, most of which are related to the opening of the two gulfs and the Red Sea during the Miocene, are also seen as either positive or negative lineaments on satellite images.

The Central Tablelands are comprised of two plateaus. El Tih ("the wandering") is the southern plateau, and El Egma is to the north. El Tih is a horseshoe-shaped (concave to the north) series of connected topographic highs (cuestas) (maximum 5,250 feet; 1,600 m) which are separated from the southern mountains by the Ramlet Himeiyir depression. The plateau is composed of Cretaceous limestones, with shales and sandstones at the base. The exposures of the lower Cretaceous (Nubia-type) sandstone at the base of the southern escarpments of El Tih are considered the principal recharge areas for this aquifer.

On El Egma plateau, massive carbonate rocks of Eocene age, containing bands of flint, are exposed in a broad zone that dips gently to the north and occupies most of the middle third of the peninsula. The plateau is separated from El Tih plateau by three tectonic depressions: Yaraqa on the west, El Aqaba on the east, and El Kuntilla on the northeast. El Egma is drained by Wadi El Arish, the tributaries of which create a well-developed and very complex network of ephemeral streams. Between the Central Tablelands and the Coastal zone (described below) lies a northeast-southwest trending belt of folded sedimentary rocks disposed in anticline/syncline pairs. These structures form prominent hills, which rise up out of the surrounding limnic, alluvial and aeolian deposits that fill the intervening synclines. The anticlines are generally accompanied by brittle faults, probably coeval with the uplift.

Three sub-belts can be defined, with the degree of deformation decreasing to the southeast. The Gebel El Maghara anticline is the principal feature in the northernmost belt and is characterized by Jurassic sandstones exposed in the core and on the northern flank, northwest-dipping axial surfaces, doubly plunging hinge lines, intense brittle deformation, and thrust faulting with overturned bedding on the southern flank. Gebel Maghara has considerable exposures of Nubian sandstones of lower Cretaceous age (Aptian-Albian) surrounding the inner core.

The Giddi-Yelleq-Halal anticlines mark the second anticlinal axis. The Nubian sandstone is also exposed in the cores of these structures, although there is a lesser amount of erosion of the core, as compared to Gebel Maghara. The syncline between the Maghara belt and the Giddi-Yelleq-Halal belt consists of Eocene rocks covered by unconsolidated Quaternary sands. The syncline to the south of this belt is occupied by lacustrine and alluvial deposits in Wadi El Bruk and the central portion of Wadi El Arish.

The southernmost belt consists of approximately 15 small hills that are much less prominent but have steeper dips than the northern folds. The hills lie in the midst of a northeast-southwest trending fault zone, the Minshera–Abu Kandu shear zone. South of the Minshera–Abu Kandu shear zone, the roughly east-west trending dextral strike-slip Zarga El Naab fault marks the southern border of the deformation associated with the Syrian Arc. Some authors refer to this fault as the Ragabet El-Naam fault. South of this fault, the plateaus are undeformed, except for brittle fracturing and faulting associated with the opening of the Red Sea and the gulfs of Suez and Aqaba. Although well defined in the east, this fault becomes obscure in the middle of the peninsula.

The synclines associated with the above-mentioned anticlines are poorly exposed. In general, the synclinal axes form wide plains that are covered by alluvium and lacustrine deposits. These are underlain by Tertiary sediments and are the location of the agricultural activity in the interior of the Sinai. The synclines receive drainage from the anticlinal blocks and in some cases are the locations of "basins of interior drainage." Drainage within these basins does not join with the exterior drainage (that which empties into the sea).

The Mediterranean Coastal Plain Region extends the entire width of northern Sinai (approximately 118 miles; 190 km). The region is bounded on the north by the Mediterranean Sea and on the south by the ridge of Gebel Maghara. The present form of the land is due to the deposition and subsequent reworking of Nile deltaic sediments during a marine transgression in the late Pliocene and Pleistocene. Several subsequent phases of regression resulted in the stranding of the paleoshores, most of which have been reworked by aeolian processes.

The region is divided into two main low areas, separated by a ridge, the Bardawil Coastal Promontory. The western area consists of low, rolling hills that have seen considerable military activity, rendering the area unsafe for travel. In the east, the towns of El Arish and Rafah lie among the coastal dunes, which tend to increase in height up to 200 feet (60 m) to the south and east.

Stratigraphy

The stratigraphy of the northern Sinai consists of three major divisions: the Lower Clastic (Pre-Cenomanian) Division, the Middle Calcareous (Cenomanian-Eocene) Division, and the Upper Clastic (Oligocene-Miocene to Recent) Division.

Paleozoic rocks of the Sinai are generally thin, (approximately 2,215 feet [675 m] total thickness), sparsely fossiliferous, and predominantly continental (fluvial and paralic), except for a lower Carboniferous marine sequence (Um Bogma formation). The exposures of the Paleozoic sequence are limited to the southern part of central Sinai, near Abu-Zenima and Um Bogma.

The Triassic (245–208 million years ago [Ma]) is exposed as a 656-foot (200-m) thick section at Gebel Arief El-Naqa. Elsewhere, the unit has been penetrated in at least two wells. At Gebel El-Halal, in a petroleum exploratory well, a thickness of 2,200 feet (671 m) was penetrated, and at Abu Hamth, only 92 feet (28 m) of Triassic is known. The Triassic sequence changes from continental sandstones, siltstones, and shales overlain by a marine transgressive sequence (argillaceous micrites, biomicrites, and sparites) exposed in the south at Arief El-Naqa to a transgressive sequence of shallow marine carbonate deposits, with the top of the section at 3,640 feet (1,109 m) at El-Halal.

Jurassic (208–144 Ma) deposits conformably overlie the Triassic and continue the trend of continental sedimentation in the southern parts of the peninsula and gradually deepening marine conditions to the north. The most significant outcrop area is at the core of Gebel Maghara, where over 6,651 feet (2,000 m) of Jurassic is exposed, while at the El-Halal well, 10,610 feet (3,234 m) of Jurassic are known. At Gebel Maghara, the sequence records a relatively stable shelf, which deepened to the north. High-energy reefs gave way southward to low-energy lagoons that interfingered with continental-fluvial environments. The exposures at Gebel Maghara record a mid-Jurassic marine regression, followed by the worldwide mid-Mesozoic transgression. The regressive sediments include erosional products derived from the southern mountainous province. These were logged in the Abu Hamth well (near Nakhl) as an approximately 3,280-foot (1,000-m) thick deposit of interbedded sand and shale.

The Cretaceous system (144–66 Ma) in the Sinai is typically separated into two principal divisions: the lower Cretaceous Malhah formation is closely related to other terrigenous sandstones found widely distributed in northern Africa (the "Nubian Sandstone"), while the upper Cretaceous (herein belonging to the Middle Calcareous Division) represents a change to marine conditions, which overspread the Sinai in the Cenomanian.

The lower Cretaceous (pre-Cenomanian) (144–98 Ma) is widely referred to as "Nubian-type Sandstone" in the Sinai. This nomenclature commonly refers to sandy facies of the entire Lower Clastic Division, especially in the central portions of the peninsula. In the fold belt, the unit is exposed in the central part of the larger anticlines, and in the south, in escarpments and tectonic depressions separating El Tih plateau from the southern mountains and from El Egma plateau. The lower Cretaceous attains a thickness of around 1,640 feet (500 m) in the southern exposures, dipping 120° to the northeast. At the Abu Hamth well in central Sinai near Nakhl, the unit is around 770 feet (235 m) thick and occurs at 2,395 feet (730 m) below land surface. The northern exposures range between 164 and 1,706 feet (50 and 520 m) in thickness, and in the deep well at El Khabra, the lower Cretaceous is 3,202 feet (976 m) thick. The lower Cretaceous trends toward finer grain sizes and decreasing age to the north. The basal beds are conglomeratic and were deposited after the uplift that produced the Jurassic/Cretaceous unconformity. In the lower part of the section, the dominant lithology is ferruginous sandstones, interbedded with shales. These were deposited in northerly flowing streams that were debauching onto a shallow carbonate platform. In the upper part of the lower Cretaceous (Aptian-Albian; 119–98 Ma), the sandstones give way to shales and sandy limestones, marking the marine transgression that continued throughout the remainder of the Mesozoic.

The Middle Calcareous (Cenomanian-Eocene) Division is comprised of predominantly carbonate rocks and includes a thin, but important shale unit (Paleocene Esna formation). There are three upper Cretaceous sub-units (Cenomanian, Turonian-Santonian, and Senonian; 98–75 Ma), along with the Paleocene and the Eocene systems (66–36 Ma). In most well logs made public by the oil industry, however, the upper Cretaceous is undifferentiated.

The Cenomanian (98–91 Ma) Halal formation follows the familiar pattern of clastics (marls and shales) to the south and carbonates (dolomites grading into deep marine deposits) to the north. At about the latitude of Nakhl (29°57' N), these are interbedded, indicating the position of the transition between a marine environment to the north and a terrigenous environment in the south.

The Turonian (91–89 Ma) Wata formation is conformable with the Cenomanian and is distinguished based on the appearance of the "Ora"; interbedded shales, marls, limestones, and sandstones. The limestone content increases toward the top of the section, until it becomes a pure, massive limestone that forms the tops of cuestas, such as those of the El Tih plateau. This indicates a continuation of the environments of the Cenomanian and has led to problems separating the units. The Coniacian (88–87 Ma) Matallah formation marks the beginning of widespread deposition of chalky limestones, a trend which continued throughout the remainder of the Cretaceous. The remaining Cretaceous units

(Santonian-Campanian [Duwwi Formation] and Maastrichtian [Sudr formation]; 88–66 Ma) are characterized by increasing chalk content and a southerly transgression of chalk-bearing deposits.

The upper Cretaceous units attain substantial thicknesses, exceeding 1,968 feet (600 m), and are widespread throughout north and central Sinai. They form many of the side slopes of the anticlinal hills. The Paleocene (66–58 Ma) is represented by the Esna formation, a gray-green shale deposit that is widespread and thin (maximum 213 feet; 65 m) at Gebel Minshera, averaging around 98–115 feet (30–35 m). The shales are usually grouped with the Maastrichtian (75–66 Ma), as they were both deposited in similar environments between structural highs.

The Eocene (58–36 Ma) is characterized by bedded and massive chalky limestones with flint bands interspersed throughout the lower part (the Egma formation). The upper Eocene (43–66 Ma) (Minya Formation) contains more clastics (marls, green-brown shales, and gypsiferous shales) and represents a regression of upper Eocene age. In northeastern Sinai, the Eocene is known as the Plateau Limestone and forms the tablelands of the El Egma plateau and the Khorasha plateau. The Eocene also occurs near the surface in the synclines within the fold belt.

The Oligocene (36–24 Ma) was a time of uplift in Sinai that resulted in the northward retreat of the sea and the deposition of regressive sediments (prograding deltas and submarine fans). These deposits are seen only in boreholes but attain great thicknesses, exceeding 5,900 feet (1,800 m) offshore. Onshore, the Oligocene is the age of basaltic intrusions, which occur as northwest-southeast olivine-bearing dikes. The dikes outcrop on the southern side of Gebel El Maghara and in the El Hasana area. The Early Miocene (24–17 Ma) is exposed south of El Arish as a thin (49 feet; 15 m) sequence of marl, marly limestone, and shale. Elsewhere in northeast Sinai, Miocene rocks are found in boreholes, reaching a maximum thickness of greater than 770 feet (235 m). Thicknesses of greater than 3,281 feet (1,000 m) have been reported in offshore exploratory wells.

Late Tertiary and Quaternary (Pliocene-Pleistocene-Holocene) (5.3 Ma–present) deposits are widespread throughout the Sinai and vary widely in depositional environment. The late Tertiary includes a conglomerate bed of Pliocene age (Al Hajj Formation) (5.3–1.6 Ma); and the Quaternary (1.6 Ma–present) includes the lower Pleistocene El Fagra calcareous sandstone (kunkur); coastal plain deposits of the Sudr and El Qaa plains; deltaic deposits, such as at Nuweiba on the Gulf of Aqaba; alluvial and pluvial terraces along the major wadis; wadi fill deposits, including colluvium in the upper reaches of Wadi El Arish and its tributaries; lake deposits, such as in Lake Bardawil and the Bitter Lakes, and those of Wadi Feiran; and eolian deposits that are widespread throughout the northern Sinai and scattered elsewhere.

St. Catherine's monastery, sitting at the base of Mount Sinai, Sinai Peninsula, Egypt *(Photo by Timothy Kusky)*

Tectonic History

The tectonic development of the Sinai began with the assemblage of several island arcs during the Pan-African orogeny in the Late Proterozoic and early Cambrian. These rocks are the igneous and metamorphic basement complex that forms the mountainous southern region. In addition to older fabrics related to the assemblage of the complex, the basement rocks have been affected by the Gulf of Suez rifting (north-north-west trend), as well as by the left-lateral strike-slip faulting in the Dead Sea–Araba Valley (north-northeast trend).

Sediments deposited in the Tethys Sea were buried and compacted to become the sedimentary rocks of the Central Tablelands of El Tih and El Egma plateaus. These rocks were deformed by the closure of the Tethys basin. Deformation in the southern portion of the central region is limited to fault-

ing and epeirogenic tilting (circa 4° N), whereas both folding and faulting are seen in the northern part of the peninsula.

Two fault zones mark the division between the two styles of deformation. The southern fault (Ragabet El Naam; also referred to as the Zarga El Naab fault) is an east-west striking, right-lateral, vertical strike-slip fault. The northern fault zone (Minshera-Abu Kandu shear zone) is composed of major faults with large (3,281-foot; 1,00-m) vertical displacements that parallel the trend of the fold belt.

The Gulf of Suez began forming as a rift no earlier than the Late Eocene (40 Ma). Three separate half-grabens (northern, central, and southern) make up the rift zone. In the northern and southern basins, the blocks are tilted eastward, whereas in the central half-graben, the blocks dip to the west. The Gulf of Suez is made up of a series of linked pull-apart basins formed from the shifting of the Arabian plate up to 65 miles (105 km) along the left-lateral Dead Sea–Araba fault. This results in a westerly and clockwise motion of the Sinai, which is acting to close the Gulf of Suez by at least 15.5 miles (25 km).

The origins of the North Sinai fold belt are less than clear but are widely regarded to be related to the Syrian arc, the origin of which is also problematic but appears to be the result of shortening of the upper crust as Arabia rotated away from Africa. This caused compression in the northern Sinai and extension in the south. The folding began in the Late Cretaceous (Senonian), as the Tethys Sea began to close, and continued to the Early Miocene and may still be active.

See also ARABIAN SHIELD.

Further Reading

Bentor, Yaacov K. "The Crustal Evolution of the Arabo-Nubian Massif with Special Reference to the Sinai Peninsula." *Precambrian Research* 28 (1985).

El-Baz, Farouk, Timothy M. Kusky, Ibrahim Himida, and Salaal Abdel-Mogheeth, eds., "Groundwater Potential of the Sinai Peninsula, Egypt, Desert Research Center, Cairo." Egypt, 1998.

El-Gaby, Samir, Franz K. List, and Resa Tehrani. "The Basement Complex of Eastern Desert and Sinai." Chapter 10 in *The Geology of Egypt*, edited by Rusdi Said. Rotterdam, Netherlands: A.A. Balkema, 1990.

Jenkins, D. A. "North and Central Sinai." Chapter 19 in *The Geology of Egypt*, edited by Rushdi Said: 361–380. Rotterdam, Netherlands: A.A. Balkema, 1990.

Kusky, Timothy M., Mohamed Abdelsalam, Robert Stern, and Robert Tucker. "The East African and Related Orogens, and the Assembly of Gondwana." *Precambrian Research* 123 (2003).

Kusky, Timothy M., and Farouk El-Baz. "Neotectonics and Fluvial Geomorphology of the Northern Sinai Peninsula." *Journal of African Earth Sciences* 31 (2000).

Stern, Robert J. "Arc Assembly and Continental Collision in the Neoproterozoic East Africa Orogen: Implications for the Consolidation of Gondwanaland." *Annual Review of Earth and Planetary Sciences* 22 (1994).

Geologists in southern Sinai, hiking on dark-colored Precambrian rocks that are overlain unconformably by light-colored sandstone and carbonates of the Egma and Tih plateaus in the background *(Photo by Timothy Kusky)*

Sinian   In Chinese literature, the Sinian Era refers to the latest Precambrian falling approximately 850–615 million years ago, and the corresponding system of rocks. The Sinian is the approximate equivalent of the Russian Riphean and in more recent international literature the Sinian would fall within the Neoproterozoic. There is considerable variation in the definitions and subdivisions of the Sinian in the scientific literature. Some classifications have the Sinian Era starting at 800 million years ago and ending at 570 million years ago, and divide the era into the Sturtian Period (800–610 Ma), and Vendian Period (610–570 Ma). In this scenario the Vendian is divided into the Varanger Epoch (610–590 Ma) and the Ediacara Epoch (590–570 Ma). Additional dating is clearly needed on the ages of the boundaries of the Sinian since the basal Cambrian rocks in China are suggested to be 615 Ma old, whereas elsewhere the basal Cambrian is dated at 544 Ma.

Rocks of the Sinian System in China rest unconformably on older deformed rocks of the Yangtze, North China, and Tarim cratons but are best developed on the Yangtze craton. Strata of the Sinian System are well exposed and have been studied extensively near the Three Gorges of the Yangtze River in Hubei Province. In these and other regions the lower part of the sequence is comprised of clastic rocks including many of glacial origin, and an upper calcareous and argillaceous succession. Deeper water rocks of the Sinian System are found around the margins of the Yangtze craton.

Sinian strata in China have abundant soft-bodied metazoan fossils similar to the Vendian fauna, and these strata are locally conformably overlain by rocks that bear shelly fauna and thus are Cambrian. Remarkable faunal assemblages including the earliest fossil records of multicellular life have been found in 585–600-million-year-old Sinian strata in southern China. These life-forms developed during global ice age conditions associated with the formation of the supercontinent of Gondwana and predate the more widespread and diverse Ediacarian fauna.

See also CAMBRIAN; PRECAMBRIAN; PROTEROZOIC; SUPERCONTINENT CYCLE; VENDIAN.

sinkhole   *See* KARST.

Skeleton Coast, Namibia   Namibia's Atlantic coastline is known as the Skeleton Coast, named for the grief and death that beset many sailors attempting to navigate the difficult waters moved by the cold Benguela current that sweeps the coast and warm winds coming off the Namib Sand Sea and Kalahari Desert. The coastline is littered with numerous shipwrecks, testifying to the difficult and often unpredictable nature of shifting winds and ocean currents. Giant sand dunes of the Namib sand-sea reach to the coast, and in places these dunes are also covered in bones of mammals that have searched in vain for water. Many dune types are present, including transverse dunes and barchans, and the winds in the region often cause a steady roar to grow from the blowing sands. The Desert Elephant lives in the region, eating and drinking in generally dry inland riverbeds, but sometimes venturing to the harsh coast. Oryx, giraffes, hyenas, springboks, ostriches, rare rhinos, and lions also roam the area, whereas Cape fur seals populate parts of the coast. Whales and dolphins swim along the coast, and occasionally, giant whale skeletons are washed up and exposed on the shore.

The cold Benguela current breaks off from the circum-Antarctica cold current and forms a cold sea breeze that often shrouds the region in mist and fog, especially during winter months. This mist sustains an unusual plant life in the desert and forms an additional navigational hazard for ships.

See also DESERT.

Slave craton   The Slave craton is an Archean granite-greenstone terrane located in the northwestern part of the Canadian shield. The Archean history of the craton spans the interval from 4.03 billion years, the age of the world's oldest rocks, known as the Acasta Gneisses exposed in a basement culmination in the Wopmay orogen, to 2.6–2.5 billion years ago, the age of major granitic plutonism throughout the province. The margins of the craton were deformed and loaded by sediments during Proterozoic orogenies, and the craton is cut by several Proterozoic mafic dike swarms.

Most of the volcanic and sedimentary rocks of the Slave craton were formed in the interval between 2.7 billion and 2.65 billion years ago. Syntectonic to post-tectonic plutons form about half of the map area of the province. The pre-late granite geology of the Slave Province shows some broadscale tectonic zonations. Greenstone belts are concentrated in a narrow northerly trending swath in the central part of the province, and the relative abundance of mafic volcanics, felsic volcanics, clastic rocks, and gneisses are different on either side of this line. The dividing line is coincident with a major Bouger gravity anomaly, and with an isotopic anomaly indicating that older crust was involved in granitoid petrogenesis in the west but not in the east. Greenstone belts west of the line comprise predominantly mafic volcanic and plutonic rocks (except at Indian Lake), whereas volcanic belts to the east contain a much larger percentage of intermediate and felsic volcanic material. This is most evident in a large belt of northwest-trending felsic volcanics extending from south and east of Bathurst Inlet toward Artillery Lake. Quartzofeldspathic gneisses older than the greenstones are rare throughout the province and are confined to a line west of the central dividing line.

In the middle 1980s, Timothy Kusky proposed that these major differences in geology across the Slave Province reflect that it is divided into a number of different tectonic terranes. These ideas were initially debated but later largely accepted and modified by further mapping, seismic surveys, and geochemical analysis. The province is divided into an older gneissic terrane in the west, known as the Anton terrane, that contains the world's oldest known rocks and is overlain by a platform type sedimentary sequence. The Contwoyto terrane and Hackett River arc represent an accretionary prism and island arc that accreted to the Anton terrane in the late Archean, uplifting the Sleepy Dragon terrane in a basement culmination.

Gneissic rocks of the Anton terrane extend from Yellowknife to the Anialik River. The name is taken from the Anton complex exposed north of Yellowknife, which consists of metamorphosed granodiorite to quartz diorite, intruded by younger granitoids. The Anton terrane dips under the Wopmay orogen in the west, and its eastern contact is marked by a several kilometer thick, nearly vertical mylonite zone, best exposed in the vicinity of Point Lake. The Anton terrane includes the oldest rocks known in the world, the 4.03-billion-year-old Acasta gneisses exposed in a basement culmination along the border with the Wopmay orogen. Also, 3.48–3.21-billion-year-old tonalitic gray gneisses are exposed in several locations, and similar undated old gray gneissic rocks are preserved as inclusions and small outcrop belts within a sea of younger granites in the western part of the craton. Several different types of gneissic rocks are present in these areas, including a variety of metamorphosed igneous and sedimentary rocks. The oldest type of gneiss recognized in most places includes tonalitic to granodioritic layers with mafic amphibolite bands that are probably deformed dikes. Younger orthogneisses have tonalitic, granodioritic, and dioritic protoliths, and migmatization is common. Locally, especially near the eastern side of the Anton terrane, the older gneisses are overlain by a shallow-water sedimentary sequence that includes quartz-pebble conglomerate, quartzite, metapelite, and metacarbonates. These rocks are likely the remnants of a thin passive margin sequence.

The Sleepy Dragon terrane extends from northeast of Yellowknife to the south shore of the south arm of Point Lake. This terrane includes intermediate to mafic quartzofeldspathic gneiss complexes such as the 2.8–2.7-billion-year-old Sleepy Dragon complex in the south, banded and migmatitic gneisses near Beniah Lake, and 3.15-billion-year-old chloritic granite on Point Lake. Isolated dioritic to gabbroic bodies are found as inclusions and enclaves. The most common protoliths to the gneisses are tonalites and granodiorites, and rock types in the Sleepy Dragon terrane are broadly similar to those in the Anton terrane. Sleepy Dragon gneisses are locally overlain unconformably by shallow water sedimentary sequences, notably along the southeastern margin of the complex near Detour Lake. Here, a basal tonalite pebble-bearing conglomerate grades up into metaquartzose and calcareous sands, and then into a metacarbonate sequence consisting of marbles and calc-silicate minerals. From base to top this sequence is only 1,600 feet (500 m)

thick, but it has been shortened considerably. Several tens of kilometers north at Beniah Lake in the Beaulieu River greenstone belt up to 3,200 feet (1,000 m) of quartzite are recognized between shear zones. There are thus several locations where shallow water sediments appear to have been deposited upon Sleepy Dragon gneisses. The similarities of the lithofacies successions in these rocks to those found in Phanerozoic rift and passive margin sequences are striking.

The Contwoyto terrane is composed of laterally continuous graywacke mudstone turbidites exposed in a series of westward-vergent folds and thrusts. Mapping in the Point Lake area revealed westward-directed thrusts placing high-grade metagraywackes over lower-grade equivalents. The graywackes are composed of matrix, rock fragments (felsic volcanics, mafic volcanics, chert, granite), and feldspars. Typically only the upper parts of the Bouma sequence are preserved. Black shales and iron formations are locally found, especially near the structural base of the sequence. In many places greenstone belts conformably underlie the sediments, but the bases of the greenstone belts are either known to be truncated by faults or are poorly defined, suggesting that they are allochthonous. Ophiolite-like stratigraphy, including the presence of sheeted dikes and cumulate ultramafics, has been recognized in several greenstone belts. Other greenstone belts of the Contwoyto terrane are composed predominantly of basaltic pillow lavas and exhibit both tholeiitic and calc-alkaline differentiation trends.

Rocks of the Contwoyto terrane thus include tectonic slivers of ophiolite-like rocks, oceanic type sediments (shales, iron formations), and abundant graywackes exhibiting both volcanogenic and flysch-like characteristics. These rocks are contained in westward-verging folds and are disrupted by westward-directed thrusts. A series of granitoids intruded this package of rocks at various stages of deformation. These relationships are characteristic of an accretionary prism tectonic setting. In such an environment graywackes are eroded from a predominantly island arc source, as well as from any nearby continents, and are deposited over ophiolitic basement capped by abyssal muds and iron formations. Advance of the accretionary prism scrapes material off the oceanic basement and incorporates it in westward-vergent fold and thrust packages. This material is accreted to the front of the arc and is intruded by arc-derived magmas during deformation. Metamorphism is of the low-pressure, high-temperature variety and is similar to that of accretionary prisms that have experienced subduction of young oceanic crust, or subduction of a ridge segment.

The Hackett River arc consists of a series of northwest-striking volcanic piles and synvolcanic granitoids, especially in the south. Felsic volcanics predominate but a spectrum of compositions including basalt, andesite, dacite, and rhyolite is present. Volcanic piles in the Hackett River arc therefore differ strongly from greenstone belts in the west, which consist predominantly of mafic volcanic and plutonic rocks. In the Back River area, cauldron subsidence features, rhyolitic ring intrusions, tuffs, breccias, flows, and domes, with well-preserved subaerial and subaquatic depositional environments, have been documented. Rhyolites from the Back River complex have been dated at 2.69 billion years old, and the volcanics are broadly contemporaneous with graywacke sedimentation because the flows overlie and interfinger with the sediments. Gneissic rocks in the area are not extensively intruded by mafic dike swarms like the gneisses of the Anton and Sleepy Dragon terranes, and they have yielded ages of 2.68 billion years, slightly younger than surrounding volcanics. Since none of the gneisses in the Hackett River arc have yielded ages significantly older than the volcanics, deformed plutonic rocks in this terrane are accordingly distinguished from gneisses in the western part of the Slave Province. These gneisses are suggested to represent subvolcanic plutons that fed the overlying volcanics. Another suite of tonalitic, dioritic, and granodioritic plutonic rocks with ages 60 to 100 million years younger than the volcanics also intrude the Hackett River arc. This suite of granitoids is equated with the late- to post-tectonic granitoids that cut all rocks of the Slave craton.

A belt of graywacke turbidites in the easternmost part of the Slave Province near Beechy Lake has dominantly eastward vergent folds and possible thrusts, with some west vergent structures. The change from regional west vergence to eastward vergence is consistent with a change from a forearc accretionary prism to a backarc setting in the Beechy Lake domain.

The Hackett River terrane is interpreted as an island arc that formed above an east-dipping subduction zone at 2.7 billion to 2.67 billion years ago. The mafic to felsic volcanic suite, development of caldera complexes, and overall size of this belt are all similar to recent immature island arc systems. The Contwoyto terrane is structurally and lithologically similar to forearc accretionary complexes; west vergent folds and thrusts in this terrane are compatible with eastward dipping subduction, as suggested by the position of the accretionary complex to the west of the arc axis. The change from west to east vergence across the arc axis into the Beechy Lake domain reflects the forearc and backarc sides of the system. Mafic volcanic belts within the Contwoyto terrane are interpreted as ophiolitic slivers scraped off the subducting oceanic lithosphere.

The Anton terrane in the western part of the Slave Province contains remnants of an older Archean continent or microcontinent including the world's oldest known continental crust. Quartzofeldspathic gneisses here are as old as 4.03 billion years, with more abundant 3.5–3.1-billion-year-old crust. These gneissic rocks were deformed prior to the main orogenic event at 2.6 billion years ago. The origin of the gneisses in the Anton terrane remains unknown; many have igneous protoliths, but the derivation of the rocks is not yet clear. The Sleepy Dragon terrane might represent a microcontinent accreted to the Anton terrane prior to collision with the Hackett River arc, but more likely it represents an eastern part of the Anton terrane uplifted and transported westward during orogenesis. Studies at the southern end of the Sleepy Dragon terrane have shown that the gneisses occupy the core of a large fold or anticlinorium, consistent with the idea that the Sleepy Dragon terrane represents a basement-cored Alpine style nappe transported westward during the main orogenic event. The distribution of pre-greenstone sediments lying unconformably over the gneisses is intriguing. At Point Lake, a few meters of conglomerates, shales, and quartzites lie with possible unconformity over the Anton terrane gneiss, whereas farther east, up to 1,600 feet (500 m) of sediments unconformably overlie Sleepy Dragon terrane gneisses. These include basal conglomerates and overlying sand, shale, and carbonate sequences, and thick quartzites with unknown relationships with surrounding rocks. These scattered bits of preserved older sediments in the Slave Province may represent remnants of an east-facing platform sequence developed on the Anton–Sleepy Dragon microcontinent.

In a simple sense, the tectonic evolution of the Slave craton can be understood in terms of a collision between an older continent with platformal cover in the west with a juvenile arc/accretionary prism in the east. The Anton terrane experienced a sequence of poorly understood tectono-magmatic events between 4.03 billion and 2.9 billion years ago, then was intruded by a set of mafic dikes probably related to lithospheric extension. After extension, the thermally subsiding Anton terrane was overlain by an eastward thickening shallow water platform sequence. To the east, the Hackett River volcanic arc and Contwoyto terrane are formed as a paired accretionary prism and island arc above an east-dipping subduction zone. Numerous pieces of oceanic crust are sliced off the subducting lithosphere, and synvolcanic plutons intrude along the arc axis. Any significant rollback of the slab or progradation of the accretionary wedge will cause arc magmas to intrude the accreted sediments and volcanics. Graywacke sediments are also deposited on the backside of the system in the Beechy Lake domain.

As the arc and continent collide at about 2.65 billion years ago, large ophiolitic sheets are obducted, and a younger set of graywacke turbidites are deposited as conformable flysch. This is in contrast to other graywackes that were incorporated into the accretionary prism at an earlier stage and then thrust upon the Anton terrane. There are thus at least two ages of graywacke sedimentation in the Slave Province. Older graywackes were deposited contemporaneously with felsic volcanism in the Hackett River arc, whereas younger graywackes were deposited during obduction of the accretionary prism onto the Anton continent. Synvolcanic plutons along the arc axis became foliated as a result of the

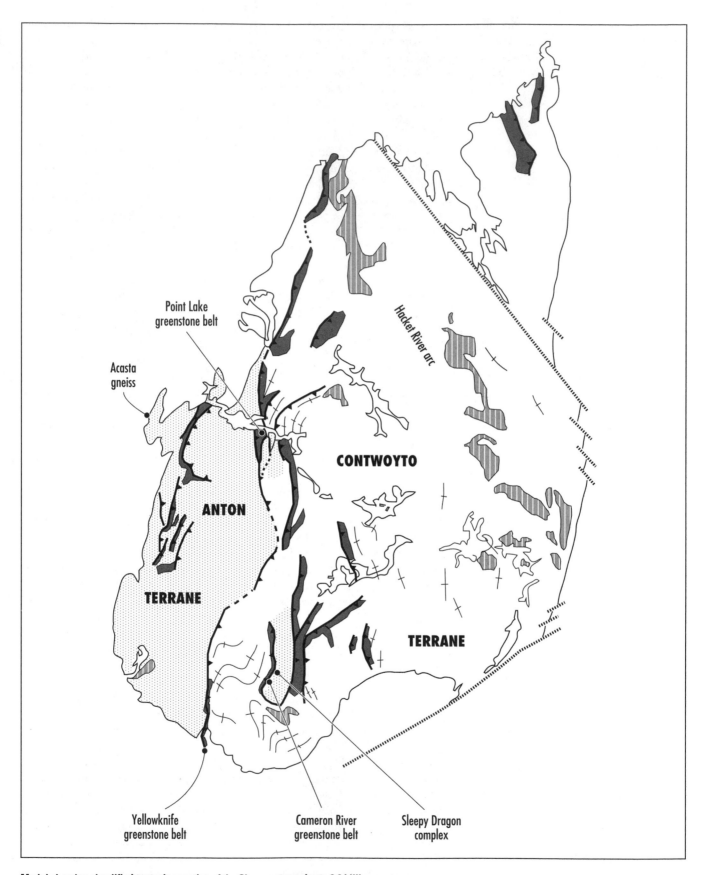

Point Lake
greenstone belt

Acasta
gneiss

Hacket River arc

CONTWOYTO

ANTON

TERRANE

TERRANE

Yellowknife
greenstone belt

Cameron River
greenstone belt

Sleepy Dragon
complex

Model showing simplified tectonic zonation of the Slave craton prior to 2.6 billion years ago

arc-continent collision, and back thrusting shortened the Beechy Lake domain. Continued convergence caused the uplift and transportation of the Sleepy Dragon terrane as a basement nappe, and strongly attenuated greenstone slivers. Numerous late- to post-tectonic granitoids represent post-collisional anatectic responses to crustal thickening, or pressure-release melts formed during post-collisional orogenic extension and collapse. These suites of granitoids have similar intrusion ages across the province.

See also ARCHEAN; CRATONS; GREENSTONE BELTS.

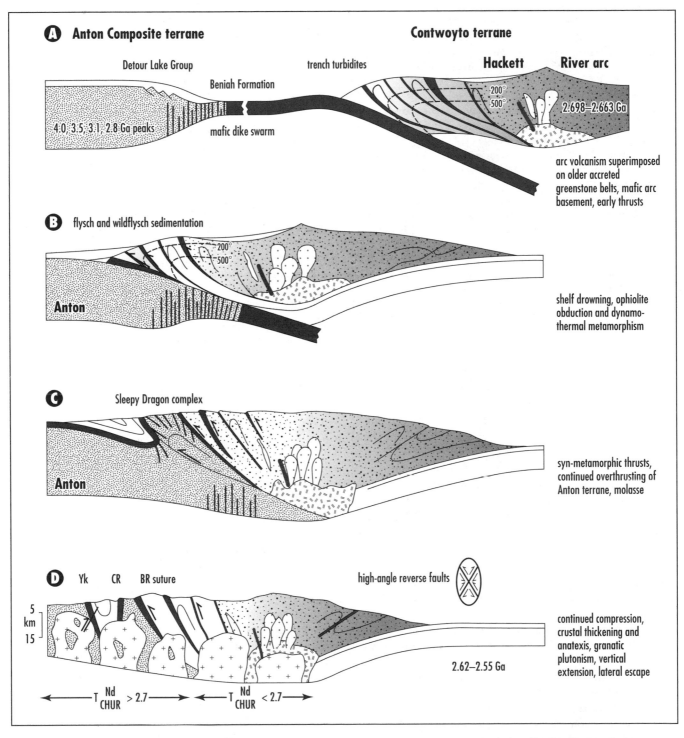

Cross-sectional tectonic evolution of the Slave craton, showing a 2.7–2.66-billion-year-old island arc terrane in the east (Hackett River arc in the Contwoyto terrane) colliding with a 4.0–2.8-billion-year-old continent and passive margin sequence in the west by 2.62 billion years ago

Further Reading

Bleeker, Wouter, and William Davis, eds. "NATMAP Slave Province Project." *Canadian Journal of Earth Sciences* 36 (1999).

Bowring, Samuel A., Ian S. Williams, and William Compston. "3.96 Ga Gneisses from the Slave Province, Northwest Territories." *Geology* 17/11 (1989b): 969–1064.

Corcoran, P. L., W. U. Mueller, and Timothy M. Kusky. "Inferred Ophiolites in the Archean Slave Province." In *Precambrian Ophiolites and Related Rocks, Developments in Precambrian Geology*, edited by Timothy M. Kusky. Amsterdam: Elsevier, 2004.

Helmstaedt, Herart, William A. Padgham, and John A. Brophy. "Multiple Dikes in the Lower Kam Group, Yellowknife Greenstone Belt: Evidence for Archean Sea Floor Spreading?" *Geology* 14 (1986): 562–566.

Henderson, John B. "Archean Basin Evolution in the Slave Province, Canada." In *Precambrian Plate Tectonics*, edited by Alfred Kroner, 213–235. Amsterdam: Elsevier, 1981.

Kusky, Timothy M. "Structural Development of an Archean Orogen, Western Point Lake, Northwest Territories." *Tectonics* 10, no. 4 (1991): 820–841.

———. "Evidence for Archean Ocean Opening and Closing in the Southern Slave Province." *Tectonics* 9, no. 6 (1990): 1,533–1,563.

———. "Accretion of the Archean Slave Province." *Geology* 17 (1989): 63–67.

Padgham, William A. "Observations and Speculations on Supracrustal Successions in the Slave Structural Province." In "Evolution of Archean Supracrustal Successions," edited by L. D. Ayrees, P. C. Thurston, K. D. Card, and W. Weber. Geological Association of Canada Special Paper 28 (1985): 133–151.

sleet *See* PRECIPITATION.

Smith, William H. (1769–1839) British *Geologist* William Smith was born in Churchill, Oxfordshire, where he became an apprentice surveyor. His experience outdoors led him to become interested in the rocks he was surveying, and he began surveying the coalfields in Somerset and then had the opportunity to survey a canal construction project, where he completed detailed geological investigations on the rocks the canal was being cut through. During this study, he mapped different sedimentary facies in the rocks and collected fossils and identified specific fossils assemblages in rocks of certain ages, in different facies. By this study, he proposed in 1796 that some strata could be identified by specific index fossils they contained, without regard to the rock type. He continued to travel and map different rock units around the United Kingdom for many years, and in 1801 Smith completed a simple geological map of England and Wales. For the next 10 years, William Smith worked incessantly on improving his maps, and in 1815 he produced the first detailed geologic map of England and Wales, showing more than 20 different units, topography, and description of the stratigraphy, and structural cross sections. During the same time, Smith produced his works on "Strata Identified by Organic Remains," in which he illustrated the fossils in the rocks through a series of wood engravings. In 1831 the Geological Society of London awarded William Smith the first Wollaston Medal, its highest honor. Since his death, William Smith has been known as the "Father of English Geology."

snow *See* PRECIPITATION.

soils All the unconsolidated material resting above bedrock. Soils are the natural medium for plant growth. Differences in soil profile and type of soil result from differences in climate, the rock type it started from, the types of vegetation and organisms, topography and time. Normal weathering produces a characteristic soil profile, marked by a succession of distinctive horizons in a soil from the surface downward. The A-Horizon is closest to the surface and usually has a gray or black color because of high concentrations of humus (decomposed plant and animal tissues). The A-Horizon has typically lost some substances through downward leaching. The B-Horizon is commonly brown or reddish, enriched in clay produced in place and transported downward from the A-horizon. The C-Horizon of a typical soil consists of slightly weathered parent material. Young soils regularly lack a B-horizon, and the B-horizon grows in thickness with increasing age.

Some unique soils form under unusual climate conditions. Polar climates are typically cold and dry, and the soils produced in polar regions are typically well-drained without an A-horizon, sometimes underlying layers of frost-heaved stones. In wetter polar climates, tundra may overlie permafrost, which prevents the downward draining of water. These soils are saturated in water and rich in organic matter. These polar soils are very important for the global environment and global warming. They have so much organic material in them that is effectively isolated from the atmosphere that they may be thought of as locking up much of the carbon dioxide on the planet. Cutting down of northern forests as is occurring in Siberia may affect the global carbon dioxide budget, possibly contributing to climate change and global warming.

Dry climates limit the leaching of unstable minerals such as carbonate from the A-horizon, which may also be enhanced by evaporation of groundwater. Extensive evaporation of groundwater over prolonged times leads to the formation of caliche crusts. These are hard, generally white carbonate minerals and salts that were dissolved in the groundwater but were precipitated when the groundwater moved up through the surface and evaporated, leaving the initially dissolved minerals behind.

In warm, wet climates, most elements (except for aluminum and iron) are leached from the soil profile, forming laterite and bauxite. Laterites are typically deep red in color and are found in many tropical regions. Some of these soils are so hard that they are used for bricks.

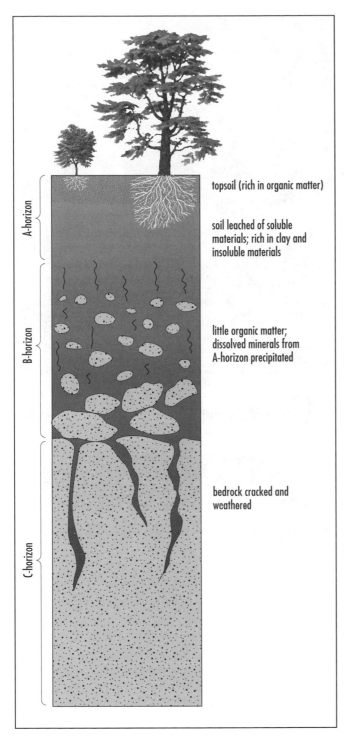

A-horizon

topsoil (rich in organic matter)

soil leached of soluble
materials; rich in clay and
insoluble materials

B-horizon

little organic matter;
dissolved minerals from
A-horizon precipitated

C-horizon

bedrock cracked and
weathered

Typical soil profile, showing organic zone and the A, B, and C horizons

Soils form at various rates in different climates and other conditions, ranging from about 50 years in moderate temperatures and wet climates to about 10,000 to 100,000 years for a good soil profile to develop in dry climates such as the desert southwest of the United States. Some soils, such as those in the tropics, have been forming for several million years and are quite mature. Deforestation causes erosion of soils, which cannot be reproduced quickly. In many places, such as parts of Madagascar, South America, and Indonesia, deforestation has led to accelerated rates of soil erosion, removing thick soils that have been forming for millions of years. These soils supported a rich diversity of life and it is unlikely that the soils will ever be restored in these regions.

See also MASS WASTING; WEATHERING.

Further Reading
Birkland, P. W. *Soils and Geomorphology.* New York: Oxford University Press, 1984.

sonar   Sonar is a process that analyzes transmitted and reflected underwater sound waves to find and locate submerged objects and measure the distances underwater. Sonar works by sending out subsurface sound waves and then waiting for the returning echoes. A computer or a human operator then analyzes the data. In 1906 Lewis Nixon invented sonar in the hopes of being able to detect underwater icebergs; interest later grew larger during World War I when sonar was identified as being able to locate submerged submarines. Various types of sonar are used today, both in science and the military, as well as special kinds of sonar found in the medical field. Most of these uses of sonar work off the same principles but have been specially adapted for use in their respected fields.

Sorby, Henry Clifton (1826–1908) British *Geologist, Biologist, Microscopist, Metallurgist* Henry Sorby was a well-known British scientist whose most influential scientific work was done from 1849 to 1864. His work was based on the application of the microscope to geology and metallurgy. In these two fields, his work included simple quantitative observation and the building and meticulous use of new experimental equipment and interpretation based on the application of elementary physicochemical principles to complex natural phenomena. His goal was to "apply experimental physics to the study of rocks." Sorby's most famous achievement was the development of the basic techniques of petrography by using the polarizing microscope to study the structure of thin rock sections. He first began his work on sedimentary rocks and by 1851 he was involved in a debate on the origin of slaty cleavage. His paper in 1853 showed that cleavage was a result of the reorientation of particles of mica accompanying the deformation flow of the deposit under anisotropic pressure. He then went on to study organisms in limestone and discussed the significance of microorganisms in chalk. Sorby then moved from slate to schist and metamorphic rocks in general. His paper on liquid inclusions in crystals, both natural and artificial, was very important. His use of the microscope helped him to find abundant smaller inclusions within the microcrystals of many metamorphic rocks.

speleothem   Any secondary mineral deposit formed in a cave by the action of groundwater. Most speleothems are made of carbonate minerals such as calcite, aragonite, or dolomite, but some are made of silicates and evaporites. Dripstone and flowstone are the most common carbonate speleothems. Yellow, brown, orange, tan, green, and red colors in dripstone and flowstone are formed through staining by organic compounds, oxides derived from overlying clays and soils, and rarely by ionic substitution in the carbonate minerals. Dripstone forms where water enters the cave through joints, bedding planes, or other structures and degasses CO_2 from water droplets, forming a small ring of calcite before each drop breaks free and falls into the cave. Each succeeding drop deposits another small ring of calcite eventually forming a hollow tube called a straw stalactite. Additional growth may occur on the outside of the straw stalactite forming a wedge-shaped hanging calcite deposit. Where the drops fall to the cave floor below they deposit additional calcite forming a mound-shaped structure known as a stalagmite. These have no central canal but consist of a series of layers deposited one over the other and typically are symmetric about a vertical axis. Flowstone is a massive secondary carbonate deposit formed by water that moves as sheet flows over cave walls and floors. The water deposits layered and terraced carbonate with complex and bizarre shapes, with shapes and patterns determined by the flow rates of the water and the shape of the cave walls, shelves, and floor. Draperies are layered deposits with furled forms intermediate between dripstone and flowstone.

Less common types of speleothems include shields, massive plate-like forms that protrude from cave walls. They are fed by water that flows through a medial crack separating two similar sides of the shield, with the crack typically parallel to regional joints in the cave.

Merged stalactite and stalagmite forming a column in Carrol Cave (Camden, Missouri)

Some speleothems have erratic forms not controlled by joints, walls, or other structures. Helictites are curved stalactite-like forms with a central canal, anthodites are clusters of radiating crystals such as aragonite and a variety of botryoidal forms that resemble beads or corals. Moonmilk is a wet powder or wet pasty mass of calcite, aragonite, or magnesium carbonate minerals. Travertine forms speleothems in some cave systems, where the waters are saturated in carbon dioxide.

Evaporite minerals form deposits in some dry dusty caves where the relative humidity drops to below 90 percent, and the waters have dissolved anions. Gypsum is the most common evaporite mineral found as a speleothem, with magnesium, sodium, and strontium sulfates being less common. Phosphates, nitrates, iron minerals, and even ice form speleothems in other less common settings.

See also GROUNDWATER; KARST.

Stalactites and stalagmites in Carrol Cave

Spitzbergen Island *See* SVALBARD.

stalactite *See* SPELEOTHEM.

stalagmite *See* SPELEOTHEM.

Steno, Nicolas (1638–1686) Danish *Anatomist, Paleontologist, Bishop* Nicolas Steno is probably the first scientist to clearly show that fossils are organic remains of formerly living organisms. He studied anatomy at Copenhagen and Leiden, and Florence. While dissecting a shark, he noted the similarity between the teeth of the modern shark and fossil shark teeth in local strata. After this revelation, Steno traveled around Tuscany collecting as many fossils as he could, becoming obsessed with understanding the origin of fossils. He produced a major work on the origin of fossils, the *Prodomus,* in 1669, that led him to ponder the difference between his observations and the history of the world as described in his version of the Bible. Steno proposed the law of stratal superposition, clearly outlining that younger rocks are deposited over older rocks, and he recognized a sequence of changing fossil forms in the stratigraphic record. He also described metalliferous mineral deposits, recognizing crosscutting veins, and he described volcanic mountain building, erosion, and faulting. Steno was converted to Catholicism, eventually being appointed bishop, in which position he completed the rest of his life.

steppe A semiarid grassy and nearly treeless upland plain found in midlatitude areas of southeastern Europe, Asia, and North America. The only Southern Hemisphere steppe is the veldt of South Africa. Most grasses on steppes are short and occur in sparse clumps, with abundant bare soil. Steppes are considered to be more arid than the prairies of the United States, but more humid than deserts. Soils in steppes are dry and have a water deficiency, with water stored only at significant depths. The grasses obtain the water they need for growth during short spring rainfall events and typically become dormant by midsummer.

Stille, Wilhem Hans (1876–1966) German *Tectonic Geologist* Wilhem Stille began his studies in stratigraphy and tectonics near his home in Hannover, Germany, and continued to do his fieldwork in this area for many years. His exploration work in Colombia introduced him to the continent of South America where he continued to do most of his research later on. Although he did not do much research abroad, he helped other students who worked on the Mediterranean region. His constant reading helped him to become a leader in synthesizing global tectonics. Stille started out as a chemistry student but soon became a geologist under the influence of Adolf von Koenen. He worked for the Prussian Geological Survey and then taught in Hannover. In 1912 he became a professor of geology and the director of the Royal Saxon Geological Survey at Leipzig. He was later named professor at the University of Berlin in 1932 and developed a reputation of being an outstanding teacher and philosopher of global tectonics. Stille was known as a leader in German geology, an outstanding investigator and collator of the history of global tectonic events, and a great teacher. He directed attention to the explanation of the relationships among large crustal features, and his studies of the eugeosynclinal belts led to the interest of their magmatic history. Stille has received many honorary doctorates and was elected an honorary member in numerous academics of science, geological societies, and other scientific organizations. He became the honorary president of the German Geological Society where he was awarded the Leopold von Buch Medal and later had the Hans Stille Medal made in his honor.

Stokes Law A physical formula describing the rate of settling of a spherical ball in a fluid, as derived by Sir George Stokes. The formula $V = Cr^2$ relates velocity (V) of the particle (in cm/sec), the particle's radius, and a constant (C) relating the relative densities of the particle, fluid, and acceleration of gravity, and the viscosity of the fluid.

storm surges Water that is pushed ahead of storms and typically moves on land as exceptionally high tides in front of severe ocean storms such as hurricanes. Storms and storm surges can cause some of the most dramatic and rapid changes to the coastal zones, and they represent one of the major, most unpredictable hazards to people living along coastlines. Storms that produce surges include hurricanes (which form in the late summer and fall) and extratropical lows (which form in the late fall through spring). Hurricanes originate in the tropics and (for North America) migrate westward and northwestward before turning back to the northeast to return to the cold North Atlantic, weakening the storm. North Atlantic hurricanes are driven to the west by the trade winds and bend to the right because the Coriolis force makes objects moving above Earth's surface appear to curve to the right in the Northern Hemisphere. Hurricane paths are further modified by other weather conditions, such as the location of high and low pressure systems, and their interaction with weather fronts. Extratropical lows (also known as coastal storms and nor'easters) move eastward across North America and typically intensify when they hit the Atlantic and move up the coast. Both types of storms rotate counterclockwise, and the low pressure at the centers of the storms raises the water several to several tens of feet. This extra water moves ahead of the storms as a storm surge that represents an additional height of water above the normal tidal range. The wind from the storms adds further height to the storm surge, with the total height of the storm surge being determined by the length, duration, and direction of wind, plus how low the pressure gets in the center of the storm. The most destructive storm surges are those that strike low-lying communities at high tide, as the effects of the storm surge and the regular astronomical tides are cumulative. Add high winds and large waves on top of the storm surge, and coastal storms and hurricanes are seen as very powerful agents of erosion. They are capable of removing entire beaches, rows of homes, causing great amounts of cliff erosion, and significantly redistributing sands in dunes and the back beach environment. Very precise prediction of the height and timing of the approach of the storm surge is necessary to warn coastal residents of when they need to evacuate and when it is not necessary to leave their homes.

Like many natural catastrophic events, the heights of storm surges to strike a coastline are statistically predictable. If the height of the storm surges is plotted on a semilogarithmic plot, with the height plotted in a linear interval and the frequency (in years) plotted on a logarithmic scale, then a linear slope results. This means that communities can plan for storm surges of certain height to occur once every 50, 100, 300, or 500 years, although there is no way to predict when the actual storm surges will occur. It must be remembered that this is long-term statistical average, and that one, two, three, or more 500-year events may occur over a relatively short period, but averaged over a long time, the events average out to once every 500 years.

Storms are known to open new tidal inlets where none were previously (without regard to whether or not any homes were present in the path of the new tidal inlet), and to close

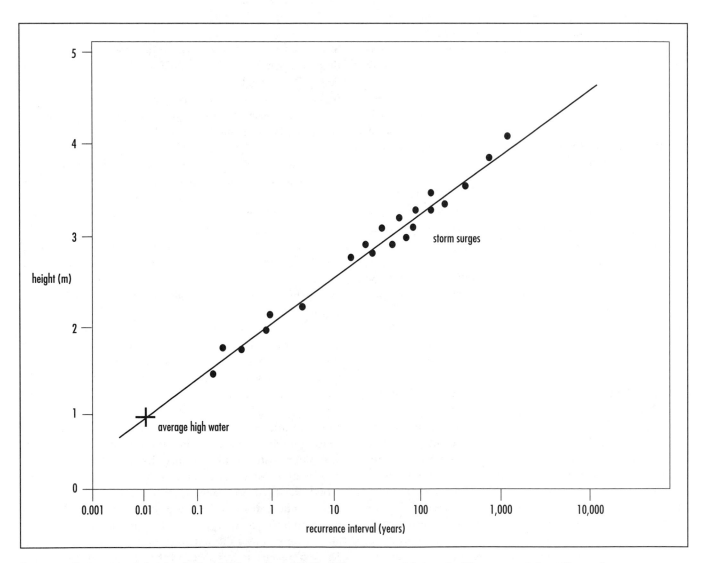

Storm surge/frequency statistics showing how it is increasingly less likely for more powerful storms to strike a segment of coast in any given year

inlets previously in existence. Storms also tend to remove large amounts of sand from the beach face and redeposit it in the deeper water offshore (below wave base), but this sand tends to gradually move back onto the beach in the intervals between storms when the waves are smaller. In short, storms are extremely effective modifiers of the beach environment, although they are unpredictable and dangerous.

See also BEACH; HURRICANE.

Strahler, Arthur N. (1918–) American *Geographer* Arthur N. Strahler received his B.A. degree in geography in 1938 from the College of Wooster in Ohio, and his Ph.D from Columbia University in 1944. He was appointed to the Columbia University faculty in 1941, serving as Professor of Geomorphology from 1958 to 1967 and as chairman of the Department of Geology from 1959 to 1962. He is a Senior Fellow of the Geological Society of America. He has been cited for his pioneering contributions to quantitative and dynamic geomorphology, contained in more than 30 major papers in leading scientific journals. He is the author of 16 textbook titles with 13 revised editions in geology, earth sciences, and related disciplines.

Further Reading
Strahler, Arthur N. *Physical Geology.* New York: Harper and Row, 1981.
———. *The Earth Sciences,* 2nd ed. New York: Harper and Row, 1971.
———. "Geomorphology and Structure of the West Kaibab Fault Zone and Kaibab Plateau, Arizona." *Bulletin of the Geological Society of America* 97 (1948): 513–540.

strain A measure of the relationship between size and shape of a body before and after deformation. Strain is one component of a deformation, a term that includes a description of the collective displacement of all points in a body. Deformation consists of three components: rigid body rotation, a rigid body translation, and a distortion known as strain. Strain is typically the only visible component of deformation, manifest as distorted objects, layers, or geometric constructs. Strain may be measured by changes in lengths of lines, changes in angles between lines, changes in shapes of objects, and changes in volume or area.

There are many measures of strain. The change in the length of lines can be quantified using several different strain measures. Extension $(\varepsilon) = (L'-L)/L$, where L = original length of line, and L' = final length of line. Stretch $(S) = L'/L = (1=\sigma)$. The quadratic elongation $(\sigma = L'/L)^2 = (1=\sigma)^2$, whereas the natural or logarithmic strain (σ) $\sigma = \log_e (L'/L) = \log_e (1=\sigma)$. The change of angles is typically measured using the angular shear $(\sigma$ angular shear, which is the change in the angle between two lines that were initially perpendicular. More commonly, structural geologists measure angular strain using the tangent of the angular shear, known as the shear strain

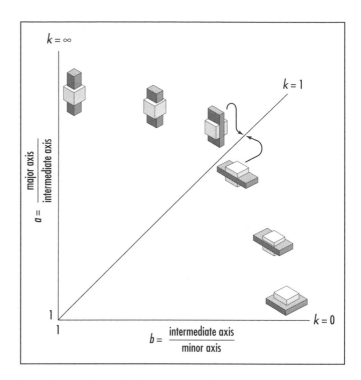

Flynn diagram showing fields of prolate, oblate, and plane strain

$(\sigma = \tan \sigma$. Volumetric strain is a measure of the change in volume of an object, layer, or region. Dilation $(\delta) = (V'-V)/V$, and measures the change of volume, whereas the volume ratio = V'/V measures the ratio of the volume after and before deformation.

Strains may be homogeneous or heterogeneous. Heterogeneous strains are extremely difficult to analyze, so the structural geologist interested in determining strain typically focuses on homogeneous domains with the heterogeneous strain field. In contrast, the geologist interested in tectonic problems involving large-scale translation and rotation often finds it necessary to focus on zones of discontinuity in the homogeneous strain field, as these are often sites of faults and high strain zones along which mountain belts and orogens have been transported. For homogeneous strains, the following five general principles hold true: straight lines remain straight and flat planes remain flat; parallel lines remain parallel and are extended or contracted by the same amount; perpendicular lines do not remain perpendicular unless they are oriented parallel to the principal strain axes; circular markers are deformed into ellipses; finally, there is one special initial ellipsoid that becomes a sphere when deformed. When these conditions are met, the strain field is homogeneous, and strain analysis of deformed objects may indicate the strain of the whole body.

Structural geologists often find it important to measure the strain in deformed rocks in order to reconstruct the history of mountain belts, to determine the amount of displace-

Strained fossil. The belemnite in this photograph was initially shorter and all of the dark parts were connected. During the deformation of the rock, the belemnite was extended, being pulled apart to form small blocks with new minerals recrystallized between the extended blocks. A measure of strain would be the ratio between the new and initial length of the belemnite. *(Photo by Timothy Kusky)*

ment across a fault or shear zone, or to accurately delineate the distribution of an ore body—this process is called strain analysis. To measure strain in deformed rocks, the geologist searches for features that had initial shapes that are known and can be quantified, such as spheres (circles), linear objects, or objects such as fossils that had initial angles between lines that are known. In most cases, geologists cannot directly see the three-dimensional shape of deformed objects in rocks. In this case, strain analysis proceeds by measuring the two-dimensional shapes of the objects on several different planes at angles to each other. The deformed shapes are graphically or algebraically fitted together to get the three-dimensional shape of the deformed object, and ultimately the three-dimensional shape and orientation of the strain ellipsoid. The strain ellipsoid has major, intermediate, and minor axes of X, Y, and Z, parallel to the principal axes of strain.

Structural geologists interested in determining the strain of a body search for appropriate objects to measure the strain. Initially spherical objects prove to be among the most suitable for estimating strain. Any homogeneous deformation transforms an initial sphere into an ellipsoid whose principal axes are parallel to the principal strains, and whose lengths are proportional to the principal stretches S1, S2, and S3. Using elliptical markers that were originally circular, it is possible to immediately tell the orientation of the principal strains on that surface, and their relative magnitudes. However, the true values of the strains are not immediately apparent, because the original volumes are not typically known. Strain markers in rocks that serve as particularly good recorders of strain, and approximate initially circular or spherical shapes, include conglomerate clasts, ooids, reduction spots in slates, certain fossils, and accretionary lapilli.

Angular strain is often measured using the change in angles of bilaterally symmetric fossils, or igneous dikes cutting across shear zones. Many fossils, such as trilobites and clams, are bilaterally symmetric, so if the line of symmetry can be found, similar points on opposite sides of the plane of symmetry can be joined, and the change in the angles from the initially right angles in the undeformed fossils can be constructed. When the fossils are deformed the right angles are also deformed, and we can use the relationships we derived above for the change in angles to determine the angular strain of the sample.

Strain represents the change from the initial to the final configuration of a body, but really it tells us very little about the path the body took to get to the final shape, known as the deformation path. The strain represents the combination of all events that occurred, but they are by no means unique. Fortunately, rocks have a memory, and there are many small-scale structures and textures in the rocks that tell us much about where the rock has been, or what its deformation path was. One of the most important attributes of the strain path to determine is whether the principal strain axes were parallel between each successive strain increment or not. A coaxial deformation is one in which the principal axes of strain are parallel with each successive increment. A noncoaxial deformation is one in which the principal axes of strain rotate with respect to the material during deformation.

Two special geometric cases of strain history are pure shear and simple shear. A pure shear is a coaxial strain (with no change in volume). Simple shear is analogous to sliding a deck of cards over itself and is a two-dimensional noncoaxial rotational strain, with constant volume, and no flattening perpendicular to the plane of slip.

In simple shear, the principal axes rotate in a regular manner. The principal strain axes start out at 45° to the shear plane, and strain S1 rotates into parallelism with it at very high (infinite) strains. The principal axes remain perpendicular, but some other lines will be lengthening with each increment, and others will be shortening. There are some orientations that experience shortening first, and then lengthening. This leads to some complicated structures in rocks deformed by simple shear; for instance folds produced by the shortening, and then extensional structures, such as faults or pull-apart structures known as boudens, superimposed on the early contractional structures.

Natural strains in rocks deform initially spherical objects into ellipsoids with elongate (prolate) or flattened (oblate) ellipsoids. All natural strains may be represented graphically on a graph known as a Flynn Diagram, which plots a = (X/Y) vs. b = (Y/Z). The number k = a−1/b−1. For k = 0, strain ellipsoids are uniaxial oblate ellipsoids or pancakes. For 0 > k > 1, deformation is a flattening deformation, forming an oblate ellipsoid. For k = 1 the deformation is plane strain if the volume has remained constant. All simple shear deformations lie on this line. For 1 > k > infinity, the strain ellipsoids are uniaxial prolate ellipsoids, or cigar shapes.

See also STRUCTURAL GEOLOGY.

strain analysis   *See* STRAIN.

stratigraphy   The study of rock strata or layers. Stratigraphy is concerned with aspects of the rock layers such as their succession, age relationships, lithologic composition, geometry, distribution, correlation, fossil content, and all other aspects of the strata. The main aim of stratigraphy is to understand and interpret the rock record in terms of paleoenvironments, mode of origin of the rocks, and the causes of similarities and differences between different stratigraphic units. Because sedimentary rocks are laid down one on top of another, we can look at a thick pile of sedimentary rocks, such as those in the Grand Canyon, and as we look lower down, we look further back in time. The time difference between the rocks at the top of the Grand Canyon and those at the base is nearly two billion years. Thus, by looking at the different layers, we can reconstruct the past conditions on the planet at this particular place.

These relationships are expressed in several laws of stratigraphy. The first, known as the Law of Original Horizontality, states that water-laid sediments form horizontal strata, parallel to the Earth's surface. So if sedimentary rocks are now found inclined, we can infer that they have been deformed. The second law, the Principle of Stratigraphic Superposition, states that the order in which the strata were deposited is from bottom to top, assuming that the strata have not since been overturned.

From the Principle of Stratigraphic Superposition, we can define the relative ages of two different sedimentary units. The older rocks are below the younger ones—this is useful for correlating geologic strata from well-exposed to poorly exposed areas, for once the relative age of a unit is known, then you know which rocks are above and below it. Where rocks are folded or tectonically deformed, some may be upside down, and simply knowing which unit you are looking at may not be enough to tell you which units might be found underground beneath this particular outcrop. One way to tell if rocks are upside down or not is to use the geometry of sedimentary features that formed when the rock was a sediment. For instance, if the original rock showed graded bedding, from coarse-grained at the bottom to fine-grained at the top, and now the rock has fine material at the base and coarse at the top, it may be upside down. If sand ripples or cross laminations are found on the bottoms of the beds instead of the tops, that would be additional evidence that the strata are now upside down.

Although the relative ages of strata can be defined by which is on top of which, the absolute ages can not be determined in this manner, nor can the intervals of time between the different units. One reason for this is that deposition is not continuous, and there may be breaks or discontinuities in the stratigraphic record, represented by unconformities.

Stratigraphic Classification
Because rocks laid down in succession each record environmental conditions on the Earth when they were deposited, experienced geologists can read the record in the stratigraphic pile like a book recording the history of time. Places like the Grand Canyon are especially spectacular because they record billions of years of history.

Classical stratigraphy is based on the correlation of distinct rock stratigraphic units, or unconformity surfaces, that are internally homogeneous and occur over large geographic areas. The formation is the basic unit of rock stratigraphy and is defined as a group of strata which constitutes a distinctive recognizable unit for geologic mapping. Thus, it must be thick enough to show up on a map, must be laterally extensive, and must be distinguishable from surrounding strata. Formations are named according to a code (the Code of North American Stratigraphic Nomenclature), using a prominent local geographic feature. Formations are divided into members and beds, according to local differences or regionally distinctive horizons. Formations may be grouped together with other formations into groups.

A more recent advance in stratigraphy is time stratigraphy, that is, the delineation of certain time-stratigraphic units. Units divided in this way have lower and upper boundaries that are everywhere the same age but may look very different and be comprised of very different rock types. Time-stratigraphic units may be identified by using fossils known to occur only during a certain period, or by correlating between unconformities (erosional surfaces) that have about the same age in different places. The primary unit of time stratigraphy is the system, which is an interval so great that it can be recognized over the entire planet. Most systems represent time periods of at least tens of millions of years. Larger groups of systems are called erathems or eras for short. Time units smaller than system are the series, and stage, which are typically used for correlation on a single continent or within a geographic province.

Time Lines and Diachronous Boundaries
In many sedimentary systems, such as the continental shelf, slope, and rise, different types of sediments are deposited in different places at the same time. We can draw time lines through these sequences to represent all the sediments deposited at a given time, or to represent the old sediment/water interference at a given time. In these types of systems, the

transition from one rock type to another, such as from the sandy delta front to the marsh facies, will be diachronous in time (it will have different ages in different places).

Correlation of Rocks

If a geologist has studied a stratigraphic unit or system in one location and figured out conditions on the Earth at that point when the rock was deposited, we may wonder how this can be related to the rest of the planet, or simply to nearby areas. In order to accomplish this task, the geologist first needs to determine the relative ages of strata in a column, then estimate the absolute ages relative to a fixed timescale. Stratigraphic units may be correlated with each other locally using various physical criteria, such as continuous exposure where a formation may be recognizable over large areas. Typically, a group of characteristics for each foundation is amassed such that each formation can be readily distinguished from each other formation. These include gross lithology or rock type, mineral content, grain size, grain shape, color, or distinctive sedimentary structures such as cross-laminations. Occasionally, key beds with characteristics so distinctive that they are easily recognized are used for correlating rock sections.

Most sedimentary rocks lie buried beneath the surface layer on the Earth, and geologists and oil companies interested in correlating different rock units have to rely on data taken from tiny drill holes. The oil companies in particular have developed many clever ways of correlating rocks with distinctive (oil rich) properties. One common method is to use well-logs, where the electrical and physical properties of the rocks on the side of the drill hole are measured, and distinctive patterns between different wells are correlated. This helps the oil companies in relocating specific horizons that may be petroleum rich.

Index fossils are those that have a wide geographic distribution, commonly occur, and are very restricted in the time interval in which they formed. Because the best index fossils should be found in many environments, most are floating organisms, which move quickly around the planet. If the index fossil is found at a certain stratigraphic level, often its age is well known, and it can be correlated with other rocks of the same age.

See also MILANKOVITCH CYCLE.

stratosphere   *See* ATMOSPHERE.

stress   The total force exerted by all the atoms on one side of an arbitrary plane upon the atoms immediately on the other side of the plane. Body forces are those that act from a distance (e.g., gravity) and are proportional to the amount of material present. Surface forces are those that act across surfaces of contact between parts of bodies, including all possible internal surfaces. There are two kinds of surface forces, including normal (compressive and ten-

sile), that act perpendicular to the surface, and shear (clockwise and anticlockwise), that act parallel to the surface. The state of stress equals the force divided by the area across which it acts.

For any applied force, it is possible to find a choice of coordinate axes such that all shear stresses are equal to zero, and only three perpendicular principal stresses have nonzero values. The principal stresses are commonly abbreviated $\sigma 1$, $\sigma 2$, and $\sigma 3$. These three principal stresses are parallel to the semimajor axes of an ellipsoid called the stress ellipsoid, parallel to the coordinate axes chosen such that they are the only nonzero stresses.

The deviatoric stress, or the difference between the principal stresses, is most important for forming most structures in rocks, because it drives the deformation. However, the mean stress (sum of $[\sigma 1 + \sigma 2 + \sigma 3] / 3$) is important for determining which deformation mechanisms operate, and the strength of materials.

Stress has dimensions of force per unit area. In the SI system, we use the Pascal (Pa) which is 1 Newton per meter squared (N/m^2). The sign convention that geologists use considers compressive stresses to have a positive sign.

See also STRUCTURAL GEOLOGY.

strike-slip fault   A fault in which movement is subhorizontal, parallel to the strike of the fault. Intense seismic activity and deformation characterize strike-slip faults, considerable differential movements between blocks, weak metamorphism, and extremely high rates of and variation in sedimentation. A strike-slip fault may be a transform (plate boundary) fault only if it penetrates the thickness of the lithosphere and ends in a subduction zone or terminates at a spreading ridge.

Many strike-slip faults occur in wide zones that are characterized by many branches of the fault that split, diverge, and rejoin in an anastomosing style. Regions where faults converge are typically associated with uplift, whereas regions where fault strands diverge are characterized by subsidence. Individual strands of strike-slip faults may end, with the displacement being taken up by another nearby fault strand. The region between faults that are linked in this way may host an accommodation zone, where minor faults, folds, and other structures may help transfer the displacement from one fault strand to another.

Strike-slip faults are rarely straight but instead are characterized by bends or steps. These steps may show extensional or contractional deformation, depending on the orientation of the step, with respect to the sense of motion along the fault. Right steps in right-slipping (also known as right lateral) faults, and left steps in left-slipping (or left lateral) faults form extensional bends. Steps in the opposite direction place the bend in compression, forming folds, thrust faults, and other contractional structures. Sedimentary basins known as pull-apart basins form at extensional bends, whereas fold-thrust belts and rapid uplift characterize contractional bends.

Many strike-slip systems are not purely strike-slip but have a component of compression or extension on a regional scale, known as transpression or transtension. Transtensional regimes are associated with normal faulting, drape folding, and volcanic activity. In contrast, thrust faulting, high-angle reverse faulting, folding, uplift, and the formation of flower structures characterize transpressive regimes. Relative plate convergence directions can change and thus change the strike-slip regime from transpressive to transtension, or vice versa.

Different types of sedimentary basins are associated with transpression and transtension. In transpression, folding, thrusting, and vertical uplift of mountains forms small fore-deep basins, cannibalistic nappes, flower structures, and shallow sedimentary basins. Most of the eroded sediments from transpression are carried and transported elsewhere, outside the mountain range, and deposited far away in cratonic or oceanic basins.

In contrast, in zones of transtension, basins associated with normal faulting tend to be deep and associated with very rapid sedimentation, with very little material transported outside the orogen. Pull-apart basins are often associated with strike-slip systems. These form very quickly, may be tens of kilometers deep, and have normal faults on two sides and strike-slip faults on the other two sides. These basins may contain virtually every type of sedimentary environment. Individual facies have a small lateral extent, and conglomerates and breccias may be locally derived. Lacustrine sediments are hallmark deposits of pull-apart basins, with the lakes being long and narrow, with thick accumulations of sediments. These are typically bounded by alluvial fans fed from one or both ends of the basin. Continual subsidence leads to facies remaining relatively stationary. After lakes are filled, they become sites of braided streams, shallow alluvial plains, or other, relatively quiet, sedimentary environments.

It is often a challenge to recognize ancient strike-slip systems in the geologic record. Different segments of a strike-slip fault system may show pure strike-slip relationships, extensional belts, or contractional belts. Also, strike-slip faults often form along boundaries between different rock terranes, making it difficult to know if there has been lateral displacement between the different belts. Thus, to recognize strike-slip orogens it is necessary to make a regional assessment of the tectonics and not examine only a small or local area.

Some characteristics of strike-slip systems are diagnostic and recognizable in old orogenic belts. In some locations it may be possible to match up geologic units, such as plutons or distinctive belts of rocks that have been laterally displaced by the strike-slip fault. In other cases, alluvial fans may be displaced from their sources, perhaps with distinctive clasts derived from one valley recognizable in a displaced fan. Strike-slip systems are characterized by the development of numerous angular unconformities in one location (such as at a restraining bend), at the same time that rapid and continuous sedimentation may be occurring in a nearby location (such as in a pull-apart basin).

The regional structural pattern may be used to determine the orientation of regional stresses, and their relationship to the main faults in the system. Predictable patterns of folds, strike-slip, normal, and thrust faults are associated with strike-slip systems, and these differ from thrust and normal fault-dominated orogens.

The types of basin filling may aid in the identification of ancient strike-slip orogens. Conglomerates, basin-interior lake sediments, and local volcanic rocks dominate early fill of pull-aparts, whereas later stages and transpressional phases may be dominated by fluvial sedimentation. All of these features must be assessed together, as a single basin could be interpreted in many different ways.

See also STRUCTURAL GEOLOGY; PLATE TECTONICS; TRANSFORM PLATE MARGIN PROCESSES.

stromatolite Columnar or mound-like structures produced by a combination of biological secretions from cyanobacteria and sediment trapping by filamentous organisms including bacteria or algae. There have been various definitions proposed for stromatolites based on a variety of models for their origin, morphology, age, and distribution. One of the most useful definitions was proposed by Stanley Awramik and

Stromatolites are produced by colonies of blue-green algae and bacteria that construct mats and cover surfaces, such as the subaquatic carbonate layer shown here. Each pinnacle is produced by a colony of microorganisms, growing toward the Sun (as indicated by the inclination of the two poles) and producing daily lamina. *(Photo by J. Vanyo)*

Lynn Margulis, and modified by M. Walter as "organosedimentary structures produced by sediment trapping, binding, and/or precipitation as a result of the growth and metabolic activity of micro-organisms, principally cyanophytes."

Stromatolites are among the oldest known organic structures on Earth, found in rocks as old as 3.5 billion years. They are the most common form of life preserved in all of the Precambrian rock record, peaking in abundance in the Proterozoic, with a few active stromatolite colonies in the recent record. Stromatolites nearly completely disappeared from the rock record at the beginning of the Phanerozoic with the first appearance of grazing Metazoans, perhaps indicating that they were low on the food chain and largely devoured by these higher organisms. However it is also possible that stromatolites declined because of competition for ecological niches by other organisms such as green and red algae and other marine organisms.

There are a bewildering number of morphological variations in different types of stromatolites which some stratigraphers, notably those in the Russian and Chinese schools, attempted to use as index fossils for certain periods. However, these attempts failed and current thinking now attributes variation in stromatolite morphology to local environmental influences and bacterial processes.

Some stromatolites form mats, crinkled mats, or gently domed layers. These grade into bulbous, nodular, and hemispherical forms that may be isolated or laterally linked. Other stromatolites form columns that may have conical or subspherical laminations and may form singular columns, branching columns, or dendritic structures. The columns are typically erect but may also be inclined, wavy, sinusoidal, or irregular. Sizes of stromatolites range from a few millimeters to several meters in height.

Most stromatolites grow by photosynthetic processes where sunlight reaches mats of filamentous microorganisms that initiate growth on some feature, such as a pebble or carbonate grain. The microorganisms secrete layers of calcium carbonate, perhaps daily, and the filamentous organisms then grow or move upward though the layer to the surface. Repeated growth over many days and years leads to a layered mound, column, or branching structure. Occasionally the filaments may trap or be buried by sedimentary particles that move over the mound, many probably moved by storms. In these cases the sticky mats may trap the particles forming a detrital layer. The microorganisms must then be able to move through this layer to the surface in order for the organic community to survive.

See also HELIOTROPISM; LIFE'S ORIGINS AND EARLY EVOLUTION; PRECAMBRIAN; PROTEROZOIC.

Further Reading

Grey, Kathleen. *Handbook for the Study of Stromatolites and Associated Structures*. Perth, Australia: Geological Survey of Western Australia, 1989.

Walter, Malcolm R. "Stromatolites." *Developments in Sedimentology,* v. 20 (1976).

structural geology   The study of the deformation of the Earth's crust or lithosphere. We know that the surface of the Earth is actively deforming from such evidence as earthquakes and active volcanism, and from rocks at the surface of the Earth that have been uplifted from great depths. The rates of processes (or timescales) of structural geology are very slow compared to ordinary events. For instance, the San Andreas fault moves only a couple of centimeters a year, which is considered relatively fast for a geological process. Even this process is discontinuous near the surface, with major earthquakes happening every 50–150 years, but perhaps with more continuous flowing types of deformation at depth. Mountain ranges such as the Alps, Himalaya, or those in the American west are uplifted at rates of a few millimeters a year, with several kilometers height being reached in a few million years. These types of processes have been happening for billions of years, and structural geology attempts to understand the current activity and this past history of the Earth's crust.

Structural geology and tectonics are both concerned with reconstructing the motions of the outer layers of the Earth. Both terms have similar roots—structure comes from the Latin *struere,* meaning "to build," whereas tectonics comes from the Latin *tektos,* meaning "builder."

Motions of the surface of the Earth can be rigid body rotations, where a unit of rock is transported from one place to another without a change in size and shape. These types of motions fall under the scope of tectonics. Alternatively, the motions may be a deformation, involving a change in the shape and size of a unit of rock, and this falls under the field of structural geology.

During motions on faults or with the uplift of mountains, rocks break at shallow levels of the crust and flow like soft plastic at deeper levels of the crust. These processes occur at all scales ranging from the scale of plates, continents, and regional maps, to what is observable only using electron microscopes.

Structural geology and tectonics have changed dramatically since the 1960s. Before 1960, structural geology was a purely descriptive science, and since then it has become an increasingly quantitative discipline, especially with applications of principles of continuum mechanics and with increasing use of laboratory experiments and the microscope to understand the mechanisms of deformation.

Tectonics has also undergone a recent revolution (since the understanding of plate tectonics in the 1960s) and the framework it provided for understanding the large-scale deformation of the crust and upper mantle. Both structural geology and tectonics have made extensive use of new tools since the 1960s, including geophysical data (e.g., seismic lines), paleomagnetism, electron microscopes, petrology, and geochemistry.

Most studies in structural geology rely on field observations of deformed rocks at the Earth's surface and proceed either downscale to microscopic observations or upscale to regional observations. None of these observations alone provides a complete view of structural and tectonic processes, so structural geologists must integrate observations at all scales and use the results of laboratory experiments and mathematical calculations to make better interpretations of our observations.

To work out the structural or tectonic history of an area, the geologist will usually proceed in a logical order. First, the geologist systematically observes and records structures (folds, fractures, contacts) in the rock, usually in the field. This consists of determining the geometry of the structures, including where they are geographically, how are they oriented, and what are their characteristics. Additionally, the structural geologist is concerned with determining how many times the rocks have been deformed and which structures belong to which deformation episode.

The term *attitude* is used to describe the orientation of a plane or line in space. Attitude is measured using two angles: one measured from geographic north, and the other from a horizontal plane. The attitude of a plane is represented by a strike and a dip, whereas the attitude of a line is represented by a trend and plunge. Strike is the horizontal angle, measured relative to geographic north, of the horizontal line in a given planar structure. The horizontal line is referred to as the strike line and is the intersection of a horizontal plane with the planar structure. It is easily measured in the field with a compass, holding the compass against the plane, and keeping the compass horizontal. Dip is the slope of the plane defined by the dip angle and the dip direction. It is the acute angle between a horizontal plane and the planar structure, measured in a vertical plane perpendicular to the strike line. It is necessary to specify the direction of the dip.

To understand the processes that occurred in the Earth, structural geologists must also understand the kinematics of formation of the structures; that is, the motions that occurred in producing them. This will lead to a better understanding of the mechanics of formation, including the forces that were applied, how they were applied, and how the rocks reacted to the forces to form the structures.

To improve understanding of these aspects of structural geology, geologists make conceptual models of how the structures form, and test predictions of these models against observations. Kinematic models describe a specific history of motion that could have carried the system from one configuration to another (typically from undeformed to deformed state). Kinematic models are not concerned with why or how motion occurred, or the physical properties of the system (plate tectonics is a kinematic model).

Mechanical models are based on continuum mechanics (conservation of mass, momentum, angular momentum, and energy), and our understanding of how rocks respond to applied forces (based on laboratory experiments). With mechanical models we can calculate theoretical deformation of a body subjected to a given set of physical conditions of forces, temperatures, and pressures (an example of this is the driving forces of tectonics based on convection in the mantle). Mechanical models represent a deeper level of analysis than kinematic models, constrained by geometry, physical conditions of deformation, and the mechanical properties of rocks.

It is important to remember that models are only models, and they only approximate the true Earth. Models are built through observations and provide predictions to test the model's relevance to the real Earth. New observations can support or refute a model. If new observations contradict predictions, models must be modified or abandoned.

Structural Geology and the Interior of the Earth

Structures at the surface of the Earth reflect processes occurring at deeper levels. We know that the Earth is divided into three concentric shells—the core, mantle, and crust. The core is a very dense iron-nickel alloy, comprised of the solid inner core and the liquid outer core. The mantle is composed of lower density, solid magnesium-iron silicates, and is actively convecting, bringing heat from the interior of the Earth to the surface. This heat transfer is the main driving mechanism of plate tectonics. The crust is the thin low-density rock material making up the outer shell of the Earth.

Temperature increases with depth in the Earth at a gradient of about 54°F per half a mile (30°C/km) in the crust and upper mantle, and with a much smaller gradient deeper within Earth. The heat of the Earth comes from several different sources, including residual heat trapped from initial accretion, radioactive decay, latent heat of crystallization of the outer core, and dissipation of tidal energy of the Sun-Earth-Moon system.

Heat flows out of the interior of the Earth toward the surface through convection cells in the outer core and mantle. The top of the mantle and the crust is a relatively cold and rigid boundary layer called the lithosphere and is about 61 miles (100 km) thick. Heat escapes through the lithosphere largely by conduction, transport of heat in igneous melts, and in convection cells of water through mid-ocean ridges.

Structural geology studies predominantly only the outer 12–18 miles (20–30 km) of the lithosphere, putting into perspective that we are inferring a great deal about the interior of the Earth by examining only its skin.

Characteristics of the Crust

Earth's crust is divisible broadly into continental crust of granitic composition and oceanic crust of basaltic composition. Continents comprise 29.22 percent of surface, whereas 34.7 percent of Earth's surface is underlain by continental crust (continental crust under submerged continental shelves accounts for the difference). The continents are in turn divided into orogens, made of linear belts of concentrated deformation, and cratons, the stable, typically older interiors of the continents.

The distribution of surface elevation is strongly bimodal, as reflected in the hypsometric diagrams. Continental freeboard is the difference in elevation between the continents and ocean floor and results from difference in thickness and density between continental and oceanic crust, tectonic activity, erosion, sea level, and strength of continental rocks.

Controls of Deformation
Deformation of the lithosphere is controlled by the strength of rocks, which in turn is most dependent on temperature and pressure. Strength increases with pressure and decreases exponentially with increasing temperature. Because temperature and pressure both increase downward, a cross section

through the crust or lithosphere will have different zones where the effects of either pressure or temperature dominate.

In the upper layers of the crust, effects of pressure dominate, rocks get stronger with depth and fail brittly throughout this region. Below about nine miles (15 km), the effects of temperature become increasingly more important, and the rocks get weaker, deforming by flowing ductile deformation.

Other important properties that determine how the lithosphere deforms are composition (e.g., quartz v. olivine in crust, mantle, continents and oceans), and strain rate. Strain rate has its greatest variations along plate boundaries, and most structures develop as a consequence of plate interactions along plate boundaries.

Structural Geology and Plate Tectonics
The surface of the Earth is divided into 12 major and several minor plates that are in motion with respect to each other. Plate tectonics describes these relative motions, which are, to a first approximation, rigid body rotations. However, deformation of the plates does occur (primarily in 10s to 100s of

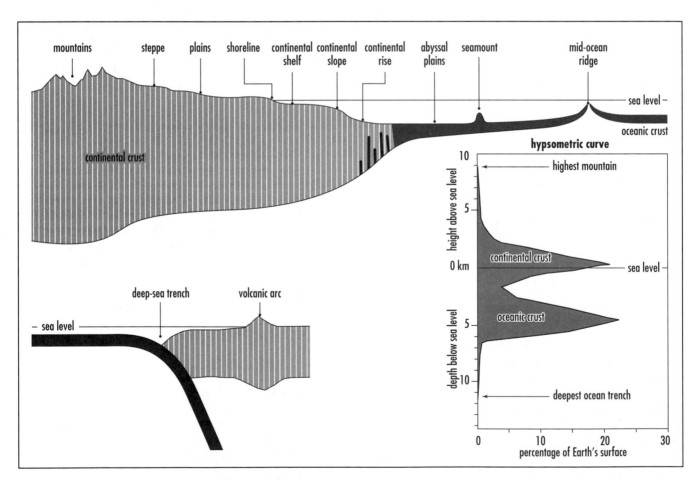

Hypsometric curve showing the distribution of land with different elevations on the planet. Note the bimodal distribution, reflecting two fundamentally different types of crust (oceanic and continental) that have different isostatic compensation levels. Cross sections show typical continental margin–ocean transition, and ocean trench–island arc boundary.

kilometers-wide belts along the plate boundaries), and in a few places, extends into the plate interiors. Structural geology deals with these deformations, which in turn give clues to the types of plate boundary motions that have occurred, and to the tectonic causes of the deformation.

Plate boundaries may be divergent, convergent, or conservative/transform. The most direct evidence for plate tectonics comes from oceanic crust, which has magnetic anomalies or stripes recording plate motions. However, the seafloor magnetic record only goes as far back as 180 million years, the age of the oldest in-place oceanic crust. Any evidence for plate tectonics in the preceding 96 percent of Earth history must come from the continents and the study of them (structural geology).

Highly deformed continental rocks are concentrated in long linear belts called orogens, comparable to those associated with modern plate boundaries. This observation suggests that these belts represent former plate boundaries. The structural geologist examines these orogens, determines the geometry, kinematics, and mechanics of these zones and makes models for the types of plate boundaries that created them. The types of structures that develop during deformation depend on the orientation and intensity of applied forces, the physical conditions (temperature and pressure) of deformation, and the mechanical properties of rocks.

The most important forces acting on the lithosphere that drive plate tectonics and cause the deformation of rocks are the gravitational "ridge push" down the flanks of oceanic ridges, gravitational "trench pull" of subducting lithosphere caused by its greater density than surrounding asthenosphere, the drag of traction exerted by the convecting mantle on the overriding lithosphere, and the resistance of trenches and mountain belts.

At low temperature and pressure and high intensity of applied forces, rocks undergo brittle deformation, forming fractures and faults. At high temperature and pressure and low intensity of applied forces, rocks undergo ductile deformation by flow, coherent changes in shape, folding, stretching, thinning, and many other mechanisms.

Different styles of deformation characterize different types of plate boundaries. For instance, at mid-ocean ridges new material is added to the crust, and relative divergent motion of the plates creates systems of extensional normal faults and ductile thinning at depth. At convergent plate boundaries, one plate is typically subducted beneath another, and material is scraped off the downgoing plate in a system of thrust faults and folds. Along transform plate boundaries, systems of strike-slip faults merge downward with zones of ductile deformation with horizontal relative displacements. All types of plate boundaries have small-scale structures in common, so it is necessary to carefully examine regional patterns before making inferences about the nature of ancient plate boundaries.

See also METAMORPHISM; PLATE TECTONICS.

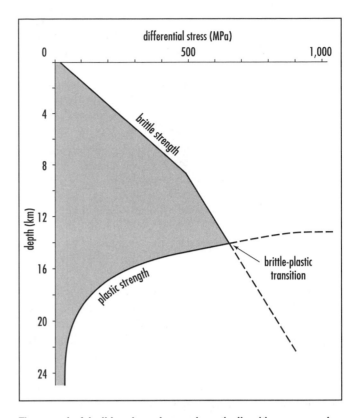

The strength of the lithosphere changes dramatically with pressure and temperature based on the mechanical properties of the minerals in the rocks at different depths. There are several weak and strong zones at different depths that vary in depth in different parts of the world based on the different rock types in different places.

Further Reading
Hatcher, Robert D. *Structural Geology, Principles, Concepts, and Problems,* 2nd ed. Englewood Cliffs, N.J.: Prentice Hall, 1995.
van der Pluijm, Ben A., and Stephen Marshak. *Earth Structure, An Introduction to Structural Geology and Tectonics.* Boston: WCB-McGraw Hill, 1997.

subduction   *See* SUBDUCTION ZONE.

subduction zone   Long narrow belts where an oceanic lithospheric plate descends beneath another lithospheric plate and enters the mantle in the processes of subduction. Subduction zones are of two basic types, the first being where oceanic lithosphere of one plate descends beneath another oceanic plate, such as in the Philippines and Marianas of the southwest Pacific. The second type of subduction zone forms where an oceanic plate descends beneath a continental upper plate, such as in the Andes of South America. Deep-sea trenches typically mark the place on the surface where the subducting plate bends to enter the mantle, and oceanic or continental margin arc systems form above subduction zones a few hundred kilometers from the trench. As the oceanic plate enters the trench it must bend, forming a flexural bulge up to few thousand feet (a couple of hundred meters) high typically about 100 miles (161 km) wide before the oceanic plate enters the trench. The outer trench slope, on the downgoing plate, is in most cases marked by a series of down-to-the-trench normal faults. Trenches may be partly or nearly entirely filled with sediments, many of which become offscraped and attached to the accretionary prism on the overriding plate. The inner trench slope on the overriding plate typically is marked by these folded and complexly faulted and offscraped sediments, and distinctive disrupted complexes known as mélanges may be formed in this environment.

In ocean–ocean subduction systems the arc develops about 100–150 miles (150–200 km) from the trench. Immature or young oceanic island arcs are dominated by basaltic volcanism and may be mostly underwater, whereas more mature systems have more intermediate volcanics and have more of the volcanic edifice protruding above sea level. The area between the arc and the accretionary prism is typically occupied by a forearc basin, filled by sediments derived from the arc and uplifted parts of the accretionary prism. Many island arcs have back arc basins developed on the opposite side of the arc, typically separating the arc from an older rifted arc or a continent.

Ocean–continent subduction systems are broadly similar to ocean–ocean systems, but the magmas must rise through continental crust so are chemically contaminated by this crust, becoming more silicic and enriched in certain sialic elements. Basalts, andesites, dacites, and even rhyolites are common in continental margin arc systems. Ocean–continent subduction systems tend to also have concentrated deformation including deep thrust faults, fold/thrust belts on the back arc side of the arc, and significant crustal thickening. Other continental margin arcs experience extension and may see rifting events that open back arc basins that may extend into marginal seas, or close. Crustal thickening in continental margin subduction systems is also aided by extensive magmatic underplating.

Oceanic plates may be thought of as conductively cooling upper boundary layers of the Earth's convection cells, and in this context subduction zones are the descending limbs of the mantle convection cells. Once subduction is initiated the sinking of the dense downgoing slabs provides most of the driving forces needed to move the lithospheric plates and force seafloor spreading at divergent boundaries where the mantle cells are upwelling.

The amount of material cycled from the lithosphere back into the mantle of the Earth in a subduction zone is enormous, making subduction zones the planet's largest chemical recycling systems. Many of the sedimentary layers and some of the upper oceanic crust are typically scraped off the downgoing slabs and added to accretionary prisms on the front of the overlying arc systems. Hydrated minerals and sediments release much of their trapped seawater in the upper few hun-

dred kilometers of the descent into the deep Earth, adding water to the overlying mantle wedge and triggering melting that supplies the overlying arcs with magma. The material that is not released or offscraped and underplated in the upper few hundred kilometers of subduction forms a dense slab that may go through several phase transitions and either flatten out at the 416-mile (670-km) mantle discontinuity, or descend all the way to the core mantle boundary. The slab material then rests and is heated at the core mantle boundary for about a billion years, after which it may rise to form a mantle plume that rises through the mantle to the surface. In this way, there is an overall material balance in subduction zone–mantle convection–plume systems.

Most continental crust has been created in subduction zone-arc systems of various ages stretching back to the Early Archean.

See also ANDES; CONVERGENT PLATE MARGIN PROCESSES; MANTLE PLUMES; MARIANAS TRENCH.

Further Reading

Kusky, Timothy M., and Ali Polat. "Growth of Granite-Greenstone Terranes at Convergent Margins and Stabilization of Archean Cratons." In *Tectonics of Continental Interiors*, edited by Stephen Marshak and Ben van der Pluijm, 43–73. Amsterdam: Elsevier, Tectonophysics, 1999.

Stern, Robert J. "Subduction Zones." *Reviews of Geophysics*, v. 40 (2002): 3-1–3-38.

subsidence Sinking of the land relative to sea level or some other uniform surface. It can be a gradual barely perceptible process, or it may occur as a catastrophic collapse of the surface. Subsidence naturally occurs along some coastlines and in areas where groundwater has dissolved cave systems in rocks such as limestone. It may occur on a regional scale, affecting an entire coastline, or it may be local in scale, such as when a sinkhole suddenly opens and collapses in the middle of a neighborhood. Other subsidence events reflect the interaction of humans with the environment and include ground surface subsidence as a result of mining excavations, groundwater and petroleum extraction, and several other processes. Compaction is a related phenomenon, where the pore spaces of a material are gradually reduced, condensing the material and causing the surface to subside. Subsidence and compaction do not typically result in death or even injury, but they do cost Americans alone tens of millions of dollars per year. The main hazard of subsidence and compaction is damage to property.

Subsidence and compaction directly affect millions of people. Coastal subsidence can affect entire cities, regions, and countries. Residents of New Orleans, Louisiana, live below sea level and are constantly struggling with the consequences of living on a slowly subsiding delta. Coastal residents in the Netherlands have constructed massive dike systems to try to keep the North Sea out of their slowly subsiding land. The city of Venice, Italy, has dealt with subsidence in a uniquely charming way, drawing tourists from around the world to their flooded streets. Millions of people live below the high-tide level in Tokyo. The coastline of Texas along the Gulf of Mexico is slowly subsiding, placing residents of Baytown and other Houston suburbs close to sea level and in danger of hurricane-induced storm surges and other more frequent flooding events. In Florida sinkholes have episodically opened up swallowing homes and businesses, particularly during times of drought.

The driving force of subsidence is gravity, with the style and amount of subsidence controlled by the physical properties of the soil, regolith, and bedrock underlying the area that is subsiding. Subsidence does not require a transporting medium, but it is aided by other processes, such as groundwater dissolution, which can remove mineral material and carry it away in solution, creating underground caverns that are prone to collapse.

Natural subsidence has many causes. Dissolution of limestone by underground streams and water systems is one of the most common, creating large open spaces that collapse under the influence of gravity. Groundwater dissolution results in the formation of sinkholes, large, generally circular depressions caused by collapse of the surface into underground open spaces.

Earthquakes may raise or lower the land suddenly, as in the case of the 1964 Alaskan earthquake where tens of thousands of square miles suddenly sank or rose 3–5 feet, causing massive disruption to coastal communities and ecosystems.

Earthquake-induced ground shaking can also cause liquefaction and compaction of unconsolidated surface sediments, leading to subsidence. Regional lowering of the land surface by liquefaction and compaction is known from the massive 1811 and 1812 earthquakes in New Madrid, Missouri, and from many other examples.

Volcanic activity can cause subsidence, as when underground magma chambers empty out during an eruption. In this case, subsidence is often the lesser of many hazards that local residents need to fear. Subsidence may also occur on lava flows, when lava empties out of tubes or underground chambers.

Some natural subsidence on the regional scale is associated with continental scale tectonic processes. The weight of sediments deposited along continental shelves can cause the entire continental margin to sink, causing coastal subsidence and a landward migration of the shoreline. Tectonic processes associated with extension, continental rifting, strike-slip faulting, and even collision can cause local or regional subsidence, sometimes at rates of several inches per year.

See also GROUNDWATER; PLATE TECTONICS.

Further Reading

Beck, B. F. *Engineering and Environmental Implications of Sinkholes and Karst.* Rotterdam: Balkema, 1989.

Dolan, Robert, and H. Grant Goodell. "Sinking Cities." *American Scientist* 74, no. 1 (1986): 38–47.

Holzer, Thomas L., ed., *Man-Induced Land Subsidence.* Geological Society of America, Reviews in Engineering Geology VI, 1984.

sulfide minerals   Many of the ores of metallic minerals occur as compounds of the sulfide ion, S^{2-}, with the metals attached as cations. Most of these are soft and look like metals and form many of the world's large ore deposits. One of the most common sulfide minerals is pyrite, FeS_2, found as a minor component in many rocks and as an accessory mineral in many ore deposits. Pyrrhotite (Fe_7S_8–FeS) is a less-common iron sulfide mineral. Most ore deposits are formed from hydrothermal fluids. Lead (Pb) and Zinc (Zn) are commonly found in sulfide compounds such as galena (PbS) and sphalerite (ZnS), and copper deposits are dominated by the two sulfide minerals chalcopyrite ($CuFeS_2$) and bornite (Cu_5FeS_4).

See also MINERALOGY.

sun dogs   *See* SUN HALOS.

sun halos   A number of unusual phenomena are related to the interaction of the Sun's rays with ice crystals in the upper atmosphere. A ring of light that circles and extends outward from the Sun, or the Moon, is known as a halo. The halo forms when the Sun's or Moon's rays get refracted by ice crystals in high-level cirriform clouds. Most sun halos form at an angle of 22° from the Sun because randomly oriented small ice particles refract the light at this angle. Occasionally a 46° halo is visible, formed when sub-horizontally oriented columnar ice crystals refract the light at this higher angle. Most sun halos are simply bright bands of light but some exhibit rainbow-like zones of color. These form when the light is dispersed by the ice crystals and light of different wavelengths (colors) is refracted by different amounts depending on its speed.

Atmospheric crystals sometimes cause hexagonal or platy ice crystals to fall slowly through the atmosphere, and this vertical motion causes the crystals to become uniformly oriented with their long dimensions in a horizontal direction. This orientation prevents light that is refracted through the ice crystals from forming a halo, but when the Sun approaches the horizon it causes two bright spots or colored bright spots to appear on either side of the sun. These spots are commonly called sundogs, or parhelia.

Sun pillars are a similar phenomenon but are formed by light that is reflected off the ice crystals instead of refracted through them. In this case, usually at sunset or sunrise, the Sun's rays reflect off the sub-horizontally oriented ice crystals and form a long column of light extending downward from the Sun.

See also RAINBOW.

sun pillars   *See* SUN HALOS.

supercontinent cycle Semi-regular grouping of the planet's landmasses into single or large continents that remain stable for a period of time, then disperse, and eventually come back together as new amalgamated landmasses with a different distribution. At several times in Earth history, the continents have joined together forming one large supercontinent, with the last supercontinent Pangea (meaning all land) breaking up approximately 160 million years ago. This process of supercontinent formation and dispersal and re-amalgamation seems to be grossly cyclic, perhaps reflecting mantle convection patterns but also influencing climate and biological evolution. Early workers noted global "peaks" in age distributions of igneous and metamorphic rocks and suggested that these represent global orogenic or mountain building episodes, related to supercontinent amalgamation.

The basic idea of the supercontinent cycle is that continents drift about on the surface until they all collide, stay together, and come to rest relative to the mantle in a place where the gravitational potential surface (geoid) has a global low. The continents are only one-half as efficient at conducting heat as oceans, so after the continents are joined together, heat accumulates at their base, causing doming and breakup of the continent. For small continents, heat can flow sideways and not heat up the base of the plate, but for large continents the lateral distance is too great for the heat to be transported sideways. The heat rising from within the Earth therefore breaks up the supercontinent after a heating period of several tens or hundreds of millions of years, the heat then disperses and is transferred to the ocean/atmosphere system, and continents move away until they come back together forming a new supercontinent.

The supercontinent cycle has many effects that greatly affect other Earth systems. First, the breakup of continents causes sudden bursts of heat release, associated with periods of increased, intense magmatism. It also explains some of the large-scale sea-level changes, episodes of rapid and widespread orogenesis, episodes of glaciation, and many of the changes in life on Earth.

Compilations of Precambrian isotopic ages of metamorphism and tectonic activity suggest that the Earth experiences a periodicity of global orogenesis of 400 million years. Peaks have been noted at time periods including 3.5, 3.1, 2.9, 2.6, 2.1, 1.8, 1.6, and 1.1 billion years ago, as well as at 650 and 250 million years ago. One hundred million years after these periods of convergent tectonism and metamorphism, rifting is common and widespread. A. H. Sutton (1963) proposed the term *chelogenic cycle,* in which continents assemble and desegregate in antipodal supercontinents.

Relationship of Supercontinents, Lower Mantle Convection, and the Geoid

Some models for the formation and dispersal of supercontinents suggest a link between mantle convection, heat flow, and the supercontinent cycle. Stationary supercontinents insulate the mantle causing it to heat up, because the cooling effects of subduction and seafloor spreading are absent. As the mantle then heats up, convective upwelling is initiated, causing dynamic and isostatic uplift of the continent, injection of melts into the continental crust, and extensive crustal melting. These crustal melts are widespread in the interiors of some reconstructed supercontinents, such as the Proterozoic anorogenic granites in interior North America, which were situated in the center of the supercontinent of Rodinia when they formed between one billion and 800 million years ago.

After intrusion of the anorogenic magmas, the lithosphere is weakened and can be more easily driven apart by divergent flow in the asthenosphere. Thermal effects in the lower mantle lag behind surface motions. So, the present Atlantic geoid high and associated hot spots represent a "memory" of heating beneath Pangea. Likewise, the circum Pangea subduction zones may have a memory in a global ring of geoid lows.

Other models for relationships between supercontinents and mantle convection suggest that supercontinents may also result from mantle convection patterns. Continental fragments may be swept toward convective downwellings, where they reaggregate as supercontinents.

Plate Tectonics, Supercontinents, and Life

Plate tectonic motions, especially the supercontinent cycle, profoundly affect the distribution and evolution of life on Earth. Plate tectonic activity such as rifting, continental collision, and drifting continents affects the distribution of lifeforms, the formation and destruction of ecological niches, and radiation and extinction blooms. Plate tectonic effects also can induce sea-level changes, initiate periods of global glaciation, change the global climate from hothouse to icehouse conditions, and affect seawater salinity and nutrient supply. All of these consequences of plate tectonics have profound influences on life on Earth.

Changes in latitude brought on by continental drift bring land areas into latitudes with better or worse climate conditions. This has different consequences for different organisms, depending on their temperature tolerance, as well as food availability in their environment. Biological diversity generally increases toward the equator, so, in general, as continents drift poleward more organisms tend to go extinct, and as they drift equatorward, diversification may increase.

Tectonics and supercontinent dispersal breaks apart and separates faunal provinces, which then evolve separately. Continental collisions and supercontinent amalgamation build barriers to migration but eventually bring isolated fauna together. One of the biggest mass extinctions (at the end of Permian) occurred with the formation of a supercontinent (Pangea), sea-level regression, evaporite formation, and global warming. At the boundary between the Permian and

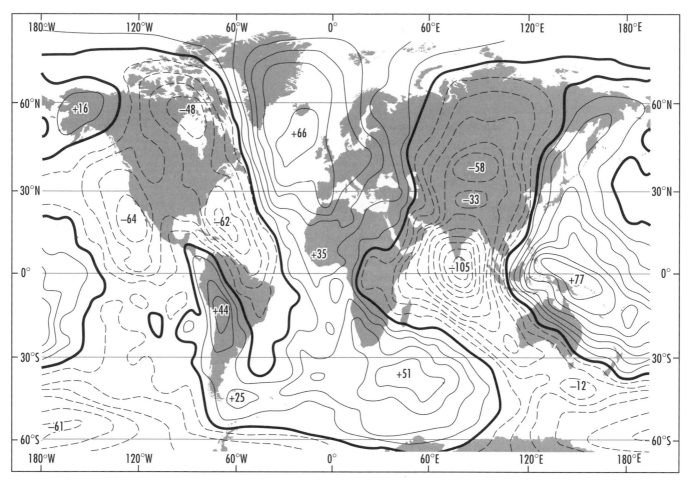

The geoid is an imaginary surface that would be sea level if it extended through the continents. The geoid surface is perpendicular to the gravity plumb lines at every location.

Triassic periods and between the Paleozoic and Mesozoic periods (245 million years ago), 96 percent of all species became extinct. Lost were the rugose corals, trilobites, many types of brachiopods, and marine organisms including many foraminifer species.

The Siberian flood basalts cover a large area of the Central Siberian Plateau northwest of Lake Baikal. They are more than half a mile thick over an area of 210,000 square miles (547,000 km²) but have been significantly eroded from an estimated volume of 1,240,000 cubic miles (3,3133,000 km³). They were erupted over a period of less than one million years (remarkably short!) 250 million years ago, at the end of the Permian at the Permian-Triassic boundary. They are remarkably coincident in time with the major Permian-Triassic extinction, implying a causal link. The Permian-Triassic boundary at 250 million years ago marks the greatest extinction in Earth history, where 90 percent of marine species and 70 percent of terrestrial vertebrates became extinct. It has been postulated that the rapid volcanism and degassing released enough sulfur dioxide to cause a rapid global cooling,

inducing a short ice age with associated rapid fall of sea level. Soon after the ice age took hold, the effects of the carbon dioxide took over and the atmosphere heated, resulting in a global warming. The rapidly fluctuating climate postulated to have been caused by the volcanic gases is thought to have killed off many organisms, which were simply unable to cope with the wildly fluctuating climate extremes.

Continental breakup may cause physical isolation of species that cannot swim or fly between the diverging continents. Physical isolation (via tectonics) produces adaptive radiation-continental dispersal and thus increases biotic diversity. For example, mammals had an explosive radiation (in 10–20 million years) in the Paleocene-Eocene, right after breakup of Pangea.

Sea-Level Changes, Supercontinents, and Life

Sea level has changed by hundreds of meters above and below current levels at many times in Earth history. In fact, sea level is constantly changing in response to a number of different variables, many of them related to plate tectonics.

The diversity of fauna on the globe is closely related to sea levels, with greater diversity during sea-level high stands and lower diversity during sea-level lows. For instance, sea level was 1,970 feet (600 m) higher than now during the Ordovician, and the sea-level high stand was associated

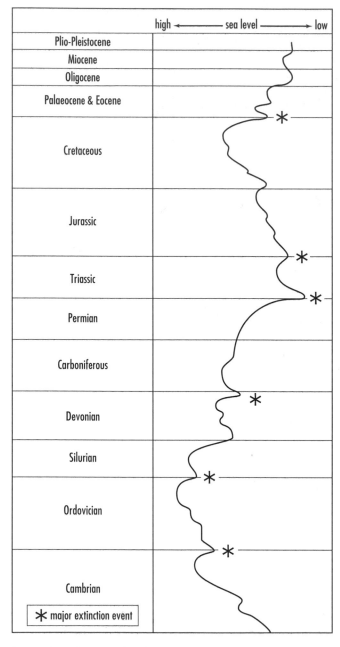

high ←——— sea level ———→ low

Plio-Pleistocene	
Miocene	
Oligocene	
Palaeocene & Eocene	
Cretaceous	✳
Jurassic	
Triassic	✳
Permian	✳
Carboniferous	
Devonian	✳
Silurian	
Ordovician	✳
Cambrian	✳

✳ major extinction event

Sea level has risen and fallen dramatically throughout Earth's history. Global sea-level rise and fall must be separated from local subsidence and uplift events along individual coastlines, by correlating events between different continents. Global eustatic sea-level curves (Vail curves) also show the height of the world's oceans, after local effects have been removed. Asterisks show the positions of six major extinction events.

with a biotic explosion. Sea levels reached a low stand at the end of the Permian, and this low was associated with a great mass extinction. Sea levels were high again in the Cretaceous.

Sea levels may change at different rates and amounts in response to changes in several other Earth systems. Local tectonic effects may mimic sea-level changes through regional subsidence or uplift, and these effects must be taken into account and filtered out when trying to deduce ancient, global (eustatic) sea-level changes. The global volume of the mid-ocean ridges can change dramatically, either by increasing the total length of ridges, or by changing the rate of seafloor spreading. The total length of ridges typically increases during continental breakup, since continents are being rifted apart, and some continental rifts can evolve into mid-ocean ridges. Additionally, if seafloor spreading rates are increased, the amount of young, topographically elevated ridges is increased relative to the slower, older topographically lower ridges that occupy a smaller volume. If the volume of the ridges increases by either mechanism, then a volume of water equal to the increased ridge volume is displaced and sea-level rises, inundating the continents. Changes in ridge volume are able to change sea levels positively or negatively by about 985 feet (300 m) from present values, at rates of about 0.4 inches (1 cm) every 1,000 years.

Continent-continent collisions, such as those associated with supercontinent formation, can lower sea levels by reducing the area of the continents. When continents collide, mountains and plateaus are uplifted, and the amount of material that is taken from below sea level to higher elevations no longer displaces seawater, causing sea levels to drop. The contemporaneous India-Asia collision has caused sea levels to drop by 33 feet (10 m).

Other things, such as mid-plate volcanism, can also change sea levels. The Hawaiian Islands are hot-spot–style mid-plate volcanoes that have been erupted onto the seafloor, displacing an amount of water equal to their volume. Although this effect is not large at present, at some periods in Earth history there were many more hot spots (such as in the Cretaceous), and the effect may have been larger.

The effects of the supercontinent cycle on sea level may be summarized as follows. Continent assembly favors regression, whereas continental fragmentation and dispersal favor transgression. Regressions followed formation of the supercontinents of Rodinia and Pangea, whereas transgressions followed the fragmentation of Rodinia, and the Jurassic-Cretaceous breakup of Pangea.

Effects of Transgressions and Regressions
During sea-level transgressions, continental shelves are covered by water, and available habitats are enlarged, increasing the diversity of fauna. Transgressions are generally associated with greater diversification of species. Regres-

sions cause extinctions through loss of environments, both shallow marine and beach. There is a close association between Phanerozoic extinctions and sea-level lowstands. Salinity fluctuations also affect diversity—the formation of evaporites (during supercontinent dispersal) causes reduction in oceanic salinity. For instance, Permian-Triassic rifts formed during the breakup of Pangea had lots of evaporites, (up to 4.4 miles, or 7 km thick), which lowered the salinity of oceans.

Supercontinents affect the supply of nutrients to the oceans, and thus, the ability of life to proliferate. Large supercontinents cause increased seasonality and thus lead to an increase in the nutrient supply through overturning of the ocean waters. During breakup and dispersal, smaller continents have less seasonality, yielding decreased vertical mixing, leaving fewer nutrients in shelf waters. Seafloor spreading also increases the nutrient supply to the ocean; the more active the seafloor spreading system, the more interaction there is between ocean waters and crustal minerals that get dissolved to form nutrients in the seawater.

See also GREENHOUSE EFFECT; ICE AGE; PLATE TECTONICS.

suspect terranes   Fault-bounded geologic entities of regional extent, each characterized by a geologic history that is different from geologic histories of contiguous terranes. The term was coined by Peter Coney and modified by Davie Jones in an attempt to explain the juxtaposition of very different fault-bounded blocks of rock in the North American Cordillera, and later modified to include seafloor topographic highs such as seamounts, oceanic plateaus, and even island arcs. The application of the term was of some use in separating units of different origin in the Cordillera and other orogenic belts. Some geologists were overzealous in the application of the term and the concept, however, and divided orogens into hundreds of terranes, many of which were merely fragments of the same terrane that were separated by strike-slip or other faults. In addition, some workers used the concept as an end in itself without attempting to understand the origin of the terranes, as done in more traditional tectonic analysis.

See also PLATE TECTONICS.

Further Reading
Coney, Peter, Davie Jones, and James W. H. Monger. "Cordilleran Suspect Terranes." *Nature*, v. 288 (1988): 329–333.
Jones, Davie, David G. Howell, Peter J. Coney, and James W. H. Monger. "Recognition, Character and Analysis of Tectonostratigraphic Terranes in Western North America." *Journal of Geologic Education*, v. 3 (1983): 295–303.
Sengör, A. M. Celal. "Lithotectonic Terranes and the Plate Tectonic Theory of Orogeny: a Critique of the Principles of Terrane Analysis." In *Terrane Analysis of China and the Pacific Rim*, edited by Thomas J. Wiley, David G. Howell, and F. L. Wong, 9–43. Houston, Tex.: Circum Pacific Council on Energy and Mineral Research, 1990.

suture zone   The structurally complex zone marking the place where two once separated tectonic plates have collided and the oceans that once separated the plate have closed. Sutures may mark the places where two continents that were previously on opposite sides of an ocean basin have collided, or they may mark the collision zones between continents, arc terranes, oceanic plateaus, or other seafloor topographic highs. Suture zones are very complex and structurally variable, typically marked by numerous shear zones, including mylonite zones, faults, and mélanges. Ophiolites are commonly found decorating the suture zones. These may range from complete ophiolite sequences to partial ophiolites, to thin mafic or ultramafic slivers in deep-level high-grade shear zones.

See also PLATE TECTONICS.

Further Reading
Dewey, John F. "Suture Zone Complexities." *Tectonophysics* (1977).

Svalbard and Spitzbergen Island   Spitzbergen is the largest island (15,000 square miles; 40,000 km²) of Svalbard, a territory of Norway located in the Arctic Ocean. The islands are located on the Barents Shelf and are bounded by the Greenland Sea on the west and the Arctic Ocean on the north. The entire Svalbard archipelago was originally referred to as Spitzbergen, but in 1940 the name was changed to Svalbard, and the name Spitzbergen was reserved for the largest island of the archipelago that also includes the islands of Nordaustlandet, Edgeoya, Barentsoya, Prins Karls Forland, and many smaller islands. About half of the island of Spitzbergen is covered by permanent ice and glaciers, and many deeply incised fiords rise to a level of about 3,200 feet (1,000 m), reflecting a peneplained erosion surface that has rebounded since the Cenozoic. Since the entire archipelago lies so far north between 76°–81°N, the Sun remains above the horizon from late April through late August, but remains below the horizon in winter months. The warm Gulf Stream current has a moderating effect on the climate.

The Svalbard archipelago is well exposed and preserves a complex history of Archean and younger events. The island chain is broken into three main terranes separated by north-south striking faults. The eastern terrane has a basement of Archean through Proterozoic gneisses and amphibolites overlain by psammitic and pelitic schists and marbles that are approximately 1,750 million years old. These are overlain by pelites, psammites, and felsic volcanics that are about 970 million years old, overlain by 900–800-million-year-old quartzites, silts, and limestones. A Vendian group of pelites and glacial tillites formed during the Varanger glaciation. These are overlain by Cambro-Ordovician carbonates, correlated with similar rocks of eastern Greenland. Mid-Paleozoic tectonism is related to the closure of the Iapetus Ocean during the Caledonian orogeny, known locally as the Friesland orogeny. West-vergent fold and thrust structures formed in the Middle and Late Ordovician, whereas late tectonic batholiths intruded in the Silurian through Early Devonian. North-south striking mylonite zones are concentrated on the western side of the terrane and indicate sinistral transpressive strains.

The central terrane contains a basement of mainly Proterozoic and possible Archean igneous gneisses, overlain by dolostones and Varanger tillites, overlain by Ediacarian phyllites. These are followed by Cambro-Ordovician carbonates. Devonian strata on Svalbard are only exposed in the central terrane and include Old Red Sandstone facies dated by identification of fossil fish remains, similar to those of Scotland. These beds are associated with sinistral transpressive tectonics with the opening of pull-apart basins, and the deposition of conglomerates, sandstones, and shales in fluvial systems in these basins. Devonian and Mesozoic strata are folded and show eastward vergence.

The western terrane has a gneissic Proterozoic basement, overlain by Varanger tillites interbedded with mafic volcanics and overlain by Ediacarian fauna. It is thought that this terrane correlates more with sequences on Ellsemere Island than in the rest of Svalbard or Greenland, so it was probably brought in later by strike-slip faulting. Deformation in Early Ordovician times in the western terrane is linked with subduction tectonics, which may have continued to the Late Ordovician. Later deformation occurred in the Devonian, possibly associated with the Ellsemerian orogeny.

Some models for the tectonic evolution of Svalbard invoke more than 600 miles (1,000 km) of sinistral strike-slip displacements in the Silurian–Late Devonian on the north-south faults, bringing the eastern terrane into juxtaposition with central Greenland. This motion is associated with the formation of the pull-apart basins filled by the Old Red Sandstone in the central terrane.

In the Carboniferous through Early Eocene, most of Svalbard was relatively stable and experienced platform sedimentation, continuous with that of northern Greenland and the Sverdrup basin of northern Canada. Early Carboniferous anhydrites, breccias, conglomerates, sabkha deposits and carbonates form the basal 3,000 feet (1,000 m) of the section, and these grade up into 1,500 feet (450 m) of fine-grained siliciclastic rocks, cherts, and glauconitic sandstones. Mesozoic strata include more than 8,000 feet (2,500 m) of interbedded deltaic and marine deposits. A Late Cretaceous period of non-deposition was followed by the deposition of nearly 5,000 feet (1,500 m) of deltaic sandstones, shales, and marine beds in the Paleocene and Early Eocene.

In the Eocene, western Spitzbergen collided in a dextral transpressional event with the northeast margin of Greenland, forming folds, thrusts, and later normal faults along the western coast of the island. Small pull-apart basins formed during this event and are filled by sediments derived from the contemporaneous uplifted fold belt. Erosion and peneplaination in the Oligocene through Holocene formed the flat sur-

face evident across the archipelago today, with Quaternary glaciations depressing the crust. Postglacial rebound plus thermal uplift were associated with the opening of the Arctic Ocean and the Norwegian and Greenland basins. Quaternary flood basalts in the northern part of Svalbard are associated with these extensional basin-forming events.

Further Reading

Harland, W. Brian, and Evelyn K. Dowdeswell, eds. *Geological Evolution of the Barents Shelf Region.* London: Graham and Trotman, 1988.
Harland, W. Brian, and Rod A. Gayer. "The Arctic Caledonides and Earlier Oceans." *Geological Magazine* 109 (1972).

syncline   *See* FOLD; STRUCTURAL GEOLOGY.

synthetic aperture radar   Space-borne remotely sensed imagery has been routinely acquired and used as a reconnaissance tool by geologists since the initial launch of the Landsat series of satellites in 1972. More recently, space-borne sensors such as Thematic Mapper (TM), Seasat Synthetic Aperture Radar (SAR), Shuttle Imaging Radar (SIR-A and SIR-B), and SPOT (Système pour l'Observation de la Terre) have scanned the Earth's surface with other portions of the electromagnetic spectrum in order to sense different features, particularly surface roughness and relief, and to improve spatial resolution. While TM and SPOT images have proved spectacularly effective at differentiating between various rock types, synthetic aperture radar (SAR) is particularly useful at delineating topographically expressed structures. Spaceborne SAR systems also play a major role in exploration of other bodies in the solar system.

Synthetic aperture radar (SAR) is an active sensor where energy is sent from a satellite (or airplane) to the surface at specific intervals in the ultrahigh frequency range of radar. The radar band refers to the specific wavelength sent by the source and may typically include X-band (4 cm), K-band (2 cm), P-band (1 meter), L-band (23.5 cm), C-band, or others. SAR allows the user to acquire detailed images at any time of day or night and also in inclement weather. Synthetic aperture radar is very complicated, but it basically works by first obtaining a two-dimensional image and then fine-tuning that image with computers and sensors to create a decisively more accurate image. It is useful in military, science, and mapping as it provides detailed resolutions of a particular area. Synthetic aperture radar is widely used by governments and militaries but is expensive for others who may wish to use it. The advancement of technology, however, is making it possible and economical in other applications.

The effectiveness of orbital SAR for structural studies depends primarily on three factors: (1) roughness contrasts; (2) local incidence angle variations (i.e., topography); and (3) look azimuth relative to topographic trends. The strength of the radar signal may also be attenuated by atmospheric or soil moisture and is affected to some degree by the types of atomic bonds in the minerals present in surface materials. Bodies of water are generally smoother than land and appear as dark, radar-smooth terrain. Structure is delineated on land by variations in local incidence angle, the precise backscatter dependence being controlled by surface roughness. Different SAR satellites have different radar incidence or look angles, and some, such as RADARSAT, are adjustable and specifiable by the user. The 20° look angle chosen for Seasat was intended to maximize the definition of sea conditions but had the incidental benefit of producing stronger sensitivity to terrain than would larger angles. Look azimuth has been shown to be an extremely important factor for low relief terrain of uniform roughness, with topography within about 20° of the normal to look azimuth being strongly highlighted.

See also REMOTE SENSING.

Further Reading

Kusky, Timothy M., Paul D. Lowman Jr., Penny Masouka, and Herb Blodget. "Analysis of Seasat L-Band Imagery of the West Bay–Indian Lake fault system, Northwest Territories." *Journal of Geology*, v. 101 (1993).

Syrian arc fold belt The term *Syrian arc fold belt* is used to denote the broad S-shaped belt of deformed Paleozoic, Mesozoic, and Cenozoic rocks that sweep across a region of complex convergent, transform, and divergent plate boundaries from North Africa, through the Sinai peninsula, into the Levant and as far as the Euphrates Valley. The Syrian arc includes several individual components with distinct geological histories such as Palmyra fold belt, Levantine fold belt, and the Cairo-Sinai fold belt, but there are enough broad-scale similarities between them to warrant a general grouping.

The main active tectonic elements surrounding the Syrian arc are the Red Sea, North African and Levantine continental margins, the Bitlis suture, and the Dead Sea transform. About 65–68 miles (105–110 km) of sinistral displacements have accumulated along the Dead Sea transform, with 37–40 miles (60–65 km) of translation in pre-Miocene times, and 25–28 miles (40–45 km) from the Miocene to present. Displacement along the Dead Sea transform diminishes northward to about 12–18 miles (20–30 km) before it merges with the Bitlis suture and East Anatolian fault. The displacement not accommodated in this northern segment of the Dead Sea transform may have been taken up in the Palmyride section of the Syrian arc fold belt.

The northern part of the Syrian arc lies about 155 miles (250 km) south of the Bitlis suture, with the area between the two occupied by the Tauride foreland basin and fold and thrust belt, which continues eastward into the Zagros. Considerable controversy surrounds the origin of the northern part of the Syrian arc: (1) contractional deformation associated with transpressional strains along the Dead Sea transform, with the fold belts representing a series of en echelon fold belts associated with northward diminution of displacements on the Dead Sea transform; (2) far-field effects of contraction across the Bitlis/Zagros suture; (3) reorganization of plate motions between the North Atlantic–Eurasian and African plates in the Santonian; or (4) combination of the different models.

The Syrian arc fold belt continues southward through the Levantine fold belt into the North Sinai fold belt and then continues with diminished intensity toward the southwest past Cairo. The numerous rugged isolated hills of the Syrian arc in northern Sinai are distinct from the other physiographic provinces of the Sinai Peninsula, which is divided into three physiographic provinces. Precambrian magmatic and metamorphic rocks of the Arabian-Nubian Shield form high, rugged mountains in the south with drainages flowing to the east and west. This basement terrane is overlain by a dissected homoclinal limestone plateau consisting of a Mesozoic and early Cenozoic continental margin sequence in the central part of the peninsula, with drainages flowing north. A sandy plain parallels the coastline in the north.

The northern part of the limestone plateau is deformed by a series of east-northeast trending folds and faults of the Syrian arc. The hills, including Gebels Maghara, Halal, and Yelleg, are the expression of doubly plunging anticlines, with axial surfaces striking northeast-southwest. These folds affect Jurassic through Cretaceous carbonates, deposited on a shallow platform that deepened toward an open basin (Tethys Sea) to the north. The rocks are exposed in breached southward-vergent anticlines with shallow fold limbs on the north side and steep to overturned southern limbs. Thrust faults pass through the southern limbs of some of the folds, and there may be east-northeast striking strike-slip faults at depth reflecting formation of the fold belt in a zone of transpression.

Contractional deformation in the northern part of the Syrian arc, and in northern Sinai, began with minor uplifts in the Late Cretaceous prior to opening of the Red Sea, followed by tectonic quiescence through the Paleogene and interrupted by minor Middle Eocene uplift. The major phase of uplift in the northern part of the Syrian arc began in the Late Oligocene to Early Miocene, concomitant with rifting of the Red Sea basin and motion along the Dead Sea transform. Recent seismicity is consistent with transpressional deformation. Thus, deformation in this part of the Syrian arc appears to be related to both far-field stresses transmitted from contraction across the Bitlis suture in the Late Cretaceous, transpressional deformation associated with Red Sea rifting and motion along the Dead Sea transform in Miocene to Recent times.

Tectonism in the north Sinai-Cairo region began in the Late Cretaceous and mostly ended by the Eocene. However, less intense deformation and uplift probably continues through the present day, as demonstrated by historical earthquakes and recent seismicity, uplifted Holocene beach terraces in the area between Bardawil Lagoon and Gebel Maghara, and fault scarps cutting recent alluvium. The North Sinai-Cairo-Faiyum area has a well-documented historical earthquake record that demonstrates ongoing activity in the fault and fold belt. On August 7, 1847, a large earthquake with an estimated magnitude of 5.5–5.9 shook the Faiyum-Cairo region. Hundreds of people were killed and injured, and thousands of structures were destroyed. The earthquake was felt in much of north Africa and across Egypt, with heavy damage reported south of Cairo as far as Assuit. A magnitude 4.8 earthquake shook the area again on January 10, 1920. On October 12, 1992, another magnitude 5.9 earthquake shook the Dahshur area, northeast of the Faiyum region, along the same belt that extends into northern Sinai. Heavy damage was reported in the region including Cairo, and hundreds of people were injured and thousands of houses were damaged in the rest of Egypt. Most recent seismicity in the Sinai Peninsula is related to rifting and strike-slip faulting along the Red Sea and Gulfs of Suez and Aqaba, although faulting related to movements along the Faiyum–northern Sinai trend is possible.

Most of the Syrian arc fold belt appears to be developed over areas that have a thicker-than-average section of Late Paleozoic–Mesozoic rocks, suggesting that the fold belt developed over a zone of older rifting. In the Palmyride fold belt, this Mesozoic rift takes the form of an aulacogen developed over an older possibly Proterozoic age suture. Further south the increased thickness of Mesozoic sediments formed along the rifted Levant–North Africa margin. The fold belt extends southwest through a fault-bounded depositional trough including the Faiyum depression that may also follow an older rift structure. The Syrian arc thus appears to mark a zone of long-standing crustal weakness. Much of the Cretaceous deformation may be due to the fact that this weak zone was the first to fail when far-field stresses were imposed by collisional events across the Bitlis/Zagros suture. Younger transpressional deformation along the Dead Sea transform related to the opening of the Red Sea and the rotation of Arabia away from Africa are also recorded.

See also CONVERGENT PLATE MARGIN PROCESSES; SINAI PENINSULA.

Further Reading

Abdel, Aal. A., and Jeffrey J. Lelek. "Structural Development of the Northern Sinai, Egypt, and Its Implications on the Hydrocarbon Prospectivity of the Mesozoic." *GeoArabia* 1 (1994).

Barazangi, Mauwia, Don Seber, Thomas Chaimov, John Best, Robert Litak, Damen Al-Saad, and Tarif Sawaf. "Tectonic Evolution of the Northern Arabian Plate in Western Syria." In *Recent Evolution and Seismicity of the Mediterranean Region,* edited by E. Boshi and E. Al.: 117–140. Amsterdam: Kluwer Academic Publishers, 1993.

Bosworth, William, Rene Guiraud, and L. G. Kessler. "Late Cretaceous (ca. 84 Ma) Compressive Deformation of the Stable Platform of Northeast Africa (Egypt): Far-field Stress Effects of the 'Santonian Event' and Origin of the Syrian Arc Deformation Belt." *Geology* 27 (1999).

Coleman, Robert G. "Geological Evolution of the Red Sea." *Oxford Monographs on Geology and Geophysics* 24 (1993).

Dewey, John F., Mark R. Hempton, William S. F. Kidd, F. Saroglu, and A. M. Celal Sengör. "Shortening of Continental Lithosphere: the Neotectonics of Eastern Anatolia—a Young Collision Zone." In "Collision Tectonics," edited by Mike P. Coward, and A. C. Ries. *Geological Society Special Publication* 19 (1986).

Garfunkel, Zvi. "Internal Structure of the Dead Sea Leaky Transform (Rift) in Relation to Plate Kinematics." *Tectonophysics* 80 (1981).

Hempton, Mark R. "Constraints on Arabian Plate Motion and Extensional History of the Red Sea." *Tectonics* 6 (1987).

Jenkins, D. A. "North and Central Sinai." In *The Geology of Egypt,* edited by Rushdi Said: 361–380. Rotterdam, Netherlands: A.A. Balkema, 1990.

Krenkel, Erich. *Geologie der Erde, Geologie Afrikas.* Berlin: Gebrüder Borntraeger, 1925.

Kusky, Timothy M., and Farouk El-Baz. "Neotectonics and Fluvial Geomorphology of the Northern Sinai Peninsula." *Journal of African Earth Sciences* 31 (2000).

Le Pichon, Xavier, and Jean M. Gaulier. "The Rotation of Arabia and the Levant Fault System." *Tectonophysics* 153 (1988): 271–294.

McBride, John H., Muawia Barazangi, John Best, Damen Al-Saad, Tarif Sawaf, Mohamed Al-Otri, and Ali Gebran. "Seismic Reflection Structure of Intracratonic Palmyride Fold-Thrust Belt and Surrounding Arabian Platform, Syria." *American Association of Petroleum Geologists Bulletin* 74 (1990).

Moustafa, Adel., and Mosbah H. Khalil. "Structural Characteristics and Tectonic Evolution of North Sinai Fold Belts." In *The Geology of Egypt,* edited by Rushdi Said: 381–389. Rotterdam, Netherlands: A.A. Balkema, 1990.

Sadek, H. "The Principal Structural Features of the Peninsula of Sinai." 14th International Geological Congress, 1926, Madrid (1928): 895–900.

Sestini, Giulliano. "Tectonic and Sedimentary History of the NE African Margin (Egypt/Libya)." In *The Geological Evolution of the Eastern Mediterranean,* edited by J. E. Dixon, and A. H. F. Robertson. Oxford: Blackwell Scientific Publishers, 1984.

Taconic Mountains, Taconic Orogeny *See* APPALACHI-
ANS.

taiga A high-latitude forest and/or swampy region that
occupies regions between the tundra and the steppe. The
taiga forest is a common feature in northern Canada and
Nunavet and Siberia. The taiga is sometimes called the cold
woodland and is characterized by low trees that are spaced
well apart, numerous shrubs, and an extensive moss and
lichen cover on the ground and rocky surfaces. Much of the
northern regions occupied by taiga were glaciated in the
Pleistocene and so have thin poorly developed and poorly
drained soils resting on bedrock. Many poorly drained
depressions develop into bogs or thick swampy accumulations of peat known as muskeg. The northern forest is inhabited by many herds of caribou, wandering black and grizzly
bears, wolverines, wolves, and billions of swarms of
mosquitoes.

tectonics *See* PLATE TECTONICS.

tectosphere Cratons are large areas of thick continental
lithosphere that have been stable for long periods of geological time, generally since the Archean. Most cratons are characterized by low heat flow and no or few earthquakes. Many
have a thick mantle root or tectosphere, characterized seismo-
logically and from xenolith studies to be cold and refractory,
having had basaltic melt extracted from it during the Archean.
Seismological studies have shown that many parts of the tectosphere are strongly deformed with most of the minerals oriented in planar or linear fabrics. Current understanding about
the origin of stable continental cratons and their roots hinges
upon recognizing which processes change the volume and
composition of continental lithosphere with time, and how

and when juvenile crust evolved into stable continental crust.
Despite decades of study, several major unresolved questions
remain concerning Archean tectosphere: (1) how is it formed?
Large quantities of melt extraction (komatiite in composition,
if melting occurred in a single event) are required from petrological observations, yet little of this melt is preserved in
Archean cratons, which are characterized by highly evolved
crust compositions. (2) In what tectonic settings is it formed?
Hypotheses range from intraplate, plume-generated settings to
convergent margin environments. (3) Finally, once formed,
does the chemical buoyancy and inferred rheological strength
of the tectosphere preserve it from disruption? Until recently,
most scientists would argue that cratonic roots last forever—
isotopic investigations of mantle xenoliths from the Kaapvaal,
Siberian, Tanzanian, and Slave cratons document the longevity of the tectosphere in these regions. However, the roots of
some cratons are now known to have been lost, including the
North China craton, and the processes of the loss of the tectosphere are as enigmatic as the processes that form the roots.

    See also ARCHEAN; CRATONS; PLATE TECTONICS.

Further Reading
Jordan, Thomas H. "Structure and Formation of the Continental
    Lithosphere." In "Oceanic and Continental Lithosphere, Similari-
    ties and Differences," edited by M. A. Menzies, and K. Cox. *Jour-
    nal of Petrology,* Special Lithosphere Issue (1988): 11–37.
Kusky, Timothy M., and Ali Polat. "Growth of Granite-Greenstone
    Terranes at Convergent Margins and Stabilization of Archean
    Cratons." *Tectonophysics,* vol. 305 (1999): 43–73.
Rudnick, Roberta L. "Making Continental Crust." *Nature,* vol. 378
    (1995): 571–578.

tektite Very unusual rounded and pitted aerodynamically
shaped clumps of black, greenish brown, or yellow glass,
usually found in groups or clusters known as tektite fields.

The glasses are typically less than a few centimeters long, and high in silica content (65–85 percent), but are of nonvolcanic origin and show no relationship to the underlying geology. Tektites were named by Eduard Suess in 1900, and their origin has been debated ever since. It is now thought that they are melts derived from hypervelocity meteorite impacts, and that the tektites formed small melt balls that shot through the hole blasted in the atmosphere by the impactor, reaching spectacular heights before cooling and falling to Earth in tektite fields. Some tektite fields can be shown to be related to specific nearby or distant impact craters such as the 15-million-year-old Reis crater in Germany, whereas specific craters have not been found to be associated with other tektite fields. Alternative models for the origin of tektites have suggested that they may be melts derived from impacts on the moon or other nearby planets.

See also IMPACT CRATER; METEORITE.

temperate climate   *See* CLIMATE.

terminal moraine   Ridge-like accumulations of drift deposited at the farthest point of travel of a glacier's terminus. Terminal moraines may be found as depositional landforms at the bases of mountain or valley glaciers marking the locations of the farthest advance of that particular glacier, or they may be more regional in extent, marking the farthest advance of a continental ice sheet. There are several different categories of terminal moraines, some related to the farthest advance during a particular glacial stage, and others referring to the farthest advance of a group of, or all, glacial stages in a region. Continental terminal moraines are typically succeeded poleward by a series of recessional moraines marking temporary stops in the glacial retreat or even short advances during the retreat. They may also mark the boundary between glacial outwash terrain toward the equator, and knob and kettle or hummocky terrain toward the pole from the moraine. The knob and kettle terrain is characterized by knobs of outwash gravels and sand separated by depressions filled with finer material. Many of these kettle holes were formed when large blocks of ice were left by the retreating glacier, and the ice blocks melted later leaving large pits where the ice once was. Kettle holes are typically filled with lakes, and many regions characterized by many small lakes have a recessional kettle hole origin.

See also GLACIER.

terrace   Any long, narrow, and relatively flat or gently inclined surface that is narrower than a plain and bounded by an ascending slope on one side and a descending slope on the other side. Terraces form in a variety of different settings. River terraces are found in many valleys as elevated surfaces on a single side or both sides of the present river and floodplain surfaces and mark a former higher level of the river sys-

Stream terraces form when a stream drops from one level to a deeper level, eroding former deposits. These photographs show stream terraces from a variety of perspectives: (A) Stream terraces that formed on both sides of a valley and were later cut by the active stream at a lower level; (B) The same terrace from a perspective that illustrates the depositional nature of the terrace and its flat upper surface; (C) The active wadi (streambed) cutting through the coarse-grained conglomerates of the ancient (Pleistocene) terrace *(Photos by Timothy Kusky)*

tem. Rivers and streams may downcut through older terraces for a variety of reasons, including climatically influenced changes in discharge, or uplift of the river valley and slopes causing a change in the river profile. Coastal or beach terraces are constructional surfaces that slope seaward (or lakeward) and are built by sedimentary deposits. Coastal terraces denote uplift of the land surface or lowering of sea levels.

See also BEACH; RIVER SYSTEM.

Tertiary The first period of the Cenozoic era, extending from the end of the Cretaceous of the Mesozoic at 66 million years ago until the beginning of the Quaternary 1.6 million years ago. The Tertiary is divided into two periods including the older Paleogene (66–23.8 Ma) and the younger Neogene (23.8–1.8 Ma), and further divided into five epochs including the Paleocene (66–54.8 Ma), Eocene (54.8–33.7 Ma), Oligocene (38.7–23.8 Ma), Miocene (23.8–5.3 Ma), and Pliocene (5.3–1.6 Ma). The term *Tertiary* was first coined by the Italian geologist Giovanni Arduino in 1758, and later adopted by Charles Lyell in 1833 for his post-Mesozoic sequences in western Europe. The term *Tertiary* is being gradually replaced by the terms *Paleogene* and *Neogene* periods.

The Tertiary is informally known as the age of mammals for its remarkably diverse group of mammals, including marsupial and placental forms that appeared abruptly after the extinction of the dinosaurs. The mammals radiated rapidly in the Tertiary while climates and seawater became cooler. The continents moved close to their present positions by the end of the Tertiary, with major events including the uplift of the Alpine-Himalayan mountain chain.

Pangea continued to break apart through the early Tertiary while the African and Indian plates began colliding with Eurasia, forming the Alpine-Himalayan mountain chain. Parts of the Cordilleran mountain chain experienced considerable amounts of strike-slip translation of accreted terranes with some models suggesting thousands of kilometers of displacement of individual terranes. The Cordillera of western North America experienced an unusual geologic event with the subduction of at least one oceanic spreading ridge beneath the convergent margin. The boundary between three plates moved rapidly along the convergent margin from about 60 million years ago in the north, to about 35 million years ago in the south, initiating a series of geological consequences including anomalous magmatism, metamorphism, and deformation. New subduction zones were initiated in the southwest Pacific (SE Asia) and in the Scotia arc in the south Atlantic. The Hawaiian-Emperor seamount chain formed as a hot-spot track with the oldest preserved record starting about 70 million years ago, and a major change in the direction of motion of the Pacific plate recorded by a bend in the track near Midway island formed 43 million years ago.

The San Andreas fault system was initiated about 30 million years ago as the East Pacific rise was subducted beneath western north America and the relative motions between the Pacific plate and the North American plate became parallel to the margin. Around 3.5 million years ago the Panama arc grew, connecting North and South America and dramatically changing the circulation patterns of the world's oceans and influencing global climate. The East African rift system began opening about 5–2 million years ago, forming the sheltered environments that hosted the first known *Homo sapiens*.

Climate records show a general cooling of ocean waters and the atmosphere from the earliest Tertiary through the Paleocene, with warming then cooling in the Eocene. The oceans apparently became stratified with cold bottom waters and warmer surface waters in the Eocene, with further cooling reflecting southern glaciations in the Oligocene. Late Oligocene through Early Miocene records indicate a period of warming, followed by additional cooling in the mid-Miocene with the expansion of the Antarctic ice sheet that continued through the end of the Miocene. Pliocene climates began fluctuating wildly from warm to cold, perhaps as a precursor to the Pleistocene ice ages and interglacial periods. The Late Pliocene climates and change into the Pleistocene ice ages were strongly influenced by the growth of the Panama arc and the closing of the ocean circulation routes between the Pacific and Atlantic oceans. The Panama isthmus blocked warm Caribbean waters from moving west into the Pacific Ocean but forced these waters into the Gulf Stream that brings warm water northward into the Arctic Ocean basin. Warm waters here cause increased evaporation and precipitation, leading to rapid growth of the northern glaciers.

Nearly all of the mammals present on the Earth today appeared in the Cenozoic, and most in the Tertiary, with the exception of a primitive group known as the pantotheres that arose in the Middle Cretaceous. The pantotheres evolved into the first marsupial, the opossum, which in turn branched into the first placental mammals that spread over much of the northern continents, India, and Africa by the Late Cretaceous. Pantotheres and earlier mammals laid eggs, whereas marsupial offspring emerge from an eggshell-like structure in the uterus early in their development. In contrast, placental mammals evolve more fully inside the uterus and emerge stronger with a higher likelihood of surviving infancy. It is believed that this evolutionary advantage led to the dominance of placental mammals and the extinction of the pantotheres.

Mammalian evolution in the Tertiary was strongly influenced by continental distributions. Some continents like Africa, Madagascar, India, and Australia were largely isolated. Connections or land bridges between some of these and other continents, such as the Bering land bridge between

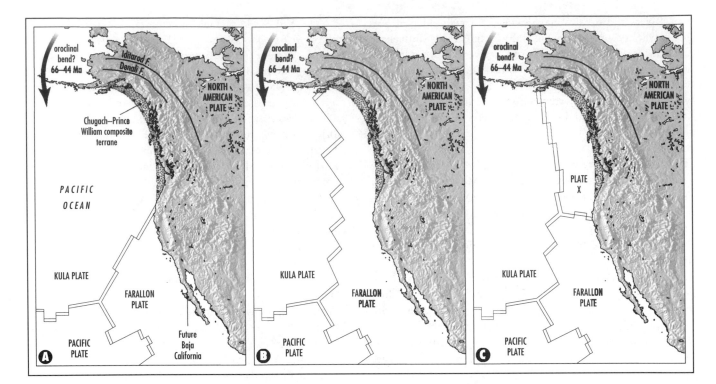

Reconstruction of part of the Tertiary world, that of the ridge subduction event of the Paleocene-Eocene of North America. Three possible plate configurations are illustrated.

Alaska and Siberia allowed communication of taxa between continents. With the land distribution patterns certain families and orders evolved on one continent, and others on other continents. Rhinoceroses, pigs, cattle, sheep, antelope, deer, cats, and related families evolved primarily in Asia, whereas horses, dogs, and camels evolved chiefly in North America with some families reaching Europe. Horses have been used as a model of evolution with progressive changes in the size of the animals, as well as the complexity of their teeth and feet.

Marine faunas included gastropods, echinoids, and pelecypods along with bryozoans, mollusks, and sand dollars in shallow water. Coiled nautiloids floated in open waters, whereas sea mammals including whales, sea cows, seals, and sea lions inhabited coastal waters. The Eocene-Oligocene boundary is marked by minor extinctions, and the end of the Pliocene saw major marine extinctions caused by changes in oceanic circulation with massive amounts of cold waters pouring in from the Arctic and from meltwater from growing glaciers.

See also CENOZOIC; NEOGENE.

Further Reading

Bradley, Dwight C., Timothy M. Kusky, Peter Haeussler, David C. Rowley, Richard Goldfarb, and Steve Nelson. "Geologic Signature of Early Ridge Subduction in the Accretionary Wedge, Fore-arc Basin, and Magmatic Arc of South-Central Alaska." In "Geology of a Transpressional Orogen Developed During a Ridge-Trench Interaction Along the North Pacific Margin," edited by Virginia B. Sisson, Sarah Roeske, and Terry L. Pavlis. *Geological Society of America,* Special Paper (2003).

Pomeral, C. *The Cenozoic Era.* New York: John Wiley and Sons, 1982.

Savage, R. J. G., and M. R. Long. *Mammal Evolution: An Illustrated Guide.* New York: Facts On File, 1986.

Tethys   The ocean basin that closed between Gondwana (Africa, Arabia, India) and Eurasia to form the Alpine-Himalayan mountain chain was named the Tethys by Eduard Suess in 1893. The history of the Tethys Ocean was complex, involving the creation and destruction of at least two major oceans. Paleo-Tethys closed in the older Cimmeride orogeny and the Neo-Tethys Ocean closed in the younger Alpine orogeny. Tethys is best known as the ocean that separated the vast U-shaped arms of the Pangean supercontinent beginning in the Permian. The Laurentian masses of North America, Baltica, Siberia, and Kazakhstan were located north of the westward tapering and generally equatorial Tethys Seaway, whereas the Gondwanan landmasses of Africa, Arabia, India, Australia, and Antarctica formed the southern arm of the supercontinent. During the Late Permian (270 million years ago) large blocks rifted from the Gondwanan margin and migrated north to collide with Eurasia forming accretionary orogens and the complex structure of the Eurasian continent. By 40 million years ago Gondwana had moved north to close

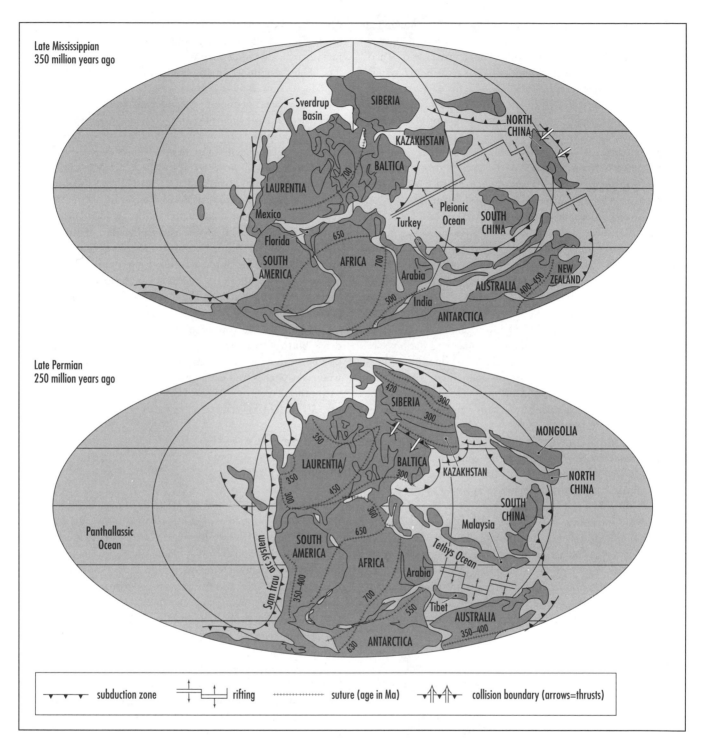

Simplified paleogeographic maps of the closure of the Tethys Ocean

the remaining parts of Neo-Tethys, forming the Alpine-Himalayan mountain chain.

See also PANGEA; PANTHALASSA.

Further Reading

Sengör, A. M. Celal. "The Story of Tethys: How Many Wives Did Okeanos Have?" *Episodes*, v. 8 (1985): 3–12.

thermodynamics Thermodynamics is the study of the transformation of heat into and from other forms of energy. It forms the basis of many principles of chemistry, physics, and earth sciences. The First Law of Thermodynamics states that energy can be neither created nor destroyed, and that heat and mechanical work are mutually convertible. The Second Law of Thermodynamics states that it is impossible for an unaided self-acting machine to transfer heat from a low-temperature body to a higher-temperature body. The Third Law of Thermodynamics states that it is impossible to reduce any system to absolute zero temperature (0°K, ⁻273°C, or ⁻459°F).

Energy is the capacity to do work, and it can exist in many different forms. Potential energy is energy of position, such as when an elevated body exhibits gravitational potential in that it can move to a lower elevation under the influence of gravity. Kinetic energy is the energy of motion and can be measured as the mean speed of the constituent molecules of a body. Einstein's Theory of Relativity showed that mass too can be converted to energy, as $E = mc^2$, where E = energy, m = mass, and c = the speed of light. This remarkable relationship forms the basis of atomic power and many mysteries of the universe.

Heat is a form of kinetic energy that manifests itself as motion of the constituent atoms of a substance. According to the laws of thermodynamics, heat may be transferred only from high-temperature bodies to lower-temperature bodies, and it does so by convection, conduction, or radiation. The specific heat of a substance is the ratio of the quantity of heat required to raise the temperature of a unit mass of the substance through a given range of temperature to the heat required to raise the temperature of an equal mass of water through the same range.

Conduction is the flow of heat through a material without the movement of any part of the material. The heat is transferred as kinetic energy of the vibrating molecules, which is passed from one molecule or atom to another. Convection is the transfer of heat through a fluid (liquid, gas, or slow-moving solid such as the Earth's mantle) by moving currents. Radiation is a heat transfer mechanism by infrared rays. All materials radiate heat, but hotter objects emit more heat energy than cold objects. Infrared radiation can pass through a vacuum and operates at the speed of light. Radiative heat can be reflected and refracted across boundaries but does not affect the medium through which it passes.

See also MANTLE.

thermohaline circulation   Vertical mixing of seawater driven by density differences caused by variations in temperature and salinity. Variations in temperature and salinity are found in waters that occupy different ocean basins, and those found at different levels in the water column. When the density of water at one level is greater than or equal to that below that level, the water column becomes unstable and the denser water sinks, displacing the deeper, less-dense water below. When the dense water reaches the level at which it is stable it tends to spread out laterally and form a thin sheet, forming intricately stratified ocean waters. Thermohaline circulation is the main mechanism responsible for the movement of water out of cold polar regions, and so it exerts a strong influence on global climate. The upward movement of water in other regions balances the sinking of dense cold water, and these upwelling regions typically bring deep water, rich in nutrients, to the surface. Thus, regions of intense biological activity are often associated with upwelling regions.

The coldest water on the planet is formed in the polar regions, with large quantities of cold water originating off the coast of Greenland and in the Weddell Sea of Antarctica. The planet's saltiest ocean water is found in the Atlantic Ocean, and this is moved northward by the Gulf Stream. As this water moves near Greenland it is cooled and then sinks to flow as a deep cold current along the bottom of the western North Atlantic. The cold water of the Weddell Sea is the densest on the planet, where surface waters are cooled to -35.4°F (-1.9°C), then sink to form a cold current that moves around Antarctica. Some of this deep cold water moves northward into all three major ocean basins, mixing with other waters and warming slightly. Most of these deep ocean currents move at a few to 10 centimeters per second.

Ocean bottom topography exerts a strong influence on dense bottom currents. Ridges deflect currents from one part of a basin to another and may restrict access to other regions, whereas trenches and deeps may focus flow from one region to another.

See also CLIMATE CHANGE; OCEAN CURRENTS.

thermosphere   *See* ATMOSPHERE.

thrust   *See* FAULT.

thunderstorms   Any storm that contains lightning and thunder may be called a thunderstorm. However, the term normally implies a gusty heavy rainfall event with numerous lightning strikes and thunder, emanating from a cumulonimbus cloud or cluster or line of cumulonimbus clouds. There is a large range in the severity of thunderstorms from minor to severe, with some causing extreme damage through high winds, lightning, tornadoes, and flooding rains.

Thunderstorms are convective systems that form in unstable rising warm and humid air currents. The air may start rising as part of a converging air system, along a frontal system, as a result of surface topography, or from unequal surface heating. The warmer the rising air is than the surrounding air, the greater the buoyancy forces acting on the rising air. Scattered thunderstorms that typically form in summer months are referred to as ordinary thunderstorms, and these typically are short-lived, only produce minor to moderate rainfall, and do not have severe winds. However, severe thunderstorms associated with fronts or combinations of unstable conditions may have heavy rain, hail, strong winds or tornadoes, and drenching or flooding rains.

Ordinary thunderstorms are most likely to form in regions where surface winds converge causing parcels of air to rise, and where there is not significant wind shear or changes in the wind speed and direction with height. These storms evolve through several stages beginning with the cumulus or growth stage, where the warm air rises and condenses into cumulus clouds. As the water vapor condenses it releases large amounts of latent heat that keeps the cloud warmer than the air surrounding it and causes it to continue to rise and build as long as it is fed from air below. Simple cumulus clouds may quickly grow into towering cumulus congestus clouds in this way. As the cloud builds above the freezing level in the atmosphere, the particles in the cloud get larger and heavier and eventually are too large to be kept entrained in the air currents, and they fall as precipitation. As this precipitation is falling drier air from around the storm is drawn into the cloud, but as the rain falls through this dry air it may evaporate, cooling the air. This cool air is then denser than the surrounding air and it may fall as a sudden downdraft, in some cases enhanced by air pulled downward by the falling rain.

The development of downdrafts marks the passage of the thunderstorm into the mature stage in which the upward and downward movement of air constitutes a convective cell. In this stage the top of the storm typically bulges outward in stable levels of the stratosphere, often around 40,000 feet (12 km), forming the anvil shape characteristic of mature thunderstorms. Heavy rain, hail, lightning, and strong, turbulent winds may come out of the base of the storms, which can be several miles in diameter. Cold downwelling air often expands out of the cloud base forming a gust front along its leading edge, forcing warm air up into the storm. Most mature storm cells begin to dissipate after half an hour or so, as the gust front expands away from the storm and can no longer enhance the updrafts that feed the storm. These storms may quickly turn into gentle rains, and then evaporate, but the moisture may be quickly incorporated into new, actively forming thunderstorm cells.

Severe thunderstorms are more intense than ordinary storms, producing large hail, wind gusts of greater than 50 knots (57.5 mi/hr, or 92.5 km/hr), more lightning, and heavy rain. Like ordinary thunderstorms, severe storms form in areas of upwelling unstable moist warm air, but severe storms tend to develop in regions where there is also strong wind shear. The high level winds have the effect of causing the rain that falls out of the storm to fall away from the region of upwelling air so that it does not have the effect of weakening the upwelling. In this way the cell becomes much longer lived and grows stronger and taller than ordinary thunderstorms, often reaching heights of 60,000 feet (18 km). Hail may be entrained for long times in the strong air currents and even thrown out of the cloud system at height, falling several kilometers from the base of the cloud. Downdrafts from severe storms are marked by bulbous mammatus clouds.

Supercell thunderstorms form where strong wind shear aloft is such that the cold downwelling air does not cut off the upwelling air, and a giant rotating storm with balanced updrafts and downdrafts may be maintained for hours. These storms may produce severe tornadoes, strong downbursts of wind, large (grapefruit-sized) hail, very heavy rains, and strong winds exceeding 90 knots (103.5 mi/hr, or 167 km/hr).

Unusual winds are associated with some thunderstorms, especially severe storms. Gust fronts maybe quite strong with winds exceeding 60 miles per hour (97 km/hr), followed by cold gusty and shifty winds. Gust fronts may be marked by lines of dust kicked up by the strong winds, or ominous-looking shelf clouds formed by warm moist air rising above the cold descending air of the gust front. In severe cases, gust fronts may force so much air upward that they generate new multicelled thunderstorms with their own gust fronts that merge, forming an intense gust front called an outflow boundary. Intense downdrafts beneath some thunderstorms spread laterally outward at speeds sometimes exceeding 90 miles per hour (145 km/hr) when they hit the ground and are termed downbursts, microbursts, or macrobursts depending on their size. Some clusters of thunderstorms produce another type of unusual wind called a straight-line wind, or derecho. These winds may exceed 90 miles per hour and extend for tens or even hundreds of miles.

Thunderstorms often form in groups called mesoscale convective systems, or as lines of storms called squall lines. Squall lines typically form along or within a zone up to a couple of hundred miles in front of the cold front where warm air is compressed and forced upward. Squall lines may form lines of thunderstorms hundreds or even a thousand miles long, and many of the storms along the line may be severe with associated heavy rain, winds, hail, and tornadoes. Mesoscale convective complexes form when many individual thunderstorm cells across a region start to act together, forming an exceedingly large convective system that may cover more than 50,000 square miles (130,000 km²). These systems move slowly and may be associated with many hours of flooding rains, hail, tornadoes, and wind.

Cumulonimbus clouds typically become electrically charged during the development of thunderstorms, although the processes that lead to the unequal charge distribution are not well known. About 20 percent of the lightning generated in thunderstorms strikes the ground, with most passing from cloud to cloud. Lightning is an electrical discharge that heats the surrounding air to 54,000°F (30,000°C) causing the air to expand explosively, causing the sound waves we hear as thunder. As the air expands along different parts of the lightning stroke the sound is generated from several different places, causing the thunder to have a rolling or echoing sound, enhanced by the sound waves bouncing off hills, buildings, and the ground. Cloud-to-ground lightning forms when negative electrical charges build up in the base of the cloud, causing positive charges to build in the ground. When the electrical potential gradient reaches three million volts per

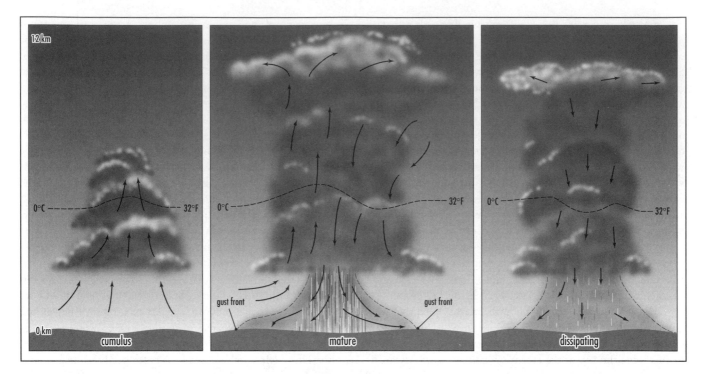

12 km

0°C — — — — — — — — — 32°F

0 km

cumulus

gust front   gust front

0°C — — — — — — — — 32°F

mature

0°C — — — — — — — — 32°F

dissipating

Schematic growth and evolution of a typical thunderstorm showing cumulus clouds, up and down drafts, gust fronts, and freezing isotherm

meter along several tens of meters, electrons rush to the cloud base and form a series of stepped leaders that reach toward the ground. At this stage, a strong current of positive charge moves up, typically along an elevated object, from the ground to the descending leader. As the two columns meet, huge numbers of electrons rush to the ground, and a several-centimeter-wide column of positively charged ions shoots up along the lightning stroke, all within a ten-thousandth of a second. The process then may be repeated several or even dozens of times along the same path, all within a fraction of a second.

See also PRECIPITATION; TORNADOES.

Further Reading

Ahrens, C. Donald. *Meteorology Today, An Introduction to Weather, Climate, and the Environment, Seventh Edition.* Pacific Grove, Calif.: Thomson Brooks/Cole, 2003.

Schaefer, Vincent, and John Day. *A Field Guide to the Atmosphere, The Peterson Field Guide Series.* Boston: Houghton Mifflin, 1981.

Tibetan Plateau   The Tibetan Plateau is the largest high area of thickened continental crust on Earth, with an average height of 16,000 feet (4,880 m) over 470,000 square miles (1,220,000 km²). Bordered on the south by the Himalayan Mountains, the Kunlun Mountains in the north, the Karakorum on the west, and the Hengduan Shan on the east, Tibet is the source of many of the largest rivers in Asia. The Yangtze, Mekong, Indus, Salween, and Brahmaputra Rivers all rise in Tibet and flow through Asia, forming the most important source of water and navigation for huge regions.

Southern Tibet merges into the foothills of the northern side of the main ranges of the Himalaya, but they are separated from the mountains by the deeply incised river gorges of the Indus, Sutlej, and Yarlung Zangbo (Brahmaputra) Rivers. Central and northern Tibet consist of plains and steppes that are about 3,000 feet (1,000 m) higher in the south than the north. Eastern Tibet includes the Transverse Ranges (the Hengduan Shan) that are dissected by major faults in the river valleys of the northwest-southeast flowing Mekong, Salween, and Yangtze Rivers.

Tibet has a high plateau climate, with large diurnal and monthly temperature variations. The center of the plateau has an average January temperature of 32°F (0°C) and an average June temperature of 62°F (17°C). The southeastern part of the plateau is affected by the Bay of Bengal summer monsoons, whereas other parts of the plateau experience severe storms in fall and winter months.

Geologically, the Tibetan Plateau is divided into four terranes, including the Himalayan terrane in the south, and the Lhasa terrane, the Qiangtang terrane, and Songban-Ganzi composite terrane in the north. The Songban-Ganzi terrane includes Triassic flysch and Carboniferous-Permian sedimentary rocks, and a peridotite-gabbro-diabase sill complex that may be an ophiolite, overlain by Triassic flysch. Another fault-bounded section includes Paleozoic limestone and marine clastics, probably deposited in an extensional basin. South of the Jinsha suture, the Qiangtang terrane contains Precambrian basement overlain by Early Paleozoic sediments that are up to 12 miles (20 km) thick. Western parts of the Qiangtang terrane contain Gondwanan tillites and Triassic-Jurassic coastal swamp and shallow marine sedimentary rocks. Late Jurassic–Early Cretaceous deformation uplifted these rocks, before they were unconformably overlain by Cretaceous strata.

The Lhasa terrane collided with the Qiangtang terrane in the Late Jurassic and formed the Bangong suture, containing flysch and ophiolitic slices, that now separates the two terranes. It is a composite terrane containing various pieces that rifted from Gondwana in the Late Permian. Southern parts of the Lhasa terrane contain abundant Upper Cretaceous to Paleocene granitic plutons and volcanics, as well as Paleozoic carbonates and Triassic-Jurassic shallow marine deposits. The center of the Lhasa terrane is similar to the south but with fewer magmatic rocks, whereas the north contains Upper Cretaceous shallow marine rocks that onlap the Upper Jurassic–Cretaceous suture.

The Himalayan terrane collided with the Lhasa terrane in the Middle Eocene forming the ophiolite-decorated Yarlungzangbo suture. Precambrian metamorphic basement is thrust over Sinian through Tertiary strata including Lower Paleozoic carbonates and Devonian clastics, overlain uncon-

formably by Permo-Carboniferous carbonates. The Himalayan terrane contains Lower Permian Gondwanan flora and probably represents the northern passive margin of Mesozoic India, with carbonates and clastics in the south, thickening to an all-clastic continental rise sequence in the north.

The Indian plate rifted from Gondwana and started its rapid (3.2–3.5 inches per year, 80–90 mm/yr) northward movement about 120 million years ago. Subduction of the Indian plate beneath Eurasia until about 70 million years ago formed the Cretaceous Kangdese batholith belt containing diorite, granodiorite, and granite. Collision of India with Eurasia at 50–30 million years ago formed the Lhagoi-Khangari of biotite and alkali granite and the 20–10-million-year-old Himalayan belt of tourmaline-muscovite granites.

Tertiary faulting in Tibet is accompanied by volcanism, and the plateau is presently undergoing east-west extension with the formation of north-south graben associated with hot springs and probably deep magmatism. Seismic reflection profiling has detected some regions with unusual characteristics beneath some of these graben, interpreted by some seismologists as regions of melt or partially molten crust.

Much research has been focused on the timing of the uplift of the Tibetan Plateau and modeling the role this uplift

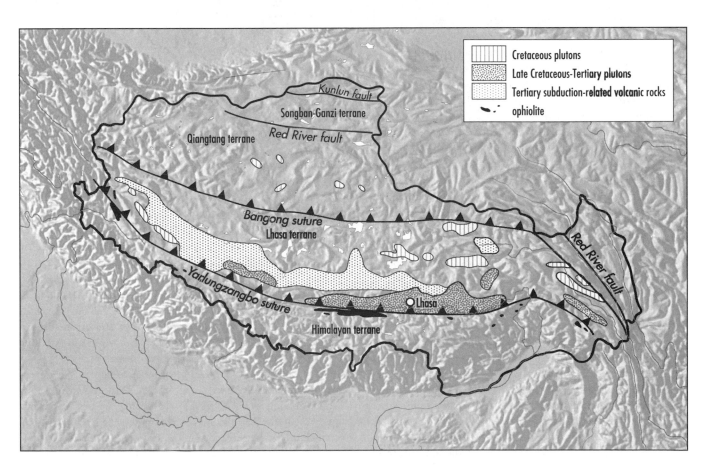

Map of tectonic terranes in Tibet

has had on global climate. The plateau strongly affects atmospheric circulation, and many models suggest that the uplift may contribute to global cooling and the growth of large continental ice sheets in latest Tertiary and Quaternary times. In addition to immediate changes to airflow patterns around the high plateau, the uplift of large amounts of carbonate platform and silicate rocks expose them to erosion. The weathering of these rocks causes them to react with atmospheric carbon dioxide, which combines these ions to produce bicarbonate ions such as $CaCO_3$, drawing down the atmospheric carbon dioxide levels and contributing to global cooling.

The best estimate of the time of collision between India and Asia is between 54 million and 49 million years ago. Since then convergence between India and Asia has continued, but at a slower rate of 1.6–2.0 inches per year (40–50 mm/yr), and this convergence has resulted in intense folding, thrusting, shortening, and uplift of the Tibetan Plateau. Timing the uplift to specific altitudes is difficult, and considerable debate has centered on how much younger than 50 million years ago the plateau reached its current height of 16,404 feet (5 km). Most geologists would now agree that this height was attained by 13.5 million years ago, and that any additional height increase is unlikely since the strength of the rocks at depth has been exceeded, and the currently active east-west extensional faults are accommodating any additional height increase by allowing to crust to flow laterally.

When the plateau reached significant heights it began to deflect regional airflow currents that in turn deflect the jet streams, causing them to meander and change course. Global weather patterns were strongly changed. In particular, the cold polar jet stream is now at times deflected southward over North America, northwest Europe, and other places where ice sheets have developed. The uplift increased aridity in central Asia by blocking moist airflow across the plateau, leading to higher summer and cooler winter temperatures. The uplift also intensified the Indian Ocean monsoon, because the height of the plateau intensifies temperature-driven atmospheric flow as higher and lower pressure systems develop over the plateau during winter and summer. This has increased the amount of rainfall along the front of the Himalaya Mountains, where some of the world's heaviest rainfalls have been reported, as the Indian monsoons are forced over the high plateau. The cooler temperatures on the plateau led to the growth of glaciers, which in turn reflect back more sunlight, further adding to the cooling effect.

Paleoclimate records show that the Indian Ocean monsoon underwent strong intensification 7–8 million years ago, in agreement with some estimates of the time of uplift, but younger than other estimates. The effects of the uplift would be different if the uplift occurred rapidly in the Late Pliocene–Pleistocene (as suggested by analysis of geomorphology, paleokarst, and mammal fauna), or if the uplift occurred gradually since the Eocene (based on lake sediment analysis).

Most geologists accept analysis of data that suggests that uplift began about 25 million years ago, with the plateau reaching its current height by 14 million or 15 million years ago. These estimates are based on the timing of the start of extensional deformation that accommodated the exceptional height of the plateau, sedimentological records, and on uplift histories based on geothermometry and fission track data.

See also ATMOSPHERE; FLOOD; HIMALAYA MOUNTAINS; SUPERCONTINENT CYCLES.

Further Reading
Chang, Chen Fa, Yu Sheng Pan, and Yi Ying Sun. "The Tectonic Evolution of Qinghai-Tibet Plateau: a Review." In *Tectonic Evolution of the Tethyan Region,* edited by A. M. Celal Sengör. Dordrecht: Kluwer Academic Publishers, 1989.
Dewey, John F., Robert M. Shakleton, Chen Fa Chang, and Yi Ying Sun. "The Tectonic Evolution of the Tibet Plateau." In *Tectonic Evolution of the Himalayas and Tibet,* edited by Robert Shakleton, John F. Dewey, and Brian F. Windley. Philosophical Transactions of the Royal Society of London Series A., 1988.
Molnar, Peter, and P. Tapponier. "Active Tectonics of Tibet." *Journal of Geophysical Research* (1978).
Raymo, Maureen E., and William F. Ruddiman. "Tectonic Forcing of Late Cenozoic Climate." *Nature* 359 (1992).
Ruddiman, William F. *Tectonic Uplift and Climate Change.* New York: Plenum Press, 1997.

tidal wave *See* TSUNAMI.

tides The periodic rise and fall of the ocean surface, and alternate submersion and exposure of the intertidal zone along coasts. Currents caused by the rise and fall of the sea surface are the strongest currents in the ocean and were attributed to the gravitational effects of the Sun and Moon since at least the times of Pliny the Elder (23–79 C.E.).

The range in sea surface height between the high and low is known as the tidal range, and this varies considerably from barely detectable to more than 50 feet (15 m). Most places have two high tides and two low tides each tidal day, a period of about 24 hours and 50 minutes, corresponding to the time between successive passes of the Moon over any point. The tidal period is the time between successive high or low tides. Places with two high and two low tides per day have semidaily or semidiurnal tides. Fewer places have only one high and one low tide per day, a cycle referred to as a diurnal or daily tide. Semidiurnal tides are often not equal in heights between the two highs and two lows.

Spring tides are those that occur near the full and new Moons and have a tidal range larger than the mean tidal range. In contrast, neap tides occur during the first and third quarters of the Moon and are characterized by lower than average tidal ranges.

Sir Isaac Newton was the first to clearly elucidate the mechanics of tides, and how they are related to the gravita-

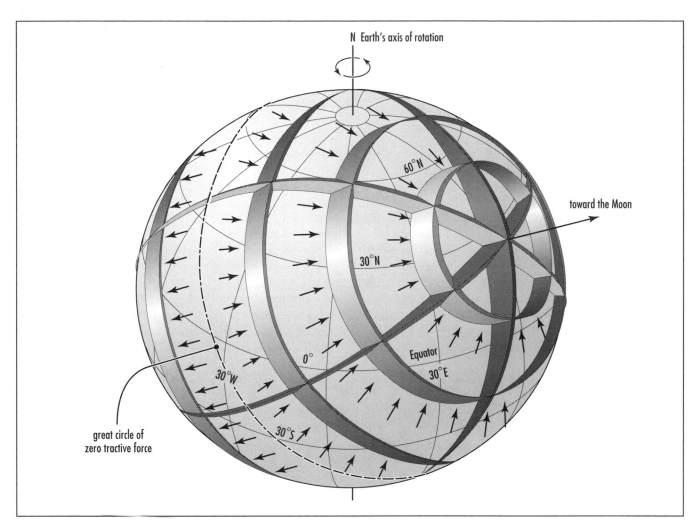

Equilibrium tide model showing the horizontal component of tide-producing forces in small arrows for when the Moon is over 30°N latitude. The large fences represent the height of the tidal bulge at any point on the Earth. The Earth rotates through these tidal bulges, which remain essentially stationary.

tional attraction of the Moon. In his equilibrium theory of tides he assumed a nonrotating Earth, covered with water and having no continents. In this simplified model aimed at understanding the origin of tides, gravitational attraction pulls the Earth and Moon toward each other, while centrifugal forces act in the opposite direction and keep them apart. Since the Moon is so much smaller than the Earth, the center of mass and rotation of the Earth-Moon system is located within the Earth 2,900 miles (4,670 km) from the Earth's center, on the side of the Earth closest to the Moon. This causes unbalanced forces since a unit of water on the Earth's surface closest to the Moon is located 59 Earth radii from the Moon's surface, whereas a unit of water on the opposite side of the Earth is located 61 Earth radii from the nearest point on the Moon. Since the force of gravity is inversely proportional to the distance squared between the two points, the Moon's gravitational pull is much greater for the unit of

water closer to the Moon. However, centrifugal forces that act perpendicular to the axis of rotation of the Earth also affect the tides and must be added with the gravitational forces to yield a vector sum that is the tide producing force. Together these forces result in the gravitational force of the Moon exceeding the centrifugal force on the side of Earth closest to the Moon, drawing water in a bulge toward the Moon. On the opposite side of the Earth the centrifugal force overbalances the gravitational attraction of the Moon so there the water is essentially dragged away from the Earth.

The interaction of the gravitational forces and centrifugal forces creates a more complex pattern of tides on Newton's model Earth. Directly beneath the Moon and on the opposite side of the Earth, both the gravitational force and the centrifugal force act perpendicular to the surface, but elsewhere the vector sum of the two forces is not perpendicular to the surface. The result of adding the centrifugal force and gravity vectors is a two-sided egg-shaped bulge that points toward and away from the Moon. Newton called these bulges the equilibrium tide. The situation, however, is even more complex, since the Sun also exerts a gravitational attraction on the Earth and its water, forming an additional egg-shaped bulge that is about 0.46 times as large as the lunar tidal bulge.

If we consider the Earth to be rotating through the tidal bulges on a water-covered planet, the simplest situation arises with two high tides and two low tides each day, since the lunar tides dominate over the effects of the solar tides. However, the Earth has continents that hinder the equal flow of water, and bays and estuaries that trap and amplify the tides in certain places, plus frictional drag slows the passage of the tidal bulge through shallow waters. In addition, the Coriolis Force must be taken into account as tides involve considerable movement of water from one place to another. These obstacles cause the tides to be different at different places on the Earth, explaining the large range in observed tidal ranges and periods.

tornadoes A rapidly circulating column of air with a central zone of intense low pressure that reaches the ground. Most tornadoes extend from the bottom of severe thunderstorms or supercells as funnel-shaped clouds that kick up massive amounts of dust and debris as they rip across the surface. The exact shape of tornadoes is quite variable, from thin rope-like funnels, to classic cylindrical shapes, to powerful and massive columns that have almost the same diameter on the ground as at the base of the cloud. Many tornadoes evolve from an immature upward swirling mass of dust to progressively larger funnels that may shrink and tilt as their strength diminishes. Funnel clouds are essentially tornadoes that have not reached the ground. More tornadoes occur in the United States than anywhere else in the world, with most of these occurring in a region known as "tornado alley," extending from Texas through Oklahoma, Nebraska, Kansas, Iowa, Missouri, and Arkansas.

Most tornadoes rotate counterclockwise (as viewed from above) and have diameters from a few hundred feet to huge systems with diameters of a mile or more. Wind speeds in tornadoes range from about 40 miles per hour to more than

300 miles per hour (65–480 km/hr) and may move forward at a few miles per hour to more than 70 miles per hour (2–115 km/hr). Most tornadoes last for only a few minutes but some last longer, with some reports of massive storms lasting for hours and leaving trails of destruction hundreds of miles long. Some supercell thunderstorms produce families or outbreaks of tornadoes with half a dozen or more individual funnel clouds produced, over the course of a couple of hours, from a single storm.

The strength of winds and potential damage of tornadoes is measured by the Fujita scale, proposed by the tornado expert Dr. Theodore Fujita. The scale measures the rotational speed (not the forward speed) and classifies the tornadoes into F0–F5 categories.

Many tornadoes form in the mid-western states' tornado alley in the springtime when warm moist air from the Gulf of Mexico is overrun by cold air from the north. Tornadic supercell thunderstorms form in front of the cold front as the warm moist air is forced upward in front of the cold air in this region. Many tornadoes are also spawned by supercell thunderstorms that have large rotating updrafts. Spinning roll clouds and vortexes may form as these storms roll across the plains, and if these horizontally spinning clouds are sucked into the storm by an updraft, the circulation may be rotated to form a tornadic condition that may evolve into a tornado. Before the supercell spawns a tornado, rotating clouds may be visible, and then a wall cloud may descend from the rotating vortex. Funnel clouds are often hidden behind the wall cloud, so these types of clouds should be eyed with caution.

Some tornadoes have formed from smaller and even non-severe thunderstorms, from squall lines, and even smaller cumulus clouds. These types of tornadoes are usually short-lived and less severe (F0–F1) than the supercell tornadoes. Waterspouts are related phenomena and include tornadoes that have migrated over bodies of water, and they may also form in fair weather over warm shallow coastal waters. These weak (F0) funnel clouds form in updrafts, usually when cumulus clouds are beginning to form above the coastal region. Their formation is aided by converging surface air, such as when sea breezes and other systems meet.

See also THUNDERSTORMS.

Fujita Tornado Scale

Scale	Category	Wind Speed	Damage Potential
F0	weak	40–72 mph	Minor; broken tree branches, damaged signs
F1	weak	73–112 mph	Moderate; broken window, trees snapped
F2	strong	113–157 mph	Considerable; large trees uprooted, mobile homes tipped, weak structures destroyed
F3	strong	158–206 mph	Severe; trees leveled, walls torn from buildings, cars flipped
F4	violent	207–260 mph	Devastating; frame homes destroyed
F5	violent	261–318 mph	Incredible; strong structures damaged, cars thrown hundreds of yards

trace fossil   *See* BIOGENIC SEDIMENT; ICHNOFOSSILS.

trade winds   Steady winds that blow from the northeast to southwest in the northern hemisphere and between 0° and 30° latitude, and from southeast to northwest in the southern hemisphere. The trade winds are formed as the cool air from Hadley Cell circulation returns to the surface at about 15°–30° latitude and then returns to the equatorial region. The Coriolis Force deflects the moving air to the right in the Northern Hemisphere, causing the air to flow from northeast to southwest, and to the left in the Southern Hemisphere, causing a southeast to northwest flow. They are named trade winds because sailors used the reliability of the winds to aid their travels from Europe to the Americas. The doldrums, an area characterized by weak stagnant air currents, bound the trade winds on high latitudes by the horse latitudes, characterized by weak winds, and toward the equator.

See also ATMOSPHERE; HADLEY CELLS.

transform plate margin processes   Processes that occur where two plates are sliding past each other along a transform plate boundary, either in the oceans or on the continents. Famous examples of transform plate boundaries on land include the San Andreas fault in California, the Dead Sea transform in the Middle East, the East Anatolian transform in Turkey, and the Alpine fault in New Zealand. Transform boundaries in the oceans are numerous, including the many transform faults that separate segments of the mid-ocean ridge system. Some of the larger transform faults in the oceans include the Romanche in the Atlantic, the Cayman fault zone on the northern edge of the Caribbean plate, and the Eltanin, Galapagos, Pioneer, and Mendocino fault zones in the Pacific Ocean.

There are three main types of transform faults, including those that connect segments of divergent boundaries (ridge-ridge transforms), offsets in convergent boundaries, and those that transform the motion between convergent and divergent boundaries. Ridge-ridge transforms connect spreading centers and develop because in this way, they minimize the ridge segment lengths and minimize the dynamic resistance to spreading. Ideal transforms have purely strike-slip motions and maintain a constant distance from the pole of rotation for the plate.

Transform segments in subduction boundaries are largely inherited configurations formed in an earlier tectonic regime. In collisional boundaries the inability of either plate to be subducted yields a long-lived boundary instability, often formed to compensate the relative motion of minor plates in complex collisional zones, such as that between Africa and Eurasia.

The development of a divergent-convergent transform boundary is best represented by the evolution of the San Andreas–Fairweather fault system. When North America overrode the East Pacific rise, the relative velocity structure

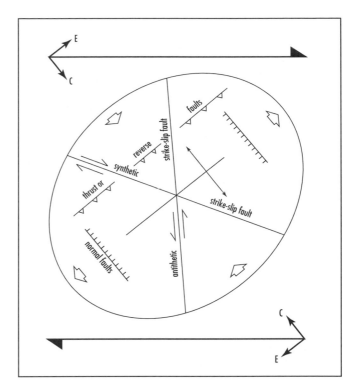

Orientation of structures in transform margins, including strike-slip faults, normal faults, thrust faults, and folds

was such that a transform resulted, with a migrating triple junction that lengthened the transform boundary.

Transform Boundaries in the Continents
Transform boundaries on the continents include the San Andreas fault in California, the North Anatolian fault in Turkey, the Alpine fault in New Zealand, and, by some definitions, the Altyn Tagh and Red River faults in Asia. Transform faults in continents show strike-slip offsets during earthquakes and are high angle faults with dips greater than 70°. They never occur as a single fault but rather as a set of subparallel faults. The faults are typically subparallel because they form along theoretical slip lines (along small circles about the pole of rotation), but the structural grain of the rocks interferes with this prediction. The differences between theoretical and actual fault orientations lead to the formation of segments that have pure strike-slip motions and segments with compressional and extensional components of motion.

Extensional segments of transform boundaries form at left steps in left-slipping (left lateral) faults and at right steps in right-slipping (right lateral) faults. Movement along fault segments with extensional bends generates gaps where deep basins known as pull-apart basins form. There are presently about 60 active pull-apart basins on the planet, including places like the Salton trough along the San Andreas fault, and the Dead Sea along the Dead Sea transform. Pull-apart

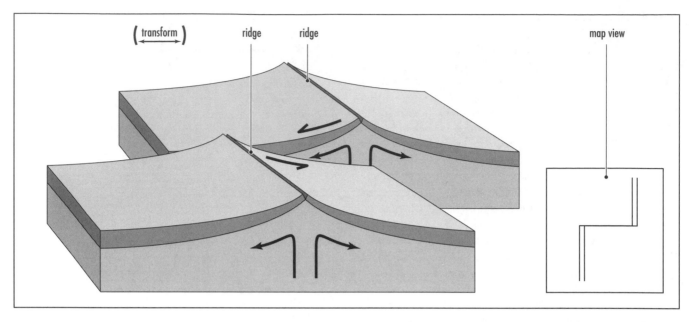

Three-dimensional view of a transform fault in the ocean basin, apparently offsetting a segment of the mid-ocean ridge. The sense of motion on the transform is opposite the apparent offset. Note that the lateral motion between the two segments of the oceanic crust ceases once the opposite ridge segment is passed. At this point, magmas from the ridge intrude the transform, and the contact becomes igneous.

basins tend form with an initially sigmoidal form, but as movement on the fault continues, the basin becomes very elongate parallel to the bounding faults. In some cases the basin may extend so much that oceanic crust is generated in the center of the pull-apart, such as along the Cayman fault in the Caribbean. Pull-apart basins have stratigraphic and sedimentologic characteristics similar to rifts, including rapid lateral facies variations, basin-marginal fanglomerate and conglomerate deposits, interior lake basins, and local bimodal volcanic rocks. They are typically deformed soon after they form, however, with folds and faults typical of strike-slip regime deformation.

Compressional bends form at right bends in left lateral faults and left steps in right lateral faults. These areas are characterized by mountain ranges and thrust-faulted terrain that uplift and aid erosion of the extra volume of crust compressed into the bend in the fault. Examples of compressional (or restraining) bends include the Transverse Ranges along the San Andreas fault, and Mount McKinley along the Denali fault in Alaska. Many of the faults that form along compressional bends have low-angle dips away from the main strike-slip fault but progressively steeper dips toward the center of the main fault. This forms a distinctive geometry known as a flower or palm tree structure, with a vertical strike-slip fault in the center and branches of mixed thrust/strike-slip faults branching off the main fault.

In a few places along compressional bends, two thrust-faulted mountain ranges may converge, forming a rapidly subsiding basin between the faults. These basins are known as

ramp valleys. Many ramp valleys started as pull-apart basins and became ramp valleys when the fault geometries changed.

A distinctive suite of structures that form in predictable orientations characterizes transform plate margins. Compressional bends form at high angles to the principal compressive stress, and at about 30°–45° from the main strike-slip zone. These are often associated with flower structures, containing a strike-slip fault at depth, and folds and thrusts near the surface. Dilational bends often initiate with their long axes perpendicular to the compressional bends, but large amounts of extension may lead to the long axis being parallel to the main fault zone. Folds, often arranged in en echelon or a stepped manner, typically form at about 45° from the main fault zone, with the fold axes developed perpendicular to the main compressive stress. The sense of obliquity of many of these structures can be used to infer the sense of shear along the main transform faults.

Strike-slip faults along transform margins often develop from a series of echelon fractures that initially develop in the rock. As the strain builds up, the fractures are cut by new sets of fractures known as Riedel fractures in new orientations. Eventually after several sets of oblique fractures have cut the rock, the main strike-slip fault finds the weakest part of the newly fractured rock to propagate through, forming the main fault.

Transform Boundaries in the Oceans
Transform plate boundaries in the oceans include the system of ridge-ridge transform faults that are an integral part of the

mid-ocean ridge system. Magma upwells along the ridge segments, cools, and crystallizes, becoming part of one of the diverging plates. The two plates then slide past each other along the transform fault between the two ridge segments, until the plate on one side of the transform meets the ridge on the other side of the transform. At this point, the transform fault is typically intruded by mid-ocean ridge magma, and the apparent extension of the transform, known as a fracture zone, juxtaposes two segments of the same plate that move together horizontally. Fracture zones are not extensions of the transform faults and are no longer plate boundaries. After the ridge/transform intersection is passed, the fracture zone juxtaposes two segments of the same plate. There is typically some vertical motion along this segment of the fracture zone, since the two segments of the plate have different ages and subside at different rates.

The transforms and ridge segments preserve an orthogonal relationship in almost all cases, because this geometry creates a least work configuration, creating the shortest length of ridge possible on the spherical Earth.

Transform faults generate very complex geological relationships. They juxtapose rocks from very different crustal and even mantle horizons, show complex structures, alteration by high-temperature metamorphism, and have numerous igneous intrusions. Rock types along oceanic transforms typically include suites of serpentinite, gabbro, pillow lavas, lherzolites, harzburgites, amphibolite-tectonites, and even mafic granulites.

Transform faults record a very complex history of motion between the two oceanic plates. The relative motion includes dip-slip (vertical) motions due to subsidence related to the cooling of the oceanic crust. A component of dip-slip motion occurs all along the transform, except at one critical point, known as the crossover point, where the transform juxtaposes oceanic lithosphere of the same age formed at the two different ridge segments. This dip-slip motion occurs along with the dominant strike-slip motion, recording the sliding of one plate past the other.

Fracture zones are also called the non-transform extension region. The motion along the fracture zone is purely dip-slip, due to the different ages of the crust with different subsidence rates on either side of the fracture zone. The amount of differential subsidence decreases with increasing distance from the ridge, and the amount of dip-slip motion decreases to near zero after about 60 million years. Subsidence decreases according to the square root of age.

Transform faults in the ocean may juxtapose crust with vastly different ages, thickness, temperature, and elevation. These contrasts often lead to the development of a deep topographic hole on the ridge axis at the intersection of the ridge and transform. The cooling effects of the older plate against the ridge of the opposing plate influence the axial rift topography all along the whole ridge segment, with the highest topographic point on the ridge being halfway between two transform segments. Near transform zones, magma will not reach its level of hydrostatic equilibrium because of the cooling effects of the older cold plate adjacent to it. Therefore, the types and amounts of magma erupted along the ridge are influenced by the location of the transforms.

Transform faults are not typically vertical planes, nor are they always straight lines connecting two ridge segments. The fault planes typically curve toward the younger plate with depth, since they tend to seek the shortest distance through the lithosphere to the region of melt. This is a least energy configuration, and it is easier to slide a plate along a vertically short transform than along an unnecessarily thick fault. This vertical curvature of the fault causes a slight change in the position and orientation of the fault on the surface, causing it to bend toward each ridge segment. These relationships cause the depth of earthquakes to decrease away from the crossover point, due to the different depth of transform fault penetration. Motion on these curved faults also influences the shape and depth of the transform-ridge intersection, enhancing the topographic depression and in many causing the ridge to curve slightly into the direction of the transform. Faults and igneous dikes also curve away from the strike of the ridge, toward the direction of the transform in the intersection regions.

Many of the features of ridge-transform intersections are observable in some ophiolite complexes (on-land fragments of ancient oceanic lithosphere), including the Arakapas transform in Troodos ophiolite in Cyprus, and the Coastal Complex in the Bay of Islands ophiolite in Newfoundland.

See also OPHIOLITES; PLATE TECTONICS; STRIKE-SLIP FAULT.

transgression A rise or landward migration of the shoreline caused by either a global sea-level rise, a fall in the land's surface, or a supply of sediment that is less than the space created for the sediment by subsidence. Transgressions may also be marked by a replacement of shallow water deposition to deep-water facies or a general landward shift in marine facies. Global sea-level rises and falls on different timescales depending on the cause. Changes in ridge volume or mantle plume activity cause slow changes to the ocean ridge volumes and slow rises or falls in sea level, whereas changes in the volume of continental glaciers may cause faster changes in the volume of water in the ocean. All of these may be related to the supercontinent cycle and cause sea-level regressions or transgressions. Local tectonic activity may cause the land surface to rise or fall relative to a stable global sea level, causing local regressions or transgressions. For instance, rapid subsidence caused by tectonic thinning or loading of the crust may cause the shoreline to migrate landward in a local transgression. To interpret patterns of global sea-level rise and fall it is necessary to isolate the effects of local tectonic subsidence or uplift, and sediment supply issues, from the global sea-level signature. This can be difficult and requires precise dating and correlation of events along different shorelines, plus a detailed understanding of the local tectonic and sedimentation history. When the local effects are isolated they can be subtracted from the global sea-level curve, and the causes of global sea-level changes can be investigated.

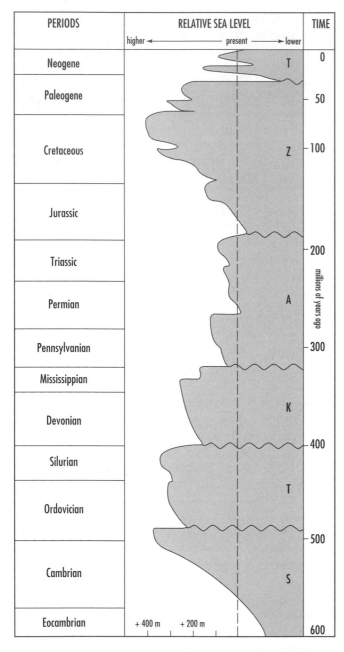

PERIODS	RELATIVE SEA LEVEL	TIME
	higher ◄————— present ————► lower	
Neogene		T — 0
Paleogene		— 50
Cretaceous		Z — 100
Jurassic		
Triassic		— 200
Permian		A
Pennsylvanian		— 300
Mississippian		
Devonian		K
Silurian		— 400
Ordovician		T
Cambrian		— 500
		S
Eocambrian	+ 400 m + 200 m	600

(vertical axis label: millions of years ago)

Plot showing the six major unconformity-bounded sequences for North America, formed when major transgressions occur. Sea levels have risen to more than 1,150 feet (350 m) above present levels and fallen to 655 feet (200 m) below present levels. The six major transgressive sequences are known as the Sauk (S), Tippicanoe (T), Kaskaskia (K), Absaroka (A), Zuni (Z), and Tejas (T).

Transgressive marine sequences record major marine advances over the land at several times in the Phanerozoic. Most transgressive sequences are preceded by an erosional unconformity and show a progressive landward shift in sedimentary facies that, according to Walther's Law, is also recorded in the vertical sequence. The base of transgressive sequences is typically marked by a beach sandstone or conglomerate unit, followed upward by an offshore muddy facies, then typically a deeper water limestone facies.

In the 1950s, using index fossils and isotopic dates of key horizons, an effort pioneered by Laurence L. Sloss and coworkers in the petroleum industry, correlated many transgressive sequences across North America and the world. Many large, laterally extensive rock units that are bounded by unconformities of regional or global significance were recognized and precisely dated in many places. Some of these unconformity-bounded sequences are so significant that they are found in almost all shallow water deposits of that age in the world. These sequences always occur where sea level has dropped from high to low, and the overlying sequence is transgressive. Index fossils were used to show that these unconformities have the same age on all continents and are clearly related to changes in sea level. Sea level has fluctuated by as much as 1,150 feet (350 m) higher than the present level and 655 feet (200 m) below the present level. Using these correlations, six major transgressive sequences have been recognized in the stratigraphic record of the continents. These transgressive sequences include the Eocambrian-Cambrian (600–500 Ma) Sauk Sequence, the Middle Ordovician–Lower Devonian (470–410 Ma) Tippicanoe Sequence, the Middle Devonian–Upper Mississippian (410–320 Ma) Kaskaskia Sequence, the Lower Pennsylvania–Lower Jurassic (320–185 Ma) Absaroka Sequence, the Middle Jurassic–Upper Cretaceous (185–30 Ma) Zuni Sequence, and the Tertiary-Recent (30–0 Ma) Tejas Sequence.

See also REGRESSION; SEQUENCE STRATIGRAPHY.

Triassic   The oldest of three Mesozoic periods, and the corresponding system of rocks. The Triassic period is bounded below by the Permian period of the Paleozoic era, and above by the Jurassic period. The time span by the Triassic ranges from 248 million years ago to 206 million years ago, divided into Early (248–242 Ma), Middle (242–227 Ma), and Late (227–206 Ma) epochs, and seven ages including, from oldest to youngest, the Induan, Olenekian, Anisian, Ladinian, Carnian, Norian, and Rhaetian.

The base of the Triassic is also the base of the Mesozoic, meaning middle life, so this boundary is marked by a profound change in the fossil biota in the stratigraphic sequence. The end of the Permian saw the extinction of 90 percent of marine organisms, followed by a re-radiation of the pelecypods, sea urchins, lobsters, and ammonoids. The ammonoids have proven very useful in subdividing the Triassic in dozens

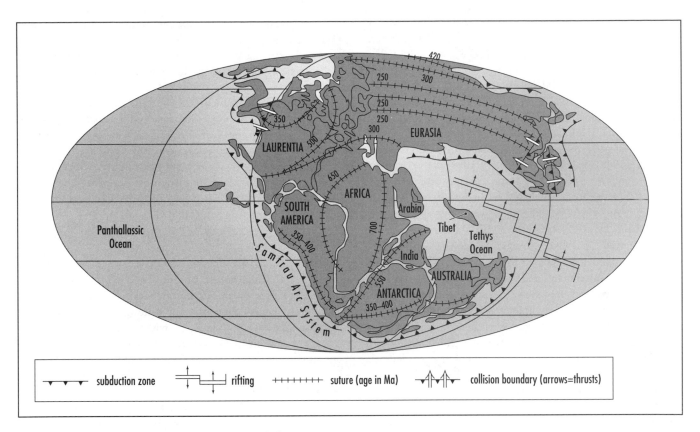

Late Triassic reconstruction of Tethys and the Panthalassic Ocean

of biozones, and the ages of the periods are under consideration for revision based on these higher resolution divisions. The first hexacorals appeared in the Early Triassic, whereas oysters appeared in the Ladinian, and the first dinosaurs are known from the Carnian. The first mammals, turtles, and crocodiles all appeared in the Late Triassic. Many extinctions occurred throughout the Triassic, including the loss of dozens of families of marine gastropods, cephalopods, brachiopods, bivalves, and sponges. Terrestrial extinctions include numerous families of insects, freshwater fish, and reptiles. Some models have suggested that these numerous extinctions created opportunities for mammals to radiate and succeed in a previously hostile world. The cause of so many extinctions is not well known, but several seem to be grouped at the end of the Early Triassic, in the Carnian, and at the end of the Triassic. The triggering mechanism for the first two events is unknown, but some models suggest that a meteorite impact, identified as the 50-mile (80-km) wide Manicuagan structure in Ontario, may have been responsible for the end-Triassic event.

The supercontinent of Pangea stretched nearly from pole to pole in the Triassic, surrounded by the Panthalassa Ocean, and partly surrounding the wedge-shaped Tethys Ocean that contained numerous reefs and carbonate platforms. The western margin of Pangea was dominated by convergent margin activity with active volcanism stretching from Alaska through the North and South American cordillera and into Antarctica.

Numerous flood basalts were erupted and swarms of diabase dikes intruded during the Triassic, including the Karoo of South Africa, the Permo-Triassic Siberian traps, and along the eastern coast of North America, western North Africa, and northeastern South America.

Triassic climates were generally warm, and sea levels started low, fluctuated in the Middle Triassic, and were generally about 300 feet (100 m) higher than the present level in the Late Triassic. Deserts covered much of inland Pangea, with extensive evaporite basins, red beds, and coal swamps forming in different locations on land, and marine carbonates deposited in much of the Tethys Ocean.

See also FLOOD BASALT; MESOZOIC; PANGEA; PANTHALLASA; TETHYS.

tropical climate The climate in the tropics depends on which definition of the tropics is used. The equatorial belt between 10° north and south latitude is characterized by upwelling of warm moist air masses, and frequent thunderstorms. Most of the deserts of the world fall between 15° and 30° latitude, because this is where cool dry air descends in Hadley Cell global circulation belts. The major deserts of

North Africa (Sahara) and South West Asia are included in this belt, as are the Kalahari of southern Africa, the Australian desert, and the Sonoron of the United States and Mexico. Thus, most of the true tropics are characterized by dry sunny conditions, with the cool downwelling surface air dramatically warmed on the surface. There is a large annual temperature cycle, especially in the continental interiors, whereas coastal regions in tropical climates show a much smaller annual temperature variation because of the moderating effects of the ocean. Rainfall is extremely variable, with some years showing no rainfall in many places, and occasional intense rains in others. Climate zones include dry tropical climate and Mediterranean-type including subdesert. These regions are characterized by a very large soil moisture deficiency and by low amounts of soil water storage. Potential evapotranspiration may be many times greater than the actual rainfall.

Tropical rainy climates and seasonal wet/dry climates characterize other parts of the tropics. Much of the Indian Ocean realm and southeast Asia are in the tropics and characterized by drenching monsoonal rains, and average temperatures in all months greater than 64°F (18°C). Tropical rainy climates have no rainy season, and large annual rainfall that exceeds the evaporation potential. Tropical rainforests, such as those in southeast Asia, the southwest Pacific islands, and parts of South and Central America receive more than 2.4 inches (6 cm) of rain in the driest season, whereas the monsoonal tropical climate has a pronounced dry season with some months receiving less than 2.4 inches of rain. Tropical savanna climates have at least one month with less than 2.4 inches of rain.

See also CLIMATE; DESERT; TROPICS.

tropics The tropics fall between 10° and 25° north and south latitude, bounded on the low-latitude side by equatorial regions, and on the high latitude side by subtropical then mid-latitude belts. Most informal usages of the term *tropics* include the entire latitudinal belt between the tropics of Cancer (23.5°N) and Capricorn (23.5°S). The tropics are characterized by a marked seasonal cycle, because the sun is directly overhead the tropic of Cancer at the summer solstice, and above the tropic of Capricorn at the winter solstice. This causes a large variation in the solar insolation at different times of the year in each tropical zone.

troposphere The atmosphere is divided into several layers, based mainly on the vertical temperature gradients that vary significantly with height. The lower 36,000 feet (11 km) of the atmosphere is characterized by circulating air known as the troposphere, where the temperature generally decreases gradually, at about 14°F (6.4°C) per kilometer, with increasing height above the surface. This is because the Sun heats the surface, which in turn warms the lower part of the troposphere. Most of the atmospheric and weather phenomena we are familiar with occur in the troposphere.

The main components of the atmosphere are nitrogen, oxygen, argon, carbon dioxide, helium, krypton, neon, and xenon. Increasing levels of photochemical smog in the troposphere is an increasing problem in many places. Smog is made mainly of the gas ozone (O_3), produced by secondary chemical reactions from automobile and other pollution such as hydrocarbons, although some is produced naturally. While ozone in the stratosphere forms a protective shield against ultraviolet radiation, low-level ozone is harmful. Ozone irritates the respiratory system and also retards tree and crop growth, and even causes rubber to break down. The formation of low-level ozone is particularly bad near major urban areas such as Los Angeles, California.

See also ATMOSPHERE.

tsunami Long wavelength seismic sea waves generated by the sudden displacement of the seafloor. The name is of Japanese origin, meaning "harbor wave." Tsunami are also commonly called tidal waves, although this is improper because they have nothing to do with tides. Tsunami may rise unexpectedly out of the ocean and sweep over coastal communities, killing hundreds of people and causing millions of dollars in damage. Such events occurred in 1946, 1960, 1964, 1992, 1993, and 1998 in coastal Pacific areas. In 1998 a catastrophic 50-foot (15-m) high wave unexpectedly struck Papua New Guinea, killing more than 2,000 people and leaving more than 10,000 homeless. The December 2004 Indian Ocean tsunami killed approximately 300,000 people, making it the most destructive tsunami known in history.

Tsunami are generated most often by thrust earthquakes along deep-ocean trenches and convergent plate boundaries. Tsunami therefore occur most frequently along the margins of the Pacific Ocean, a region characterized by numerous thrust-type earthquakes. About 80 percent of all tsunami strike circum-Pacific shorelines, with the most being generated in and striking southern Alaska. Volcanic eruptions, giant submarine landslides, and the sudden release of gases from sediments on the seafloor may also generate tsunami. Tsunami are not rare on Pacific islands including Hawaii and Japan, which now have extensive warning systems in place to alert residents when they are likely to occur. Before these warning systems were in place, residents would have no warning when the tsunami, in some cases reaching 50 feet or more in height. would occasionally strike coastal areas.

Some historical tsunami have been absolutely devastating to coastal communities, wiping out entire populations with little warning. One of the most devastating tsunami in recent history was generated by the eruption of the Indonesian volcano Krakatau in 1883. When Krakatau erupted, it blasted a large part of the center of the volcano out, and seawater rushed in to fill the hole. This seawater was immediately heated and it exploded outward in a steam eruption and a huge wave of hot water. The tsunami generated by this eruption reached more than 120 feet (37 m) in height and killed more than 36,500 people in nearby coastal regions. Another famous tsunami was also generated by a volcanic eruption of Santorin (now called Thira) on the Mediterranean island of Crete. In 1600 B.C.E., this volcano was the site of the most powerful eruption in recorded history, and it generated a

tsunami that destroyed many Mediterranean coastal areas and probably led to the eventual downfall of the Minoan civilization on Crete. The tsunami deposited volcanic debris at elevations of up to 800 feet (245 m) above the mean ocean level on the nearby island of Anaphi, and the wave was still more than 20 feet (6 m) high when it ran up the shorelines on the far side of the Mediterranean in Israel.

December 26, 2004: Indian Ocean Earthquake and Tsunami

One of the worst natural disasters of the 21st century unfolded on December 26, 2004, following a magnitude 9.0–9.2 earthquake off the coast of Sumatra in the Indian Ocean. During this catastrophic earthquake, a segment of the seafloor the size of California suddenly moved upward and seaward by several tens of feet, releasing more energy than all the earthquakes on the planet in the last 25 years combined. The sudden displacement of this volume of undersea floor displaced huge volumes of water and generated the most destructive tsunami ever recorded.

Within minutes of the earthquake, a mountain of water more than 100 feet high was ravaging northern Sumatra, sweeping into coastal villages and resort communities with a fury that crushed all in its path, removing buildings and vegetation, and eroding shoreline areas down to bedrock. Similar scenes of destruction and devastation rapidly moved up the coast of nearby Indonesia, where residents and tourists were enjoying a holiday weekend. In some cases, the sea retreated to unprecedented low levels before the waves struck, drawing people to the shore to investigate the phenomena—in other cases, the sea waves simply came crashing inland without warning. Buildings, vehicles, trees, boats, and other debris were washed along with the ocean waters, forming projectiles that smashed at speeds of up to 30 miles per hour into other structures, leveling all in its path, killing approximately 300,000 people.

The displaced water formed a deepwater tsunami that moved at speeds of 500 miles per hour across the Indian Ocean, smashing within an hour into Sri Lanka and India, wiping away entire fishing communities and causing widespread destruction of the shore environment. South of India are many small islands, including the Maldives, Chagos, and Seychelles, many with maximum elevations only tens of feet above sea level. As the tsunami approached these islands, many wildlife species and tribal residents fled to the deep forest, perhaps sensing the danger as the sea retreated and the ground trembled with the approaching wall of water. As the tsunami heights were higher than many of the maximum elevations of some of these islands, the forest was able to protect and save many lives in places where the tsunami rose with less force than in places where the shoreline geometry caused large breaking waves to crash ashore.

Several hours later the tsunami reached the shores of Africa and Madagascar, and though distance diminished its height to less than 10 feet, several hundred people were killed by the waves and high water. Kenya and Somalia were hit severely, with harbors experiencing rapid and unpredictable rises and falls in sea level, and many boats and people washed to sea.

The tsunami traveled around the world, being measured as minor (inches) changes in sea level more than 24 hours later in the North Atlantic and Pacific. Overall, more than 300,000 people perished in the December 26th Indian Ocean tsunami, though many could have been saved, if a tsunami warning system had been in place in the Indian Ocean. Tsunami warning systems are capable of saving lives by alerting residents of coastal areas that a tsunami is approaching their location. These systems are most effective for areas located more than 500 miles (750 km), or one hour away from the source region of the tsunami, but may also prove effective at saving lives in areas closer to a tsunami. The National Oceanographic and Atmospheric Administration operates the Pacific Tsunami Warning Center in Honolulu, Hawaii, integrating data from several different sources, including seismic stations that record earthquakes and quickly sort out those earthquakes that are likely to be tsunamogenic. A series of tidal gauges placed around the Pacific monitors the passage of any tsunamis past their location, and if these stations detect a tsunami, warnings are quickly issued for local and regional areas likely to be affected. Analyzing all of this information takes time, however, so this Pacific-wide system is most effective for areas located far from the earthquake source.

Tsunami warning systems designed for shorter-term local warnings are also in place in many communities, including Japan, Alaska, Hawaii, and some Pacific islands. These warnings are based mainly on estimating the magnitude of nearby earthquakes and on the ability of public authorities to rapidly issue warnings so that the population has time to respond. For local earthquakes, the time between the shock event and the tsunami hitting the shoreline may be only a few minutes. So if you are in a coastal area and feel a strong earthquake, you should take that as a natural warning that a tsunami may be imminent and leave low-lying coastal areas.

U.S. scientists detected the magnitude of the Sumatra earthquake and tried to warn countries in soon-to-be-affected regions that a tsunami might be approaching. However, despite efforts by some scientists over the past few years, no systematic warning system was in place in the Indian Ocean. Initial cost estimates for a crude system were about $20 million, deemed too expensive by poor nations that needed funds for more obviously pressing humanitarian causes. When the earthquake struck on a Sunday, scientists who tried contacting countries and communities surrounding the Indian Ocean to warn them of the impending disaster typically found no one in the office and no systematic list of phone numbers of emergency response personnel. Having such a simple phone-pyramid list could potentially have saved tens of thousands of lives. Indian Ocean communities are now planning to establish a tsunami warning system before the next tsunami strikes.

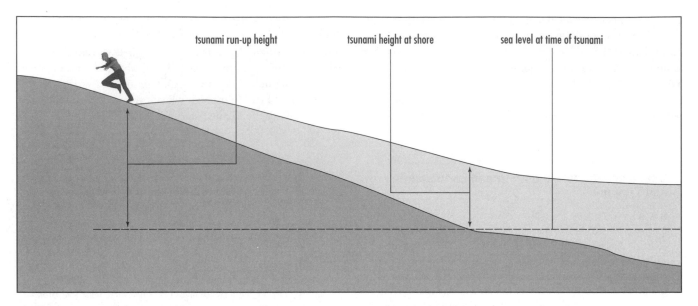

Tsunami run-up is a measure of the height of the tsunami above sea level at the farthest point it reaches from the shore.

Movement of Tsunami

Tsunami are waves with exceptionally large distances between individual crests, and they move like other waves across the ocean. We define wavelength as the distance between crests, wave-height as the vertical distance from the crest to the bottom of the trough, and the amplitude as one-half of the wave height. Most ocean waves have wavelengths of 300 feet (100 m) or less; tsunami are exceptional in that they have wavelengths that can be 120 miles (200 km) or greater. When tsunami are traveling across deep ocean water, their amplitudes are typically less than three feet (1 m). You would probably not even notice even the largest of tsunami if you were on a boat in the deep ocean. Circular or elliptical paths that decrease in size with depth describe the motion of water in waves. All motion from the waves stops at a depth equal to one-half the distance of the wavelength. Tsunami therefore are felt at much greater depths than ordinary waves, and this effect may be used with deep ocean bottom tsunami detectors to help warn coastal communities when tsunami are approaching.

Waves with long wavelengths travel faster than waves with short wavelengths. Since the longer the wavelength the faster the wave in deep open water, tsunami travel extremely fast across the ocean. Normal ocean waves travel at less than 55 miles per hour (90 km/hr), whereas many tsunami travel at 500 to 600 miles per hour (800 to 950 km/hr), faster than most commercial airliners!

When waves encounter shallow water the friction of the seafloor along the base of the wave causes them to slow down dramatically, and the waves effectively pile up on themselves as successive waves move into shore. This causes the wave height or amplitude to increase dramatically, sometimes 15 to 150 feet (4.5–45 m) above the normal still water line for tsunami.

When tsunami strike the coastal environment, the first effect is sometimes a significant retreat or drawdown of the water level, whereas in other cases the water just starts to rise quickly. Since tsunami have long wavelengths, it typically takes several minutes for the water to rise to its full height. Also, since there is no trough right behind the crest of the wave, on account of the very long wavelength of tsunami, the water does not recede for a considerable time after the initial crest rises onto land. The rate of rise of the water in a tsunami depends in part on the shape of the seafloor and coastline. If the seafloor rises slowly, the tsunami may crest slowly, giving people time to outrun the rising water. In other cases, especially where the seafloor rises steeply, or the shape of the bay causes the wave to be amplified, tsunami may come crashing in huge walls of water with breaking waves that pummel the coast with a thundering roar and wreaking utmost destruction.

Because tsunami are waves, they travel in successive crests and troughs. Many deaths in tsunami events are related to people going to the shoreline to investigate the effects of the first wave, or to rescue those injured or killed in the initial crest, only to be drowned or swept away in a succeeding crest. Tsunami have long wavelengths, so successive waves have a long lag time between individual crests. The period of a wave is the time between the passage of individual crests, and for tsunami the period can be an hour or more. Thus, a tsunami may devastate a shoreline area, retreat, and then another crest may strike an hour later, then another, and another in sequence.

Tsunami Run-Up

Run-up is the height of the tsunami above sea level at the farthest point it reaches on the shore. This height may be considerably different from the height of the wave where it first hits the shore. Many things influence the run-up of tsunami, including the size of the wave, the shape of the shoreline, the profile of the water depth, and other irregularities particular to individual areas. Some bays and other places along some shorelines may amplify the effects of waves that come in from a certain direction, making run-ups higher than average. These areas are called wave traps, and in many cases the incoming waves form a moving crest of breaking water, called a bore. Tsunami magnitudes are commonly reported using the maximum run-up height along a particular coastline.

Origin of Tsunami

Tsunami may be generated by any event that suddenly displaces the seafloor, which in turn causes the seawater to move suddenly to compensate for the displacement. Most tsunami are earthquake induced or caused by volcanic eruptions, although giant submarine landslides have initiated others. It is even possible that gases dissolved on the seafloor may suddenly be released, forming a huge bubble that erupts upward to the surface, generating a tsunami.

EARTHQUAKE-INDUCED TSUNAMI Earthquakes that strike offshore or near the coast have generated most of the world's tsunami. In general, the larger the earthquake, the larger the potential tsunami, but this is not always the case. Some earthquakes produce large tsunami, whereas others do not. Earthquakes that have large amounts of vertical displacement of the seafloor result in larger tsunamis than earthquakes that have predominantly horizontal movements of the seafloor. This difference is approximately a factor of 10, probably because earthquakes with vertical displacements are much more effective at pushing large volumes of water upward or downward, generating tsunami. Another factor that influences how large a tsunami may be generated by a specific earthquake is the speed at which the seafloor breaks during the earthquake—slower ruptures tend to produce larger tsunami.

Tsunami earthquakes are a special category of earthquakes that generate tsunami that are unusually large for the earthquake's magnitude. Tsunami earthquakes are generated by large displacements that occur along faults near the seafloor. Most are generated on steeply dipping seafloor surface penetrating faults that have vertical displacements along them during the earthquake, displacing the maximum amount of water. These types of earthquakes also frequently cause large submarine (underwater) landslides or slumps, which also generate tsunami. In contrast to tsunami generated by vertical slip on vertical faults, which cause a small region to experience a large uplift, other tsunami are generated by movement on very shallowly dipping faults. These are capable of causing large regions to experience minor uplift, displacing large volumes of water and generating a tsunami. Some of the largest tsunami may have been generated by earthquake-induced slumps along convergent tectonic plate boundaries. For example, in 1896 a huge 75-foot (23-m) high tsunami was generated by an earthquake-induced submarine slump in Sanriku, Japan, killing 26,000 people in the wave. Another famous tsunami generated by a slump from an earthquake is the 1946 wave that hit Hilo, Hawaii. This tsunami was 50 feet (15 m) high, killed 150 people, and caused about $25 million in damage to Hilo and surrounding areas. The amazing thing about this tsunami is that it was generated by an earthquake-induced slump off Unimak Island in the Aleutian Chain of Alaska 4.5 hours earlier! This tsunami traveled at 500 miles per hour (800 km/hr) across the Pacific, hitting Hawaii without warning.

Another potent kind of tsunami-generating earthquake occurs along subduction zones. Sometimes, when certain kinds of earthquakes strike in this environment, the entire forearc region above the subducting plate may snap upward by up to a few tens of feet, displacing a huge amount of water. The tsunami generated during the 1964 magnitude 9.2 Alaskan earthquake formed a tsunami of this sort, and it caused numerous deaths and extensive destruction in places as far away as California. The 2004 Indian Ocean tsunami was likewise generated by a magnitude 9.0–9.2 subduction zone earthquake in the Sumatra trench-subduction zone system.

VOLCANIC ERUPTION-INDUCED TSUNAMI Some of the largest recorded tsunamis have been generated by volcanic eruptions. These may be associated with the collapse of volcanic slopes, debris and ash flows that displace large amounts of water, or by submarine eruptions that explosively displace water above the volcano. The most famous volcanic eruption-induced tsunami include the series of huge waves generated by the eruption of Krakatau in 1883, which reached run-up heights of 120 (36.5 m) feet and killed 36,500 people. The number of people that perished in the eruption of Santorin in 1600 B.C.E. is not known, but the toll must have been huge. The waves reached 800 feet (245 m) in height on islands close to the volcanic vent of Santorin. Flood deposits have been found 300 feet (91 m) above sea level in parts of the Mediterranean Sea and extend as far as 200 miles (320 km) up the Nile River to the south. Several geologists suggest that these were formed from a tsunami generated by the eruption of Santorin. The floods from this eruption may also, according to some scientists, account for some historical legends such as the great biblical flood, the parting of the Red Sea during the exodus of the Israelites from Egypt, and the destruction of the Minoan civilization of the island of Crete.

LANDSLIDE-INDUCED TSUNAMI Many tsunamis are generated by landslides that displace large amounts of water. These may be from rock falls and other debris that falls off cliffs into the water, such as the huge avalanche that triggered a 200-foot (61-m) high tsunami in Lituya Bay, Alaska. Submarine landslides tend to be larger than avalanches that originate above the waterline, and they have generated some of

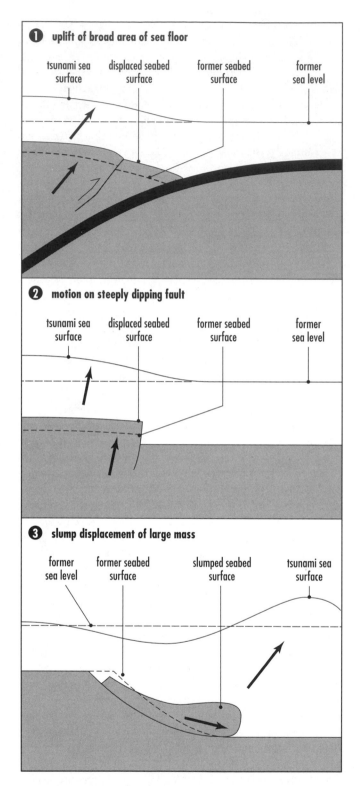

① uplift of broad area of sea floor

tsunami sea surface | displaced seabed surface | former seabed surface | former sea level

② motion on steeply dipping fault

tsunami sea surface | displaced seabed surface | former seabed surface | former sea level

③ slump displacement of large mass

former sea level | former seabed surface | slumped seabed surface | tsunami sea surface

Causes of tsunami, including vertical displacements of the seafloor and slumping of large blocks into the water

the largest tsunami on record. Many submarine landslides are earthquake-induced, but tsunami are thought to be landslide-induced when the earthquake is not large enough to produce

the observed size of the associated tsunami. Sediments near deep-sea trenches are often saturated in water, and close to failure. When an earthquake strikes these areas, large parts of the submarine slopes may give out simultaneously, displacing water and generating a tsunami. The 1964 magnitude 9.2 earthquake in Alaska generated more than 20 tsunami, and these were responsible for most of the damage and deaths from this earthquake.

Some steep submarine slopes that are not characterized by earthquakes may also be capable of generating huge tsunami. Recent studies along the east coast of North America, off the coast of Atlantic City, New Jersey, have revealed significant tsunami hazards. A thousands of feet thick pile of unconsolidated sediments on the continental slope is so porous and saturated with water that is on the verge of collapsing under its own weight. A storm or minor earthquake may be enough to trigger a giant submarine landslide in this area, possibly generating a tsunami that could sweep across the beaches of Long Island, New Jersey, Delaware, and much of the east coast of the United States.

GAS HYDRATE ERUPTION-INDUCED TSUNAMI Decaying organic matter on the seafloor releases large volumes of gas. Under some circumstances, including cold water at deep depths, these gases may coagulate forming gels called gas hydrates. It has recently been recognized that these gas hydrates occasionally spontaneously release their trapped gases in giant bubbles that rapidly erupt to the surface. Such catastrophic degassing of gas hydrates poses a significant tsunami threat to regions not previously thought to have a significant threat, such as along the east coast of the United States.

OTHER TSUNAMI Giant tsunami may be generated by the impact of asteroids with the Earth. These types of events do not happen very often, but when they do they are cataclysmic. Geologists are beginning to recognize deposits of impact-generated tsunami and now estimate that they may reach several thousand feet in height. One such tsunami was generated about 66 million years ago by an impact that struck the shoreline of the Yucatán Peninsula, producing the Chicxulub impact structure. This impact produced a huge crater and sent a 3,000-foot (915-m) high tsunami around the Atlantic, devastating the Caribbean and the U.S. Gulf coast. Subsequent fires and atmospheric dust that blocked the Sun for several years killed off much of the planet's species, including the dinosaurs.

See also BEACH; EARTHQUAKES; PLATE TECTONICS.

Further Reading

Bernard, E. N., ed., *Tsunami Hazard: A Practical Guide for Tsunami Hazard Reduction.* Dordrecht, The Netherlands: Kluwer Academic Publishers, 1991.

Dawson, A. G., and S. Shi. "Tsunami Deposits." *Pure and Applied Geophysics* 157 (2000): 493–511.

Driscoll, N. W., J. K. Weissel, and J. A. Goff. "Potential for Large-Scale Submarine Slope Failure and Tsunami Generation along the United States Mid-Atlantic Coast." *Geology* 28 (2000): 407–410.

Dvorak, J., and T. Peek. "Swept Away." Earth 2, no 4 (1993): 52–59.

Latter, J. H. "Tsunami of Volcanic Origin, Summary of Causes, with Particular Reference to Krakatau, 1883." *Journal of Volcanology* 44 (1981): 467–490.

McCoy, F., and G. Heiken. "Tsunami Generated by the Late Bronze Age Eruption of Thera (Santorini), Greece." *Pure and Applied Geophysics* 157 (2000): 1,227–1,256.

Minoura, K., F. Inamura, T. Nakamura, A. Papadopoulos, T. Takahashi, and A. Yalciner. "Discovery of Minoan Tsunami Deposits." *Geology* 28 (2000): 59–62.

Revkin, A. C. "Tidal Waves Called Threat to East Coast." *The New York Times* (2000): A18.

Satake, K. "Tsunamis." *Encyclopedia of Earth System Science* 4 (1992): 389–397.

Steinbrugge, K. V. *Earthquakes, Volcanoes, and Tsunamis, An Anatomy of Hazards.* New York: Skandia America Group, 1982

Tsuchiya, Y., and N. Shuto, eds., *Tsunami: Progress in Prediction Disaster Prevention and Warning.* Boston: Kluwer Academic Publishers, 1995.

U.S. Geological Survey. "Surviving a Tsunami—Lesson from Chile, Hawaii, and Japan." *U.S. Geological Survey Circular* 1187, 1987.

tundra   Treeless plains of the Arctic and subarctic regions, characterized by a marshy surface with abundant growth of mosses and lichens, and a dark organic-rich soil overlying permafrost. When heated in the summer months, the permafrost layer remains impermeable and develops a layer of meltwater on or below the surface forming wet swampy or mucky conditions. Bacterial action is slow in these high-latitude swamps, so organic material accumulates, adding to the soil layer. The size of most plants in the tundra is limited by the depth to which roots can penetrate and by the mechanical abrasion by wind-driven snow and dry winds in the winter months.

Some unique and unusual landforms characterize the tundra environment. Frost action slowly brings rock fragments to the surface, as ice crystals form below the rocks and force them upward through the soil. Once on the surface many of these rock particles may be pushed laterally by ice to form stone polygons. Patterned ground may form through permafrost cracking into polygons by thermal contraction, then the cracks fill with water that freezes, forming ice wedges. Eventually these subsurface ice wedges may form polygonal or patterned ground where the surface is broken into regular polygons separated by narrow ice wedges, and partly covered by thaw lakes. Pingos are large, sometimes several hundred feet high, mounds of soil and rock cored by ice that grows as more and more ice accumulates under the surface.

Tundra regions are also susceptible to solifluction and slumping of melt-ice layers particularly during the summer months when a melt layer is well developed. The slightest incline can cause soil layers to move downhill in solifluction lobes. Downhill creep of permafrost layers is also common, as is intense weathering and shattering of any exposed bedrock. Bedrock exposures are commonly shattered into fields of angular boulders called blockfields or kurums.

See also GLACIER; MASS WASTING.

Tunguska, Siberia   The location of what is thought to be the site of a collision of a comet with the Earth. On June 30, 1908, a huge explosion rocked the Tunguska area of Siberia and devastated more than 1,158 square miles (3,000 km²) of forest. The force of the blast is estimated to have been equal to 15 megatons and is thought to have been produced by the explosion six miles (10 km) above the surface of Earth of an asteroid with a diameter of 200 feet (61 m). Shock waves were felt thousands of miles away, and people located closer than 60 miles (97 km) to the site of the explosion were knocked unconscious, and some were thrown into the air by the force of the explosion. Fiery clouds and deafening explosions were heard more than 300 miles (480 km) from Tunguska. After years of study and debate it is now thought that this huge explosion was produced by a fragment of Comet Encke that broke off the main body and exploded in the air about five miles (8 km) above the Siberian Plains.

The sequence of events associated with the Tunguska event has been reconstructed as follows. About 7:00 A.M. on June 30, a huge fireball was seen moving westward across Siberia. Next, an explosion was heard that was centered on the remote Tunguska region, with reports of the explosion and pressure waves being felt more than 600 miles (965 km) away. People who were hundreds of miles from the site of the explosion were knocked down, and a huge 12-mile (19-km) high column of fire was visible for more than 400 miles (640 km). Seismometers recorded the impact, and barometers around the world recorded the air pressure wave as it traveled two times around the globe. The impact caused a strange, bright, unexplained glow that lit up the night skies in Scotland and Sweden.

Several years after the impact, a scientific expedition to the remote region discovered that many trees were charred and knocked down in a 2,000 square mile (5,180 km²) area, with near total destruction in the center 500 square miles (1,295 km²). Many theories were advanced to explain the strange findings, and more than 50 years later, in 1958, a new expedition to Tunguska found small melted globules of glass and metal, identified as pieces of an exploded meteorite or asteroid.

One of the biggest puzzles at Tunguska is the absence of an impact crater, despite all other evidence that points to an impact origin for this event. It is now thought that a piece of a comet, Comet Encke, broke off the main body as it was orbiting nearby Earth, and this fragment entered Earth's atmosphere and exploded about 5 miles (8 km) above the Siberian plains at Tunguska. Comets are weaker than metallic or stony meteorites, and they more easily break up and explode in the atmosphere before they hit the Earth's surface.

turbidite A deposit of a submarine turbidity current consisting of graded sandstone and shale, typically deposited in a thick sequence of similar turbidites. Most turbidites are thought to be deposited in various sub-environments of submarine fans, in shallow to deepwater settings. Typically, a water-saturated sediment on a shelf or shallow water setting is disturbed by a storm, earthquake, or some other mechanism that triggers the sliding of the sediment downslope. The sediment-laden sediment/water mixture then moves rapidly downslope as a density current and may travel tens or even hundreds of miles at tens of miles per hour until the slope decreases, and the velocity of the current decreases. As the velocity of the current decreases, the ability of the current to hold coarse material in suspension decreases, so the current drops first its coarsest load, then progressively finer material as the current decreases further. In this way, the coarsest material is deposited closest to the channel or slope that the turbidity current flowed down, and the finest material is deposited further away. The same sequence of coarse to fine material is deposited upward in the turbidite bed as the current velocity decreases with time at any given location. This is how graded beds are formed, with the coarsest material at the base and finer material at the top.

Classical complete turbidite beds consist of a sequence of sedimentary structures divided into a regular A–E sequence known as the Bouma sequence, after the sedimentologist Arnold Bouma who first described the sequence. The A horizon consists of coarse to fine-grained graded sandstone beds, representing material deposited rapidly from suspension. The B horizon consists of parallel-laminated sandstones deposited by material that moved in traction on the bed, whereas division C contains cross-laminated sands deposited in the lower flow regime. The D and E horizons represent the transition from material deposited from the waning stages of the turbidity current and background pelagic sedimentation.

Variations in the thickness and presence or absence of individual horizons of the Bouma sequence have been related to where on the submarine fan or slope the turbidite was deposited. Turbidites with more of the A-B-C horizons are

Graywacke and shale of a turbidite sequence, Chugach Mountains, southern Alaska *(Photo by Timothy Kusky)*

interpreted to have been deposited closer to the slope or channel, whereas turbidite sequences with more of the C-D-E horizons are interpreted as more distal deposits.

Many turbidite sequences are deposited in foreland basins and in deep-sea trench settings. These environments have steep slopes in the source areas, a virtually unlimited source of sedimentary material, and many tectonic triggers to initiate the turbidity current.

See also CONVERGENT PLATE MARGIN PROCESSES; FORELAND BASIN; SEDIMENTARY ROCKS.

Further Reading

Bouma, Arnold, H. *Sedimentology of Some Flysch Deposits.* Amsterdam: Elsevier, 1962.

Kuenen, Phillip H., and Carlo I. Migliorini. "Turbidite Currents as a Cause of Graded Bedding." *Journal of Geology,* v. 58 (1950): 91–127.

Walker, Roger G. *Facies Models.* Toronto: Geoscience Canada Reprint Series 1, Geological Association of Canada, 1983.

turbidity current *See* TURBIDITE.

typhoon *See* CYCLONE.

U

unconformity A substantial break or gap in a stratigraphic sequence that marks the absence of part of the rock record. These breaks may result from tectonic activity with uplift and erosion of the land, sea-level changes, climate changes, or simple hiatuses in deposition. Unconformities normally imply that part of the stratigraphic sequence has been eroded but may also indicate that part of the sequence was not ever deposited in that location.

There are several different types of unconformities. Angular unconformities are angular discordances between older and younger rocks. Angular unconformities form in places where older layers were deformed and partly eroded before the younger layers were deposited. Disconformities represent a significant erosion interval between parallel stra-

ta. They are typically recognized by their irregular surfaces, missing strata, or large breaks between dated strata. Nonconformities are surfaces where strata overlies igneous or metamorphic rocks. Unconformities are significant in that they record an unusual event, such as tectonism, erosion, sea-level change, or climate change.

Unconformities are typically overlain by a progradational marine sequence, starting with shallow-water sandstone, conglomerate or quartzite, and succeeded by progressively deeper water deposits such as sandstone, shale, and limestone. Unconformities are often used by stratigraphers and other geologists to separate different packages of rocks deposited during different tectonic, climatic, or time systems.

See also PLATE TECTONICS; STRATIGRAPHY.

Unconformity, Scotland. Steeply dipping beds in lower half of photograph are overlain with a sharp angular discordance (at nose level of person) by younger horizontal beds

uniformitarianism The doctrine that proposes that geologic processes and natural laws that are operating at present on the Earth have acted in a similar way and with similar intensity throughout geologic time. The principle is commonly used by geologists to interpret ancient rocks, although contrasting or opposed views exist, including catastrophism and other non-uniformitarian views. The term was coined by William Whewell to explain James Hutton's approach to geology, best summarized in a famous quote from his 1795 book *The Theory of the Earth:* "The present is the key to the past, and the past is the key to geological processes." In 1830 another famous geologist named Charles Lyell announced support for James Hutton's approach, and the two became the two main progenitors of uniformitarianism philosophy.

Uniformitarianism is one of two main ways to think about old rocks and orogens. The uniformitarian view aims to apply plate tectonic principles to old rocks to see how similar they are to younger rocks and orogens. In this approach, the scientist notes differences between then and now and searches for causes and secular trends. Some differences between the present and early Earth may be expected because of higher earlier heat production and heat flow in the early part of Earth history. In the opposing, non-uniformitarian view, the scientist assumes a model for the earlier history of the Earth, based on theoretical considerations of heat flow, biological evolution, etc., and then applies this model to old rocks and orogens to determine if the model is compatible with the observations. The two approaches can yield very different solutions to explain observations, and there is currently widespread debate among scientists that work on the early history of the Earth as to which paradigm may be more appropriate.

See also ARCHEAN; PLATE TECTONICS; PROTEROZOIC.

United States Geological Survey (USGS) The USGS (http://www.usgs.gov/) was established in 1879 and has become a world leader in the natural sciences through scientific excellence and responsiveness to society's needs. It focuses into four major areas: natural hazards, resources, the environment, and information and data management. The USGS stands as the sole science agency for the Department of the Interior. The USGS serves the nation by providing reliable scientific information to describe and understand the Earth; minimize loss of life and property from natural disasters; manage water, biological, energy, and mineral resources; and enhance and protect the quality of life.

Ural Mountains The boundary between Europe and Asia is typically taken to be the Ural Mountains, a particularly straight mountain range that stretches 1,500 miles (2,400 km) from the Arctic tundra to the deserts north of the Caspian Sea. Naroda (6,212 feet; 1,894 m) and Telpos-Iz (meaning "nest of winds," 5,304 feet; 1,617 m) are the highest peaks, found in the barren rocky and tundra-covered northern parts of the range. Southern parts of the mountain range rise to 5,377 feet (1,639 m) at Yaman-Tau, in the Mugodzhar Hills. The southern parts of the range are densely forested, whereas the northern parts are barren and covered by tundra or bare rock.

The Ural River flows out of the southern Urals into the Caspian Sea, and the western side of the range is drained by the Kama and Belaya Rivers, which are tributaries that also feed into the Caspian Sea, providing more than 75 percent of the water that flows into this shallow, closed basin. The eastern side of the range is drained by the Aob-Irtysh drainage system that flows into the Ob Gulf on the Kara Sea.

The Urals are extremely rich in mineral resources, including iron ore in the south, and large deposits of coal, copper, manganese, gold, aluminum, and potash. Ophiolitic rocks in the south are also rich in chromite and platinum, plus deposits of bauxite, zinc, lead, silver, and tungsten are mined. Basins on the western side of the Urals produce large amounts of oil, and regions to the south in the Caspian are yielding many new discoveries. The Urals are also very rich in

rare minerals and gems, yielding many excellent samples of emeralds, beryl, and topaz.

The Urals form part of the Ural-Okhotsk mobile belt, a Late Proterozoic to Mesozoic orogen that bordered the Paleoasian Ocean. The Ural Mountains section of this orogen saw a history that began with Early Paleozoic, probably Cambrian rifting of Baikalian basement, and Late Ordovician spreading to form a back arc or oceanic basin that was active until the Mid-Carboniferous. Oceanic arcs grew in this basin, but by Middle Devonian they began colliding with the East European continent, forming flysch basins. The Kazakhstan microcontinent collided with the Laurussian continent in the Permian, forming a series of foredeep basins on the Russian and Pechora platforms. These foredeeps are filled with molasse and economically important Middle to Late Permian coal deposits, as well as potassium salts.

The Urals show a tectonic zonation from the Permian flysch basins on the East European craton to the Permian molasse basins on the western slopes of the Urals, then into belts of allochthonous carbonate platform rocks derived from the East European craton and thrust to the west over the Permian foredeeps. These rocks are all involved in westward-vergent fold-thrust belt structures, including duplex structures, indicating westward tectonic transport in the Permian. The axial zone of the Urals includes a chain of anticlinoria bringing up Riphean rocks, whose eastern contact is known as the Main Uralian fault. This major fault zone brings oceanic and island arc rocks in large nappe and klippen structures, placing them over the passive margin sequence.

The eastern slope of the Urals consists of a number of Ordovician to Carboniferous oceanic and island arc synformal nappes, imbricated with slices of the Precambrian crystalline basement. It is uncertain if these Precambrian gneisses are part of the East European craton, part of the accreted Kazakhstan microcontinent, or an exotic terrane. The eastern slopes of the Urals are intruded by many Devonian-Permian granites.

See also CASPIAN SEA.

Further Reading
Coleman, Robert G., ed. *Reconstruction of the Paleo-Asian Ocean.* The Netherlands: VSP International Science Publishers, 1992.

uranium A radioactive mineral that spontaneously decays to lighter "daughter" elements by loosing high-energy particles at a predictable rate known as a half-life. The half-life specifically measures how long it takes for half of the original or parent element to decay to the daughter element. ^{238}U decays to ^{236}Pb, and ^{235}U decays to ^{207}Pb, and radium, through a long series of steps with a cumulative half-life of 4.4 billion years. During these steps, intermediate daughter products are produced, and high-energy particles including alpha particles, consisting of two protons and two neutrons, are released—this produces heat.

The main ore of uranium is uraninite (UO_2), a black, brown, or steel gray mineral that has cubic or octahedral crystal forms. Uraninite typically occurs in veins with tin, lead, and copper minerals, and as a detrital mineral in sandstones. It is also a primary mineral in many granites and pegmatites.

See also MINERALOGY; RADIOACTIVE DECAY.

Uranus The seventh planet from the Sun, Uranus is a giant gaseous sphere with a mass 15 times that of the Earth, and a diameter four times as large as Earth (51,100 km). The equatorial plane is circled by a system of rings, some associated with the smaller of the 15 known moons circling the planet. Uranus orbits the Sun at a distance of 19.2 astronomical units (Earth–Sun distance) with a period of 84 Earth years and has a retrograde rotation of 0.69 Earth days. Its density is only 1.2 grams per cubic centimeter, compared to the Earth's average density of 5.5 grams per cubic centimeter. The density is higher, however, than Jupiter's or Saturn's, suggesting that Uranus has a proportionally larger rocky core than either one of these giant gaseous planets.

Most of the planets in the solar system have their rotational axis roughly perpendicular to the plane of the ecliptic, the plane that the planets approximately orbit the Sun within. However, the rotational axis of Uranus is one of the most unusual in the solar system as it lies roughly within the plane of the ecliptic, as if it is tipped over on its side. The cause of this unusual orientation of the planet is not known, but some astronomers have speculated that it may be a result of a large impact early in the planet's history. As a consequence of this unusual orientation, as Uranus orbits the Sun, it goes through seasons where the north and south poles are alternately pointing directly at the Sun, and periods in between (spring and autumn) when the pole aligned in between these extremes. The poles experience very long summers and winters because of the long orbital period of Uranus and are alternately plunged into icy cold darkness for 42 years, then exposed to the distant Sun for 42 years. With the rapid rotation rate and

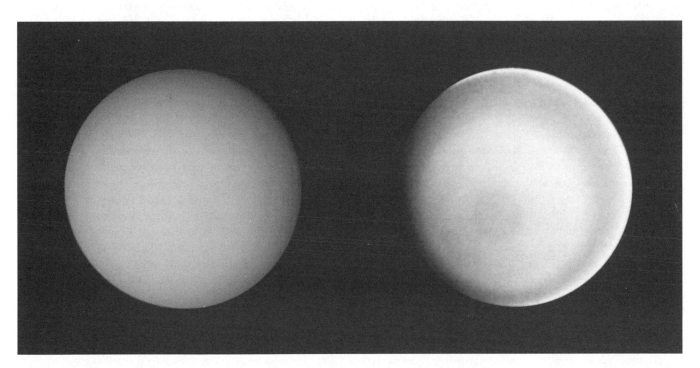

These two pictures of Uranus—one in true color (left) and the other in false color—were compiled from images returned January 17, 1986, by the narrow-angle camera of *Voyager 2*. The spacecraft was 5.7 million miles (9.1 million km) from the planet, several days from closest approach. The picture at left has been processed to show Uranus as human eyes would see it from the vantage point of the spacecraft. The picture is a composite of images taken through blue, green, and orange filters. The darker shadings at the upper right of the disk correspond to the day-night boundary on the planet. Beyond this boundary lies the hidden northern hemisphere of Uranus, which currently remains in total darkness as the planet rotates. The blue-green color results from the absorption of red light by methane gas in Uranus's deep, cold, and remarkably clear atmosphere. The picture at right uses false color and extreme contrast enhancement to bring out subtle details in the polar region of Uranus. Images obtained through ultraviolet, violet, and orange filters were respectively converted to the same blue, green, and red colors used to produce the picture at left. The very slight contrasts visible in true color are greatly exaggerated here. In this false-color picture, Uranus reveals a dark polar hood surrounded by a series of progressively lighter concentric bands. One possible explanation is that a brownish haze or smog, concentrated over the pole, is arranged into bands by zonal motions of the upper atmosphere. The bright orange and yellow strip at the lower edge of the planet's limb is an artifact of the image enhancement. In fact, the limb is dark and uniform in color around the planet. The *Voyager* project is managed for NASA by the Jet Propulsion Laboratory. *(Photo by NASA)*

unusual orientation of the planet, an observer on the pole experiencing the change from winter to spring would first observe the distant Sun rising above the horizon and tracing out part of a circular path, then sinking below the horizon. Eventually the Sun would finally emerge totally and trace out complete circle paths every 17 hours. The position of the Sun would progressively change over the next 42 years until it sunk below the horizon for the following 42 years.

The atmosphere of Uranus is roughly similar to Jupiter's and Saturn's, consisting mostly of molecular hydrogen (84 percent), helium (14 percent), and methane (2 percent). Ammonia seems to be largely absent from the atmosphere of Uranus, part of a trend that has ammonia decreasing in abundance outward in the outer solar system, with less at lower temperatures. The reason is that ammonia freezes into ammonia ice crystals at $-335°F$ ($70°K$), and Jupiter's and Saturn's upper atmospheres are warmer than this, whereas Uranus's is $-355°F$ ($58°K$). Any ammonia therefore would have crystallized and fallen to the surface. The atmosphere of Uranus appears a blue-green color because of the amount of methane, but so far relatively few weather systems have been detected on the planet. Enhancement of imagery, however, has revealed that the atmosphere is characterized by winds that are blowing around the planet at 125 to 310 miles per hour (200–500 km/hr) in the same sense as the planet's rotation, with detectable channeling of the winds into bands. The winds are responsible for transporting heat from the warm to the cold hemisphere during the long winter months.

The magnetic field of Uranus is surprisingly strong, about 100 times as strong as the Earth's. However, since the rocky core of the planet is so far below the cloud level, the strength of the magnetic field at the cloud tops on Uranus is actually similar to that on the surface of the Earth. Like the rotational axis, the orientation of the magnetic poles and field on Uranus are highly unusual. The magnetic axis is tilted at about 60° from the spin axis and is not centered on the core of the planet but is displaced about one-third of the planetary radius from the planet's center. The origin of these unusual magnetic field properties is not well understood but may be related to a slurry of electrically conducting ammonia clouds near the planet's rocky surface. Whatever the cause, a similar unusual field exists on nearby Neptune.

Uranus has 15 known large moons with diameters over 25 miles (40 km), orbiting between 31,070 miles (50,000 km) from the planet (for the smallest moon) to 362,260 miles (583,000 km) for the largest moon. From largest to smallest, these moons include Oberon, Titania, Umbriel, Ariel, Miranda, Puck, Belinda, Rosalind, Portia, Juliet, Desdemona, Cressida, Bianca, Ophelia, and Cordelia. The 10 smallest moons all orbit inside the orbit of Miranda and are associated with the ring system around Uranus, and all of the moons rotate in the planet's equatorial plane, not the ecliptic of the solar system. Most of these moons are relatively dark, heavily cratered, and

geologically inactive bodies, with the exception of Miranda, which shows a series of ridges, valleys, and different morphological terrains. One of the most unusual are a series of oval wrinkled or faulted terrains of uncertain origin, but perhaps related to subsurface magmatism, impacts, or volcanism.

The ring system around Uranus was only discovered recently, in 1977, when the planet passed in front of a bright star and the rings were observed by astronomers studying the planet's atmosphere. There are nine known rings at 27,340–31,690 miles (44,000–51,000 km) from the planet's center, and each of these rings appears to be made of many much smaller rings. The rings are generally dark and narrow with widths of up to six miles (10 km), with wide spaces between the main rings ranging at 125–620 miles (200–1,000 km). The rings are only a few tens of meters thick. Most of the particles that make up the rings are dust to boulder-sized, dark-colored, and trapped in place by the gravitational forces between Uranus and its many moons.

urbanization and flash flooding   Urbanization is the process of building up and populating a natural habitat or environment, such that the habitat or environment no longer responds to input the way it did before being altered by humans. When heavy rains fall in an unaltered natural environment, the land surface responds to accommodate the additional water. Desert regions may experience severe erosion in response to the force of falling raindrops that dislodge soil, and also by overland flow during heavy rains. This causes upland channel areas to enlarge, becoming able to accommodate larger floods. Areas that frequently receive heavy rains may develop lush vegetative cover, which helps to break the force of the raindrops and reduce soil erosion, and the extensive root system holds the soil in place against erosion by overland flow. Stream channels may be large so that they can accommodate large volume floods.

When the natural system is altered in urban areas, the result can be dangerous. Many municipalities have paved over large parts of drainage basins, and covered much of the recharge area with roads, buildings, parking lots, and other structures. The result is that much of the water that used to seep into the ground and infiltrate into the groundwater system now flows overland into stream channels, which may themselves be modified or even paved over. The net effect of these alterations is that flash floods may occur much more frequently than in a natural system since more water flows into the stream system than before the alterations. The floods may occur with significantly lower amounts of rainfall as well, and since the water flows overland without slowly seeping into the ground, the flash floods may reach urban areas more quickly than the floods did before the alterations to the stream system. Overall the effect of urbanization is faster, stronger, and bigger floods, which have greater erosive power and do more damage. It is almost as if the natural environment responds to urban growth by increasing its ability to return the environment to its natural state.

Many examples of the effects of urbanization on flood intensity have been documented from California and the desert southwest. Urban areas such as Los Angeles, San Diego, Tucson, Phoenix, and other cities have documented the speed and severity of floods from similar rainfall amounts along the same drainage basin. What these studies have documented is that the floodwaters rise much more quickly after urbanization, and they rise up to four times the height they reached before urbanization, depending on the amount of paving over of the surface. The increased speed at which the floodwaters rise and the increased height to which they rise are directly correlated with the amount of land surface that is now covered over by roads, houses, and parking lots, blocking infiltration.

In natural systems, floods gradually wane after the highest peak passes, and the slow fall of the floodwaters is related to the stream system being recharged by groundwater that seeped into the shallow surface area during the heavy rainfall event. However, in urbanized areas the floodwaters not only rise quickly but also recede faster than in the natural environment. This is attributed to the lack of groundwater continuing to recharge the stream after the flood peak in urbanized areas.

Many other modifications to stream channels have been made in urbanized areas, with limited success in changing nature's course to suit human needs. Many stream channels have been straightened, which only causes the water to flow faster and have more erosive power. Straightening the stream course also shortens the stream length and thereby steepens the gradient. The stream may respond to this by aggrading and filling the channel with sediment, in an attempt to regain the natural gradient.

See also DESERT; GEOMORPHOLOGY; GROUNDWATER.

Further Reading

Arnold, J. G., P. J. Boison, and P. C. Patton. "Sawmill Brook—An Example of Rapid Geomorphic Change Related to Urbanization." *Journal of Geology* 90 (1982): 115–166.

valley breeze   *See* KATABATIC WINDS.

Valley of Ten Thousand Smokes   The eruption of Novarupta volcano in the Aleutian Ranges on the Alaska Peninsula in June 1912 produced a huge ash flow, hundreds of feet thick covering more than 40 square miles (100 km²). A foot of ash covered the distant port of Kodiak on Kodiak Island. The eruption was about 100 times greater than the 1982 eruption of Mount St. Helens and produced a Plinian ash cloud that probably rose 20 miles (32 km) in the atmosphere. As the ash from the eruption cooled, vapors were released and made their way to the surface. The result was that a huge area with thick ash developed thousands of steaming fumaroles, with still hot gasses rising from the fumaroles when an expedition mounted by the National Geographic Society in 1916 explored the site of the eruptions. During the expedition, botanist Robert Griggs gave the valley its name, and the Valley of Ten Thousand Smokes was declared a National Monument by President Woodrow Wilson soon after the National Geographic expedition. The landscape in the Valley of Ten Thousand Smokes is still barren, almost a century after the catastrophic eruptions.

The Valley of Ten Thousand Smokes is at the base of Katmai and Novarupta volcanoes, in Alaska's Aleutian chain of subduction-related convergent margin volcanoes. On the Alaskan Peninsula, the main andesitic volcanoes are adjacent to a subsidiary chain of rhyolite domes, and the 1912 eruptions were unusual in that magma from the main andesitic system beneath Mount Katmai migrated to and mixed with magma from the rhyolitic domes beneath Novarupta. Major activity began on June 6, 1912, when a five-day-long swarm of earthquakes yielded to a series of spectacular fissure eruptions around Novarupta, the opening of explosion craters, and the formation of a collapse caldera nearly two miles (3 km) across

and 4,000 feet (1,200 m) deep on the top of Mount Katmai. Magma from Katmai migrated underground to Novarupta, mixed with the magma there, and erupted from Novarupta, draining the magma from beneath Katmai, leading to the formation of the giant caldera. Eruptions from the fissures filled the valley with more than 650 feet (200 m) of ash flows and other pyroclastic material. An estimated 4.5 cubic miles (7 km³) of material was erupted from the volcanic vents.

Studies of the deposits in the Valley of Ten Thousand Smokes have led to the recognition that many huge sheets of siliceous volcanic rocks are not lava flows but are pyroclastic or ash flows. Some of the flows from Novarupta are dominantly white rhyolitic ash and pumice, whereas others include brown andesitic pumice. Some flows show strongly banded alternations between the andesitic and rhyolitic types, and there are numerous blocks of solid andesite and exotic sedimentary rocks. The ash flows are dominantly unsorted and unstratified (except at large distances from the vents) and are composed largely of sand and dust-sized pumice and lapilli particles. Some of the ash flows are reworked by rivers and show cross-beds and other fluvial features. Most of the ash flows are only weakly indurated, but some are welded and exhibit columnar jointing.

For such a large eruption, the eruption of Novarupta was preceded by only a few days of earthquakes. Early eruptions were minor pumice falls, followed rapidly by extremely violent eruptions that gradually over several days became more gas poor and less explosive, leading to the extrusion of the Novarupta dome. A late-stage eruption was particularly violent and may have been related to the injection of a new batch of magma into the chamber beneath Novarupta.

Mount Katmai, Novarupta, and the Valley of Ten Thousand Smokes are located in the remote reaches of the Alaskan Peninsula, and a few days warning was enough for the local residents to evacuate, so no deaths resulted from this huge eruption. However, if such a huge eruption were to occur in populated parts of the country, the results would be different. If Mount Katmai were located in St. Louis, Missouri, then Chicago would be buried under nearly a foot of ash, and the eruption would be heard as far as New York City and Los Angeles. The gases from the eruption would reach New York and Boston, tarnishing brass doorknobs and causing breathing problems in these distant cities.

See also CONVERGENT PLATE MARGIN PROCESSES; VOLCANO.

Further Reading

Curtis, G. H. "The Stratigraphy of the Ejecta from the 1912 Eruption of Mount Katmai and Novarupta, Alaska." *Geological Society of America Memoir* 116 (1968).

Fenner, Clarence N. "The Origin and Mode of Emplacement of the Great Tuff Deposit of the Valley of Ten Thousand Smokes." *National Geographic Society,* Contribution Technical Papers, Katmai Series, no. 1 (1923).

Vendian

In some classifications the latest Precambrian period in the Neoproterozoic is called the Vendian, stretching from 610–570 million years ago. The period is divided into the Varanger (610–590 million years ago) and Ediacara (590–570 million years ago) epochs, in turn divided into the Smalfjord, Mortensnes, Wonokan, and Poundian ages. The Vendian is one of two periods divided from the Sinian sub-era—the Sturtian (800–610 million years ago) and the Vendian.

The Vendian is an extremely important time in Earth history as it preserves some of the earliest known animal fossils. The Varanger was a time of widespread and even global glaciation reaching to low latitudes and leaving glacial deposits on many continents. This glaciation may have been induced by the breakup of Rodinia and the amalgamation of the supercontinent of Gondwana at the end of the Proterozoic. Carbonate platforms on continental margins involved in collisions were uplifted above sea level and croded, causing eroded silicates and carbonates to combine with and draw down atmospheric carbon dioxide, lowering global temperatures.

In many places around the world the Vendian is recognized as containing the oldest multicelled organisms. In south China, well-preserved assemblages of early multicelled metazoan fossils overlie rocks of the Varanger glaciation, making them the oldest known animal fossils. Other Vendian rocks and Ediacaran fossils are known from Russia, Ukraine, Northwest Canada, Australia, Norway, England, Newfoundland, Namibia, China, the southwest United States, and North Carolina.

See also LIFE'S ORIGINS AND EARLY EVOLUTION; PRECAMBRIAN; PROTEROZOIC; SINIAN; SUPERCONTINENT CYCLES.

Further Reading

Glaessner, Martin F. *The Dawn of Animal Life: A Biohistorical Study.* Cambridge: Cambridge University Press, 1984.

Lipps, Jere H., and Phillip W. Signor, eds., *Origin and Early Evolution of the Metazoa.* New York: Plenum, 1992.

Venus

The second planet from the Sun, Venus is the planet in our solar system that most closely resembles Earth with a planetary radius of 3,761 miles (6,053 km), or 95 percent of that of the Earth's radius. Venus orbits the Sun in a nearly circular path at 0.72 astronomical units (Earth-Sun distance) and has a mass equal to 81 percent of Earth, and density of 5.2 grams per cubic centimeter, very similar to Earth's 5.5 grams per cubic centimeter. The orbital period (year) of Venus is 0.62 Earth years, but it has a retrograde rotation about its axis of 243 days, with its north pole turned essentially upside down so that its equatorial pole is inclined at 177.4° from the orbital plane. One result of this tilt and the slow retrograde rotation is that the two effects to combine such that each day on Venus takes the equivalent of 117 Earth days. Another effect is that the slow rotation has not set up a geodynamo current in the planet's core, so Venus has no detectable mag-

Mosaic of Magellan radar images showing the surface of Venus, with Aphrodite Terra at the center of the image. This global view of the surface of Venus is centered at 180° east longitude. Magellan synthetic aperture radar mosaics from the first cycle of Magellan mapping are imposed onto a computer-simulated globe to create this image. Data gaps are filled with *Pioneer Venus orbiter* data, or a constant mid-range value. Simulated color is used to enhance small-scale structure. The simulated hues are based on color images recorded by the Soviet *Venera 13* and *14* spacecraft. The image was produced by the Solar System Visualization project and the Magellan science team at the Jet Propulsion Laboratory (JPL) Multimission Image Processing Laboratory and is a single frame from a video released at the October 29, 1991, JPL news conference. *(Photo by NASA)*

netic field. Thus, it lacks a magnetosphere to protect it from the solar wind so it is constantly bombarded by high-energy particles from the Sun. These particles lead to constant ionization of the upper levels of the atmosphere. Venus is usually one of the brightest objects in the sky (excepting the Sun and Moon) and is usually visible just before sunrise or just after sunset, since its orbit is close to the Sun.

The atmosphere of Venus is very dense and is composed mostly (96.5 percent) of carbon dioxide and is nearly opaque to visible radiation, so most observations of the planet's surface are based on radar reflectivity. Spacecraft and earth-based observations of Venus show that the atmospheric and cloud patterns on the planet are more visible in ultraviolet wavelengths, since some of the outer clouds seem to be made of mostly sulfuric acid which absorbs UV radiation, whereas other clouds reflect this wavelength, producing a highly contrasted image. These observations show that the atmosphere contains many large fast-moving clouds that are moving around 250 miles per hour (400 km/hr) and rotate around the planet on average once every four days. The atmospheric patterns on Venus resemble the jet stream systems on Earth. Aside from carbon dioxide, the atmosphere contains nitrogen plus minor or trace amounts of water vapor, carbon monoxide, sulfur dioxide, and argon.

Although many basic physical Venutian properties are similar to Earth's, the atmosphere on Venus is about 90 times more massive and extends to much greater heights than Earth's atmosphere. The mass of the Venutian atmosphere causes pressures to be exceedingly high at the surface, a value of 90 bars, compared to Earth's one bar. The troposphere, or region in which the weather occurs, extends to approximately 62 miles (100 km) above the surface. The upper layers of the Venutian atmosphere between about 75–45-mile (120–70-km) height are composed of sulfuric acid cloud layers, underlain by a mixing zone that is underlain by a layer of sulfuric acid haze at 30–20 miles (50–30 km). Below about 30 kilometers the air is clear.

The carbon-dioxide–rich and water-poor nature of the Venutian atmosphere has several important consequences for the planet. First of all, these gases are greenhouse gases that trap solar infrared radiation, an effect that has raised the surface temperature to an astounding 750°K (900°F, or 475°C, compared to 273°K for Earth). Second, surface process are much different on Venus than on Earth because of the elevated temperatures and pressures. There is no running water, rock behaves differently mechanically under high temperatures and pressures, and heat flow from the interior is different with such a drastically different surface temperature.

Earth and Venus had essentially the same amounts of gaseous carbon dioxide, nitrogen, and water in their atmospheres soon after the planets formed. However, since Venus is closer to the Sun (it is 72 percent of the distance from the Sun that Earth is), it receives about two times as much solar radiation as the Earth, which is enough to prevent the oceans from condensing from vapor. Without the oceans, the carbon dioxide does not dissolve in the seawater or combine with other ions to form carbonates, so the CO_2 and water stayed in the atmosphere. As the water vapor was lighter than the CO_2 it rose to high atmospheric levels and was dissociated into H and O ions, and the hydrogen escaped to space, whereas the O combined with other ions. Thus, the oceans never formed on Venus, and the water that could have formed them dissociated and is lost in space. Earth is only slightly further from the Sun, but conditions here are exactly balanced to allow water to condense, the atmosphere to remain near the equilibrium (triple) point of solid, liquid, and vapor water. These conditions allowed life to develop on Earth, and life further modified the atmosphere-ocean system to maintain its ability to support further life. Venus never had a chance.

With Venus's thick atmosphere, the surface must be mapped with cloud-penetrating radar from Earth and spacecraft. The surface shows many remarkable features including a division into a bimodal crustal elevation distribution reminiscent of Earth's continents and oceans. Most of the planet is topographically low, including about 27 percent flat volcanic lowlands and about 65 percent relatively flat plains, probably basaltic in composition, surrounded by volcanic flows. The plains are punctuated by thousands of volcanic structures including volcanoes and elongate narrow flows, including one that stretches 4,225 miles (6,800 km) across the surface. Some of the volcanoes are huge, with more than 1,500 having diameters of more than 13 miles (20 km), and one (Sapas Mons) more than 250 miles (400 km) across and almost one mile (1.5 km) high. About 8 percent of the planet consists of highlands made of elevated plateaus and mountain ranges. The largest continent-like elevated landmasses include the Australian-sized Ishtar Terra in the southern hemisphere, and African-sized Aphrodite Terra in equatorial regions. Ishtar Terra has interior plains rimmed by what appears to be folded mountain chains, and Venus's tallest mountain, Maxwell Mons, reaching 7 miles (11 km) in height above surrounding plains. Aphrodite Terra also has large areas of linear folded mountain ranges, many lava flows, and some fissures that probably formed from lava upwelling from crustal magma chambers, then collapsing back into the chamber instead of erupting.

The surface of Venus preserves numerous impact structures and unusual circular to oval structures and craters that are most likely volcanic in origin. Some of the most unusual appearing are a series of rounded pancake-like bulges that overlap each other on a small northern hemisphere elevated terrane named Alpha Regio. These domes are about 15.5 miles (25 km) across, and probably represent lava domes that filled and then had the magma withdrawn from them, forming a flat, cracked lava skin on the surface. There are many basaltic shield volcanoes scattered about the surface and some huge volcanic structures known as coronae. These are hundreds of kilometers across and are characterized by a series of circular fractures reflecting a broad upwarped dome, probably formed as a result of a plume from below. Many volcanoes dot the surface in and around coronae, and lava flows emanate and flow outward from some of them. Impact craters are known from many regions on Venus, but their abundance is much less than expected for a planet that has had no changes to its surface since formation. No impacts less than 2 miles (3 km) across are known since small meteorites burn up in the thick atmosphere before hitting the surface. The paucity of other larger impacts reflects the fact that the surface has been reworked and plated by basalt in the recent history of the planet, as confirmed by the abundant volcanoes, lava flows, and the atmospheric composition of the planet. It is likely that volcanism is still active on Venus.

Many of the surface features on Venus indicate some crustal movements. For instance, the folded mountain ranges show dramatic evidence of crustal shortening, and there are many regions of parallel fractures. Despite these features there have not been any features found that are indicative of plate tectonic types of processes operating. Most of the structures could be produced by crustal downsagging or convergence between rising convective plumes, in a manner similar to that postulated for the early Earth before plate tectonics was recognized.

viscosity   The resistance of a material to flow, sometimes measured as the ratio of tangential frictional force per unit area (stress) to the velocity gradient perpendicular to the direction of flow. The study of the flow of matter and the relationships between stress, strain, strain rate, and material and applied properties is known as rheology. Many geologic materials behave viscously, where strain accumulates as a function of time. For instance, water that flows downhill exhibits viscous behavior because it continuously accumulates strain with time as it flows. Likewise, many magmas and rocks continuously flow under an applied stress such that strain continues to build up with time under the applied stress. For viscous materials stress is proportional to strain rate, such that increasing the stress causes the rate of flow to increase. Viscous materials may exhibit Newtonian viscous behavior where a plot of stress versus strain rate is linear, or they may exhibit nonlinear viscous behavior where a stress-strain rate plot is nonlinear or curved. Some materials exhibit more complex nonviscous rheologies including some time-dependent behavior where strain accumulation and strain recovery are delayed in visco-elastic behavior. Materials that are distorted under applied stress but return to their undeformed state after the stress is removed exhibit elastic behavior, whereas plastic behavior describes a response of a material to stress where the material develops a strain without the loss of continuity, or without the development of fractures.

See also DEFORMATION OF ROCKS; IGNEOUS ROCKS; STRUCTURAL GEOLOGY.

volcanic bomb   Clots of magma that are more than 2.5 inches (64 mm) long and were partly or entirely plastic when erupted from a volcanic vent. In contrast, volcanic blocks are of the same size but were entirely solid when erupted. Pyroclastic fragments smaller than blocks and bombs are known as lapilli and ash. The sizes and shapes of volcanic bombs are extremely variable, with some bombs being more than 20 feet (6 m) in diameter Shapes of these pyroclastic bombs are determined by the rheology of the magma when it is erupted as well as the length and speed of the flight path the bomb took when it was ejected from the volcano, and any deformation that occurred during impact with the surface. Some bombs appear to show significant shape modification through the expansion of bubbles or gas vesicles in the magma. Expansion of gases often occurs after the bombs have landed and cause the skin of the bomb to expand and crack.

Basaltic magmas typically show spheroidal bombs or lapilli, whereas less-fluid basaltic eruptions often produce almond or spindle-shaped bombs. Ribbon-shaped bombs typically have vesicles that are elongate parallel to the length of the bomb and are often broken or shattered upon impact with the surface. Some bombs have formed around older material included in the magma and are referred to as cored bombs.

See also PYROCLASTIC.

Further Reading
Fisher, Richard V., and Hans U. Schminke. *Pyroclastic Rocks*. Berlin: Springer-Verlag, 1984.
Williams, Howell, and Alexander McBirney. *Volcanology*. San Francisco: Freeman, Cooper and Company, 1979.

volcaniclastic   *See* PYROCLASTIC.

volcano   A mountain or other constructive landform built by a singular eruption or a sequence of volcanic eruptions of molten lava and pyroclastic material from a volcanic vent. Volcanoes have many forms ranging from simple vents in the Earth's surface, through elongate fissures that erupt magma, to tall mountains with volcanic vents near their peaks. Volcanic landforms and landscapes are as varied as the volcanic rocks and eruptions that produce them. Shield volcanoes include the largest and broadest mountains on Earth. Mauna Loa is the largest shield volcano on the planet, more than 100 times as large as Mount Everest, a nonvolcanic mountain. Shield volcanoes have surface slopes of only a few degrees, produced by basaltic lavas that flow long distances before cooling and solidifying. Stratovolcanoes, in contrast, are the familiar steep-sided cones like Mount Fuji of Japan, or Mount Rainier of Washington. Stratovolcanoes are made of stickier lavas such as andesites and rhyolites, and they may have slopes of 30°. Other volcanic constructs include cinder or tephra cones, including the San Francisco Peaks in Arizona, which are loose piles of cinder and tephra. Calderas, like Crater Lake in Oregon, are huge circular depressions, often many kilometers in diameter, that are produced when deep magma chambers under a volcano empty out during an eruption, and the overlying land collapses inward producing a topographic depression. Yellowstone Valley occupies one of the largest calderas in the United States. This caldera contains many geysers, hot springs, and fumaroles related to groundwater circulating to great depths, being heated by shallow magma, and mixing with volcanic gases, which escape through minor cracks in the crust of the earth.

There is tremendous variety in the style of volcanic eruptions, both between volcanoes and from a single volcano during the course of an eruptive phase. This variety is related to the different types of magma produced by partial melting beneath the volcano, and by the amount of dissolved gases in the magma. Geologists have found it useful to classify volcanic eruptions based on the explosive characteristics of the eruption, the materials erupted, and by the type of landform produced by the volcanic eruption.

Tephra is material that comes out of a volcano during an eruption, and it may be thrown through the air or transported over the land as part of a hot moving flow. Tephra includes both new magma from the volcano and older broken rock fragments that got caught in the eruption. The term includes pyroclasts, which are rocks ejected by the volcano, and ash. Large pyroclasts are called volcanic bombs, smaller fragments are lapilli, and the smallest grade into ash.

While the most famous volcanic eruptions produce huge explosions, many eruptions are relatively quiet and nonexplosive. Nonexplosive eruptions have magma types that have low amounts of dissolved gases, and they tend to be basaltic in composition. Basalt flows easily and for long distances and tends not to have difficulty flowing out of volcanic necks. Nonexplosive eruptions may still be spectacular, as any visitor to Hawaii lucky enough to witness the fury of Pele, the Hawaiian goddess of the volcano, can testify. Mauna Loa, Kilauea, and other nonexplosive volcanoes produce a variety of eruption styles including fast-moving flows and liquid rivers of lava, lava fountains that spew fingers of lava trailing streamers of light hundreds of feet into the air, and thick sticky lava flows that gradually creep downhill. The Hawaiians devised clever names for these flows, including aa lava for blocky rubbly flows because walking across these flows in bare feet makes one exclaim "a! a!" in pain. Pahoehoes are ropy textured flows, after the Hawaiian term for rope.

Explosive volcanic eruptions are among the most dramatic of natural events on Earth. With little warning, long-

Volcanoes and Plate Tectonics

The types of volcanoes and associated volcanic eruptions are different in different tectonic settings because each tectonic setting produces a different type of magma. Mid-ocean ridges and intraplate "hot spot" types of volcanoes typically produce nonexplosive eruptions, whereas convergent tectonic margin volcanoes may produce tremendously explosive and destructive eruptions. Much of the variability in the eruption style may be related to the different types of magma produced in these different settings and to the amount of dissolved gases in these magmas. Magmas with large amounts of gases tend to be highly explosive, whereas magmas with low contents of dissolved gases tend to be nonexplosive.

Eruptions from mid-ocean ridges are mainly basaltic flows, with low amounts of dissolved gases. These eruptions are relatively quiet, with basaltic magma flowing in underwater tubes and breaking off in bulbous shapes called pillow lavas. The eruption style in these underwater volcanoes resembles toothpaste being squeezed out of a tube. Eruptions from mid-ocean ridges may be observed in the few rare places where the ridges emerge above sea level, such as Iceland. Eruptions there include lava fountains, where basaltic cinders are thrown a few hundred feet in the air and accumulate as cones of black, glassy fragments, and they also include long streamlike flows of basalt.

Hot-spot volcanism tends to be much like that at mid-ocean ridges, particularly where the hot spots are located in the middle of oceanic plates. The Hawaiian Islands have the most famous hot-spot type of volcano in the world, with the active volcanoes on the island of Hawaii known as Kilauea and Mauna Loa. Mauna Loa is a huge shield volcano, characterized by a gentle slope of a few degrees from the base to the top. This gentle slope is produced by basaltic lava flows that have a very low viscosity and can flow and thin out over large distances before they solidify. Magmas with high viscosity are much stickier and solidify in short distances, producing volcanoes with steep slopes. Measured from its base on the Pacific Ocean seafloor to its summit, Mauna Loa is the tallest mountain in the world, a fact attributed to the large distances that its low-viscosity lavas flow and to the large volume of magma produced by this hot-spot volcano.

Volcanoes associated with convergent plate boundaries produce by far the most violent and destructive eruptions. Recent convergent margin eruptions include Mount Saint Helens in Washington State, and Mount Pinatubo volcano in the Philippines. The magmas from these volcanoes tend to be much more viscous, higher in silica content, and they have the highest concentration of dissolved gases. Many of the dissolved gases such as water are released from the subducting oceanic plate as high mantle temperatures heat it up as it slides beneath the convergent margin volcanoes. These gases build up pressure beneath the volcano as it prepares to erupt, much like dissolved gas in a carbonated soda that is shaken before it is opened. When the gas and magma pressure in these volcanoes exceeds the weight of overlying rocks the volcano may suddenly explode, emitting enormous quantities of magma, pulverized rocks, and gases.

dormant volcanoes can explode with the force of hundreds of atomic bombs, pulverizing whole mountains and sending the existing material together with millions of tons of ash into the stratosphere. Explosive volcanic eruptions tend to be associated with volcanoes that produce andesitic or rhyolitic magma and have high contents of dissolved gases. These are mostly associated with convergent plate boundaries. Volcanoes that erupt magma with high contents of dissolved gases often produce a distinctive type of volcanic rock known as pumice, which is full of bubble wholes, in some cases making the rock light enough to float on water.

When the most explosive volcanoes erupt they produce huge eruption columns known as Plinian Columns (named after Pliny the Elder, the Roman statesman who died in 79 C.E. sampling volcanic gas during an eruption of Mount Vesuvius). These eruption columns can reach 28 miles (45 km) in height, and they spew hot turbulent mixtures of ash, gas, and tephra into the atmosphere where winds may disperse them around the planet. Large ash falls and tephra deposits may be spread across thousands of square kilometers. These explosive volcanoes also produce one of the scariest and most dangerous clouds on the planet. Nuée ardentes are hot glowing clouds of dense gas and ash that may reach temperatures of nearly 1,832°F (1,000°C), rush down volcanic flanks at 435 miles per hour (700 km/hr),

and travel more than 62 miles (100 km) from the volcanic vent. Nuée ardentes have been the nemesis of many a volcanologist and curious observer, as well as thousands upon thousands of unsuspecting or trusting villagers who previously lived on the flanks of volcanoes. Nuée ardentes are but one type of pyroclastic flow, which include a variety of mixtures of volcanic blocks, ash, gas, and lapilli that produce volcanic rocks called ignimbrites.

Most volcanic eruptions emanate from the central vents at the top of volcanic cones. However, many flank eruptions have been recorded, where eruptions blast out of fissures on the side of the volcano. Occasionally volcanoes blow out their sides, forming a lateral blast like the one that initiated the 1980 eruption of Mount Saint Helens in Washington State. This blast was so forceful that it began at the speed of sound, killing everything in the initial blast zone.

See also NUÉE ARDENTE; PLATE TECTONICS.

Further Reading

Fisher, R. V. *Out of the Crater: Chronicles of a Volcanologist.* Princeton: Princeton University Press, 2000.

Fisher, R. V., G. Heiken, and J. B. Hulen. *Volcanoes: Crucibles of Change.* Princeton: Princeton University Press, 1998.

Simkin, T., and R. S. Fiske. *Krakatau 1883: The Volcanic Eruption and Its Effects.* Washington D.C.: Smithsonian Institution Press, 1993.

W

wadi A term used for a dry streambed in desert regions of North Africa, the Middle East, and southwest Asia. Most wadis have gravelly bases, and many have steep to vertical sides that cut through mountainous regions, older fluvial deposits, whereas others have more gentle sides in places where the wadi cuts across plains and flatter regions. Wadis may be dry for many months or years and then be ravaged by flash floods during brief but intense rainfalls somewhere in the watershed, then quickly lose water to underlying porous sediments. In desert regions it is often possible to find water at shallow depths beneath wadi channels because the gravels and sands typically are very porous and may host a significant amount of subsurface base flow of water, not visible from the surface. Thus, many small villages and settlements in rural arid regions are located adjacent to wadis, where

Wadi Dayqa, northern Oman, showing small amount of water flowing over cobbles with more extensive subsurface base flow *(Photo by Timothy Kusky)*

wells are dug and water is carried to villages by pumps or other gravity-fed irrigation systems.

See also DESERT; GEOMORPHOLOGY; RIVER SYSTEM.

wash *See* WADI.

watershed *See* DRAINAGE BASIN.

water table The depth or level at which groundwater below the surface fills all of the pore spaces between individual grains of the regolith. The distribution of water in the ground can be divided into the unsaturated and the saturated zones. The top of the water table is defined as the upper surface of the saturated zone. Below this surface, all openings are filled with water. After a rainfall, much of the water stays near the surface, since clay in the near-surface horizons of the soil retains much water because of its molecular attraction. This forms a layer of soil moisture in many regions and is able to sustain seasonal plant growth. Some of this near surface water evaporates, and plants use some of the near surface water; other water runs directly off into streams. The remaining water seeps into the saturated zone, or into the water table. Once in the saturated zone it moves by percolation, gradually and slowly, from high areas to low areas, under the influence of gravity. These lowest areas are usually lakes or streams. Many streams form where the water table intersects the surface of the land. Once in the water table, the paths that individual particles follow vary, the transit time from surface to stream may vary from days to thousands of years along a single hillside! Water can flow upward because of high pressure at depth and low pressure in the stream. The level of the water table changes with different amounts of precipitation—in humid regions, it reflects the topographic variation, whereas in dry times or

454

places it tends to flatten out to the level of the streams of lakes. Water flows faster when the slope is greatest, so groundwater flows faster during wet times. The fastest rate of groundwater flow yet observed in the United States is 800 feet per year (250 m/yr).

See also ARTESIAN WELL; GROUNDWATER.

waves Geometrically regular and repeating undulations on the surface of water that move and transport energy from one place to another. Waves are generated by winds that blow across the water surface, and the frictional drag of the wind on the surface transfers energy from the air to the sea, where it is expressed as waves. The waves may travel great distances across entire ocean basins, and they may be thought of as energy in motion. This energy is released or transferred to the shoreline when the waves crash on the beach. It is this energy that is able to move entire beaches and erode cliffs grain-by-grain, slowly changing the appearance of the beach environment.

When waves are generated by winds over deep water, often from distant storms, they develop a characteristic spacing and height, known as the wavelength and height. The wave crest is the highest part of the wave, and the wave trough is the low point between waves. Wavelength is the distance between successive crests or troughs, the wave height is the vertical distance between troughs and crests, and amplitude is one-half of the wave height. Wave fronts are (imaginary) lines drawn parallel to the wave crests, and the wave moves perpendicular to the wave fronts. The time (in seconds) that it takes successive wave crests to pass a point is known as the wave period.

The height, wavelength and period of waves is determined by how strong the wind is that generates the waves, how long it blows, and the distance over which the wind blows (known as the fetch). As any sailor can tell you, the longer and stronger and the greater the distance the wind blows over, the longer the wavelength, the greater the wave height, and the longer the period.

It is important to remember that waves are energy in motion, and the water in the waves does not travel along with the waves. The motion of individual water particles as a wave passes is roughly a circular orbit that decreases in radius with depth below the wave. You have probably experienced this motion sitting in waves in the ocean, feeling yourself moving roughly up and down, or in a circular path, as the waves pass you.

The motion of water particles in waves changes as the water depth decreases, and the waves approach shore. At a depth equal to roughly one-half of the wavelength, the circular motion induced by the wave begins to feel the sea bottom, which exerts a frictional drag on the wave. This changes the circular particle paths to elliptical paths and causes the upper part of the wave to move ahead of the deeper parts. Eventually, the wave becomes oversteepened and begins to break, as the wave crashes into shore in the surf zone. In this surf zone, the water is actually moving forward, causing the common erosion, transportation, and deposition of sand along beaches.

Most coastlines are irregular and have many headlands, bays, bends, and changes in water depth from place to place. These variables cause waves that are similar in deep water to approach the shoreline at different angles in different places. You may have noticed on beaches, how waves may come ashore gently in one place, yet form nice breakers down the beach. These changes can be attributed to changes in water depth, steepness of the underwater slope, and the shape of the beach. Wave refraction occurs when a straight wave front approaches a shoreline obliquely. The part of the wave front that first feels shallow water (with a depth of less than one-half of the wavelength, known as the wave base) begins to slow down while the rest of the wave continues at its previous velocity. This causes the wave front to bend, or be refracted. Refraction tends to cause waves to bend toward

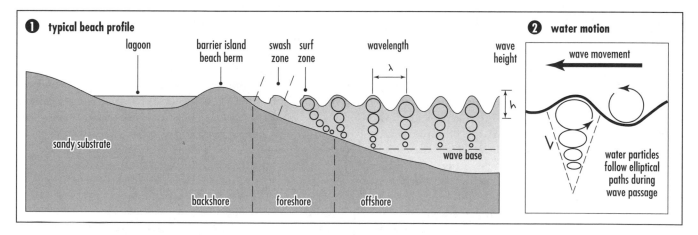

Wave motion showing orbital paths followed by water particles. Note that wave motion dies out a distance equal to half of the wavelength of the waves.

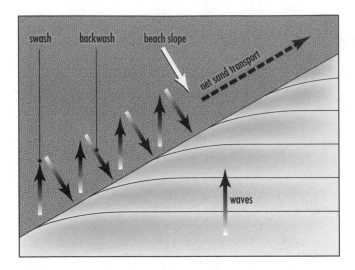

Waves that crash obliquely on beaches wash water and sand particles obliquely, parallel to the wave direction, up the beach face. When the backwash moves the water and sand particles back down the beach face, gravity moves the particles parallel to the beach face slope. The net effect is that the wash/backwash action on beaches moves sand particles down the beach face, sometimes many hundreds of meters per day.

headlands concentrating energy in those places, and to concentrate less energy in bays. Material is eroded from the headlands and transported to and deposited in the bays.

When waves approach a beach obliquely, a similar phenomenon occurs. Even though the waves are refracted, they may still crash onto the beach obliquely, moving sand particles sideways up the beach with each wave. As the wave returns to the sea in the backwash, the wave energy has been transferred to the shoreline (and has moved the sand grains), and gravity is the driving force moving the water and sand, which then moves directly downhill. The net result is that sand particles move obliquely up and straight down the beach slope, moving slightly sideways along the beach with each passing wave. It is common for individual sand grains to move almost a mile per day along beaches, through this process known as longshore drift. If the supply or transportation route of the sand particles is altered by human activity, such as the construction of seawalls or groins, the beach will respond by dramatically changing in some way.

See also BEACH; CONTINENTAL MARGIN; TSUNAMI.

weathering The process of mechanical and chemical alteration marked by the interaction of the lithosphere, atmosphere, hydrosphere, and biosphere. The resistance to weathering varies with climate, composition, texture, and how much a rock is exposed to the elements of weather. Weathering processes occur at the lithosphere/atmosphere interface. This is actually a zone that extends down into the ground to the depth that air and water can penetrate—in some regions this is a few meters, in others it is a kilometer or more. In this zone, the rocks make up a porous network,

with air and water migrating through cracks, fractures, and pore space. The effects of weathering can often be seen in outcrops on the side of the roads, where they cut through the zone of alteration into underlying bedrock. These roadcuts and weathered outcroppings of rock show some similar properties. The upper zone near the surface is made of soil or regolith in which the texture of the fresh rock is not apparent, a middle zone in which the rock is altered but retains some of its organized appearance, and a lower zone, of fresh unaltered bedrock.

Processes of Weathering

There are three main types of weathering. Mechanical weathering is the disintegration of rocks, generally by abrasion. Mechanical weathering is common in the talus slopes at the bottom of the mountains, along beaches, and along river bottoms. Chemical weathering is the decomposition of rocks through the alteration of individual mineral grains and is a common process in the soil profile. Biological weathering involves the breaking down of rocks and minerals by biological agents. Some organisms attack rocks for nutritional purposes; for instance, chitons bore holes through limestone along the seashore, extracting their nutrients from the rock.

Generally, mechanical and chemical weathering are the most important, and they work hand-in-hand to break down rocks into the regolith. The combination of mechanical, chemical, and biological weathering produces soils, or a weathering profile.

MECHANICAL WEATHERING There are several different types of mechanical weathering which may act separately or together to break down rocks. The most common process of mechanical weathering is abrasion, where movement of rock particles in streams, along beaches, in deserts, or along the bases of slopes causes fragments to knock into each other. These collisions cause small pieces of each rock particle to break off, gradually rounding the particles and making them smaller, and creating more surface area for processes of chemical weathering to act upon.

Some rocks develop joints or parallel sets of fractures, from differential cooling, the pressures exerted by overlying rocks, or tectonic forces. Joints are fractures along which no observable movement has occurred. Joints promote weathering in two ways: they are planes of weakness across which the rock can break easily, and they act as passageways for fluids to percolate along, promoting chemical weathering.

Crystal growth may aid mechanical weathering. When water percolates through joints or fractures, it can precipitate minerals such as salts, which grow larger and exert large pressures on the rock along the joint planes. If the blocks of rock are close enough to a free surface such as a cliff, large pieces of rock may be forced off in a rockfall, initiated by the gradual growth of small crystals along joints.

When water freezes to form ice, its volume increases by 9 percent. Water is constantly seeping into the open spaces provided by joints in rocks. When water filling the space in a joint freezes, it exerts large pressures on the surrounding rock. These forces are very effective agents of mechanical weathering, especially in areas with freeze-thaw cycles. They are responsible for most rock debris on talus slopes of mountains.

Heat may also aid mechanical weathering, especially in desert regions where the daily temperature range may be extreme. Rapid heating and cooling of rocks sometimes exerts enough pressures on the rocks to shatter them to pieces, thus breaking large rocks into smaller fragments.

Plants and animals may also aid mechanical weathering. Plants grow in cracks and push rocks apart. This process may be accelerated if plants such as trees become uprooted, or blown over by wind, exposing more of the underlying rock to erosion. Burrowing animals, worms, and other organisms bring an enormous amount of chemically weathered soil to the surface and continually turn the soils over and over, greatly assisting the weathering process.

CHEMICAL WEATHERING Minerals that form in igneous and metamorphic rocks at high temperatures and pressures may be unstable at temperatures and pressures at the Earth's surface, so they react with the water and atmosphere to produce new minerals. This process is known as chemical weathering. The most effective chemical agents are weakly acidic solutions in water. Therefore, chemical weathering is most effective in hot and wet climates.

Rainwater mixes with CO_2 from the atmosphere and from decaying organic matter, including smog, to produce carbonic acid according to the reaction:

$$H_2O + CO_2 \rightarrow H_2CO_3$$

Water + carbon dioxide $\rightarrow$ carbonic acid

Carbonic acid ionizes to produce the hydrogen ion (H^+), which readily combines with rock-forming minerals to produce alteration products. These alteration products may then rest in place and become soils, or be eroded and accumulate somewhere else.

Hydrolysis is a process that occurs when the hydrogen ion from carbonic acid combines with K-feldspar to produce kaolinite, a clay mineral, according to the reaction:

$$2\ KAlSi_3O_8 + 2\ H_2CO_3 + H_2O \rightarrow Al_2Si_2O_5(OH)_4 + 4\ SiO_2 + 2\ K^{+1} + 2\ HCO_3$$

feldspar + carbonic acid + water $\rightarrow$ kaolinite + silica + potassium + bicarbonate ion

This reaction is one of the most important reactions in chemical weathering. The product, kaolinite, is common in soils and is virtually insoluble in water. The other products, silica, potassium, and bicarbonate are typically dissolved in water and carried away during weathering.

Much of the material produced during chemical weathering is carried away in solution and deposited elsewhere, such as in the sea. The highest-temperature minerals are leached the easiest. Many minerals combine with oxygen in the atmosphere to form another mineral, by oxidation. Iron is very easily oxidized from the Fe+2 state to the Fe+3 state, forming goethite or with the release of water, hematite.

$$2FeO + OH \rightarrow Fe_2O_3 + H_2O$$

Different types of rock weather in distinct ways. For instance, granite contains K-feldspar and weathers to clays. Building stones are selected to resist weathering in different climates, but now, increasing acidic pollution is destroying many old landmarks. Chemical weathering results in the removal of unstable minerals and a consequent concentration of stable minerals. Included in the remains are quartz, clay, and other rare minerals such as gold and diamonds, which may be physically concentrated in placer deposits.

On many boulders, weathering only penetrates a fraction of the diameter of the boulder, resulting in a rind of the altered products of the core. The thickness of the rind itself is useful for knowing the age of the boulder, if rates of weathering are known. These types of weathering rinds are useful for determining the age of rock slides and falls, and the time interval between rockfalls in any specific area.

Exfoliation is a weathering process where rocks spall off in successive shells, like the skin of an onion. Exfoliation is caused by differential stresses within a rock formed during chemical weathering processes. For instance, feldspar weathers to clay minerals, which take up a larger volume than the original feldspar. When the feldspar minerals turn to clay, they exert considerable outward stress on the surrounding rock, which is able to form fractures parallel to the rock's surface. This need for increased space is accommodated by the minerals through the formation of these fractures, and the rocks on the hillslope or mountain are then detached from their base and are more susceptible to sliding or falling in a mass wasting event.

If weathering proceeds along two or more sets of joints in the subsurface, it may result in shells of weathered rock, which surround unaltered rocks, looking like boulders. This is known as spheroidal weathering. The presence of the several sets of joint surfaces increases the effectiveness of chemical weathering, because the joints increase the available surface area to be acted on by chemical processes. The more subdivisions within a given volume, the greater the surface area.

Factors That Influence Weathering

The effectiveness of weathering processes is dependent upon several different factors, explaining why some rocks weather one way at one location, and a different way in another location. Rock type is an important factor in determining the weathering characteristics of a hillslope, because different minerals react differently to the same weathering conditions. For instance, quartz is resistant to weathering, and quartz-rich rocks typically form large mountain ridges. Conversely, shales readily weather to clay minerals, which are easily washed away by water, so shale-rich rocks often occupy the bottoms of valleys. Examples of topography being closely related to the underlying geology in this manner are abundant in the Appalachians, Rocky Mountains, and most other mountain belts of the world.

Rock texture and structure is important in determining the weathering characteristics of a rock mass. Joints and other weaknesses promote weathering by increasing the surface area for chemical reactions to take place on, as described above. They also allow water, roots, and mineral precipitates to penetrate deeply into a rock mass, exerting outward pressures that can break off pieces of the rock mass in catastrophic rockfalls and slides.

The slope of a hillside is important for determining what types of weathering and mass wasting processes occur on that slope. Steep slopes let the products of weathering get washed away, whereas gentle slopes promote stagnation and the formation of deep weathered horizons.

Climate is one of the most important factors in determining how a site weathers. Moisture and heat promote chemical reactions, so chemical weathering processes are strong, fast, and dominant over mechanical processes in hot wet climates. In cold climates, chemical weathering is much less important. Mechanical weathering is very active during freezing and thawing, so mechanical processes such as ice wedging tend to dominate over chemical processes in cold climates. These differences are exemplified by two examples of weathering. In much of New England, a hike over mountain ridges will reveal fine, millimeter-thick striations that were formed by glaciers moving over the region more than 10 thousand years ago. Chemical weathering has not removed even these one-millimeter thick marks in 10 thousand years. In contrast, new construction sites in the tropics, such as roads cut through mountains, often expose fresh bedrock. In a matter of 10 years these road cuts will be so deeply eroded to a red soil-like material called gruse, that the original rock will not be recognizable.

As in most things, time is important. It takes tens of thousands of years to wash away glacial grooves in cold climates, but in tropics, weathered horizons that extend to hundreds of meters may form over a few million years.

See also SOILS.

Wegener, Alfred Lothar (1880–1930) German *Meteorologist, Geophysicist* Alfred Wegener is well known for his studies in meteorology and geophysics and is considered by many to be the father of continental drift. He completed his studies in Berlin and presented a thesis on astronomy. His interest in meteorology and geology led him on a Danish expedition to northeastern Greenland in 1906–1908. This was the first of four Greenland expeditions he would make, and this area remained one of his dominant interests. Wegener is famous for being the first person to come up with the idea for continental drift. He studied the apparent correspondence between the shapes of the coastlines of western Africa and eastern South America. Later on he learned that evidence of paleontological similarities was being used to support the theory of the "land bridge" that had connected Brazil to Africa. He continued to study the paleontological and geological evidence, concluded that these similarities demanded an explanation, and wrote an extended account of his continental drift theory in his book *Die Entstehung der Kontinente und Ozeane* (The Origin of the Continents and Oceans). As a meteorologist he began to look at ancient climates and used paleoclimatic evidence he found to strengthen his theory of continental drift. Wegener was by no means the first to think of the theory of continental drift. However, he was the first to go to great lengths to develop and establish the theory. He is also known for his work on dynamics and thermodynamics of the atmosphere, atmospheric refraction and mirages, optical phenomena in clouds, acoustical waves, and the design of geophysical instruments.

Werner, Abraham Gottlob (1749–1817) Polish *Geologist, Mineralogist* Abraham Gottlob Werner was enormously influential in the field of geology. Werner developed techniques for identifying minerals using human senses, and this appealed to a broad audience interested in learning more about geology. Werner also proposed a new classification for certain geologic formations. In the 18th century, rocks were explained and were classified into three categories with accordance to the "biblical flood," including Primary for ancient rocks without fossils (believed to precede the flood), Secondary for rocks containing fossils (often attributed to the flood itself) and Tertiary for sediments believed to have been deposited after the flood. Werner did not dispute the commonly held belief in the biblical flood, but he did discover a different group of rocks that did not fit this classification: rocks with a few fossils that were younger than primary rocks but older than secondary rocks. He called these "transition" rocks. Geologists of succeeding generations classified these rocks into the geologic periods still accepted today.

white smoker chimneys *See* BLACK SMOKER CHIMNEYS.

wildcat Oil and gas wells that are drilled on structures, formations, depths, or regions not yet known to contain hydrocarbons, or that have not yet yielded any oil or gas, are known as wildcat wells. These risky wells are also called outpost wells, deeper-pool or shallower-pool wells, and exploratory wells. Wildcat wells are routinely drilled by large companies but may cost hundreds of thousands or even millions of dollars to drill, so the risk is high. Some small petroleum exploration companies have been made or broken by wildcat wells, and investors have gotten rich or bankrupt by investing fortunes in wildcatters.

See also HYDROCARBON; PETROLEUM.

Witwatersrand basin   The Witwatersrand basin on South Africa's Kaapvaal craton is one of the best known of Archean sedimentary basins and contains some of the largest gold reserves in the world, accounting for more than 55 percent of all the gold ever mined in the world. Sediments in the basin include a lower flysch-type sequence, and an upper molassic facies, both containing abundant silicic volcanic detritus. The strata are thicker and more proximal on the northwestern side of the basin that is at least locally fault bounded. The Witwatersrand basin is a composite foreland basin that developed initially on the cratonward side of an Andean arc, similar to retroarc basins forming presently behind the Andes. A continental collision between the Kaapvaal and Zimbabwe cratons 2.7 billion years ago caused further subsidence and deposition in the Witwatersrand basin. Regional uplift during this later phase of development placed the basin on the cratonward edge of a collision-related plateau, now represented by the Limpopo Province. There are many similarities between this phase of development of the Witwatersrand basin and basins such as the Tarim and Tsaidam north of the Tibetan Plateau.

The Witwatersrand basin is an elongate trough filled predominantly by 2.8–2.6-billion-year-old clastic sedimentary rocks of the West Rand and Central Rand groups, together constituting the Witwatersrand supergroup. These are locally, in the northwestern parts of the basin, underlain by the volcanosedimentary Dominion group. The structure trends in a northeasterly direction parallel with, but some distance south of, the high-grade gneissic terrane of the Limpopo Province. The high-grade metamorphism, calc-alkaline plutonism, uplift, and cooling in the Limpopo are of the same age as and closely related to the evolution of the Witwatersrand basin. Strata dip inward with dips greater on the northwestern margin of the basin than on the southeastern margin. The northwestern margin of the basin is a steep fault that locally brings gneissic basement rocks into contact with Witwatersrand strata to the south. Dips are vertical to overturned at depth near the fault, but only 20° near the surface, demonstrating that this is a thrust fault. A number of folds and thrust faults are oriented parallel to the northwestern margin of the basin.

The predominantly clastic fill of the Witwatersrand basin has been divided into the West Rand and the overlying Central Rand groups, which rest conformably on the largely volcanic Dominion group. The Dominion group was deposited over approximately 9,000 square miles (15,000 km²), but it is correlated with many similar volcanic groups along the northern margin of the Kaapvaal craton. The Dominion group and its correlatives, and a group of related plutons, has been interpreted as the products of Andean arc magmatism, formed above a 2.8-billion-year-old subduction zone that dipped beneath the Kaapvaal craton. The overlying West Rand and Central Rand groups were deposited in a basin at least 50,000 square miles (80,000 km²) in area. Stratigraphic thicknesses of the West Rand group generally increase toward the fault-bounded northwestern margin of the basin, whereas thicknesses of the Central Rand group increase toward the center of the basin. Strata of both groups thin considerably toward the southeastern basin margin. The northeastern and southwestern margins are poorly defined, but some correlations with other strata (such as the Godwan formation) indicate that the basin was originally larger than the present basin. Strata that were originally deposited north of Johannesburg are buried, removed by later uplift, omitted by igneous intrusion, and cut out by faulting.

The West Rand group consists of southeastward-tapering sedimentary wedges that overlie the Dominion group, and onlap granitic basement in many places. The maximum thickness of the West Rand group, 25,000 feet (7,500 m), occurs along the northern margin of the basin, and the group thins southeast to a preserved thickness of 2,700 feet (830 m) near the southern margin. Shale and sandstone in approximately equal proportions characterize the West Rand group, and a thin horizon of mafic volcanics is locally present. This volcanic horizon thickens to 800 feet (250 m) near the northern margin of the basin but is absent in the south. The West Rand group contains mature quartzites, minor chert, and sedimentation patterns indicating both tidal and aeolian reworking. Much of the West Rand group is an ebb-dominated tidal deposit later influenced by beach-swash deposition. More shales are preserved near the top of the group. Overall, the West Rand group preserves a transition from tidal flat to beach then deeper water deposition, which indicates a deepening of the Witwatersrand basin during deposition. Upper formations in the West Rand group contain magnetic shales and other fine-grained sediments suggestive of a distal shelf or epicontinental sea environment of deposition.

The lower West Rand group records subsidence of the Witwatersrand basin since the sediments grade vertically from beach deposits to a distal shallow marine facies. This inundation of shallow water sedimentary environments suggests that the subsidence was rapid, and the absence of coarse, immature, fanglomerate type sediments suggests that the subsidence was accommodated by flexure and not faulting. A decreasing rate of subsidence and/or a higher rate of clastic sediment supply is indicated by the progressively shallower water facies deposits in the upper West Rand group. Numerous silicic volcanic clasts in the West Rand group indicate that a volcanic arc terrane to the north was contributing volcanic detritus to the Witwatersrand basin. Additionally, the presence of detrital ilmenite, fuchsite, and chromite indicate that an ultramafic source such as an elevated greenstone belt was also contributing detritus to the basin.

The Central Rand group was deposited conformably on top of the West Rand group and attains a maximum preserved thickness of 9,500 feet (2,880 m) northwest of the center of the basin, and north of the younger Vredefort impact structure. Sediments of the Central Rand group consist of coarse-grained graywackes and conglomerates along with subordinate quartz arenite interbedded with local lacustrine or shallow marine shales and siltstones. The conglomerates are typically poorly sorted and larger clasts are well rounded, while the smaller pebbles are angular to subangular. Paleocurrent indicators show that the sediments prograded into the basin from the northwestern margin in the form of a fan-delta complex. This is economically important because numerous goldfields in the Central Rand group are closely associated with major entry points into the basin. Some transport of sediments along the axis of the basin is

indicated by paleocurrent directions in a few locations. A few tuffaceous horizons and a thin mafic lava unit are found in the Central Rand group in the northeast part of the basin. The great dispersion of unimodal paleocurrent directions derived from most of the Central Rand group indicates that these sediments were deposited in shallow braided streams on coalescing alluvial fans. The paleorelief is estimated at 20 feet (6 m) in areas proximal to the source, and 1–2 feet (0.5 m) in more distal areas. Some of the placers in the Central Rand group have planar upper surfaces, commonly associated with pebbles and heavy placer mineral concentrations, which may be attributed to reworking by tidal currents. Clasts in the conglomerates include vein quartz, quartz arenite, chert, jasper, silicic volcanics, shales and schists, and other rare rocks.

The Central Rand group contains a large amount of molassic type sediments disposed as sand and gravel bars in coalesced alluvial fans and fluvial systems. The West Rand/Central Rand division of the Witwatersrand basin into a lower flysch-type sequence and an upper molasse facies is typical of foreland basins. Extensive mining of paleoplacers

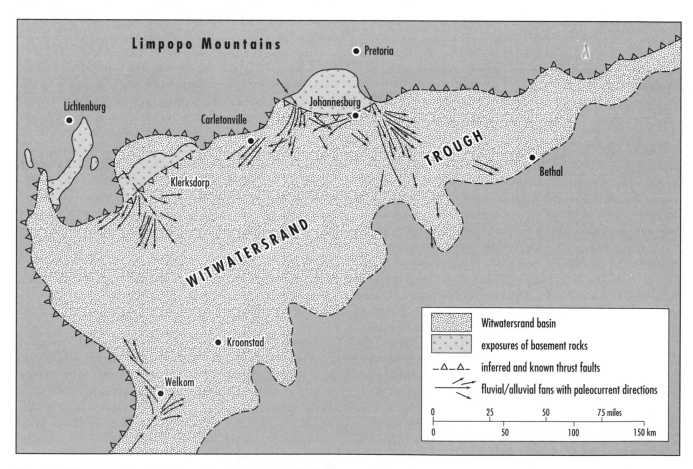

Map of the Witwatersrand basin showing paleocurrent directions indicating several discrete sources for detrital sediments in the basin, located mostly north of the basin

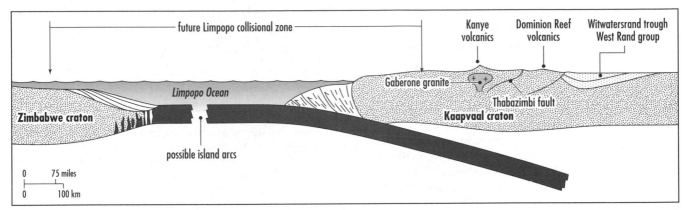

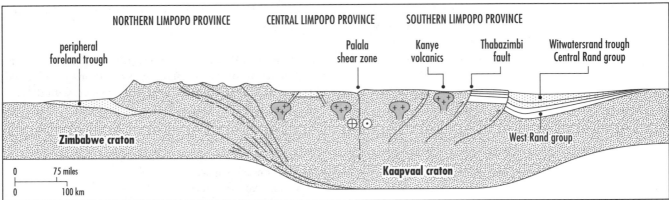

Two stages in the tectonic evolution of the Witwatersrand basin showing early formation as a foreland trough that developed behind an Andean arc complex built on the Kaapvaal craton. A later stage shows the collision between the Kaapvaal and Zimbabwe cratons, forming an uplifted Tibetan-style plateau, with further deposition in the Witwatersrand basin in a collisional foredeep.

for gold and uranium has enabled the dendritic paleodrainage patterns to be mapped, and the points of entry into the basin to be determined. The source of the Central Rand sediments was a mountain range located to the northwest of the basin, and this range contained a large amount of silicic volcanic material.

The growth of folds parallel to the basin margin during sedimentation and the preferential filling of synforms by some of the mafic lava flows in the basin indicate that folding was in progress during Central Rand group sedimentation. Deformation of this kind is diagnostic of flexural foreland basins, and studies show that the depositional axis of the basin migrated southeastward during sedimentation, with many local unconformities related to tilting during flexural migration of the depositional centers.

The Witwatersrand basin exhibits many features that are characteristic of foreland basins including an asymmetric profile with thicker strata and steeper dips toward the mountainous flank, a basal flysch sequence overlain by molassic-type sediments, and thrust faults bounding one side of the basin. Compressional deformation was in part synsedimentary, and associated folds and faults trend parallel to the basin margins, and the depositional axis migrated away from the thrust

front with time, as in younger foreland basins. Stratigraphic relationships within the underlying Dominion group, the presence of silicic volcanic clasts throughout the stratigraphy, and minor lava flows within the basin suggest that the foreland basin was developed behind a volcanic arc, partly preserved as the Dominion group. Sediments of the West Rand group are interpreted as deposited in an actively subsiding foreland basin developed adjacent to an Andean margin and fold thrust belt.

Deformation in the Limpopo Province and northern margin of the Kaapvaal craton are related to a collision between the northern Andean margin of the Kaapvaal craton with a passive margin developed on the southern margin of the Zimbabwe craton that began before 2.64 billion years ago, when Ventersdorp rifting, related to the collision, began. It is possible that some of the rocks in the Witwatersrand basin, particularly the molasse of the Central Rand group, may represent erosion of a collisional plateau developed as a consequence of this collision. The plateau would have been formed in the region between the Witwatersrand basin and the Limpopo Province, a region characterized by a deeply eroded gneiss terrane. A major change in the depositional style occurs in the Witwatersrand basin between the Central

Rand and West Rand groups, and this break may represent the change from Andean arc retroarc foreland basin sedimentation to collisional plateau erosion-related phases of foreland basin evolution.

Paleoplacers in the Witwatersrand basin have yielded more than 850 million tons of gold, dwarfing all the world's other gold placer deposits put together. Many of the placer deposits (called reefs in local terminology) preserved detrital gold grains on erosion surfaces, along foreset beds in cross-laminated sandstone and conglomerate, in trough cross-beds, in gravel bars, and as detrital grains in sheet sands. Most of the gold is located close to the northern margin of the basin in the fluvial channel systems. Some of the gold flakes in more distal areas were trapped by stromatolite-like filamentous algae, and some appears to have even have been precipitated by some types of algae, although it is more likely that these are fine recrystallized grains that were trapped by algal filaments. Besides gold there are more than 70 ore minerals recognized in the Witwatersrand basin, and most of these are detrital grains, and others are from metamorphic fluids. The most abundant detrital grains include pyrite, uraninite, brannerite, gold, arsenopyrite, cobaltite, chromite, and zircon. Gold mining operations in the Witwatersrand employ more than 300,000 people and have led to the economic success of South Africa.

See also ARCHEAN; CONVERGENT PLATE MARGIN PROCESSES; CRATONS; KAAPVAAL CRATON.

Further Reading

Antrobus, E. S. A., ed., *Witwatersrand Gold—100 Years*. Johannesburg: Geological Society of South Africa, 1986.

Burke, Kevin, William S. F. Kidd, and Timothy M. Kusky. "Is the Ventersdorp Rift System of Southern Africa Related to a Continental Collision between the Kaapvaal and Zimbabwe Cratons at 2.64 Ga ago?" *Tectonophysics* 11 (1985).

———. "Archean Foreland Basin Tectonics in the Witwatersrand, South Africa." Tectonics 5 (1986).

Tankard, Anthony J., M. P. A. Jackson, Ken A. Eriksson, David K. Hobday, D. R. Hunter, and W. E. L. Minter. *Crustal Evolution of Southern Africa: 3.8 Billion Years of Earth History*. New York: Springer-Verlag, 1982.

wollastonite A calc-silicate mineral of the pyroxenoid group found in contact metamorphosed carbonate rocks, and in the invading igneous rocks near their contacts with the carbonates as a result of contamination. Named after Wollaston, Massachusetts, wollastonite may appear as tabular twinned crystals or in cleavable masses and has colors that range from white to gray, brown, red and yellow. The chemical formula for wollastonite is $CaSiO_3$.

See also METAMORPHISM.

X

xenolith   A foreign inclusion in an igneous rock, typically a fragment of the local surrounding country rock. Xenoliths are commonly found near the margins of igneous intrusions and are typically flattened parallel to the margins of the pluton, reflecting high strains associated with emplacement and perhaps expansion of the pluton. Many studies of plutons use their compositions to learn about the types of rocks that the magma passed through from its source region to its emplacement location. In most cases the xenoliths are from within a few kilometers below the pluton, but in other cases the xenoliths come from much deeper and contain important information about deep crustal or even mantle petrography and petrology.

Suites of deep xenoliths found in continental and oceanic alkalic basalts are compositionally and mineralogically diverse and include some mantle rocks including dunite, lherzolite, and spinel lherzolite. Kimberlites typically contain large amounts of xenoliths including many deep crustal and mantle xenoliths, including diamonds. Diamonds are derived from depths of approximately 93 miles (150 km), making kimberlites the host of the most deep-level xenoliths, and an important source of information about the deep mantle roots beneath Archean cratons, and the upper mantle.

See also IGNEOUS ROCKS; KIMBERLITE; PETROLOGY.

xenotime   A tetragonal mineral that is isostructural with zircon and has the formula YPO_4. Xenotime may be yellow, brown, or even red and typically has high concentrations of rare earth elements, including uranium, thorium, beryllium, and zirconium, as well as aluminum and calcium. Xenotime is a moderately rare mineral but occurs as an accessory in some granites and pegmatites.

See also MINERALOGY; ZIRCON.

X-ray fluorescence   A geochemical technique that is commonly used to determine the concentration of major and trace elements in minerals and rocks, at concentrations ranging from less than one part per million to high concentrations. Samples of the rocks and minerals are cleaned and powdered, then melted and fused into a glass disc or pressed powder pellet. The sample is then irradiated with primary X rays and reemits secondary X rays with wavelengths and energies that are characteristic of the minerals present in the sample. To determine the concentrations of the elements in the sample, the peaks from the analysis are compared to similar peaks from a standard sample with known concentrations of the same elements.

See also GEOCHEMISTRY.

Y

yardangs Yardangs are elongate streamlined wind-eroded ridges, which resemble an overturned ship's hull sticking out of the water. These unusual features are formed by abrasion, by the long-term sandblasting along specific corridors. The sandblasting leaves erosionally resistant ridges but removes the softer material which itself will contribute to sandblasting in the downwind direction and eventually contribute to the formation of sand, silt, and dust deposits.

See also DESERT; GEOMORPHOLOGY.

Yellowstone National Park The northwest corner of Wyoming, and adjacent parts of Idaho and Montana, was established as a national park in 1872 by President Ulysses S. Grant, and it remains the largest national park in the conterminous United States. The park serves as a large nature preserve and has large populations of moose, bear, sheep, elk,

bison, numerous birds, and a diverse flora. The park sits on a large upland plateau resting at about 8,000 feet (2,400 m) elevation straddling the continental divide. The plateau is surrounded by mountains that range 10,000–14,000 feet (3,000–4,250 m) above sea level. Most of the rocks in the park formed from a massive volcanic eruption that occurred 600,000 years ago, forming a collapse caldera that is 28 miles (45 km) wide and 46 miles (74 km) long. The deepest part of the caldera is now occupied largely by Yellowstone Lake. The region is still underlain by molten magma that contributes heat to the groundwater system, which boasts more than 10,000 hot springs, 200 geysers, and numerous fumaroles, vents, and hot mud pools. The most famous geyser in the park is Old Faithful, which erupts an average of once every 64.5 minutes blowing 11,000 gallons (41,500 l) of water 150 feet (46 m) into the air. The most famous hot springs include

Yellowstone Falls on the Yellowstone River in Yellowstone National Park *(Photo by Timothy Kusky)*

Fumaroles and geysers in a geyser basin, Yellowstone National Park *(Photo by Timothy Kusky)*

Mammoth hot springs, Yellowstone National Park, showing thick terraces and pools of travertine mineral deposits *(Photo by Timothy Kusky)*

Mammoth hot springs on the northern side of the park, where giant travertine and mineral terraces have formed from the spring, and where simple heat-loving (thermophyllic) organisms live in the hot waters. Other remarkable features of the park include the petrified forests buried and preserved by the volcanic ash, numerous volcanic formations including black obsidian cliff, and waterfalls and canyons including the spectacular Lower Falls in the Grand Canyon of the Yellowstone.

The massive eruption from Yellowstone caldera 600,000 years ago covered huge amounts of the western United States with volcanic ash, and if such an eruption were to occur today, the results would be devastating. There has been some concern recently about some increase in some of the thermal activity in Yellowstone, although it is probably related to normal changes within the complex system of heated groundwater and seasonal or longer changes in the groundwater system. First, Steamboat geyser, which had been quiet for two decades, began erupting in 2002. New lines of fumaroles formed around Nymph Lake, including one line 250 feet (75 m) long that forced the closure of the visitor trail around the geyser basin. Other geysers have seen temperature increases from 152°F (67°C) to 190°F (88°C) over a several-month period. Other changes include a greater discharge of steam from some geysers, changes in the frequency of eruptions, and a greater turbidity of thermal pools. Perhaps most worrying is the discovery of a large bulge beneath Yellowstone Lake, although its age and origin are uncertain. Fears are that the bulge may be related to the emplacement of magma to shallow crustal levels, a process that sometimes precedes eruptions. However, the bulge was recently discovered because new techniques are being used to map the lake bottom. The feature has an unknown age and may have been there for decades or even hundreds of years.

Yellowstone Park is underlain by a hot spot, the surface expression of a mantle plume. As the North American plate has migrated 280 miles (450 km) southwestward in the past 16 million years with respect to this hot spot, the volcanic effects migrated from the Snake River Plain to the Yellowstone Plateau. There is currently a parabolic shaped area of seismicity, active faulting, and centers of igneous intrusion that is centered around the parabolic area, all of which are migrating northeastward. Heat and magma from this mantle plume has emplaced as much as 7.5 miles (12 km) of mafic magma into the continental crust overlying the plume along this trace, causing the surface eruptions of the massive Snake River Plain flood basalts and the Yellowstone volcanics. It is likely that, on geological timescales, massive volcanism and other effects of this hot spot will continue and also will slowly move northeast.

See also GEYSER.

Further Reading

Leeman, William P. "Development of the Snake River Plain—Yellowstone Plateau Province, Idaho and Wyoming: An Overview and Petrologic Model." In *Cenozoic Geology of Idaho, Bulletin 26,* edited by Bil Bonnichsen and R. M. Breckenridge. Moscow: Idaho Bureau of Mines and Geology, 1982.

Morgan, Lisa A., David J. Doherty, and William P. Leeman. "Ignimbrites of the Eastern Snake River Plain: Evidence for Major Caldera Forming Eruptions." *Journal of Geophysical Research* 89 (1984).

Morgan, W. Jason. "Deep Mantle Convection Plume and Plate Motions." *American Association of Petroleum Geologists Bulletin* 56 (1972).

Rogers, David W., William R. Hackett, and H. Thomas Ore. "Extension of the Yellowstone Plateau, Eastern Snake River Plain, and Owyhee Plateau." *Geology* 18 (1990).

Yosemite Valley   Yosemite Valley and National Park are located in eastern California in the Sierra Nevada Mountains. The main feature of the park is a beautiful glacially scoured valley, surrounded by peaks, huge monoliths, and pinnacles

Full moon over the Half Dome, Yosemite Park, California *(Photo by Timothy Kusky)*

including the famous Half Dome, reaching heights of 4,800 feet (1,465 m). The park was established in 1890 largely as a result of the efforts of the conservationist John Muir and was designated a World Heritage Site in 1984. Yosemite Falls flows out of a hanging valley into the main Yosemite Valley, dropping 2,425 feet (740 m), making it the tallest waterfall in North America. The park includes abundant stands of giant and other sequoias, meadows, and acts as a preserve for many other fauna and flora, preserving great biological diversity.

See also GLACIER; ROCKY MOUNTAINS.

Z

Zagros and Makran Mountains The Zagros are a system of folded mountains in western and southern Iran, extending about 1,100 miles (177 km) from the Turkish-Russian-Iranian border, to Zendam fault north of the Straits of Hormuz. The Makran Mountains extend east from the Zagros, through the Baluchistan region of Iran, Pakistan, and Afghanistan. The mountains form the southern and western borders of the Iranian plateau and Dasht-e-Kavir and Dasht-e-Lut Deserts. The northwestern Zagros are forested and snowcapped and include many volcanic cones, whereas the central Zagros are characterized by many cylindrical folded ridges and interridge basins. The southwest Zagros and Makran ranges are characterized by more subdued topography with bare rock, sand dunes, and lowland salt marshes. Many major oil fields are located in the western foothills of the Central Zagros, where many salt domes have punctured through overlying strata creating many oil traps.

Southwestern Central Iran has been an active continental margin since the Mesozoic, with at least three main phases of magmatic activity related to subduction of Tethyan oceanic crust beneath the mountain ranges. Late Cretaceous magmatism in the Makran formed above subducting oceanic crust related to the Oman ophiolite preserved on the Arabian continental margin. In the late Eocene, the axis of active magmatism shifted inland away from the Mesozoic magmatic belt, but then shifted back during the Oligocene-Miocene. The Oligocene-Miocene magmas are also related to subduction of oceanic crust, suggesting that the Arabian-Iranian collision did not begin until the Miocene. Most of the southern Zagros consist of folded continental margin sediments of the Arabian platform deformed since the Miocene, and mostly since the Pliocene. In contrast, the Makran is an oceanic accretionary wedge consisting of folded Cretaceous to Eocene flysch and ribbon chert-bearing mélange that is resting above subducting oceanic crust of the Gulf of Oman. A large ophiolitic sheet is thrust over the ophiolitic mélange and flysch and is part of a large ophiolitic belt that stretches the length of the Makran-Zagros ranges, falling between the Cenozoic volcanics and accretionary wedge/folded platform rocks of the Makran and Zagros. The main differences between the Zagros and the Makran can be attributed to the fact that continent/continent collision has begun in the Zagros but has not yet begun in the Makran.

Iran is seismically active, as shown by the devastating magnitude 6.7 earthquake that destroyed the ancient walled fortress city Bam on December 26, 2003, killing an estimated 50,000 people. The Zagros belt is extremely active, where thrust-style earthquakes occur beneath a relatively ductile layer of folded sedimentary rocks on the surface. The Makran accretionary wedge is also seismically active, where subduction zone and upper plate accretionary wedge earthquakes occur. The boundary between the Makran and Zagros is a structurally complex region where many strike-slip faults, including the Zendan fault and related structures, rupture to the surface. The Bam earthquake was a strike-slip earthquake, related to this system of structures. The Central Iranian plateau is also seismically active and experiences large magnitude earthquakes that rupture to the surface.

See also CONVERGENT PLATE MARGIN PROCESSES; KUWAIT.

Further Reading

Berberian, F., and Manuel Berberian. "Tectono-plutonic Episodes in Iran." In *Zagros-Hindu Kush-Himalaya Geodynamic Evolution*, edited by Harsh K. Gupta and Frances M. Delany. American Geophysical Union Geodynamics Series 3, 1981.

Berberian, Manuel. "Active Faulting and Tectonics of Iran." In *Zagros-Hindu Kush-Himalaya Geodynamic Evolution*, edited by Harsh K. Gupta and Frances M. Delany. American Geophysical Union Geodynamics Series 3, 1981.

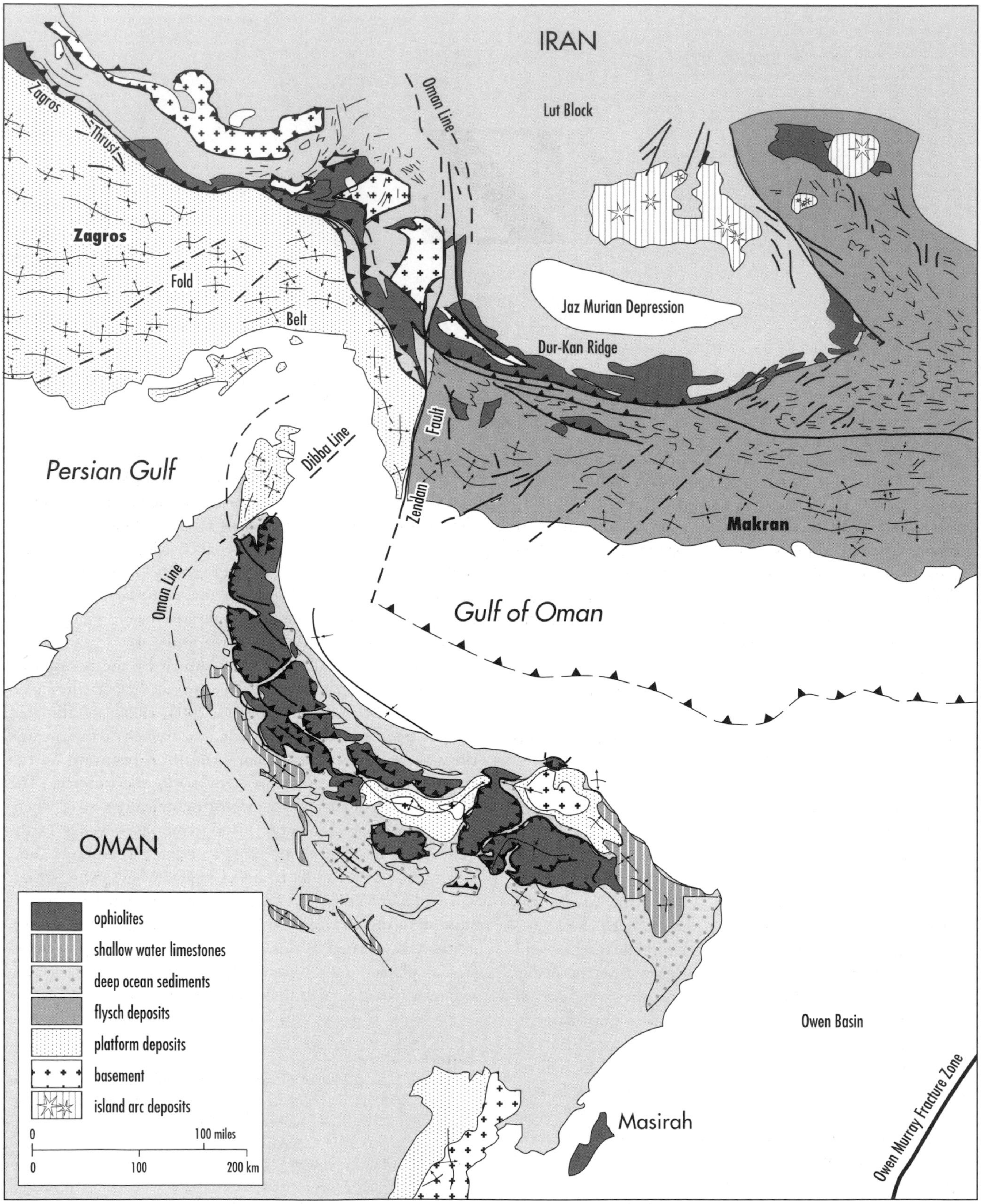

IRAN

Zagros Thrust

Lut Block

Zagros

Fold

Belt

Oman Line

Jaz Murian Depression

Dur-Kan Ridge

Persian Gulf

Dibba Line

Zendan Fault

Oman Line

Makran

Gulf of Oman

OMAN

ophiolites
shallow water limestones
deep ocean sediments
flysch deposits
platform deposits
basement
island arc deposits

0			100 miles
0	100		200 km

Owen Basin

Masirah

Owen Murray Fracture Zone

Tectonic map of the Zagros, Makran, and Northern Oman Mountains

Glennie, Ken W., M. W. Hughes-Clarke, M. G. A. Boeuf, W. F. H. Pilaar, and B. M. Reinhardt. "Interrelationship of Makran-Oman Mountains belts of convergence." In *The Geology and Tectonics of the Oman Region,* edited by A. H. F. Robertson, Mike P. Searle, and Alison C. Ries. Geological Society Special Publication 49, 1990.

zeolite Any of a group of white to colorless hydrous aluminosilicates of alkali and alkaline earth metals, characterized by easy and reversible loss of water of hydration, and fusion and swelling when strongly heated. Zeolites have a composition similar to feldspars, with sodium, potassium, and calcium or barium or strontium as their chief metals, and a ratio of (Al + Si) to non-hydrous oxygen of 1:2. Many different varieties of zeolites are found as secondary minerals in cavities in basalt and other vuggy rocks, in secondary hydrothermal deposits, in beds of volcanic tuff, and in sedimentary layers in saline lakes and even deep-sea sediments. Some of the common zeolite minerals include natrolite, heulandite, analcime, thomsonite, stilbite, laumontite, harmonite, and many others. Many of these zeolites are thought to form by reaction of pore waters with solid aluminosilicate minerals such as feldspar and clay minerals, or with volcanic glass. Many zeolites have been found over a narrow range of low-grade metamorphism and diagenesis, resulting in the definition of a zeolite metamorphic facies. There are numerous industrial applications for zeolites that utilize their properties of absorption and loss of water, including water softening, drying agents, or gas absorbers.

See also DIAGENESIS; METAMORPHISM; MINERALOGY.

Zimbabwe craton The Zimbabwe craton is a classic granite greenstone terrane. In 1971 Clive W. Stowe proposed a division of the Zimbabwean (then Rhodesian) craton into four main tectonic units. His first unit includes remnants of older gneissic basement in the central part of the craton, including the Rhodesdale, Shangani, and Chilimanzi gneissic complexes. Stowe's second (northern) unit includes mafic and ultramafic volcanics overlain by a mafic/felsic volcanic sequence, iron formation, phyllites, and conglomerates of the Bulawayan group all overlain by sandstones of the Shamvaian group. The third or southern unit consists of mafic and ultramafic lavas of the Bulawayan group, overlain by sediments of the Shamvaian Group. The southern unit is folded about east-northeast axes. Stowe defined a fourth unit in the east, including remnants of schist and gneissic rocks, enclosed in a sea of younger granitic rocks. In 1979 James F. Wilson proposed a regional correlation between the greenstone belts in the craton. His general comparison of the compositions of the upper volcanics in the greenstone belts resulted in a distinction between the greenstone belts located in the western part of the craton from those in the eastern section. The greenstone belts to the west of his division are composed of dominantly calc-alkaline rock suites including basalt, andesite, and dacite flows and pyroclastic rocks. This western section includes bimodal volcanic rocks consisting of tholeiite and magnesium-rich pillow basalt and massive flows, with some peridotitic rocks alternating with dacite flows, tuffs, and agglomerates. The eastern section of the Zimbabwe craton is characterized by pillowed and massive tholeiitic basalt flows and less-abundant magnesium-rich basalts and their metamorphic equivalents. The eastern section contains a number of phyllites, banded iron formations, local conglomerate, and grit and rare limestone. Wilson identified an area of well-preserved circa 3.5-billion-year-old gneissic rocks and greenstones in the southern part of the province and named this the Tokwe segment. He suggested that this may be a "mini-craton," and that the rest of the Zimbabwe craton stabilized around this ancient nucleus.

Despite these early hints that the Zimbabwe craton may be composed of a number of distinct terranes, much of the work on rocks of the Zimbabwe craton has been geared toward making lithostratigraphic correlations between these different belts and attempting to link them all to a single supergroup style nomenclature. Many workers attempted to pin the presumably correlatable 2.7-billion-year-old stratigraphy of the entire Zimbabwe craton to an unconformable relationship between older gneissic rocks and overlying sedimentary rocks exposed in the Belingwe greenstone belt. More recently, Timothy Kusky, Axel Hoffman, and others have emphasized that the sedimentary sequence unconformably overlying the gneissic basement may be separated from the mafic/ultramafic magmatic sequences by a regional structural break, and that the presence of a structural break in the type stratigraphic section for the Zimbabwe craton casts doubt on the significance of any lithostratigraphic correlations across Stowe's divisions of the craton.

Central Gneissic Unit (Tokwe Terrane)

Three and a half billion-year-old gneissic and greenstone rocks are well-exposed in the area between Masvingo (Fort Victoria), Zvishavane (Shabani), and Shurugwi (Selukwe), in the Tokwe segment. The circa 3.5–3.6-billion-year-old Mashaba tonalite forms a relatively central part of this early gneissic terrane, and other rocks include mainly tonalitic to granodioritic, locally migmatitic gneissic units such as the circa 3.475-billion-year-old Tokwe River gneiss, Mushandike granodiorite (2.95 billion years old), the 3.0-billion-year-old Shabani gneiss, and the 3.5-billion-year-old Mount d'Or tonalite. Similar rocks extend in both the northeast and southwest directions, but they are less well exposed and intruded by younger rocks in these directions. The Tokwe terrane probably extends to the northeast to include the area of circa 3.5 Ga greenstones and older gneissic rocks southeast of Harare. The Tokwe segment represents the oldest known portion of the Tokwe terrane, which was acting as a coherent terrane made up of 3.5–2.95-billion-year-old tectonic elements by circa 2.9 Ga.

The 3.500–2.950-billion-year-old Tokwe terrane also contains numerous narrow greenstone belt remnants, which are typically strongly deformed and multiply-folded along with interlayered gneiss. The area in northeastern-most part of the central gneissic terrane southeast of Harare best exhibits this style of deformation, although it continues southwest through Shurugwi. In the Mashava area west of Masvingo, ultramafic rocks, iron formations, quartzites, and mica schist are interpreted as 3.5-billion-year-old greenstone remnants tightly infolded with the ancient gneissic rocks. The 3.5-billion-year-old Shurugwi (Selukwe) greenstone belt was the focus of Clive W. Stowe's classic studies in the late 1960s, in which he identified Alpine-type inverted nappe structures and proposed that the greenstone belt was thrust over older gneissic basement rocks, forming an imbricated and inverted mafic/ultramafic allochthon. This was subsequently folded and intruded by granitoids during younger tectonic events.

The Tokwe terrane is in many places unconformably overlain by a heterogeneous assemblage of volcanic and sedimentary rocks known as the Lower Greenstones. In the Belingwe greenstone belt this Lower Greenstone assemblage is called the Mtshingwe group, composed of mafic, ultramafic, intermediate, and felsic volcanic rocks, pyroclastic deposits, and a wide variety of sedimentary rocks. Isotopic ages on these rocks range from 2.9–2.83 billion years old, and the rocks are intruded by the 2.83-billion-year-old Chingezi tonalite. The Lower Greenstones are also well-developed in the Midlands (Silobela), Filabusi, Antelope–Lower Gwanda, Shangani, Bubi, and Gweru-Mvuma greenstone belts. The upper part of the Lower Greenstones have yielded U-Pb ages of 2.8 billion years in the Gweru greenstone belt, and 2.79 billion years in the Filabusi belt.

The Buhwa and Mweza greenstone belts contain the thickest section of three-billion-year-old shallow water sedimentary rocks in the Zimbabwe craton. The Buhwa belt contains a western shelf succession and an eastern deeper-water basinal facies association. The shelf sequence is up to 2.5 miles (4 km) thick and includes units of quartzite and quartz arenite, shale, and iron formation, whereas the eastern deepwater association consists of strongly deformed shales, mafic-ultramafic lavas, chert, iron formation, and possible carbonate rocks. The Buhwa greenstone belt is intruded by the Chipinda batholith, which has an estimated age of 2.9 billion years. Shelf-facies rocks may have originally extended along the southeastern margin of the Tokwe terrane into Botswana, where a similar assemblage is preserved in the Matsitama greenstone belt. Rocks of the Matsitama belt include interlayered quartzites, iron formations, marbles and metacarbonates, and quartzofeldspathic gneisses in a 6–12-mile (10–20 km) thick structurally imbricated succession. The strong penetrative fabric in this belt may be related to deformation associated with the formation of the Limpopo belt to the south, but early nappes and structural imbrication that predates the regional cleavage forming event are also recognized. The Matsitama belt (Mosetse complex) is separated from the Tati belt to the east by an accretionary gneiss terrane (Motloutse complex) formed during convergence of the two crustal fragments. The Tati and Vumba greenstone belts (Francistown granite-greenstone complex) were overturned prior to penetrative deformation, possibly indicating that they represent lower limbs of large regional nappe structures. The mafic, oceanic-affinity basalts of the Tati belt are overlain by andesites and other silicic igneous rocks and intruded by syntectonic granitoids, typical of magmatic arc deposits. Similar arc-type rocks occur in the lower Gwanda greenstone belts to the east. The Lower Gwanda and Antelope greenstone belts are allochthonously overlain by basement gneisses that were thrust over the greenstones prior to granite emplacement.

A second sequence of sedimentary rocks lies unconformably over the Lower Greenstone assemblage and overlaps onto basement gneisses in several greenstone belts, most notably in the Belingwe belt where the younger sequence is known as the Manjeri formation. The Manjeri formation contains conglomerates and shallow water sandstones and locally carbonates at the base, and ranges stratigraphically up into cherts, argillaceous beds, graywacke, and iron formation. The top of the Manjeri formation is marked by a regional fault.

The Manjeri formation is between 2,000 and 800 feet (600 and 250 m) thick along most of the western side of the Belingwe belt except where it is cut out by faulting, and it thins northward to zero meters north of Zvishavane. It is considerably thinner on the western edge of the belt. On the scale of the Belingwe belt, the Manjeri formation thickens toward the southeast, with some variation in structural thickness attributed to either sedimentary or tectonic ramping. The age of the Manjeri formation is poorly constrained and may be diachronous across strike. However, the Manjeri formation must be younger than the unconformably underlying circa 2.8-billion-year-old Ga Lower Greenstones, and it must be older than or in part contemporaneous with the thrusting event that emplaced circa 2.7-billion-year-old magmatic rocks of the Upper Greenstones over the Manjeri formation. The Manjeri formation overlaps onto gneissic basement of the Tokwe terrane on the eastern side of the Belingwe belt, and at Masvingo, and rests on older (3.5-billion-year-old Sebakwian group) greenstones at Shurugwi. Regional stratigraphic relationships suggest that the Manjeri formation forms a southeast thickening sedimentary wedge that prograded onto the Tokwe terrane.

Northern Belt (Zwankendaba Arc)

The northern volcanic terrane includes the Harare (Salisbury), Mount Darwin, Chipuriro (Sipolilo), Midlands (Silobela and Que Que), Chegutu (Gatoma), Bubi, Bulawayo, and parts of the Filabusi and Gwanda greenstone belts. These contain a lower volcanic series overlain by a calc-alkaline suite of basalts, andesites, dacites, and rhyolites. Pyroclastic, tuffaceous, and volcaniclastic horizons are common. Also common are iron formations and other sedimentary rocks including slates, phyllites, and conglomerate. In the Bulawayo-Silobela area (Mulangwane Range), the top of the upper volcanics include a series of porphyritic and amygdaloidal andesitic and dacitic agglomerate and other pyroclastic rocks.

U-Pb ages from felsic volcanics of the northern volcanic belt include 2.696, 2.698, 2.683, 2.702, and 2.697 billion years. Isotopic data from the Harare-Shamva greenstone belt and surrounding granitoids suggests that the greenstones evolved on older continental crust between 2.715 billion and 2.672 billion years ago. The age of deformation is constrained by a 2.667-billion-year-old syntectonic gneiss, a 2.664-billion-year-old late syntectonic intrusion, and 2.659-billion-year-old shear zone-related gold mineralization. Other post-tectonic granitoids yielded U-Pb zircon ages of 2.649, 2.618, and 26.01 billion years. Isotopic data for the felsic volcanics suggest that the felsic magmas were derived from a melt extracted from the mantle 200 million years before volcanism and saw considerable interaction between these melts and older crustal material.

Southern Belt

The southern belt of tholeiitic mafic-ultramafic dominated greenstones structurally overlies shallow water sedimentary sequences and gneissic rocks in parts of the Belingwe, Mutare (Umtali), Masvingo (Fort Victoria), Buhwa, Mweza, Antelope, and Lower Gwanda belts. The most extensively studied of these is the Belingwe belt, which many workers have used as a stratigraphic archetype for the entire Zimbabwe craton. The allochthonous Upper greenstones are here discussed separately from the structurally underlying rocks of the Manjeri formation that rest unconformably on Tokwe terrane gneissic rocks.

The Negezi group of the Belingwe greenstone belt rests allochthonously over the Manjeri formation, and it contains ultramafic and mafic volcanic and plutonic rocks of the Reliance and Zeederbergs formations. The Reliance formation is composed of variably altered high-magnesium basalts, komatiites, and their intrusive equivalents. High strain zones are common within the Reliance formation especially in the lower 650 feet (200 m). The Zeederbergs formation is composed almost entirely of 2.7-billion-year-old extrusive volcanic rocks, which are typically pillowed tholeiitic basalts.

Geochemical studies have suggested that the komatiites of the Reliance formation in Belingwe could not have been erupted through continental crust, but rather that they are similar to intra-plate basalts and distinct from mid-ocean ridge and convergent margin basalts. The geochemistry of the Ngezi group in the Belingwe greenstone belt suggests that it could be a pre-served oceanic plateau and that there was no evidence for them to have been derived from a convergent margin.

Craton-Wide Overlap Assemblage (Shamvaian Group)

The Shamvaian group consists of a sequence of coarse clastic rocks that overlie the Upper Greenstones in several locations. These conglomerates, arkoses, and graywackes are well-known from the Harare, Midlands, Masvingo, and Belingwe greenstone belts. The Cheshire formation is the top unit of the Belingwe greenstone belt. It is a heterogeneous succession of sedimentary rocks that contain a number of various lithofacies including conglomerate, sandstone, siltstone, argillite, limestone, cherty limestone, stromatolitic limestone, and minor banded iron formation. The Shamvaian group is intruded by the circa 2.6-billion-year-old Chilimanzi suite granites, providing an upper age limit on deposition. In the Bindura-Shamva greenstone belt the Shamvaian group is 1.2 miles (2 km) thick, beginning with basal conglomerates and grading up into a thick sandstone sequence. Tonalitic clasts in the basal conglomerate have yielded igneous ages of 3.2, 2.9, 2.8, and 2.68 billion years. Felsic volcanics associated with the Shamvaian group in several greenstone belts have ages of 2.66 billion to 2.64 billion years.

Chilimanzi Suite

The Chilimanzi suite of K-rich granitoids is one of the last magmatic events in the Zimbabwe craton, with reported ages of 2.57 billion to 2.6 billion years. These granites appear to be associated with a system of large intracontinental shear zones that probably controlled their position and style of intrusion. These relatively late structures are related to north-northwest to south-southeast shortening and associated southwestward extrusion of crust during the continental accretion and collision as recorded in the Limpopo belt.

Accretion of the Archean Zimbabwe Craton

The oldest part of the Zimbabwe craton, the Tokwe terrane, preserves evidence for a complex series of tectonomagmatic events ranging in age of 3.6–2.95 billion years ago. These events resulted in complex deformation of the Sebakwian greenstones and intervening gneissic rocks. This may have involved convergent margin accretionary processes that led to the development of the Tokwe terrane as a stable continental nucleii by 2.95 billion years ago.

A widespread unit of mixed volcanic and sedimentary rocks was deposited on the Tokwe terrane at circa 2.9 billion years ago. These lower greenstones include mafic and felsic volcanic rocks, coarse conglomerates, sandstones, and shales. The large variation in volcanic and sedimentary rock types, along with the rapid and significant lateral variations in stratigraphic thicknesses that typify the Lower Greenstones, are characteristic of rocks deposited in continental rift or rifted arc settings. The Tokwe terrane was subjected to rifting at 2.9 Ga leading to the formation of widespread graben in which the Lower Greenstones were deposited. It appears that the southeastern margin of the Tokwe terrane may have been rifted from another, perhaps larger fragment at this time, along a line extending from the Buhwa-Mweza greenstone belts to the Mutare belt, allowing a thick sequence of passive margin-type sediments (preserved in the Buhwa greenstone belt) to develop on this rifted margin. Age constraints on the timing of the passive margin development are not good, but they appear to fall within in the range of 3.09 billion to 2.86 billion years ago. By 2.7 billion years ago, a major marine transgression covered much of the southern half of the Tokwe terrane, as recorded in shallow water sandstones, carbonates, and iron formations of the Manjeri-type units preserved in several greenstone belts. The Manjeri-type units overlap basement of the Tokwe terrane in several places (e.g., Belingwe, Masvingo) and lie unconformably over the circa 3.5-billion-year-old and 2.9-billion-year-old greenstones. Regional stratigraphic relationships suggest that the Manjeri formation forms a southeast thickening sedimentary wedge that prograded onto the Tokwe terrane, in a manner analogous to the Ocoee-Chilhowee and correlative Sauk Sequence shallow-water progradational sequence of the Appalachians, and similar sequences in other

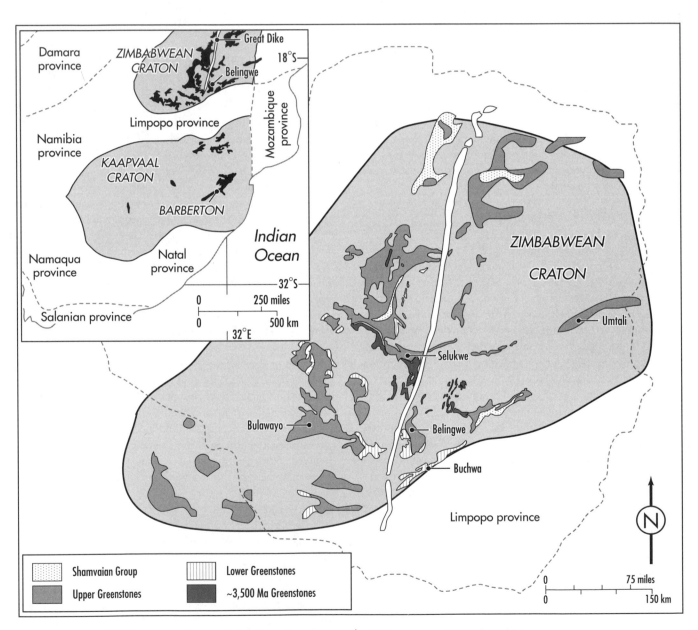

Tectonic map of the Zimbabwe craton showing the locations of greenstone belts of different ages, plus the Great Dike

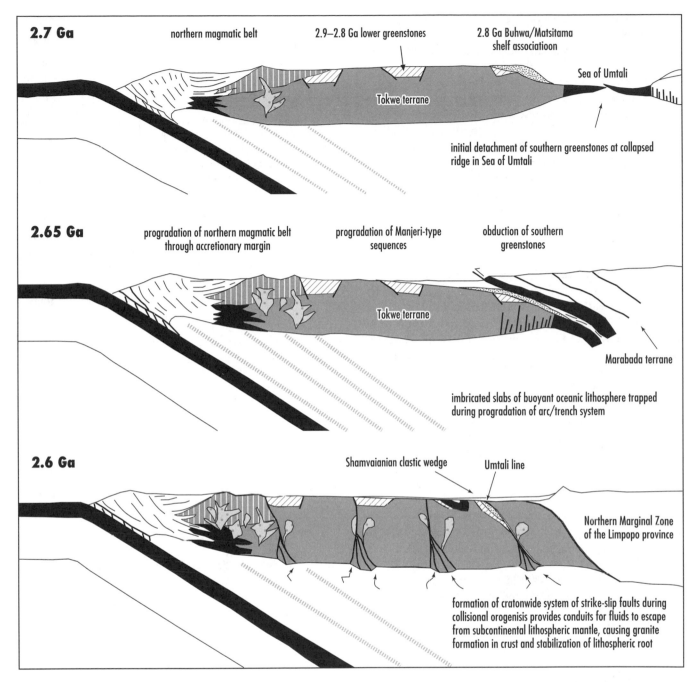

2.7 Ga   northern magmatic belt   2.9–2.8 Ga lower greenstones   2.8 Ga Buhwa/Matsitama shelf associatioon

Sea of Umtali

Tokwe terrane

initial detachment of southern greenstones at collapsed ridge in Sea of Umtali

2.65 Ga   progradation of northern magmatic belt through accretionary margin   progradation of Manjeri-type sequences   obduction of southern greenstones

Tokwe terrane

Marabada terrane

imbricated slabs of buoyant oceanic lithosphere trapped during progradation of arc/trench system

2.6 Ga   Shamvaianian clastic wedge   Umtali line

Northern Marginal Zone of the Limpopo province

formation of cratonwide system of strike-slip faults during collisional orogenisis provides conduits for fluids to escape from subcontinental lithospheric mantle, causing granite formation in crust and stabilization of lithospheric root

Tectonic cross sections showing tectonic evolution of the Zimbabwe craton 2.7–2.6 billion years ago

mountain belts. The progradation could have been driven by sedimentary or tectonic flexural loading of the margin of the Tokwe terrane, but most evidence points to the latter cause. The top of the Manjeri-type units represents a regional detachment surface, upon which allochthonous units of the southern greenstones were emplaced. Loading of the passive margin by these thrust sheets would have induced flexural subsidence and produced a foreland basin that migrated onto the Tokwe terrane.

The 2.7-billion-year-old greenstones are divided into a northwestern arc-like succession, and a southeastern allochthonous succession. The northwestern arc succession contains lavas with strong signatures of eruption though older continental crust, and the arc appears to be a continental margin type of magmatic province. In contrast, the southern greenstones are allochthonous and were thrust in place along a shear zone that is well exposed in several places, including the Belingwe belt. These southern greenstones have a stratigraphy reminiscent of thick oceanic crust, suggesting that they may represent an oceanic plateau that was obducted onto the Tokwe terrane 2.7 billion years ago. All of the southern greenstones are distributed in a zone confined to about 100 miles (150 km) from the line of passive margin-type sediments extending from Buhwa-Mweza to Mutare. This "Umtali line" may represent the place where a ocean or back arc basin opened between 2.9 billion and 2.8 billion years ago, then closed at 2.7 billion years, and forms the root zone from which the southern greenstones were obducted. This zone contains numerous northeast-striking mylonitic shear zones in the quartzofeldspathic gneisses. Closure of the Sea of Umtali at circa 2.7 billion years ago deposited a flysch sequence of graywacke-argillite turbidites that forms the upper part of the Manjeri formation and formed a series of northeast-striking folds.

The latest Archean tectonic events to affect the Zimbabwe craton are associated with deposition of the Shamvaian group, and intrusion of the Chilimanzi suite granitoids at circa 2.6–2.57 billion years ago. These events appear to be related to collision of the now-amalgamated Zimbabwe craton with the northern Limpopo province, as the Zimbabwe and Kaapvaal cratons collided. Interpretations of the Limpopo orogeny suggest that the Central Zone of the Limpopo province collided with the Kaapvaal craton at circa 2.68 billion years ago, and that this orogenic collage collided with the southern part of the Zimbabwe craton at 2.58 billion years ago. Deposition of the Shamvaian group clastic sediments occurred in a foreland basin related to this collision, and the intrusion of the Chilimanzi suite occurred when this foreland became thickened by collisional processes and was cut by sinistral intracontinental strike-slip faults. Late folds in the Zimbabwe craton are oriented roughly parallel to the collision zone and appear contemporaneous with this collision. The map pattern of the southern Zimbabwe craton shows some interference between folds of the early generation (related to the closure of the Sea of Umtali) and these late folds related to the Limpopo orogeny.

See also ARCHEAN; BELINGWE GREENSTONE BELT; CRATONS; KAAPVAAL CRATON.

Further Reading

Bickle, Mike J., and Euan G. Nisbet, eds. *The Geology of the Belingwe Greenstone Belt, Geological Society of Zimbabwe Special Publication* 2. Rotterdam: A.A. Balkema, Rotterdam, 1993(a).

Coward, Mike P., and P. R. James. "The Deformation Patterns of two Archean Greenstone Belts in Rhodesia and Botswana." *Precambrian Research* 1 (1974): 235–258.

Fedo, Christopher M., and Kenneth A. Eriksson. "Stratigraphic Framework of the ~ 3.0 Ga Buhwa Greenstone Belt: a Unique Stable Shelf Succession in the Zimbabwe Archean Craton." *Precambrian Research* 77 (1996): 161–178.

Hoffman, A., and Timothy M. Kusky. "The Belingwe Greenstone Belt: Ensialic or Oceanic?" In *Precambrian Ophiolites and Related Rocks,* edited by Timothy M. Kusky. Elsevier Amsterdam: Elsevier, 2004.

Jelsma, Hielke A., Michael L. Vinyu, Peter J. Valbracht, G. Davies, Jan R. Wijbrans, and Ed A. T. Verdurmen. "Constraints on Archean Crustal Evolution of the Zimbabwe Craton: a U-Pb Zircon, Sm-Nd and Pb-Pb Whole Rock Isotope Study." *Contributions to Mineralogy and Petrology* 124 (1996): 55–70.

Kusky, Timothy M., and William S. F. Kidd. "Remnants of an Archean Oceanic Plateau, Belingwe Greenstone Belt, Zimbabwe." *Geology* 20, no. 1 (1992): 43–46.

Kusky, Timothy M., and Pamela A. Winsky. "Structural Relationships along a Greenstone/ Shallow Water Shelf Contact, Belingwe Greenstone Belt, Zimbabwe." *Tectonics* 14, no. 2 (1995): 448–471.

Kusky, Timothy M. "Tectonic Setting and Terrane Accretion of the Archean Zimbabwe Craton." *Geology* 26 (1998): 163–166.

Stowe, Clive W. "Alpine Type Structures in the Rhodesian Basement Complex at Selukwe." Journal of the Geological Society of London (1974): 411–425.

———. "The Structure of a Portion of the Rhodesian Basement, South and West of Selukwe." Ph.D. thesis, University of London, 1968(a).

———. "The Geology of the Country South and West of Selukwe." *Bulletin of the Geological Survey of Rhodesia* 59 (1968b).

Taylor, P. N., Jan D. Kramers, Stephen Moorbath, J. F. Wilson, J. L. Orpen, and A. Martin. "Pb/Pb, Sm-Nd, and Rb-Sr Geochronology in the Archean Craton of Zimbabwe." *Chemical Geology (Isotope Geosciences)* 87 (1991): 175–196.

Wilson, James F., Robert W. Nesbitt, and C. Mark Fanning. "Zircon Geochronology of Archean Felsic Sequences in the Zimbabwe Craton: a Revision of Greenstone Stratigraphy and a Model for Crustal Growth." In *Early Precambrian Processes,* edited by Mike P. Coward and Alison C. Ries, 109–126. Geological Society Special Publication 95, 1995.

Wilson, James F. "A Preliminary Reappraisal of the Rhodesian Basement Complex." Special Publication of *the Geological Society of South Africa* 5 (1979): 1–23.

zinc A native metallic element that forms a blue white mineral, commonly called zinc blende or sphalerite. Lead and zinc typically form in ore deposits together and may occur with minerals including copper and other base sulfides. There is a large industrial demand for lead and zinc for use in batteries, ammunition, electrical components, as a galvanizing agent, as a precipitating agent for gold extraction, and in medicines. Zinc forms six common minerals found in many lead-zinc deposits, including sphalerite (ZnS), smithsonite ($ZnCO_3$), Hemimorphite ($Zn_4Si_2O_7[OH]2 \cdot H_2O$), Zincite ($ZnO$), Willimenite ($Zn_2SiO_4$), and Franklinite (Fe, Zn, Mn, $[Fe, Mn]_2O_4$).

Lead-zinc ore deposits are of several different types, including stratabound deposits of syngenetic origin including deposits in Dzhezkazgan, stratabound deposits of secondary or epigenetic origin such as the deposits of southeast Missouri. Many lead-zinc deposits are known from volcanosedimentary terranes such as Archean greenstone belts and island arc associations, including those of Kuroko, Japan, and Kidd Creek,

Canada. Replacement deposits include those of Cerro de Pasco in Peru, whereas veins and contact metasomatic or skarn deposits are known from many places throughout the world.

zircon A common accessory mineral in many rock types that typically forms small tetragonal prisms, with the formula $ZrSiO_4$. Zircon is the chief ore of zirconium, used as a refractory for smelting, and for gemstones that resemble diamonds. Studies of zircon morphology and growth can reveal whether it crystallized from the magma that formed igneous rocks that it is found in, if it grew during metamorphism, or if it is a detrital mineral or xenolith incorporated in the rock from an older source. Since zircon may contain significant quantities of radiogenic uranium, thorium, and lead, it is commonly used to obtain U-Pb geochronologic ages of igneous rocks that it grew in, or to obtain ages of metamorphism of the host rock. Studies of zircon populations in sedimentary rocks can also reveal an enormous amount about the ages of rocks in the source terrane for that sedimentary deposit.

See also GEOCHRONOLOGY; MINERALOGY.

APPENDIX I

Periodic Table of Elements

1 H 1.008																	2 He 4.003
3 Li 6.941	4 Be 9.012											5 B 10.81	6 C 12.01	7 N 14.01	8 O 16.00	9 F 19.00	10 Ne 20.18
11 Na 22.99	12 Mg 24.31											13 Al 26.98	14 Si 28.09	15 P 30.97	16 S 32.07	17 Cl 35.45	18 Ar 39.95
19 K 39.10	20 Ca 40.08	21 Sc 44.96	22 Ti 47.88	23 V 50.94	24 Cr 52.00	25 Mn 54.94	26 Fe 55.85	27 Co 58.93	28 Ni 58.69	29 Cu 63.55	30 Zn 65.39	31 Ga 69.72	32 Ge 72.59	33 As 74.92	34 Se 78.96	35 Br 79.90	36 Kr 83.80
37 Rb 85.47	38 Sr 87.62	39 Y 88.91	40 Zr 91.22	41 Nb 92.91	42 Mo 95.94	43 Tc (98)	44 Ru 101.1	45 Rh 102.9	46 Pd 106.4	47 Ag 107.9	48 Cd 112.4	49 In 114.8	50 Sn 118.7	51 Sb 121.8	52 Te 127.6	53 I 126.9	54 Xe 131.3
55 Cs 132.9	56 Ba 137.3	57-71* 	72 Hf 178.5	73 Ta 180.9	74 W 183.9	75 Re 186.2	76 Os 190.2	77 Ir 192.2	78 Pt 195.1	79 Au 197.0	80 Hg 200.6	81 Tl 204.4	82 Pb 207.2	83 Bi 209.0	84 Po (210)	85 At (210)	86 Rn (222)
87 Fr (223)	88 Ra (226)	89-103‡ 	104 Rf (261)	105 Db (262)	106 Sg (263)	107 Bh (262)	108 Hs (265)	109 Mt (266)	110 Ds (271)	111 Uuu (272)	112 Uub (285)	113 Uut (284)	114 Uuq (289)	115 Uup (288)			

atomic number — 1
symbol — H
atomic weight — 1.008

Numbers in parentheses are the atomic mass numbers of radioactive isotopes.

*lanthanide series	57 La 138.9	58 Ce 140.1	59 Pr 140.9	60 Nd 144.2	61 Pm (145)	62 Sm 150.4	63 Eu 152.0	64 Gd 157.3	65 Tb 158.9	66 Dy 162.5	67 Ho 164.9	68 Er 167.3	69 Tm 168.9	70 Yb 173.0	71 Lu 175.0
‡actinide series	89 Ac (227)	90 Th 232.0	91 Pa 231.0	92 U 238.0	93 Np (237)	94 Pu (244)	95 Am (243)	96 Cm (247)	97 Bk (247)	98 Cf (251)	99 Es (252)	100 Fm (257)	101 Md (258)	102 No (259)	103 Lr (260)

The Chemical Elements

element	symbol	a.n.	element	symbol	a.n.	element	symbol	a.n.	element	symbol	a.n.
actinium	Ac	89	erbium	Er	68	molybdenum	Mo	42	selenium	Se	34
aluminum	Al	13	europium	Eu	63	neodymium	Nd	60	silicon	Si	14
americium	Am	95	fermium	Fm	100	neon	Ne	10	silver	Ag	47
antimony	Sb	51	fluorine	F	9	neptunium	Np	93	sodium	Na	11
argon	Ar	18	francium	Fr	87	nickel	Ni	28	strontium	Sr	38
arsenic	As	33	gadolinium	Gd	64	niobium	Nb	41	sulfur	S	16
astatine	At	85	gallium	Ga	31	nitrogen	N	7	tantalum	Ta	73
barium	Ba	56	germanium	Ge	32	nobelium	No	102	technetium	Tc	43
berkelium	Bk	97	gold	Au	79	osmium	Os	76	tellurium	Te	52
beryllium	Be	4	hafnium	Hf	72	oxygen	O	8	terbium	Tb	65
bismuth	Bi	83	hassium	Hs	108	palladium	Pd	46	thallium	Tl	81
bohrium	Bh	107	helium	He	2	phosphorus	P	15	thorium	Th	90
boron	B	5	holmium	Ho	67	platinum	Pt	78	thulium	Tm	69
bromine	Br	35	hydrogen	H	1	plutonium	Pu	94	tin	Sn	50
cadmium	Cd	48	indium	In	49	polonium	Po	84	titanium	Ti	22
calcium	Ca	20	iodine	I	53	potassium	K	19	tungsten	W	74
californium	Cf	98	iridium	Ir	77	praseodymium	Pr	59	ununbium	Uub	112
carbon	C	6	iron	Fe	26	promethium	Pm	61	ununpentium	Uup	115
cerium	Ce	58	krypton	Kr	36	protactinium	Pa	91	ununquadium	Uuq	114
cesium	Cs	55	lanthanum	La	57	radium	Ra	88	ununtrium	Uut	113
chlorine	Cl	17	lawrencium	Lr	103	radon	Rn	86	unununium	Uuu	111
chromium	Cr	24	lead	Pb	82	rhenium	Re	75	uranium	U	92
cobalt	Co	27	lithium	Li	3	rhodium	Rh	45	vanadium	V	23
copper	Cu	29	lutetium	Lu	71	rubidium	Rb	37	xenon	Xe	54
curium	Cm	96	magnesium	Mg	12	ruthenium	Ru	44	ytterbium	Yb	70
darmstadtium	Ds	110	manganese	Mn	25	rutherfordium	Rf	104	yttrium	Y	39
dubnium	Db	105	meitnerium	Mt	109	samarium	Sm	62	zinc	Zn	30
dysprosium	Dy	66	mendelevium	Md	101	scandium	Sc	21	zirconium	Zr	40
einsteinium	Es	99	mercury	Hg	80	seaborgium	Sg	106			

a.n. = atomic number

Appendix II

The Geologic Timescale

Era	Period	Epoch	Age (millions of years)	First Life-forms	Geology
		Holocene	0.01		
	Quaternary				
		Pleistocene	3	Humans	Ice age
Cenozoic		Pliocene	11	Mastodons	Cascades
		Neogene			
		Miocene	26	Saber-toothed tigers	Alps
	Tertiary	Oligocene	37		
		Paleogene			
		Eocene	54	Whales	
		Paleocene	65	Horses, Alligators	Rockies
	Cretaceous		135		
				Birds	Sierra Nevada
Mesozoic	Jurassic		210	Mammals	Atlantic
				Dinosaurs	
	Triassic		250		
	Permian		280	Reptiles	Appalachians
	Pennsylvanian		310		Ice age
				Trees	
	Carboniferous				
Paleozoic	Mississippian		345	Amphibians	Pangaea
				Insects	
	Devonian		400	Sharks	
	Silurian		435	Land plants	Laursia
	Ordovician		500	Fish	
	Cambrian		544	Sea plants	Gondwana
				Shelled animals	
			700	Invertebrates	
Proterozoic			2500	Metazoans	
			3500	Earliest life	
Archean			4000		Oldest rocks
			4600		Meteorites

Classification of Species

Group	Characteristics	Geologic Age
Vertebrates	Spinal column and internal skeleton. About 70,000 living species. Fish, amphibians, reptiles, birds, mammals.	Ordovician to recent
Echinoderms	Bottom dwellers with radial symmetry. About 5,000 living species. Starfish, sea cucumbers, sand dollars, crinoids. Cambrian to recent	
Arthropods	Largest phylum of living species with more than 1 million known. Insects, spiders, shrimp, lobsters, crabs, trilobites.	Cambrian to recent
Annelids	Segmented body with well-developed internal organs. About 7,000 living species. Worms and leeches.	Cambrian to recent
Mollusks	Straight, curled, or two symmetrical shells. About 70,000 living species. Snails, clams, squids, ammonites.	Cambrian to recent
Brachiopods	Two asymmetrical shells. About 120 living species.	Cambrian to recent
Bryozoans	Moss animals. About 3,000 living species.	Ordovician to recent
Coelenterates	Tissues composed of three layers of cells. About 10,000 living species. Jellyfish, hydra, coral.	Cambrian to recent
Porifera	The sponges. About 3,000 living species.	Proterozoic to recent
Protozoans	Single-celled animals. Foraminifera and radiolarians.	Precambrian to recent

Summary of Solar System Data

Body	Orbit in millions of miles	Radius miles	Mass	Density	Axis tilt	Rotation	Year	Temp. (°C)	Atmospheric composition
Mercury	36	1,500	0.1	5.1	10	58.6 days	88 days	425	Carbon dioxide
Venus	67	3,760	0.8	5.3	6	242.9 days	225 days	425	Carbon dioxide minor water
Earth	93	3,960	1.0	5.5	23.5	24 hours	365 days		78% nitrogen 21% oxygen
Mars	141	2,110	0.1	3.9	25.2	24.5 hours	687 days	−42	Carbon dioxide minor water
Jupiter	483	44,350	318	1.3	3.1	9.9 hours	11.9 years	2000	60% hydrogen, 36% helium, 3% neon, 1% methane and ammonia
Saturn	886	37,500	95	0.7	26.7	10.2 hours	29.5 years	2000	Same as Jupiter
Uranus	1,783	14,500	14	1.6	98	10.8 hours	84 years		Similar to Jupiter, no ammonia
Neptune	2,793	14,450	18	2.3	29	15.7 hours	165 years		Same as Uranus
Pluto	3,666	1,800	0.1	1.5		6.4 days	248 years		

Evolution of Life and the Atmosphere

Evolution	Origin (millions of years)	Atmosphere
Humans	2	Nitrogen, oxygen
Mammals	200	Nitrogen, oxygen
Land animals	350	Nitrogen, oxygen
Land plants	400	Nitrogen, oxygen
Metazoans	700	Nitrogen, oxygen
Sexual reproduction	1,100	Nitrogen, oxygen, carbon dioxide
Eukaryotic cells	1,400	Nitrogen, carbon dioxide, oxygen
Photosynthesis	2,300	Nitrogen, carbon dioxide, oxygen
Origin of life	3,800	Nitrogen, methane, carbon dioxide
Origin of Earth	4,600	Hydrogen, helium

INDEX

Note: Page numbers in **boldface** indicate main entries; *italic* page numbers indicate photographs and illustrations.

A

aa lava **1**
abrasion *456, 464*
Absaroka Sequence **434**
absolute dating system **165**
Abu Hamth, Egypt *388, 389*
abyssal hills **1**
abyssal plains **1**, *299*
Acadian orogeny, Appalachians **13**, **14–15**
Acasta gneisses, Canada **22**, **100**, *392, 394*
accretionary prisms *94, 95*
accretionary processes, continental crust formation **100–103**
accretionary wedges **1–3**, *2*, **94–95**
achondrite **276**
acid rain **32**
acritarchs **60–61**, *73*
active arcs *94, 95*
Adirondack Mountains **3–6**, *4*, **194**, **195–197**
advection **208**
advection fog **155**
aerial photographs *352*
Afar Depression, Ethiopia *6, 7*, **124–125**
Afghanistan, Makran Mountains **467–469**, *468*

Afif terrane, Arabian shield **19**, *20*
Africa *See also specific countries*
    Atlas Mountains **31**
    desertification **115**, *116*, *117*
    East African orogen **59**, *343*
    East African rift system *6, 7*, **62**, *355*
    flood basalt **153**
    Kaapvaal craton *25*, *45, 46*, *100, 101*, **236–238**, *237*, *459–462, 460, 461*, *472, 474*
    Lake Victoria **246**
    Limpopo Province **461**, *461*, *472, 474*
    mantle plumes **261**
    Nile River **246**, **293–295**, *294*
    Sahara Desert *116*, **360–362**, *367*
    Sahel *116, 117*, **362–363**
    sand seas *367*
African craton **20**
agates **6–7**
agglomerates **346**
AGI (American Geological Institute) **11**
AGU (American Geophysical Union) **11**
Agulhas-Falkland fracture **148**
Ahaggar massif, Sahara *360*

Ahmadi ridge, Kuwait **243**
A-horizon sequences **442**
A-horizon soil profiles **396**, *397*
air pressure **7–8**, *31*
    Venus **450**
Airy isostatic model **230**
Alabama
    sinkhole *240*
Al-Amar-Idsas suture zone, Arabian shield **19**, *20*
Alaska
    Aleutian Islands and trench **8**
    braided stream, Mount McKinley *54*
    Brooks Range **54**
    Chugach Mountains **74–76**, *75, 77*, *207*, *286*, *442*
    cirques *77*
    conjugate joints *160*
    glaciers *176*
    McCarty fiord *151*
    McCarty tidewater glacier *176*
    mica schist *373*
    Mount McKinley *286*, **286–287**
    North Slope and Arctic National Wildlife Refuge (ANWR) **297–298**
    oxbow lakes, Kuskokwim River *312*

Alaska *(continued)*
    permafrost 325–326
    permafrost zone *326*
    seiche waves 381
    tsunami 436, 437, 439
    Valley of Ten Thousand Smokes
        448–449
Aleutian Islands and trench 8
algae 405
Algeria, Atlas Mountains 31
Algoma-type banded iron formation
    38, 39
Algonquin terrane, Grenville
    province 192
Al-Jazirah, Sudan 293
Alleghenian orogeny 13, 16
alluvial fans **8–9**, *9*
alluvium 113, 147, 154
almandine 164
alpha decay 348
alpha particles 348
Alpha Regio, Venus 450
Alpine fault, New Zealand **292–293**
Alps **9–10**, *111*
alstonite 62
altimeters 10
Altiplano, Andes 10
altocumulus clouds 84
altostratus clouds 84
Amazon River 10
amber **10–11**
AMCG (anorthosites, mangerites,
    charnockites, granitic gneisses)
    suite, Adirondack Mountains 3–5
American Geological Institute (AGI)
    11
American Geophysical Union (AGU)
    11
American Meteorological
    Association (AMS) **11**
ammonia 446
ammonite 234
amniotes 66
amphibians 65–66
amphiboles **11–12**, 273, 281
amphibolite 12
AMS (American Meteorological
    Association) **11**
Amu Darya River 22
analcime 469
Ancient Gneiss complex, South
    Africa 236

Andean-style arc systems 94, *95*
Anderson, Don 260
Anderson, E. M. 160
Andes 10, **12**
andesite **12**
andesitic magma 227
andradite 164
Andreanof Islands, Alaska 8
aneroid barometers 10
angle of repose 266
angular shear 401
angular strain 402
angular unconformities 443
angular velocity 332
anorthosite 345
Antananarivo block, Madagascar
    255, *256, 257*
Antarctica
    Dry Valleys **127**
    icebergs 222, 376
    ice cap 223, 375, 376–377
    ice sheet 175, 223, 376
    ozone hole 313
    polynyas 340
    Ross Ice Shelf 222–223, **359**,
        376
    sea ice 375
    thermohaline circulation 424
    Weddell Sea 424
Antarctic Circumpolar Current 300
Antarctic ridge 278
antecedent streams 126
anticlines *12*, **12–13**, 111
anticlinoria 12
antiforms 156
antigorite 385
Antler orogeny 357
Anton complex, Slave craton 392
Antongil block, Madagascar 255, *256*
Anton terrane, Slave craton 392,
    393, 394, 395
ANWR (Arctic National Wildlife
    Refuge) **297–298**
Apex chert, Australia 26–27
aphanite 225
Aphrodite Terra, Venus *449*, 450
Apollo objects 276
Appalachian-Caledonide orogen 58
Appalachians **13–16**, *14, 16*
    Caledonides and 58
    Penobscottian orogeny
        321–325, *323, 324*

apparent polar wandering (APW)
    335–336
Aqiq-Tuluhah orogeny, Arabian
    shield 18
aquatic organisms
    benthos 46–47, 50, 136
    nektons **289**
    plankton **329–330**
aquicludes 17
aquifers **16–17**, 200
    artesian wells 27
    cones of depression 86
    fracture zone aquifers **161–162**
    Saharan 361
Arabian folds 243
Arabian-Nubian Shield 17, 59, 343
Arabian plate 6, 304
Arabian shield **17–22**, *22*
    intrusive rocks 18–19
    Najd fault system 19, 20
    ophiolite belts 19
    rock unit classification 18
    tectonic evolution 19–20
    tectonic models, history of
        17–18
aragonite 56, 62
Aral Sea **22**
arc/continent collisions 96, 98
Archean (Archaean) era **22–27**, *26*
    cratonic basins 25–26
    cratons 22–23, 38, 99–103, 419
    gneiss terranes 24–25, *26*
    granite-greenstone terranes
        23–24, *25*
    life 26–27
    ophiolites 307
Archimedes' Principle 90
arcs 94–96, *97*
    Arabian shield 17–18, 19–20
    and crustal growth 100, 101,
        333
    subduction zones 410
Arctic National Wildlife Refuge
    (ANWR) **297–298**
Arctic Ocean
    icebergs 222
    ice cap 375, 376–377
    sea ice 375
    Svalbard and Spitzbergen Island
        415–416
arenites 369
aretes 77

Argand, Emily 91
Argentina, Patagonia **320–321**
Arizona
    Grand Canyon **181**
    Meteor Crater 228
    San Francisco Peaks 452
    urbanization and flash flooding
      447
arkoses 369
Aroostook-Matapedia trough,
    Appalachians 15, 16
Ar-Rayn terrane, Arabian shield 19,
    20
ARS (Attitude Reference System)
    371
artesian systems 17, 27, 200
artesian wells **27**
Arthropods 106
Aruma Group, Oman Mountains
    304
asbestos **27–28**, 171
Asbestos Hazard Emergency
    Response Act (1986) 27
asbestosis 27–28
Asia
    Aral Sea **22**
    Caspian Sea **66–67**
    Gobi Desert **179**
    Himalaya Mountains 98–99,
      210, **210–211**, 285–286, 428
    monsoons 285
    Tibetan Plateau **426–428**
    Ural Mountains **444**
Asia-India convergence 98–99, 210,
    211, 229, *334*, 428
Asir terrane, Arabian shield 19, 20
assimilation 226, 339
asteroids **28–30**, *29*, 171, 276, 440
asthenosphere **30**
Atacama Desert **30–31**
Athollian orogeny 374
Atlantic Ocean
    abyssal plains **1**
    currents 300–301
    evaporites 144
    Mid-Atlantic ridge 51,
      223–224, 278
    North Atlantic cold bottom
      water 80–82
    ridges 125
    thermohaline circulation 424
Atlas Mountains **31**

atmosphere **31–35**, 130 *See also*
    winds
    air pressure **7–8**
    Archean 23
    biosphere 49–50
    carbon cycle and 63, *64*
    climate and 34–35
    clouds **83–84**
    Coriolis effect 99
    formation and evolution of
      32–34, 203–204
    gases 32–34
    general circulation 31, *33*, 80,
      138, **165**, 205–206, *206*, *431*
    geostrophic currents 174
    greenhouse effect **185–187**, *186*
    Hadley cells 31, *33*, 80,
      205–206, *206*, 431
    inversions **229–230**
    ionosphere **230**
    Jupiter 234
    layers of 31, 34
    life's origins and evolution
      248–250, 480
    Mars 262–263
    mesosphere **271–272**
    meteorology 277, 349, 458
    ozone hole **313**
    precipitation *See* precipitation
    Saturn 371–372
    troposphere 31, *34*, **436**
    Uranus 446
    Venus 450
atolls **35**, 351
atomic weight 231
attitude 407
Attitude Reference System (ARS)
    371
augen gneiss 179
aulacogens 159
Aurora Australis 35–36
Aurora Borealis 35–36
auroras (northern lights) **35–36**, *36*
Australia
    Apex chert 26–27
    Ayres Rock 229
    Ediacarian fauna 276, 345
    Great Barrier Reef **184**
    Great Sandy Desert 367
    greenstone belts 187, 188, 189,
      190, 191
    Lake Eyre **246**

    willy-willys 127
    Wittenoom asbestos disaster
      28
australopithecines 338
Austro-Asian feedback system 138
avalanches 265, **269–270**
Avalon Composite terrane,
    Appalachians 15
Avalonia 58
avulsion 248
Awramik, Stanley 405–406
axes of strain 402
axis of rotation
    Earth 332, 378
    Uranus 445–446
    Venus 449
Ayres Rock, Australia 229

B

back arc basins 95, 96, 102
backarc regions 94
backshore *43*
backwash 456, *456*
bacteria
    biosphere and 49–50
    chemosynthetic 51, 71, 249
    stromatolites 405, *405*
badlands *37*, **37–38**
Bahra anticline, Kuwait 243
Baja California, Mexico **38**
bajada 9
Balcones escarpment, Texas 151
Baltica 58
Baltic Sea 10–11
Baltic shield **38**, 241–242
Bam earthquake, Iran 467
Bancroft terrane, Grenville province
    193
banded iron formations (BIFs)
    38–40, *40*, 70, 249, 250
Bangladesh, cyclones 217
banner clouds 84
Barberton greenstone belt, Belingwe,
    Zimbabwe 188, 189, 236
barchan dunes 367, *367*
Barents shelf 415
barometric altimeters 10
barrier beaches **40–41**
barrier chains 40
barrier islands 40, 43, 164, 214
barrier reefs 351

Barringer Meteor Crater, Arizona
228
basalt metamorphism 273
basalts, flood **153–154,** *154, 301*
    Deccan flood basalts, India 68,
        109, 153, 261
    Siberian 317
basic rocks **41**
basin and dome fold interference
    patterns *156, 158*
Basin and Range Province, United
    States **41–42,** 358
basins **41**
batholiths **42**
bauxite **42**
beaches **42–44,** *43,* 147–148
    barrier beaches **40–41**
    benthic environments 46–47,
        49, 50
    groins **198**
    littoral zones 251
    sand dunes 365, 366
    storm surges and 400–401
    terraces 421
    waves and *455,* 455–456, *456*
beach terraces *213, 214*
Beaufort Gyre 300
Beaumont, Leonce Elie de 91
bedding **44** *See also* stratigraphy
    cross-bedding 44, **105,** *106*
    sedimentary structures 380
bed load 251, 355
Beechy Lake, Canada 393
Beekmantown Group, Appalachians
    13
Bekily block, Madagascar *256, 257*
belemnite fossil *402*
Belingwe greenstone belt, Zimbabwe
    44–46, *45, 46,* 189, 236, 470
Belomorian mobile belt, Baltic shield
    38
Bemarivo block, Madagascar 255,
    256
bench marks **46**
Benguela current 391
Benioff, Hugo 336
Benioff Zone 336
benthic environments **46–47,** 49, 50
benthos organisms 46–47, 50, 136
Bergeron, Tor 79–80
Bering land bridge 421–422
Bermuda triangle 359

Bertrand, Marcel 91
beta decay 348
beta particles 348
Betsimisiraka suture, Madagascar
    255
B-horizon sequences 442
B-horizon soil profiles 396, *397*
Big Bang Theory **47–49**
Big Crunch 48
biofacies 147
biogenic sediments 49
biological weathering 456
biomass, carbon cycle and 63–64
biosphere **49–50,** 130
biotite **50**
bioturbation 49
Bi'r Umq suture zone, Arabian shield
    19, 20
Bishah-Rimmah orogeny, Arabian
    shield 18
Bishop Tuff 253
Bitlis suture 417, 418
bituminous coal 85
black smoker chimneys 50, **50–52,**
    51, 71, 125, 249, 307
blockfields 441
Blue Nile 293
body forces 404
body waves (earthquakes) 132–133
bog iron ore (limonite) 250
Boil Mountain ophiolite, Maine
    321–325, *323, 324*
Bolivia
    Altiplano **10**
    Lake Titicaca **246**
Border Ranges ultramafic-mafic
    complex (BRUMC), Alaska 75
bores 439
Borneo
    Sarawak Chamber 68
bornite 411
boudinage *52,* **52–53**
boudins 52–53
Bouguer gravity anomalies 183
Bouma sequences 183–184, 442
Bowen, Norman Levi **53,** 227
brachiopods **53**
Bradley, Dwight 92
braid-deltas 113
braided streams 53, *54,* 154–155
breakup basalts 153
breccias **53–54,** 150

Bright Angel Shale, Grand Canyon
    181
brittle deformation 110, 111–112,
    149, 159, 409
Bronson Hill anticlinorium,
    Appalachians 14
Bronze Age 211–212
Brooks Range, Alaska **54,** 297
Brown, John S. 5
brown ochre (limonite) 250
bryozoa **54–55**
Buhwa greenstone belt, Zimbabwe
    craton 470
Burbidge, Geoff and Margaret 48
Burgess Shale, Canada 60, *60,* 61
burial metamorphism 273–274
Burke, Kevin C. 55
Bushveld complex, Kaapvaal craton
    238

C

Cairo-Sinai fold belt, Syrian arc 417
Calaveras fault, California 364
calcareous oozes 1, 46–47
calcite **56,** 62
calderas **56,** *57,* 452
    Crater Lake, Oregon *56, 57*
    Krakatau eruption 436–437,
        439
    Long Valley caldera, California
        252–253
    Yellowstone National
        Park/Yellowstone Valley 56,
        227, 261, 452, *464,* **464–465,**
        465
Caledonian orogen 58, 373,
    415–416
Caledonides **58**
California
    alluvial fans 9, *9*
    Death Valley 8, *8, 143, 144,*
        337
    fog 155
    Loma Prieta earthquake 132
    Long Valley caldera **252–253**
    Mount Whitney **287**
    mudflows 269
    northers 369
    San Andreas Fault 130–131,
        132, **364–365,** *365,* 421, 431
    Santa Ana winds **369**

urbanization and flash flooding 447

Yosemite Valley and National Park *465*, 465–466

Callisto 234

calving 176, 222

Cambrian period **58–61**, *59, 60*

Cambrian system 58

Cameron River greenstone belt *394*

Canada

Acasta gneisses 22, 100, 392, *394*

banded iron formation, Ungava Peninsula *40*

Burgess Shale 60, *60*, 61

esker *142*

Great Lakes **184**

Great Slave Lake, Nunavet *185*, **185**

Grenville province 3–5, **192–198**, *196, 197*

Hoffman, Paul Felix *195*, **211**

Hudson Bay *213*, **213–214**

Huronian Supergroup 344–345

Penobscottian orogeny **321–325**, *323, 324*

permafrost 325–326

Slave craton **391–396**, *394, 395*

Slave Province 101, 187, 189

taiga **419**

Ungava Peninsula banded iron formation *40*

Yellowknife greenstone belt *25*, *329, 392, 394*

Canadian Rockies *12*, 357

Canadian shield 192, 214, 385, 391

Cancer, tropic of **436**

capacity, river 355

Cape Cod, Massachusetts **61**

Cape Horn and the Strait of Magellan **61**

Capricorn, tropic of 436

carbon-14 dating **61**, 165–166

carbonate compensation depth (CCD) 299

carbonate iron formations 39

carbonate minerals **62**

carbonate mounds 62

carbonate platforms 62, 320, 351

carbonate ramps 62

carbonates **62**, 282

diagenesis 120

dolomite 126

carbonatite **62–63**

carbon cycle **63–64**, *64*

life's origins and 249

Siberian Taiga Forest and Global Carbon Sink **385**

carbon dioxide

carbon cycle 63

greenhouse effect 32, 34, 186, 377

life's origins and 248–249

stored in oceans 379

Carboniferous period **64–66**, *66*

Carboniferous System 64–65

carbon sink 385

Caribbean Ocean 318

Caribbean oceanic plateau 153–154

Carlsbad Cavern 68

Carrol Cave, Missouri *398*

Caspian Sea **66–67**

Cassini spacecraft 233

cataclasis 54, **67**

cataclasites 67

catazonal plutons 42

cathodoluminescence 67

Catskill Mountains 14

caves **67–68**

karst terrains 238–240

speleothems *398*, **398**

subsidence and 410, 411

cavitation 141–142

C-band 416

CCD (carbonate compensation depth) 299

CCT (computer compatible tapes) 369

Cenozoic era **68–70**, *69*

Cretaceous-Tertiary (K-T) extinction 68, 105, 109, 264–265

ice age (Late Cenozoic) 221

Neogene period **289–290**, 421

Quaternary period **347**

Tertiary period 69, **421–422**, *422*

Central Rand group, Witwatersrand basin 459, 460–461

centrifugal forces 429–430

Ceres (asteroid) 30

cerussite 62

CFCs (chlorofluorocarbons) 32, 313, 377

Chain Lakes massif, Maine 321–325, *323*

chalcopyrite 411

chalk 104–105

Challenger, H.M.S. 70, 259, 299, 302

Challenger Deep 261

Challenger Mission 41-G 371

Charon (moon) *339*, 339

chemical sediments **70**, 380

chemical weathering 140, 456, 457

chemosynthesis 51–52, **71**

chert 70, **71**

Chicxulub crater, Yucatan Peninsula, Mexico 30, 68, **71**, 440

Chile

Atacama Desert **30–31**

Patagonia **320–321**

Chilimanzi suite, Zimbabwe craton 471

China

chromite *73*, 74

Dabie Shan ultra-high pressure metamorphic belt **71**

Dongwanzi ophiolite **71–73**, *296–297*, 307

Ediacaran fossils 449

encroaching sand dunes 367

fracture/joint *159*

Gobi Desert 179, 367

loess, Shanxii Province *252*

Ming Dynasty ocean exploration 302

North China craton 71–72, 101, **295–297**, 307

Shaanxi Province earthquake 128

Sinian era **391**

Sinian fauna **73**

Wutal Shan slump *267*

Yellow River 248

Chinook winds **73**

chlorite 281–282

chlorofluorocarbons (CFCs) 32, 313, 377

chocolate block boudins 53

chondrite 276

chondrule 276

C-horizon sequences 442

C-horizon soil profiles 396, *397*

chromite 73, **73–74**

chromitite 73

chrysotile 28, 385

Chugach accretionary wedge, Alaska 75

Chugach Mountains, Alaska **74–76,** *75, 77, 207, 286, 442*

Chugach-Prince William superterrane, Alaska 74

cinder cones 452

circumdenudation 229

cirques **76–77,** *77,* 177

cirrocumulus clouds 84

cirrostratus clouds 84

cirrus clouds 84

classical stratigraphy 384, 403

classification of species 479

clastic rocks 77

    conglomerates 87

    flysch **155**

clastic sediments 380

clathrates (gas hydrates) 63, **164,** 440

clay **77–78,** 238, 281–282

cleavage **78,** *79,* 282, 397

climate **78–80** *See also* climate change

    atmosphere and 34–35

    carbon cycle and 63–64

    classification of 79–80

    climatology 79

    clouds and 84

    dendrochronology 114

    dust storms and 128

    El Niño and the Southern Oscillation (ENSO) **138,** 285, 295, 377

    monsoons 428

    paleoclimatology **314–315**

    precipitation **342**

    seasons **378**

    soils and 396–397

    steppe **399**

    Tibetan Plateau and 428

    tropical 42, 396–397, **435–436,** 458

        bauxite 42

        Everglades, Florida 145

        sea breezes 375

        soil of 396–397

        weathering and 458

    tundra **441**

    weathering and 458

climate change **80–83**

    carbon cycle and 64

    desertification 116

    El Niño and the Southern Oscillation (ENSO) 138

    Gaia hypothesis 82

    glaciers and 175

    greenhouse effect **185–187,** *186*

    ice ages **220–222**

    loess and 252

    Milankovitch cycles **278–280,** *279*

    paleoclimatology **314–315**

    Panama isthmus and 318

    permafrost and 326

    Sahara Desert, Africa 116, 361

    Sahel, Africa 116, 117, 362–363

    sea-level rise and fall *See* sea-level rise and fall

climatology 79

clinometers **83**

Cloos, Hans **83**

closed universe model 48

closing salts 144

Cloud, Preston **83**

clouds **83–84**

    fog 155

    Neptune 291, *292*

    precipitation 342

    thunderstorms 424–426, *426*

    tornado 430

    Venus 450

coal **85**

    biogenic rocks 49

    Carboniferous period and 65

    diagenesis 120

coastal deserts 118

coastal downwelling 85

coastal subsidence 410

coastal upwelling **85**

coaxial deformation 402

coccoliths 104–105

Coconino Sandstone, Grand Canyon 181

cold bottom waters 80–82

collapse structures 239–240

collisional foreland basins 158–159

collision zones 86, 94, 96, 98–99

Colorado, Southern Rockies 356–357

Colorado Plateau 358

Colorado River 181, 340

Columbia, Nevada del Ruiz lahars 245, 266

Columbia River flood basalts 153

columnar joints 160

Comet Encke 441

Comet Halley *85,* 86

comets 29–30, *85,* **85–86,** 276–277, 441

compaction 411

compasses **86,** 258

competence, stream 355

compression 404–405, 432

compressional arcs 96

compressional bends 432

compressional waves (primary waves, P-waves) 132–133, 260, 383, *384*

computer compatible tapes (CCT) 369

concentric folds 157

concordant plutons 42, 339

concordia curve 167, *167*

concretions **86**

condensation 155

condensation theory (origin of the solar system) 203, *204*

condensation trails (contrails) 84

conduction 208, 424

cones of depression **86,** *87*

Coney, Peter 415

confined aquifers 17, 27

conglomerates **87**

    flysch 155

    gravel 183

    molasse 284

conjugate joints 160, *160*

Connecticut Valley–Gaspe trough, Appalachians 14, *15,* 16

conodonts **87–88**

consequent steams 126

contact metamorphism 273

continental crust *88,* **88–89,** 106 *See also* crust

    cratons **99–103,** *102*

    hypsometric curve (hypsographic curve) 407–408, *408*

continental divide 122

continental drift **89–91,** 412, 458

continental drift and plate tectonics: historical development of theories **91–92**

continental freeboard 408
continental interior deserts 117
continental margins **93**
  continental shelves 88, 93, 147
  convergent plate margin
    processes 93, **94–99**, *97, 98,*
    333
      accretionary wedges **1–3,** *2*
      foreland basins 158–159
      Mount Whitney/Sierra
        Nevada mountain ranges
        287
      subduction zones **409–410**
      volcanic eruptions and 452
    divergent or extensional
      boundaries 93, **122–125,** *123,*
      124, 333
    passive margins 93, *319,*
      319–320, *320,* 327
    transform plate margin processes
      93, 333, 346, *431,* **431–433,**
      432
continental rifts 23, 122–124, *124*
continental shelves 88, 93, 147
continental shields *See* shields
continent-continent collisions
  98–99, 414
continuum mechanics 407
contour currents 174
contrails (condensation trails) 84
Contwoyto terrane, Slave craton
  392, 393, *394, 395*
convection and the Earth's mantle
  93–94, *94*
  Archean era 23
  asthenosphere 30
  heat flow 208
  Holmes, Arthur 90–91
  mantle plumes **260–261,** *261*
  supercontinent cycles 412
convergent plate margin processes
  94–99, *97, 98,* 333
  accretionary wedges **1–3,** *2*
  andesite 12
  collisions 86, 98–99
  collision zones 86, 94, 96,
    98–99
  continental margins 93
  foreland basins 158–159
  Mount Whitney/Sierra Nevada
    mountain ranges 287

subduction zones 1–3, 94–96,
  333, **409–410**
  volcanic eruptions and 452
coral reefs 99, 351
  atolls **35**
  benthic environment 47
  Great Barrier Reef **184**
corals 99, 108
Cordillera, North America 357,
  415, 421 *See also* Rocky
  Mountains
Cordilleran orogeny 357
core 129
core-mantle boundary 383–384
Coriolis effect 80, **99,** 299–300
coronae 451
correlation of rocks 404
cosmic background radiation
  48–49
cosmology 49
country rocks 42
cover-collapse sinkholes 240
cover-subsidence sinkholes 240
Cox, Allan 337
cracks *See* fractures
Crater Lake, Oregon *56, 57*
craters *See* impact craters
cratonic basins 25–26
cratonization 103
cratons **99–103,** *102 See also*
  greenstone belts; shields
  Archean era and 22–23, 38,
    99–103, 419
  Baltic shield **38**
  continental crust and 88–89
  deformation of rocks and 111
  epeirogeny **140**
  Kaapvaal craton, South Africa
    25, *45, 46,* 100, 101,
    236–238, *237*
  North China craton 71–72
  Precambrian (Archean and
    Proterozoic) 341
  Slave craton **391–396,** *394, 395*
  tectosphere **419**
  world distribution *341*
  Zimbabwe craton *See*
    Zimbabwe craton
creep *103,* **103–104,** 169, 441
Cretaceous period *104,* **104–105**
Cretaceous-Tertiary (K-T) extinction
  68, 71, 105, 109, 264–265

Crete, Santorin eruption and tsunami
  437, 439
crinoids *65,* **105**
cross-bedding 44, **105,** *106*
cross-lamination 105
cross sections **105**
cross-stratification 105
crust **105–106,** 129 *See also*
  continental crust; oceanic crust
  formation of 89, 202–203,
    309–310
  hypsometric curve (hypsographic
    curve) *219,* **219,** 407–408,
    408
  measuring 383–384
  Mohorovicic discontinuity (the
    Moho) 250, **284,** 383, 384
  Venus 450–451
crustaceans **106,** 108, 330
crustal heat flow 208
cryptobauxites 42
crystals **106–107**
  cleavage 78
  geodes 168
  mineralogy 282
  weathering and 456
ctenophores 330
C-type asteroids 29
cuestas 211
cumulonimbus clouds 84, 424, 425
cumulus clouds 84, 425, *426*
currents *See* ocean currents
cyanobacteria 405, *405*
cyclones 214, 217 *See also*
  hurricanes
cyclothems 65

D

Dabie Shan ultra-high pressure
  metamorphic belt, China 71
dacite **108**
Dactyl 29
daily tides 428
Dalradian Supergroup, Scottish
  Highlands 373
Dalziel, Ian 195
Dana, James Dwight 91, **108**
Daraina-Milanoa Group,
  Madagascar 255
Darcy (measure of permeability) 326
Darcy's Law **108**

Darwin, Charles **108–109**
 on atolls 35
 evolution 145, 166
 on Lyell, Charles 253
 on Patagonia 320–321
Das Antlitz der Erde (Suess) 91, 312
DASI (Degree Angular Scale
 Interferometer) experiment 48–49
dating systems *See* geochronology
daughter elements 444
Dead Sea **109**
Dead Sea transform 417
Death Valley, California *143, 144,*
 337
debris avalanche 217, 269–270
debris falls 269
debris flows 245, 268–269
debris slides 269
decay series 349
Deccan flood basalts, India 68, **109,**
 153, 261
declination **109,** 172
decollement *110,* **110**
deep marine facies 148
deep-sea oozes 1, 46–47
deep-sea trenches *See* trenches
deflation 140
deflation basins 140
deformation of rocks **110–113** *See
 also* mass wasting
 boudinage *52,* **52–53**
 brittle 110, 111–112, 149, 159,
 409
 cataclasis **67**
 cleavage **78,** *79*
 ductile 110, 409
 fault-block mountains 89, 111,
 312
 faults *See* faults
 fold and thrust mountain ranges
 89, 111, 312
 folds *See* folds
 fractures *159,* **159–161**
 orogeny 88–89, 312
 regional 112
 strain 110, *401,* **401–403,** *402,*
 451
 stress 110, **404,** 451
 structural geology 406–407
 viscosity **451**
 volcanic mountain ranges 89,
 111–112, 312

deformation path 402
Degree Angular Scale Interferometer
 (DASI) experiment 48–49
deltas **113,** 147, 352
DEMs (digital elevation models)
 121–122
Denali (Mount McKinley) *286,*
 286–287
dendrite **113**
dendritic drainage **1**
dendrochronology **114**
dendroclimatology **114**
density 282
Der Bau der Erde (Kober) 91
derecho 425
desalination **114–115**
Descent of Man (Darwin) 145
desertification 115–117, 138, 171,
 362–363
desert pavements 140
deserts **115–119**
 alluvial fans 9, *9*
 Atacama Desert **30–31**
 coastal 118
 continental interior/mid-latitude
 deserts 117
 Dry Valleys, Antarctica **127**
 dust devils **127**
 dust storms *127,* **127–128**
 El-Baz, Farouk 137
 eolian sediments *139,* **139–140**
 Gobi Desert **179,** *252,* 367
 horse latitudes 212
 inselbergs 229
 Kuwait 242
 map of *118*
 monsoon 118
 pediments *321,* **321**
 playas 242, **337**
 polar 119, 127
 rainshadow 117–118
 Sahara Desert 116, **360–362,**
 367
 sand dunes 115, 139, *139,* 242,
 360, **365–367,** *366, 367, 368,*
 368, 391
 sand seas **367–367,** *368*
 Santa Ana winds **369**
 Sinai Peninsula, Egypt 386,
 386–390, *390,* 417–418
 Skeleton Coast, Namibia **391**

trade wind (Hadley cell) deserts
 117
 tropical climates and 435–436
 urbanization and flash flooding
 446–447
 wadis 386–387, 389, *454,* **454**
 yardangs 139–140, **464**
desiccation cracks 160
deviatoric stress 404
Devonian Period **119**
dewatering 120
Dewey, John F. 55, **119–120**
de Wit, Maarten 101, 189
Dharwar craton 228–229
D-horizon sequence 442
diachronous boundaries 403–404
diagenesis 120, 272
diamond anvil press **121**
diamonds **120–121,** 241
diapirism 226
diapirs **121**
diatoms **121**
diatremes 241
Dibdibba arch, Kuwait 243
Die Entstehung der Alpen (Suess)
 91, 312
Die Entstehung der Kontinente
 (Wegener) 91, 312
*Die Entstehung der Kontinente und
 Ozeane* (The Origin of the
 Continents and Oceans, Wegener)
 91, 312, 458
digital elevation models (DEMs)
 121–122
dike injection extension model 122
dikes 184, 225, 308–309, 310, 339
dilational bends 432
dinosaurs 68, 234–235, 264–265
dip 407
dip-slip faults *148,* 149, 433
discharge areas 200
disconfomities 443
discordant plutons 42, 339
discordia curve *167*
dissolution 238–239
dissolved load 251, 355
diurnal tides 428
divergent or extensional boundaries
 93, **122–125,** *123, 124,* 333
divides 122
D" layer 260
dogtooth spar 56

doldrums **125–126**
dolomite 62, **126**
dolostone 70, 126
Dominion group, Witwatersrand basin 459, 461
Dongwanzi ophiolite, China **71–73**, 296–297, 307
Doppler radar **126**
downbursts 425
downdrafts 425
downslope currents 174
downwelling, coastal **85**
drainage basins 41, **126–127**
    alluvial fans 9
    Amazon River **10**
    dendritic drainage **113–114**
    floods and 151–153
    stream categories 126
draperies 398
dripstone 398
drizzle 342
dropstones 177
droughts
    deserts and 115, 116, 117
    El Niño and the Southern Oscillation (ENSO) and 138
    as geological hazard 171
    Hadley cells and 206
    Sahel 362–363
Dry Valleys, Antarctica **127**
ductile deformation 110, 409
dunes *See* sand dunes
dunites **127**
Durness sequence, Scottish Highlands 373
dust devils **127**
dust storms *127*, **127–128**
Du Toit, Alexander L. 90, 91, **128**, 247, 335

E
early Mesozoic truncation event 358
Earth **129–130**, *130,* 202–205
earthflows 269
earthquakes **130–136**, 169
    Bam, Iran 467
    body waves 132–133
    elastic rebound theory 131
    epicenter 132, *133*
    Faiyum-Cairo region 418
    focus 132

Indian Ocean earthquake. *See* Sumatra earthquake
Loma Prieta earthquake, California 132
    magnitude 135–136
    origins of 132–134
    Richter scale 135, **353–354**
    San Andreas Fault, California 130–131, 132, 364–365, *365*
    seismic waves 132
    seismographs *134,* 134–135, **381–382**
    subsidence and 411
    surface waves 133–134
    tsunami 436, 439
    warning systems 131
    worst in history 131
    Zagros and Makran Mountains 467
earthquake warning systems 131, 437
Earth Resources Technology Satellite (ERTS-1) 369
Earth's orbit 378
East African orogen 59, 343
East African rift system 6, *7, 62, 355*
Eastern and Western Ghats, India **228–229**
East Pacific Rise *51, 94,* 278
echinoderms 105, **136**
ecliptic plane 378
eclogites **136,** 164
economic geology 353
economic resource assessments 353
ecosystems **136**
Ediacara epoch 276, 345, 391, 449
Egypt
    Abu Hamth 388, 389
    dust storm *127*
    Nile River 293–294, *294*
    Sinai Peninsula **386–390,** *390,* 417–418
E-horizon sequence 442
Einstein, Albert 48, 49, 424
Ekman spirals 85, **136**
elasticas 157
elastic behavior 451
elastic deformation 110
elastic rebound theory 131
El-Baz, Farouk **136–137**
Eldredge, Niles 145–146
electrodialysis 114–115

electromagnetic spectrum 348
    infrared radiation **229**
    remote sensing and 352–353, *353,* 369–371
electron microprobes **137**
electron microscopes **137–138**
El Egma plateau, Egypt 387, 389, 390, *391*
elements
    daughter 444
    electron microprobes **137**
    isotopes of **230–231**
    mineral forming 280, 281
    periodic table 477–478
    radioactive decay 348–349
    in seawater 378, 379
elevations 46
El Niño and the Southern Oscillation (ENSO) **138,** 285, 295, 377
Elsasser, Walter M. 172–173
Elsinore fault, California 364
El Tih plateau, Egypt 387, 389, 390, *3931*
Elzevir terrane, Grenville province 193, 195
energy 424
ENSO (El Niño and the Southern Oscillation) 138, **138,** 285, 295, 377
Environmental Protection Agency, asbestos regulation 28
Eocene epoch **139,** 421, 422, *422*
eolian sediments **139–140**
eons **140**
    Archean (Archaean) **22–27,** *26*
    Hadean 23, **202–205**
    Phanerozoic **328**
    Proterozoic **342–345,** *344*
epeirogeny **140,** 377
epicenters 132, *133*
epicontinental seas **140**
epifauna 47
epizonal plutons 42
epochs **140**
equilibrium theory of tides 429, *429,* 430
eras **140**
Erastothenes 168
erathems (time stratigraphy) 140, 403
ergs 367
Eriksson, Leif 302

erosion 140–142, *141*
    alluvial fans 8–9, *9*
    badlands *37*, **37–38**
    beach 42–43
    pediments and *321*, **321**
    waves and 454–455
erratics, glacial 177
eskers *142*, **142**
Eskola, Pentti **142–143**, 274
Espenschied, Lloyd 10
estuaries **143**, 147
Ethiopia
    Afar Depression **6**, 7, 124–125
    Nile River 293, 294
Ethiopian rift 6, 7
eukaryotes 250, 275, 345
Euler's theorem 332
Europa 234
Europe
    Alps **9–10**
    Caspian Sea **66–67**
    little ice age 220
    Ural Mountains **444**
eustatic (global) sea-level rise 377
eutrophication **143**
euxinic basins **143**
euxinic environments **143**
evaporation fog 155–156
evaporites *143*, **143–144**, *144*
    Messinian salt crisis 290
    playas 337
    sediments 70
    speleothems 398
Everglades, Florida **145**
evolution **145–146** *See also* fossils;
    life, origins and evolution of
        Cloud, Preston **83**
        Darwin, Charles **108–109**
        faunal succession 150
        Lyell, Charles **253**
The Evolution of Igneous Rocks
    (Bowen) 53
Ewing, Maurice 336
exfoliation **146**, 457
exfoliation domes 146
expansive clays 78
explosion tubes 314
extension 401
extensional back arcs 96
extensional bends 404–405, 431–432
extensional boundaries *See* divergent
    or extensional boundaries

extensional foreland basins 159
extension joints 160
extinctions *See* mass extinctions
extratropical cyclones **146**
extratropical lows 399–400
extrusive rocks *See* volcanic rocks

F

facies **147–148**
Falkland Islands 148
Falkland Plateau 148
falls (mass movements) 267
fan-deltas 113
fault-block mountains 89, 111, 312
fault blocks 212–213
faults *148*, **148–150**, *149*
    Alpine fault, New Zealand
        292–293
    Basin and Range Province,
        United States 41–42
    decollement *110*, **110**
    deformation of rocks general
        111–112, *112*
    dip-slip faults *148*, 149, 433
    earthquakes and 132
    horsts **212–213**
    San Andreas Fault, California
        130–131, 132, **364–365**, *365*,
        421
    strike-slip faults 111, *148*, 149,
        190, **404–405**, 431, 432
    structural (tectonic) breccias 54
    thrust faults 2, 2–3, 149,
        189–190
    transform faults 277–278, *331*,
        333, 337, 431–433
fault scarps 149, *149*
fault zone aquifers 161
fault zones 150
faunal succession **150**
Federal Emergency Management
    Agency (FEMA) **150**
feedback mechanisms 82, 138
feldspar **150**, 282
felsic rocks 226
FEMA (Federal Emergency
    Management Agency) **150**
Ferrel cells *206*
ferricrete 42
ferro-carbonatites 62
Fertile Crescent 290, *291*

fertilization effect 64
Filabusi greenstone belt, Zimbabwe
    craton 471
Finger Lakes, New York 150
fiords 150, *151*
fireballs *See* meteors/meteorites
First Law of Thermodynamics 423
firths 373
fish 289
fission **151**
fission track dating technique 168
flash distillation 114
flash floods 151, 446–447
flood basalts **153–154**, *154, 301*
    Deccan flood basalts, India 68,
        109, 153, 261
    Siberian 317
floodplains 152, **154**
floods **151–153**, 354–355
    erosion 141
    floodplains **154**
    as geological hazards 169
    hurricanes and 151, 217
    levees and 247–248
    urbanization and flash flooding
        446–447
Florida
    artesian system 27
    Everglades **145**
    Hurricane Andrew 217
    sinkholes 239, 411
flower structures 432
flows 267
flowstone 398
fluorite 207
flute marks 381
fluvial sediments 8–9, **154–155**
fluvial systems 147, 154
Flynn diagrams *401*, 402–403
flysch **155**
focus 132
Foehn winds 73
fog **155–156**, 342
fold and thrust mountain ranges 89,
    111, 312
fold axial surfaces 12
fold facing 157
fold generations 157
fold hinges 12, 111
fold interference patterns *156*,
    157–158, *158, 374*

folds 111, *111*, *156*, **156–158**, *158*
    Adirondack Mountains *4, 5*
    anticlines *12*, **12–13**
    cleavage and 78
    greenstone belts 189–190
    Syrian arc fold belt **417–418**
fold trains 156
foliation 78, *79*, 273
footwalls 149
foraminifera 65, 345
forearc basins 94, 96
foreland basins 41, **158–159**
    Indo-Gangetic Plain **229**
    Witwatersrand **459–462**, *460,*
    461
foreshocks *See* earthquakes
foreshore *43*
formations **159**, 404
fossil fuels 168, 218, 289, 327–328
    See also coal; petroleum/petroleum
    industry
fossils **159**
    amber 10–11
    Archean 26–27
    belemnite *402*
    brachiopods **53**
    bryozoa **54–55**
    Cambrian 61
    carbon-14 dating **61**
    conodonts **87–88**
    corals 99, 108
    crinoids 65, **105**
    diatoms **121**
    echinoderms 105, **136**
    evolution and 145
    faunal succession 150
    gastropods **164**
    graptolites *182*, **182**
    ichnofossils (trace fossils) 49,
    61, **225**
    index fossils 121, 136, *182,*
    182, **228**, 404
    Metazoa **275–276**, 345, 449
    nektons 289
    paleontology **315**
    Protozoa 345
    sedimentary rocks 381
    Sinian fauna, China 73
    Smith, William H. 396
    Steno, Nicolas 399
    strain and 402, *402*
Fowler, William 48

Fox Islands, Alaska 8
fracture porosity 340
fractures *159*, **159–161**
fracture zone aquifers **161–162**
fracture zones 433
France, Système pour l'Observation
    de la Terre (SPOT) 370
franklinite 474
Fredrickton trough, Appalachians
    15
free-air gravity anomalies 183
freeboard, continental 408
freeze-thaw cycles 104, 268, 457
freezing drizzle 342
freezing rain 342
freshwater 114, *363*, **363–364**
frictional sliding 159
Friesland orogeny **415–416**
fringing reefs 351
Frontenac terrane, Grenville province
    193, 195
frost heaving 104, 268, 441
Fujita Tornado Scale 430
fumaroles *162*, **162**, 465
    geysers **174–175**, 464, *464*, 465
    hydrothermal vents 47, 50, 125,
    307
        black smoker chimneys 50,
        50–52, *51*, 71, 125, 249,
        307
        white smoker chimneys 50,
        51
    Valley of Ten Thousand Smokes
    448–449
    Yellowstone National Park 464,
    464, 465
funnel clouds 430
fusion **162**
fusulinid foraminifera 65

G

gabbro **163**
Gabel Maghara, Syrian arc fold belt
    417
Gaia hypothesis 82, **163–164**
Galileo (spacecraft) *29, 130*
Galveston Island, Texas 40–41, 151,
    164, **164**, 214
gamma rays 348
Gamow, George 48
Ganges River 229

Ganymede 234
Garlock fault, California 364
garnet 164, 281, *373*
gas, natural 218, **289**, 458
gas hydrates (clathrates) 63, **164,**
    440
gas law 8
Gaspra (asteroid) *29*
gastropods **164**
Gebel Arief El-Naqa 388
Gebel Halal 388, 417
Gebel Maghara 387, 388, 389
Gebel Yelleg 417
Geiger counters **164–165**
general circulation models **165**
General Theory of Relativity 48
generations (folds) 157
geobarometry **165**
geochemistry **165**
    Goldschmidt, Victor Moritz
    179–180
    X-ray fluorescence **463**
geochronology **165–168**
    carbon-14 dating 61, **165–166**
    ion microprobes 230
    zircon and 475
geodes **168**
geodesy **168–169**
geodynamics **169**
geodynamo theory **172–173**
Geographic Information Systems
    (GIS) **169**
geoid 168–169, **169**, 412, *413*
geological hazards **169–172**
Geological Society of America (GSA)
    172
geologic timescale 479
geologists
    Bowen, Norman Levi 53, **227**
    Cloos, Hans **83**
    Cloud, Preston **83**
    Dana, James Dwight 91, **108**
    Darwin, Charles 35, **108–109**,
    145, 166, 320–321
    Du Toit, Alexander L. 90, 91,
    128, 247, 335
    El-Baz, Farouk **136–137**
    Eskola, Pentti **142–143**, 274
    Gilbert, Grove Karl 113, 140,
    175, 321
    Grabau, Amadeus William
    180–181

geologists *(continued)*
    Hess, Harry Hammond
        209–210, 336
    Hoffman, Paul Felix 195, **211**
    Hutton, James 166, **217–218**,
        443
    Lawson, Andrew Cowper 247
    Lyell, Charles 166, 212, **253**,
        289, 443
    Pettijohn, Francis John **328**
    Powell, John Wesley **340–341**
    Sedgwick, Adam **379–380**
    Smith, William H. **396**
    Sorby, Henry Clifton **397–398**
    Stille, Wilhem Hans **399**
    Werner, Abraham Gottlob **458**
geomagnetic reversals 173
geomagnetism **172–173**, *173*
    auroras and 35–36, *36*
    compasses 86, 258
    declination **109**
    magnetosphere **258–259**
geomorphology **173–174**
    Gilbert, Grove Karl **175**
    Strahler, Arthur N. **401**
geophysics **174**
    geodynamics **169**
    techniques 352
    Wegener, Alfred Lothar **458**
geostrophic currents **174**
geosynchronous satellite orbit 178
geothermal energy **174**
    fumaroles *162*, **162**, 448, 464,
        464, 465
      geysers **174–175**, 464, *464*, 465
      hot springs 213, **464–465**, *465*
geothermal gradient 226, 227
geothermal wells **174**
Germany
    Reis crater 420
    Stille, Wilhem Hans **399**
geyser cones 174
geysers **174–175**, 464, *464*, 465
Ghats, Eastern and Western, India
    228–229
Giddi-Yelleq-Halal, Egypt 388
Gilbert, Grove Karl 113, 140, **175**,
    321
GIS (Geographic Information
    Systems) **169**
glacial erratics 177
glacial marine drift 177

glacial striations 176–177
glaciers/glaciation **175–177**, *176*
    calving 176
    Cenozoic 69
    cirques **76–77**, *77*
    erosion and 142
    eskers *142*, 142
    Finger Lakes, New York 150
    fiords 150, *151*
    formation of 175–176
    as geologic hazard 171
    hanging valley *207*, **207**
    ice ages 175, **220–222**,
        249–250
    icebergs 176, 177, 222,
        222–223, 359, 375, 376
    ice caps **223**
    ice sheets 175, 223, 278, 337
    ice shelves 359
    kames **238**
    landforms 176–177
    load 252
    Milankovitch cycles 278–279
    moraines 177, *285*, **285**, *285*,
        285
    movement of 176
    Pangean 318–319
    Pleistocene 337–338
    sea ice 376
    sea-level rise and fall 376–377
    transport and deposit 177
    types of 175
Global Carbon Sink, Siberia **385**
global positioning systems (GPS)
    168, 169, **177–179**
gneiss **179**
gneiss terranes
    Archean 24–25, *26*
    central gneiss belt, Grenville
        province 192–193
    greenstone belts and 188–189
    Tokwe terrane, Zimbabwe
        craton 469–470, 471–473
Gobi Desert **179**, *252*, 367
Golconda allochthon 357
gold **179**
    Patagonia 321
    placer 329
    Witwatersrand basin, South
        Africa 461
Goldschmidt, Victor Moritz
    179–180

Gondwana (Gondwanaland) *58*,
    180
    Cambrian period 59
    Carboniferous period *65*, *66*
    ice ages and 220
    Proterozoic era 343
    Tethys ocean 422–423
Gould, Steven J. 145–146
GPS (global positioning systems)
    168, 169, **177–179**
Grabau, Amadeus William **180–181**
graben **181**
graded bedding 44
graded streams 175
Grand Canyon **181**, 340, 403
granite **181–182**
granite-anorthosite association 345
granite-gneiss terranes 24–25, *26*
granite-greenstone terranes 23–24,
    100, 103, 190, 469
granite tectonics 83
granitic magma 227
granular flows 269
granulite *182*, **182**
graptolites *182*, **182**
gravel **183**
gravimeters **183**
gravity
    anomalies **183**
    geoid 168–169, **169**, 412, *413*
    mass wasting 266
    Mercury 271
    tides and 428–429
gravity unit (g.u.) 183
graywacke **183–184**
Great Barrier Reef **184**
Great Bitter Lake, Egypt 387
Great Dark Spot (GDS) *292*
Great Dike, Zimbabwe **184**
Great Glen fault, Scottish Highlands
    373
Great Lakes, North America **184**
Great Red Spot, Jupiter *233*, 234
Great Salt Lake, Utah **185**
Great Sandy Desert, Australia 367
Great Slave Lake, Nunavet, Canada
    185, **185**
greenhouse effect **185–187**, *186*
    gases contributing to 32, 34–35
    Late Paleocene global hothouse
        221
    life's origins and 248, 249

sea-level rise and 376–377
  Siberian Taiga Forest and Global
    Carbon Sink **385**
  thermohaline circulation and
    81–82
  Venus 450
Greenland
  Caledonides **58**
  ice sheet 175, 223, 376–377
  Isua belt 22, 26, 38
  thermohaline circulation 424
  Wegener, Alfred Lothar 458
greenschist **187**
greenstone belts **187–192**, *191*
  Archean era and 23–24, *25*,
    100–101, 103
  Belingwe greenstone belt,
    Zimbabwe **44–46**, *45, 46,*
    189, 236, 470
  geometry of 187–188, *191*
  gneiss terrains and 188–189
  Kaapvaal craton, South Africa
    236–238
  Slave craton 391, *394*
  structural elements of 189–191
  structural *vs.* stratigraphic
    thickness of 188
  Zimbabwe craton 469–474,
    472
Greenville plutonic belt,
  Appalachians 14
Grenville province, Canada 3–5,
  192–198, *196, 197*
  central gneiss belt 192–193
  central granulite terrane
    193–194
  central metasedimentary belt
    193
  Grenville front 194
  tectonic evolution of 194–198
groins **198**
ground moraines 285
groundwater **198–201**, *199*
  aquifers **16–17**, 27, 86,
    161–162, 200, 361
  artesian wells 27
  cones of depression 86
  contamination of 200–201
  drinking water standards 200
  fracture zone aquifers **161–162**
  karst terrains **238–240**
  meteoric 277

  saltwater intrusion of *363,*
    363–364
  speleothems *398,* **398**
  subsidence and 410, 411
  water table 17, 86, 200,
    454–455
groundwater basins 41
growth habit 282
GSA (Geological Society of America)
  172
g.u. (gravity unit) 183
Gulf of Aden 6
Gulf of Aqaba 386, 387
Gulf of Suez 386, 387, 390
gust fronts 425
Guth, Alan 48, 49
guyots 35, 209, 299, 377
Gwanda greenstone belt, Zimbabwe
  craton 471
gypsum 398

H

Hackett River arc, Slave craton
  392–393, *394, 395*
Hadean era 23, **202–205**
Hadhramaut Group, Oman
  Mountains 304
Hadley, George 80
Hadley cell deserts 117
Hadley cells 31, *33*, 80, **205–206**,
  206, 431
hail 342, 425
Hajar Mountains (Oman Mountains)
  303–306, *305*
Hajar Supergroup, Oman Mountains
  306
Hajar Unit, Oman Mountains 304
half-life 165, 348–349
halide minerals **207**
halite 207
Halley's comet *85, 86*
halocline 346
halogen 207
hammada 360
Hamrat Duru Group, Oman
  Mountains 306
hanging valleys *207,* 207
hanging walls 149
hardness 282
hard water 201
Harker, Alfred **207**

harmonite 469
hartzburgite **207–208**, 308
Hawaii **208**, 377–378, 439
  aa lava 1
  pahoehoe lava 314
  tsunami 436, 437, 439
  volcanoes 208, 451, 452
Hawaiian-Emperor seamount chain
  208, 260, *261,* 377–378
Hawasina nappes, Oman Mountains
  304
Hayward fault, California 364
heat flow **208**, 407, 424 *See also*
  convection and the Earth's mantle
heavy minerals **208–209**
Heezen, Bruce 336
Heinrich Events 81
helictite 398
heliotropism **209**
hemimorphite 474
Hengshan high-pressure granulite
  belt, North China craton 295–296
Hermit Shale, Grand Canyon 181
Hess, Harry Hammond **209–210**,
  336
Hess Rise 301
heterogeneous strain 401
heterosphere 31, *34*
heulandite 469
high-constructive deltas 113
high-destructive deltas 113
high-grade metamorphism 272
Highland Boundary fault, Scottish
  Highlands 373
high-pressure regions 8
Al-Hijaz terrane, Arabian shield 19,
  20
Himalaya Mountains *210,* **210–211**
  convergent plate margin
    processes 98–99
  Mount Everest **285–286**
  rainfall 428
Himalayan terrane, Tibetan Plateau
  426–427, *427*
Hindu Kush, Indo-Gangetic Plain
  229
hinge lines 156, 157
hinterlands **211**
Hoffman, Paul Felix 195, **211**
hogbacks **211**, *212*
Holmes, Arthur 90, 334–335
Holocene epoch 69, **211–212**, 347

holoplankton 329–330
homeostasis 163
homoclines **212**
Homo erectus 338
homogeneous strain 401
Homo habilis 338
Homo sapiens neanderthalensis 338
Homo sapiens sapiens (humans) **338**
    Holocine 69, 211–212, 347
    Neolithic 290
    Sahara Desert 361–362
homosphere 31, *34*
Honduras, Hurricane Mitch 217
Hooke's Law 110
horizontal aquifers 161–162
horizontal plane 407
hornfels 273
horns 77, 177
horse latitudes **212**
horsts **212–213**
hothouses 221, 250
hot spots **213**, 452
    Hawaii **208**, 260, *261*
    Iceland 224
    mantle plumes **260–261**, *261*
    Yellowstone National Park 465
hot springs **213**, 464–465, *465*
Howard, Luke 83
Hoyle, Fred 48
Hubble, Edwin 48
Hubble Space Telescope *263, 339, 372*
Hudson Bay, Canada *213*, **213–214**
humans *(Homo sapiens sapiens)* **338**
    Holocine 69, 211–212, 347
    Neolithic 290
    Sahara Desert 361–362
Huronian division **214**
Huronian Supergroup, Canada 344–345
Hurricane Alice 151
Hurricane Andrew 217
Hurricane Camille 217
Hurricane Claudette 151, 153
Hurricane Manoun *215*
Hurricane Mitch 217
hurricanes **214–217**, *215, 216*
    floods and 151, 217
    Galveston Island, Texas 40–41, 151, **164**, 214

polar lows **340**
Saffir-Simpson scale 216–217
storm surges 217, 399
Hutton, James 166, **217–218**, 443
hydrocarbons **218**
    coal 85
    gas hydrates **164**
    natural gas **289**
    petroleum *See* petroleum/petroleum industry
hydroelectric power **218**
hydrofracturing **218**
hydrologic cycle **218**, 378
hydrology 108, **218–219**
hydrolysis 457
hydrosphere 130, **219**
    biosphere 49–50
    hydrologic cycle 218
hydrostatic equilibrium, principle of (isostasy) 230
hydrothermal deposits **219**, 275
hydrothermal fluids **219**
hydrothermal solutions **219**
hydrothermal systems 249, 275
hydrothermal vents 47, 50, 125, 307
    black smoker chimneys 50, **50–52**, *51*, 71, 125, 249, 307
    white smoker chimneys 50, 51
hypabyssal rocks 225
hypsometric curve (hypsographic curve) *219*, **219**

I

Iapetus Ocean
    Scottish Highlands and 374
ice ages 175, **220–222**, 249–250
    See also glaciers/glaciation
icebergs 177, **222–223**, 359
    calving 176, 222
    sea ice 375, 376
ice caps 175, **223**, 375
ice dams 153
icehouses 250
Iceland **223–225**, *224*, 261
Iceland spar 56
ice sheets 175, 223, 278, 337
ice shelves 359
ichnofossils (trace fossils) 49, 61, **225**
Ida (asteroid) *29*

Idaho
    Columbia River flood basalts 153
    Lake Bonneville 337
    Northern Rockies 357
    Yellowstone National Park/Yellowstone Valley *56*, 227, 261, 452, *464*, **464–465**, *465*
igneous rocks **225–227** *See also* plutonic rocks (intrusive rocks); volcanic rocks (extrusive rocks)
    amphiboles 11–12
    basic rocks **41**
    batholiths **42**
    Bowen, Norman Levi 53
    breccia 53–54
    carbonatite **62–63**
    diapirs 121
    feldspar **150**
    granite **181–182**
    Ontong-Java plateau, Pacific Ocean 154, 301–302, **306–307**
    pegmatite **321**
    S-type asteroids 29
    viscosity **451**
    xenoliths **463**
ignimbrites 298
impact craters **227–228**
    Chicxulub crater, Yucatan Peninsula, Mexico 30, 68, **71**, 440
    Mercury 271
    Venus 451
impactogens 159
Imperial fault, California 364
inclination 172
inclinometers 83
index fossils **228**
    diatoms **121**
    echinoderms **136**
    graptolites *182*, **182**
    stratigraphy and 404
India
    Cherrapunji region rainfalls 153
    Deccan flood basalts 68, **109**, 153, 261
    Eastern and Western Ghats **228–229**
    monsoons 285

India-Asia convergence 98–99, 210, 211, 229, *334*, 428
Indian Ocean
    currents 300
    ridge 278
    tsunami 303, 436, 437, 439
Indian plate 427
Indo-Australian plate 292
Indo-Gangetic Plain **229**
Indus River 229
Indus Valley, Indo-Gangetic Plain 229
infauna 47
inflationary theory of the universe 48
infrared imagery 352–353
infrared radiation **229**, 348, 424
Inner Mongolia 295
Inner Mongolia–Northern Hebei orogen 297
insects 65
inselbergs **229**
interfluves 122
intermediate rocks 41, 226
intertidal zones 47
Intremo Group (Sèries Quartzo-Schisto-Calcaire, QSC), Madagascar 255, 257
intrusive rocks *See* plutonic rocks (intrusive rocks)
inversions **229–230**
invertebrates, Cambrian period 58, 60–61
Io 234
ion microprobe **230**
ionosphere 31–32, *34*, **230**
Iran, Zagros and Makran Mountains **467–469**, *468*
iron, banded iron formations (BIFs) **38–40**, *40*
iron meteorites 276
ironstone sediments 70
Irving, Earl 335–336
Ishtar Terra, Venus 450
island arcs 94–96, 97
    Arabian shield 17–18, 19–20
    crustal growth and 100, 101, 333
    subduction zones 410
isograd *See* metamorphism
isostasy **230**
    continental drift theory and 90
    Lawson, Andrew Cowper 247

isostatic anomalies 230
isostatic correction 183
isostatic models 230
isotopes **230–231**, 348–349
isotopic dating systems 166–168
Israel, Dead Sea **109**
Isua belt, Greenland 22, 26, 38
Italy, Venice subsidence 411

J

Jabal Akhdar, Oman 305–306
jade **232**
Japan
    Mount Fuji **286**
    paired metamorphic belts **232–233**
    Tokyo subsidence 411
    tsunami 436, 439
jet streams 205, **233**
Ji'balah Group, Arabian shield 19
jifn 127
Johnson, Peter 18
Joides Resolution 233
joints 111, 159, *160*
    types of 160–161
    weathering and 456, 457, 458
joint systems 160
Jones, Davie 415
Jordan, Dead Sea **109**
Jupiter 28, *233*, **233–234**, 480
Jurassic Period **234–235**

K

Kaapvaal craton, South Africa 25, *45*, 46, 100, 101, **236–238**, *237*, 459–462, *460*, *461*, *472*, *474*
Kaibab Limestone, Grand Canyon 181
kames **238**
kaolinite **238**, *457*
Karakoram Range, Indo-Gangetic Plain 229
Karelian province, Baltic shield 38
Karoo Sequence, Kaapvaal craton 238
karst terrains **238–240**
    caves 67–68
    Everglades, Florida 145
Kaskaskia Sequence 434
katabatic winds **240–241**

Kazakhstan, Aral Sea **22**
K-band 416
Kearsarge–Central Maine basin, Appalachians 14–15, 15–16
kelp 47
Kelvin, Lord 166
Kentucky, Mammoth Cave 68
kerogens 26
kettle holes 420
Keweenawan rift, North America 343–344
K-feldspar 457
Kidd, Bill *55*
Kilauea, Hawaii 452
kimberlites 120–121, **241**, *463*
kinematics 407
kinetic energy 424
Kirkwood Gaps 28
klippen (klippe) **241**
knob and kettle terrains 420
Kober, Leopold 91
Kola Peninsula, Russia **241–242**
Kollsman, Paul 10
komatiite 23, 101, **242**
Koppen, Wladimir 79
Krakatau eruption 56, 302, 436–437, 439
K-T (Cretaceous-Tertiary) extinction 68, 105, 109, 264–265
Kuiper belt 85
kurums 441
Kuskokwim River, Alaska *312*
Kusky, Timothy 45–46, 92, 391–392
Kuwait **242**
Kuwait Bay **242–243**

L

lagoons **245**
Lagrange Points 28
lahars **245**, 266
Lake Bardawil, Egypt 387
Lake Bomoseen, Vermont *111*
Lake Bonneville, Utah, Nevada, Idaho 337
lake breezes 374
Lake Eyre, Australia **246**
lakes **245–246**
    Aral Sea **22**
    eutrophication **143**
    Finger Lakes, New York 150

lakes *(continued)*
    Great Lakes, North America **184**
    Great Salt Lake, Utah **185**
    Great Slave Lake, Nunavet *185,*
        185
    Lake Eyre, Australia **246**
    Lake Titicaca 10, **246**
    Lake Victoria **246**
    pack ice **375**
    seiche waves **381**
Lake Titicaca 10, **246**
Lake Victoria, Africa **246**
land bridges 421–422
Landsat *368,* 369, 370, 416
landslides 169, 266, 267, 439–440
Langmuir Circulation **247**
La Niña 138
Lapland-Kola province, Baltic shield
    38
Laramide orogeny **358**
Large Format Camera (LFC) **371**
laser altimeter 10
latent heat 32
lateral moraines **285**
laterite **247**
Laurasia **247**
Laurentia 58, 247, 422
laurmontite **469**
lava 247, 451–453
    aa lava **1**
    pahoehoe lava **314**
    pillow lavas *25, 324,* **328–329,**
        329
lava fountains **452**
lava tubes **314**
Law of Faunal Succession **150**
Law of Original Horizontality **403**
Lawson, Andrew Cowper **247**
L-band 416
lead-zinc deposits **474**
left-lateral strike-slip fault *148*
Lemaitre, Georges 48
lenticular clouds 84
Levantine fold belt, Syrian arc 417
levees **247–248**
Lhasa terrane, Tibetan Plateau
    426–427, *427*
lherzolite **307**
Libby, Willard F. 61, 165–166
life, origins and evolution of *64,*
    248–250
    Archean (Archaean) 26–27
    atmosphere and 480

black smoker chimneys and
    51–52
Cambrian 58, 60–61
Carboniferous period 65–66
Cenozoic 68, 69
chemosynthesis and 71
classification of species 480
Cloud, Preston **83**
Cretaceous 105
Devonian 119
faunal succession **150**
Hadean Eon 204–205
Mesozoic 272
Ordovician 311
Ordovician period 311
Proterozoic 345
Silurian 386
stromatolites *405,* **405–406**
supercontinent cycle and
    412–415
Tertiary 421–422
Triassic 434–435
lightning 425–426
light scattering 379
lignite 85
limestone
    caves 68
    metamorphism of 273
    sediments 70
limonite **250**
Limpopo Province, Africa 461, *461,*
    472, 474
Linde, Andrei 49
linear dunes 367
liquefaction 411
lithification **250**
lithology 159
lithosphere 49–50, 129, **250–251**
    See also crust
Little Bitter Lake, Egypt 387
littoral drift (longshore drift) 40,
    198, **251,** 456
littoral zones **251**
lizardite 385
Llandoverian age 386
load 53, **251–252,** 356
Loch Marce, Scotland *374*
lochs 373
lode deposits 179
loess 128, 139, *252,* **252**
Logan, William 214
Loma Prieta earthquake, California
    132

longshore drift (littoral drift) 40,
    198, **251,** 456
Long Valley caldera, California
    252–253
look azimuth 416, 417
lophophores 53
Louisiana
    Hurricane Andrew **217**
    New Orleans subsidence 410
Lovelock, James 82, 163
low-grade metamorphism 272
low-pressure regions 8
low velocity zone 30
Ludlovian age 386
luster 282
Lyell, Charles 166, 212, **253,** 289,
    443

M
macrobursts 425
macroevolution 145
Madagascar **254–258,** *256*
    Hurricane Manoun *215*
    rainbow *351*
mafic dikes 184
mafic rocks (basic rocks) 41, **41,**
    226
magma *See also* volcanoes
    crust, formation of *332,* 333
    explosive *vs.* nonexplosive
        eruptions and 452–453
    igneus rocks 225–227
    plutons 339
    volcanic bombs **451**
magma chamber *332,* 333
magmatic differentiation 226, 227
magmatic tectonics 83
magnesio-carbonatites 62
magnesite 62
magnetic anomalies 315, *316,* 335
magnetic compasses 86, **258**
magnetic fields
    Earth 172–173, *173, 335, 335*
        auroras and 35–36, *36*
        compasses 86, 258
        declination **109**
        magnetosphere *36, 36,*
            258–259
        paleomagnetism 315, *316,*
            335
    Uranus 446

magnetic poles 86, 109, 315, 335–336

magnetosphere 36, *36*, **258–259**

magnetotail 259

Maine
boudins *52*
Penobscottian orogeny **321–325**, *323, 324*
rock cleavage 79

Main Ethiopian Rift 6

Majlis Al Jinn Cave, Oman 67

Makran Mountains, Iran **467–469**, *468*

mammals 66, 421–422

mammatus clouds 84

Mammoth Cave, Kentucky 68

Mammoth Hot Springs, Yellowstone National Park 465, *465*

manganese nodules **259**

Manihiki Plateau 301

Manjeri Formation, Belingwe greenstone belt, Zimbabwe 44–46, *46*, 470, 472–473

Manso River 320

mantle **259–260**
asthenosphere **30**
convection and 23, 90–91, **93–94**, *94*, 208, 260–261, *261*, 412

mantle plumes 94, 213, **260–261**, *261*, 301

mantle root (tectosphere) 250, **419**

marble **261**

Margulis, Lynn 406

Marianas Islands 261

Marianas-style arc systems 94, *95*

Marianas trench **261**

Mars **261–263**, *263*, 480

marsupials 421

Massachusetts
Cape Cod **61**
wollastonite 462

mass extinctions 170, 171, **263–265**
Cretaceous-Tertiary (K-T) boundary 68, 71, 105, 109, **264–265**
Devonian 119
evolution and 146
Permian-Triassic extinction 264, 317, 412–413
supercontinent cycle and 412–413, *414*

mass spectrometers **265**

mass spectrometry **265**

mass wasting **265–270**
creep *103*, **103–104**, 169, 268, 441
debris avalanche 217, 269–270
debris flows 245, 268–269
erosion 142
granular flows and earth flows 269
mudflows 217, 269
rockfalls 269
slides 267, 269
slumps *267*, **267–268**, *439, 440*
slurry flows 268
solifluction 268, 441
subaqueous 270

mathematicians
Milankovitch, Milutin M. **278**
Sedgwick, Adam **379–380**

Matrix theory (M-theory) of the universe 49

Matsitama greenstone belt, Zimbabwe craton 470

Matthews, D. H. 336–337

Matuyama 337

Mauna Loa, Hawaii 451, 452

Mazinaw terrane, Grenville province 193

McCarty fiord, Alaska *151*

McCarty tidewater glacier, Alaska *176*

McHugh complex, Alaska 74–75

McLelland, Jim 5

meanders 154, **270**, 312

mean stress 404

mechanical weathering 456–457

mechanics 407

medial moraines 285

Mediterranean Coastal Plain Region, Sinai 388

Mediterranean region, Messinian salt crisis 290

medusa 99

mélanges *95*, **270–271**

membrane desalination 114–115

membrane distillation 115

Mercalli, Giuseppi 135

Mercury *271*, **271**, 480

meroplankton 330

Merrimack trough, Appalachians 15

mesoproterozoic interval 343

mesoscale convective systems 425

mesosphere 31, *34*, **271–272**

Mesozoic era **272**
Cretaceous period *104*, **104–105**
Jurassic period **234–235**
Triassic period **434–435**, *435*

Mesozoic truncation event 358

mesozonal rocks 42

Messinian salt crisis 290

metabauxite 42

metamorphic facies 147, 274, *274*

metamorphism/metamorphic rocks **272–273**, 273
amphiboles 12
C-type asteroids 29
Dabie Shan ultra-high pressure metamorphic belt, China 71
eclogites **136**
Eskola, Pentti 142–143
garnets **164**
gneiss **179**
granulites *182*, **182**
greenschist **187**
Japan's paired metamorphic belts **232–233**
marble **261**
metasomatic **275**
schist **372–373**, *373*
serpentinite **385**
wollastonite **462**

metasomatic processes **275**

metasomatism **275**

Metazoa **275–276**, 345, 449

Meteor Crater, Arizona 228

meteoric water **277**

meteorites *See* meteors/meteorites

meteorology **277**
radiosondes 349
Wegener, Alfred Lothar **458**

meteor showers 86, 276

meteors/meteorites 29–30, **276–277**
See also asteroids; comets
Chicxulub crater, Yucatán Peninsula, Mexico 30, 68, **71**, 440
as geological hazard 171
Hadean Eon 203
impact craters **227–228**
Neogene impact event 290
tektites and 420

methane 32, 164, 289, 377

Mexico
Baja California 38
Chicxulub crater, Yucatán
Peninsula 30, 68, **71**, 440
Miami Oolite 145
mica 50, **277**, 281–282
mica schist 372–373, *373*
microbursts 425
microscopes
electron microscopes **137–138**
polarizing microscope **339–340**
Sorby, Henry Clifton **397–398**
microwave remote sensing 353
Mid-Atlantic ridge 51, 223–224,
278 *See also* mid-ocean ridge
systems
mid-latitude cyclones *See*
extratropical cyclones
mid-latitude deserts 117
Mid-Ocean Ridge Basalt (MORB)–
type ophiolites 100–101
mid-ocean ridge systems **277–278**
black smoker chimneys 50,
50–52, *51*, 71, 125, 249, 307
crust formation at *332*
divergent margins *124*,
124–125
heat loss by mantle convection
and 93–94.*94*
Iceland 223–224
life's origins 249
transform faults 277–278, 433
volcanic eruptions and 452
Mid-Pacific Mountains 301
Midyan terrane, Arabian shield 19,
20
Milankovitch, Milutin M. **278**
Milankovitch cycles **278–280**, *279*
millidarcy 326
milligal 183
mineral cleavage 78
mineralogy **280–282**
diamond anvil press and multi-
anvil press **121**
Goldschmidt, Victor Moritz
179–180
Werner, Abraham Gottlob **458**
Miocene epoch 421
Miramichi massif, Appalachians 15,
16
Miranda 446
mirror plane crystal symmetry 107

Mississippi
Hurricane Camille 217
Washington County flood
(1927) 152
Mississippian Period 65
Mississippi River 152, 248, *283*,
283–284
Missouri
Carrol Cave *398*
New Madrid earthquakes 131,
411
St. Louis floods 152
Missouri River 283, *283*
Miyashiro, Akiho 232
mobilist school 91
Modified Mercalli Intensity Scale
135, 135–136, 354
Mohorovicic discontinuity (the
Moho) 250, **284**, 383, 384
Moh's Hardness Scale 282
Moine thrust zone 374
Moinian Assemblage, Scottish
Highlands 373, *374*
Mojave segment, San Andreas Fault
364
molasse **284**
Mollusca 164
Mongolia, Gobi Desert 179
monoclines 111
Mono Lake-Inyo Craters 253
monsoon deserts 118
monsoons **284–285**, 428, 436
Montana
Rocky Mountains 357
W. R. Grace and Co. asbestos
disaster 28
Yellowstone National Park *464*,
464–465, *465*
Moon *130*
age of 166
origin of 203, *205*
tides and 428–430
Moon halos 411
moonmilk 398
Moores, Eldridge 195
moraines 177, *285*, **285**
MORB (Mid-Ocean Ridge Basalt)-
type ophiolites 100–101
Morocco, Atlas Mountains 31
mountain breezes *See* katabatic
winds
mountain glaciers 175

mountains 129 *See also* mid-ocean
ridge systems; orogeny/orogens
Adirondack Mountains **3–6**, *4*,
194, 195–196
alluvial fans **8–9**, *9*
Alps **9–10**, *111*
Andes 10, **12**
Appalachians **13–16**, *14*, *16*,
58, 321–325
Atlas Mountains **31**
Basin and Range Province,
United States **41–42**, 358
Brooks Range, Alaska **54**
Chugach Mountains, Alaska
74–76, *75*, *77*, 207, 286, 442
exfoliation **146**
fault-block ranges 89, 111, 312
fault-blocks 112
fold and thrust belts 89, 111,
312
Himalaya Mountains **98–99**,
210, **210–211**, 285–286, 428
hinterlands **211**
inselbergs **229**
Mount Everest 54, **285–286**
Mount Fuji, Japan **286**
Mount Logan, Canada *286*
Mount McKinley, Alaska *286*,
286–287
Mount Washington, New
Hampshire **287**
Mount Whitney, California **287**
Oman Mountains (Hajar
Mountains) **303–306**, *305*
rainshadow deserts 117–118
Rocky Mountains **356–358**
Brooks Range, Alaska **54**
Canadian Rockies *12*, 357
Chinook winds **73**
upslope fog 155–156
Ural Mountains **444**
Venus 450
volcanic ranges 89, 111–112,
312
Mount Elbert 356
Mount Everest **285–286**
Mount Fuji, Japan **286**
Mount Katmai, Alaska 448–449
Mount Logan, Canada *286*
Mount Marcy massif, Adirondack
Mountains 3–4, *5*
Mount McKinley, Alaska *54*, *286*,
286–287

Mount Pinatubo, Philippines 245
Mount Saint Helens, Washington
  245, 453
Mount Sinai, Egypt *390*
Mount Washington, New Hampshire
  287
Mount Whitney, California **287**
Mozambique Belt 59, 343
Mozambique Ocean 59
MSS (Multi-Spectral Scanner) 369
M-theory (Matrix theory) of the
  universe 49
Mtshingwe Group (Lower
  Greenstones), Belingwe greenstone
  belt, Zimbabwe 44–46, 470
mud cracks 381
mudflows 217, 269
mudstone 273
multi-anvil press **121**
multi-effect distillation 114
Multi-Spectral Scanner (MSS) 369
multiverses 49
Murray, John 70
muskegs **287**
Mweza greenstone belt, Zimbabwe
  craton 470
mylonite *287*, **287**
mylonitic gneiss 179

N

Nabitah suture zone, Arabian shield
  19, 20
NAGT (National Association of
  Geoscience Teachers) **288**
nailhead spar 56
Najd fault system, Arabian shield
  19, 20
Namib Desert, Africa 367
Namibia, Skeleton Coast **391**
nappes 91, 241, **288**
NAS (National Academy of Science)
  288
NASA *See* National Aeronautics and
  Space Administration (NASA)
Natal, Boris 102
National Academy of Science (NAS)
  288
National Aeronautics and Space
  Administration (NASA) **288**
  asteroids and comets 30
  Cassini spacecraft 233

Galileo photos *29, 130*
  Hubble Space Telescope *263,
    339, 372*
  Voyager 2 292, 445
National Association of Geoscience
  Teachers (NAGT) **288**
National Oceanographic and
  Atmospheric Association (NOAA)
  288–289, 437
National Weather Service (NWS) **289**
natrolite 469
natural gas 218, **289**, 458
The Natural History of Rocks
  (Harker) 207
natural selection 145
Neandertals 338
Near Islands, Alaska 8
Nebraska Sand Hills 367
nebular theory 203
nektons **289**
Neogene period **289–290**, 421
Neolithic (New Stone Age) 211, **290**
Neoproterozoic era **290**, 343
Neo-Tethys Ocean 304–305, 422
Neptune **290–292**, *292*, 480
net convergence 8
Netherlands, subsidence in 410–411
Nevada
  Lake Bonneville 337
  thrust fault *111*
Nevado del Ruiz, Columbia 245,
  266
Newfoundland
  decollement *110*
  fault scarp 149, *149*
  ophiolites 324
New Hampshire, Mount Washington
  287
New Madrid, Missouri 131
New Mexico, Southern Rockies
  356–357
Newton, Sir Isaac 428–429
New York
  Finger Lakes 150
  Long Island seawater intrusion
    364
  Potsdam Sandstone 13
New Zealand, Alpine fault and
  Otago Schist **292–293**
Ngezi Group (Upper Greenstones),
  Belingwe greenstone belt,
  Zimbabwe 44–46

Nicaragua, Hurricane Mitch 217
Nile River 246, **293–295**, *294*
nimbostratus clouds 84
Nipissing terrane, Grenville province
  192
nitrogen oxide 377
nitrous oxide 32
Nixon, Lewis 397
NOAA (National Oceanographic
  and Atmospheric Association)
  288–289, 437
noncoaxial deformation 402
nonconforminties 443
nonelastic deformation 110
non-uniformitarian view 444
nor'easters 146, 399–400
normal fault 149
normal surface forces 404
Norseman-Wiluna greenstone belt,
  Australia 188, 189, 190
North America *See also specific
  countries and states*
  Adirondack Mountains **3–6**, *4*,
    194, 195–197
  Appalachians **13–16**, *14, 16*,
    58, 321–325, *323, 324*
  Cenozoic era 68–69
  Cordillera 357, 415, 421
  east coast passive margin 320,
    320
  Great Lakes **184**
  Keweenawan rift 343–344
  Laurentia 58, 247, 422
  Lawson, Andrew Cowper 247
  Ordovician period 310–311
  Paleocene-Eocene subduction
    event *422*
  Panama isthmus **317–318**
  Rocky Mountains *12*, 54,
    68–69, 73, **356–358**, *357*
    Brooks Range **54**
  Tertiary ridge subduction event
    421, *422*
North American plate 364, 465
North Atlantic cold bottom water
  80–82
North Atlantic Igneous Province
  153
North China craton 71–72, 101,
  295–297, 307
northern lights (auroras) **35–36**, *36*
northers 369

North Korea 295
North Pole 315, 378
North Sinai fold belt 390
North Slope, Alaska and Arctic
National Wildlife Refuge (ANWR)
297–298
Norway, Svalbard and Spitzbergen
Island **415–416**
Novarupta eruption, Alaska 448
Nubian-type Sandstone 389
nuclear accelerators 151
nuclear fission 151
nuclear fusion 162
nuclear reactors 151
nuée ardentes **298**, 453
Nunavet, Canada 25, 329, 419
NWS (National Weather Service) 289

O

Occupational Safety and Health
Administration (OSHA), asbestos
regulation 28
ocean basins/oceans **299** *See also*
beaches; sea-level rise and fall
abyssal plains **1**
atolls **35**, 351
benthic environments **46–47**,
49, 50
carbon cycle and 63
desalination **114–115**
evaporites 144
lagoons **245**
layers of 63
littoral zones **251**
manganese nodules 259
mid-ocean ridges *See* mid-ocean
ridge systems
oceanography **302–303**
plate tectonics and *336*
pycnocline **346**
reefs 47, 62, 99, 184, **351**
seawater, composition of
378–379
tides **428–430**, *429*
transform faults *432*, 432–433
tsunami 302, **436–441**, *438,
440*
waves *455*, 455–456, **455–456**,
456
beaches and 43, *455*,
455–456, *456*

rogue waves **358–359**
tsunami **436–441**, *438, 440*
wind and *455*
ocean-continent subduction systems
410
ocean currents **299–301**, *300*, 302
Benguela current 391
coastal downwelling **85**
coastal upwelling **85**
Coriolis effect **99**
Ekman spirals **136**
El Niño and the Southern
Oscillation (ENSO) 138
geostrophic currents 174
Langmuir Circulation **247**
Skeleton Coast, Namibia 391
themohaline circulation 80–82,
300, **424**
tides and 428
waves and **358–359**
Ocean Drilling Program (ODP) 233,
302, 310
oceanic crust 106, 299, **301**
Caspian Sea **66–67**
formation of 309–310, *332,
333*
gabbro 163
Hess, Harry Hammond 209
hypsometric curve (hypsographic
curve) 407–408, *408*
ophiolites and 101, 125
seismic layers 125
oceanic migrations 302
oceanic plateau **301–302**
accretion 101
flood basalts 153–154
Ontong-Java plateau 154,
301–302, **306–307**
oceanic plates 409–410
ocean-ocean subduction systems 410
oceanography 70, **302–303**
Odenbach, Frederick L 382
ODP (Ocean Drilling Program) 233,
302, 310
offscraping 2, *2*
offshore *43*
Ohio River 283, *283*
oil *See* petroleum/petroleum
industry
oil traps 327
Old Red Sandstone 374, 380, 416
Oligocene epoch 421, 422

olistostromes 95
olivine 281
Oman
hogback mountain *212*
Majlis Al Jinn Cave 67
Oman Mountains (Hajar
Mountains) **303–306**, *305,
467*
sinkhole 239, *240*
Wadi Dayqa *454*
Wahiba Sand Sea 368, *368*
Oman Mountains (Hajar Mountains)
303–306, *305, 467*
On the Origin of Species (Darwin)
109, 145
Oort cloud 85, 203
opal 6–7
ophiolites **307–310**, *308, 309*
accretion 100–101
Arabian shield 19
Boil Mountain ophiolite, Maine
321–325, *323, 324*
Dongwanzi ophiolite, China
71–73, 296–297, 307
formation 309–310
North China craton 296–297
oceanic crust and 100, 125
oldest 307
Oman Mountains 304–305
podiform chromites and 74
Opisthobranchia 164
optical imagery 352–353
Ordovician extinction 264
Ordovician period **310–311**, *311*
Oregon
Columbia River flood basalts
153
Crater Lake 56, *57*
Original Horizontality, Law of 403
*The Origin of Continents and
Oceans* (Wegener) 90
origin of universe *See* Big Bang
Theory
orogeny/orogens 88–89, 112,
311–312 *See also* mountains
Penobscottian orogeny
321–325, *323, 324*
Proterozoic 343
Turkic (accretionary-type) 102
orthobauxite 42
orthogneiss 179

orthopyroxene-bearing granulite 182, *182*
Otago Schist, New Zealand **292–293**
Ottawan orogeny 3, 5, 192, 197–198
outflow boundaries 425
outwash 177
outwash plains 177
overland flow 141
Oweineat Mountains 360
oxbow lakes 154, *312,* **312**
oxide iron formations 38–39
oxide minerals 282, **313**
oxygen 32, 280, 328
ozone 32, 313, 436
ozone hole **313**

P

Pacific Ocean
  abyssal hills 1
  currents 300
  East Pacific Rise 278
  guyots 35, 209, 299, 377
  Hess, Harry Hammond 209
  Marianas trench **261**
  oceanic plateaus 154
  Ontong-Java plateau 154, 301–302, **306–307**
  Panama isthmus 318
  plateaus 301
  ridges 125
  seafloor magnetic stripes *316*
  tsunami 436, 437
Pacific plate 292, 364
pack ice 214, 222, 340, 375
pahoehoe lava **314**
paired metamorphic belts **232–233**
Pakistan, Makran Mountains **467–469,** *468*
Paleocene epoch 221, 421, *422*
paleoclimatology **314–315**
Paleogene period 139, 421
Paleolithic division **315**
paleomagnetism **315–316,** *316,* 335
paleontology **315**
  faunal succession 150
  Grabau, Amadeus William **180–181**
  Steno, Nicolas **399**
Paleoproterozoic interval 342–343

Paleozoic era 315, 317
  Cambrian period **58–61,** *59, 60*
  Carboniferous period **64–66,** *66*
  Devonian period 119
  ice age 220
  Ordovician period **310–311,** *311*
  Permian period **326–327**
  Silurian period **385–386**
palm tree structures 432
Palmyride fold belt, Syrian arc fold belt 417, 418
Palymra fold belt, Syrian arc 417
Pan-African orgenic belts *59,* 343
Panama Canal 318
Panama isthmus 221, **317–318**
Pangea **318–319,** *319*
  Mesozoic era 272
  Paleozoic era 247, 317
  Panthalassa **319**
  Permian period 326–327
  supercontinent cycle 412
  Triassic period 435
  Wegener, Alfred 90, *90*
Panthalassa 319, 435, *435*
pantotheres 421
parabolic dunes 367
paragneiss 179
parallel drainage patterns 126
parallel folds 157
parallel strata 380
parhelia 411
Parry Sound terrane, Grenville province 192–193, *197*
partial melting 226–227, *227*
passive margins 93, *319,* **319–320,** *320,* 327
Patagonia **320–321**
P-band 416
peat 85
pediments *321,* **321**
pegmatite **321**
Pennsylvanian period 65
Penobscottian orogeny **321–325,** *323, 324*
Penrose-type ophiolites 307–309
Penzias, Arno 48
perched water tables 17, 200
Percival, William 48
peridotite **325**
periodic table of the elements 477–478

permafrost 325–326, **325–326,** *326,* 441
permeability 199, **326**
Permian period **326–327**
Permian-Triassic extinction 264, 317, 412–413
Peru
  Altiplano 10
  El Niño and the Southern Oscillation (ENSO) 138
  Lake Titicaca **246**
petrogenesis *See* petrology
petrography 327, 397
petroleum/petroleum industry 218, **327–328**
  anticlines and 12–13
  Caspian Sea 66–67
  conodonts 88
  correlation of rocks 404
  hydrofracturing **218**
  Kuwait 243
  natural gas **289**
  wildcat wells **458–459**
  Zagros and Makran Mountains, Iran 467
petrology **328**
  Bowen, Norman Levi **53**
  Harker, Alfred **207**
Pettijohn, Francis John **328**
phanerites 225
Phanerozoic eon **328**
Philippines, Mount Pinatubo eruption 245
phosphates 70
photochemical smog 436
photosynthesis 49, **328,** 406
phototropism 209
physical weathering 140
phytoplankton 329
Piazzi, Giuseppe 30
piedmont glaciers 175
Pietersburg greenstone belt, Kaapvaal craton 237
Pilbara craton, Australia 26–27, 190
pillow lavas *25, 324,* **328–329,** *329*
pingos 441
pipes 241
Piscataquis volcanic arc, Appalachians 14, 15, 16
Pitman, Walter 336
placer deposits 179, **329,** 461
placer mining 329

plagioclase feldspars 282
plane of the ecliptic 378
planets *See* solar system
plankton **329–339**
plastic behavior 451
plate boundaries, map of *336*
plate tectonics **330–337**, *336*
    continental margins **93**
    convergent plate margins *See*
      convergent plate margin
      processes
    divergent or extensional
      boundaries **122–125**, *123,*
      124, 333
    earthquakes **130–136**
    geological hazards and 170
    heat loss by mantle convection
      and 93–94.*94*
    historical development of
      theories **91–92**, 333–337
    metamorphism and 275, *275*
    passive margins 93, *319,*
      319–320, *320,* 327
    Proterozoic 343
    pull-apart basins 41, 109, **346**,
      404–405, 431–432
    seafloor magnetic stripes 315,
      316
    structural geology and
      406–407, 407–408
    supercontinent cycles **412–415**
    suture zones **415**
    transform plate margin processes
      333, *431,* **431–433**, *432*
    volcanoes and 452
platforms, carbonate 62, 320, 351
playas 242, **337**
Pleistocene epoch **337–339**, 347
plieus clouds 84
Plinian columns 453
Pliny the Elder 11, 453
Pliocene epoch 421, 422
plumose structures 160, *160*
Pluto 277, *339,* **339**, 480
plutonic rocks (intrusive rocks) 225,
  225
    breccias *53*
    dunites **127**
    gabbro **163**
    greenstone belts 187
    harzburgites **207–208**
    peridotite **325**

plutons 42, 225, *225,* **339**
podiform chromites 73
Point Lake greenstone belt, Slave
  craton 392, 393, *394*
polar cells *206*
polar deserts 119, 127
polar glaciers 175
polar ice caps 223
polarizing microscope **339–340**, 397
polar jet stream 233
polar lows **340**
polar soils 396
pole of rotation *331, 332*
Polynesians 302
polynyas **340**
polyps 351
Pongola Supergroup, Kaapvaal
  craton 237
population growth 171–172
porosity 120, 199, 267, 326, **340**
porphyries 225
porphyroblasts 287
porphyroclasts 287, *287*
Portugal, ocean exploration 302
potassium-argon dating technique
  168
potential energy 424
Potsdam Sandstone, Appalachians
  13
Powell, John Wesley **340–341**
Pratt isostatic model 230
Precambrian eon *341,* **341–342** *See*
  also Archean era; Proterozoic era
    diamonds 121
    Hoffman, Paul Felix 211
    Vendian period **449**
precession of the equinoxes
  279–280
precipitation **342**
    erosion and 141
    floods 151
    fog **155–156**
    hydrologic cycle **218**
    monsoons 436
    rainbows *351,* 351
    rainfall pattern changes 83
    thunderstorms 425
    urbanization and flash flooding
      446–447
    water table and 454–455
Pridolian age 386
primary porosity 340

primary waves (P-waves,
  compressional waves) 132–133,
  260, 383, *384*
primitive (P-type) asteroids 29
principle of hydrostatic equilibrium
  (isostasy) 230
principal stress 404
Principle of Stratigraphic
  Superposition 399, 403
Principles of Geology (Lyell) 253,
  289
Prodomus (Steno) 399
prograde metamorphism 272–273
prokaryotes 249, 345
Prosobranchia 164
Proterozoic era **342–345**, *344*
    Huronian division **214**
    ice age 220
    Neoproterozoic **290**
protistids 345
protomylonites 287
Protozoa **345**
pseudorandom code 178
pseudotachylite 150
P-type (primitive) asteroids 29
pull-apart basins 41, 109, **346**,
  404–405, 431–432
Pulmonata 164
pumice 453
punctuated equilibrium 145–146
pure shear extension model 122,
  123
pure shear strain 402
P-waves (primary waves,
  compressional waves) 132–133,
  260, 383, *384*
P-wave shadow 383, *384*
pycnocline **346**
pyrite 411
pyroclastic flow 298
pyroclastic rocks **346**, 451
pyrope 164
pyroxene 281
pyrrhotite 411

Q

Qiangtang terrane, Tibetan Plateau
  426–427, *427*
Qinglong foreland basin and fold
  thrust belt, North China craton
  296

QSC (Sèries Quartzo-Schisto-Calcaire, Intremo Group), Madagascar 255, 257
quadratic elongation 401
quartz 282
quartzo-feldspathic gneiss terranes 100
Quaternary period 347
    Holocene epoch 69, **211–212**, 347
    Pleistocene epoch **337–339**

R
radar 370–371, *449*
    Doppler radar **126**
    synthetic aperture radar (SAR) **416–417**
radar altimeter 10
radar band 416
radar remote sensing 353
Radarsat 371, 417
radiation **348**
    Geiger counters **164–165**
    infrared **229**, 348, 424
    solar 278–280, *279*, 348
radiation fog 155
radiation temperature inversions 230
radioactive decay 94, 166–168, **348–349**, *349*
radiocarbon dating *See* carbon-14 dating
radio communications 230
radiolara 345
radiometric dating 165
radiosondes 349
radium **349–350**
radon 171, **349–351**
rain 342
rainbows *351*, **351**
rainshadow deserts 117–118
ramp valleys 432
Ranotsara fault zone, Madagascar 254
Rat Islands, Alaska 8
RBV (Return Beam Vidicon) 369–370
Recent *See* Holocene epoch
recessional moraines 285
recharge areas 200
rectangular drainage patterns 126

recumbent synclines 111, *111*
Red Sea *319*, 320, 417
Redwall Limestone, Grand Canyon 181
reefs 35, 47, 62, 99, 184, **351**
Rees, Martin 49
regional metamorphism 274
regolith *103*, 103, 265, 456
regressions **351–352**, 414–415
Reis crater, Germany 420
relative dating systems 165
Reliance formation, Belingwe greenstone belt, Zimbabwe 471
remote sensing **352–353**
    infrared radiation 229
    satellite imagery **369–371**
    synthetic aperture radar (SAR) **416–417**
reptiles 65–66
Resolution, H.M.S. 233
resources **353**
retroarc foreland basins 159
Return Beam Vidicon (RBV) 369–370
reverse faults 112
reverse osmosis 115
rheology 451
Rhodesian craton *See* Zimbabwe craton
rhodochrosite 62
Richter, Charles 354
Richter scale 135, **353–354**
ridge push 409
ridge-ridge transform faults 431, 432–433
ridges *See* mid-ocean ridge systems
ridge-transform intersections 432–433
Riedel fractures 432
rift basins/rifting 41, **354**, *355*
    Afar Depression, Ethiopia 6
    continental *23*, 122–124, *124*
    graben 181
    Proterozoic 343–344
rift-rift-rift (RRR) triple junctions 125, *355*
rift system 354, *355*
rift valley 122, *125*
right-lateral strike-slip fault *148*
rimmed shelves 62
rinds 457
Ringwood, A. E. 260

ripple marks 381
river systems **354–356** *See also* drainage basins
    alluvial fans **8–9**, *9*
    Amazon River **10**
    braided streams **53**, *54*, 154–155
    deltas **113**, 147, 352
    erosion 141–142
    estuaries **143**, 147
    floodplains 152, **154**
    floods 151, 152, 354–355, 446–447
    fluvial sediments **154–155**
    Gilbert, Grove K. 175
    levees **247–248**
    load **251–252**, 356
    meanders **270**
    Mississippi River 152, 248, *283*, **283–284**
    Nile River 246, **293–295**, *294*
    oxbow lakes *312*, **312**
    sedimentary facies 147–148
    terraces *420*, **420–421**
    urbanization and flash flooding **446–447**
    wadis 386–387, 389, *454*, **454**
rivulets 140
roche moutonnée 177
rock correlation 404
rockfalls 269
rock flour 177
rockslides 269
Rocky Mountains **356–358**
    Brooks Range, Alaska **54**
    Canadian Rockies *12*, 357
    Chinook winds **73**
Rocky Mountain trench 357
rocky shore environments 47
Rodinia 59, 180, 195, 197, 343, *344*
Rogers, W. B. and H. D. 91
rogue waves **358–359**
Rossby Waves 205–206
Ross Ice Shelf, Antarctica 222–223, **359**, 376
rotation (deformation of rocks) 110, 401
rotational axis
    Earth 332, 378
    Uranus 445–446
    Venus 449

roto inversion crystal symmetry 107
rounding 369, 381
RRR (rift-rift-rift) triple junctions
    125, *355*
Rub' al-Khali, Arabia 367
Runcorn, Stanley K. 335–336
Russia, Kola Peninsula **241–242**
Ryoke-Abukuma metamorphic belt,
    Japan 232

S

sabkhas 144, 243
Saffir-Simpson hurricane scale
    216–217
Sahara Desert **360–362**
    desertification 116
    sand seas 367
Sahel, Africa 116, 117, **362–363**
Saih Hatat, Oman 305–306
St. Louis University Earthquake
    Center 382
saltation 140, 251, 355, 367
salt deposits 143, 144, 290
salt diapirs 121
salt marshes 47
Salton Sea, California 364
saltwater intrusion 201, *363*,
    363–364
Sambirano Group, Madagascar 255
San Andreas Fault, California
    130–131, 132, **364–365**, *365*, 421,
    431
Sanbagawa metamorphic belt, Japan
    232
sand avalanches 367
sand dunes **365–367**, *366*, *367*
    Kuwait 242
    Sahara 360
    Skeleton Coast, Namibia 391
    Wahiba sand sea, Oman 368,
        368
    wind and 115, 139, *139*, *366*,
        367
sand ripples 367, 368, *368*
sand seas **367–368**, *368*
    Skeleton Coast, Namibia 391
sandstone
    graywacke **183–184**
    metamorphism of 273
    turbidite 442
sandstorms 115

San Francisco Peaks, Arizona 452
San Jacinto fault, California 364
Santa Ana winds **369**
Santa Cruz Mountains, California
    132
Santorin (Thira), eruption of, Crete
    437, 439
SAR (synthetic aperture radar) 371,
    416–417
Sarawak Chamber, Borneo 68
satellites
    global positioning systems (GPS)
        177–179
    imagery 352–353, **369–371**
    synthetic aperture radar (SAR)
        416–417
saturated zone 200
Saturn **371–372**, *372*, 480
Sauk Sequence 434
scaly pahoehoe lava 314
Scandinavia, Caledonides 58
scanning electron microscopes (SEM)
    137–138
scattering of light 379
schist **372–373**, *373*
schistose 187
schistosity 273, 372
Scotland **373–374**, *374*
    basin and dome fold interference
        pattern *156*, *158*
    Caledonides 58
    Sedgwick, Adam **379–380**
    unconformity *443*
SDTS (Spatial Data Transfer
    Standard) 122
sea breezes **374–375**
seafloor magnetic stripes 315, *316*
sea ice 222, 340, **375–376**
sea-level rise and fall **376–377**
    coastal cities and 81, 171
    polar ice caps and
        223–376–377
    regressions 351–352, 414–415
    subsidence **410–411**
    supercontinent cycle and
        413–415, *414*
    transgressions 414–415,
        433–434
sea lilies (crinoids) **105**
seamounts 299, **377–378**
Sea of Umtali, Zimbabwe craton
    474

seasons
    Earth **378**
    Saturn 372
    Uranus 445–446
seawater **378–379**
    color of 379
    composition of 378–379
    intrusion 201, *363*, **363–364**
secondary porosity 340
secondary waves (S-waves, shear
    waves) 132–133, 259, 260, 383,
    384
Second Law of Thermodynamics
    423–424
Sedgwick, Adam **379–380**
sedimentary basins
    Altiplano, Andes **10**
    Dead Sea **109**
    foreland basins 41, **158–159**
    Indo-Gangetic Plain **229**
    pull-apart basins 41, 109, **346**,
        404–405, 431–432
    strike-slip systems and 404–405
    transtension and transpression
        and 405
    Witwatersrand **459–462**, *460*,
        461
    Witwatersrand basin **459–462**,
        460, *461*
    Witwatersrand basin, South
        Africa 179, 238, 329,
        459–462, *460*, *461*
sedimentary facies 147
sedimentary rocks **380–381** *See also*
    sediments
    banded iron formations (BIFs)
        38–40, *40*, 70, 249, 250
    biogenic **49**
    breccias 53, **53–54**, 150
    carbonates **62**
    carbon cycle and 63
    Carboniferous period 65
    chert **71**
    clastic rocks **77**
    concretions **86**
    conglomerates **87**
    cross-bedding 105, *106*
    diapirs **121**
    dolomite 126
    flysch **155**
    gravel **183**
    graywacke **183–184**

heavy minerals 208–209
lithification 250
mélanges 270–271
Milankovitch cycles and 279
Pettijohn, Francis John 328
sandstone 368–369
stratification (bedding) 44
stratigraphy 384, 403–404
turbidite 442, *443*
Sedimentary Rocks (Pettijohn) 328
sedimentary strata 384
sedimentary structures 380–381
sediment flows 268
sediment load 251–252, 356
sediments 380
    abyssal plains 1
    alluvial fans 8–9, *9*
    beaches 42
    biogenic 49
    chemical 70, 380
    clastic 380
    clay 77–78, 238, 281–282
    clays 77–78
    deltas 113, **113**, 147, 352
    diagenesis **120**, 272
    eolian 139–140
    evaporites *143*, 143–144, *144*
    fluvial 154–155, **154–155**
    molasse **284**
seiche waves 381
seismic discontinuities 259–260,
    383, 384
seismic moment 354
seismic waves 132
seismograms *134*, 381, *382*
seismographs *134*, 134–135,
    381–382
seismology **383–384**
seismometers *See* seismographs
selective scattering 379
Selukwe (Shurugwi) greenstone belt,
    Zimbabwe craton 470
SEM (scanning electron microscopes)
    137–138
Semail nappe, Oman Mountains
    304–305
semidiurnal tides 428
Sengör, A. M. Celal 55, 91, 102
sequence stratigraphy **384–385**
series (time stratigraphy) 403
Sèries Quartzo-Schisto-Calcaire
    (QSC, Intremo Group),
    Madagascar 255, *257*

serpentinite **385**
Seychelles 68
Shabani-Tokwe gneiss complex,
    Zimbabwe 44, *46*
shadow zones 383, *384*
shale 273, 443, *443*
shale diapirs 121
Shamvaian group, Zimbabwe craton
    471, *474*
Sharbot Lake terrane, Grenville
    province 193
Shatsky Rise 301
shear joints 160
shear rupture 160
shear strain 401
shear surface forces 404
shear waves (S-waves, secondary
    waves) 132–133, *259*, 260, 383,
    384
shear zones 67
sheeted dikes 308–309, *310*
shelly pahoehoe lava 314
shields **385** *See also* cratons
    Arabian shield **17–22**, *22*
    Baltic shield **38**, 241–242
    continental crust 88, *89*
    deformation of rocks 111
shields (speleothem) 398
shield volcanoes 451
shoestring sands 155
shooting stars *See*
    meteors/meteorites
shrink/swell potential 78
Shurugwi (Selukwe) greenstone belt,
    Zimbabwe craton 470
Shuttle Imaging Radar (SIR-A, SIR-
    B, SIR-C) 370
Siberia
    comet impact, Tunguska 86,
        441
    flood basalts 317, 413
    Global Carbon Sink **385**
    permafrost 325–326
    Taiga Forest **385**, **419**
siderite 62
Sierra Nevada mountain ranges 287,
    358, 465–466
silica-content rock classification 41,
    226
silicate iron formations 39
silicate minerals 280–282
silicate tetrahedra 280–281, *281*

siliceous oozes 1
silicic rocks 41, 226
silicoflagellates 345
silicon 280
sills 225, 339
Silurian period **385–386**
Silurian System 386
similar folds 157
simple shear extension model 122,
    123
simple shear strain 402
Sinai Peninsula, Egypt **386–390**,
    390, 417–418
Sinian era **391**
Sinian fauna, China **73**
Sinian System 391
sinkholes 171, 238–240, *240*, 410,
    411
siphonophores 330
SIR (Shuttle Imaging Radar) 370
Sitter, William de 48
Skeleton Coast, Namibia **391**
slabby pahoehoe lava 314
slaty cleavage 273, 397
Slave craton, Canada **391–396**, *394*,
    395
Slave Province, Canada 101, 187,
    189
Sleepy Dragon terrane, Slave craton
    392, 393, *394*, *395*
sleet 342
slickenlines 150
slickensides 150
slides (mass movements) 267, 269
Sloss, Laurence L. 434
slumps *267*, **267–268**, 439, *440*
slurry flows 268
Smith, William H. **396**
smithsonite 474
smog 436
snow 342
Snowball Earth 175
soil profiles 396, *397*
soils **396–397**
    clay 78
    laterites 247
    permafrost and 325–326
    rivers and 354–355
    weathering and 456
solar flares 36, 259
solar radiation 278–280, *279*, 348

solar system
    asteroids **28–30**, *29*
    comets *29–30, 85*, **85–86**,
        276–277, 441
    data summary 480
    Earth **129–130**, *130*
    formation of 202–205, *204*
    Jupiter *233*, **233–234**
    Mars **261–263**, *263*
    Mercury *271*, **271**
    Neptune **290–292**, *292*
    Pluto 277, *339*, **339**
    Saturn **371–372**, *372*
    Uranus *445*, **445–446**
    Venus *449*, **449–457**
solar wind 35–36, 258–259
solifluction 268, 441
solution sinkholes 240
sonar **397**
Songban-Ganzi terrane, Tibetan
    Plateau 426–427, *427*
Sonoma orogeny 357
Sorby, Henry Clifton **397–398**
sorting 368–369, 380–381
South Africa
    Du Toit, Alexander Logie 128
    Kaapvaal craton 25, *45*, 46,
        100, 101, **236–238**, *237, 472*,
        474
    kimberlites and diamonds 241
    Limpopo Province 460, 461,
        461, 472, 474
    veldt 399
    Witwatersrand basin 179, 238,
        329, **459–462**, *460, 461*
South America *See also specific*
    countries
    Altiplano **10**
    Amazon River **10**
    Andes **12**
    Atacama Desert **30–31**
    Cape Horn and the Strait of
        Magellan **61**
South Dakota badlands *37*, 38
southern lights 35–36
Southern Oscillation *See* El Niño
    and the Southern Oscillation
    (ENSO)
South Pole 315
Space Shuttle orbiters 371
Spatial Data Transfer Standard
    (SDTS) 122

species, classification of 480
specific gravity, measuring 183
speleothems *398*, **398**
spessartine 164
sphalerite 411, 474
spheroidal weathering 457
spinel 313
Spitzbergen Island, Norway **415–416**
SPOT (Système pour l'Observation
    de la Terre) 370
springs 200
spring tides 428
squall lines 425
St. Louis floods, Missouri 152
stages 403
stalactites *398*, 398
stalagmites *398*, 398
standard air pressure 8
standard model of the universe 48
star dunes 367
steady state theory of the universe
    48
Steamboat Geyser, Yellowstone
    National Park 465
steam fog *155–156*
Steinman, G. 307
Steno, Nicolas 91, **399**
steppe **399**
steps 404
stereo aerial photographs 352
stilbite 469
Stille, Wilhem Hans **399**
stinky black shales 143
Stokes Law **399**
Stone Age 211, 290, 315
stony-irons 276
stoping 226, 339
storm surges 217, **399–401**, *401*
Stowe, Clive W. 469, 470
Strahler, Arthur N. **401**
straight-line winds 425
strain 110, *401*, **401–403**, *402*
    analysis 402
    viscosity and 451
strain path 402
Strait of Magellan **61**
"Strata Identified by Organic
    Remains" (Smith) 396
stratification (bedding) **44**
    cross-bedding 44, **105**, *106*
    formations **159**
    unconformities *443*, **443**

Stratigraphic Superposition, Principle
    of 399, 403
stratigraphy **403–404**
    Grabau, Amadeus William
        180–181
    sequence stratigraphy 384–385,
        403–404
    time stratigraphy 403–404
stratocumulus clouds 84
stratosphere 31, *34*
stratovolcanoes 451–452
stratus clouds 84
stream capture 126–127
stream flow 141
stream order 114
streams *See* river systems
stream terraces *420*, 420–421
stress 110, **404**, 451
stretch 401
strike 407
strike line 407
strike-slip faults 112, *148*, 149,
    404–405
    greenstone belts and 190
    transform faults and 431, *431*,
        432
string theory 49
stromatolites 209, *209, 345, 405*,
    405–406
strontianite 62
structural geology **406–409**, *408*,
    409
    cross sections **105**
    Dewey, John F. **119–120**
    plate tectonics and 406–407,
        407–408
sturgeon 67
Sturtian period 391
stylolites 120
S-type asteroids 29
subaqueous mass wasting 270
subduction zones 1–3, 94–96, 333,
    409–410
subsequent streams 126
subsidence **410–411**
subsidence events 410
subtropical jet stream 233
Sudan
    Al-Jazirah 293
    drought 116
Suess, Eduard 91, 312
sulfide iron formations 40

sulfide minerals 282, **411**
Sumatra earthquake 135, 303, 354, 437, 439
sun dogs 411
sun halos **411**
sun pillars 411
sunspots 36, 259
Supai Formation, Grand Canyon 181
supercell thunderstorms 425, 430
supercontinent cycles **412–415**
  epicontinental seas **140**
  evolution and 146
  Gondwana (Gondwanaland) **180**
    Cambrian period 59
    Carboniferous period 65, 66
    ice ages and 220
    Proterozoic era 343
    Tethys ocean 422–423
  Laurasia **247**
  Pangea **318–319**, *319*, 412
    Mesozoic era 272
    Paleozoic era 247, 317
    Panthalassa **319**
    Permian period 326–327
    Triassic period 435
    Wegener, Alfred 90, *90*
  Proterozoic era 343, *344*
  Tethys ocean **422–423**, *423*
Superior Province 187, 189, 190
Superior-type banded iron formations 38, 39
superposed streams 126
superswell 94
surface forces 404
surface waves (earthquakes) 132, 133–134
surf zone 455, *455*
suspect terranes **415**
suspended load 251, 355
Sutton, A. H. 412
suture zones 19–20, **415**
Svalbard, Norway **415–416**
Svecofennian orogeny, Baltic shield 38
swash zone *455*
S-waves (shear waves, secondary waves) 132–133, 259, 260, 383, *384*
S-wave shadow 383, *384*
Swaziland Supergroup, Kaapvaal craton 236

Sykes, Lynn 337
synclines 111, *111*
synforms 156
synthetic aperture radar (SAR) 371, **416–417**
Syr Darya River 22
Syrian arc fold belt **417–418**
Système pour l'Observation de la Terre (SPOT) 370

T

Taconic orogeny 13–14, 324
TAG hydrothermal mound 51
taiga 385, **419**
Tanzania carbonatite 62
Tapeats Sandstone, Grand Canyon 181
tarns 77
Tati greenstone belt, Zimbabwe craton 470
tectonicists
  Burke, Kevin C. **55**
  Dewey, John F. **119–120**
  Hoffman, Paul Felix 195, **211**
  Stille, Wilhem Hans **399**
tectonics *See* plate tectonics
tectosphere 250, **419**
Tejas Sequence 434
tektite **419–420**
temperate glaciers 175
temperature inversions 229–230
Tendaho rift 7
tensile cracking 159–160
tephra 452
tephra cones 452
Terai 229
terminal moraines 285, **420**
terraces *420*, **420–421**
Tertiary period **421–422**, *422 See also* Cretaceous-Tertiary (K-T) extinction
  Neogene period **289–290**
  Paleogene period 139, 421
Tethys Ocean **422–423**, *423*
  Alps and 9–10
  Cenozoic era 69
  Triassic period 435, *435*
Texas
  Balcones escarpment rainfalls 151
  Galveston Island 40–41, 151, 164, **164**, 214

Teyq sinkhole, Oman *240*
thalweg 154
thecamoebians 345
Thematic Mapper (TM) imaging system 370
Theory of Earth (Hutton) 217
Theory of Relativity 421–422
The Theory of the Earth (Hutton) 443
thermal distillation 114
thermal metamorphism 273
thermobarometry 165
thermocline 63, 346
thermodynamics **423–424**
thermohaline circulation 80–82, 300, **424**
thermoluminescence 168
thermophyllic organisms 49
thermosphere 31, *34*
Third Law of Thermodynamics 424
Thomson, Charles Wyville 70
thomsonite 469
thorium-lead dating technique 168
thrust faults 2–3, 149, 189–190
thunderstorms **424–426**, *426*, 430
Tibesti massif, Sahara 360
Tibetan Plateau **426–428**
  continent-continent collision zone 98–99
  permafrost 325–326
tidal bulges 430
tidal inlets 400–401
tidal period 428
tidal waves *See* tsunami
tide-dominated deltas 113
tide pools 47
tides 251, **428–430**, *429*
Tierra del Fuego 61, 320–321
time lines 403–404
time stratigraphy 403–404
tintinnids 345
Tippicanoe Sequence 434
Titan 372
Titicaca, Lake 10, **246**
TM (Thematic Mapper) imaging system 370
Tokwe terrane, Zimbabwe craton 469–470, 471–473
Tomiko terrane, Grenville province 192
tonalite 323–324

tonalitic suite, Adirondack
    Mountains 5
topographic elevations 121–122
tornadoes 430
Tornquist Sea 58
trace fossils (ichnofossils) 49, 61,
    225
trade winds 117, **431**
trailing margins *See* passive margins
transcurrent plate boundaries *See*
    transform plate margin processes
transform faults 277–278, *331*, 333,
    337, 431–433
transform plate margin processes
    93, 333, 346, *431*, 431–433, *432*
transgressions 414–415, **433–434**
transgressive sequences 434, *434*
translation (deformation of rocks)
    110, 401
transmission electron microscopes
    (TEM) 137–138
transverse dunes 367
Transverse Ranges, California 364
traps *See* flood basalts
travertine 398
tree rings 114
trellis drainage patterns 126
trenches 94–95, 333
    Aleutian 8
    Marianas **261**
    subduction zones 409–410
trench pull 409
triangulation 169
Triassic-Jurassic extinction 264
Triassic period **434–435**, *435*
trilaterating 178
trilobites 61
Triton 291–292, *292*
Trojans (asteroids) 28
tropical climates **435–436**
    Everglades, Florida 145
    sea breezes 375
    soil of 396–397
    weathering and 42, 458
tropics 378, **436**
tropopause 31, *34*
troposphere 31, *34*, **436**
trubidite 442, *443*
trubidity currents 442
Tsarantana sheet, Madagascar 255
tsunami 302, **436–441**, *438*, *440*

tsunami run-up *438*, 438–439
Tucker, Robert 257
tundra **441**
Tunguska, Siberia 86, **441**
tunicates 330
Tunisia, Atlas Mountains 31
turbidite 60
turbidity currents 174, 442
twist hackle 160
typhoons 214

U

ultra-high pressure metamorphic
    belts 71
ultramafic rocks 41
ultramylonites 287
ultraviolet radiation 230, 313, 348
Umtali line, Zimbabwe craton 474
unconfined aquifer 17
unconformities *443*, **443**
unconformity-bounded sequences
    434, *434*
underplating 2, *2*
Ungava Peninsula banded iron
    formation, Canada *40*
uniformitarianism **443–444**
United Arab Emirates
    barchan dunes *367*
    Oman Mountains (Hajar
        Mountains) 303–306, *305*,
        467
United Kingdom
    Harker, Alfred 207
    Scotland
        basin and dome fold
            interference pattern *156*,
            158
        Caledonides **58**
        Scottish Highlands
            373–374, *374*
        Sedgwick, Adam **379–380**
        unconformity *443*
    Smith, William H. **396**
United States Department of Defense
    Navstar satellites 177–178
United States *See also specific states*
    Basin and Range Province
        41–42, 358
    Great Lakes **184**
    Mississippi River 152, *283*,
        283–284

Powell, John Wesley **340–341**
Southwest desertification 116
Yellowstone National
    Park/Yellowstone Valley 56,
    227, 261, 452, *464*, **464**, 464–465,
    465
United States Coast and Geodetic
Survey
    bench marks 46
United States Exploring Expedition
    108
United States Geological Survey
    (USGS) **444**
    bench marks 46
    collection and analysis of
        geochemical data 165
    digital elevation models (DEMs)
        121–122
    economic resource assessments
        353
    tsunami mapping 437
United States Public Health Service
200
universe, theories of **47–49**
unloading joints 160
updrafts 425
upslope fog 155–156
upwelling, coastal **85**
Ural Mountains **444**
Ural-Okhotsk mobile belt 444
uraninite 445
uranium 349, **444–445**
uranium-lead dating 166–167
Uranus 445, **445–446**, 480
urbanization and flash flooding
    446–447
USGS *See* United States Geological
Survey
Utah
    cross bedding *106*
    Great Salt Lake **185**
    Lake Bonneville 337
    Middle Rockies 357
uvarovite 164
Uzbekistan, Aral Sea **22**

V

Vail curves *414*
Valdez group, Alaska 75
Valles Marineris, Mars 262
valley breeze *See* katabatic winds

Valley of Ten Thousand Smokes, Alaska **448–449**
Van Allen radiation belts 35
vapor compression condensation 114
Varanger epoch 391, **449**
varves 380
vaterite 56, 62
veldt 399
Vendian period 391, **449**
Vendoza fauna 276
Venice, Italy 411
vents *See* hydrothermal vents
Venus 449, **449–457**, 480
Vermont, Lake Bomoseen syncline *111*
Vikings 302
Vine, John 336–337
viscosity 94, **451**
Vishnu Schist, Grand Canyon 181
volcanic blocks 451
volcanic bombs **451**
volcanic islands 35
volcaniclastic sediments and rock 346
volcanic mountain ranges 89, 111–112, 312
volcanic necks 42, 225, 339
volcanic rocks (extrusive rocks) 225
    andesite **12**
    breccias 53–54, 150
    concretions **86**
    dacites **108**
    flood basalts **153–154**, *154, 301*
    greenstone belts 187
    kimberlites 120–121, **241**, 463
    mélanges 95, **270–271**
volcanoes **451–453**
    Andes 12
    atolls 35
    calderas 56, 57, 452
        Crater Lake, Oregon 56, 57
        Krakatau eruption 436–437, 439
        Long Valley caldera, California **252–253**
        Yellowstone National Park/Yellowstone Valley 56, 227, 261, 452, *464,* **464–465,** *465*
    climate change and 82

eruptions, explosive *vs.* nonexplosive 452–453
    as geological hazard 169
    graben 181
    Hawaii 208, 451, 452
    hot spots **213**, 452
    Iceland 223
    lahars 245, 269
    lava **247**
        aa lava **1**
        pahoehoe lava **314**
        pillow lavas *25, 324, 328–329, 329*
    map of *336*
    Mars 262
    mid-ocean ridges 125
    Novarupta eruption, Alaska 448
    nuée ardente **298**, 453
    plate tectonics and 452
    pyroclastic rocks 346
    seamounts (guyots) **377–378**
    subsidence and 411
    tsunami and 439
    Valley of Ten Thousand Smokes, Alaska **448–449**
    Venus 450–451
    volcanic bombs **451**
volcanogenic massive sulfide (VMS) deposits 51–52
Volga River 66
volumetric strain 401
Voyager 2
    Neptune images *292*
    Uranus images *445*
Vumba greenstone belt, Zimbabwe craton 470

W

W. R. Grace and Co. asbestos disaster 28
WAAS (Wide Area Augmentation System) 179
wackes 183–184, 369
Wadi Al-Batin, Kuwait 243
Wadi Dayqa, Oman *454*
Wadi El Arish, Egypt 386
Wadi El Bruk, Egypt 387
Wadi Geraia, Egypt 387
wadis 386–387, 389, *454,* **454**
Wahiba Sand Sea, Oman 368, *368*
Walcott, Charles D. 60

Walter, M. 406
Walther's Law 147
Walvis–Rio Grande Ridge 144, 261
washes *See* wadis
Washington
    Columbia River flood basalts 153
    Mount Saint Helens 245, 453
    Northern Rockies 357
watershed *See* drainage basins
waterspouts 430
water table 200, **454–455**
    aquifers and 17
    cones of depression 86
water vapor 32
wave crests 455
wave-dominated deltas 113
wave fronts 455
wavelengths 438, 455, *455*
wave periods 455
wave refraction 455–456
waves *455,* **455–456,** *456*
    beaches and 43, *455,* 455–456, *456*
    rogue waves **358–359**
    seiche waves **381**
    tsunami **436–441,** *438, 440*
    wind and 455
wave traps 439
weathering 140, **456–458**
    bauxite 42
    chemical 456, 457
    exfoliation **146**
    factors that influence 457–458
    mechanical 456–457
    mountains 41–42
    yardangs **464**
weathering profiles 456
Weddell Sea, Antarctica 424
Wegener, Alfred Lothar 89, 90, 91, 312, 318, 334, **458**
Wentworth scale 77
Werner, Abraham Gottlob **458**
Western Ghats *See* Eastern and Western Ghats, India
West Rand group, Witwatersrand basin 459–460
Whewell, William 443
White Nile River 293
white smoker chimneys 50, 51
Wide Area Augmentation System (WAAS) 179

Wiechert, E. 381–382
wildcat wells **458–459**
willimenite 474
willy-willys 127
Wilson, J. Tuzo 333, 337
Wilson, James F. 469
Wilson, Robert 48
winds 80
    Chinook winds **73**
    coastal downwelling and 85
    coastal upwelling and 85
    doldrums **125–126**
    eolian sediments **139–140**
    general circulation models **165,**
        205–206, *206*
    horse latitudes **212**
    hurricane 216–217
    jet streams 205, **233**
    katabatic winds **240–241**
    lake breezes 374
    land breezes 375
    Langmuir Circulation and 247
    load 252
    loess *252*, **252**
    sand dunes and 360, *366, 367*
    Santa Ana winds **369**
    sea breezes **374–375**
    thunderstorms 425
    tornadoes 430
    trade winds 117, **431**
    waves 455
witherite 62
Wittenoom asbestos disaster,
   Australia 28
Witwatersrand basin, South Africa
   179, 238, 329, **459–462**, *460, 461*
wollastonite **462**
Wopmay orogen, Slave craton 392

Wosnesinski glacier, Alaska *176*
Wrangellia superterrane, Alaska 74
Wyoming
    Rocky Mountains 356–357
    Yellowstone National Park *464,*
    464–465, *465*

X

X-band 416
xenoliths 463
xenotime 463
X-ray fluorescence **463**
X rays 348

Y

Yafikh-Ragbah orogeny, Arabian
   shield 18
Yanbu suture zone, Arabian shield
   19, 20
Yangtze craton 391
Yangtze River 391
yardangs 139–140, **464**
Yellowknife greenstone belt *25, 329,*
   394
Yellowknife greenstone belt, Slave
   craton 392
Yellow River, China 248
Yellowstone National
   Park/Yellowstone Valley 56, 227,
   261, 452, *464*, **464–465**, *465*
Yilgarn Province, Australia 187
Yosemite Valley and National Park
   465, **465–466**
Yucatán Peninsula, Mexico 440
Yucatán Pennisula, Mexico 30, 68,
   71

Z

Zagros Mountains, Iran 121,
   467–469, *468*
Zeederbergs formation, Belingwe
   greenstone belt, Zimbabwe 471
zeolite **469**
Zimbabwe craton *45*, 45–46, 461,
   461, **469–474**, *472, 473*
    accretion of 471–474
    Belingwe greenstone belt,
      Zimbabwe **44–46**, *45, 46,*
      189, 236, 470
    Chilimanzi suite 471
    Great Dike **184**
    Northern Belt (Zwankendaba
      Arc) 470–471
    Shamvaian group 471, 474
    Southern belt 471
    Tokwe terrane *26*, 469–470,
      471–473
zinc **474–475**
zincite 474
zircon 463, **475**
zirconium 475
zone of ablation (glaciers *176*
zone of accumulation (glaciers) *176*
zooplankton 329
Zoroaster Granite, Grand Canyon
   181
Zunhua structural belt, North China
   craton 72, 296–297
Zuni Sequence 434
Zwankendaba Arc (Northern Belt),
   Zimbabwe craton 470–471